I0235407

NASA SP-2003-4409

The Wind and Beyond: A Documentary Journey into the History of Aerodynamics in America

Volume 1: The Ascent of the Airplane

James R. Hansen, Editor
with D. Bryan Taylor, Jeremy Kinney, and J. Lawrence Lee

The NASA History Series

National Aeronautics and Space Administration
NASA History Office
Office of External Relations
Washington, D.C.   2003

Published by Books Express Publishing
Copyright © Books Express, 2012
ISBN 978-1-78039-691-0

Books Express publications are available from all good retail and online booksellers. For publishing proposals and direct ordering please contact us at: info@books-express.com

Table of Contents

Foreword xi
Acknowledgments xv
Series Introduction: Talking with the Wind, Collaborating with Genius ... xvii
Significant Aircraft List xxxi
Series Bibliographic Essay: Days on the Wing xxxv
Biographies of Volume 1 Contributors lxxi

CHAPTER ONE: THE ACHIEVEMENT OF FLIGHT 1

Essay: From Cayley's Triple Paper to Orville Wright's Telegram 1
The Documents 33
    1-1(a)   George Cayley, "On Aerial Navigation," Part One, 1809 34
    1-1(b)   George Cayley, "On Aerial Navigation," Part Two, 1810 42
    1-1(c)   George Cayley, "On Aerial Navigation," Part Three, 1810 47
    1-2   Thomas Jefferson, letter to William D. B. Lee, 1822 57
    1-3(a)   Francis H. Wenham, "On Aerial Locomotion," 1866 59
    1-3(b)   Report of the first wind tunnel built by Francis H. Wenham, 1871 ... 80
    1-3(c)   Minutes of the Royal Aeronautical Society, 1872 84
    1-4   Samuel P. Langley, "Langley's Law," 1891 89
    1-5(a)   Otto Lilienthal, "The Problem of Flying," 1893 92
    1-5(b)   Otto Lilienthal, "Practical Experiments in Soaring," 1893 ... 95
    1-5(c)   Otto Lilienthal, "The Best Shapes for Wings," 189 99
    1-5(d)   Vernon, "The Flying Man," 1894 101
    1-6   Hiram Maxim, "Natural and Artificial Flight," 1896 107
    1-7   B. Baden-Powell, "Present State of Aeronautics," 1896 126
    1-8(a)   Samuel P. Langley, "Story of Experiments in Mechanical Flight," 1897 ... 131
    1-8(b)   Samuel P. Langley, "The New Flying Machine," 1897 141
    1-8(c)   Alexander Graham Bell, "The Aerodromes in Flight," 1897 145
    1-9(a)   Octave Chanute, "Conditions of Success in the Design of Flying Machines," 1898 ... 149
    1-9(b)   Augustus M. Herring, "A Solution to the Problem of the Century," 1897 ... 156
    1-9(c)   Octave Chanute, "Experiments in Flying," 1900 171
    1-9(d)   Octave Chanute, "Aerial Navigation," 1901 179
    1-10(a) Orville Wright on their interest in flight, 1920 188
    1-10(b) Wilbur Wright, letter to the Smithsonian Institution, 1899 .. 188
    1-10(c) Wilbur Wright, letter to Octave Chanute, 1900 189
    1-11(a) Wilbur Wright, letter to Milton Wright, 1900 193
    1-11(b) Wilbur Wright, "Some Aeronautical Experiments," 1901 193
    1-12(a) Octave Chanute, remarks at a meeting of the Aéro-Club, 1903 ... 208

1-12(b)	Wilbur Wright, "Experiments and Observations in Soaring Flight," 1903	218
1-13(a)	Orville Wright, diary entry for 17 December 1903	222
1-13(b)	Wilbur and Orville Wright, statement to the Associated Press, 5 January 1904	223
1-14(a)	Wilbur and Orville Wright, letter to the Secretary of War, 1905	226
1-14(b)	U.S. Army specifications for heavier-than-air flying machine, 1907	226
1-14(c)	Wilbur and Orville Wright, letter to General James Allen, 1908	229
1-14(d)	Octave Chanute, "Recent Aeronautical Progress in the United States," 1908	229
1-15	Simon Newcomb, "Aviation Declared a Failure," 1908	233

CHAPTER TWO: BUILDING A RESEARCH ESTABLISHMENT 237
Essay: From the Wright Bicycle Shop to the Langley Full-Scale Tunnel 237
The Documents 273

2-1(a)	Wilbur Wright, letter to George A. Sprat, 15 December 1901	274
2-1(b)	Wilbur Wright, letter to Octave Chanute, 15 December 1901	275
2-1(c)	Wilbur Wright, letter to Octave Chanute, 23 December 1901	282
2-1(d)	Wilbur Wright, letter to Octave Chanute, 19 January 1902	284
2-2(a)	W. I. Chambers, "Report on Aviation," 1912	288
2-2(b)	Jerome C. Hunsaker, *Report on Facilities for Aeronautical Research in England, France, and Germany, Part III*, 1913	296
2-2(c)	Albert F. Zahm, *Report on European Aerodynamical Laboratories*, 1914	301
2-3(a)	U.S. Congress, Act Establishing an Advisory Committee for Aeronautics, 3 March 1915	306
2-3(b)	George P. Scriven, Chairman's Letter for the first Annual Report of the National Advisory Committee for Aeronautics, 1915	307
2-3(c)	George P. Scriven, "Existing Facilities for Aeronautic Investigation in Government Departments," 1915	309
2-4	Jerome C. Hunsaker, "The Wind Tunnel of the Massachusetts Institute of Technology," 1916	311
2-5	G. I. Taylor, "Pressure Distribution Over the Wing of an Aeroplane in Flight," 1916	318
2-6	Henry. T. Tizard, "Methods of Measuring Aircraft Performances," 1917	340
2-7(a)	Jerome C. Hunsaker, letter to H. M. Williams, 1918	359
2-7(b)	"Education in Advanced Aeronautical Engineering," *NACA Annual Report for 1920*	360
2-8(a)	U.S. Army, "Full Flight Performance Testing," 1918	362
2-8(b)	"Special Aerodynamic Investigations," *NACA Annual Report for 1919*	373

2-9(a)	"Office of Aeronautical Intelligence," *NACA Annual Report for 1918*	376
2-9(b)	John J. Ide, "Report on Visit to England," 1921	377
2-9(c)	John J. Ide, "Wind Tunnel at Issy-les-Moulineaux," 1921	380
2-10(a)	D. W. Taylor, letter to Frederick C. Hicks, 1919	384
2-10(b)	Edward P. Warner, "Report on German Wind Tunnels and Apparatus," 1920	386
2-11	"Free Flight Tests," *NACA Annual Report for 1918*	397
2-12(a)	Edward P. Warner, F. H. Norton, and C. M. Hebbert, "Design of Wind Tunnels and Wind Tunnel Propellers," 1919	398
2-12(b)	F. H. Norton and Edward P. Warner, *The Design of Wind Tunnels and Wind Tunnel Propellers, II*, 1920	426
2-13(a)	"Report of Committee on Aerodynamics," *NACA Annual Report for 1919*	440
2-13(b)	"Report of Committee on Aerodynamics," *NACA Annual Report for 1920*	444
2-13(c)	"Report of Committee on Aerodynamics," *NACA Annual Report for 1921*	454
2-14	Jerome C. Hunsaker, "Recommendations for Research Program—Comparison of Wing Characteristics in Models and Free Flight," 1920	463
2-15(a)	Joseph S. Ames, letter to A. H. [sic] Zahm, L. J. Briggs, E. B. Wilson, W. F. Durand, E. N. Fales, J. G. Coffin, H. Bateman, and F. H. Norton, 23 August 1920	468
2-15(b)	F. H. Norton, letter to Joseph S. Ames, 23 August 1920	468
2-15(c)	A. F. Zahm, letter to Joseph S. Ames, 14 September 1920	469
2-15(d)	Joseph G. Coffin, letter to Joseph S. Ames, 18 September 1920	470
2-15(e)	W. F. Durand, letter to Joseph S. Ames, 24 September 1920	472
2-15(f)	Ludwig Prandtl, letter to William Knight, 1920	474
2-15(g)	"International Standardization of Wind-Tunnel Results," *NACA Annual Report for 1922*	475
2-15(h)	G.W. Lewis, memorandum to Langley Memorial Aeronautical Laboratory, 1925	476
2-15(i)	Aerodynamics Department, The National Physical Laboratory, "A Comparison between Results for R.A.F. 15 in the N.P.L. Duplex Tunnel and in the N.A.C.A Compressed Air Tunnel," [1925]	476
2-16	"Summary of General Recommendations," *NACA Annual Report for 1921*	478
2-17	F. H. Norton, "The National Advisory Committee's 5-Ft. Wind Tunnel," 1921	481
2-18(a)	Max M. Munk, letter to J. C. Hunsaker, 1920	500
2-18(b)	J. C. Hunsaker, letter to Lester D. Gardner, 1921	501
2-18(c)	F. H. Norton, memorandum to NACA Executive Officer [G.W. Lewis], 1921	502

2-18(d) W. Margoulis, "A New Method of Testing Models in Wind
　　　　　Tunnels," 1920 502
2-18(e) Edward P. Warner, letter, to G. W. Lewis, 1921 506
2-18(f) W. Margoulis, "A New Method of Testing Models in Wind
　　　　　Tunnels," [1921] 507
2-18(g) Max M. Munk, "On a New Type of Wind Tunnel," 1921 511
2-18(h) F. H. Norton, memorandum to G. W. Lewis,
　　　　　"Design of Compressed Air Wind Tunnel," 1921 521
2-18(i) "Compressed Air Wind Tunnel,"
　　　　　NACA Annual Report for 1921 522
2-18(j) Max M. Munk and Elton W. Miller,
　　　　　*The Variable Density Wind Tunnel of the National Advisory
　　　　　Committee for Aeronautics*, 1925 523
2-19(a) Ludwig Prandtl, letter to Jerome C. Hunsaker, 1916 537
2-19(b) Joseph S. Ames, letter to Jerome C. Hunsaker, 1920 537
2-19(c) Ludwig Prandtl, *Applications of Modern Hydrodynamics to
　　　　　Aeronautics*, 1921 538
2-20(a) Max M. Munk, memorandum to G. W. Lewis, 1925 557
2-20(b) Fred E. Weick and Donald H. Wood, *The Twenty-Foot
　　　　　Propeller Research Tunnel of the National Advisory Committee
　　　　　for Aeronautics*, 1928 560
2-20(c) Fred E. Weick, *From the Ground Up*, 1988 576
2-21(a) Elliot G. Reid, "Memorandum on Proposed Giant Wind
　　　　　Tunnel," 1925 585
2-21(b) Arthur W. Gardiner, "Memorandum on Proposed Giant
　　　　　Wind Tunnel," 1925 588
2-21(c) Smith J. DeFrance, *The NACA Full-Scale Wind Tunnel*, 1933 591
2-22   Max M. Munk, memorandum on "Recommendations for
　　　　　New Research," 1926 602
2-23(a) Daniel Guggenheim, letter to Herbert Hoover, 1926 605
2-23(b) California Institute of Technology, "Development of
　　　　　Aeronautics," 1926 608
2-24(a) C. G. Grey, "On Research," 1928 611
2-24(b) C. G. Grey, "On Research" continued, 1928 614
2-25(a) Frank A. Tichenor, "Air—Hot and Otherwise: Why the
　　　　　N.A.C.A.?" 1930 617
2-25(b) Frank A. Tichenor, "Air—Hot and Otherwise: The N.A.C.A.
　　　　　Counters," 1931 624
2-25(c) Edward P. Warner, "Speaking of Research," 1931 628
2-26   Joseph S. Ames and Smith J. DeFrance, "Report of
　　　　　Proceedings of the Sixth Annual Aircraft Engineering
　　　　　Research Conference," 1931 631
2-27   Minutes of the Second Technical Committee Meeting,
　　　　　United Aircraft and Transport Corporation, 1929 636
2-28(a) A. L. Klein, "The Wind Tunnel as an Engineering
　　　　　Instrument," 1930 643

2-28(b) A. L. Klein, letter to V. E. Clark, 1934 650
2-29   Starr Truscott, memorandum to [Henry J. E. Reid],
         "Work in connection with special aerodynamic tests
         for Bureau of Aeronautics which has been requested
         by Mr. Lougheed," 1932 652
2-30(a) Edward P. Warner, "Research to the Fore," 1934 657
2-30(b) "Research Symphony," 1935 659

NASA History Series 667

Index 677

x

Foreword

Airplane travel is surely one of the most significant technological achievements of the last century. The impact of the airplane goes far beyond the realm of the history of technology and touches upon virtually every aspect of society, from economics to politics to engineering and science. While space exploration often claims more public glory than aeronautics research, many more individuals have been able to fly within the Earth's atmosphere than above it. Thus, aeronautics and air travel have had an enormous practical impact on many more individuals. For this reason, if no other, it is certainly an appropriate time to document the rich legacy of aeronautical achievements that has permeated our society. It is especially timely to do so during the centennial anniversary of the Wright brothers' historic flight of 1903.

Dr. James R. Hansen and his collaborators do more than just document the last century of flight. They go back and expertly trace the historical origins of what made the first heavier-than-air, controlled, powered airplane flight possible on 17 December 1903. Some names covered in this volume, such as Isaac Newton and Leonardo da Vinci, are familiar to even the most casual reader. Other heralded, but less well-known, early pioneers of flight such as George Cayley, Otto Lilienthal, Theodore von Kármán, and Theodore Theodorsen will come alive to readers through their original letters, memos, and other primary documents as they conjoin with the authors' insightful and elegantly written essays.

This first volume, plus the succeeding five now in preparation, covers the impact of aerodynamic development on the evolution of the airplane in America. As the six-volume series will ultimately demonstrate, just as the airplane is a defining technology of the twentieth century, aerodynamics has been the defining element of the airplane. Volumes two through six will proceed in roughly chronological order, covering such developments as the biplane, the advent of commercial airliners, flying boats, rotary aircraft, supersonic flight, and hypersonic flight.

This series is designed as an aeronautics companion to the *Exploring the Unknown: Selected Documents in the History of the U.S. Civil Space Program* (NASA SP-4407) series of books. As with *Exploring the Unknown*, the documents collected during this research project were assembled from a diverse number of public and private sources. A major repository of primary source materials relative to the history of the civil space program is the NASA Historical Reference Collection in the NASA Headquarters History Office. Historical materials housed at NASA field centers, academic institutions, and Presidential libraries were other sources of documents considered for inclusion, as were papers in the archives of private individuals and corporations.

The format of this volume also is very similar to that of the *Exploring the Unknown* volumes. Each section in the present volume is introduced by an overview essay that is intended to introduce and complement the documents in the section and to place them in a chronological and substantive context. Each essay contains references to the documents in the section it introduces, and many also contain references to documents in other sections of the collection. These introductory essays are the responsibility of Dr. Hansen, the series' author and chief editor, and the views and conclusions contained therein do not necessarily represent the opinions of either Auburn University or NASA.

The documents included in each section were chosen by Dr. Hansen's project team from a much longer list initially assembled by the research staff. The contents of this volume emphasize primary documents, including long-out-of-print essays and articles as well as material from the private recollections of important actors in shaping aerodynamic thinking in the United States and abroad. Some key legislation and policy statements are also included. As much as possible, the contents of this volume (and the five volumes to come) in themselves comprise an integrated historical narrative, though Dr. Hansen's team encourages readers to supplement the account found herein with other sources that have already or will come available.

For the most part, the documents included in each section are arranged chronologically. Each document is assigned its own number in terms of the section in which it is placed. As a result, for example, the fifteenth document in the second chapter of this volume is designated "Document 2-15." Each document is accompanied by a headnote setting out its context and providing a background narrative. These headnotes also provide specific information and explanatory notes about people and events discussed. Many of the documents, as is the case with Document 2-15, involve document "strings," i.e., Document 2-15(a–e). Such strings involve multiple documents—in this case, five of them (a through e) that have been grouped together because they relate to one another in a significant way. Together, they work to tell one documentary "story."

The editorial method that has been adopted seeks to preserve, as much as possible, the spelling, grammar, and language usage as they appear in the original documents. We have sometimes changed punctuation to enhance readability. We have used the designation [. . .] to note where sections of a document have not been included in this publication, and we have avoided including words and phrases that had been deleted in the original document unless they contributed to an understanding of the writer's thought process in making the record. Marginal notations on the original documents are inserted into the text of the documents in brackets, each clearly marked as a marginal comment. Page numbers in the original document are noted in brackets internal to the document text.

Copies of all documents in their original form are available for research by any interested person at the NASA History Office or Auburn University.

While the *Exploring the Unknown* series has been a good model in many ways, this volume indeed represents the beginning of a yet another new undertaking into uncharted waters. I am confident that Dr. Hansen and his team have crafted a landmark work that will not only be an important reference work in the history of aeronautics, but will be interesting and informative reading as well. We hope you enjoy this useful book and the forthcoming volumes.

Stephen Garber
NASA History Office
October 2003

ACKNOWLEDGMENTS

This volume represents the collected efforts of many members of an outstanding team. At Auburn University, a number of individuals provided generous assistance to Dr. James R. Hansen's project team. Dr. Paul F. Parks, former University Provost, strongly encouraged and supported the project from its inception, as did Dr. Michael C. Moriarty, Vice President for Research. To undertake his leadership of the project, Dr. Hansen gave up his job as Chair of the Department of History, something he would not have felt comfortable doing without being certain that the administration of his department would be in the capable hands of worthy successors—first, Dr. Larry Gerber, and then Dr. William F. Trimble. Both Gerber and Trimble gave hearty and vocal support to Auburn's NASA history project. A number of colleagues in aerospace history gave help to the project, including Distinguished University Professor Dr. W. David Lewis and Dr. Stephen L. McFarland. Dr. Roy V. Houchin, who earned a Ph.D. under Hansen, lent aid and comfort to the project team from his vantage point inside the U.S. Air Force. A number of Hansen's current graduate students helped the project in various ways, notably Andrew Baird, Amy E. Foster, and Kristen Starr, as did Dr. David Arnold, also of the USAF, who earned a Ph.D. in aerospace history during the time period when this project was being conducted.

Historians and archivists at a number of other facilities also aided the project. Most of these are acknowledged in the "Series Bibliographic Essay," which appears early in this volume.

At NASA Headquarters, a number of people in the NASA History Office deserve credit. M. Louise Alstork painstakingly edited the essays and proofread all the documents. Jane Odom, Colin Fries, and John Hargenrader helped track down documents from our Historical Reference Collection. Nadine Andreassen provided much valuable general assistance and helped with the distribution. In the Office of Aerospace Technology, Tony Springer served as an invaluable sounding board on technical aeronautics issues. We also owe a special debt to Roger D. Launius, the former NASA Chief Historian, who provided the initial impetus and guidance for this worthy project.

At Headquarters Printing and Design Office, several individuals deserve praise for their roles in turning a manuscript into a finished book. Anne Marson did a careful job in copyediting a lengthy and detailed manuscript. Melissa Kennedy, a graphic designer, performed her craft in an exemplary manner. Jeffrey McLean and David Dixon expertly handled the physical printing of this book. Thanks are due to all these devoted professionals.

The Wind and Beyond
Series Introduction

Talking with the Wind, Collaborating with Genius

Aerodynamics is not, strictly speaking, about the wind. Rather, it concerns the motion of the air (or other gaseous fluid) plus the forces acting on a body in motion relative to the airflow.[1] Still, in choosing a title for this multivolume documentary history of aerodynamics development in the United States, we ultimately concluded that no title would be more suggestive than *The Wind and Beyond*.

Historically, the most basic instrument of aerodynamic research, by far, has proved to be the wind tunnel. Many of the scientists and engineers who achieved vital discoveries in aerodynamics described their breakthroughs in terms of a special capacity to "talk with the wind." By this, they meant visualizing what the air was doing—using the mind's eye as part of a creative process by which the air virtually told them what needed to be done in order to make an airplane fly effectively. "Beyond the wind" suggests many salient themes: the profound human curiosity that gave birth to the science of aerodynamics and, before that, to the dream of flight itself; the countless puzzles and mysteries hidden deep within all natural phenomena, both in macrocosm and microcosm, plus the myriad theories and concepts devised to fathom the forces at work; the countless technological forms resulting from inventiveness as human ingenuity moved toward mastering those natural forces, enabling humanity to leave the ground and soar above the clouds; and the potential for flight above the atmosphere and into space, where the ambitions of humankind may someday result in colonization of other worlds and perhaps contact with extraterrestrial—mythically speaking, even a sort of divine—intelligence. All this, and much more, is evoked by the title *The Wind and Beyond*.

The poetry in the title is not original to this publication. In 1967, writer Lee Edson selected it for the autobiography of Theodore von Kármán, one of the most brilliant thinkers to make the sojourn into aerodynamics during the twentieth century. (Von Kármán had just died, in 1963, at age eighty-two; Edson became his co-author.) Recounting in the first person the amazing life story of the colorful Hungarian-born

---

[1] Webster's *Third New International Dictionary* defines "aerodynamics" as "the branch of dynamics that deals with the motion of air and other gaseous fluids and with the forces acting on bodies in motion relative to such fluids."

scientist who moved to the United States in 1930, only to tower over much of its aeronautical science for the next four decades, *The Wind and Beyond: Theodore von Kármán, Pioneer in Aviation and Pathfinder in Space* (Boston and Toronto: Little, Brown and Company) quickly became a classic. Behind its popularity, which lasts to this day, was von Kármán's restless intellectual dynamism; his humor and eccentricities (which were at least as awesome as his intellect); and his ability to express himself with extreme cogency (even amid his thick accent), whether mathematically or verbally. He made a number of fundamental breakthroughs in aerodynamic theory and, from his post at the Guggenheim Aeronautical Laboratory at the California Institute of Technology (GALCIT), he exerted massive influence over American aeronautical and early space research and development (R&D), especially during the adolescence of the United States Air Force. These facts encouraged readers to care about his remarkably storied career. As American aerodynamicist and aerospace historian John Anderson has written, "The name von Kármán has been accorded almost godlike importance in the history of aerodynamics in the United States, and that aura existed well before his death in 1963 and continues today because of the magnitude of his contributions to, and advocacy for, aerodynamic research and development. Also, his influence is seen today in the contributions of his students, many of whom have gone on to leadership roles in the field."[2] Besides his autobiography, two biographies of von Kármán have been written.[3] This coverage reveals his powerful place in the field, because certainly no other aerodynamicist has enjoyed such a spotlight. Most are still in the figurative dark, not yet having been chosen by a single biographer.

What audacity, then, we have in redeploying the title The Wind and Beyond! When we first suggested to colleagues in aerospace history and to the NASA History Office that we wanted to use it for our documentary study of aerodynamic development in the United States, thoughtful people raised eyebrows and asked serious questions about the propriety of using a title so closely identified with another book, especially one as adored as von Kármán's autobiography. Even though nearly forty years had passed since its original publication, reverence for the book was too great, some said, to adopt its title.

In the end, however, we decided that we could pay no greater respect to von Kármán's memory and to the vibrant intellectual passions of his extraordinary life than by using his title once again. In this way, we hope to refresh the memories of those already aware of his book and his other legacies—for some individuals

---

[2] John D. Anderson, *A History of Aerodynamics and Its Impact on Flying Machines* (Cambridge, MA: Cambridge University Press, 1997), p. 420.

[3] See Paul Hanle, *Bringing Aerodynamics to America* (Cambridge, MA: MIT Press, 1982) and Michael H. Gorn, *The Universal Man: Theodore von Kármán's Life in Aeronautics* (Washington, DC: Smithsonian Institution Press, 1992).

Series Introduction: Talking with the Wind, Collaborating with Genius

now in their golden years, memories of a teacher that are quite personal. But even more importantly, we thought it desirable to bring *The Wind and Beyond* to the attention of a new generation of students now that we find ourselves in the twenty-first century—some forty years since von Kármán's death. We feel that the great aerodynamicist, a very generous spirit, would approve.

Dr. Theodore von Kármán (1881–1963), shown here at his blackboard at the California Institute of Technology, personified the restless intellectual dynamism that turned the study of aerodynamics into one of the classic expressions of 20th century science and technology. NASA Image #P30570B (JPL)

Anyone with the curiosity, and the fortitude, to venture into a serious study of aerodynamics—or, as in this publication, into its marvelously intricate historical development—will certainly keep wonderful company. For somewhere deep in the heart of the adventure moves the spirit not just of von Kármán but also of many of the greatest geniuses of all time. Roughly 500 years ago, during the period of great intellectual and cultural growth in Europe known as the Renaissance, the prototypical "universal man," Leonardo da Vinci, became obsessed with the problems of flight. No subject fascinated him more. Not his study of the circulation of the blood and the action of the eye; not his discoveries in meteorology and geology; not his insights into the effects of the moon on tides; not his conceptions of continent formation; not his schemes for the canalization of rivers; not any of his other myriad inventive ideas; not even the brilliant artistic compositions for which people most remember him today—*Mona Lisa, The Adoration of the Magi, The Last Supper*. Art historian Kenneth Clark once called Leonardo "undoubtedly the most curious man who ever lived." And his greatest curiosity—and the curiosity left most unfulfilled—centered on the preconditions of flight.[4]

As we shall see in Chapter 1 of this work, what da Vinci, the quintessential artist-engineer, lacked was hardly imagination or ingenuity. He made amazing sketches of birds in flight and, on paper, designed remarkable flying machines, including a flapping-wing "ornithopter," a parachute, and a helicopter. What the great maestro and all other aeronautical conjurers of the next four centuries lacked was a *science* of flight that could turn fantasy into reality. As von Kármán would explain in his autobiography: "For advances in aviation in any real sense it was necessary for science to catch up with the dreamers and experimenters and set a basis for further development in a rational manner."[5]

In critical ways the Scientific Revolution of the seventeenth and eighteenth centuries primed the pump for the emergence of the science of flight; but in at least one key respect, that great intellectual fermentation, one of the greatest in all human history, also retarded it. With his monumental writings of the late 1600s, Isaac Newton, the culminating figure of the Scientific Revolution, gave birth to the science of mechanics. But in doing so he made calculations and expressed ideas about "upward forces" (what later scientists would call "lift") and "forces preventing motion" (later called "drag") that many would read to mean that human flight by means of a supporting wing was impossible. And, of course, everyone took Sir Isaac's conclusions very seriously, almost as divine guidance.

No real breakthrough occurred until fellow Englishman George Cayley, more than a hundred years later, showed that the amount of lift attained by a forward-moving plate was directly proportional not to the square of the angle, as Newton

---

[4] On Leonardo's fascination for flight see Serge Bramley's biography *Discovering the Life of Leonardo da Vinci* (New York, NY: Harper Collins, 1991) as well as Irma Richter, ed., *The Notebooks of Leonardo da Vinci* (Oxford: Oxford University Press, 1952; "World Classics Edition," 1980). Readers interested in a psychological treatment of Leonardo's dreams of flight, see Sigmund Freud, *Leonardo da Vinci: A Study in Psychosexuality* (New York, NY: Vintage Books, 1961). Perhaps surprisingly, Freud's analysis does not try to reduce Leonardo's genius to any sort of pathology; rather, it provides a sensitive and respectful inquiry into da Vinci's creative greatness.

[5] Theodore von Kármán with Lee Edson, *The Wind and Beyond: Theodore von Kármán, Pioneer in Aviation and Pathfinder in Space* (Boston and Toronto: Little, Brown and Company, 1967), p. 58.

Series Introduction: Talking with the Wind, Collaborating with Genius

reasoned, but to the angle made by the plate. An airplane wing need not be impossibly large and heavy as Newton had posited; its weight could be sustained by the force of lift, and its drag compensated for by a propulsion device actuated by a powerful enough engine. Although it would be several decades before Cayley's principle of the modern airplane achieved the status of a paradigm, one can argue, as von Kármán himself did, that it was at this point, during the first half of the nineteenth century, that science moved "off the sharp horns of the dilemma created by Newton."[6] In this sense, Cayley's notions were truly revolutionary.

If anyone can be known as the "father of modern aeronautics," it is Sir George Cayley (1773–1857). Not only did he set down the fundamental concepts that defined the airplane in its present-day form, but he was also the first to link a scientific study of aerodynamic principles with the actual design and operation of flying machines. SI Negative No. 75-16335

Cayley built and flew a few gliders of his own design, some of them manned, and so did several other intrepid experimenters during the 1800s. The Industrial Revolution was sweeping across Europe, America, and other parts of the globe, bringing with it greater and greater fascination for machines of all kinds. The special interest in new forms of faster transportation spurred all kinds of efforts to develop aviation into a practical technology. Piloted balloons like those originally built by the Montgolfier brothers in France had been aloft since the 1780s, and even military establishments were finding ways of deploying them. Another form of lighter-than-air craft, the steerable airship or "dirigible," proved even more practical. The impetus for learning more about what it took to fly grew ever stronger.

What brought experimenters, finally in the late nineteenth century, to the birth of a formal study of "aerodynamics" was the matter of lift—specifically the question, "Why do curved surfaces, in particular, perform so well in flight?" Cayley had shown that the curved shape of trout enjoyed low drag in water. This prompted applications of the shape to airplanes. In the 1880s and early 1890s, the great "Bird Man," German engineer Otto Lilienthal, built and flew a number

---

[6] von Kármán, *The Wind and Beyond*, p. 58.

of effective hang gliders with wings having large curvatures. His flight experiments showed that, compared to a flat plate of the same size, a large curved surface provided significantly greater lift for the same speed. But Lilienthal could not show why. As von Kármán himself noted, the search for this why "led to the study of what we now call aerodynamics, the science of flight."[7]

It is the birth of this science and, even more, the unfolding of the century of rapid development in flight technology that followed it that form the storyline of the volumes to come.

Quickly, and quite necessarily, aerodynamics became highly mathematical. But, as readers of our volumes will perceive, the story involves much more than abstruse theories and differential equations. Unlike other modern technological developments, in which theory mostly preceded practice (often by several decades), aeronautical practice moved largely in parallel with theory, and even occurred before the discovery of many fundamental laws and theories. It had not happened this way in other science-based fields like thermodynamics, electricity, or atomic power, but, for some reason, it did with aerodynamics. Von Kármán recognized this as "an extraordinary thing," the fact that "scientific knowledge and its technological application proceeded almost in parallel."[8] He did not begin to try to explain the phenomenon historically—nor has anyone else. Perhaps it can be explained simply by an age-old quest: we were too anxious to fly. We had dreamed about it for too long, from time immemorial, to wait around for detailed explanations. We were plunging ahead, into the wind and beyond, even if we did not yet understand in detail how it worked.

But that excited energy took us only a little way up. Not until aerodynamicists existed as professional scientists and discovered the why behind lift—and a small but growing community of aeronautical practitioners could truly benefit from "its first real understanding of what makes flight possible"—did conditions become ripe for the "amazingly swift progress" to come.[9] We are fully aware of this synergy, and a leitmotif of our study will be the critical relationship between theory and experiment. Without a fruitful interplay between the two, progress in virtually all science is spotty and can grind to a halt. That certainly proved to be true over and over again with flight technology. In aerodynamics, specifically, we will see this most clearly in the matter of wing design where designers often, especially early in the century, shaped their airfoils independent of theory, sometimes with fortunate results but often with extremely poor—even fatal—ones. As one of the documents

---

[7] von Kármán, *The Wind and Beyond*, pp. 58–59.

[8] von Kármán, *The Wind and Beyond*, p. 59.

[9] von Kármán, *The Wind and Beyond*, p. 59.

(from 1933) in a later volume suggests, "A large number of investigations are carried out with little regard for the theory, and much testing of airfoils is done with insufficient knowledge of ultimate possibilities."[10] How theory came to inform experiment in constructive ways over the course of the twentieth century should become apparent in our study. But the back-and-forth between theory and experiment remains a central concern today; to discern the most helpful balance between the two approaches to gathering knowledge will perhaps forever be an issue for scientists and engineers to resolve.

In conceiving the nature of our study, we determined very early on that our documentary history of aerodynamics should not, and could not, dwell only in the realm of ideas, concepts, equations, principles, and the like. As absolutely critical as they were to the development of a science of flight, their historical evolution—apart from the social drive to turn them into the physical realities of actual flight hardware, e.g., airplanes, airships, helicopters, autogiros, missiles, and spacecraft—made no sense. Our study would then only concern "the wind," but not what lay "beyond" it in terms of technological forms or their socio-economic, political, military, and cultural contexts.

In adopting a broader approach, we felt secure knowing that John D. Anderson in his 1997 *History of Aerodynamics . . .* took exactly the same tact, as can be seen in the rest of his title *. . . and Its Impact on Flying Machines*. In the preface to his book, Anderson wrote: "In addition to examining the history of aerodynamics per se and assessing the state of the art of aerodynamics during various historical periods, the book seeks to answer an important question: How much of the contemporary state of the art in aerodynamics at any given time was incorporated into the actual design of flying machines at that time? That is, what was the impact of aerodynamic knowledge on contemporary designs of flying machines?"[11]

From the start we determined that we wanted to show through our flow of documents not a linear development of technology but rather a matrix of dynamic relationships involving not just ideas, but also organizations, social and economic forces, political influences, and more. With this in mind, and recognizing that the general perception of aviation history is focused on the evolution of aircraft, we began work, ironically, on a list of aerodynamically significant American aircraft. We consulted numerous sources, notably *Milestones of the Air: Jane's 100 Significant Aircraft* (1969), a compilation that included thirty-two American aircraft; *Progress*

---

[10] Theodore Theodorsen, "Theory of Wing Sections of Arbitrary Shape," NACA Technical Report No. 411, printed in *Eighteenth Annual Report of the National Advisory Committee for Aeronautics, 1932* (Washington, DC: the NACA, 1933), p. 29.

[11] Anderson, *A History of Aerodynamics and Its Impact on Flying Machines* (Cambridge, MA: Cambridge University Press, 1997), p. xi.

in Aircraft Design Since 1903 (1973), prepared by the NASA Langley Research Center, which covered ninety aircraft, seventy-six of which were American designs; and *Quest for Performance: The Evolution of Modern Aircraft* (1985), which evaluated ninety-three American aircraft in addition to a large number of foreign designs.

The list of fifty-two aerodynamically significant U.S. aircraft we came up with early in our project is appended to this introduction. We used the list not as a definitive categorization that defined the contents and structure of our narrative, but as a flexible working outline that helped us organize and focus our research effort as we moved out into aerospace archives and records centers across the United States in search of significant documents. Our objective in highlighting certain aircraft was to use them as subjects in a "pedigree chart," to use a genealogical analogy, illustrating the vital interactions between people, institutions, and ideas in the historical unfolding of American aviation technology. The fifty-two aircraft we chose represented different developmental stages of particular technologies in this "family history" of the evolution of aerodynamics. Some highlighted the primogeniture of certain technological concepts, while others illustrated developments in maturity. Many of the aircraft were significant because of the knowledge and techniques engendered through their use in research, rather than simply because of their specific performance capabilities. We treated some closely related sequences of aircraft (such as the DC-1/2/3, or Convair XF92/F102/F106) as one subject because they represented a continuum of technological developments. We hoped that the research lines emanating from each identified aircraft would provide links to an equal number of *nonaircraft* highlights, such as identification of the key people involved, the development of testing facilities and programs, and the broader social impact and meaning of the machine's operation.

Throughout our research and document-collecting phase, we kept this list of aerodynamically significant American aircraft close by our side. But it never defined or handcuffed our approach to the subject. We did not want our work to become just "another wearisome airplane book."[12] We hoped to contribute to an emerging historiography committed to fostering "Aviation History in the Wider View."[13] This approach moved the scholarship beyond narrow "gee-whiz" fascination with aircraft types to the "people, ideologies, and organizations" as well as the deeper social and intellectual roots of the technology of flight.[14] This became our

---

[12] See Richard K. Smith's review of James Sinclair's *Wings of Gold: How the Aeroplane Developed in New Guinea* (Sydney, Australia, 1980) in *Technology and Culture* 22 (1981): 641–643. Smith praises Sinclair's book for working in economic, engineering, and social history, and for not being just "another wearisome airplane book."

[13] See James R. Hansen, "Aviation History in the Wider View," *Technology and Culture* 30 (1989): 643–656.

[14] See Joseph Corn's review of Roger Bilstein, *Flight in America 1900–1983: From the Wrights to the Astronauts* (Baltimore, MD, 1984) in *Technology and Culture* 26 (1985): 872.

Series Introduction: Talking with the Wind, Collaborating with Genius

goal—to the extent we could accomplish a significant historical treatment in a documentary study involving something as technical as aerodynamics. A careful study of aircraft themselves was, without question, critically important to a serious understanding of aerodynamic development, but aircraft, as complex and dynamic as they are, are not alive. They never, once, have made history on their own. In our study, the primary roles are played by the people who provided the aerodynamic concepts; those who designed, built, and then used the airplane; and all the institutions fundamental to those activities.

We apologize up front, especially to our international readers, for our focus on American topics. But the magnitude of the task of adequately covering aerodynamic development in the United States alone was so profound that we simply could not hope to relate the global saga that aerodynamic development actually experienced. The American story was epic enough. And one certainly does not need to apologize for the American record. From the Wright brothers to the present-day pioneers, no national community has contributed more to aerodynamics and to resulting flight technology than the United States.

Still, the American experience has always connected to a larger world, and has been influenced in many critical and formative ways, especially in the early decades of the century, by foreign—particularly European—developments. We do our best in this study to relate the most essential connections, and to give all the credit that is due to the scientists, engineers, and institutions of the many other countries that have participated in the study of flight. In our volumes, readers will find numerous references to international events and developments, relating not only to countries whose aeronautical achievements are well known, such as Great Britain, France, Germany, Italy, Russia (and the former Soviet Union), and Japan, but also to the People's Republic of China, Sweden, the Netherlands, Poland, Israel, Brazil, and several others. Certainly, the cast of featured characters in American aerodynamics will itself be international: von Kármán from Hungary; Igor Sikorsky, Wladimir Margoulis, and Alexandre Seversky from Russia; Max Munk and Adolf Busemann from Germany; Alexander Lippisch from Switzerland; Theodore Theodorsen from Norway; Antonio Ferri from Italy; Hsue-tsen Tsien from China; among others. In embarking on their new lives and careers in America, many of them brought ideas, values, and techniques learned while studying or otherwise associating with some of the world's greatest aerodynamic thinkers: Germany's Ludwig Prandtl, Switzerland's Jakob Ackeret, England's Frederick William Lanchester, Russia's Dimitri Riabouchinsky and Nicolai Joukowski, and many more. In addition to coming into contact with the work of these foreign thinkers—either directly or indirectly—native-born American aerodynamicists engaged in a professional discipline that was, like most scientific fields, increasingly international in character. They encountered books, articles, and certainly hundreds and hundreds of foreign technical reports, both translated into English and in their foreign tongues. Many of them also traveled to international meetings and symposia where they met the world's leading lights face to face. Some of these forums, such as the Volta Congress on High-Speed Aeronautics, held in Italy in 1935, proved historic in their significance for future technological developments in America and elsewhere.

Our focus had one other major limiting factor. Because the project had been sponsored by the history program of the National Aeronautics and Space Administration (NASA), we set out to highlight as much as possible the role of NASA—and its predecessor organization, the National Advisory Committee for Aeronautics (NACA)—in the development of aerodynamic theory and practice. In exploring this, we hoped to contribute additional insights into the issue of NACA/NASA's overall significance in America's "progress" into air and space. In league with a growing body of historical scholarship that has mostly affirmed the positive significance of these two related organizations, our study confirms that the NACA and NASA were a vital part of the broader spectrum of aerodynamic research activities involving both civilian and military aviation and aerospace organizations in the United States. Without direct government involvement in the form of well-equipped and rather expansive research laboratories dedicated to "the scientific study of the problems of flight with a view to their practical solution" (Law establishing the NACA: Public Law 271, 63d Congress, approved 3 March 1915), many fundamental aspects of aerospace science and technology, most involving aerodynamics, would not have been addressed as quickly or as comprehensively. In the early decades of the century, the U.S. aircraft industry was simply too fledgling to deal with a wide range of fundamental questions. Even after that industry had exploded in size and capability, the short-term financial interests of modern corporate capitalism would not engender the kind of long-term investment in research and development that NACA/NASA laboratories provided. The proper and most effective role of government in aerodynamic R&D remains a vital issue today, and we hope the historical insights provided by our study will inform contemporary discussion of it.

Within the framework of our contract with the NASA History Office, we tried to branch out from the NACA/NASA focus to include sufficient information about aerodynamic developments related to the U.S. military, universities, the aircraft manufacturing industry, the airlines, and other pertinent government agencies such as the U.S. Bureau of Standards, the U.S. Department of Transportation, and the Federal Aviation Administration (and its precursor, the Civil Aeronautics Authority). But we concede that the attention we paid to these entities is too limited. Nonetheless, we hope that, overall, readers will come away with a feel for the whole scope of aerodynamic history—NACA/NASA and otherwise, American and otherwise.

Series Introduction: Talking with the Wind, Collaborating with Genius

We would like to explain how even the admittedly confined scope of our study ramified beyond anything anyone had originally intended.

In 1997, the NASA History Office contracted with us (through the Science Technology Corporation) to produce a documentary history of American aerodynamic development with a focus on NACA/NASA contributions. There were at least two major reasons for the request. First, the History Office's publication of the first two volumes of *Exploring the Unknown: Selected Documents in the History of the U.S. Civil Space Program* (NASA SP-4407; Vol. 1, 1995; Vol. 2, 1996), edited by John M. Logsdon, et al., were well received both inside and outside of NASA. These extensive documentary volumes (three more volumes have since been published) quickly became an essential reference for anyone interested in the history of the U.S. space program. Given that NASA's space activities had been covered much more extensively in NASA history books than had its aeronautical activities, Dr. Roger Launius, then the NASA chief historian, wisely felt that a complementary series of documentary histories devoted to aeronautics would prove valuable. Launius chose to start with aerodynamics, the central component of all aeronautical technology. Other volumes, dedicated to propulsion, stability and control, structures and materials, supersonics, and hypersonics, will follow.

Another inspiration for what became our project was the upcoming Wright centennial. In 2003, the United States and the rest of the world will be celebrating the 100th anniversary of the historic first powered flight by the Wright brothers at Kitty Hawk, North Carolina, on 17 December 1903. Although perhaps not comparable in magnitude to the United States Bicentennial in 1976 or the fifty-year commemoration of the end of World War II celebrated in 1995, the Wright centennial will certainly be a major national and international event. All over the world scholars and aviation experts will be discussing the significance of flight, assessing the course of its historical development, and projecting aviation's future. Major conferences will undoubtedly take place on the centennial of powered flight. In October 1998 Wright State University in Dayton, Ohio, the Wright's hometown, launched the anniversary events with a National Aerospace Conference. Its theme, "The Meaning of Flight in the 20th Century," attracted some 400 historians, policy planners, aerospace professionals, and just plain interested people.[15] The following month, on 13 November 1998, President William J. Clinton signed into law the Centennial of Flight Commemoration Act (PL 105-389). This act established a commission responsible for planning all national events involved with the Wright centennial celebration.

From the start of our project, we kept the Wright centennial in mind—and not just as a deadline for publication. We believed, as NASA did, that a documentary history

---

[15] *National Aerospace Conference Proceedings*, Wright State University, April 1999.

of aerodynamics, one which started with "The Achievement of Flight" and that took the reader through the entire twentieth century, to the present, and then beyond to our future in flight, would be a worthy contribution to the national celebration of the Wright centennial.

Originally, our contract called for a single volume incorporating between 150 and 250 documents. Initially, we planned eight chapters. Early into the project, however, we knew that this framework was too confining. In producing rough drafts of our first three chapters, we were forced to address two facts: that our resulting book would grow to become extremely large, much larger than first anticipated, and that some important topics in the overall history of aerodynamics could not be well covered in a single volume structure. For the first three chapters alone, it seemed we would need to publish in the neighborhood of 150 documents. However, the latter point became the key issue for us. Expanding the work into two volumes with a total of twelve or more chapters would result in a significantly better history of aerodynamics—as well as two volumes that were much handier for the reader.

No celebration of the Wright centennial would be complete without fully documenting the epic story of how aerodynamics progressed with such incredible speed and import, not just to the drama at Kitty Hawk on 17 December 1903, but also for the 100 years of systematic development following the historic first achievement of powered flight. NASA Image #65-H-611

Series Introduction: Talking with the Wind, Collaborating with Genius

At this point, a key concept emerged to guide our organization of the two volumes. The volumes would cover, respectively, "The Aerodynamics of Propeller Aircraft," and "The Aerodynamics of Jets and Rockets." In the first volume we would then have room not only to cover the mainline stories but also give adequate coverage to important aerodynamic developments in lighter-than-air research, flying boats, and rotorcraft, among others. It was evident to us, as we were sure it would be to any informed student of aviation history, that each of these areas possessed tremendous historical value and interest, and were vital components of technological developments related to propeller-driven aircraft.

The second volume would begin with the dramatic story of the high-speed research that led to the development of dedicated research aircraft and the breaking of the sound barrier. It would include chapters on transonic research facilities and supersonic flight, plus chapters covering the first generations of military and civil jet aircraft, the quest for a commercial supersonic transport, hypersonics, and the development of today's most advanced wind tunnels and other aerodynamic research tools. It would conclude with a special chapter looking ahead to the future of flight.

As bundles of photocopied documents continued to grow (and fill up our file cabinets) as a result of our research trips into archives and depositories around the country, a truly radical idea for our publication surfaced. What we were producing, whether we liked it or not (and whether or not the NASA History Office could find the funding to publish it in such an expansive format) were two multi-volume boxed sets: one, with the general introduction, series reference material, and chapters devoted to propeller aircraft; and the other, with chapters covering the aerodynamics of jets and rockets. All told, the various chapters included over 350 document strings (groups of related documents) comprising almost 1,000 individual documents—or nearly eight times as many as anticipated in the initial concept. In keeping what was in fact the original format for the one-volume concept, each volume would be introduced by a full-length analytical essay and substantial introductory headers for each document string.

At this point, the practical demands and costs of publishing intervened, and our view of the project's organization evolved into an outline of six separately bound volumes, each containing two chapters. The plan became for the first three volumes to cover the aerodynamics of propeller-driven aircraft, while volumes four through six covered the aerodynamics of jets and rockets. Also planned was a general introduction with bibliographic information included as a third section in the first volume and an epilogue included as a third section in the sixth volume.

The more we thought about this revised and greatly expanded format, the more we liked it, and came to consider it absolutely vital to the success of our publication. Besides offering much more historical content, it seemed to us most "user-friendly." We wanted the students of aerodynamics history to know by looking just at the physical makeup of our work that its authors intended for it to be *read*, that each volume in the series tells a story that can be appreciated independently of the others. A reader can start with the first volume to follow the document narrative we put together in "Chapter 1: The Achievement of Flight," or turn to the fourth volume to read "Chapter 8: The Supersonic Design Revolution." In sum, though we offer a reference work to those who want it, we also offer an integrated series of independent volumes. Each tells a story worth reading. (We were flattered beyond description when one of our referees, after reading our draft of Chapter 1, told us "it was the best single shorthand telling of the invention of the airplane" that he had ever read.) We chose the documents for every chapter to serve as critical stepping-stones through an important, often complex, narrative that would be compelling to the reader every step of the way.

Because, in the end, we wished for our readers, too, to "talk with the wind," as we felt we had been doing these past few years while engaging the history of aerodynamics. As suggested earlier in this introduction, it is our dearest hope that Dr. von Kármán himself would heartily endorse our use of his title, *The Wind and Beyond*, as a token of our great affection, and that he would warmly welcome this collection of books into his library. There, cigar in hand and with a wryly quizzical look on his face, he would spend evening after evening reading through our volumes word for word, realizing that each tells a great story full of remarkable ideas and people—and that together our volumes represent an epic intellectual journey, a crucial episode in humankind's quest to conquer nature and fly. We flatter ourselves to imagine him nodding frequently in tacit recognition and approval, perhaps even learning a few things here and there that he himself never knew—though this seems hardly possible, knowing the universal experience of the man. Optimistically, we believe he would excuse our minor errors and mistakes as instructively as he did the scientific and technical papers written by the prized pupils at Caltech who came to love him so much.

Writer Lee Edson began von Kármán's autobiography with a tribute entitled, "Collaboration with a Genius." In sum, that is what we see ourselves doing in this documentary history: collaborating not only with the genius of von Kármán but with all the other tremendously inventive minds that played a part in the development of aerodynamic theory and practice from before the Wright brothers up to the present. May there be many more in the new century at hand.

Significant Aircraft List

The list of Aerodynamically Significant American Aircraft was compiled to aid the development of a research plan for the AU/NASA History Project. As such, the list is not intended to be a definitive categorization, but rather a flexible working document to aid in organizing and focusing our research effort. The objective of highlighting certain aircraft in this way is to use them as subjects in a "pedigree chart" illustrating the dynamic relationship of people, institutions, and ideas in the history of technology.

The aircraft themselves represent different developmental stages of particular technologies in this "family history" of aerodynamics. Some highlight the primogeniture of certain technological concepts, while others illustrate developments in maturity. Many of the aircraft are more significant because of the knowledge and techniques engendered through their use in research rather than their specific performance capabilities. Along these development lines there may be other aircraft that will serve just as well, if not better, than the ones listed. Some closely related sequences of aircraft, like the Douglas DC-1/2/3, or the Convair XF92/F102/F106, are treated as one subject on the list because they represent a continuum of technological developments.

The listings on the accompanying page were prepared from a number of different sources, each representing slightly different perspectives. The different sources were compared to each other both to develop and to filter the list. Over 155 aircraft "candidates" were considered in drawing up the list, fifty-two of which appear on the following page. One of the considerations during this initial filtering was to reduce the list to a more manageable proportion.

The first source used is *Milestones of the Air: Jane's 100 Significant Aircraft* (1969). This compilation includes thirty-two American aircraft and, of this number, twenty-four are included on our list. The eight aircraft not included were deleted because there were no collaborating entries from the other sources. Two of the aircraft not listed, however, should perhaps still be considered for their aerodynamic significance: the Sikorsky VS-300 and the Lockheed U-2.

The next two sources are NASA publications: *Quest for Performance: The Evolution of Modern Aircraft*, by Laurence K. Loftin, Jr. (1985); and *Progress in Aircraft Design Since 1903*, prepared by the Langley Research Center (1974). Loftin's study of the period from 1914 to 1980 covers ninety-three American aircraft, in addition to a large number of foreign designs. One category of aircraft accorded considerable attention by Loftin, but not well represented in the other sources, is that of flying boats and amphibians. Few of these types appear on the following list, but they deserve, and will get, more attention in a later volume. The Langley publication covers ninety aircraft, seventy-six of which are American designs. Of interest is the fact that the Langley listing does not duplicate Loftin's, and includes a number of aircraft not considered in *Quest for Performance*. The aircraft from these two sources that appear on the following list are those that were also included in other lists.

The final listing of twenty-four aircraft is based on the previous lists and represents an attempt to identify flight vehicles associated with specific NACA/NASA aerodynamic research programs. The intent of this filter was to focus further recognition of the aircraft as technological representatives of a dynamic interrelated train of research and ideas. A second consideration was to help direct initial research efforts to document trails that are most likely accessible in NASA collections.

Significant Aircraft List

Aircraft	Year	Jane's	Loftin	LRC	NACA/NASA Aerodynamics
Wright Flyer	1903	✓		✓	
Curtiss JN-4	1916		✓	✓	Flight tests
Curtiss NC-4	1919	✓	✓	✓	
Dayton-Wright RB-1	1920	✓	✓		
Curtis R2C-1	1923	✓	✓		
Sperry Messenger	1923				Airfoils, Full-scale wind tunnel
Ford Tri-Motor/5-AT	1926	✓	✓	✓	
Ryan NYP	1927	✓	✓	✓	
Lockheed Vega	1927	✓	✓	✓	
Curtiss AT-5	1927				NACA Cowling
Lockheed Air Express	1927				
Pitcairn PCA (*Cierva)	1931	✓		✓	Rotary-wing tests
Taylor/Piper Cub	1931	✓	✓	✓	
Boeing 247	1933	✓	✓		
Douglas DC-2/3	1933	✓	✓	✓	
Sikorsky S-42	1934	✓	✓		
Boeing B-17	1935	✓	✓	✓	Flaps, Cowls, Airfoil

Aircraft	Year	Jane's	Loftin	LRC	NACA/NASA Aerodynamics
Brewster XF2A	1938	✓			Drag clean-up
Consolidated B-24	1939		✓	✓	
Lockheed P-38	1939		✓	✓	Compressibility
North American P-51	1940	✓	✓	✓	Laminar flow
Republic P-47	1941			✓	
Hughes H-1	1941				
Boeing B-29	1942		✓	✓	Scale model testing, etc.
Lockheed L 1049/C-69	1943		✓	✓	
Lockheed P-80	1944		✓	✓	
Bell X-1	1946	✓		✓	Supersonic flight
North American F-86	1947		✓	✓	Swept wings
Boeing B-47	1947			✓	
Convair XF-92/102/106	1948	✓	✓	✓	Delta wing/area rule
Northrup N-1M	1950				
Bell X-5	1951	✓			Variable geometry wing
North American F-100	1953		✓	✓	
Lockheed F-104	1954	✓	✓		High-speed tests
Boeing 707	1954	✓	✓	✓	
Vought F-8	1955		✓	✓	Supercritical wing
Convair B-58	1956	✓		✓	
Bell Model 204/UH-1	1956			✓	Ogee tip
McDonnell F-4	1958	✓	✓	✓	
North American X-15A	1959	✓		✓	Hypersonic upper atmosphere
Sikorsky S-60/64	1959	✓		✓	
Lockheed YF-12/SR-71	1963			✓	Hypersonic flight
General Dynamics F-111	1964		✓	✓	TACT, Transonic research
North American XB-70	1964		✓		Supersonic cruise

Aircraft	Year	Jane's	Loftin	LRC	NACA/NASA Aerodynamics
Northrup HL-10	1966	✓			Lifting body
General Dynamics F-16	1974		✓		
Rockwell B-1	1974		✓		
Rockwell STS	1977				Space Shuttle
Bell XV-15	1978				Tilt-rotor
Gates Learjet 55	1980				Whitcomb winglet
Grumman X-29	1984				HiMAT, High maneuverability
Lockheed F-117	1988				

Series Bibliographic Essay: Days on the Wing

Journeys to America's Aviation Archives

Many historians and other scholars would admit that, next to seeing their work finally appear in print, research is the most enjoyable experience of their enterprise. That was the case for all of us with this project. The comprehensive nature and scope of our documentary project required the members of the project team to travel to and spend extended periods of time in those archives and libraries across the United States that held significant materials on the development of aerodynamics. These repositories included public institutions such the National Archives and Records Administration (NARA) and the National Air and Space Museum (NASM) as well as private facilities such as the Boeing Company's Historical Archives. The "days on the wing" that members of the project team spent at these places resulted in our accumulating the bulk of the material readers will see in the following volumes. Those days of research and photocopying also provided us with some unprecedented impressions not only of the state of aerodynamics archives in the United States, but also of aeronautics archives in general.[16]

Represented in the volumes of our documentary history is a cornucopia of historical materials, of various types and in various formats. In the archives we found diverse manuscript material in the form of correspondence, journal and diary entries, meeting minutes, personal scrapbooks, and pilot reports. We also found many relevant published primary sources such as periodical and newspaper articles, autobiographies, and transcripts of oral histories. Our primary criterion in the research phase was to find documents that, either alone or in combination, could illuminate some significant point or historical moment that contributed to the development of aerodynamics in the United States. Project researchers, which included three Ph.D. students in the history of technology at Auburn University working under the supervision of project director Professor James R. Hansen, also sought related materials to support the documents—photographs, negatives, blueprints, technical manuals, motion pictures, video records, sound recordings, electronic data, etc. Though very few of these ancillary materials

---

[16] This section's title has been inspired by the title of the World War I memoirs of Belgian fighter ace Willy Coppens de Houthulst, *Days on the Wing* (London: J. Hamilton, 1934).

could be included in the documentary volumes to follow, given the medium of textual presentation, everything we collected helped us to see the historical development of aerodynamics in a fuller context.

We went about identifying and locating significant documents in the history of aerodynamics through several methods. First, we looked to the previous historical record by searching through the footnotes and endnotes of notable works in the field of aeronautical history. (An annotated list of significant secondary works related specifically to the history of aerodynamics is at the end of this essay.) Such a process allowed us to build a foundation for the evaluation of key documents as well as improve the research team's general knowledge of the subject. We went one step further and asked leading historians to suggest archives and document collections that they felt had been important to their research as well as to exploration of the subject in general. In recognition of their vital contribution, we have listed these individuals in our acknowledgements.

In a clear indication of the growing importance of digital information, we found the Internet to be a highly valuable resource and general research tool. The readily accessible technology in the form of our desktop PCs facilitated the search and acquisition of information pertinent to the project. This ranged from locating specific documents, to inventory listings for specific archives, to e-published articles related to aeronautical history and aerodynamic developments of the present and very recent past. The use of mainstream search engines and of keywords such as "aerodynamics," "wind tunnels," "laminar flow," "helicopters" brought us into a global matrix of information concerning the past, present, and future of aerodynamics and its social, economic, political, and military contexts. Much of this was in an electronic format that was instantaneously available. The project's specific use of Internet and other types of digital resources is mentioned at appropriate points through this essay as well as at various points in the text. E-mail also facilitated correspondence with archivists and librarians and kept us in our offices at home working at a pace that greatly helped the progress of the project.

Another valuable research tool for our process of document identification, location, and acquisition were two published descriptions of American aerospace archives. One of these was *A Guide to Sources for Air and Space History; Primary Historical Collections in United States Repositories* (1994 edition, sponsored by the National Air and Space Museum and edited by Cloyd Dake Gull), an extensive listing available in both published and digital form.[17] This guide highlights approximately 2,250 collections in over 370 repositories across the United States,

---

[17] Cloyd Dake Gull, ed., *A Guide to Sources for Air and Space History; Primary Historical Collections in United States Repositories* (Washington, DC: National Air and Space Museum, Smithsonian Institution, 1994), and at *http://www.nasm.edu/nasm/arch/ARCH_REPOS/gtsash1.html*

which span from the flight experiments of the nineteenth century to the space flight missions of the 1980s. First appearing in 1976, the last major edition of this bibliography, from 1994, was available to us on the Internet. Gull's listings are organized alphabetically, meaning that it takes some time for users of the bibliography to search and read through the descriptions. Also, we found that the descriptions were in no way comprehensive. Too often, Gull details a particular collection of papers thoroughly but neglects to describe other potentially significant collections found at the very same repository. Perspective researchers need to contact target archives for more specific information about their collections.

The second major bibliographical work, Catherine Scott's edition of *Aeronautics and Space Flight Collections* (1985) details the state of aerospace archives into the 1980s.[18] Scott opens the volume with a guide to regional aerospace archives across the United States; the guide is organized by subject (e.g., Wright Brothers, Charles A. Lindbergh) and by region (e.g., Northeast, Southeast). The volume includes specific essays describing the aeronautics collections in the Library of Congress, New York Public Library, United States Air Force Historical Collection, Gimble Collection at the United States Air Force Academy, National Air and Space Museum Library, and the History of Aviation Collection at The University of Texas at Dallas. Most of these essays are written by the archivists responsible for them.

There were limitations to using both Gull's and Scott's work. Neither one mentions aerodynamics specifically, meaning that further inquiry by the project team was necessary. In neither work is there a specific listing of aerodynamics archives per se. The closest thing we found to an aerodynamics archive cited in Scott's bibliography involved the volumes of wind tunnel test reports housed in the GALCIT Aero Library of the California Institute of Technology in Pasadena. But this facility was closed in the early 1990s and most of its holdings turned over to the Caltech Archives (although some went to aerospace corporations). As a result of such institutional transitions, along with corporate downsizing and loss of funding for records keeping, the archival landscape surveyed by Gull and by Scott had changed significantly by the time we started our project in 1997. Not just for our purposes but also generally speaking, by the mid-1990s both guides were woefully out-of-date and in need of revision and updating. Despite their limitations, however, these guides still qualify as points of departure for any individual starting a research project in aerospace history in the United States.

The research methodology of utilizing the previous historical record, suggestions from leading historians, and published guides to American aerospace archives

---

[18] Catherine D. Scott, ed., *Aeronautics and Space Flight Collections* (New York, NY: Haworth Press, 1985).

provided the preliminary foundation for building what would become a vast documentary collection related to the development of aerodynamics in America. From it we successfully moved forward to collecting many documents of known value and significance and many others that were forgotten or completely unknown. Our objective was not only to make the most important of them available to historians and present them to readers in an accessible form, but also to make sense of them, to contextualize them, to interpret them, and to place them within a meaningful narrative—telling the story of the development of aerodynamics in America, from before the Wright brothers up to the present.

Members of the project team visited private, corporate, and both civilian and military government archives to amass the documents that made the completed volumes possible. Virtually all of the repositories we visited employed enthusiastic caretakers dedicated to the preservation of aviation history, even though they all represent different interests, outlooks, and agendas in their records keeping. We found that the governmental, military, and corporate memory of American aerodynamics was principally alive and well. No one in the project experienced the typical horror stories about tyrannical and paranoid archivists hindering historical research. All of the archivists and specialists we met were open to the promise of our project. At each archive, the project team not only found many important documents but we also benefited from conversations with archivists and, in the process, learned considerably more about the aviation history we were researching.

Many of these repositories are located near historic places related to U.S. aeronautics. This enhanced our overall research experience greatly. We visited Huffman Prairie where the Wright brothers developed their aircraft after going back to Dayton from Kitty Hawk. We toured the dry lakebeds of Edwards Air Force Base in the Mojave Desert, where the Bell X-1 and so many other important experimental aircraft had test flown. We explored Boeing Field in Seattle, and we stood in the vast test section of the thirty- by sixty-foot Full-Scale Tunnel at Langley Research Center in Virginia. All of these first-hand experiences made the importance of the project all the more apparent and committed us anew to documenting the history of aerodynamics in America.

Although we specify the bibliographic origin of the many hundreds of documents reproduced and analyzed in our study, we have included a general exposition on the state of those American archives containing collections pertinent to the development of aerodynamics. We believe this would be helpful to scholars wishing to expand upon the themes and topics presented in our volumes. The following descriptive essay therefore discusses the primary institutions holding significant collections of these records. It begins by taking a look at the NACA/NASA materials we examined and is followed by summary inventories of what we found in military, corporate, and academic libraries and repositories.

NASA

NACA/NASA collections predominate in this documentary history. The very nature of our work's overarching theme—the development of aerodynamics in the United States—meant that considerable time would be spent in archival repositories maintained by the National Aeronautics and Space Administration (NASA). Throughout the twentieth century, NASA and its predecessor agency, the National Advisory Committee for Aeronautics (NACA), founded in 1915, have been the constant in aerodynamics research in the United States.

The NASA History Office at NASA Headquarters in Washington, D.C., directs an extensive history and archival program dedicated to the preservation of both NACA and NASA involvement in the development of flight.[19] NASA created its history office in 1959 and assigned it the duty of widely disseminating information about the agency's activities and about aerospace developments generally, under the mandate of the 1958 National Air and Space Act. A commitment to serving both governmental and public interests has ensured that NACA/NASA involvement in the development of aerodynamics has been preserved. The Headquarters History Office perpetuates this mission through its direction of history offices at all of NASA's field installations. This includes offices at Langley Research Center in Hampton, Virginia; Ames Research Center at Moffett Field, California; Glenn Research Center at Lewis Field in Cleveland, Ohio; and Dryden Flight Research Center at Edwards Air Force Base in California. These are the NASA facilities that have been most directly involved in aeronautics research. For a detailed synopsis of the NASA History Program and its archival holdings across the United States, see *Research in NASA History: A Guide to the NASA History Program* (1997), which is also accessible via the Internet.[20]

As an archival repository, the NASA History Office itself maintains extensive biographical and subject files. The biography files cover approximately 5,000 individuals from the 1800s to the present and comprise 180 linear feet of archival space. Arranged alphabetically, these files contain speeches, correspondence, articles, clippings, press releases, and photographs. The aeronautics collection is thirty-two linear feet and focuses on NACA/NASA activities from 1945 to the present. These records contain photographs, newspaper and *Congressional Record* clippings, articles, speeches, news releases, reports, studies, pamphlets, and brochures on specific topics such as SST, V/STOL, X-aircraft, aerodynamics, wind tunnels, and NACA contributions to historic aircraft.

---

[19] NASA History Office, Code IQ, NASA Headquarters, 300 E Street SW, Washington, DC 20546-0001, *http://history.nasa.gov/*

[20] *Research in NASA History: A Guide to the NASA History Program* (Washington, DC: NASA History Office, NASA HHR-64, June 1997), and at *http://www.hq.nasa.gov/office/pao/History/hhrhist.pdf*

The archival core of our documentary project, however, proved to be the substantial collections housed in the Historical Archives at Langley, the original and the oldest NACA/NASA laboratory.[21] The collections housed at Langley are immensely valuable to the study of the development of aerodynamics and are the closest thing to a dedicated "aerodynamics archive" in the United States. The collections possess materials dating back to the creation of the laboratory in 1917 and include technical reports and data logs, internal correspondence and memoranda, personal papers of leading engineers, programs and minutes of technical conferences, transcripts of oral histories, and numerous special files, such as the one compiled on the flight testing of the North American XP-51 Mustang. Members of our project team concentrated most of their efforts in four important collections: the NACA research authorization (RA) files; the Milton Ames collection; and the papers of research engineers John Stack and Fred E. Weick. Detailed finding aids for these and other collections at Langley are available in the archives and also appear in bibliographies at the end of James R. Hansen's two works on the history of NACA/NASA Langley, both of which are referenced later in the "Secondary Sources" portion of this essay.

In our view, the over 2,000 research authorization (RA) files housed in Langley's Historical Archives compose the single most valuable collection for the study of the NACA's involvement in the development of aerodynamics—and one of the most important intact collections in all of American aeronautics history. Through the voluminous materials in these RA files, historians are able to trace the evolution of a research program from the original idea to the completed project featured in a NACA publication. A clerk began an RA file whenever a research authorization was signed by the NACA chairman in Washington, D.C., authorizing one of the laboratories to pursue a specific research project. An RA file thus should exist for each and every authorized research program the NACA ever conducted. A typical RA file consisted of the RA sheet itself, which officially authorized the research, specified its title and sequential number, and delivered a statement indicating the reason and method of research. Following this cover sheet in every RA file one finds copies of all correspondence, memoranda, blueprints, drawings, and reports that NACA file clerks designated as related to the RA. Copies of a large number of original items such as this would also be placed in other related RA or correspondence files. This records-keeping scheme, invented by the NACA in the early 1920s, resulted in a truly historic and extremely extensive documentary collection. Researching the NACA RA files allows the historian to gain both a broad perspective and a very detailed picture of the NACA's process of selecting research problems

---

[21] Historical Program Manager, Mail Stop 446, Langley Research Center, Hampton, VA 23681-0001.

and then exploring ways to solve them. The RA files also illuminate the interaction between the NACA committee structure and the NACA executive office in Washington, D.C., and between those bodies and the field centers such as Langley. They also provide an almost day-to-day picture of the research life of the laboratory and of the social, intellectual, and professional dynamics taking place within it. Typically, there are also indications in the RA files of which RA contributed to a particular NACA publication. Still, quite a bit of time needs to be spent in the RA files before a historical researcher can become adept at following the many intricacies involved in the administration and contents of the RA files. Many documents that follow in our volumes come from the RA files at Langley. Many more, which we approximate to be in the neighborhood of some two million,[22] might have been included had we time to examine them all.

Another critical assembly of documents we fully utilized in the Historical Archives at NASA Langley was the Milton Ames Collection. These files resulted from an attempt made in the 1970s by the former Langley research engineer (1936–1941) and staff member in the NACA's Washington Office (1949–1958) to write a technical history of NACA Langley laboratory, from its establishment in 1917 to the birth of NASA in 1958. By this time Ames was serving as the NACA's chief of aerodynamics. What made this collection so important to us was that it embodied dozens of documents that Ames, a veteran researcher and aerodynamics specialist, had personally identified as historically significant. Organized into seven boxes, the Ames Collection spans from the Wright-brother era to Langley laboratory's involvement in the Apollo program and includes hundreds of significant aerodynamics documents. Specific themes that the Ames Collection highlights include the establishment of government aeronautical research establishments in Europe and North America; the construction of the Variable Density Wind Tunnel (VDT); the contributions of significant individuals to NACA research; NACA research programs related to low-drag engine cowlings, the design and operation of wind tunnels and other aerodynamic research facilities, airfoils and propellers, boundary-layer control, and the potential of laminar flow; the design of airships, flying boats, and rotary wing aircraft; NACA research contributions to the aircraft that fought in World War II; the establishment of other NACA laboratories and field stations, notably what became NACA/NASA Ames, Lewis, Wallops Island, and Dryden installations; post–World War II developments in high-speed flight, including the breaking of the mythical sound barrier; and Langley researcher Richard T. Whitcomb's conception of the so-called "area

---

[22] For more information about the holdings of the Langley Research Center Historical Archives, specifically the RA files, see James R. Hansen's *Engineer in Charge: A History of the Langley Aeronautical Laboratory, 1917–1958* (Washington, DC: NASA SP-4305, 1987), pp. 567–588.

rule." Virtually all of the papers in the Ames Collection deserved our attention as potential items for publication in our documentary history.'

A third major collection we reviewed systematically in the Historical Archives at NASA Langley involved the business papers of Langley research engineer and supersonics pioneer John Stack. Divided into six major sections, the Stack Collection documents Stack's deep and diverse involvement in major aerodynamic research projects from the 1930s to the 1970s. A pioneering specialist in high-speed aerodynamics, Stack is considered by all aeronautical historians to be one of the world's most significant aerodynamic researchers. Many of his activities proved crucial to the achievement of flight at transonic and supersonic speeds. He was one of the driving forces behind the Bell X-1 and North American X-15 research airplanes; the slotted-throat transonic wind tunnel; the national SST program; and the General Dynamics F-111 (TFX) variable-sweep wing program. Stack received both the prestigious Collier Trophy (twice) and the Wright Brothers Memorial Trophy. Topics in the Stack papers cover wind tunnel design, operation, and test techniques; research problems related to airfoils, compressible flow, and the boundary layer; reports of important meetings such as NACA-sponsored conferences on aerodynamics in the late 1940s; memoranda and correspondence related to high-speed aerodynamics during World War II and the immediate postwar period; and aircraft development projects such as the North American P-51, Republic P-47B, SST, V/STOL, and the TFX (F-111). Regrettably, no biography of Stack has ever been attempted. When it is, the papers in the Stack Collection will be an important starting place. Unfortunately, the collection includes almost no papers of a personal nature.

Another collection we reviewed at NASA Langley held the papers of American aeronautics pioneer Fred E. Weick, a prolific NACA Langley researcher well known for his work at the laboratory in the 1920s and 1930s. After leaving government employment permanently in 1936, Weick went on to a highly successful career in academia and industry. The thousands of items in this collection (donated by the Weick family to the Langley archives after Weick's death in 1994) include extensive correspondence, reports, memos, articles, newspaper clippings, speeches, and oral histories. These papers document his work with the navy's Bureau of Aeronautics, the NACA, the United Aircraft and Transport Corporation, the Engineering Research Corporation (ERCO), Texas A&M University, and the Piper Aircraft Corporation. In searching through the Weick papers, we gained numerous critical insights into major aeronautical developments from the 1920s to the 1970s, including propeller design; the drag-reduction experiments that resulted in the NACA's famous Collier Trophy–winning program of the late 1920s; the design of pioneering general aviation aircraft such as the Weick W-1A and the Ercoupe; and the refinement of agricultural aerial application (i.e., crop-dusting) aircraft such as the Piper Pawnee. Occasionally in the volumes to follow, we have also reproduced excerpts from Weick's published autobiography, *From the Ground Up: The Autobiography of an Aeronautical Engineer* (Washington, DC, and London: Smithsonian Institution Press, 1988), which was coauthored by project director James R. Hansen.

On the grounds of the Langley Research Center we also reviewed the contents of the Old Dominion University Experimental Aerodynamics Library. Housed within Langley's Full-Scale Wind Tunnel building, this library is the repository for the bound volumes on the history of wind tunnels compiled by Langley aerospace engineer and cryogenic wind tunnel pioneer Robert Kilgore, an individual whose work is much discussed in our study.[23] Kilgore's volumes contain countless manuscripts and published primary documents, arranged chronologically and devoted to specific subjects related both to his own career and to the development of wind tunnels in general. The titles of these volumes are *Development of Cryogenic Wind Tunnels* (Black Volumes 1969–1989) and the *ETB Weekly Briefs* (1980s–1990s). The latter documents Langley's Experimental Techniques Branch, the organization that Kilgore directed while working to develop the cryogenic wind tunnel concept. Kilgore also compiled a similar multivolume general history of wind tunnels that he donated to Langley's Floyd L. Thompson Technical Library.

Without question, we have included more documents from Langley than from any other archives, repository, or library. This is due both to the integrity of the collections at this particular NASA center, and to the extraordinary historical significance of the aerodynamics research activities that have taken place at this, the oldest of the NACA/NASA facilities, over ten decades of operation.

But documents from many other major facilities also play a critical role in our research story. One of the most important of these was the NASA Dryden Flight Research Facility History Office located on Edwards Air Force Base in the Mojave Desert of California, northeast of Los Angeles, which serves as the central repository documenting NACA/NASA's involvement in high-speed flight research.[24] Since fall 1946, the NACA and NASA have maintained a collaborative presence at Edwards with the Air Force and the Navy in the experimental testing of research aircraft. NACA/NASA pilots, engineers, and managers played leading roles in the success of state-of-the-art R&D projects that extended humankind's presence into both the air and space. Dryden's history office maintains thirty linear feet of materials

---

[23] Old Dominion University, Department of Aeronautical Engineering, The Langley Full-Scale Tunnel, P.O. Box 65309, Langley Air Force Base, VA 23665-5309, and at *http://www.lfst.com/*

[24] Office of External Affairs, Mail Stop TR-42, NASA Dryden Flight Research Center, Edwards, CA 93523, and at *http://www.dfrc.nasa.gov/History/*

concerning the X-series aircraft (X-1 through X-43) and D-558 program; Century series fighters (F-100, F-101, F-102, F-104, F-105, F-106, F-107); lifting bodies; the North American XB-70 supersonic research platform; the supercritical wing on the modified F-8; and the Highly Maneuverable Aircraft Technology (HiMAT) research vehicle. The office also sponsors an extensive oral history program covering important pilots and engineers in the history of flight research at Dryden.

To the north of Dryden at the mouth of the San Francisco Bay is NASA Ames Research Center. Founded in 1939, the Ames laboratory established the NACA's presence on the West Coast, which contributed greatly to the success of the aircraft manufacturing industry during World War II and the Cold War era. The Ames Research Center Library maintains a history collection documenting the activities of the center from December 1939 through August 1998, when the collection was formally created.[25] The collection measures twenty-eight linear feet and consists chiefly of published primary materials, including promotional brochures, newspaper clippings, and programs from official ceremonies. Other components of the collection are the manuscript and research materials used in the writing of Elizabeth A. Muenger's *Searching the Horizon: A History of Ames Research Center, 1940–1976*.[26] The library also maintains extensive Web-based resources describing the location, scope, and content of other archival collections concerning the history of Ames.[27] The majority of the Ames records are held by the National Archives and Records Administration, specifically at the Pacific Region Office at San Bruno, California. There are also materials about Ames housed at the NASA Headquarters History Office in Washington, D.C.

National Archives and Records Administration

A major part of our research necessarily involved the National Archives and Records Administration (NARA), which at its various facilities around the country maintains and preserves the largest collection of aerospace history archival records in the United States.[28] The federal government's extensive civilian and military involvement in aeronautical research and development since the late

---

[25] Research Information Resources, Library, Mail Stop 202-3, NASA Ames Research Center, CA 94035-1000, and at *http://mainlib.arc.nasa.gov/*

[26] Elizabeth A. Muenger, *Searching the Horizon A History of Ames Research Center, 1940–1976* (Washington, DC: NASA SP-4304, 1985).

[27] Sources for Researchers of Ames History at *http://history.arc.nasa.gov/research.htm*

[28] National Archives and Records Administration, 700 Pennsylvania Avenue NW, Washington, DC 20408, and at *http://www.nara.gov/research/*

nineteenth century has ensured the survival of significant collections. Organized into record groups (RG), these collections document the activities of particular federal agencies and departments involved in the development of American aeronautics. For our purposes, the most important record group was RG 255, which holds documents of the NACA and NASA from 1915 to 1988. RG 255 measures a gargantuan 5,182 cubic feet; however, the files are not located in one place but are distributed among various NARA facilities across the United States.[29] RG 255 consists of correspondence, technical and administrative reports, blueprints and drawings, photographs, meeting minutes, and memoranda from NACA and NASA committees, subcommittees, research units and facilities, and national and international offices. The primary NARA facilities holding files of interest to our project were the NARA Mid-Atlantic Region Office (City Center Philadelphia) in Philadelphia, and the NARA Pacific Region (San Francisco) Office at San Bruno, California.

An effort to integrate portions of the archival collections of NACA/NASA into the National Archives in the early 1990s resulted in, among other things, the move of NASA Langley's correspondence files (dating from 1917 to 1958) to the NARA facility in Philadelphia.[30] These files are also very significant historically. The NACA's manner of "correspondence control"—wherein all memoranda, letters, and notices went through Langley's engineer-in-charge before being sent out to NACA Headquarters or to any outside addresses—led to a very effective historical archive, one that enables historians to follow the business of the laboratory as it routinely unfolded, virtually day by day. In our view, however, it was a mistake for NARA to remove these correspondence files from Langley, where they not only were being well preserved but also fit into a complete network of archival materials involving the NACA research authorization (RA) files and a vast collection of library holdings. Documents in RA files, for example, routinely include cross-references to items in Langley's correspondence files. Previous to NARA's acquisition of the correspondence files, researchers could simply move from one part of the Langley historical archives to another to check the cross reference; now, unfortunately, they must travel from Hampton, Virginia, to Philadelphia, Pennsylvania.

The other NARA facility that contributed to our research was the NARA Pacific Region (San Francisco) Office at San Bruno, California, the central repository for the records of the Ames Research Center from 1939 to 1988.[31] Finding

---

[29] Guide to Federal Records in the National Archives of the United States—Record Group 255, *http://www.nara.gov/guide/rg255.html#255.2*

[30] NARA Mid Atlantic Region (Center City Philadelphia), 900 Market Street, Philadelphia, PA 19107-4292, and at *http://www.nara.gov/regional/philacc.html*. This facility contains a total of 347 cubic feet of NACA records, and finding aids are available. A portion of the collection is indexed by subject.

[31] NARA Pacific Region, 1000 Commodore Drive, San Bruno, CA 94066-2350, and at *http://www.nara.gov/regional/sanfranc.html*

aids are available for most of the 1,025-cubic-foot collection. Interspersed in the center's central files are correspondence, data sheets, meeting minutes, memoranda, and technical reports pertaining to aerodynamics, high-performance aircraft technology, and wind-tunnel tests of such notable World War II aircraft as the P-51 Mustang and P-38 Lightning. The materials from 1939 to the early 1960s are processed, well organized, and readily accessible while those from 1970 have not been fully integrated into the collection.

United States Air Force

The inextricable link between the technical development of flight and the military's participation in that quest has been a persistent theme throughout the twentieth century—and through our documentary history as well. The Air Force History Support Office (AFHSO), at Bolling Air Force Base, Washington, D.C., oversees the writing and publication of books, monographs, and professional studies and reports that document Air Force history.[32] These efforts are primarily for the benefit of Air Force leadership, but the program also exists for scholars of the military's role in aerospace. The History Support Office also directs the history of the various facilities of the Air Force as well as an extensive museum program across the nation. We reviewed several of these works in our research and they helped provide an important context for the development of aerodynamics.

Wright-Patterson Air Force Base (1948–present)—and its predecessors, McCook Field (1917–1927) and Wright Field (1927–1948), near Dayton, Ohio—has been a major center for the technical development of American military aviation. In addition to its importance as a research facility, Wright-Patterson Air Force Base is also the home of two important archival repositories. Since 1917, the successive incarnations of the Air Force Aeronautical Systems Center (ASC), have worked to enhance the performance of military aircraft and their abilities to achieve successful completion of military missions. The ASC History Office maintains and preserves not only the records of that organization, but also the individual papers of important Wright Field engineers who worked at the Air Force Flight Dynamics Laboratory.[33] Of particular importance to aerodynamics researchers are the William Lamar Files, which include many documents regarding the development of supersonic and hypersonic research aircraft and spacecraft between 1949 and 1978. Some finding aids are available. We reviewed the contents

---

[32] Air Force History Support Office, AFHSO/HOS, Reference and Analysis Division, 200 McChord Street, Box 94, Bolling Air Force Base, Washington, DC 20332-1111, and at *http://www.airforcehistory.hq.af.mil/*

[33] United States Air Force Aeronautical Systems Center History Office, 2275 D Street, Suite 2, Wright-Patterson Air Force Base, OH 45433-7219.

of these collections, though admittedly we could have done much more with them. Two of the primary problems of working in military archives, of course, are document classification and security clearance.

The Research Division of the United States Air Force Museum in Dayton maintains an extensive collection of archival materials supporting the exhibition of Air Force history.[34] The museum's collection approximates 200,000 documents. It is comprised of photographs, technical manuals, drawings, reports, and other materials, most of them dealing with specific artifacts in the museum's collection. Many of the documents are organized by subject airplanes, which facilitates a researcher's study of milestone aircraft such as the Martin B-10 of the 1930s and the Century series fighters of the 1950s and 1960s.

While Wright-Patterson AFB has served the Army, and later the Air Force, as its major center for aircraft research and development, Edwards AFB in the Mojave Desert of California has, since the early years of the Cold War, been the more broadly based center of high-performance flight research and testing. The Air Force, Navy, and NACA/NASA have used this remote location to perform the stunning aerospace trials related primarily to high-speed flight. The Air Force, through its Flight Test Center, has also used the facility to evaluate experimental and prototype military aircraft. The United States Air Force Air Force Flight Test Center History Office (AFFTC/HO) is responsible for documenting the activities of this organization.[35] The history office preserves important material on the X-series aircraft, with a particularly strong collection documenting the Bell X-1, fighter aircraft from the air force inventory, biographical files on leading test pilots and military officers, and general documents on high-speed flight research.

Because it is located so close to Auburn University and holds such a major collection of aerospace archives and library materials, Maxwell Air Force Base in Montgomery, Alabama, the home of the Air Force's Air University (which trains the service's officers for higher command and staff duties), was a mainstay of our research project. On base at Maxwell is the Air Force Historical Research Agency (AFHRA), which is the central repository for Air Force historical documents.[36] Unit histories, command and theater of operations records, and personal papers of leading military engineers and personnel document the air force's continued involvement in the aerodynamic development of the airplane.

---

[34] United States Air Force Museum/MUA, Research Division, 2601 E Street, Wright-Patterson Air Force Base, OH 45433-7609, and at *http://www.wpafb.af.mil/museum/mua.htm*

[35] United States Air Force Air Force Flight Test Center History Office, 305 E. Popson Avenue, Edwards Air Force Base, CA 93524-6595, and at *http://www.edwards.af.mil/history/index.html*

[36] Air Force Historical Research Agency, 600 Chennault Circle, Bldg 1405, Maxwell Air Force Base, AL 36112-6424, and at *http://www.maxwell.af.mil/au/afhra/*

National Air and Space Museum

Of course, no research project related to aeronautical development in the United States would be complete without examination of the collection at the National Air and Space Museum (NASM) in Washington, D.C., one of the premiere aerospace archives and libraries in the world.[37] NASM collections represent government, military, corporate, and academic involvement in humankind's journey through air and space. Overall, NASM possesses over 1,400 archival collections amounting to some 10,000 cubic feet of material. The different types of collections include personal and professional papers, corporate and institutional records, and "artificial" collections created from published sources to serve as reference files for a particular subject. A good point of departure before embarking upon any research at the museum is Paul E. Silbermann and Susan E. Ewing's *Guide to the Collections of the National Air and Space Archives*,[38] which describes 250 of the individual, corporate, government, military, and artificial collections acquired by the museum through 1989. Readers should be aware that this guide is not exhaustive, however, as NASM holds more than the 250 collections described therein.

One of the more useful collections consulted by the project team was the NASM Technical Files, which consists of 1,300 cubic feet of aviation and space-related materials arranged as a vertical file. Initially organized for the use of the museum's curatorial staff, these records are organized by subject, aircraft, individuals, organizations, events, and objects to include correspondence, reports, brochures, press releases, clippings, and photographs. A particularly helpful document was the original 1934 Douglas Aircraft Company report on the development of the DC-1 and DC-2, the first truly modern aircraft to emerge from the airplane design revolution of the 1920s and 1930s. In addition to its huge collection of manuscript material, the NASM archives also maintains an estimated 1.5 million photographs; 700,000 feet of motion picture film; and 2 million technical drawings.

United States Navy

Members of the project team also visited the Naval Historical Center at the Washington Navy Yard in Washington, D.C. Here, for example, we researched the

---

[37] Smithsonian Institution, National Air and Space Museum, National Air and Space Archives, MRC 322, Seventh Street and Independence Avenue SW, Washington, DC 20560, and at http://www.nasm.edu/nasm/arch/archdiv.htm or http://www.nasm.edu/nasm/arch/ARCH_REPOS/ GUIDE.PT7.html#NASM

[38] Paul E. Silbermann and Susan E. Ewing, *Guide to the Collections of the National Air and Space Archives* (Washington, DC: National Air and Space Museum, April 1991).

Navy's participation in the development of early wind tunnels.[39] Records of U.S. Navy aviation are located in two buildings within the Navy Yard. The main History Center, located in Building 57, contains Navy organizational archives and transcripts of interviews with several captains and admirals before their retirements, including individuals prominent in naval aviation. The Aviation History Office in Building 157 maintains other naval aviation records. These include documents, published and unpublished manuscripts, and photographs concerning Navy aircraft programs and facilities. Finding aids are available.

Industrial and Corporate

A primary objective of our documentary research was to visit archival centers based in the aircraft industry and to fully integrate the industry's contribution to aerodynamics development into our larger study. Unfortunately, records-keeping realities within the private sector made it very difficult for us to reach this goal. Economic and financial transitions within the aerospace industry have doomed some important collections to be closed forever. Fortunately, some of the succeeding corporate entities have been able to secure the integrity of some vital archival collections.[40] Most notably, through a series of mergers in the 1990s, the Boeing Company of Seattle, Washington, has been able to integrate into its own historical memory the archives of North American Rockwell, Inc., as well as that of the McDonnell-Douglas Corporation. As a result, Boeing now maintains the preeminent aerospace corporate archives in the United States.

The Boeing Company's Historical Archives represents an unprecedented commitment at the corporate level to preserve aviation history.[41] The program maintains archival facilities in Seattle; Long Beach, California; and St. Louis, Missouri. These facilities are principally organized and maintained to serve the company's needs rather than those of the general historian—and the company must give its permission for access to and use of its materials. Boeing's Historical Archives are not actually equipped to handle and provide service to general researchers, though its directors will do their best to respond to outside inquiries

---

[39] Naval Historical Center, Naval Aviation History Branch, Washington Navy Yard, 805 Kidder Breese, SE, Washington, DC 20374-5060, and at http://www.history.navy.mil/index.html or http://www.history.navy.mil/branches/nhcorg4.htm

[40] The following aerospace manufacturers that had no archival program or did not open their doors to this scholarly study included Lockheed-Martin, Northrop-Grumman, members of the United Technologies conglomerate (Sikorsky, Vought), Piper, and Cessna.

[41] Historical Archives, The Boeing Company, MS4H-0Z, P.O. Box 3707, Seattle, WA 98124, and at http://www.boeing.com/history/

and research objectives. Overall, the Boeing Company was very generous in its assistance to our conduct of research, and we are happy to have the company's permission to reproduce some of the historically significant documents we found in its collection. If our documentary history does nothing more than inspire the aerospace industry to provide a greater commitment to archival collection and to scholarly research, we shall have provided a critically important service to the history of aeronautics and space in America.

Because Boeing's materials were so helpful to us, more about its archival collections should be said here. Seattle lumber magnate William E. Boeing started manufacturing aircraft in 1916. By the 1920s, his company produced passenger and mail airplanes as well as military fighters and bombers. Boeing's enterprise was one of the first manufacturing concerns to benefit from the innovations brought forth by the airplane design revolution when the company moved toward specialization in multi-engine aircraft. The Boeing Company pioneered the all-metal "modern" airplane, the Model 247, as well as the Boeing B-17 Flying Fortress and B-29 Superfortress of World War II. At the dawn of the jet age, the company produced the first successful jet bombers, the B-47 Stratojet and B-52 Stratofortress plus the prototype Model 367-80, the ancestor of the long running 707 series of modern commercial jetliners. In the 1950s and 1960s, Boeing won the contract for the ill-fated SST and introduced the "jumbo-jet" 747. In the 1990s, Boeing innovated the "paper-less" 777 airliner, the first aircraft to be designed entirely with digital imaging technology. All of these aircraft developments are detailed in our study.

Boeing made important business acquisitions over the course of the twentieth century that greatly enhanced its prominence as an aircraft manufacturer. In the 1920s, Boeing's control of the Stearman Company ensured a strong position in the general aviation segment of the aviation industry. In 1960, Boeing acquired the Vertol Aircraft Company of Philadelphia, Pennsylvania, manufacturers of the CH-46 Sea Knight and CH-47 Chinook military helicopters. This made Boeing a power in the manufacture of rotary-wing aircraft. These company developments, too, make the Boeing archives critically important to any research into the history of American aeronautics.

The Boeing Company Historical Archives facility in Seattle is the main repository for the company's archival collections. Ninety percent of the archival and microform material held there is organized around specific aircraft designated by the Boeing model number; all of these numbers are indexed on computer. Researchers need to be aware of the Boeing model number in order to access the records successfully. For example, if a researcher wants to find records on the famous B-17 Flying Fortress, one must know that Boeing's designation for the aircraft was Model 299. Materials that the project team consulted at Boeing included its extensive collection of aircraft engineering reports; personal and company correspondence from the 1920s to the present; subject and biography files; aircraft project files; personal and professional papers; and oral history interviews of leading engineers and managers such as Claire Egtvedt, Maynard Pennell, Ed Wells, and William Cook.

From 1929 to 1934, Boeing Aircraft was part of United Aircraft and Transport Corporation, which included America's leading airlines and manufacturers and was one of the three major aviation conglomerates to emerge.[42] The corporation sponsored a technical advisory committee that included many leading aeronautical engineers such as Igor Sikorsky, Jack Northrop, and Fred E. Weick. In the papers of Claire Egtvedt, one of Boeing's leading managers and engineers, survives the only available transcriptions of the immensely valuable meeting minutes from 1929 to 1934. Nowhere else can a historian have the opportunity to see the verbatim discussion of the technical development of the products from America's most powerful aviation corporation. Readers will find an outstanding example of the importance of these meeting transcripts in Volume 1.

The Seattle location also maintains the records of North American Rockwell Corporation. In 1934, North American Aviation, Inc., emerged from the nucleus of the General Aviation Manufacturing Corporation of Dundalk, Maryland. The resulting company specialized initially in the design of small, single-engine trainers—notably the NA-16, BT-9, and AT-6 airplanes. By the early 1940s, North American had set its sights higher and manufactured its highly successful P-51 Mustang fighter and B-25 Mitchell medium bomber. The corporation's highly profitable transition into the design, production, and manufacture of high-performance military jet aircraft included the F-86 Sabre, FJ-1 Fury (the navy's first swept-wing jet), and the F-100 Super Sabre. North American became the corporate partner with the air force and NACA/NASA in the hypersonic X-15 research airplane and the experimental supersonic, delta-wing XB-70 Valkyrie. In 1966, North American Aviation merged with the Rockwell Standard Corporation and became North American Rockwell Corporation. In 1973, North American Rockwell became Rockwell International to reflect the company's widening range of businesses. It introduced the B-1 Bomber with variable-geometry wings. Rockwell's pioneering work on flight research and experimental design continued with its work on the tail-less X-31 enhanced-fighter maneuverability demonstrator. In December

---

[42] Member companies of the United Aircraft and Transport Corporation included airframe manufacturers Avion (Northrop), Boeing, Sikorsky, Chance Vought, and Stearman; engine and propeller makers Pratt and Whitney and Hamilton Standard; and the commercial operators United Airlines and Boeing Air Transport. The airline component of the corporation included United Airlines, Boeing Air Transport, National Air Transport, Pacific Air Transport, and Varney Air Lines. The corporation also controlled the Boeing School of Aeronautics, United Aircraft Exports, and the United Airports Company.

1996, the space and defense divisions of the Boeing Company and Rockwell International merged and the Rockwell contingents operated as a subsidiary under the name Boeing North American.

The overwhelming majority of North America's archival collection is not catalogued and remains unprocessed due to the relatively recent acquisition of the materials. As in the case of the Boeing files, effective evaluation of this collection requires extensive knowledge of North American's model number system, which can be acquired from secondary sources. The largest component of the collection is project correspondence recorded on microfilm—especially of late 1930s and 1940s projects such as the first trainer, the NA-16; Model NA-73 (P-51 Mustang); and the prototype XFJ-1 Fury. The collection also includes photographs, blueprints, and a small number of promotional materials from North American's earliest ancestor, General Aviation.

The Boeing Company, along with its North American component, merged with the McDonnell Douglas Corporation in August 1997. Itself the product of a 1967 merger of two eminent aerospace companies, McDonnell Douglas was a world leader in the design and production of high-performance jet fighter aircraft and commercial airliners. Due to the proprietary and military nature of its products, the records documenting the aerodynamic development of McDonnell-Douglas's fighter aircraft could not be made accessible to our project, and their size and importance is thus unknown.[43] As a result, we were able to research only the former Douglas Company's involvement in the aerodynamic development of its products.

Donald W. Douglas founded the company that bore his name in 1920, and the small enterprise quickly rose to prominence when it helped initiate the onset of the airplane design revolution with its introduction of the pioneering DC series of transports in 1933. Douglas employed such leading aeronautical engineers as James H. "Dutch" Kindelberger, John K. Northrop, and Edward H. Heinemann.

---

[41] It is unfortunate that the documentary project was unable to access the archival records of McDonnell Aircraft, due to its important position in aerospace history. After working for various aircraft manufacturers, such as the Aviation Division of the Ford Motor Company and the Glenn L. Martin Company, McDonnell established his own company in St. Louis, Missouri, in 1939. An early project McDonnell pursued was his Doodlebug of 1927, which the NACA latter used as a research vehicle for STOL experiments. Primarily a component manufacturer during World War II, the company submitted pioneering designs such as the twin-engine, long-range XP-67 fighter. The postwar period and the emergence of jet aircraft, specifically fighters, would catapult McDonnell to a prominent position within the world aerospace industry. The FH-1 Phantom was the first military jet aircraft to operate from an American aircraft carrier and was quickly followed by the F2H Banshee. The highly versatile F-101 Voodoo and F-4 Phantom II of the 1950s and 1960s led to the even more adaptable F-15 Eagle and F/A-18 Hornet. Hughes Helicopters became part of McDonnell Douglas in January 1984 and produced the important OH-6A Cayuse light observation helicopter and the AH-64 Apache advanced attack helicopter.

During World War II, the company produced the A-20 Havoc and A-26 Invader light bombers, the C-47 Skytrain and C-54 Skymaster transports, and the TBD Devastator and SBD Dauntless naval aircraft. The Cold War era saw it manufacture the military aircraft A4D Skyhawk and C-17 Globemaster III, the experimental high-speed research aircraft of the D-558 series and the X-3 Stiletto, and the commercial airliners MD-90 and MD-11.

The Boeing Company's Historical Archives at Long Beach, California, maintain primarily the corporate memory of the Douglas Aircraft Company.[44] Archival materials housed there include internal correspondence, engineering data, photographs, press releases, GALCIT wind tunnel reports, personal papers, the company newspaper *Airview*, and files organized by model number and subject. A collection that proved crucial to the success of the documentary history was the group of nine boxes containing the professional papers of Richard T. Cathers, who served as chief of the Advanced Design Group at McDonnell Douglas from the 1970s to the 1980s. Cathers's materials included documents related to developmental work on commercial airliners such as the DC-10, proposed SST and HSCT aircraft, and STOL transports.

The N. Paul Whittier Aviation Library and Archives of the San Diego Aerospace Museum preserves the history of the various incarnations of two other significant West Coast aircraft manufacturers, Convair and Teledyne Ryan.[45] The museum holds the surviving papers of Consolidated-Vultee and Convair from the 1930s to the 1960s and Ryan Aircraft and Teledyne Ryan from 1925 to 1960. These records are organized by aircraft project and are in the form of scrapbooks, technical reports, project correspondence, letters, memoranda, and GALCIT tests. The Convair files were especially helpful in illuminating the aerodynamic development of high-lift devices and big aircraft, primarily the B-24 and the B-36. There were also records related to the development of the jet-powered XF2Y Sea Dart seaplane interceptor; the VTOL XFY-1 Pogostick; and the Convair 240, 340, 440, 880, and 990 series airliners. The Ryan scrapbooks cover the development of the NYP *Spirit of St. Louis*, the Fireball FR-1 hybrid jet-piston engine aircraft, the Dragonfly YO-51 STOL observation plane, the X-13 Vertijet and Vertiplane V/STOL projects, and the ducted-fan SV-5A and SV-5B. The museum also maintains extensive subject files, which include entries on "aerodynamics" and "wings."

Another useful collection found at the San Diego Aerospace Museum was of the papers of Edward H. Heinemann (1908–1991), a prolific aircraft designer

---

[44] Historical Archives, The Boeing Company, 3855 Lakewood Boulevard, Mail Code D036-0038, Long Beach, CA 90846.

[45] San Diego Aerospace Museum Archives, 2001 Pan American Plaza, Balboa Park, San Diego, CA 92101-1636, and at *http://www.aerospacemuseum.org/library.htm*

who primarily worked for the Douglas Aircraft Company. Heinemann was personally responsible for the design of over twenty aircraft, which included the SBD Dauntless, A-20 and A-26 bombers, A-1 Skyraider, F4D Skyray, A4 Skyhawk, and D-558 Skystreak. Twelve boxes, ranging from 1926 to 1986, include correspondence; biographical articles; and materials related to Heinemann's designs, patents, speeches, and involvement in specific aircraft projects.

Back in the country's heartland, the Henry Ford Museum and Greenfield Village Archives in Dearborn, Michigan, also contributed to our project. There, we found records documenting Henry Ford's and the Ford Motor Company's involvement in aeronautics from 1925 to 1936.[46] Interested in doing for aviation what he had already succeeded in doing for the automobile industry, Henry Ford bought the Stout Metal Airplane Company and its idea for an all-metal airplane; he hoped this would be the foundation for his new aviation enterprise. The Stout Metal Airplane Company Collection, with some ninety linear feet of records in 169 boxes, documents Ford's earliest activities and their precursors. The internal correspondence, technical reports, and photographs pertaining to the highly successful Tri-Motor transport proved the most illustrative from the standpoint of aerodynamic development during the airplane design revolution. Another valuable resource found at the Henry Ford Museum and Greenfield Village Archives was the oral history of Harold Hicks, the engineer behind the development of the Ford Tri-Motor.

We also visited the Missouri Historical Society Archives in St. Louis, Missouri. It maintains records pertaining to regional involvement in aviation, some of which relates to aerodynamic development.[47] Under the Graeco-Byzantine dome of the renovated Historic United Hebrew Temple at Forest Park, researchers can evaluate the papers of Lloyd Engelhardt (1905–1973), who served from 1924 to 1970 as an aeronautical engineer for the Curtiss-Wright Corporation, McDonnell Aircraft Corporation, and McDonnell Douglas Corporation. The approximately seven linear feet of materials include Engelhardt's engineering papers for specific airplanes; inter-office memoranda; and published primary articles on V/STOL, supersonic and hypersonic design, and commercial airliners.

Academic Sources

Along with the governmental, military, and industrial archives we visited in the course of our project, we also sought to identify significant materials located

---

[46] Henry Ford Museum, Archives and Research Library, Dearborn, MI 48121, and at *http://www.hfmgv.org/research/index.html*

[47] Missouri Historical Society Archives, Jefferson Memorial Building/Forest Park, St. Louis, MO 63112-1099, and at *http://www.mohistory.org/*

in the libraries and archives of America's universities. We soon realized that one of the most important academic collections to examine was at the California Institute of Technology (Caltech) in Pasadena, California.

The philanthropic Guggenheim Fund for the Promotion of Aeronautics catapulted the Caltech to a preeminent position within the aeronautical community in the late 1920s, a position the institution retains today. The institute's archives maintain materials related to key individuals associated with the university's aeronautical engineering program and the operation of the GALCIT.[48] Organized by individual collection, these papers include fifteen linear feet of the personal and professional correspondence, government files, manuscripts, technical papers, and personal diaries of Dr. Clark B. Millikan; a twenty-page oral history along with the papers for the 1928–1934 period of Dr. Arthur E. Raymond, equaling approximately one linear foot; and the two linear feet of oral history and papers from 1928–1974 of Dr. Arthur L. Klein. Perhaps the most meaningful collection at Caltech is the eighty-one linear feet of Theodore von Kármán's 145,000 pages of letters, scientific manuscripts, reports, unpublished speeches, and lecture notes that span from the 1880s to 1960. GALCIT and its aeronautical library closed in the early 1990s. The library's voluminous collection of wind tunnel test reports of specific aircraft was distributed among members of the aerospace industry in lieu of complete disposal.

Our research also took us to The University of Texas at Dallas History of Aviation Collection at The University of Texas at Dallas Eugene McDermott Library. There we found an archive containing over 200 private collection items and 2.5 million items pertaining to aviation history.[49] Of particular interest to aeronautical researchers is the library's Lighter-Than-Air Collection. Included in this are the Rosendahl Collection, with almost 350 boxes of dirigible and blimp materials, and the Robinson Collection, which contains forty-five boxes of materials dealing with dirigible operations and, surprisingly, also with B-58 bomber development. The General Aviation Collection includes two boxes of personal papers and documents regarding Edwin A. Link and Link flight trainers, as well as other documents dealing with general aviation, but few of these are directly related to aerodynamics. Good finding aids are available.

---

[48] California Institute of Technology, Institute Archives, Mail Code 015A-74, Pasadena, CA 91125, and at *http://broccoli.caltech.edu/~archives/*

[49] University of Texas at Dallas, History of Aviation Collection, Eugene McDermott Library, Special Collections, P.O. Box 830643, Richardson, TX 75083-0643, and at *http://www.utdallas.edu/library/special/aviation/*

Primary Sources—Published

In addition to the primary archival and manuscript material that was central to our documentary history, our project team also consulted a large number of published primary sources in the form of articles, reports, and memoranda from national and international agencies, trade groups, and the popular aviation media. The team found these materials at three major libraries. The Floyd L. Thompson Technical Library at NASA Langley Research Center holds complete series of the major NACA and NASA publications—Technical Report (TR), Technical Note (TN), Technical Memorandum (TM), Advanced Confidential Report (ACR), and the Wartime Confidential Report (WCR).[50] Several of these publications, or selections from them, are reproduced in the documentary collection to follow. The Thompson library also maintains a comprehensive card-file index to all types of aeronautical literature that dates back to the early days of the NACA; much of this has not been computerized and is therefore not available online. Most of these publications, such as professional papers generated by the American Institute of Aeronautics and Astronautics (AIAA) and the North Atlantic Treaty Organization's (NATO's) Advisory Group for Aircraft Research and Development (AGARD), are accessible through the library system.

An enormous amount of NACA/NASA information has become increasingly available via the Internet. The NASA Center for Aerospace Information (CASI) Technical Report Server,[51] the NACA Report Server,[52] and the NASA Technical Report Server[53] provide full-text FTP reproductions of major technical documents. At a certain point, all NACA/NASA publications should be available online and will be an extraordinary resource of research into the history of aeronautics and space exploration.

The Air University Library (AUL) at Maxwell Air Force Base, Alabama,[54] and the Auburn University Library, Auburn, Alabama,[55] are both outstanding repositories for primary aeronautical periodicals and secondary works. These libraries possess full or partial series of the various incarnations of *Aviation Week and Space*

---

[50] The Floyd L. Thompson Technical Library, Langley Research Center, Hampton, VA 23681-0001.

[51] NASA Center for Aerospace Information Technical Report Server, *http://www.sti.nasa.gov/RECONselect.html*

[52] NACA Report Server, *http://naca.larc.nasa.gov/*

[53] NASA Technical Report Server, *http://techreports.larc.nasa.gov/cgi-bin/NTRS*

[54] Air University Library, 600 Chennault Circle, Building 1405, Maxwell Air Force Base, AL 36112-6424, and at *http://www.au.af.mil/au/aul/aulv2.htm*

[55] Auburn University Libraries, 231 Mell Street, Auburn University, AL 36849-5606, and at *http://www.lib.auburn.edu/*

Technology, *Journal of the Royal Aeronautical Society*, *Aero Digest*, *Flight*, and *The Aeroplane* as well as publications from professional societies such as the Society of Automotive Engineers, the American Society of Mechanical Engineers, and the AIAA. The Auburn University Interlibrary Loan Office provided access to materials that were not otherwise available in a timely and efficient manner.

Naturally, our project benefited from considerable documentation pertaining to the achievement of the Wright Brothers. In much the same spirit of this documentary history, Marvin W. McFarland edited *The Papers of Wilbur and Orville Wright, Including the Chanute-Wright Letters and Other Papers of Octave Chanute*. This work from 1953 reproduced seminal documents from the development of the first practical heavier-than-air aircraft.[56]

An individual who was intimately connected with several of the successful aerodynamics programs at Langley was Fred E. Weick. His memoirs, *From the Ground Up: The Autobiography of an Aeronautical Engineer*, co-written by James R. Hansen, provide remarkable insights into the story of early aeronautical development in the United States that are not just associated with the NACA or Langley, but with the navy and the aircraft industry as well.[57] Weick went on to successful careers in academia and industry and his autobiography illuminates those aspects of his career as well.

One of the most prolific and important aerodynamicists of the twentieth century was Theodore von Kármán. As noted in our introduction, von Kármán's autobiography, *The Wind and Beyond: Theodore von Kármán, Pioneer in Aviation and Pathfinder in Space* (co-written with Lee Edson) inspired our study in many ways, including our choice of an overall title.[58]

Because our project focused on the history of aerodynamics in America and not the world—and because our budget would not have allowed it anyway—we did very little research outside of the United States. Wanting to place American developments in global context, we did review a significant amount of secondary literature related to the worldwide story and also located a number of non-American (mostly European) reports and articles, some of which are reproduced in this study. We did take the opportunity to explore the contents of one European library, however. The library of the Royal Aeronautical Society (RAeS) in London provided valuable information on the development of the early wind

---

[56] Marvin W. McFarland, ed., *The Papers of Wilbur and Orville Wright, Including the Chanute-Wright Letters and Other Papers of Octave Chanute* (New York, NY: McGraw-Hill, 1953).

[57] Fred E. Weick and James R. Hansen, *From the Ground Up: The Autobiography of an Aeronautical Engineer* (Washington, DC: Smithsonian Institution Press, 1988).

[58] Theodore von Kármán with Lee Edson, *The Wind and Beyond: Theodore von Kármán, Pioneer in Aviation and Pathfinder in Space* (Boston and Toronto: Little, Brown and Company, 1967).

tunnel, primarily the activities of society members Francis Wenham and Horatio Phillips in the late nineteenth century. The RAeS is one of the oldest institutions devoted to aeronautical and aerospace engineering, and it maintains the oldest library in the world dedicated to the study of aeronautics and astronautics. Its collection includes all of the society's earliest records, dating back to its first meeting in 1866. The early Annual Reports are a goldmine of papers containing the thoughts and ideas of prominent European researchers in the field. Finding aids and excellent assistance are available, but access for non-RAeS members is expensive, 10£ per day.[59]

Secondary Sources

The project team looked to the work of other historians and scholars not only to formulate our research plan but also to inform our overall perspective. The topics, themes, and interpretations offered in these volumes represent our collective reflection upon what has become a vigorous field of aerospace history. As a result, a growing body of substantial research into the history of flight and its technical development has guided us through our deliberations. Much of this work has been produced with the help of the NASA History Office. Since establishment in 1959, this office has sponsored a steady stream of important historical works discussing NACA/NASA's involvement in the development of aeronautics and space. Many of these works are now out of print, but are available online.[60] Information on how to access them and other resources created by the NASA History Office records are provided in footnotes at appropriate points within this bibliographical essay.

But many fine works in aerospace history have come from places other than NASA. Products of rigorous independent scholarship, these works have appeared in books and articles published by academic and commercial presses. One of the most fundamental studies that we utilized in our project was John D. Anderson, Jr.'s *A History of Aerodynamics and Its Impact on Flying Machines*, published by Cambridge University Press in 1997. Anderson's was the first work of its kind to directly address the emerging practice of aerodynamics from the nineteenth century to the present.[61] Heavily footnoted and written in a context that benefits both aerodynamic

---

[59] The Royal Aeronautical Society, 4, Hamilton Place, London, W1V 0BQ, United Kingdom, and at *http://www.raes.org.uk/*

[60] NASA Histories Online, *http://www.hq.nasa.gov/office/pao/History/online.html*

[61] John D. Anderson, Jr., *A History of Aerodynamics and Its Impact on Flying Machines* (Cambridge, MA: Cambridge University Press, 1997).

practitioners and historians, we viewed this work as our main point of departure. It was our goal that our volumes would read profitably side-by-side with Anderson's.

Lawrence K. Loftin, Jr.'s *Quest for Performance: The Evolution of Modern Aircraft*[62] helped us enormously to understand some of the arcane aspects of aircraft engineering and aerodynamic design. His book also aided us in identifying, classifying, and describing the major eras in heaver-than-air powered flight. Similarly, *The Technical Development of Modern Aviation*,[63] by Ronald Miller and David Sawers, assisted us by addressing aerodynamic issues within the overall system of the airplane. The essays by John D. Anderson, Jr., James R. Hansen, Richard K. Smith, and Terry Gwynn-Jones that appeared in John T. Greenwood's edition of *Milestones of Aviation* elaborated on different technological trends in aeronautics crucial to aerodynamic development.[64] Walter G. Vincenti's *What Engineers Know and How They Know It: Analytical Studies from Aeronautical History*,[65] challenged us to apply epistemological and sociological notions to what we were discovering in the empirical evidence. A general work that was immensely valuable to the less technically literate in our group (and thus to any nonspecialist's understanding of the mechanics of flight) was Theodore A. Talay's *Introduction to the Aerodynamics of Flight*.[66]

In our study of the Wrights, we placed tremendous value on the work of the National Air and Space Museum's Dr. Tom D. Crouch, the world's leading scholar of the early flight period. In *A Dream of Wings: Americans and the Airplane, 1875–1905* and *The Bishop's Boys: A Life of Wilbur and Orville Wright*, Crouch has brilliantly recreated the international and personal environment in which the Wrights developed their aircraft.[67] Another major work that deserves special consideration is NASM's Peter L. Jakab's book, *Visions of a Flying Machine: The Wright Brothers and the Process of Invention*, which successfully tackles the Wrights' creative engineering approach to solving the design challenges they faced.[68]

---

[62] Lawrence K. Loftin, Jr., *Quest for Performance: the Evolution of Modern Aircraft* (Washington, DC: NASA SP-468, 1985), http://www.hq.nasa.gov/office/pao/History/SP-468/cover.htm

[63] Ronald Miller and David Sawers, *The Technical Development of Modern Aviation*, (New York, NY: Praeger, 1970).

[64] John T. Greenwood, ed., *Milestones of Aviation* (New York, NY: Macmillan, 1989).

[65] Walter G. Vincenti, *What Engineers Know and How They Know It: Analytical Studies from Aeronautical History* (Baltimore, MD: Johns Hopkins University Press, 1990).

[66] Theodore A. Talay, *Introduction to the Aerodynamics of Flight* (Washington, DC: NASA SP-367, 1975), http://history.nasa.gov/SP-367/cover367.htm

[67] Tom D. Crouch, *A Dream of Wings: Americans and the Airplane, 1875–1905* (Washington, DC: Smithsonian Institution Press, 1989) and *The Bishop's Boys: A Life of Wilbur and Orville Wright* (New York, NY: W.W. Norton and Co., 1989).

[68] Peter L. Jakab, *Visions of a Flying Machine: The Wright Brothers and the Process of Invention* (Shrewsbury, England: Airlife, 1990).

General histories of the NACA and NASA proved valuable to the conceptualization of our documentary project. A fine overall survey of the two organizations, including coverage of their research commitment to aeronautics, is Roger E. Bilstein's *Orders of Magnitude: A History of the NACA and NASA, 1915–1990*.[69] There are two important histories that focus on the NACA specifically as well. Alex Roland's history, *Model Research: The National Advisory Committee for Aeronautics, 1915–1958*, provides a useful and provocative critique of the NACA style of research.[70] The fact that our work questions the validity of many of Roland's interpretations does not detract from the overall esteem we have for his insightful, shrewdly argued book.

A work that we include here as a secondary source but in truth serves more usefully as a primary document is George W. Gray's 1948 book, *Frontiers of Flight: The Story of NACA Research*. Readers will find that we have used it primarily as the latter, reproducing selections from it as actual documents. The book does celebrate the NACA's achievements, especially in regard to the Committee's contributions to the outcome of World War II.[71] But Gray's technical discussions are so exceptional in communicating difficult engineering concepts and activities to a lay readership that we have incorporated them as a reference—not as propaganda in the service of the NACA's historical reputation.

No single body of work underpins our entire study more than the histories of NACA/NASA Langley Research Center provided by James R. Hansen. Langley has been one of the epicenters—some might even argue the epicenter—of American aerodynamics research since the 1920s, and thus its history informed our study throughout. Hansen's two books, *Engineer in Charge: A History of the Langley Aeronautical Laboratory, 1917–1958* and *Spaceflight Revolution: NASA Langley Research Center from Sputnik to Apollo*,[72] examine the nature of aerodynamics research activities at Langley. The books cover the laboratory's establishment in 1917 through the transition to space and into the 1960s—with thoughts on what has happened beyond and into the 1990s. Hansen's goal was to present more than just an institutional history of an important aeronautics research facility.

---

[69] Roger E. Bilstein, *Orders of Magnitude: A History of the NACA and NASA, 1915–1990* (Washington, DC: NASA SP-4406, 1989), http://www.hq.nasa.gov/office/pao/History/SP-4406/cover.html

[70] Alex Roland, *Model Research: The National Advisory Committee for Aeronautics, 1915–1958*, 2 vols. (Washington, DC: NASA SP-4103, 1985).

[71] George W. Gray, *Frontiers of Flight: The Story of NACA Research* (New York, NY: Alfred A. Knopf, 1948).

[72] James R. Hansen, *Engineer in Charge: A History of the Langley Aeronautical Laboratory, 1917–1958* (Washington, DC: NASA SP-4305, 1987) and *Spaceflight Revolution: NASA Langley Research Center from Sputnik to Apollo* (Washington, DC: NASA SP-4308, 1995).

In these books, he more broadly interprets the events within a government research laboratory devoted to a technological development. As director of our documentary project, he worked to expand—and to correct—some of the coverage he provided in his two-volume study. For example, he delved into the history of NACA/NASA research on helicopters and rotary-wing aircraft, mainly to correct his previous neglect of that topic. An excellent illustrated history of Langley based largely on Hansen's books is James Schultz's *Winds of Change: Expanding the Frontiers of Flight, Langley Research Center's Seventy-Five Years of Accomplishment, 1917–1992*.[73]

After establishment in 1940, NACA Ames Aeronautical Laboratory (later renamed NASA Ames Research Center) became an important center for aerodynamics research, one that directly benefited from the high concentration of aviation manufacturers on the West Coast. Over the past thirty years, considerable effort has been made to document the Ames contribution to aeronautics. Edwin P. Hartman, a NACA/NASA official and leading aeronautical engineer and educator, contributed the first appraisal of the center's activities in *Adventures in Research: A History of Ames Research Center, 1940–1965*.[74] Elizabeth A. Muenger provided another detailed account in *Searching the Horizon: A History of Ames Research Center, 1940–1976*.[75] The latest update, *Atmosphere of Freedom: Sixty Years at the NASA Ames Research Center*, by Glenn E. Bugos, focuses on the history of the center during the last quarter of the twentieth century.[76]

One of the more important research programs conducted at the Ames Research Center has concerned Vertical and Short Takeoff and Landing (V/STOL) aircraft. In the last section of Volume 3 of our documentary history, we study the history of V/STOL research as it has related to helicopter and other rotary-wing aircraft development, as well as the NACA/NASA role in that history. Readers will find that one type of modern rotary-wing aircraft, the tilt rotor, has been a pet subject for NASA research, especially at Ames laboratory. We will make significant use of a 2000 NASA publication on tilt rotor development entitled *The History of the XV-15 Tilt Rotor Research Aircraft: From Concept to Flight*, authored by three Ames researchers who have worked prominently in this field.[77] Selections from this book will also be used as part of our documentary story. A recent book

---

[73] James Schultz, *Winds of Change: Expanding the Frontiers of Flight, Langley Research Center's Seventy-Five Years of Accomplishment, 1917–1992* (Washington, DC: NASA, 1992).

[74] Edwin P. Hartman, *Adventures in Research: A History of Ames Research Center, 1940–1965* (Washington, DC: NASA SP-4302, 1970).

[75] Elizabeth A. Muenger, *Searching the Horizon: A History of Ames Research Center, 1940–1976* (Washington, DC: NASA SP-4304, 1985).

[76] Glenn E. Bugos, *Atmosphere of Freedom: Sixty Years at the NASA Ames Research Center* (Washington, DC: NASA SP-2000-4314).

that proved tremendously useful for our discussion of helicopter development was Eugene K. Liberatore's *Helicopters Before Helicopters*.[78] So taken were we with the conceptually stimulating epilogue to this book that we reproduced it as the final document in that section.

The highly successful flight research programs at the NACA's High-Speed Flight Research Station at Muroc, California, (later renamed NASA Dryden Flight Research Center) have produced many of the most memorable moments in aviation history, including the flight of the Bell XS-1, which broke the mythical "sound barrier" in October 1947. Historian Richard P. Hallion provides an exhaustive survey of the high-flying, high-speed activities over the Mojave Desert in *On the Frontier: Flight Research at Dryden, 1946–1981*.[79] A helpful companion to Hallion's work is Lane E. Wallace's *Flights of Discovery: An Illustrated History of the Dryden Flight Research Center*.[80] Michael H. Gorn documents the career of a former director of the center and the man for whom the center is now named in *Hugh L. Dryden's Career in Aviation and Space*.[81] An important part of Gorn's biography is the inclusion of facsimile reproductions of key documents from Dryden's career.

Some of the flight research programs conducted at Dryden have garnered their own histories. J. D. Hunley's edition of *Toward Mach 2: The Douglas D-558 Program* documents the joint NACA–navy quest to break the sound barrier in the late 1940s.[82] Two notable histories of lifting bodies come from former NASA employees. R. Dale Reed, a former Dryden engineer, and Milton O. Thompson recount their involvement in *Wingless Flight: The Lifting Body Story* and *Flying Without Wings: NASA Lifting Bodies and the Birth of the Space Shuttle*, respectively.[83]

---

[77] Martin D. Maisel, Demo J. Giulianetti, and Daniel C. Dugan, *The History of the XV-15 Tilt Rotor Research Aircraft: From Concept to Flight* (Washington, DC: NASA SP-2000-4517, NASA Monographs in Aerospace History, No. 17, 2000).

[78] E. K. Liberatore, *Helicopters Before Helicopters* (Malabar, FL: Krieger Publishing Co., 1998).

[79] Richard P. Hallion, *On the Frontier: Flight Research at Dryden, 1946–1981* (Washington, DC: NASA SP-4303, 1984), http://www.dfrc.nasa.gov/History/Publications/SP-4303/

[80] Lane E. Wallace, *Flights of Discovery: An Illustrated History of the Dryden Flight Research Center* (Washington, DC: NASA SP-4309, 1996).

[81] Michael H. Gorn, *Hugh L. Dryden's Career in Aviation and Space* (Washington, DC: NASA, NASA Monographs in Aerospace History, No. 5, 1996), http://www.dfrc.nasa.gov/History/Publications/Monograph_5/

[82] J.D. Hunley, ed., *Toward Mach 2: The Douglas D-558 Program* (Washington, DC: NASA SP-4222, 1999), http://www.dfrc.nasa.gov/History/Publications/D-558/

[83] R. Dale Reed, *Wingless Flight: The Lifting Body Story* (Washington, DC: NASA SP-4220, 1998), http://www.dfrc.nasa.gov/History/Publications/WinglessFlight/; Milton O. Thompson and Curtis Peebles, *Flying Without Wings: NASA Lifting Bodies and the Birth of the Space Shuttle* (Washington, DC: Smithsonian Institution Press, 1999).

Computers Take Flight: A History of NASA's Pioneering Digital Fly-by-Wire Project documents an important episode in the refinement of aircraft control systems.[84] Albert L. Braslow's *A History of Suction-Type Laminar-Flow Control with Emphasis on Flight Research* addresses an ongoing effort at Dryden to further refine aerodynamic efficiency through the application of new technology.[85] We found Braslow's insights extremely useful as we endeavored to accept the esoteric nature of the subtle aerodynamic conditions in the so-called "boundary layer."

Often overlooked is the fact that women played (and continue to play) a part in the success of aerodynamics in the United States. Sheryll Goecke Powers, in *Women in Flight Research*, tells the story of the unnamed women who in the early days of Dryden—and at other NACA laboratories before them—worked as "human computers" and were responsible for processing and analyzing wind tunnel and flight research data.[86]

Several important works addressing the development of hypersonic aerodynamics in the United States informed our study. *In Hypersonics Before the Shuttle: A Concise History of the X-15 Research Airplane*, Dennis R. Jenkins analyzes the highly successful flight research program of the late 1950s and 1960s that, in key respects, led technologically to the Space Shuttle program of the 1970s and beyond.[87] Wendell H. Stillwell's *X-15 Research Results with a Selected Bibliography* provides an analysis and exhaustive listing of reports and other "lessons learned" materials drawn from the X-15 program.[88] In "Transiting from Air to Space: The North American X-15," which appeared in *The Hypersonic Revolution: Case Studies in the History of Hypersonic Technology*, Robert S. Houston, Richard P. Hallion, and Ronald G. Bostonis detail many crucial aspects of our country's prototype transatmospheric vehicle or "spaceplane."[89] A definitive source listing for the Space Shuttle appears in Roger D. Launius and Aaron K. Gillette's *Toward a History of the Space*

---

[84] *Computers Take Flight: A History of NASA'a Pioneering Digital Fly-by-Wire Project* (Washington, DC: NASA SP-2000-4224), http://www.dfrc.nasa.gov/History/Publications/f8ctf/

[85] Albert L. Braslow, *A History of Suction-Type Laminar-Flow Control with Emphasis on Flight Research* (Washington, DC: NASA, *NASA Monographs in Aerospace History*, No. 13, 1999).

[86] Sheryll Goecke Powers, *Women in Flight Research* (Washington, DC: NASA, *NASA Monographs in Aerospace History*, No. 6, 1997), http://www.dfrc.nasa.gov/History/Publications/WIFR/contents.html

[87] Dennis R. Jenkins, *Hypersonics Before the Shuttle: A Concise History of the X-15 Research Airplane* (Washington, DC: NASA SP-2000-4518, *NASA Monographs in Aerospace History*, No. 18).

[88] Wendell H. Stillwell, *X-15 Research Results with a Selected Bibliography* (Washington, DC: NASA SP-60, 1965), http://www.hq.nasa.gov/office/pao/History/SP-60/cover.html

[89] Bolling Air Force Base, *The Hypersonic Revolution: Case Studies in the History of Hypersonic Technology*, 3 vols. (Washington, DC: Air Force History and Museums Program, 1998), http://www.hq.nasa.gov/office/pao/History/hyperrev-x15/cover.html

Shuttle: An Annotated Bibliography.[90] All of these works gave us a framework for our profile of development in hypersonic aerodynamics.

One of the keys to the success of NASA, and the NACA before it, has been the ability to design and construct world-class facilities in support of pioneering research programs. In *The Wind Tunnels of NASA*, veteran NACA/NASA wind tunnel engineer Donald D. Baals and technical writer William R. Corliss provide the standard reference for NASA's preeminent research tools.[91] Those looking for detailed technical information on the operation and use of wind tunnels may find Alan Pope's *Wind-Tunnel Testing* useful.[92] Pope later collaborated with Jewel B. Barlow and William H. Rae, Jr., on *Low-Speed Wind Tunnel Testing*,[93] and he co-authored *High-Speed Wind Tunnel Testing* with Kenneth L. Goin.[94] These books contain a wealth of information on the ways engineers have used wind tunnels to develop aircraft designs over several decades.

Two other histories highlight the correlation between advanced research tools and successful research programs. *The High Speed Frontier: Case Histories of Four NACA Programs, 1920–1950*, was written by NACA/NASA engineer John V. Becker, one of John Stack's chief associates and the man who many regard as the "father of the X-15." In his book, Becker offers both first-hand and historical insights into the story of the NACA's work in the 1930s, 1940s, and 1950s on high-speed airfoils, propellers, cowlings, internal-flow systems, and transonic wind tunnels.[95] An anthology of scholarly articles entitled *From Engineering Science to Big Science: The NACA and NASA Collier Trophy Research Project Winners*, edited by Clemson University historian of technology Pamela Mack, cover the processes by which NACA and NASA scientists and engineers successfully employed wind tunnels, aerodynamics, and other types of experimental tools and theoretical methods to generate successful research programs.[96]

---

[90] Roger D. Launius and Aaron K. Gillette, *Toward a History of the Space Shuttle: An Annotated Bibliography* (Washington, DC: NASA, *NASA Monographs in Aerospace History*, No. 1, 1992), http://www.hq.nasa.gov/office/pao/History/Shuttlebib/cover.html

[91] Donald D. Baals and William R. Corliss, *The Wind Tunnels of NASA* (Washington, DC: NASA SP-440, 1981), http://www.hq.nasa.gov/office/pao/History/SP-440/cover.htm

[92] Alan Pope, *Wind-Tunnel Testing*, 2d ed. (New York, NY: John Wiley & Sons, 1954).

[93] Jewel B. Barlow, William H. Rae, Jr., and Alan Pope, *Low-Speed Wind Tunnel Testing*, 3d ed. (New York, NY: John Wiley & Sons, 1999).

[94] Eugene E. Bauer, *Boeing in Peace and War* (Enumclaw, WA: TABA Publishing, 1991).

[95] John V. Becker, *The High Speed Frontier: Case Histories of Four NACA Programs, 1920–1950* (Washington, DC: NASA SP-445, 1980), http://www.hq.nasa.gov/office/pao/History/SP-445/cover.htm

[96] Pamela Mack, ed., *From Engineering Science to Big Science: The NACA and NASA Collier Trophy Research Project Winners* (Washington, DC: NASA SP-4219, 1998).

Two biographical studies explore von Kármán's role in the American academic and scientific community. The first of these, Paul A. Hanle's *Bringing Aerodynamics to America*, examines the dramatic steps taken by proponents of American aeronautics to transfer European knowledge of aerodynamics from Ludwig Prandtl and his talented group of students at the University of Göttingen to the United States. This initiative led to the importation of two of Prandtl's prize students: von Kármán to Caltech, and Max M. Munk to the NACA.[97] The work of both of these individuals is covered in our text and documents to come. The second of the biographies, *The Universal Man: Theodore von Kármán's Life in Aeronautics*, by Michael H. Gorn, provides another important picture of von Kármán's career as an aerodynamicist and aeronautical research leader.[98] Whereas Hanle's book concentrates on the era of the 1920s and 1930s, Gorn's study spends more time looking at von Kármán's career during and after World War II to his death in the early 1960s.

No single definitive history of the U.S. Air Force's commitment to aeronautical research exists; one is badly needed. Works featuring research from the bountiful Sarah Clark collection in the National Archives have started to appear, but none of them discuss the American military's involvement in aerodynamic development. In *From Huffman Prairie to the Moon: The History of Wright-Patterson Air Force Base*, Lois E. Walker and Shelby E. Wickam provide a valuable introduction to the aeronautical research activities that have taken place at the army and—later—air force installations near Dayton, Ohio.[99]

Project members found value in several secondary works on the history of the Boeing Company. Boeing Historical Services, the unit that manages the firm's historical archives, published *A Brief History of the Boeing Company*, which provides a useful and condensed introduction to the individuals, aircraft, and programs of Boeing, North American Rockwell, and McDonnell Douglas.[100] In *Boeing's Ed Wells*, Mary Wells Geer presents insightful biographical reminiscences of her engineer father, which highlight Boeing's important activities in aircraft design and manufacture during the 1940s.[101] Two popular histories, Robert J. Serling's *Legend and Legacy: The Story of Boeing and Its People* and Harold Mansfield's *Vision: The Story of Boeing*, present the company in a very favorable light while providing a

---

[97] Paul A. Hanle, *Bringing Aerodynamics to America* (Cambridge, MA: MIT Press, 1982).

[98] Michael H. Gorn, *The Universal Man: Theodore von Kármán's Life in Aeronautics* (Washington, DC: Smithsonian Institution Press, 1992).

[99] Lois E. Walker and Shelby E. Wickam, *From Huffman Prairie to the Moon: The History of Wright-Patterson Air Force Base* (Washington, DC: General Printing Office, 1986).

[100] *A Brief History of the Boeing Company* (Seattle, WA: Boeing Historical Services, 1998).

[101] Mary Wells Geer, *Boeing's Ed Wells* (Seattle, WA: University of Washington Press, 1992).

basic survey of the company's history.[102] The standard reference for Boeing aircraft over the decades is *Boeing Aircraft Since 1916* by Peter M. Bowers.[103] Another work we consulted was Eugene E. Bauer's *Boeing in Peace and War*.[104]

North American Aviation and Rockwell have yet to find their historians. Norm Avery's two-volume *North American Aircraft 1934–1998* assists readers mostly in determining the model numbers of individual aircraft and exciting them with stunning photographs.[105] Wagner's *Mustang Designer: Edgar Schmued and the P-51* focuses on the design environment at North American in which the famous World War II fighter gestated. Wagner also highlights the careers of contemporary fellow designers, such as Don Berlin of Curtiss-Wright, to illustrate the existence of an American design community.[106]

Two works were especially helpful in studying the history of the Douglas Company. Douglas J. Ingells, a prolific historian of the aviation industry, in *The McDonnell Douglas Story*, traces the development of two separate companies and the path they followed after their historic merger in 1967.[107] In the same format as Bowers's book on Boeing, Rene Francillon's *McDonnell Douglas Aircraft Since 1920* rates as the major point of departure for any serious study of the corporation's aircraft.[108] Fully half of Henry M. Holden's *The Legacy of the DC-3* delves into the early history of Douglas and the development of the DC-1/2/3 airplanes.[109]

For our epilogue on thoughts about the future of flight, Joseph J. Corn's *The Winged Gospel: America's Romance with Aviation, 1900–1950* offered critical social and cultural insights.[110] Corn's classic little book focuses on America's popular infatuation with the airplane rather than on technological enthusiasm and concepts within the aircraft industry and the aeronautical engineering community.

---

[102] Robert J. Serling, *Legend and Legacy: The Story of Boeing and Its People* (New York, NY: St. Martin's Press, 1992); Harold Mansfield, *Vision: The Story of Boeing* (New York, NY: Popular Press, 1966).

[103] Peter M. Bowers, *Boeing Aircraft Since 1916* (Annapolis, MD: Naval Institute Press, 1989; reprint, New York, NY: Funk and Wagnalls, 1966).

[104] Eugene E. Bauer, *Boeing in Peace and War* (Enumclaw, WA: TABA Publishing, 1991).

[105] Norm Avery, *North American Aircraft 1934–1998*, 2 vols. (Santa Ana, CA: Narkiewisc/Thompson, 1998).

[106] Ray Wagner, *Mustang Designer: Edgar Schmued and the P-51* (Washington, DC: Smithsonian Institution Press, 2000).

[107] Douglas J. Ingells, *The McDonnell Douglas Story* (Fallbrook, CA: Aero Publishers, Inc., 1979).

[108] Rene Francillon, *McDonnell Douglas Aircraft Since 1920*, 2 vols. (London: Putnam Aeronautical Books, 1990).

[109] Henry M. Holden, *The Legacy of the DC-3* (Niceville, FL: Wind Canyon Publishing, 1997).

[110] Joseph J. Corn, *The Winged Gospel: America's Romance with Aviation, 1900–1950* (New York, NY: Oxford University Press, 1983).

Nonetheless, it provides an overarching thesis illuminating the dynamics between American culture and ideology and American technological belief and outlook.

Points of Departure

Despite what we feel was a rather comprehensive examination of aeronautics libraries and archives countrywide, without question we left many stones unturned in our endeavor to document the history of aerodynamic development in the United States, in terms of both topics and archival facilities. There was simply no way to visit every potentially interesting repository or engage every related topic. For example, we recognize that more needs to be done to cover general aviation aircraft's aerodynamic design and to explore aerodynamics in university research. It is our hope that our documentary collection will provoke others to add to our research and to create collections of their own. Readers should remember, however, that the NASA History Office contracted with Auburn University to document the history of aerodynamics in America as it related most specifically to the contributions made by the NACA and NASA. Such a focus demanded that we emphasize certain archives and topics, and minimize others.

As mentioned above, one area we regret to have neglected involves general—or private—aviation. A leading general aviation manufacturer during the period of the airplane design revolution of the 1920s and 1930s, well known for its "Mystery Ship" and Model R air racers, was The Travel Air Company. We had hoped to visit the special collections and archives of Wichita State University Library, where the papers of Walter E. Burnham and Herb Rawdon are stored. These materials document the engineering development of Travel Air general aviation and racing aircraft during the 1920s and 1930s.[111]

Given the enormous scope of our project we were not even able to cover relevant NACA/NASA materials as exhaustively as we had hoped. One set of records we regret to have minimized involves the activities of NASA John H. Glenn Research Center at Lewis Field in Cleveland, Ohio, and the original NACA Aircraft Engine Research Laboratory (AERL), which later became NASA Lewis Research Center. While primarily concerned with problems pertaining to propulsion systems, the Glenn center has sponsored numerous projects pertinent to the relationship between aerodynamics and propulsions systems. Work in this facility's historic icing tunnel plus many other issues certainly relate to compromised aerodynamic integrity, a problem that has had an important impact on the study

---

[111] Special Collections and University Archives, Wichita State University, 1845 Fairmount, Wichita, KS, 67260-0068, and at *http://www.twsu.edu/library/specialcollections/sc.html*

of aerodynamics for both commercial and military aircraft.[112] Given our focus on NACA/NASA contributions, we could have done more to illuminate the importance of the work done at this installation, whose work dates back to 1941. Many of NASA Glenn's retired administrative files rest within six boxes at NARA's facility at Suitland, Maryland, or are still in storage at the center's Plum Brook Station, located on Lake Erie near Sandusky. There were also nineteen boxes of documents (mostly speeches, talks, and papers by center authors) accessioned from NASA Glenn to the NARA center in Dayton, Ohio, plus four additional boxes of center records stored at the federal records center in Chicago, Illinois. For more description of NASA Lewis's records situation, readers should consult Virginia P. Dawson's *Engines and Innovation: Lewis Laboratory and American Propulsion Technology*.[113]

Even within NARA, the project team concentrated on documenting NACA and NASA involvement in the development of aerodynamics as it related to Langley, Ames, and Dryden centers. This brought us to the NARA facilities at Philadelphia and San Bruno but not to the other NARA facilities where the Glenn center records are kept. Our focus also took us to the NARA facility at College Park, Maryland, where a significant, if not the largest, portion of the previously mentioned NACA Record Group 255 exists.[114]

Another important collection that NARA College Park maintains is RG 342, the Records of the United States Air Force Commands, Activities, and Organizations. Within that organization rests the well-known Sarah Clark Collection, named after the clerk who processed the documents in preparation for their transfer from the air force to NARA. Officially known as the research and development project files of the Engineering Division, Material Command of Wright-Patterson Air Force Base in Ohio, this collection features an extensive record of the military's pursuit of aeronautical excellence from 1917 to 1945. The post-1945 version of the Sarah Clark collection is located at NARA's National Personnel Records Center in St. Louis.[115]

Other facilities where our project could have profitably spent more time include the NASA Marshall History Office in Huntsville, Alabama;[116] the NARA Southeast

---

[112] History Office, Mail Stop 21-8, Glenn Research Center, 21000 Brookpark Road, Cleveland, Ohio, 44135, and at *http://www.grc.nasa.gov/WWW/PAO/html/history.htm*

[113] Virginia P. Dawson, *Engines and Innovation: Lewis Laboratory and American Propulsion Technology* (Washington, DC: NASA SP-4306, 1991), and at *http://history.nasa.gov/SP-4306/sp4306.htm*

[114] National Archives at College Park, 8601 Adelphi Road, College Park, MD 20740-6001, and at *http://www.nara.gov/nara/dc/Archives2_directions.html*

[115] NARA's National Personnel Records Center, Military Personnel Records, 9700 Page Avenue, St. Louis, Missouri 63132-5100, and at *http://www.nara.gov/regional/mpr.html*

Regional Office at East Point, Georgia, near Atlanta;[117] and the Arnold Engineering Development Center (AEDC) History Office at Tullahoma, Tennessee.[118] At each of these facilities, we are sure we could have found additional materials related to the development of hypersonic aircraft and high-speed wind tunnels.

The preceding bibliographic essay includes references from archives, libraries, printed materials, primary sources, and secondary literature that proved most influential to our study. We apologize to authors of other major works in aerospace history if we have neglected to include references to them in this essay. Not mentioning them does not mean that we were not aware of them, or failed to take them into account in preparation of our study. We have, in fact, cited many other works in our footnotes. But in a project of this magnitude, there was simply not enough time and energy on our part to locate, read, and completely digest every last document, every last article, and every last book that may have been relevant to the development of aerodynamics in America. As detailed and voluminous as this overall study became, just in terms of documents, it still only scratches the surface of what is available out there. Somewhere, on some dusty shelf, in someone's file drawer, hidden in some crumpled box, or some garage container, there are works waiting for other curious historians and scholars to discover, analyze, and interpret them.

What we offer, in the end, is not the final or penultimate "documentary journey through the history of aerodynamics in America," but our own unique journey along the rivers and streams running through our massive country. We invite you to retrace our steps and see what we found, what we mapped, the detours we took, and the locales we found most engaging. We hope you enjoy the adventure as much as we did, and that you are equally stimulated and informed by it, even though you may have chosen to follow a different route. We are sure to revisit this one incredible trip over and over again as we look back through these volumes, like a family browsing over the photo albums of their memorable cross-country vacation to the Grand Canyon. We are relieved the journey is over, and we feel

---

[116] George C. Marshall Space Flight Center History Office, CN22, Marshall Space Flight Center, Alabama 35812, and at *http://history.msfc.nasa.gov/*

[117] NARA Southeast Region (Atlanta), 1557 St. Joseph Avenue, East Point, Georgia 30344-2593, and at *http://www.nara.gov/regional/atlanta.html*

[118] Arnold Engineering Development Center History Office, 100 Kindel Drive, Suite B-330, Arnold Air Force Base, Tennessee 37389-2330, and at *http://candice.arnold.af.mil/aedc/history.htm*; E-mail: *history@hap.arnold.af.mil*

thankful that our own "family," the Auburn University project team, made it home in one piece. We are immensely proud of the contributions we made individually and collectively to what we hope will serve, for many years to come, as a standard reference work in the history of aeronautics.

Biographies of Volume 1 Contributors

James R. Hansen, Professor of History at Auburn University, has written about aerospace history for the past twenty-three years. His two-volume study of NASA Langley Research Center—*Engineer in Charge* (NASA SP-4305, 1987) and *Spaceflight Revolution* (NASA SP-4308, 1995)—earned critical acclaim, as did his *From the Ground Up* (Smithsonian Institution Press, 1988), the life story of aviation pioneer Fred E. Weick. His newest book, *The Bird Is on the Wing* (Texas A&M University Press, 2003), explores the role of aerodynamics in the American progress of the airplane from before the Wright brothers to the present. Currently, Hansen is at work on the authorized biography of Neil A. Armstrong.

D. Bryan Taylor is Instructional Technology (IT) Coordinator for the College of Liberal Arts at Auburn University. He holds an M.A. in history from Brigham Young University and is working on his Ph.D. in the history of technology at Auburn.

Jeremy R. Kinney is a curator in the Aeronautics Division, National Air and Space Museum, Smithsonian Institution. He holds a Ph.D. in the history of technology from Auburn University, with a specialization in aerospace history. Prior to joining NASM, he served as the American Historical Association NASA Fellow in Aerospace History.

J. Lawrence Lee holds B.M.E., M.A., and Ph.D. degrees from Auburn University. Following a lengthy career in mechanical engineering, Lee changed his focus to the history of technology, where his primary interests are transportation technology and engineering issues.

Chapter 1

The Achievement of Flight

Success four flights Thursday morning all against twenty-one mile wind started from level with engine power alone average speed through air thirty-one miles longest 57 seconds inform press home Christmas.

Orville Wright telegram sent on 17 December 1903 from the lifesaving station in Kitty Hawk, North Carolina, to his father, Bishop Milton Wright, in Dayton, Ohio. This document announced humankind's first successful powered flight of a heavier-than-air flying machine.

From Cayley's Triple Paper to Orville Wright's Telegram

Even in a study of NASA's—and its predecessor agency, the NACA's—contributions to aerodynamic development, one must begin by looking at a fruitful pairing, not of two government agencies but of two remarkable individuals, the Wright brothers. It was the inventive genius of Wilbur and Orville Wright, rather than an institutional imperative, that set the stage for aviation's astonishing technological progress in the United States and abroad during the twentieth century. Without retracing the path from the earliest concept of the airplane to the pioneering achievement of the Wrights, readers of this volume will not be able to truly appreciate the formative ideas from which research and development in the field of modern aerodynamics blossomed.

The Wrights accomplished what no one before them had the insight to recognize and tackle—they worked out a means of controlling an aircraft aerodynamically by deflecting the wing and tail surfaces themselves, as Wilbur had seen birds do. In this respect, they were far ahead of their predecessors and contemporaries, including some of the world's top scientists. The secret of their success was that they learned to fly: before they ever attempted to put an engine on their glider, and thus complete the "invention" of the airplane, they learned how to control a flying machine in the dynamic element of the air. They learned about the use of aerodynamic surfaces to control the airplane in a satisfactory manner and, together with the experience they developed as pilots of their kites and gliders, were able to fly their "powered" plane without difficulty. They were marvelously clear and pioneering thinkers, a rare combination. As such, they deserve a place of distinction at the beginning of any history of modern aerodynamics, especially one published during the centennial of their great achievement.

For this first chapter in the first volume of this series, then, there is no more fitting destination than the wind-blown sands of Kitty Hawk, North Carolina, where on 17 December 1903, the Wrights made the world's first successful powered flight of a heavier-than-air craft. The flight of the Wright airplane in 1903 culminated a long, faltering search for basic aerodynamic knowledge, a winding and irregular path that runs far back into the dawn of human experience, back even to our ancestors' prehistoric fascination with the flight of birds. No doubt, humankind dreamed about flying as soon as humans began dreaming. When civilizations arose, men and women imagined gods in the sky and gave them wings—no matter what ancient civilization one looks at, its gods could fly. The power of flight, denied to mortals, was envisioned as an ability of immortal deities. They lived in the heavens above; they moved across the sky in golden chariots; they soared majestically in and above the clouds.

While the materials needed to fabricate a craft resembling a modern hang glider were available thousands of years ago, unquestionably one of the major reasons that mechanical flight did not happen for millennia rested in the basic intuitive misconception that, in order for humans to fly, we needed to imitate birds. But exactly how did birds accomplish flight? Before modern scientific observations with high-speed cameras and, more importantly, before the dawn of a mathematically derived and rational theory of aerodynamics in the early nineteenth century, naked-eye observation of bird flight fostered a wrongheaded belief in flapping wings. From the wings crafted in ancient Greek mythology by Daedalus and Icarus to the ornithopter designed by Leonardo da Vinci during the Renaissance, the best idea anyone could come up with was to mimic a flapping wing structure and adapt it for human use. This path to heavier-than-air flight proved not only ignominiously unsuccessful but also often fatal. Donning makeshift wings of wood, feathers, and cloth, a handful of foolish men in the Middle Ages, flapping away without a whit of aerodynamic understanding, jumped to their deaths from the tops of cathedral towers and castle walls.

Not just popular imagination but also esoteric intellectual conceptions of the age mystified flight rather than rationalized it. Various medieval and Renaissance thinkers ascribed the ability of the wing to produce lift to psychic and occult forces possessed by the birds. This view held that for humans to fly required not some understanding and application of natural laws but rather their circumvention through the invocation of magical powers.[1]

This is not to say that, from ancient times to da Vinci, there was no one who expressed sound aerodynamic ideas. In certain basic respects, aerodynamics

---

[1] Charles H. Gibbs-Smith covers the pre-history of flight in *Aviation: An Historical Survey* (London: HMSO, 1970).

began in classical Greece during the fourth century B.C. with the first "scientific" observer of nature, the philosopher Aristotle. He developed a theory of motion that a projectile such as an arrow stayed in motion because the air separated in front of the arrow and then rapidly filled in the region behind it. According to this concept, the arrow kept flying because of the constant application of the medium's force in its wake. Despite being fundamentally wrong (though it was a widely accepted theory until Galileo's time), Aristotle's "medium theory" represented an attempt to explain the phenomena of flight in rational terms with the help of mental pictures of physical operations. This type of visual thinking was a crucial first pathway in the maze leading to mature understanding.

With other aerodynamic concepts, Aristotle proved even more insightful. He was the first thinker to elaborate on the notion of a "continuum." In a radically altered form, this became a basic concept of modern science, the foundation of aerodynamic flow applications today. In his writings, one can also find physical reasoning consistent with a concept of aerodynamic "resistance," inversely expressed through the appreciation that a body in motion, in Aristotle's words, "will move indefinitely unless some obstacle comes into collision with it."[2]

None of Aristotle's ideas prompted anyone to consider designing an airplane, however, nor did any of the concepts developed by Archimedes, the greatest engineering mind of the Hellenistic world. Archimedes, who developed the mechanical lever, built ingenious catapults, and created a screw device for raising water, is also recognized as founder of the science of fluid statics. But not even Archimedes, the author of the world's first "Eureka" experience, ever came close to the idea of mechanical flight. It is one thing to say that the brilliant minds of antiquity produced concepts in the raw, like the continuum or hydrostatic pressure that contributed to the development of aerodynamics, but quite another to say that any of them had technological import.

If science did not stimulate technology, maybe technology stimulated science? One might think that something like the invention of the kite, which took place in China around 1000 A.D., and which appeared in Europe three centuries later in the form of the windsock, would have provoked aerodynamic thinking. Or that

---

[2] Aristotle expressed most of the major tenets of his natural philosophy in his *Treatise on the Heavens (De Caelo)*, Book I. An excellent English translation of this entire treatise (in two books) was done by Stuart Leggat in 1995. His translation and commentary help the reader to appraise Aristotle's ideas in relation to the cosmologies of his predecessors. Leggat makes it clear that, while tied to the thinking of his day, Aristotle placed natural philosophy on a new footing, with his use of mathematics-style demonstration and his appeal to observation rather than reliance on what were then more standard forms of argument. John D. Anderson does a good job of interpreting Aristotle's significance in aerodynamics in *A History of Aerodynamics and Its Impact on Flying Machines* (Cambridge, England, and New York, NY: Cambridge University Press/Cambridge Aerospace Series, 1997), pp. 14-17.

the understanding of basic aerodynamics improved with the design of windmills, a powerful new machine that migrated into Western Europe from the Near East around the late thirteenth century. But neither seems to have been the case. History records no serious attempt to bridge the gap between the dream of flight, the actual world of invention, and basic scientific understanding of aerodynamic forces until the work of Leonardo da Vinci in the late fifteenth and early sixteenth centuries. Even then nothing came of da Vinci's thoughts. Leonardo's precious notebooks, encoded with his curious "mirror-like" reverse handwriting, stayed locked up and unknown in the hands of different private collectors for 300 years.

One wonders what might have germinated if da Vinci's ideas had been published and widely discussed during the Scientific Revolution of the sixteenth through eighteenth centuries. Certainly no one before da Vinci, or for three centuries after him, offered anything close to his astonishing vision of flying machines or considered so many of the principles affecting flight. In the same era that Christopher Columbus sailed across the Atlantic Ocean and Vasco da Gama circumnavigated Africa, Leonardo explored his own New World: diagramming the wing structures of birds and bats, developing streamlined shapes and aerodynamically more efficient artillery projectiles, and studying complex flow patterns in streams of water. He drew plans for an ornithopter, a parachute, and a vague sort of helicopter. He meticulously studied the mechanics of bird flight, and shortly before his death in 1519 he was apparently close to the breakthrough concept that a flying machine could be made with fixed instead of flapping wings.

Perhaps most significantly, da Vinci realized that "a bird is an instrument working according to mathematical law, an instrument which is within the capacity of man to reproduce with all its movements." In other words, he realized that applied mathematics was the key not only to understanding basic problems in nature but also to designing advanced technology. "Let no man who is not a mathematician read the elements of my work," he wrote. Immersed in Euclid's geometry and other mathematical texts, and coupling this knowledge with his meticulous observation of natural phenomena, Leonardo came up with the first quantitative expressions for fluid flow conditions, and then applied the same analytical thinking about flow conditions to aerodynamic phenomena. In 1513, he wrote:

> What quality of air surrounds birds in flight? The air surrounding birds is above thinner than the usual thinness of the other air, as below it is thicker than the same, and it is thinner behind than above in proportion to the velocity of the bird in its motion forwards, in comparison with the motion of its wings towards the ground; and in the same way the thickness of the air is thicker in front of the bird than below, in proportion to the said thinness of the two said airs.[3]

Considering this passage, modern aerodynamic engineer and historian John D. Anderson concluded that Leonardo was "three centuries ahead of his time." In his book *A History of Aerodynamics and Its Impact on Flying Machines* (1997), Anderson notes that what Leonardo contributed was "a valid description, expressed in the technical language of the early sixteenth century, of the sources of lift as well as pressure drag (form drag) on an aerodynamic body." Da Vinci's ideas about lift and drag, however, were not disseminated until the late 1790s when the Italian physicist Giovanni Battista Venturi discovered twelve volumes of Leonardo's notes in the Institute of Paris, where they had been moved by Napoleon's troops after the looting of the Ambrosian Library in Milan.

By that time Leonardo's undeniable genius was no longer critical, for other scientists between his life and Venturi's had advanced the understanding of fluid dynamic phenomena beyond da Vinci's own thinking. This started with the work of Galileo Galilei (1564–1642), who developed concepts of inertia and momentum, offered insights into the parabolic trajectory and aerodynamics of ballistic projectiles, and deduced that aerodynamic resistance was proportional not just to velocity, as da Vinci had understood, but also to the air density. Along with Galileo's work there were the pioneering hydrodynamic theories and experiments of the Italian physicist and mathematician Evangelista Torricelli (1608–1647), the inventor of the barometer and the first to measure air pressure, as well as the work of Frenchman Blaise Pascal (1625–1662), who advanced new fundamental principles about the actions of atmospheric pressure on fluid behavior.[4]

Later in the seventeenth century came what is referred to as "the first major breakthrough in the evolution of aerodynamics." This was the formulation of the "velocity-squared law," which states that the force on an object varies as the square of the flow velocity; or in other words, if velocity doubles, then the force acting on it goes up by a factor of four, not of two as Galileo and others had believed. Derived experimentally by Edme Mariotte (1620–1684) in France and Christian Huygens (1625–1695) in Holland, and confirmed theoretically by Isaac Newton (1642–1727) in his *Principia* of 1687, the velocity-squared law lifted scientific understanding of aerodynamics to a new level by adding a crucial, quantifiably exact variable to the mechanical relationships defining aerodynamic force.[5]

---

[3] Leonardo da Vinci, Codex Atlanticus Volume E, translated by R. Giacomelli, "The Aerodynamics of Leonardo da Vinci," *Journal of the Royal Aeronautical Society* 34 (1930): 1016–1038, cited in John D. Anderson, Jr., A History of Aerodynamics (New York, NY: Cambridge University Press, 1997), p. 24.

[4] On Galileo, Torricelli, Pascal, and Newton as well, see Richard S. Westfall, *The Construction of Modern Science: Mechanisms and Mechanics* (New York and London: John Wiley & Sons, Inc., 1971).

[5] Anderson, *A History of Aerodynamics*, pp. 32–37.

On the heels of the velocity-squared law came Newton's own "sine-squared law." Though later shown to be flawed, this suggested a means by which to calculate the pressure distribution over a body in motion and make predictions about its lift and drag. Newton's three laws of motion were another major contribution to aerodynamic understanding, especially his second law, which stated that the time rate of change for a moving body (momentum being the product of velocity and mass) was proportional to the force. Applied to fluid flow, this Newtonian law provided a basic equation on which to build further aerodynamic understanding. Other major blocks in the foundation were laid by the great Swiss mathematicians Daniel Bernouilli (1700–1782) and Leonhard Euler (1707–1783), who provided differential equations that, once their solutions could be worked out in the mid-1800s, provided accurate pressure distributions over aerodynamic bodies and, thus, a way of eventually analyzing low-speed flows and reliably calculating lift.[6]

But all this simply provided analytical tools for future use, for none of these scientists and mathematicians of the seventeenth and eighteenth centuries had any interest whatsoever in applying them to the design of a flying machine. Bernouilli and Euler may have developed the mathematics that would be used to show how the pressure differential on the upper and lower surfaces of a wing developed the phenomena of lift, but they themselves never attempted to design an effective shape to achieve this. This, the problem of designing an effective wing or "airfoil," was a fundamental problem that needed a solution if heavier-than-air flight was ever to be achieved.

All that came before the early 1800s served only as prologue to the story leading to the invention of heavier-than-air flight by the Wright brothers. The genuine narrative begins in 1804 when Sir George Cayley (1773–1857), a twenty-one-year-old scientifically educated English country squire, designed and hand-launched a small glider. Not more than a meter long, Cayley's innocent-looking glider of 1804 represented a revolutionary breakthrough, as it essentially incorporated all the elements of the modern airplane. This simple yet profound design effectively combined the three essential features recognized today in the configuration of a modern airplane: a fixed wing, a body or fuselage, and a tail with both horizontal

---

[6] A number of fine new books and translations have added considerably to our historical and technical understanding of Isaac Newton's contributions, notably *The Principia: Mathematical Principles of Natural Philosophy*, a translation of Newton's great work of 1687 by Julia Budenz, Anne Whitman, and I. Bernard Cohen (Berkeley, CA: University of California Press, 1999). Other important studies that have recently appeared include Francois De Gandt and Curtis Wilson, *Force and Geometry in Newton's Principia* (Princeton, NJ: Princeton University Press, 1995); Gale E. Christianson, *Isaac Newton and the Scientific Revolution* (Oxford, England: Oxford University Press, 1996); A. Rupert Hall, *Isaac Newton: Adventurer in Thought* (Cambridge, MA, and New York, NY: Cambridge University Press/Cambridge Science Biographies, 1996); Jed. Z. Buchwald and I. Bernard Cohen, *Isaac Newton and Natural Philosophy* (Cambridge, MA: MIT Press/Dibner Institute Studies in the History of Science and Technology, 2000); and David Berlinski, *Newton's Gift: How Sir Isaac Newton Unlocked the System of the World* (New York, NY: Free Press, 2000).

and vertical surfaces. While producing such a glider may seem like child's play today, at the time it was totally at variance with existing flapping wing, or ornithopter, concepts of how to achieve heavier-than-air flight.

Cayley's configuration was the product of some bold new thinking about what it was going to take to fly. For example, Cayley understood how to produce the lift necessary for flight, because he recognized that curved surfaces produced more lift and less drag than flat surfaces. He also understood that the airplane was going to have to be stabilized and controlled, thus the need for tail surfaces. And he understood mathematically what kind of power it was going to take to achieve flight—and this was well in advance of the invention of practical lightweight engines themselves. While he benefited from his knowledge of existing scientific theory, such as Newton's sine-squared law, Cayley's 1804 glider design was essentially "his own." On his own he performed laboratory experiments; conducted flight testing, and analyzed his results. What was so remarkably new about Cayley was the way he broke the elements necessary for flight into what would later be called "systems," i.e., a system to produce lift, a system for control of the craft in the air, and a system to produce propulsion. Before Cayley, people had thought not in terms of separate systems that could be effectively integrated, but of one organic system, like that of a bird, that somehow could do everything.[7]

Though Cayley first arrived at his concept of the airplane in 1799 and integrated it into his 1804 glider, his radically new approach to understanding the airplane is best exemplified in a three-part paper "On Aerial Navigation," written in 1809 and published in three parts in *Nicholson's Journal of Natural Philosophy, Chemistry, and the Arts*.

Without question, this paper, reproduced in its entirety as the inaugural document of this volume, represents "the highlight in the history of aerodynamics"—not just for the beginning of the nineteenth century, but for all human history up to that time. (See Document 1-1.) As John Anderson writes in his history, "the work of all previous investigators pale" compared to Cayley's treatise.[8] For its relationship to the subsequent development of the airplane, this document by George Cayley compares to such seminal publications as Copernicus' 1543 *On the Revolution of Heavenly Bodies*, Darwin's 1859 *Origin of Species*, and Watson and Crick's 1953 paper on their discovery of the structure of DNA. Cayley himself kept working on flying machines up to his death in 1857.

---

[7] Anderson does a very complete job of analyzing Sir George Cayley's seminal contributions in *A History of Aerodynamics*, pp. 64–80. There is no monographic treatment of Cayley, though one is much needed. In 1968, the Science Museum of London published a "Science Museum Booklet" entitled *Sir George Cayley, 1773-1857*, authored by C. H. Gibbs-Smith.

[8] Anderson, *A History of Aerodynamics*, p. 79.

Curiously, after the publication of the "Triple Paper," Cayley shifted his attention for four decades to the development of balloons and airships. Most likely he did this because of the vexations of developing an engine for airplane applications. Steam engines with huge boilers simply weighed too much, and innovative new engines, such as gas-fueled, internal combustion engines, were still very experimental and quite limited in their power output. But even without the right engine, in 1843 he refocused again on the airplane and began designing full-scale gliders anew. One of them carried a ten-year-old boy without incident a few yards down a hill in 1849. Four years later another carried his reluctant coachman across a small valley, possibly flying as far as 500 yards. (Unaware of the historic auspices of this 1853 flight, the coachman on landing allegedly remarked, "Please, Sir George, I wish to give notice. I was hired to drive and not to fly.")

With its roots in the scientific revolution, Cayley's work provided the trunk for a growing tree of aerodynamic knowledge and invention that in half a century led to the Wright brothers' achievement in December 1903. From that solid trunk, however, at least three major branches of experimental heavier-than-air flight technology grew in the nineteenth century, and no one at the time could be certain which branch would ultimately bear the most fruit.

The first branch involved trying to learn about flight by experimenting with small-scale models. This proved to be a very useful approach for testing aerodynamic concepts, one that cost little money and did not risk life or limb. The most important individual taking this approach was the young French marine engineer, Alphonse Pénaud (1850–1880), who in the 1860s and 1870s experimented with a number of little flying models powered by twisted rubber bands, much like the toys children still play with today. His most successful model flew 131 feet across the Tuileries Gardens in Paris in August 1871. Pénaud based his model on Cayley's 1804 glider design, but improved it in key respects. He set its horizontal tail to a negative angle of eight degrees relative to the chord line of the wing, which gave his airplane greater longitudinal (or pitching) stability. He also bent its wing tips up in a dihedral angle, thereby providing lateral (or roll) stability. Changes like this, which would have been difficult, expensive, and potentially dangerous if made with a full-scale, manned aircraft, were relatively simple to make and flight test with a small-scale model.[9]

What Pénaud sought was a mechanism that would guarantee absolute inherent stability, so he ignored the possibility of active pilot control. In this he shared the

---

[9] Pénaud is discussed in most surveys of the development of flight, but there is no single exhaustive study devoted to his work. Anderson covers Pénaud adequately in *A History of Aerodynamics*, pp. 193–194. See also Tom D. Crouch, *A Dream of Wings: Americans and the Airplane, 1875–1905* (Washington, DC, and London: Smithsonian Institution Press, 1981), pp. 36, 46, 57–59, 64, and 192; as well as Crouch's *The Bishop's Boys: A Life of Wilbur and Orville Wright* (New York and London: W. W. Norton & Co., 1989), pp. 56–57, 161, 164, 168–169, 249, and 342.

mindset of many aeronautical thinkers at that time, who believed that their first task was to demonstrate how to fly in a simple straight line with a passenger on board and that the issue of control could be dealt with later. Pénaud's emphasis on automatic stability was also a limitation of the scale-model approach. Working with models powered by twisted rubber bands, and without any means of piloting them once launched, Pénaud had little choice but to go after inherent stability. With their inefficient power source, his best flights lasted only thirteen to fourteen seconds, and were characterized by Professor Langley of the Smithsonian Institution as being, "so erratic, and so short, that it was possible to learn very little from them."

At the end of the nineteenth century many people regarded Samuel Pierpont Langley (1834–1906) as the unofficial chief scientist of the United States, which makes it even more significant that Langley, though aware of the limitations of Pénaud's experiments, also focused his own work on scale models. In fact, Langley subscribed to the erroneous and ultimately tragic assumption that results from scale models could be directly scaled up to design a full-scale airplane. In

Although he ultimately failed in his effort to build a successful man-carrying airplane, one should not minimize Professor Samuel P. Langley's many contributions to aerodynamic understanding, especially those embodied in the flights of his large steam-powered models in 1896. In this photograph from December 1895, Langley's *Aerodrome No. 6* (which evolved from *No. 4*) had not yet matured into its final configuration. SI Negative No. A-2854-J

1887, Langley left his work in astronomy and solar science, for which he was very distinguished, to take the prestigious position of Assistant Secretary of the Smithsonian Institution (a few years later he became Secretary). There he embarked on a new course of remarkable aeronautical experiments. He followed the approach of Pénaud but with models that quickly grew much larger, more complicated, and were powered by small steam and gasoline engines rather than rubber bands. His research paid off, up to a point. On 6 May 1896, his *Aerodrome No. 5*, a steam-powered model weighing thirty pounds and with tandem wings spanning over thirteen feet, made two long flights from a launching platform atop a houseboat in the Potomac River south of Washington, D.C. The best of the flights that day covered 4,200 feet in forty-five seconds. Celebrated inventor Alexander Graham Bell, a close friend of Langley's, observed these flights firsthand and published ecstatic accounts of what Langley had achieved as a major step on the way toward powered flight.[10] (See Document 1-8.)

In terms of the history of aerodynamics, Langley made other major contributions. His work in the late 1880s and 1890s has been called "the first meaningful aerodynamic research" in America, a systematic program of tests involving whirling arms and other sophisticated experimental instruments of his own design, the purpose of which was to measure the various aerodynamic forces at work on different surfaces, including propeller shapes. Interestingly, however, this program of basic research "contributed little of practical value" to the design of either his or any other flying machine of the day. Moreover, contemporary scientists, including British giants Lord Kelvin and Lord Rayleigh, condemned his most provocative aerodynamic conclusion, the so-called "Langley's law," which stated that the power required for a vehicle to fly through the air decreased as the velocity increased.[11] (See Document 1-4.) Lilienthal, the Wrights, and others trying to fly believed that this counterintuitive idea had to be wrong—which it was, except for in the narrow range of Langley's test velocities (20 meters per second or less).

A difficult personality at best, Langley today is chiefly remembered as the one who failed where the Wrights succeeded. Yet perhaps the best commentary on his role and achievements came from the Wrights themselves. Writing to a friend shortly after Langley's death, Wilbur Wright said of him:

---

[10] No comprehensive biography has yet been written of Dr. Samuel P. Langley. The books by Tom Crouch cover Langley's importance in some significant detail, as does Anderson's *A History of Aerodynamics*. But as much as any other figure in the early history of aviation that has not yet received one, Langley deserves an exhaustive, singular treatment.

[11] For an analysis of "Langley's law" and how it erred, see Anderson, *A History of Aerodynamics*, pp. 179–181.

The successful launch of Langley's *Aerodrome No. 5* from atop a houseboat in the Potomac River, on 6 May 1896. SI Negative No. A-18870

The knowledge that the head of the most prominent scientific institution of America believed in the possibility of human flight was one of the influences that led us to undertake the preliminary investigation that preceded our active work. He recommended to us the books which enabled us to form sane ideas at the outset. It was a helping hand at a critical time and we shall always be grateful. . . . When scientists in general considered it discreditable to work in the field of aeronautics he possessed both the discernment to discover possibilities there and the moral courage to subject himself to the ridicule of the public and the apologies of his friends. He deserves more credit for this than he has yet received. . . . Though we have rarely followed his lead, we have always found a study of his writings very profitable, especially at the time when we were trying to find out what the real sticking points of flying were.[12]

---

[12] Wilbur Wright, letter to Octave Chanute from Dayton, Ohio, 8 November 1906, found in *The Papers of Wilbur and Orville Wright*, ed. Marvin W. McFarland, Vol. 2 (New York, NY: McGraw-Hill Book Company, Inc., 1953), pp. 736–738.

Langley's contemporary impact on the development of aerodynamics extended beyond his influence on the Wrights; as the culmination of the nineteenth century scale-model approach to aeronautical knowledge, Langley's experiments also had a much broader significance. First, his publications, beginning with his 1891 book *Experiments in Aerodynamics*, inspired others to step up their experimental efforts and to be systematic in carrying them out. Second, and perhaps even more importantly, the fact that one of the greatest scientists in the country had decided to devote his efforts to the problems of flight convinced a great number of people that "aeronautics was no longer the past-time of fools."

The notion that only fools tried to fly was certainly exacerbated by the second major branch of technological experimentation in heavier-than-air flight during the nineteenth century. This method involved venturesome attempts to build the real thing, full size, and then try to fly it with a person on board. This approach offered the advantage of more meaningful, immediate, and sensational results—not to mention the possibility of great notoriety for the successful inventor. But full-scale flying machines also meant considerably greater cost in construction, dramatically increased danger in testing, and the inability to experimentally vary design parameters quickly and easily.

Given the general impatience of human nature, though, it is not surprising that a number of individuals pursued this course anyway. In the 1840s, two Englishmen, William Henson and John Stringfellow, designed an "aerial steam carriage" with a huge 150-foot wingspan, powered by a thirty-horsepower steam engine they themselves planned to build. Their objective was to develop an "Aerial Transport Company" to haul goods commercially worldwide, an idea far ahead of its time. Their ambitious machine, however, never flew—in fact, it was never built, and all they constructed was a twenty-foot model that never made it into the air. Yet so many fanciful pictures of their proposed machine appeared in contemporary newspapers and magazines that many people believed the Henson-Stringfellow airplane flew over the Tower of London, the Pyramids, and even the Taj Mahal. These drawings cemented in the public consciousness the image of what an airplane should look like, with rectilinear wings, an enclosed cabin fuselage directly under the wing, twin screw propellers, and pilot-controlled tail surfaces. Thanks to the publicity surrounding this "machine that never flew," George Cayley's basic formula for configuring an airplane became the firmly established technological norm. After about 1845, when anyone thought of an airplane, they pictured it in terms resembling the Henson-Stringfellow vision of the basic Cayley design.[13]

---

[13] On the aerodynamics involved with the "aerial steam carriage" conceived by Henson and Stringfellow in the 1840s, see Anderson, *A History of Aerodynamics*, p. 194. See also Crouch, *A Dream of Wings*, pp. 28 and 89.

Maxim's huge four-ton biplane, with its eighteen-foot propeller, sits on its track ready for flight testing (unsuccessful) in 1894. Ultimately the monster rose slightly from its guardrails, but only after traveling some 600 feet down the track. It possessed no truly redeemable features and represented, in its two very efficient steam engines, a brute force approach to getting an airplane aloft. SI Negative A-212-A

In the second half of the nineteenth century there were other full-scale attempts at flying machines. Two naval officers, Felix du Temple in France in 1874 and Alexander Mozhaiski in Russia ten years later, sent subordinates in steam-engined machines down inclined planes to achieve short powered hops, but neither of these attempts did much to advance either aerodynamics or the practical development of the airplane. The most elaborate full-scale attempt was made by Hiram Maxim (1840–1916), the American engineer in London famous for his invention of the machine gun. In the early 1890s Maxim spent a good part of his new fortune on the three-year development of a large biplane called the *Leviathan*, with a wingspan of 104 feet, powered by two 180 horsepower steam engines driving eighteen-foot propellers, and weighing nearly four tons. The design, however, was structurally weak, aerodynamically unsound, and lacked

---

[14] See Iain McCallum, *Blood Brothers: Hiram and Hudson Maxim—Pioneers of Modern Warfare* (Chatham Publishers, 1999). Though this book focuses on weapons development, notably the "Maxim gun," it also covers Hiram S. Maxim's unsuccessful attempts to fly. For first-hand accounts of his experiments in flight, see Hiram S. Maxim, "Natural and Artificial Flight," *Aeronautical Annual* (1896); and "Aerial Navigation by Bodies Heavier Than Air," *The Aeronautical Journal* 6 (January 1902): 2–7. On Maxim's views of the early work of the Wright brothers, see "The Recent Experiments Conducted by the Wright Brothers," *The Aeronautical Journal* 10 (January 1906): 37–39.

effective controls. In 1894 the *Leviathan* crashed into a guide rail after briefly rising mere inches from the ground.[14] (See Document 1-6.)

Perhaps the closest thing to a successful powered flight of a full-scale heavier-than-air machine took place in France four years prior to Maxim's debacle, when Clement Ader's bat-winged *L'Eole* rose twelve inches off the ground in a "flight" of possibly 165 feet. But Ader's achievement was not in the same league as the historic Wright flight of 1903, no matter what French enthusiasts, then and now, may claim about it. All *L'Eole* really managed was a short, uncontrolled, powered hop and nothing more, which was fortunate given that the craft could not have sustained itself, as it had no tail and no method of lateral control, a formula for disaster.

A second Ader machine, the *Avion II*, was never completed, and a third in 1897, the *Avion III*, refused to leave the ground. Neither Du Temple, Mozhaiski, Maxim, nor Ader gave serious thought to how to control their aircraft in flight or what sort of skill the pilot would need to have. All they wanted to do was simply get a machine carrying a person into the air—and they did not truly accomplish even that.[15]

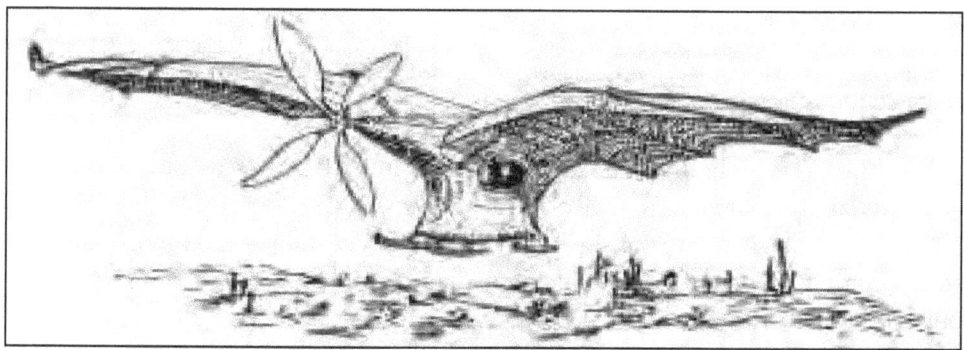

This is Ader's own perspective sketch of the *Éole*. The craft's wing spanned fourteen meters, with the length of the fuselage only 6.5 meters. The airframe weighed just over 175 kilograms. The bat-shaped wings could be folded for storage. Notice that at this point in time Ader considered fitting his plane with narrow caterpillar-like tracks rather than wheels. NASM File AA-006502-01

---

[15] One can divide the works on Clement Ader into those that support the claim that his *L'Eole* successfully flew and those who refute or seriously qualify it. Among the works favorable to Ader are: Jacques May, *Ader* (Librairie Aeronautique, 1910); Georges de Manthe, *Clement Ader, sa vie, son oeuvre* (Edition Privat, 1936); Louis Castex, *L'homme qui donna des ailes au monde* (Editions Plon, 1947); Louis Castex, *Clément Ader, ou l'homme qui voulait se faire oiseau* (Presse de la cite, 1967); Raymond Cahisa, *L'aviation d'Ader et des temps héroiques* (Albin Michel, 1950); and Pierre Lissaargue, *Clément Ader, Inventeur d'avions* (Edition Privat, collection Bibliotheque Historique, 1990). For a more critical analysis, see Charles H. Gibbs-Smith, *Clément Ader, his claims and his place in history* (London/Science Museum, Her Majesty's Stationery Office, 1968). There are a number of Web sites that are devoted to Ader, notably that administered by Le musée Clement Ader at *http://le-village.ifrance.com/eole/squeletteader.html* There is information about the *Avion III* at *www.cham.fr/museum/revue/ref/r13a04_a.html*, which is maintained by the Musée des arts et métiers in Paris. Restored in the 1980s, the *Avion III* serves as one of the centerpieces of this museum.

Fortunately, there existed a third approach, one pursued earlier by Cayley himself, which investigated the problems of flight with full-scale manned gliders. This approach may have seemed the most foolhardy to some, involving jumping off hills and crashing nose first into the dirt; or the least sophisticated to others, because it did not involve steam engines or other kinds of power plants. But learning what type of wing shape and aerodynamic configuration best lifted a man into the air, and what sort of system and piloting maneuvers then allowed that man to control his flight in three dimensions, proved absolutely essential to solving what came to be recognized in the late 1800s, not just by aeronautical enthusiasts but by the engineering community as a whole, as "The Problem of the Century."[16]

Certainly, if there was one secret to the Wrights' success, it was that they learned to fly and control their airplane in glider form before they ever put power into it. But these brothers from Dayton, Ohio, were not the first to understand the fundamental importance of this human/machine interface as part of the invention of powered, heavier-than-air flight. Before the Wrights, there were two key individuals who provided technological breakthroughs in this regard that built significantly on the work started by Cayley and laid out a clear path to the future. They were two professional engineers: Otto Lilienthal (1848–1896) and Octave Chanute (1832–1910).

German mechanical engineer Otto Lilienthal made his living manufacturing small steam engines and marine foghorns in a little factory he operated on the outskirts of Berlin. With an unbridled passion for flying machines, Lilienthal conducted laboratory experiments and gathered data throughout the 1880s, trying to grasp and measure the way wings generated lift and generally searching for some real understanding of aerodynamic principles. Feeling that he had learned all he could from unmanned gliders, in 1891 he built and flew his first full-scale manned hang glider. From then until his death, the result of a gliding accident in October 1896, Lilienthal made about 2,000 glides.[17] (See Document 1-5.)

Lilienthal was not the first human being to fly. At least five other individuals had flown in gliders, beginning with Cayley's two hapless "test pilots." In France in the

---

[16] This phrase was used by Augustus M. Herring as the title of a 1897 paper about his flying experiments with Octave Chanute.

[17] There is no satisfactory biography of Lilienthal in English, nor have any of the monographic treatments written in German ever been translated. In German, one should start with the works by Gerhard Halle: *Otto Lilienthal und seine Flugzeug-Konstruktionen* (Dusseldorf: VDI-Verlag, 1962) and *Otto Lilienthal: die erste Flieger* (Dusseldorf: VDI-Verlag, 1976). Other studies of Lilienthal have been done in German by Werner Schwipps, Jutta Wegener, Karl-Dieter Seifert, and Stephan Nitsch. Anyone interested in Lilienthal's life who does not read German should start with the British translation of his autobiographical *Birdflight As the Basis of Aviation: A Contribution Towards a System of Aviation* (London, 1911). Tom Crouch's *A Dream of Wing* also offers an excellent chapter on Lilienthal's legacy in America, and the Smithsonian Institution Press/National Air and Space Museum published a booklet on *Otto Lilienthal and Octave Chanute: Pioneers of Gliding*, 1980.

1860s Jean Marie le Bris and Louis Mouillard both flew in gliders, while American John Joseph Montgomery flew three different hang gliders of his own design off a hill near San Diego harbor in the 1880s.[18] Universally, however, these predecessors became so frightened after making a few flights that they either refused to go up again or went back to the drawing board. Lilienthal was the first to persist. He continued making gliding flights, incorporating what he learned in his flights into the design of subsequent gliders. He even arranged for a cone-shaped hill to be built for him so he could fly his gliders no matter which direction the wind came from. He became the first man to really know how to fly, earning the nickname "The Flying Man" in accounts of his exploits. His machines were very well made and included both monoplane and biplane configurations. Some of his flights covered over 1,000 feet and lasted as long as fifteen seconds. Stories about his wonderful flights and piloting abilities appeared in newspapers and magazines around the world, and Lilienthal himself wrote a book and published a number of articles explaining his techniques and understanding of aerodynamic principles. These publications educated people about flight and motivated countless people, including the Wright brothers, to consider taking on the problem of flight for themselves.

On 8 October 1896, Lilienthal was gliding about fifty feet in the air when a strong gust of air caused his craft to nose up and stall. He was unable to regain control and his hang glider entered a terminal spin and crashed to the ground. Lilienthal suffered a broken back in the crash and died the next day in a Berlin hospital. The tragedy came as a terrible blow to all would-be aviators, for if not even the great "Flying Man" could avoid a fatal accident, how could anyone else expect to do it?

As much as Lilienthal had worked to address the matter of aerodynamic control in his gliders, what killed him was nonetheless insufficient control. In essence, Lilienthal's method for control depended solely on the pilot shifting his weight under his glider in a type of "negative guidance" whereby he countered any unwanted aircraft movement by using his own body to shift the center of gravity of the whole machine. If the craft started to stall, all Lilienthal could do was throw his legs forward to try to pull the nose down and regain flying speed. Such a control system was fatally flawed because if the pilot did not shift his weight in just the right way and at just the right time his movements could actually make things worse. Furthermore, this system held little promise for purposes of subsequent development, because the ability of a pilot to control an airplane by shifting his own weight diminished rapidly as the size of the airplane increased.[19]

---

[18] See Crouch, *A Dream of Wings*, p. 90 on Le Bris; pp. 20–21, 65–72, 83, 90, 176–178, and 228 on Mouillard; and pp. 87–100 and 307–309 on Montgomery.

[19] For a more detailed technical analysis of Lilienthal's aerodynamics and the reasons behind his fatal crash in 1896, see Anderson, *A History of Aerodynamics*, pp. 138–164.

At the time of his death at age forty-eight, Lilienthal was thinking about propulsion systems and the problems that needed to be solved if he was to move forward to a powered airplane. There was no telling what he might have accomplished if he had survived.

Aviation's first martyrdom proved catalytic. Although pessimists took it as proof that humankind was not meant to fly, more optimistic individuals reacted to news of Lilienthal's demise with even greater determination to get up into the air. In America, Octave Chanute had just finished a successful summer season test flying a series of different gliders at the Indiana Dunes on the south shore of Lake Michigan, east of Chicago. By the time he heard the horrible news of Lilienthal's death, Chanute and a small group of young colleagues had evolved an advanced biplane glider incorporating a new approach to the problem of airplane stability. This glider proved to be "the most significant and influential aircraft of the pre-Wright era."[20]

Given the depth of Chanute's talents as an engineer and his encylopedic awareness of what was being tried aeronautically in America and around the world, it is not surprising that his 1896 glider represented the most advanced state of the art of that time. He emigrated to the United States with his family from France in 1838 when he was six, by the mid-1870s Chanute was one of the best known and most respected civil engineers in the country, renowned for his bridges and railway structures. At the peak of his career, Chanute was chief engineer of the Erie Railroad and president of the American Society of Civil Engineers and the engineering section of the American Association for the Advancement of Science. Interested in flight since his childhood, in the 1880s Chanute entered a period of semiretirement that allowed him to devote his energies to the problems of "aeronautical navigation." He initiated a regular correspondence with dozens of people

An accomplished railroad engineer, when Octave Chanute (1832–1910) turned his attention to the design of flying machines in the 1880s and 1890s and began corresponding with others with similar interests, he inadvertently created the first international aviation community, a creation essential to the successful invention of the airplane. SI Negative A-21147-B

---

[20] See Crouch, *A Dream of Wings*, pp. 175–202, and Anderson, *A History of Aerodynamics*, pp. 192–197.

all over the world who were interested in flying machines, including prominent members of the Aeronautical Society of Great Britain, the most important of the early organizations devoted to the study of flight. He also published what he learned, starting with a series of critically important essays on "Progress in Flying Machines," that appeared in twenty-seven installments in the *Railroad and Engineering Journal*, beginning in October 1891, and was reprinted three years later as a single volume.[21]

Not satisfied with the important role he filled in the aeronautics community, by serving as an active clearinghouse for aeronautical information and by making aeronautics a respectable concern for engineers, Chanute also wanted to test his own ideas about flying. In 1888, he laid out an idea for the design of a flying machine combining features of an airship and an airplane. Abandoning this hybrid, he quickly turned to manned gliders, unveiling a bold new design in late 1894 that was to serve as a test bed for his own novel concepts. Chanute's original glider design featured an approach to control problems that was significantly different from Lilienthal's negative or reactive stability, in which the pilot shifted his body weight. The veteran engineer envisioned instead a system of automatic stability that did not require much from a pilot—in essence, he wanted the structure itself to stabilize the airplane. To do this he proposed an arrangement of multiple wings set in tandem, suggesting as many as four pairs of tandem biplane wings, or sixteen in all. Such a redundant structure, he calculated, would reduce the movement of the center of pressure on each surface by adjusting automatically to shifts in the airplane's center of gravity. The "operator," using Chanute's term, "need only intervene when he wants to change direction, either up or down, or sideways." Furthermore, Chanute added yet another element of automatic stability by designing each one of his tandem wings so that they could rotate slightly. When hit by a strong gust of wind, explained Chanute, "the wings are blown backward . . . the aeroplane tips slightly to the front, thus decreasing the angle of incidence,

---

[21] It is truly amazing that there is yet no published biography of Chanute, considering how important he was to not only the development of flight technology but also of American engineering generally. The seminal figure in Crouch's *A Dream of Wings* is unquestionably Chanute rather than Langley or the Wrights. The main collection of Chanute Papers rests in the Manuscript Division of the Library of Congress. The Denver (CO) Public Library holds a large collection of Chanute materials assembled by Ms. Pearl I. Young, who for many years, starting in the early 1920s, worked for the National Advisory Committee for Aeronautics (NACA) at Langley Field, Virginia. Young grew fascinated with the early history of flight, particularly Chanute's contributions. At the Denver Public Library, one will find Young's *Bibliography of Items About Octave Chanute, Complete Writings of Octave Chanute*, the *Chanute-Mouillard Corespondence*, and two essays by Young: "Octave Chanute and New England Aeronautics" and on "The Contributions of Octave Chanute, 1832–1910." Anyone reconstructing Chanute's story would also want to see Marvin McFarland, ed., *The Papers of Wilbur and Orville Wright, Including the Chanute-Wright Letters and Other Papers of Octave Chanute*, recently reissued by McGraw Hill, in January 2001.

so that the aggregate 'lift' is diminished." When the wind subsided, a coil spring pulled the wings forward to return to "their normal adjusted position."

Flight trials with different gliders at the Indiana Dunes in the summer of 1896 showed Chanute the error of at least some of his concepts and stimulated a major re-rigging of the multiplane glider, dubbed the *Katydid* for its insect-like appearance. The redundant wings caused far too much lift, and with some surfaces creating more lift than others this caused dangerous imbalances. No one, including the sixty-four-year-old Chanute, wanted to risk flying it at first, so the glider was initially flown tethered like a kite. After numerous design changes, a young fellow engineer named Augustus M. Herring made a series of flights in the glider, the longest a glide of eighty-two feet. Word of the flights spread to neighboring Chicago and reporters from a number of newspapers and the national wire services trekked to the southeast shore of Lake Michigan to observe the action. Not wanting the attention, Chanute and his men packed up and went back to a workshop in Chicago. There they applied what they had learned to the design of a brand-new glider.

One should view the work of Chanute and Herring as collaborative. In the truss bracing used on the gliding machines, one detects Chanute's engineering skills; in the cruciform tail and overall design schemes lay Herring's major contributions. In terms of the fundamental aerodynamics, it is unclear whose understanding was superior. Chanute knew much more mathematics, but Herring's physical intuition was in many ways sharper. Here, Herring stands ready to take off in his two-surface gliding machine in 1896. SI Negative A-30907-H2

The result was a simple, dramatic improvement: a far simpler, triplane glider made with a single rigid box frame. Much less cumbersome than the *Katydid*, it flew at the Indiana Dunes in 1896 as the first modern aeronautical structure, of which Chanute's team had evaluated the forces, figured the strengths, and precisely calculated the performance. Looking at its external bracing of crossed diagonal wires and upright struts, one could see the face of airplane structures to come, beyond the World War I era. The new glider also featured an effective combination of horizontal and vertical tail surfaces put into a cruciform (or cross) shape. That such a tail, appropriately positioned, enhanced the pitch stability of an airplane was understood as far back as Cayley. But Chanute and his men were not satisfied with just the advantages of a fixed tail. Under young Herring's direction, they made their tail moveable (or "regulating") so that it could react positively to gusts, keeping the airplane straight and level. At the suggestion of one of Chanute's men, carpenter and electrician William Avery, the triplane soon metamorphosed into a biplane simply by removing the bottom wings. The resulting performance of this "two-surface" machine was phenomenal—the equal of anything flown by Lilienthal and, in its design features, significantly more progressive. (See Document 1-9.)

One would think that Chanute and his men would have followed up immediately on their sensational success; after all, it was the season-after-season buildup of knowledge and know-how that proved so essential to the Wright brothers' eventual achievement at Kitty Hawk. But it did not happen this way with Chanute. When the summer of 1896 ended, so did his experimental flying. The following summer he and his men did not return to the Indiana Dunes. Herring went to explore the possibility of powered hang gliders in the vicinity of St. Joseph, Michigan.[22] Chanute kept collecting aeronautical information, writing articles for publication and corresponding with dozens of others in pursuit of the dream. Perhaps due to his age, he did not persist with flight trials. It was certainly clear from his 1897 article, "Recent Experiments in Gliding Flight," that he was not sure that powered flight was imminent. In fact, without Herring around to push him in bold new technological directions and describe what it was like to pilot a plane, Chanute reverted to old ideas, notably the absolute priority of automatic stability. He set aside the world's most advanced glider and reverted to the design of yet another monstrously complex multiplane.

---

[22] On Herring, see Crouch, *A Dream of Wings*, pp. 203–223. His ancestors to this day continue to believe that Augustus Herring made a powered flight in an airplane before the Wrights; see, for example, Lou Mumford, "Family claim historical accounts of first flight aren't 'Wright,'" *South Bend Tribune*, 11 October 1998. This story can be accessed at *http://www.southbendtribune.com/98/oct/101198/local_ar/117176.htm*

The year 1896 was thus a pivotal time in the history of aeronautics, and in that year two brothers in Ohio, self-trained designers without high-school diplomas, turned their attention away from their print shop and bicycle business in order to take a crack at inventing the airplane. These two unique individuals, Wilbur Wright (1867–1912) and Orville Wright (1871–1948), brought not only fresh perspectives and new energy to the fledgling field of flight research, but also one of the most remarkable collaborations of genuine talent in the history of invention. Writing shortly before his death, Wilbur Wright offered the following insight into the fruitful relationship that sustained them in their work and jointly magnified their respective abilities:

> From the time we were little children my brother Orville and myself lived together, played together, worked together and, in fact, thought together. We usually owned all of our toys in common, talked over our thoughts and aspirations so that nearly everything that was done in our lives has been the result of conversations, suggestions and discussions between us.[23]

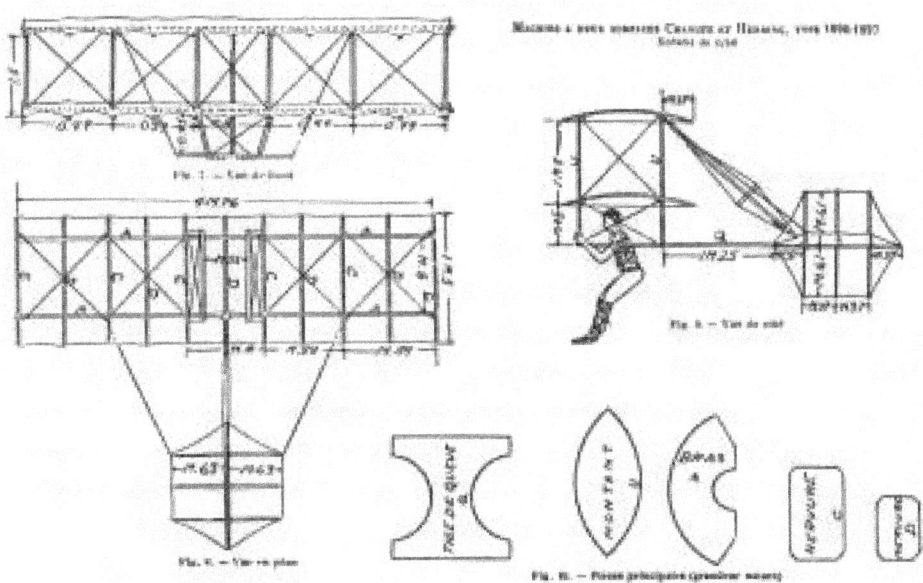

Views of Herring's three-surface glider from 1896. Smithsonian Institution

---

[23] Wilbur Wright, *Papers of Wilbur and Orville Wright*, 3 April 1912.

Without this creative synergism between them, it is impossible to imagine them inventing the airplane.

Another thing that is very important to understand about the Wright brothers is how they followed their own path in pursuit of the invention of the airplane. Not that they ignored all the work of others. In fact, one of the first things they did after deciding to address the problems of flight in 1896 was collect all the aeronautical information that was available. First, they exhausted the collections at the Dayton Public Library, then they wrote to Chanute, who was kind enough to write back and advise them to learn how to glide before they tried a powered flight. Aware of Langley's work, they also wrote to the Smithsonian Institution, which replied by sending them a number of articles, including works by both Langley and Lilienthal. (See Document 1-8.) But once they had digested this knowledge, they were not to be boxed in by it. Instead they approached technical problems from their own unique perspective and came up with equally unique solutions.

The most significant example of this crucial characteristic of the Wrights concerned the decisive matter of control. They did something that no airplane pioneer up to that time had done: they isolated the control of an airplane as the main problem that needed attention. Lilienthal, Chanute, and Langley had obviously designed wings that were able to lift machines into the air, and while there were many aerodynamic improvements yet to come regarding lift, the invention of the airplane did not have to wait for them. Neither did the problem of propulsion overly concern the Wrights, because all over the country mechanics were developing small, lightweight internal combustion engines for automobiles and motorcycles. When they needed an engine, the Wrights thought, the technology would be there. That left the issue of pilot control, the very problem that had killed Lilienthal and the aspect on which nobody else had placed such clear priority.[24]

The Wrights' thought process about control ran against the grain of current thinking, and in hindsight it appears that this was precisely why they eventually succeeded where others failed. Because Lilienthal, the great "Flying Man," had died because not even he was able to control his aircraft, conventional wisdom reasoned that what was needed was an inherently stable machine—for example, the line of thinking followed by both Chanute and Langley. The Wrights turned this logic on its head. They cared little about a flying machine's stability, but focused from the start on the critical aspect of making it controllable. No doubt,

---

[24] For an extremely insightful analysis of how the Wrights defined their problems and solved them, see Peter L. Jakab, *Visions of a Flying Machine: The Wright Brothers and the Process of Invention* (Washington, DC: Smithsonian History of Aviation Series, 1990). Jakab takes his reader step by step through the thought processes that led the Wright brothers to their successful invention of the airplane.

their intimacy with bicycle technology influenced them greatly in this matter, for a bicycle is an inherently unstable machine. The Wrights felt that the airplane would need to be the same sort of dynamically interactive device as a bicycle: unstable on its own but completely controllable and virtually automatic in the hands of an experienced operator. The key was designing control features into the airplane that a pilot could easily and effectively manage.

In 1899, they took their first important step toward that goal by experimenting in Dayton with what came to be known as their "wing-warping kite." The basic idea was a mechanism that allowed the operator to control the kite in its roll axis, considered the most challenging axis of motion to master in a flying machine. The Wrights developed a set of controls that enabled them to induce a helical twist across the wing surface that increased the lift on one side and decreased it on the other, thereby changing the aerodynamic balance of the machine. They did not realize at the time that others, like Yale physics instructor Edson Fessendon Gallaudet, had flown large wing-warping kites already. Instead they came up with the idea on their own, based on their observations of buzzards employing a similar technique, regaining "their balance, when partly overturned by a gust of wind, by a torsion of the tips of the wings." Starting with the wing-warping kite of 1899, one can follow the links in an evolutionary chain of ever more sophisticated flying machines, right up to the historic Wright *Flyer* of 1903.[25]

In 1900, the brothers built their first full-scale machine, a biplane glider large enough to carry a person. Though it resembled the Chanute-Herring glider of 1896 in many of its structural features, the Wright glider of 1900 differed in that it had a large elevator (or canard) set directly in front of the lower wing. This horizontal surface moved up and down and was the Wrights' way of managing pitch control and preventing nose dives in case of stalls. They built this machine in accordance with the tables of aerodynamic coefficients compiled by Otto Lilienthal, and estimated the amount of wing surface needed to lift the weight of their fifty-pound machine plus a pilot into the air. It was this machine that first flew at Kitty Hawk, on the windy Outer Banks of North Carolina, for two weeks in the early autumn of 1900, during what was to the Wrights a short "scientific vacation." (See Document 1-10.)

What concerned the Wright brothers most about their first season of flight tests at Kitty Hawk was the fact that their glider simply did not generate the lift

---

[25] The analysis presented in this chapter of the evolutionary technological development from the Wrights's kites to their historic 1903 *Flyer* is based largely on the narrative and interpretations of Tom Crouch, the world's preeminent Wright brothers' scholar and author of the prizewinning *The Bishop's Boys*. As with other sections of this chapter, a more detailed analysis of the Wright's aerodynamics relied on Anderson's *A History of Aerodynamics*, pp. 201–243. Of course, there are many more works on the Wrights, some of which were consulted in preparation of this chapter.

that their calculations indicated it should. Something was obviously wrong with their calculations. While they did manage to fly their machine as a kite, and finally on the last day to make a few very short free glides, overall, the results were deflating. They tried the glider "with tail in front, behind, and every other way," but with no real success. After their return to Dayton, Orville Wright noted of his brother's mood that, "When we got through Will was so mixed up he couldn't even theorize. It has been with considerable effort that I have succeeded in keeping him in the flying business at all."

The flying season in the next year proved equally frustrating. The machine they took to Kitty Hawk in the summer of 1901 was a good deal larger than their previous glider—in fact, with a twenty-two-foot wingspan, it was the largest glider anyone had ever tried to fly. Basically, they hoped to solve the problem of insufficient lift by creating more wing surface and by increasing the curvature or camber of the wings, but none of these changes worked. The glider produced much less lift than predicted, and was plagued with serious control problems. The disappointment of this season was a critical point for the Wrights, for they recognized that if they failed to solve these lift problems, sooner or later one of them was going to be killed.

Rather than giving up, though, the two brothers reevaluated their basic assumptions and realized that something was seriously wrong with the scientific tables on which they were basing their calculations. Back in Dayton they

The Wrights experienced frustration after frustration in their 1901 glider flying season at Kitty Hawk. But in battling through their many problems and failures they learned lessons extremely valuable to their successful invention of the airplane two years later. Assistants Dan Tate and Edward Huffaker launch Wilbur in his 1901 glider. SI Negative 84-12143

embarked on a new course to conduct their own basic research and find out what was wrong with the Lilienthal data. They constructed a small six-foot-long wind tunnel in the back room of their bicycle shop and tested a whole new range of airfoil shapes, about two hundred in all. They also experimented with changes in aspect ratio, with different wing tip shapes, and with varying gaps between the wings of a biplane. At the conclusion of this intense period of original investigation they sat on a wealth of new aerodynamic information that would quickly enable them to solve the problems that had hitherto blocked their progress. (For material related to the invention of the wind tunnel by British engineer Francis H. Wenham in the early 1870s, see Document 1-3.)

The resulting 1902 Wright glider flew beautifully. Besides incorporating new wing shapes that gave them the lift they wanted, the airplane also featured the Wright's most recent significant discovery, the benefits of a moveable tail rudder linked directly to the wing-warping mechanism, enabling the pilot to manage both controls easily and in concert. With this machine the Wrights really learned to fly. They experienced a magnificent season of soaring and making banks, turns, and recoveries in every direction. Appropriately, it was this machine that these intrepid aviation pioneers patented, not their powered airplane of 1903. They had set out to invent an airplane that was controllable in the air, and by the end of 1902 they had done it. (See Document 1-11.) All that was left was to add a propulsion system to the airplane.

They did encounter one other aerodynamic problem related to propulsion in the design of an effective propeller. In the beginning this was a problem that the Wrights had not anticipated, since they believed it would be sufficient to find a good marine propeller and simply adapt it for use in the air. After the surprises of 1900 and 1901, however, they did not want to take any chances, and determined to calculate precisely the performance requirements of an aviation propeller. In what proved to be another case of their marvelous talent for applying clear engineering logic to a problem that others had either overlooked or ignored, they reasoned that the propeller was no more than a rotating wing—and as a result of their wind tunnel tests they knew a lot about wings. For a propeller, therefore, they selected an appropriate airfoil shape from their own aerodynamic tables, with the camber best suited for the speed at which their machine would be flying through the air, and carved their own propeller blades. "The result," according to aviation historian and Wright biographer Tom Crouch, "was the world's first true aircraft propeller, a device whose performance could be precisely calculated."[26]

The story of the design of the homemade engine for their airplane, although interesting, does not concern aerodynamic development, save perhaps in one

---

[26] Crouch, *A Dream of Wings*, p. 294.

This photo from 24 October 1902 shows Wilbur maneuvering the glider (fitted with a single moveable vertical rudder) through a gentle right turn. By the end of the 1902 season the brothers had taken their machine through a remarkable series of banks, turns, and recoveries. SI Negative A-43395-A

respect. The Wrights wanted an engine that provided just enough power to meet their calculated requirements, with no extraneous weight or extra resources devoted to redundant capability. They knew exactly from their calculations how much power it would take to get them into the air, and once they had an engine that could deliver that precise amount of power—a mere $12\frac{1}{2}$ horsepower from a 200-pound motor—they concluded engine development right there. This efficient approach contrasted sharply with the design overkill of Professor Langley, who had no practical idea how much power it was going to take to get his aerodrome up into the air, and wasted inordinate time, money, and energy on the development of an overly powerful fifty-two horsepower engine while neglecting the more critical areas of aerodynamic structure and control.

The Wrights spent nearly three months at Kitty Hawk in late 1903 before they were ready to try their first powered flight. Besides assembling, testing, and rigging their new machine, they also practiced extensively with the 1902 glider. After a false start on 14 December using a downhill launch rail that produced a take-off speed too high for effective control, Orville and Wilbur were ready to take

The secret of the Wrights' success was that they learned to fly and control the airplane in glider form before they put power into it. They learned a great deal about the use of aerodynamic surfaces to control the airplane in a satisfactory manner, and together with the experience they had as pilots, they were able to fly the plane without too much difficulty when they did put power into it. That did not preclude a false start or two, however. In attempting to fly on 14 December 1903, pictured here, Wilbur nosed up too rapidly and spun back into the ground, damaging the front horizontal rudder, lower rear wing spar, and skid. SI Negative A-38618-A

to the air with their airplane from a level launch rail on 17 December 1903. The brothers made a total of four flights that day. The first flight, made by Orville, covered only 120 feet in a little over twelve seconds. The last, piloted by Wilbur, covered 850 feet and lasted nearly a full minute. The telegram sent by Orville to his father telling the story of this historic achievement is appropriately the destination document for this chapter, marking the realization of humankind's long-sought goal of sustained powered flight. (See Document 1-13.)

The Wrights went home to Ohio and in 1904 and 1905 built two new airplanes and tested them as they flew over a cow pasture known as Huffman Prairie, today a part of Wright-Patterson Air Force Base. Their airplanes got better, as did their piloting skills, and by the end of 1905 they were making flights of forty-five minutes, flying repeated circles and other maneuvers over their "air field." The world surprisingly took little notice of these truly revolutionary achievements. News of their initial flights produced only a scattering of stories, and those at

Huffman Prairie a brief flurry of interest. The press soon forgot about the Wrights and allowed them to continue their refinement of the airplane undisturbed.

The Wright brothers were masters at thinking visually, using what is sometimes called "the mind's eye," and translating abstractions into hardware. Their flying machine had to "look right," in both an aesthetic and a practical sense, before they were convinced that they had everything right in their invention of the airplane. Certainly everything looks right in this picture of the 1905 *Flyer* in flight above Huffman Prairie near Dayton. SI Negative No. A-317-B

Part of the explanation for the lack of greater publicity was due to the ignominious public failure of Professor Langley's full-size manned *Great Aerodrome* in 1903. Using the familiar launch platform atop the houseboat on the Potomac River, in October Langley tried to fly his craft with an assistant aboard, but the machine crashed in the attempt. A second climactic public attempt to fly the machine on 8 December, a mere nine days before the Wright's epochal achievement in isolated Kitty Hawk, also resulted in utter failure. These tests were covered by a corps of newspapermen and photographers who scathingly reported the failure of a government program that had spent $75,000 (the equivalent of about $1.5 million today) to produce an aircraft whose flying characteristics were compared to "a handful of mortar" or "a block of cement." The nation's leading scientist was humiliated and became the target of public scorn and ridicule. With Langley's disgrace capturing the headlines, the press took little notice of the work of two obscure inventors from Dayton.

The coincidence of the Wrights' success and Langley's failure in the month of December 1903 highlights a factor that proved to be critically important, not just to the invention of the airplane, but for the entire course of aerodynamic research and development in the twentieth century. The Wrights were engineers, not scientists—practical-minded and realistic men, who time and time again found simple solutions to what turned out to be key problems. Langley, the scientist, on the other hand, proved an inept technologist because he could not turn what he knew into an accomplishment of his goal. The Wrights had clear vision and demonstrated their genius as creative technologists par excellence.

What they also did, once the world and particularly the different fields of engineering became aware of what the Wrights achieved through their systematic

What the public acknowledged about powered flight after December 1903 was the ignominious, highly publicized failure of Samuel P. Langley's *Great Aerodrome*, not the successful flights of the Wrights, which few knew about and even fewer believed had actually happened. Here, Langley's *Great Aerodrome* rests on its catapult atop a houseboat, ready for flight, on 8 December 1903. SI Negative No. A-18789

engineering approach, was help prepare the ways and means for the future evolution of the airplane. The form of the airplane as the Wrights conceived it in 1903 was only the beginning—a fact they themselves perhaps did not appreciate enough. In all respects, save the decisive one of control, the performance of their 1903 airplane was highly marginal. As one analyst has observed, it really amounted to "the first aeronautical 'proof of concept' design, and could not be used for anything else, including repeated flying."[27] (The historic 1903 *Flyer* was never flown again after the 17 December flights.) If the airplane was to ever become truly practicable, its technology had to improve dramatically and become much more capable and versatile.

Over the course of the next decades, the airplane would experience a number of "reinventions" in many ways as remarkable as the original Wright invention. And in all of them, a systematic engineering approach similar to that of the Wrights proved critical. As much as the invention of the original airplane itself, this was the Wrights' legacy.

Ironically, the country's first civilian aeronautical research facility, under the auspices of the National Advisory Committee for Aeronautics (NACA), was named after Langley, not the Wrights. The NACA Langley Memorial Aeronautical Laboratory

---

[27] E. K. Liberatore, *Helicopters before Helicopters* (Malabar, FL: Krieger Publishing Co., 1998), pp. 158–159.

Observers compared the aerodynamic qualities of Langley's *Great Aerodrome*, seen here crashing into the Potomac an instant after launch, to "a sackful of mortar." SI Negative No. A-18853

was founded near Hampton, Virginia, in 1917, eleven years after the death of the discredited Smithsonian scientist. Fortunately, for the sake of the aeronautical research investigations that took place at the NACA and for the positive impact that NACA research would have on the nascent U.S. aircraft industry, the influence of Langley's devoted colleagues and friends in the Smithsonian Institution (enough to get the first NACA laboratory named after him) was not compelling enough to stamp the character of the once-fledgling organization with Langley's unsuccessful approach. The researchers at Langley followed the technological lead and the model of the Wrights. For the study of aerodynamics to affect flying machines, and for flying machines to indeed change the world, the ways of the Wrights ruled the day.

European skeptics watched in awe as Wilbur Wright unveiled what his airplane could do during his trip to France in 1908. In one sensational flight made on 13 August 1908 above the Les Hanaudières race track near Le Mans, France, Wilbur made seven circles of the track in 8 minutes, 13 $\frac{2}{5}$ seconds. Four months later, in December, he made two stunning nonstop flights across France—one a distance of 99 kilometers in 1 hour, 54 minutes, 22 seconds, and the other of 124.7 kilometers in 2 hours, 20 minutes. SI Negative No. A-42962-A

The Documents

Document 1-1(a–c)

(a) George Cayley, "On Aerial Navigation," Part One, *Nicholson's Journal*, November 1809.

(b) George Cayley, "On Aerial Navigation," Part Two, *Nicholson's Journal*, February 1810.

(c) George Cayley, "On Aerial Navigation," Part Three, *Nicholson's Journal*, March 1810.

Note: All three parts were republished in James Mead, ed., *The Aeronautical Annual* 1, (Boston: W.B. Clarke & Co., 1895): 16–48.

There is perhaps no more important individual paper in the entire history of aeronautics than Sir George Cayley's famous three-part treatise "On Aerial Navigation," written in 1809. In this paper, reproduced here in its entirety, Cayley reported all his findings on airplane aerodynamics and provided a thorough explanation of the potential of a fixed-wing flying machine. In addition to outlining a systems approach to solving the problems of lift, control, and power, he demonstrated a clear understanding of the advantages of wing camber, the first person ever to appreciate the subtleties of the effects of curvature on lift. He did not fail to address the matter of drag, and in various passages of his treatise he even expressed a modern concept of aerodynamic "streamlining," before the term was invented. The total effect of his presentation was momentous, although it may not come across as such to the modern reader who takes for granted the basic operating principles of the airplane. Cayley's delineation of the form and function of the airplane, based on his intuitive genius and solid experimental approach, is classic.

Cayley's document was written in three parts for *Nicholson's Journal*, and is often referred to as the "Triple Paper."

Document 1-1(a), George Cayley, "On Aerial Navigation," Part One, 1809.

BROMPTON, Sept. 6, 1809.

SIR, I observed in your Journal for last month, that a watchmaker at Vienna, of the name of Degen, has succeeded in raising himself in the air by mechanical means. I waited to receive your present number, in expectation of seeing some farther account of this experiment, before I commenced transcribing the following essay upon aerial navigation, from a number of memoranda which I have made at various times upon this subject. I am induced to request your publication of this essay, because I conceive, that, in stating the fundamental principles of this art, together with a considerable number of facts and practical observations, that have arisen in the course of much attention to this subject, I may be expediting the attainment of an object, that will in time be found of great importance to mankind; so much so, that a new era in society will commence, from the moment that aerial navigation is familiarly realized.

It appears to me, and I am more confirmed by the success of the ingenious Mr. Degen, that nothing more is necessary, in order to bring the following principles into common practical use, than the endeavours of skilful artificers, who may vary the means of execution, till those most convenient are attained.

Since the days of Bishop Wilkins the scheme of flying by artificial wings has been much ridiculed; and indeed the idea of attaching wings to the arms of a man is ridiculous enough, as the pectoral muscles of a bird occupy more than two-thirds of its whole muscular strength, whereas in man the muscles, that could operate upon wings thus attached, would probably not exceed one-tenth of his whole mass. There is no proof that, weight for weight, a man is comparatively weaker than a bird; it is therefore probable, if he can be made to exert his whole strength advantageously upon a light surface similarly proportioned to his weight as that of the wing to the bird, that he would fly like the bird, and the ascent of Mr. Degen is a sufficient proof of the truth of this statement.

The flight of a strong man by great muscular exertion, though a curious and interesting circumstance, in as much as it will probably be the first means of ascertaining this power and supplying the basis whereon to improve it, would be of little use. I feel perfectly confident, however, that this noble art will soon be brought home to man's general convenience, and that we shall be able to transport ourselves and families, and their goods and chattels, more securely by air than by water, and with a velocity of from 20 to 100 miles per hour.

To produce this effect, it is only necessary to have a first mover, which will generate more power in a given time, in proportion to its weight, than the animal system of muscles.

The consumption of coal in a Boulton and Watt's steam engine is only about 5 ½ lbs. per hour for the power of one horse[power]. The heat produced by the combustion of this portion of inflammable matter is the sole cause of the power generated; but it is applied through the intervention of a weight of water expanded into steam, and a still greater weight of cold water to condense it again. The engine itself likewise must be massy enough to resist the whole external pressure of the atmosphere, and therefore is not applicable to the purpose proposed. Steam engines have lately been made to operate by expansion only, and those might be constructed so as to be light enough for this purpose, provided the usual plan of a large boiler be given up, and the principle of injecting a proper charge of water into a mass of tubes, forming the cavity for the fire, be adopted in lieu of it. The strength of vessels to resist internal pressure being inversely as their diameters, very slight metallic tubes would be abundantly strong, whereas a large boiler must be of great substance to resist a strong pressure. The following estimate will show the probable weight of such an engine with its charge for one hour.

	lb.
The engine itself from 90 to	100
Weight of inflamed cinders in a cavity presenting about 4 feet surface of tube:	25
Supply of coal for one hour:	6
Water for ditto, allowing steam of one atmosphere to be 1/1800 the specific gravity of water:	32
[Total weight in pounds:]	163

I do not propose this statement in any other light than as a rude approximation to truth, for as the steam is operating under the disadvantage of atmospheric pressure, it must be raised to a higher temperature than in Messrs. Boulton and Watt's engine; and this will require more fuel; but if it take twice as much, still the engine would be sufficiently light, for it would be exerting a force equal to raising 550 lb. one foot high per second, which is equivalent to the labour of six men, whereas the whole weight does not much exceed that of one man.

It may seem superfluous to inquire farther relative to first movers for aerial navigation; but lightness is of so much value in this instance, that it is proper to notice the probability that exists of using the expansion of air by the sudden combustion of inflammable powders or fluids with great advantage. The French have lately shown the great power produced by igniting inflammable powders in close vessels; and several years ago an engine was made to work in this country in a similar manner, by the inflammation of spirit of tar. I am not acquainted with the name of the person who invented and obtained a patent for this engine, but from some minutes with which I was favoured by Mr. William Chapman, civil engineer in Newcastle, I find that 80 drops of the oil of tar raised eight hundred weight to the height of 22 inches; hence a one horse power many consume from 10 to 12 pounds per hour, and the engine itself need not exceed 50 pounds weight. I am informed by Mr. Chapman, that this engine was exhibited in a working state to Mr. Rennie, Mr. Edmund Cartwright, and several other gentlemen, capable of appreciating its powers; but that it was given up in consequence of the expense attending its consumption being about eight times greater than that of a steam engine of the same force.

Probably a much cheaper engine of this sort might be produced by a gas-light apparatus, and by firing the inflammable air generated, with a due portion of common air, under a piston. Upon some of these principles it is perfectly clear, that force can be obtained by a much lighter apparatus than the muscles of animals or birds, and therefore in such proportion may aerial vehicles be loaded with inactive matter. Even the expansion steam engine doing the work of six men, and only weighing equal to one, will as readily raise five men into the air, as Mr. Degen can elevate himself by his own exertions; but by increasing the magnitude of the engine, 10, 50, or 500 men may equally well be conveyed; and convenience alone, regulated by the strength and size of materials, will point out the limit for the size of vessels in aerial navigation.

Having rendered the accomplishment of this object probable upon the general view of the subject, I shall proceed to point out the principles of the art itself. For the sake of perspicuity I shall, in the first instance, analyze the most simple action of the wing in birds, although it necessarily supposes many previous steps. When large birds, that have a considerable extent of wing compared with their weight, have acquired their full velocity, it may frequently be observed, that they extend their wings, and without waving them, continue to skim for some time in a horizontal path. Fig. 1, in the Plate, represents a bird in this act.

Let $a\,b$ be a section of the plane of both wings opposing the horizontal current of the air (created by its own motion) which may be represented by the line $c\,d$, and is the measure of the velocity of the bird. The angle $b\,d\,c$ can be increased at the will of the bird, and to preserve a perfectly horizontal path, without the wing being waved, must continually be increased in a complete ratio, (useless at present to enter into) till the motion is stopped altogether; but at one given time the position of the wings may be truly represented by the angle $b\,d\,c$. Draw $d\,e$ perpendicular to the plane of the wings, produce the line $e\,d$ as far as required, and from the point e, assumed at pleasure in the line $d\,e$, let fall $e\,f$ perpendicular to $d\,f$. Then d e will represent the whole force of the air under the wing; which being resolved into the two forces $e\,f$ and $f\,d$, the former represents the force that sustains the weight of the bird, the latter the retarding force by which the velocity of the motion, producing the current $c\,d$, will continually be diminished. $e\,f$ is always a known quantity, being equal to the weight of the bird, and hence $f\,d$ is also known, as it will always bear the same proportion to the weight of the bird, as the sine of the angle $b\,d\,e$ bears to its cosine, the angles $d\,e\,f$, and $b\,d\,c$, being equal. In addition to the retarding force thus received is the direct resistance, which the bulk of the bird opposes to the current. This is a matter to be entered into separately from the principle now under consideration; and for the present may be wholly neglected, under the supposition of its being balanced by a force precisely equal and opposite to itself.

Before it is possible to apply this basis of the principle of flying in birds to the purposes of aerial navigation, it will be necessary to encumber it with a few practical observations. The whole problem is confined within these limits, viz. To make a surface support a given weight by the application of power to the resistance of air. Magnitude is the first question respecting the surface. Many experiments have been made upon the direct resistance of air by Mr. Robins, Mr. Rouse, Mr. Edgeworth, Mr. Smeaton, and others. The result of Mr. Smeaton's experiments and observations was, that a surface of a square foot met with a resistance of one pound, when it travelled perpendicularly to itself through air at a velocity of 21 feet per second. I have tried many experiments upon a large scale to ascertain this point. The instrument was similar to that used by Mr. Robins, but the surface used was larger, being an exact square foot, moving round upon an arm about five feet long, and turned by weights over a pulley. The time was measured by a stop watch, and the distance travelled over in each experiment was 600 feet. I shall for the present only give the result of many carefully repeated experiments, which is, that a velocity of 11.538 feet per second generated a resistance of 4 ounces; and that a velocity of 17.16 feet per second gave 8 ounces resistance. This delicate instrument would have been strained by the additional weight necessary to have tried the velocity generating a pressure of one pound per square foot; but if the resistance be taken to vary as the square of the velocity, the former will give the velocity necessary for this purpose at 23.1 feet, the latter 24.28 per second. I shall therefore take 23.6 feet as somewhat approaching the truth.

Having ascertained this point, had our tables of angular resistance been complete, the size of the surface necessary for any given weight would easily have been determined. Theory, which gives the resistance of a surface opposed to the same current in different angles, to be as the squares of the sine of the angle of incidence, is of no use in this case; as it appears from the experiments of the French Academy, that in acute angles, the resistance varies much more nearly in the direct ratio of the sines, than as the squares of the sines of the angles of incidence. The flight of birds will prove to an attentive observer, that, with a concave wing apparently parallel to the horizontal path of the bird, the same support, and of course resistance, is obtained. And hence I am inclined to suspect, that, under extremely acute angles, with concave surfaces, the resistance is nearly similar in

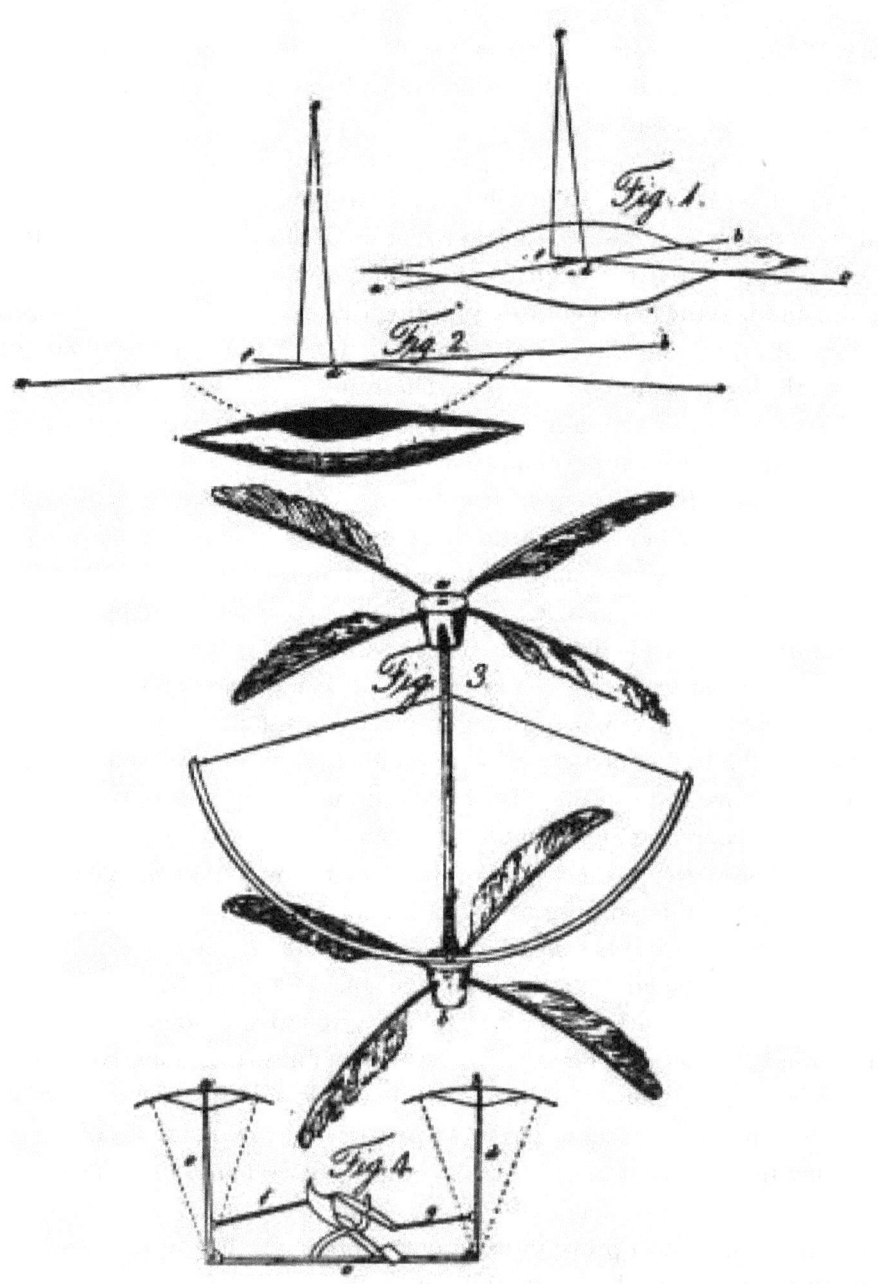

them all. I conceive the operation may be of a different nature from what takes place in larger angles, and may partake more of the principle of pressure exhibited in the instrument known by the name of the hydrostatic paradox, a slender filament of the current is constantly received under the anterior edge of the surface, and directed upward into the cavity, by the filament above it, in being obliged to mount along the convexity of the surface, having created a slight vacuity immediately behind the point of separation. The fluid accumulated thus within the cavity has to make its escape at the posterior edge of the surface, where it is directed considerably downward; and therefore has to overcome and displace a portion of the direct current passing with its full velocity immediately below it; hence whatever elasticity this effort requires operates upon the whole concavity of the surface, excepting a small portion of the anterior edge. This may or may not be the true theory, but it appears to me to be the most probable account of a phenomenon, which the flight of birds proves to exist.

Six degrees was the most acute angle, the resistance of which was determined by the valuable experiments of the French Academy; and it gave $\frac{4}{10}$ of the resistance, which the same surface would have received from the same current when perpendicular to itself. Hence then a superficial foot, forming an angle of six degrees with the horizon, would, if carried forward horizontally (as a bird in the act of skimming) with a velocity of 23.6 feet per second, receive a pressure of $\frac{4}{10}$ of a pound perpendicular to itself. And, if we allow the resistance to increase as the square of the velocity, at 27.3 feet per second it would receive a pressure of one pound. I have weighed and measured the surface of a great many birds, but at present shall select the common rook (*corvus frugilegus*) because its surface and weight are as nearly as possible in the ratio of a superficial foot to a pound. The flight of this bird, during any part of which they can skim at pleasure, is (from an average of many observations) about 34.5 feet per second. The concavity of the wing may account for the greater resistance here received, than the experiments upon plain surfaces would indicate. I am convinced, that the angle made use of in the crow's wing is much more acute than six degrees; but in the observations, that will be grounded upon these data, I may safely state, that every foot of such curved surface, as will be used in aerial navigation, will receive a resistance of one pound, perpendicular to itself, when carried through the air in an angle of six degrees with the line of its path, at a velocity of about 34 or 35 feet per second.

Let *a b*, fig. 2, represent such a surface or sail made of thin cloth, and containing about 200 square feet (if of a square form the side will be a little more than 14 feet); and the whole of a firm texture. Let the weight of the man and the machine be 200 pounds. Then if a current of wind blew in the direction *c d*, with a velocity of 35 feet per second, at the same time that a cord represented by *c d* would sustain a tension of 21 pounds, the machine would be suspended in the air, or at least be within a few ounces of it (falling short of such support only in the ratio of the sine of the angle of 94 degrees compared with radius; to balance which defect, suppose a little ballast to be thrown out) for the line $d\,e$ represents a force of 200 pounds, which, as before, being resolved into $d\,f$ and $f\,e$, the former will represent the resistance in the direction of the current, and the latter that which sustains the weight of the machine. It is perfectly indifferent whether the wind blow against the plane, or the plane be driven with an equal velocity against the air. Hence, if this machine were pulled alone by a cord $c\,d$, with a tension of about 21 pounds, at a velocity of 35 feet per second, it would be suspended in a horizontal path; and if in lieu of this cord any other propelling power were generated in this direction, with a like intensity, a similar effect would be produced. If therefore the waft of surfaces advantageously moved, by any force generated within the machine, took place to the extent required, aerial navigation would be accomplished. As the acuteness of the angle between the plane and current increases, the propelling power required is less and less. The principle is similar to that of the inclined plane, in which theoretically one pound may be made to sustain all but an infinite quantity; for in this case, if the magnitude of the surface be increased ad infinitum, the angle with the current may be diminished, and consequently the propelling force, in the same ratio. In practice, the extra resistance of the car and other parts of the machine, which consume a considerable portion of power, will regulate the limits to which this principle, which is the true basis of aerial navigation, can be carried; and the perfect ease with which some birds are suspended in long horizontal flights, without one waft of their wings, encourages the idea, that a slight power only is necessary.

As there are many other considerations relative to the practical introduction of this machine, which would occupy too much space for any one number of your valuable Journal, I propose, with your approbation, to furnish these in your subsequent numbers; taking this opportunity to observe, that perfect steadiness, safety, and steerage, I have long since accomplished upon a considerable scale of magnitude; and that I am engaged in making some farther experiments upon a machine I constructed last summer, large enough for aerial navigation, but which I have not had an opportunity to try the effect of, excepting as to its proper balance and security. It was very beautiful to see this noble white bird sail majestically from the top of a hill to any given point of the plane below it, according to the set of its rudder, merely by its own weight, descending in an angle of about 18 degrees with the horizon. The exertions of an individual, with other avocations, are extremely inadequate to the progress, which this valuable subject requires. Every man acquainted with experiments upon a large scale well knows how leisurely fact follows theory, if ever so well founded. I do therefore hope, that what I have said, and have still to offer, will induce others to give their attention to this subject; and that England may not be backward in rivalling the continent in a more worthy contest than that of arms.

As it may be an amusement to some of your readers to see a machine rise in the air by mechanical means, I will conclude my present communication by describing an instrument of this kind, which any one can construct at the expense of ten minutes labour.

a and *b*, fig. 3, are two corks, into each of which are inserted four wing feathers from any bird, so as to be slightly inclined like the sails of a windmill, but in opposite directions in each set. A round shaft is fixed in the cork *a*, which ends in a sharp point. At the upper part of the cork *b* is fixed a whalebone bow, having a small pivot hole in its centre, to receive the point of the shaft. The bow is then to be strung equally on each side to the upper portion of the shaft, and the little machine is completed. Wind up the string by turning the flyers different ways, so that the spring of the bow may unwind them with their anterior edges ascending; then place the cork with the bow attached to it upon a table, and with a finger on the upper cork press strong enough to prevent the string from unwinding, and taking it away suddenly, the instrument will rise to the ceiling. This was the first experiment I made upon this subject in the year 1796. If in lieu of these small feathers large planes, containing together 200 square feet, were similarly placed, or in any other more convenient position, and were turned by a man, or first mover of adequate power, a similar effect would be the consequence, and for the mere purpose of ascent this is perhaps the best apparatus; but speed is the great object of this invention, and this requires a different structure.

P. S. In lieu of applying the continued action of the inclined plane by means of the rotative motion of flyers, the same principle may be made use of by the alternate motion of surfaces backward and forward; and although the scanty description hitherto published of Mr. Degen's apparatus will scarcely justify any conclusion upon the subject; yet as the principle above described must be the basis of every engine for aerial navigation by mechanical means, I conceive, that the method adopted by him has been nearly as follows. Let A and B, fig. 4, be two surfaces or parachutes, supported upon the long shafts C and D, which are fixed to the ends of the connecting beam E, by hinges. At E, let there be a convenient seat for the aeronaut, and before him a cross bar turning upon a pivot in its centre, which being connected with the shafts of the parachutes by the rods F and G, will enable him to work them alternately backward and forward, as represented by the dotted lines. If the upright shafts be elastic, or have a hinge to give way a little near their tops, the weight and resistance of the parachutes will incline them so, as to make a small angle with the direction of their motion, and hence the machine rises. A slight heeling of the parachutes toward one side, or an alteration in the position of the weight, may enable the aeronaut to steer such an apparatus tolerably well; but many better constructions may be formed, for combining the requisites of speed, convenience and steerage. It is a great point gained, when the first experiments demonstrate the practicability of an art; and Mr. Degen, by whatever means he has effected this purpose, deserves much credit for his ingenuity.

Document 1-1(b), George Cayley, "On Aerial Navigation," Part Two, 1810.

HAVING, in my former communication, described the general principle of support in aerial navigation, I shall proceed to show how this principle must be applied, so as to be steady and manageable.

Several persons have ventured to descend from balloons in what is termed a parachute, which exactly resembles a large umbrella, with a light car suspended by cords underneath it.

Mr. Garnerin's descent in one of these machines will be in the recollection of many; and I make the remark for the purpose of alluding to the continued oscillation, or want of steadiness, which is said to have endangered that bold aeronaut. It is very remarkable, that the only machines of this sort, which have been constructed, are nearly of the worst possible form for producing a steady descent, the purpose for which they are intended. To render this subject more familiar, let us recollect, that in a boat, swimming upon water, its stability or stiffness depends, in general terms, upon the *weight* and distance from the centre of the section elevated above the water, by any given heel of the boat, on one side; and on the *bulk*, and its distance from the centre, which is immersed below the water, on the other side; the combined endeavour of the one to fall and of the other to swim, produces the desired effect in a well-constructed boat. The centre of gravity of the boat being more or less below the centre of suspension is an additional cause of its stability.

Let us now examine the effect of a parachute represented by A B, fig. 1, Pl. III. When it has heeled into the position *a b*, the side *a* is become perpendicular to the current, created by the descent, and therefore resists with its greatest power; whereas the side b is become more oblique, and of course its resistance is much diminished. In the instance here represented, the angle of the parachute itself is 144°, and it is supposed to heel 18°, the comparative resistance of the side *a* to the side *b*, will be as the square of the line *a*, as radius, to the square of the sine of the angle of *b* with the current; which, being 54 degrees, gives the resistances nearly in the ratio of 1 to 0.67; and this will be reduced to only 0.544, when estimated in a direction perpendicular to the horizon. Hence, so far as this form of the sail or plane is regarded, it operates directly in opposition to the principle of stability; for the side that is required to fall resists much more in its new position, and that which is required to rise resists much less; therefore complete inversion would be the consequence, if it were not for the weight being suspended so very much below the surface, which, counteracting this tendency, converts the effort into a violent oscillation.

On the contrary, let the surface be applied in the inverted position, as represented at C D, fig. 2, and suppose it to be heeled to the same angle as before, represented by the dotted lines *c d*. Here the exact reverse of the former instance takes place; for that side, which is required to rise, has gained resistance by its new position, and that which is required to sink has lost it; so that as much power oper-

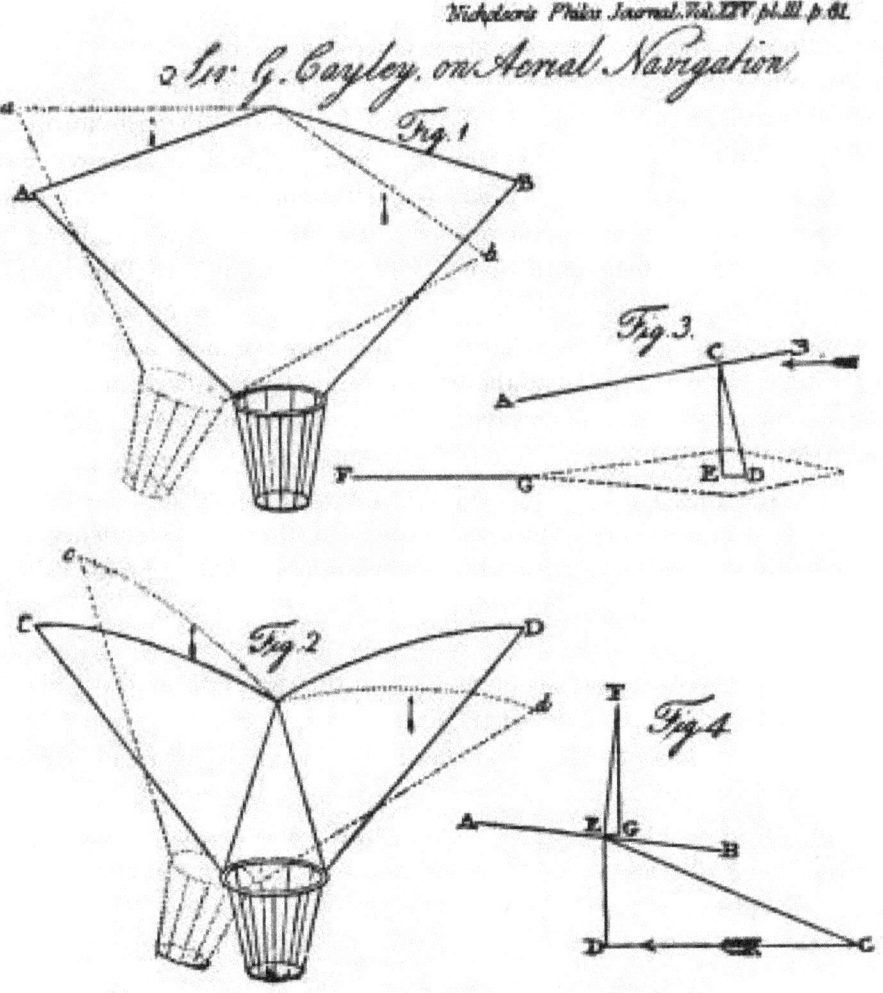

ates to restore the equilibrium in this case, as tended to destroy it in the other: the operation very much resembling what takes place in the common boat. (A very simple experiment will show the truth of this theory. Take a circular piece of writing paper, and folding up a small portion, in the line of two radii, it will be formed into an obtuse cone. Place a small weight in the apex, and letting it fall from any height, it will steadily preserve that position to the ground. Invert it, and, if the weight be fixed, like the life boat, it rights itself instantly.)

This angular form, with the apex downward, is the chief basis of stability in aerial navigation; but as the sheet which is to suspend the weight attached to it, in its horizontal path through the air, must present a slightly concave surface in a small angle with the current, this principle can only be used in the lateral extension of the sheet; and this most effectually prevents any rolling of the machine from side to side. Hence the section of the inverted parachute, fig. 2, may equally well represent the cross section of a sheet for aerial navigation.

The principle of stability in the direction of the path of the machine, must be derived from a different source. Let A B, fig. 3, be a longitudinal section of a sail, and let C be its centre of resistance, which experiment shows to be considerably more forward than the centre of the sail. Let C D be drawn perpendicular to A B, and let the centre of gravity of the machine be at any point in that line, as at D. Then, if it be projected in a horizontal path with velocity enough to support the weight, the machine will retain its relative position, like a bird in the act of skimming; for, drawing C E perpendicular to the horizon, and D E parallel to it, the line C E will, at some particular moment, represent the supporting power, and likewise its opponent the weight; and the line D E will represent the retarding power, and its equivalent, that portion of the projectile force expended in overcoming it: hence, these various powers being exactly balanced, there is no tendency in the machine but to proceed in its path, with its remaining portion of projectile force.

The stability in this position, arising from the centre of gravity being below the point of suspension, is aided by a remarkable circumstance that experiment alone could point out. In very acute angles with the current it appears, that the centre of resistance in the sail does not coincide with the centre of its surface, but is considerably in front of it. As the obliquity of the current decreases, these centres approach, and coincide when the current becomes perpendicular to the sail. Hence any heel of the machine backward or forward removes the centre of support behind or before the point of suspension; and operates to restore the original position, by a power, equal to the whole weight of the machine, acting upon a lever equal in length to the distance the centre has removed.

To render the machine perfectly steady, and likewise to enable it to ascend and descend in its path, it becomes necessary to add a rudder in a similar position to the tail in birds. Let F G be the section of such a surface, parallel to the current;

and let it be capable of moving up and down upon G, as a centre, and of being fixed in any position. The powers of the machine being previously balanced, if the least pressure be exerted by the current, either upon the upper or under surface of the rudder, according to the will of the aeronaut, it will cause the machine to rise or fall in its path, so long as the projectile or propelling force is continued with sufficient energy. From a variety of experiments upon this subject I find, that, when the machine is going forward with a superabundant velocity, or that which would induce it to rise in its path, a very steady horizontal course is effected by a considerable depression of the rudder, which has the advantage of making use of this portion of sail in aiding the support of the weight. When the velocity is becoming less, as in the act of alighting, then the rudder must gradually recede from this position, and even become elevated, for the purpose of preventing the machine from sinking too much in front, owing to the combined effect of the want of projectile force sufficient to sustain the centre of gravity in its usual position, and of the centre of support approaching the centre of the sail.

The elevation and depression of the machine are not the only purposes, for which the rudder is designed. This appendage must be furnished with a vertical sail and be capable of turning from side to side in addition to its other movements, which effects the complete steerage of the vessel.

All these principles, upon which the support, steadiness, elevation, depression, and steerage of vessels for aerial navigation, depend, have been abundantly verified by experiments both upon a small and a large scale. Last year I made a machine, having a surface of 300 square feet, which was accidentally broken before there was an opportunity of trying the effect of the propelling apparatus; but its steerage and steadiness were perfectly proved, and it would sail obliquely downward in any direction, according to the set of the rudder. Even in this state, when any. person ran forward in it, with his full speed, taking advantage of a gentle breeze in front, it would bear upward so strongly as scarcely to allow him to touch the ground; and would frequently lift him up, and convey him several yards together.

The best mode of producing the propelling power is the only thing, that remains yet untried toward the completion of the invention. I am preparing to resume my experiments upon this subject, and state the following observations, in the hope that others may be induced to give their attention towards expediting the attainment of this art.

The act of flying is continually exhibited to our view; and the principles upon which it is effected are the same as those before stated. If an attentive observer examines the waft of a wing, he will perceive, that about one third part, toward the extreme point, is turned obliquely backward; this being the only portion, that has velocity enough to overtake the current, passing so rapidly beneath it, when in this unfavourable position. Hence this is the only portion that gives any propelling force.

To make this more intelligible, let A B, fig. 4, be a section of this part of the wing. Let C D represent the velocity of the bird's path, or the current, and E D that of the wing in its waft: then C E will represent the magnitude and direction of the compound or actual current striking the under surface of the wing. Suppose E F, perpendicular to A B, to represent the whole pressure; E G being parallel to the horizon, will represent the propelling force; and G F, perpendicular to it, the supporting power. A bird is supported as effectually during the return as during the beat of its wing; this is chiefly effected by receiving the resistance of the current under that portion of the wing next the body where its receding motion is so slow as to be of scarcely any effect. The extreme portion of the wing, owing to its velocity, receives a pressure downward and obliquely forward, which forms a part of the propelling force; and at the same time, by forcing the hinder part of the middle portion of the wing downward, so increases its angle with the current, as to enable it still to receive nearly its usual pressure from beneath.

As the common rook has its surface and weight in the ratio of a square foot to a pound, it may be considered as a standard for calculations of this sort; and I shall therefore state, from the average of many careful observations, the movements of that bird. Its velocity, represented by C D, fig. 4, is 34.5 feet per second. It moves its wing up and down once in flying over a space of 12.9 feet. Hence, as the centre of resistance of the extreme portion of the wing moves over a space of 0.75 of a foot each beat or return, its velocity is about 4 feet per second, represented by the line E D. As the wing certainly overtakes the current, it must be inclined from it in an angle something less than 7°, for at this angle it would scarcely be able to keep parallel with it, unless the waft downward were performed with more velocity than the return; which may be and probably is the case, though these movements appear to be of equal duration. The propelling power, represented by E G, under these circumstances, cannot be equal to an eighth part of the supporting power G F, exerted upon this portion of the wing; yet this, together with the aid from the return of the wing, has to overcome all the retarding power of the surface, and the direct resistance occasioned by the bulk of the body.

It has been before suggested, and I believe upon good grounds, that very acute angles vary little in the degree of resistance they make under a similar velocity of current. Hence it is probable, that this propelling part of the wing receives little more than its common proportion of resistance, during the waft downward. If it be taken at one-third of the whole surface, and one-eighth of this be allowed as the propelling power, it will only amount to one twenty-fourth of the weight of the bird; and even this is exerted only half the duration of the flight. The power gained in the return of the wing must be added, to render this statement correct, and it is difficult to estimate this; yet the following statement proves, that a greater degree of propelling force is obtained, upon the whole, than the foregoing observations will justify. Suppose the largest circle that can be described in the breast of a crow, to be 12 inches in area. Such a surface, moving at the velocity of 34.5 feet per second, would meet a resistance of 0.216 of a pound, which, reduced by the proportion of the resistance of a sphere to its great circle (given by

Mr. Robins as 1 to 2.27) leaves a resistance of 0.095 of a pound, had the breast been hemispherical. It is probable however, that the curve made use of by Nature to avoid resistance, being so exquisitely adapted to its purpose, will reduce this quantity to one half less than the resistance of the sphere, which would ultimately leave 0.0475 of a pound as somewhat approaching the true resistance. Unless therefore the return of the wing gives a greater degree of propelling force than the beat, which is improbable, no such resistance of the body could be sustained. Hence, though the eye cannot perceive any distinction between the velocities of the beat and return of the wing, it probably exists, and experiment alone can determine the proper ratios between them.

From these observations we may, however, be justified in the remark—that the act of flying, when properly adjusted by the Supreme Author of every power, requires less exertions than, from the appearance, is supposed.

Document 1-1(c), George Cayley, "On Aerial Navigation," Part Three, 1810.

BROMPTON, Dec. 6, 1809.

NOT having sufficient data to ascertain the exact degree of propelling power exerted by birds in the act of flying, it is uncertain what degree of energy may be required in this respect in vessels for aerial navigation: yet, when we consider the many hundred miles of continued flight exerted by birds of passage, the idea of its being only a small effort is greatly corroborated. To apply the power of the first mover to the greatest advantage in producing this effect, is a very material point. The mode universally adopted by nature is the oblique waft of the wing. We have only to choose between the direct beat overtaking the velocity of the current, like the oar of a boat; or one, applied like the wing, in some assigned degree of obliquity to it. Suppose 35 feet per second to be the velocity of an aerial vehicle, the oar must be moved with this speed previous to its being able to receive any resistance; then, if it be only required to obtain a pressure of 1/10th of a pound upon each square foot, it must exceed the velocity of the current 7.5 feet per second. Hence its whole velocity must be 42.5 feet per second. Should the same surface be wafted downward, like a wing, with the hinder edge inclined upward in an angle of about 50° 40' to the current, it will overtake it at a velocity of 3.5 feet per second; and as a slight unknown angle of resistance generates a pound pressure per square foot at this velocity, probably a waft of little more than 4 feet per second would produce this effect; one tenth part of which would be the propelling power. The advantage in favour of this mode of application, compared with the former, is rather more than ten to one.

In combining the general principles of aerial navigation for the practice of the art many mechanical difficulties present themselves, which require a considerable course of skilfully applied experiments, before they can be overcome. But to a certain extent the air has already been made navigable; and no one, who has seen the steadiness with which weights to the amount of ten stone (including four stone, the weight of the machine) hover in the air, can doubt of the ultimate accomplishment of this object.

The first impediment I shall take notice of is the great proportion of power, that must be exerted previous to the machine's acquiring that velocity, which gives support upon the principle of the inclined plane; together with the total want of all support during the return of any surface used like a wing. Many birds, and particularly water fowl, run and flap their wings for several yards before they can gain support from the air. The swift (*hirundo apus Lin.*) is not able to elevate itself from level ground. The inconvenience under consideration arises from very different causes in these two instances. The supporting surface of most swimming birds does not exceed the ratio of $4/10$ths of a square foot to every pound of their weight: the swift, though it scarcely weighs an ounce, measures eighteen inches in extent of wing. The want of surface in the one case, and the inconvenient length of wing in the other, oblige these birds to aid the *commencement* of their flight by other expedients; yet they can both fly with great power, when they have acquired their full velocity.

A second difficulty in aerial navigation arises from the great extent of lever, which is constantly operating against the first mover, in consequence of the distance of the centre of support in large surfaces, if applied in the manner of wings.

A third and general obstacle is the mechanical skill required to unite great extension of surface with strength and lightness of structure; at the same time having a firm and steady movement in its working parts, without exposing unnecessary obstacles to the resistance of the air. The first of these obstacles, that have been enumerated, operates much more powerfully against aerial navigation upon a large scale, than against birds; because the small extent of their wings obliges them to employ a very rapid succession of strokes, in order to acquire that velocity which will give support; and during the small interval of the return of the wing, their weight is still rising, as in a leap, by the impulse of one stroke, till it is again aided by another. The large surfaces that aerial navigation will probably require, though necessarily moved with the same velocity, will have a proportionably longer duration both of the beat and return of the wing; and hence a greater descent will take place during the latter action, than can be overcome by the former.

There appears to be several ways of obviating this difficulty. There may be two surfaces, each capable of sustaining the weight, and placed one above the other, having such a construction as to work up and down in opposition when they are moved, so that one is always ready to descend, the moment the other ceases. These surfaces may be so made, by a valvelike structure, as to give no opposition in rising up, and only to resist in descent.

The action may be considered either oblique, as in rotative flyers; alternately so, without any up and down waft, as in the engine I have ascribed to Mr. Degen; by means of a number of small wings in lieu of large ones, upon the principle of the flight of birds, with small intervals of time between each waft; and lastly by making use of light wheels to preserve the propelling power both of the beat and the return of the wings, till it accumulates sufficiently to elevate the machine, upon the principle of those birds which run themselves up. This action might be aided by making choice of a descending ground like the swift.

With regard to another part of the first obstacle I have mentioned, viz. the absolute quantity of power demanded being so much greater at first than when the full velocity has been acquired; it may be observed, that, in the case of human muscular strength being made use of, a man can exert, for a few seconds, a surprising degree of force. He can run up stairs, for instance, with a velocity of from 6 to 8 feet perpendicular height per second, without any dangerous effort; here the muscles of his legs only are in action; but, for the sake of making a moderate statement, suppose that with the activity of his arms and body, in addition to that of his legs, he is equal to raising his weight 8 feet per second; if in this case he weighs 11 stone, or 154 pounds, he will be exerting, for the time, an energy equal to more than the ordinary force of two of Messrs. Boulton and Watt's steam horses; and certainly more than twelve men can bestow upon their constant labour.

If expansive first movers be made use of, they may be so constructed, as to be capable of doing more than their constant work; or their power may be made to accumulate for a few moments by the formation of a vacuum, or the condensation of air, so that these expedients may restore at one time, in addition to the working of the engine, that which they had previously absorbed from it.

With regard to the second obstacle in the way of aerial navigation, viz. the length of leverage to which large wing-like surfaces are exposed, it may be observed, that, being a constant and invariable quality, arising from the degree of support such surfaces give, estimated at their centres of resistance, it may be balanced by any elastic agent, that is so placed as to oppose it. Let A and B, Pl. IV, fig. 1, be two wings of an aerial vehicle in the act of skimming; then half the weight of the vessel is supported from the centre of resistance of each wing; as represented by the arrows under them. If the shorter ends of these levers be connected by cords to the string of a bow C, of sufficient power to balance the weight of the machine at the points A and B, then the moving power will be left at full liberty to produce the waft necessary to bend up the hinder edge of the wing, and gain the propelling power. A bow is not in fact an equable spring, but may be made so by using a spiral fusee. I have made use of it in this place merely as the most simple mode of stating the principle I wished to exhibit. Should a counterbalancing spring of this kind be adopted in the practice of aerial navigation, a small well polished cylinder, furnished with what may be termed a bag piston (upon the principle made use of by nature in preventing the return of the blood to the heart, when it has been driven into the aorta, by the intervention of the semilunar valves) would, by a vacuum being excited each stroke of the wing, produce the desired effect, with scarcely any loss by friction. These elastic agents may likewise be useful in gradually stopping the momentum of large surfaces when used in any alternate motion, and in thus restoring it during their return.

(I have made use of several of these pistons, and have no scruple in asserting, that for all blowing engines, where friction is an evil, and being very nearly airtight is sufficient, there is no other piston at all comparable with them. The most irregular cylinder, with a piston of this kind, will act with surprising effect. To give an instance; a cylinder of sheet tin, 8 inches long and $3\frac{1}{2}$ in diameter, required 4 pounds to force the piston down in 15 minutes; and in other trials became perfectly tight in some positions, and would proceed no farther. The friction, when the cylinder was open at both ends, did not exceed $\frac{1}{2}$ an ounce.)

Another principle, that may be applied to obviate this leverage of a wing, is that of using such a construction as will make the supporting power of the air counterbalance itself. It has been before observed, that only about one third of the wing in birds is applied in producing the propelling power; the remainder, not having velocity sufficient for this purpose, is employed in giving support, both in the beat and return of the wing.

Let A and B, fig. 2, be two wings continued beyond the pole or hinge upon which they turn at C. If the extreme parts at A and B be long and narrow, they may be balanced, when in the act of skimming, by a broad extension of less length on their opposite sides; this broad extension, like the lower part of the wing, will always give nearly the same support, and the propelling part of the surface will be at liberty to act unincumbered by the leverage of its supporting power. This plan may be modified many different ways; but my intention, as in the former case, is still the principle in its simplest form.

A third principle upon which the leverage of a surface may be prevented is by giving it a motion parallel to itself, either directly up and down, or obliquely so. The surface A 1, fig. 3, may be moved perpendicularly, by the shaft which supports it, down to the position K C: or, if it be supported upon two shafts with hinges at D and E, it may be moved obliquely parallel to itself into the position B L.

A fourth principle upon which the leverage may be greatly avoided, where only one hinge is used, is by placing it considerably below the plane of the wing, as at the point D, fig. 3, in respect to the surface A. It may be observed in the heron, which is a weak bird with an extended surface, that its wings curve downward considerably from the hinge to the tip; hence the extreme portion, which receives the chief part of the stroke, is applied obliquely to the current it creates; and thus evades in a similar degree the leverage of that portion of the supporting power, which is connected with the propelling power. These birds seldom carry their waft much below the level of the hinge of the wing, where this principle, so far as respects the supporting power, would vanish.

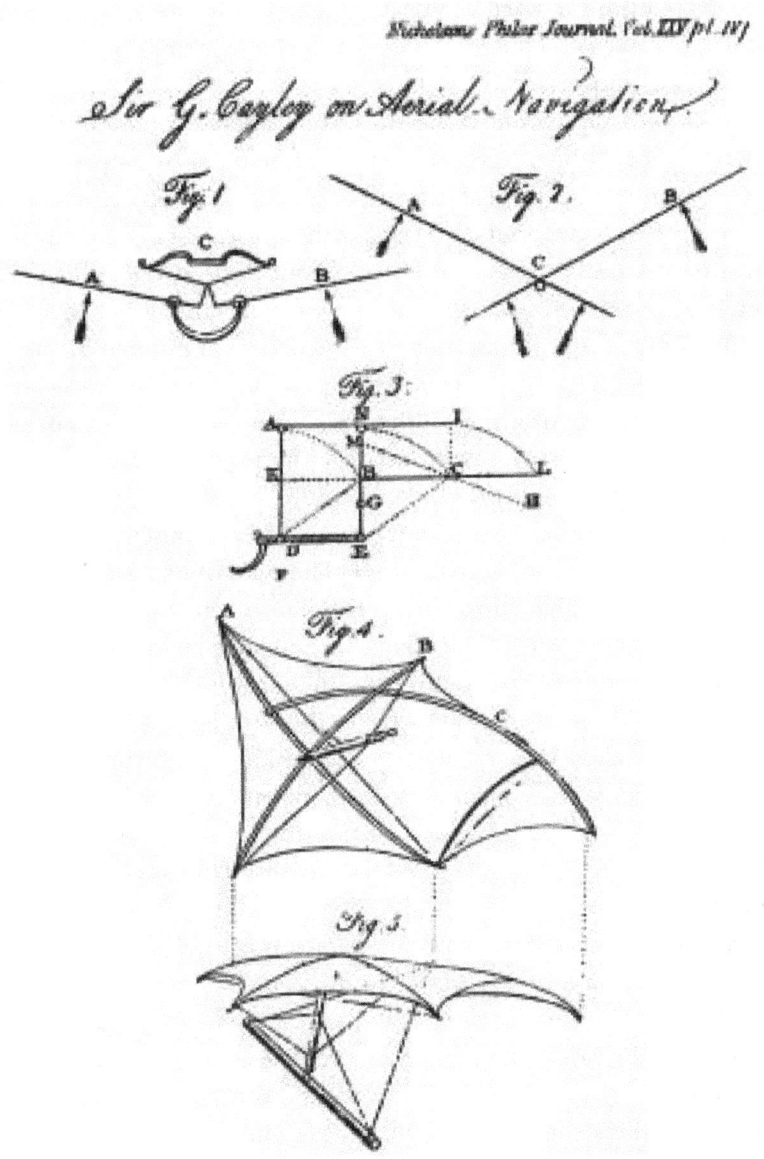

By making use of two shafts of unequal length, the two last mentioned principles may be blended to any required extent. Suppose one hinge to be at F, and the other at G, fig. 3, then the surface, at the extent of its beat, would be in the position of the line H M. If the surface A 1, fig. 3, be supported only upon one shaft, N E, be capable of being forced in some degree from its rectangular position in respect to the shaft, and be concave instead of flat as here represented; then the waft may be used alternately backward and forward, according to the principles of the machine I have ascribed to Mr. Degen. This construction combines the principles of counterpoising the supporting power of one part of the surface, by that of an opposite part, when the machine is in the act of skimming; and likewise the advantages of the low hinge, with the principle of leaving little or no interval without support.

All that has hitherto appeared respecting Mr. Degen's apparatus is, that it consisted of two surfaces, which were worked by a person sitting between them. This statement communicates no real information upon the subject; for scarcely anyone would attempt to fly without *two* wings; without these being equally poised by placing the weight *between* them; and also, without these surfaces being capable of receiving motion from his muscular action. I may be altogether mistaken in my conjecture; my only reason for ascribing this structure of mine to Mr. Degen's machine is, that, if it were properly executed upon this principle, it would be attended with success. The drawing, rather diagram, which is given of this machine in the first part of my essay, is only for the purpose of exhibiting the principle in a form capable of being understood. The necessary bracings, etc., required in the actual execution of such a plan, would have obscured the simple nature of its action; and were therefore omitted. The plan of its movement is also simply to exhibit, in a tangible form, the possibility of effecting the intended alternate motion of the parachutes. The seat is fronted lengthwise for the purpose of accommodating the mode of communicating the movement.

A fifth mode of avoiding leverage is by using the continued action of oblique horizontal flyers, or an alternate action of the same kind, with surfaces so constructed as to accommodate their position to such alternate motion; the hinge or joint being in these cases vertical. In the construction of large vessels for aerial navigation, a considerable portion of fixed sail will probably be used; and no more surface will be allotted, towards gaining the propelling power, than what is barely necessary, with the extreme temporary exertion of the first mover, to elevate the machine and commence the flight. In this case the leverage of the fixed surface is done away.

The general difficulties of structure in aerial vehicles, (arising from the extension, lightness, and strength required in them; together with great firmness in the working parts, and at the same time such an arrangement as exposes no unnecessary obstacles to the current,) I cannot better explain than by describing a wing, which has been constructed with a view to overcome them.

Fig. 4 represents the shape of the cloth, with a perspective view of the poles upon which it is stretched with perfect tightness. Upon the point where the rods A and B intersect is erected an oval shaft; embracing the two cross poles by a slender iron fork; for the purpose of preserving their strength uninjured by boring. To this shaft are braced the ends of the pole B, so as to give this pole any required degree of curvature. The pole A is strung like a common bow to the same curve as the pole B; and is only connected with the upright shaft by what may be called a check brace; which will allow the hinder end of this pole to heel back to a certain extent, but not the fore end. The short brace producing this effect is shown in fig. 4. Fig. 5 exhibits the fellow wing to that represented in fig. 4, erected upon a beam, to which it is so braced, as to convert the whole length of it into a hinge. The four braces coming from the ends of this beam are shown: two of them terminate near the top of the centre of the other shaft; the others are inserted into the point C, fig. 4, of the bending rod. A slight bow, not more than three-eighths of an inch thick, properly curved by its string, and inserted between the hinder end of the pole A, and the curved pole C, completes the wing.

This fabrick contained 54 square feet, and weighed only eleven pounds. Although both these wings together did not compose more than half the surface necessary for the support of a man in the air, yet during their waft they lifted the weight of nine stone. The hinder edge, as is evident from the construction, being capable of giving way to the resistance of the air, any degree of obliquity, for the purpose of a propelling power, may be used.

I am the more particular in describing this wing, because it exemplifies almost all the principles that can be resorted to in the construction of surfaces for aerial navigation. Diagonal bracing is the great principle for producing strength without accumulating weight; and, if performed by thin wires, looped at their ends, so as to receive several laps of cordage, produces but a trifling resistance in the air, and keeps tight in all weathers. When bracings are well applied, they make the poles, to which are attached, bear endwise. The hollow form of the quill in birds is a very admirable structure for lightness combined with strength, where external bracings cannot be had; a tube being the best application of matter to resist as a lever; but the principle of bracing is so effectual, that, if properly applied, it will abundantly make up for the clumsiness of human invention in other respects; and should we combine both these principles, and give diagonal bracing to the tubular bamboo cane, surfaces might be constructed with a greater degree of strength and lightness, than any made use of in the wings of birds.

The surface of a heron's wing is in the ratio of 7 square feet to a pound. Hence, according to this proportion, a wing of 54 square feet would weigh about $7\frac{3}{4}$ pounds: on the contrary the wings of water fowl are so much heavier, that a surface of 54 square feet, according to their structure, will weight 18 ½ lb. I have in these instances quoted nearly the extreme cases among British birds; the wing I have described may therefore be considered as nearly of the same weight in proportion to its bulk as that of most birds.

Another principle exhibited in this wing is that of the poles being couched within the cloth, so as to avoid resistance. This is accomplished by the convexity of the frame, and the excessive lightness of the cloth. The poles are not allowed to form the edge of the wing, excepting at the extreme point of the bow, where it is very thin, and also oblique to the current. The thick part of this pole is purposely conveyed considerably within the edge. In birds, a membrane covered with feathers is stretched before the thick part of the bone of the wing, in a similar manner, and for the same purpose. The edge of the surface is thus reduced to the thickness of a small cord, that is sown to the cloth, and gives out loops whenever any fastening is required. The upright shaft is the only part that opposes much direct resistance to the current, and this is obviated in a great degree by a flat oval shape, having its longest axis parallel to the current.

The joint or hinge of this wing acts with great firmness, in consequence of its being supported by bracings to the line of its axis, and at a considerable distance from each other; in fact the bracings form the hinge.

The means of communicating motion to any surfaces must vary so much, according to the general structure of the whole machine, that I shall only observe at present, that where human muscular action is employed, the movement should be similar to the mode of pulling oars; from which any other required motion may be derived; the foot-board in front enables a man to exert his full force in this position. The wings I have described were wafted in this manner; and when they lifted with a power of 9 stone, not half of the blow, which a man's strength could have given, was exerted, in consequence of the velocity required being greater than convenient under the circumstances. Had these wings been intended for elevating the person who worked them, they should have contained from 100 to 150 square feet each; but they were constructed for the purpose of an experiment relative to the propelling power only.

Avoiding direct resistance is the next general principle, that it is necessary to discuss. Let it be remembered as a maxim in the art of aerial navigation, that every pound of direct resistance, that is done away, will support 30 pounds of additional weight without any additional power. The figure of a man seems but ill calculated to pass with ease through the air, yet I hope to prove him to the full as well made in this respect as the crow, which has hitherto been our standard of comparison, paradoxical as it may appear.

The principle, that surfaces of similar bodies increase only as the squares of their homologous lines, while their weights, or rather solid contents, increase as the cubes of those lines, furnishes the solution. This principle is unanimously in favour of large bodies. The largest circle that can be described in a crow's breast is about 12 square inches in area. If a man exposes a direct bulk of 6 square feet, the ratio of their surfaces will be as 1 to 72; but the ratio of their weight is as 1 to 110; which is 1 ½ to 1 in favour of the man, provided he were within a case as well constructed for evading resistance, as the body of the crow; but even supposing him to be exposed in his natural cylindric shape, in the foreshortened posture of sitting to work his oars, he will probably receive less resistance than the crow.

It is of great importance to this art, to ascertain the real solid of least resistance, when the length or breadth is limited. Sir Isaac Newton's beautiful theorem upon this subject is of no practical use, as it supposes each particle of the fluid, after having struck the solid, to have free egress; making the angles of incidence and reflection equal; particles of light seem to possess this power, and the theory will be true in that case; but in air the action is more like an accumulation of particles, rushing up against each other, in consequence of those in contact with the body being retarded. The importance of this subject is not less than the difficulties it presents; it affects the present interests of society in its relation to the time occupied in the voyages of ships; it will have still more effect when aerial navigation, now in its cradle, is brought home to the uses of man. I shall state a few crude hints upon this point, to which my subject has so unavoidably led, and on which I am so much interested, and shall be glad if in so doing I may excite the attention of those, who are competent to an undertaking greatly beyond my grasp.

Perhaps some approach toward ascertaining the actual solid of least resistance may be derived from treating the subject in a manner something similar to the following. Admit that such a solid is already attained (the length and width being necessarily taken at pleasure). Conceive the current intercepted or disturbed, by the largest circle that can be drawn within the given spindle, to be divided into concentric tubular laminae of equal thickness. At whatever distance from this great circle the apex of the spindle commences, on all sides of this point the central lamina will be reflected in diverging pencils, (or rather an expanding ring,) making their angles of incidence and reflection equal. After this reflection they rush against the second lamina and displace it: this second lamina contains three times more fluid than the first; consequently each pencil in the first meets three pencils in the second; and their direction, after the union, will be one fourth of the angle, with respect to the axis, which the first reflection created. In this direction these two laminae proceed till they are themselves reflected, when they (considered as one lamina of larger dimensions) rush against the third and fourth, which together contain three times the fluid in the two former laminae, and thus reduce the direction of the combined mass to one fourth of the angle between the axis and the line of the second reflection. This process is constant, whatever be the angles formed between the surface of the actual solid of least resistance at these points of reflection, and the directions of the currents thus reflected.

From this mode of reasoning, which must in some degree resemble what takes place, and which I only propose as a resemblance, it appears, that the fluid keeps creeping along the curved surface of such a solid, meeting it in very acute angles. Hence, as the experiments of the French Academy show, that the difference of resistance between the direct impulse, and that in an angle of six degrees, on the same surface, is only in the ratio of 10 to 4, it is probable, that in the slight difference of angles that occur in this instance, the resistances may be taken as equal upon every part, without any material deviation from truth. If this reasoning be correct, it will reduce the question, so far as utility is concerned, within a strictly abstract mathematical inquiry.

It has been found by experiment, that the shape of the hinder part of the spindle is of as much importance as that of the front, in diminishing resistance. This arises from the partial vacuity created behind the obstructing body. If there be no solid to fill up this space, a deficiency of hydrostatic pressure exists within it, and is transferred to the spindle. This is seen distinctly near the rudder of a ship in full sail, where the water is much below the level of the surrounding sea. The cause here, being more evident, and uniform in its nature, may probably be obviated with better success; in as much as this portion of the spindle may not differ essentially from the simple cone. I fear however, that the whole of this subject is of so dark a nature, as to be more usefully investigated by experiment, than by reasoning; and in the absence of any conclusive evidence from either, the only way that presents itself is to copy nature; accordingly I shall instance the spindles of the trout and woodcock, which, lest the engravings should, in addition to the others, occupy too much valuable space in your Journal, must be reserved to a future opportunity.

Document 1-2

Thomas Jefferson, letter to William D. B. Lee, 27 April 1822.

The remarkable farsightedness of George Cayley's understanding of aerodynamics can be gauged by the lack of confidence many of the leading intellectuals and scientists of his day had in the future of "aerial navigation" with flying machines. Responding to this issue just four years before his death, an elderly Thomas Jefferson, principal author of the Declaration of Independence (1776) and third President of the United States (from 1801 to 1809), took the following cautiously pessimistic view about the possibility of mechanical flight. Jefferson possessed a wide-ranging curiosity about all things, especially science and technology. But here in 1822, in his advanced years, he seems to be admitting that they have progressed well beyond his capacity to understand them fully or to assess the potential of such radically new concepts as mechanical flight. (The spelling in this document is Jefferson's.)

Document 1-2, Thomas Jefferson, letter to William D. B. Lee, 27 April 1822.

Your letter of the 15th is received, but age has long since obliged me to withdraw my mind from speculations of the difficulty of those of your letter. That there are means of artificial buoyancy by which man may be supported in the air, the balloon has proved, and that means of directing it may be discovered is against no law of nature and is therefore possible as in the case of birds. But to do this by mechanical means alone in a medium so rare and unresisting as air must have the aid of some principle not yet generally known. However I can really give no opinion understandingly on the subject and with more goodwill than confidence wish you success.

Document 1-3(a–c)

(a) Francis H. Wenham, "On Aerial Locomotion and the Laws by Which Heavy Bodies Impelled Through Air are Sustained," *First Annual Report of the Aeronautical Society of Great Britain for the Year 1866*, pp. 10–47.

(b) Report on the first wind tunnel built by Francis H. Wenham, *Sixth Annual Report of the Aeronautical Society of Great Britain, for the Year 1871*, pp. 75–78.

(c) Minutes of Aeronautical Society, *Seventh Annual Report of the Aeronautical Society of Great Britain, for the Year 1872*, pp. 5–12.

A crucial step forward for aeronautics happened in January 1866 when a group of serious aviation enthusiasts met for the first time, creating the Aeronautical Society of Great Britain. Later this organization of gentleman amateurs and professional engineers interested in flight became the Royal Aeronautical Society. Through its regular periodic meetings and publications, the study of human flight received an important early measure of scholarly credence and viability.

The first document below, from the year 1866, is one of the most interesting of the very early papers presented at a meeting of the Aeronautical Society of Great Britain. Its author, Francis H. Wenham (1824–1908), the son of an army surgeon with no college education, rose to great prominence in the emerging field of aeronautics; some even came to consider him "the father of aeronautics" in Britain. In this 1866 paper, published in the society's first annual report, Wenham focused on a principle that its author called "sustainment," by which he meant principally aerodynamic lift. Although his paper contributed an important idea in the form of his favorable analysis of what we would call a "high aspect-ratio" wing (with multiplane layout and propeller propulsion), most of Wenham's presentation suffered from a serious lack of clarity and depth regarding the basic sources of aerodynamic force. One of the reasons for this was Wenham's weakness in higher mathematics. As readers will see, his 1866 paper contained no mathematical equations; his basic inclination was always to solve technical problems through experimental means.

This experimental approach can be clearly seen in the second document below, from the Aeronautical Society of Great Britain's report of 1871, which discusses Wenham's design of what historians now recognize to be the world's first wind tunnel. Though Wenham used his primitive, steam-engine-driven wind tunnel (maximum velocity was forty miles per hour) solely to explore the lift and drag characteristics of flat surfaces, the type of device he pioneered showed great promise. In the following decade, in 1884, another member of the Aeronautical Society of Great Britain, Horatio Phillips, built a second, improved tunnel to demonstrate the improved lifting qualities of mildly cambered surfaces. Together, these two tunnels paved the way for what, by the Wright brother era, became the most basic and versatile tool in modern aeronautical experimentation.

Document 1-3(a), Francis H. Wenham, "On Aerial Locomotion," 27 June 1866.

The resistance against a surface of a defined area, passing rapidly through yielding media, may be divided into two opposing forces. One arising from the cohesion of the separated particles; and the other from their weight and inertia, which, according to well-known laws, will require a constant power to set them in motion.

In plastic substances, the first condition, that of cohesion, will give rise to the greatest resistance. In water this has very little retarding effect, but in air, from its extreme fluidity, the cohesive force becomes inappreciable, and all resistances are caused by its weight alone; therefore, a weight, suspended from a plane surface, descending perpendicularly in air, is limited in its rate of fall by the weight of air that can be set in motion in a given time.

If a weight of 150 lbs. is suspended from a surface of the same number of square feet, the uniform descent will be 1,300 feet per minute, and the force given out and expended on the, air, at this rate of fall, will be nearly six horse-power; and, conversely, this same speed and power must be communicated to the surface to keep the weight sustained at a fixed altitude. As the surface is increased, so does the rate of descent and its accompanying power, expended in a given time, decrease. It might, therefore, be inferred that, with a sufficient extent of surface reproduced, or worked up to a higher altitude, a man might by his exertions raise himself for a time, while the surface descends at a less speed.

A man, in raising his own body, can perform 4,250 units of work—that is, this number of pounds raised one foot high per minute—and can raise his own weight—say, 150 lbs.—twenty-two feet per minute. But at this speed the atmospheric resistance is so small that 120,000 square feet would be required to balance his exertions, making no allowance for weight beyond his own body.

We have thus reasons for the failure of the many misdirected attempts that have, from time to time, been made to raise weights perpendicularly in the air by wings or descending surfaces. Though the flight of a bird is maintained by a constant reaction or abutment against an enormous weight of air in comparison with the weight of its own body, yet, as will be subsequently shown, the support upon that weight is not necessarily commanded by great extent of wing-surface, but by the direction of motion.

One of the first birds in the scale of flying magnitude is the pelican. It is seen in the streams and estuaries of warm climates, fish being its only food. On the Nile, after the inundation, it arrives in flocks of many hundreds together, having migrated from long distances. A specimen shot was found to weigh twenty-one pounds, and measured ten feet across the wings, from end to end. The pelican rises with much difficulty, but, once on the wing, appears to fly with very little exertion, notwithstanding its great weight. Their mode of progress is peculiar and graceful. They fly after a leader, in one single train. As he rises or descends, so his followers do the same in succession, imitating his movements precisely. At a distance, this gives them the appearance of a long undulating ribbon, glistening under the cloudless sun of an oriental sky. During their flight they make about seventy strokes per minute with their wings. This uncouth-looking bird is somewhat whimsical in its habits. Groups of them may be seen far above the earth, at a distance from the river-side, *soaring*, apparently for their own pleasure. With outstretched and motionless wings, they float serenely, high in the atmosphere, for more than an hour together, traversing the same locality in circling movements. With head thrown back, and enormous bills resting on their breasts, they almost seem asleep. A few easy strokes of their wings each minute, as their momentum or velocity diminishes, serves to keep them sustained at the same level. The effort required is obviously slight, and not confirmatory of the excessive amount of power said to be requisite for maintaining the flight of a bird of this weight and size. The pelican displays no symptom of being endowed with great strength, for when only slightly wounded it is easily captured, not having adequate power for effective resistance, but heavily flapping the huge wings, that should, as some imagine, give a stroke equal in vigour to the kick of a horse.

During a calm evening, flocks of spoonbills take their flight directly up the river's course; as if linked together in unison, and moved by the same impulse, they alter not their relative positions, but at less than fifteen inches above the water's surface, they speed swiftly by with ease and grace inimitable, a living sheet of spotless white. Let one circumstance be remarked,—though they have fleeted past at a rate of near thirty miles an hour, so little do they disturb the element in which they move, that not a ripple of the placid bosom of the river, which they almost touch, has marked their track. How wonderfully does their progress contrast with that of creatures who are compelled to drag their slow and weary way against the fluid a thousand fold more dense, flowing in strong and eddying current beneath them.

Our pennant droops listlessly, the wished-for north wind cometh not. According to custom we step on shore, gun in hand. A flock of white herons, or "buffalo-birds," almost within our reach, run a short distance from the pathway as we approach them. Others are seen perched in social groups upon the backs of the apathetic and mud-begrimed animals whose name they bear. Beyond the ripening dhourra crops which skirt the river-side, the land is covered with immense numbers of blue pigeons, flying to and fro in shoals, and searching for food with restless diligence. The musical whistle from the pinions of the wood-doves sounds cheerily, as they dart past with the speed of an arrow. Ever and anon are seen a covey of the brilliant, many-coloured partridges of the district, whose *long and pointed wings* give them a strength and duration of flight that seems interminable, alighting at distances beyond the possibility of marking them down, as we are accustomed to do with their plumper brethren at home. But still more remarkable is the spectacle which the sky presents. As far as the eye can reach it is dotted with birds of prey of every size and description. Eagles, vultures, kites and hawks, of manifold species, down to the small, swallow-like, insectivorous hawk common in the Delta, which skims the surface of the ground in pursuit of its insect prey. None seem bent on going forward, but all are soaring leisurely round over the same locality, as if the invisible element which supports them were their medium of rest as well as motion. But mark that object sitting in solitary state in the midst of yon plain: what a magnificent eagle! An approach to within eighty yards arouses the king of birds from his apathy. He partly opens his enormous wings, but stirs not yet from his station. On gaining a few feet more he begins to walk away, with half-expanded, but motionless wings. Now for the chance fire! A charge of No. 3 from 11 bore rattles audibly but ineffectively upon his densely feathered body; his walk increases to a run, he gathers speed with his slowly-waving wings, and eventually, leaves the ground. Rising at a gradual inclination he mounts aloft and sails majestically away to his place of refuge in the Lybian range, distant at least five miles from where he rose. Some fragments of feathers denote the spot where the shot had struck him. The marks of his claws are traceable in the sandy soil, as, at first with firm and decided digs, he forced his way, but as he lightened his body and increased his speed with the aid of his wings, the imprints of his talons gradually merged into long scratches. The measured distance from the point where these vanished, to the place where he had stood, proved that with all the stimulus that the shot must have given to his exertions, he had been compelled to run full twenty yards before he could raise himself from the earth.

Again the boat is under weigh, though the wind is but just sufficient to enable us to stem the current. An immense kite is soaring overhead, scarcely higher than the top of our lateen yard, affording a fine opportunity for contemplating his easy and unlaboured movements. The cook has now thrown overboard some offal. With a solemn swoop the bird descends and seizes it in his talons. How easily he rises again with motionless expanded wings, the mere force and momentum of his *descent* serving to raise him again to more than half-mast high. Observe him next, with lazy flapping wings, and head turned under his body; he is placidly devouring the pendant morsel from his foot, and calmly gliding onwards.

The Nile abounds with large aquatic birds of almost every variety. During a residence upon its surface for nine months out of the year, immense numbers have been seen to come and go, for the majority of them are migratory. Egypt being merely a narrow strip of territory, passing through one of the most desert parts of the earth, and rendered fertile only by the periodical rise of the waters of the river, it is probable that these birds make it their grand thoroughfare into the rich districts of Central Africa.

On nearing our own shores, steaming against a moderate head-wind, from a station abaft the wheel the movements of some half-dozen gulls are observed, following in the wake of the ship, in patient expectation of any edibles that may be thrown overboard. One that is more familiar than the rest comes so near at times that the winnowing of his wings can be heard; he has just dropped astern, and now comes on again. With the axis of his body exactly at the level of the eyesight, his every movement can be distinctly marked. He approaches within ten yards, and utters his wild plaintive note, as he turns his head from side to side, and regards us with his jet black eye. But where is the angle or upward rise of his wings, that should compensate for his descending tendency, in a yielding medium like air? The incline cannot be detected, for, to all appearance, his wings are edgewise, or parallel to his line of motion, and he appears to skim along a *solid* support. No smooth-edged rails, or steel-tired wheels, with polished axles revolving in well oiled brasses, are needed here for the purpose of diminishing friction, for Nature's machinery has surpassed them all. The retarding effects of gravity in the creature under notice, are almost annulled, for he is gliding forward upon a *frictionless* plane. There are various reasons for concluding that the direct flight of many birds is maintained with a much less expenditure of power, for a high speed, than by any mode of progression.

The first subject for consideration is the proportion of surface, to weight, and their combined effect in descending perpendicularly through the atmosphere. The datum is here based upon the consideration of *safety*, for it may sometimes be needful for a living being to drop passively, without muscular effort. One square foot of sustaining surface, for every pound of the total weight, will be sufficient for security.

According to Smeaton's table of atmospheric resistances, to produce a force of *one pound* on a square foot, the wind must move against the plane (or, which is the same thing, the plane against the wind), at the rate of twenty-two feet per second, or 1,320 feet per minute, equal to fifteen miles per hour. The resistance of the air will now balance the weight on the descending surface, and, consequently, it cannot exceed that speed. Now, twenty-two feet per second is the velocity acquired at the end of a fall of eight feet—a height from which a well-knit man or animal may leap down without much risk of injury. Therefore, if a man with parachute weigh together 143 lbs., spreading the same number of square feet of surface contained in a circle fourteen and a half feet in diameter, he will descend at perhaps an unpleasant velocity, but with safety to life and limb.

It is a remarkable fact how this proportion of wing-surface to weight extends throughout a great variety of the flying portion of the animal kingdom, even down to hornets, bees, and other insects. In some instances, however, as in the gallinaceous tribe, including pheasants, this area is somewhat exceeded, but they are known to be very poor flyers. Residing as they do chiefly on the ground, their wings are only required for short distances, or for raising them or easing their descent from their roosting-places in forest trees, the *shortness* of their wings preventing them from taking extended flights. The wing-surface of the common swallow is rather more than in the ratio of *two* square feet per pound, but having also great length of pinion, it is both swift and enduring in its flight. When on a rapid course this bird is in the habit of furling its wings into a narrow compass. The greater extent of surface is probably needful for the continual variations of speed and instant stoppages requisite for obtaining its insect food.

On the other hand, there are some birds, particularly of the duck tribe, whose wing-surface but little exceeds *half* a square foot, or seventy-two inches per pound, yet they may be classed among the strongest and swiftest of flyers. A weight of one pound, suspended from an area of this extent, would acquire a velocity due to a fall of 16 feet—a height sufficient for the destruction or injury of most animals. But when the plane is urged forward horizontally, in a manner analogous to the wings of a bird during flight, the sustaining power is greatly influenced by *the form and arrangement* of the surface.

In the case of *perpendicular* descent, as a parachute, the sustaining effect will be much the same, whatever the figure of the outline of the superficies may be, and a circle perhaps affords the best resistance of any. Take for example a circle of 20 square feet (as possessed by the pelican) loaded with as many pounds. This, as just stated, will limit the rate of perpendicular descent to 1,320 feet per minute. But instead of a circle 61 inches in diameter, if the area is bounded by a parallelogram 10 feet long by 2 feet broad, and whilst at perfect freedom to descend perpendicularly, let a force be applied exactly in a horizontal direction, so as to carry it edgeways, with the long side foremost, at a forward speed of 30 miles per hour—just double that of its passive descent: the rate of fall under these conditions will be decreased most remarkably, probably to less than one-fifteenth part, or 88 feet per minute, or one mile per hour.

The annexed line represents transversely the plane 2 feet wide and 10 feet long, moving in the direction of the arrow with a forward speed of 30 miles per hour, or 2,640 feet per minute, and descending at 88 feet per minute, the ratio being as 1 to 30. Now, the particles of air, caught by the forward edge of the plane, must be carried down eight-tenths of an inch before they leave it. This stratum, 10 feet wide and 2,640 long, will weigh not less than 134 lbs.; therefore the weight has continually to be moved downwards, 88 feet per minute, from a state of absolute rest. If the plane, with this weight and an upward rise of eight-tenths of an inch, be carried forward at a rate of 30 miles per hour, it will be maintained at the same level without descending.

The following illustrations, though referring to the action of surfaces in a denser fluid, are yet exactly analogous to the conditions set forth in air:—

Take a stiff rod of wood, and nail to its end at right angles a thin lath or blade, about two inches wide. Place the rod square across the thwarts of a rowing-boat in motion, letting a foot or more of the blade hang perpendicularly over the side into the water. The direct amount of resistance of the current against the flat side of the blade may thus be felt. Next slide the rod to and fro thwart ship, keeping all square; the resistance will now be found to have increased enormously; indeed, the boat can be entirely stopped by such an appliance. Of course the same experiment may be tried in a running stream.

Another familiar example may be cited in the lee-boards and sliding keels used in vessels of shadow draught, *which act precisely on the same principle as the plane or wing-surface of a bird when moving in air*. These surfaces, though parallel to the line of the vessel's course, enable her to carry a heavy press of sail without giving way under the side pressure, or making lee-way, so great is their resistance against the rapidly passing body of water, which cannot be deflected sideways at a high speed.

The succeeding experiments will serve further to exemplify the action of the same principle. Fix a thin blade, say one inch wide and one foot long, with its plane exactly midway and at right angles, to the end of a spindle or rod. On thrusting this through a body of water, or immersing it in a stream running in the direction of the axis of the spindle, the resistance will be simply that caused by the water against the mere superficies of the blade. Next put the spindle and blade in rapid rotation. The retarding effect against direct motion will now be increased near *tenfold*, and is equal to that due *to the entire area of the circle of revolution*. By trying the effect of blades of various widths, it will be found that, for the purpose of effecting the maximum amount of resistance, the more rapidly the spindle revolves the narrower may be the blade. There is a specific ratio between the *width* of the blade and its *velocity*. It is of some importance that this should be precisely defined, not only for its practical utility in determining the best proportion of width to speed in the blades of screw-propellers, but also for a correct demonstration of the principles involved in the subject now under consideration; for it may be remarked that the swiftest-flying birds possess extremely long and *narrow* wings, and the slow, heavy flyers short and wide ones.

In the early days of the screw-propeller, it was thought requisite, in order to obtain the advantage of the utmost extent of surface, that the end view of the screw should present no opening, but appear as a complete disc. Accordingly, some were constructed with one or two threads, making an entire or two half-revolutions; but this was subsequently found to be a mistake. In the case of the two blades, the length of the screw was shortened, and consequently the width of the blades reduced, with increased effect, till each was brought down to considerably less than *one-sixth* of the circumference or area of the entire circle; the maximum speed was then obtained. Experiment has also shown that the effective propelling area of the two-bladed screw is tantamount to its entire circle of revolution, and is generally estimated as such.

Many experiments tried by the author, with various forms of screws, applied to a small steam-boat, led to the same conclusion—that the two blades of one-sixth of the circle gave the best result.

All screws reacting on a fluid such as water, must cause it to yield to some extent; this is technically known as "slip," and whatever the ratio or percentage on the speed of the boat may be, it is tantamount to *just so much loss of propelling power*—this being consumed in giving motion to the water instead of the boat.

On starting the engine of the steam-boat referred to, and grasping a mooring-rope at the stern, it was an easy matter to hold it back with one hand, though the engine was equal in power to five horses, and the screw making more than 500 revolutions per minute. The whole force of the steam was absorbed in "slip," or in giving motion to the column of water; but let her go, and allow the screw to find an abutment on a fresh body of water, not having received a gradual motion, and with its *inertia undisturbed* when running under full way, the screw worked almost as if in a solid nut, the "slip" amounting to only eleven per cent.

The laws which control the action of inclined surfaces, moving either in straight lines or circles in air, are identical, and serve to show the inutility of attempting to raise a heavy body in the atmosphere by means of rotating vanes or a screw acting vertically; for unless the ratio of surface compared to weight is exceedingly extensive, the whole power will be consumed in "slip," or in giving a downward motion to the column of air. Even if a sufficient force is obtained to keep a body suspended by such means, yet, after the desired altitude is arrived at, *no further ascension* is required; there the apparatus is to remain stationary as to level, and its position on the constantly yielding support can only be maintained at an enormous expenditure of power, for the screw cannot obtain a hold upon a *fresh and unmoved* portion of air in the same manner as it does upon the body of water when propelling the boat at full speed; its action under these conditions is the same as when the boat is held fast, in which case, although the engine is working up to its usual rate, the tractive power is almost annulled.

Some experiments made with a screw, or pair of inclined vanes acting vertically in air, were tried, in the following manner. To an upright post was fixed a frame, containing a bevil wheel and pinion, multiplying in the ratio of three to one. The axle of the wheel was horizontal, and turned by a handle of five-and-a-half inches radius. The spindle of the pinion rotated vertically, and carried two driving-pins at the end of a cross-piece, so that the top resembled the three prongs of a trident. The upright shaft of the screw was bored hollow to receive the middle prong, while the two outside ones took a bearing against a driving-bar, at right angles to the lower end of the shaft, the top of which ended in a long iron pivot, running in a socket fixed in a beam overhead; it could thus rise and fall about two inches with very little friction. The top of the screw-shaft carried a cross-arm, with a blade of equal size at each extremity, the distance from end to end being six feet. The blades could be adjusted at any angle by clamping-screws. Both their edges, and the arms that carried them, were bevilled away to a sharp edge to diminish the effects of atmospheric resistance. A wire stay was taken from the base of each blade to the bottom of the upright shaft, to give rigidity to the arms, and to prevent them from springing upwards. With this apparatus experiments were made with weights attached to the upright screw-shaft, and the blades set at different pitches, or angles of inclination. When the vanes were rotated rapidly, they rose and floated on the air, carrying the weights with them. Much difficulty was experienced in raising a heavy weight by a comparatively small extent of surface, moving at a high velocity; the "slip" in these cases being so great as to absorb all the power employed. The utmost effect obtained in this way was to raise a weight of six pounds on one square foot of sustaining surface, the planes having been set at a coarse pitch. To keep up the rotation, required about half the power a man could exert.

The ratio of weight to sustaining surface was next arranged in the proportion approximating to that of birds. Two of the experiments are here quoted, which gave the most satisfactory result. Weight of wings and shaft, 17 $\frac{1}{2}$ oz.; area of two wings, 121 inches—equal to 110 square inches per pound. The annexed figures are given approximately, in order to avoid decimal fractions:—

	No. of revolutions per minute.	Mean sustaining speed. Miles per hour.	Feet per minute.	Pitch or angle of rise in one revolution. Inches.	Ratio of pitch to speed.	Slip per cent.
1st experiment	210	38	3,360	26	$\frac{1}{8}$ nearly	12 $\frac{1}{2}$
2nd Do.	240	44	3,840	15	$\frac{1}{13}$ Do.	8

The power required to drive was nearly the same in both experiments—about equal to one-sixteenth part of a horse-power, or the third part of the strength of a man, as estimated by a constant force on the handle of twelve pounds in the first experiment, and ten in the second, the radius of the handle being five-and-a-half inches, and making seventy revolutions per minute in the first case, and eighty in the other.

These experiments are so far satisfactory in showing the small pitch or angle of rise required for sustaining the weight stated, and demonstrating the principle before alluded to, of the slow descent of planes moving horizontally in the atmosphere at high velocities; but the question remains to be answered, concerning the disposal of the excessive power consumed in raising a weight not exceeding that of a carrier pigeon, for unless this can be satisfactorily accounted for, there is but little prospect of finding an available power, of sufficient energy in its application to the mechanism, for raising apparatus, either experimental or otherwise, in the atmosphere. In the second experiment, the screw-shaft made 240 revolutions, consequently, one vane (there being two) was constantly passing over the *same spot* 480 times each minute, or eight times in a second. This caused a descending current of air, moving at the rate of near four miles per hour, almost sufficient to blow a candle out placed three feet underneath. This is the result of "slip," and the giving both a downward and rotary motion to this column of air, will account for a great part of the power employed, as the whole apparatus performed the work of a blower. If the wings, instead of travelling in a circle, could have been urged continually forward in a straight line in a fresh and unmoved body of air the "slip" would have been so inconsiderable, and the pitch consequently, reduced to such a small angle, as to add but little to the direct forward atmospheric resistance of the edge.

The small flying screws, sold as toys, are well known. It is an easy matter to determine approximately the force expended in raising and maintaining them in the atmosphere. The following is an example of one constructed of tin-plate with three equidistant vanes. This was spun by means of a cord, wound round a wooden spindle, fitted into a forked handle as usual. The outer end of the coiled string was attached to a small spring steelyard, which served as a handle to pull it out by. The weight, or degree at which the index had been drawn, was *afterwards* ascertained by the mark left thereon by a pointed brass wire. It is not necessary to know the time occupied in drawing out the string, as this item in the estimate may be taken as the duration of the ascent; for it is evident that if the same force is re-applied at the descent, it would rise again, and a repeated series of these impulses will represent the power required to prolong the flight of the instrument. It is, therefore, requisite to know the length of string, and the force applied in pulling it out. The following are the data:

Diameter of screw	81 inches
Weight of ditto	396 grains

Length of string drawn out	2 feet
Force employed	8 lbs.
Duration of flight	16 seconds

From this it may be computed that, in order to maintain the flight of the instrument, a constant force is required of near sixty foot-pounds per minute—in the ratio of about three horse-power for each hundred pounds raised by such means. The force is perhaps over-estimated for a larger screw, for as the size and weight is increased, the power required would be less than in this ratio. The result would be more satisfactory if tried with a sheet-iron screw, impelled by a descending weight.

Methods analogous to this have been proposed for attempting aerial locomotion; but experiment has shown that a screw rotating in the air is an imperfect principle for obtaining the means of flight, and supporting the needful weight, for the power required is enormous. Suppose a machine to be constructed, having some adequate supply of force, the screw rotating vertically at a certain velocity will raise the whole. When the desired altitude is obtained, nearly the same velocity of revolution, and the same excessive power, must be continued, and consumed *entirely in "slip,"* or in drawing down a rapid current of air.

If the axis of the screw is slightly inclined from the perpendicular, the whole machine will travel forward. The "slip," and consequently the power, is somewhat reduced under these conditions; but a swift forward speed cannot be effected by such means, for the resistance of the inclined disc of the screw will be very great, far exceeding any form assimilating to the edge of the wing of a bird. But, arguing on the supposition that a forward speed of thirty miles an hour might thus be obtained, even then nearly all the power would be expended in giving an unnecessary and rapid revolution to an immense screw, capable of raising a weight, say of 200 pounds. The weight alone of such a machine must cause it to fail, and every revolution of the screw is a subtraction from the much-desired direct forward speed. A simple narrow blade, or inclined plane, propelled in a direct course at *this* speed—which is amply sufficient for sustaining heavy weights—is the best, and, in fact, the only means of giving the maximum amount of supporting power with the least possible degree of "slip," and direct forward resistance. Thousands of examples in Nature testify its success, and show the principle in perfection;—apparently the only one, and therefore beyond the reach of amendment, the wing of a bird, combining a propelling and supporting organ in one, each perfectly efficient in its mechanical action.

This leads to the consideration of the amount of power requisite to maintain the flight of a bird. Anatomists state that the pectoral muscles for giving motion to the wings are excessively large and strong; but this furnishes no proof of the expenditure of a great amount of force in the act of flying. The wings are hinged to the body like two powerful levers, and some counteracting force of a *passive* nature, acting like a spring under tension, must be requisite merely to balance the weight of the bird. It cannot be shown that, while there is no active motion, there is any real exertion of muscular force; for instance, during the time when a bird is soaring with motionless wings. This must be considered as a state of equilibrium, the downward spring and elasticity of the wings serving to support the body; the muscles, in such a case, performing like stretched india-rubber springs would do. The motion or active power required for the performance of flight must be considered exclusive of this.

It is difficult, if not impossible, by any form of dynamometer, to ascertain the precise amount of force given out by the wings of birds; but this is perhaps not requisite in proof of the principle involved, for when the laws governing their movements in air are better understood, it is quite possible to demonstrate, by isolated experiments, the amount of power required to sustain and propel a given weight and surface at any speed.

If the pelican referred to as weighing twenty-one pounds, with near the same amount of wing-area in square feet, were to descend perpendicularly, it would fall at the rate of 1,320 feet per minute, being limited to this speed by the resistance of the atmosphere.

The standard generally employed in estimating power is by the rate of descent of a weight. Therefore, the weight of the bird being 21 pounds, which, falling at the above speed will expend a force on the air set in motion nearly equal to one horse (.84 HP.) or that of 5 men; and conversely, to raise this weight again perpendicularly upon a yielding support like air, would require even more power than this expression, which it is certain that a pelican does not possess; nor does it appear that any *large* bird has the faculty of raising itself on the wing *perpendicularly* in a still atmosphere. A pigeon is able to accomplish this nearly, mounting to the top of a house in a very narrow compass; but the exertion is evidently severe, and can only be maintained for a short period. For its size, this bird has great power of wing; but this is perhaps far exceeded in the humming-bird, which, by the extremely rapid movements of its pinions, sustains itself for more than a minute in still air in one position. The muscular force required for this feat is much greater than for any other performance of flight. The body of the bird at the time is nearly vertical. The wings uphold the weight, not by striking vertically downwards upon the air, but as inclined surfaces reciprocating horizontally like a screw, but wanting in its continuous rotation in one direction, and, in consequence of the loss arising from rapid alternations of motion, the power required for the flight will exceed that specified in the screw experiment before quoted, viz.: three horse-power for every 100 pounds raised.

We have here an example of the exertion of enormous animal force expended in flight, necessary for the peculiar habits of the bird, and for obtaining its food; but in the other extreme, in large heavy birds, whose wings are merely required for the purposes of migration or locomotion, flight is obtained with the least possible degree of power, and this condition can only be commanded by a rapid straightforward course through the air.

The sustaining power obtained in flight must depend upon certain laws of action and reaction between relative weights; the weight of a bird, balanced, or finding an abutment, against the fixed inertia of a far greater weight of air, continuously brought into action in a given time. This condition is secured, not by extensive surface, but by great length of wing, which, in forward motion, takes a support upon a wide stratum of air, extending transversely to the line of direction.

The pelican, for example, has wings extending out 10 feet. If the limits of motion imparted to the substratum of air, acted upon by the incline of the wing, be assumed as one foot in thickness, and the velocity of flight as 30 miles per hour, or 2,640 feet per minute, the stratum of air passed over in this time will weigh nearly one ton, or 100 times the weight of the body of the bird, thus giving such an enormous supporting power, that the comparatively small weight of the bird has but little effect in deflecting the heavy length of stratum downwards, and, therefore, the higher the velocity of flight the less the amount of "slip," or power wasted in compensation for descent.

As noticed at the commencement or this paper, large birds may be observed to skim close above smooth water without ruffling the surface; showing that during rapid flight the air does not give way beneath them, but approximates towards a solid support.

In all inclined surfaces, moving rapidly through air, the whole sustaining power approaches toward the front edge; and in order to exemplify the inutility of surface alone, without proportionate length of wing, take a plane, ten feet long by two broad, impelled with the narrow end forward, the first twelve or fifteen inches will be as efficient at a high speed in supporting a weight as the entire following portion of the plane, which may be cut off, thus reducing the effective wing-area of a pelican, arranged in this direction, to the totally inadequate equivalent of two-and-a-half square feet.

One of the most perfect natural examples of easy and long-sustained flight is the wandering albatross. "A bird for endurance of flight probably unrivalled. Found over all parts of the Southern Ocean, it seldom rests on the water. During storms, even the most terrific, it is seen now dashing through the whirling clouds, and now serenely floating, without the least observable motion of its outstretched pinions." The wings of this bird extend fourteen or fifteen feet from end to end, and measure only eight-and-a-half inches across the broadest part. This conformation gives the bird such an extraordinary sustaining power, that it is said to *sleep* on the wing during stormy weather, when rest on the ocean is impossible. Rising high in the air, it skims slowly down, with absolutely motionless wings, till a near approach to the waves awakens it, when it rises again for another rest.

If the force expended in actually sustaining a long-winged bird upon a wide and unyielding stratum of air, during rapid flight, is but a small fraction of its strength, then nearly the whole is exerted in overcoming direct forward resistance. In the pelican referred to, the area of the body, at its greatest diameter, is about 100 square inches; that of the pinions, eighty. But as the contour of many birds during flight approximates nearly to Newton's solid of least resistance, by reason of this form, acting like the sharp bows of a ship, the opposing force against the wind must be reduced down to one third or fourth part; this gives one-tenth of a horse-power, or about half the strength of a man, expended during a flight of thirty miles per hour. Judging from the action of the living bird when captured, it does not appear to be more powerful than here stated.

The transverse area of a carrier pigeon during flight (including the outstretched wings) a little exceeds the ratio of twelve square inches for each pound, and the wing-surface, or sustaining area, ninety square inches per pound.

Experiments have been made to test the resisting power of conical bodies of various forms, in the following manner:—A thin lath was placed horizontally, so as to move freely on a pivot set midway; at one end of the lath a circular card was attached, at the other end a sliding clip traversed, for holding paper cones, having their bases the exact size of the opposite disc. The instrument acted like a steelyard; and when held against the wind, the paper cones were adjusted at different distances from the centre, according to their forms and angles, in order to balance the resistance of the air against the opposing flat surface. The resistance was found to be diminished nearly in the ratio that the height of the cone exceeded the diameter of base.

It might be expected that the pull of the string of a flying kite should give some indication of the force of inclined surfaces acting against a current of air; but no correct data can be obtained in this way. The incline of the kite is far greater than ever appears in the case of the advancing wing-surface of a bird. The tail is purposely made to give steadiness by a strong pull backwards from the action of the wind, which also exerts considerable force on the suspended cord, which for more than half its length hangs nearly perpendicularly. But the kite, as a means of obtaining unlimited lifting and tractive power, in certain cases where it might be usefully applied, seems to have been somewhat neglected. For its power of raising weights, the following quotation is taken from Vol. XLI. of the *Transactions of the Society of Arts*, relating to Captain Dansey's mode of communicating with a lee-shore. The kite was made of a sheet of holland exactly nine feet square, extended by two spars placed diagonally, and as stretched spread a surface of fifty-five square feet. "The kite, in a strong breeze, extended 1,100 yards of line five-eighths in circumference, and would have extended more had it been at hand. It also extended 360 yards of line, one and three-quarters of an inch in circumference, weighing sixty pounds. The holland weighed three and a half pounds; the spars, one of which was armed at the head with iron spikes, for the purpose of mooring it, six and three-quarter pounds; and the tail was five times its length, composed of eight pounds of rope and fourteen of elm plank, weighing together twenty-two pounds."

We have here the remarkable fact of ninety-two and a quarter pounds carried by a surface of only fifty-five square feet.

As all such experiments bear a very close relation to the subject of this paper, it may be suggested that a form of kite should be employed for reconnoitring and exploring purposes, in lieu of balloons held by ropes. These would be torn to pieces in the very breeze that would render a kite most serviceable and safe. In the arrangement there should be a smaller and upper kite, capable of sustaining the weight of the apparatus. The lower kite should be as nearly as practicable in the form of a circular flat plane, distended with ribs, with a car attached beneath like a parachute. Four guy-ropes leading to the car would be required for altering the angle of the plane—vertically with respect to the horizon, and laterally relative to the direction of the wind. By these means the observer could regulate his altitude, so as to command a view of a country in a radius of at least twenty miles; he could veer to a great extent from side to side, from the wind's course, or lower himself gently, with the choice of a suitable spot for descent. Should the cord break, or the wind fail, the kite would, in either case, act as a parachute, and as such might be purposely detached from the cord, which then being sustained from the upper kite, could be easily recovered. The direction of descent could be commanded by the guy-ropes, these being hauled taut in the required direction for landing.

The author has good reasons for believing that there would be less risk associated with the employment of this apparatus, than the reconnoitring balloons that have now frequently been made use of in warfare.

The wings of all flying creatures, whether of birds, bats, butterflies, or other insects, have this one peculiarity of structure in common. The front, or leading edge, is rendered rigid by bone, cartilage, or a thickening of the membrane; and in most birds of perfect flight, even the individual feathers are formed upon the same condition. In consequence of this, when the wing is waved in air, it gives a persistent force in one direction, caused by the elastic reaction of the following portion of the edge. The fins and tails of fishes act upon the same principle. In most rapid swimmers these organs are termed "lobated and pointed." The tail extends out very wide transversely to the body, so that a powerful impulse is obtained against a wide stratum of water, on the condition before explained. The action is imitated in Macintosh's screw-propeller, the blade of which his made of thin steel, so as to be elastic. While the vessel is stationary, the blades are in a line with the keel, but during rotation they bend on one side, more or less, according to the speed and degree of propulsion required, and are thus self-compensating; and could practical difficulties be overcome, would prove to be a form of propeller perfect in theory.

In the flying mechanism of beetles there is a difference of arrangement. When the elytra, or wing-cases, are opened, they are checked by a stop, which sets them at a fixed angle. It is probable that these serve as "aeroplanes," for carrying the weight of the insect, while the delicate membrane that folds beneath acts more as a propelling than a supporting organ. A beetle cannot fly with the elytra removed.

The wing of a bird, or bat, is both a supporting and a propelling organ, and flight is performed in a rapid course, as follows:—During the down-stroke it can be easily imagined how the bird is sustained; but in the up-stroke, the weight is also equally well supported, for in raising the wing, it is slightly inclined upwards against the rapidly passing air, and as this angle is somewhat in excess of the motion due to the raising of the wing, the bird is sustained as much during the up as the down-stroke—in fact, though the wing may be rising, the bird is still pressing against the air with a force equal to the weight of its body. The faculty of turning up the wing may be easily seen when a large bird alights; for after gliding down its aerial gradient, on its approach to the ground it turns up the plane of its wing against the air; this checks its descent, and it lands gently.

It has before been shown how utterly inadequate the mere perpendicular impulse of a plane is found to be in supporting a weight, when there is no horizontal motion at the time. There is no material weight of air to be acted upon, and it yields to the slightest force, however great the velocity of impulse may be. On the other hand, suppose that a large bird, in full flight, can make forty miles per hour, or 3,520 feet per minute, and performs one stroke per second. Now, during every fractional portion of that stroke, the wing is acting upon and obtaining an impulse from a fresh and undisturbed body of air; and if the vibration of the wing is limited to an arc of two feet, this by no means represents the small force of action that would be obtained when in a stationary position, for the impulse is secured upon a stratum of fifty-eight feet in length of air at each stroke. So that the conditions of weight of air for obtaining support equally well apply to weight of air, and its reaction in producing forward impulse.

So necessary is the acquirement of this horizontal speed, even in commencing flight, that most heavy birds, when possible, rise against the wind, and even run at the top of their speed to make their wings available, as in the example of the eagle, mentioned at the commencement of this paper. It is stated that the Arabs, on horseback, can approach near enough to spear these birds, when on the plain, before they are able to rise: their habit is to perch on an eminence, where possible.

The tail of a bird is not necessary for flight. A pigeon can fly perfectly with this appendage cut short off: it probably performs an important function in steering, for it is to be remarked, that most birds that have either to pursue or evade pursuit are amply provided with this organ.

The foregoing reasoning is based upon facts, which tend to show that the flight of the largest and heaviest of all birds is really performed with but a small amount of force, and that man is endowed with sufficient muscular power to enable him also to take individual and extended flights, and that success is probably only involved in a question of suitable mechanical adaptations. But if the wings are to be modelled in imitation of natural examples, but very little consideration will serve to demonstrate its utter impracticability when applied in these forms. The annexed diagram, fig. 1, would be about the proportions needed for a man of medium weight. The wings, *a a*, must extend out sixty feet from end to end, and measure four feet across the broadest part. The man, *b*, should be in a horizontal position, encased in a strong framework, to which the wings are hinged at *c c*. The wings must be stiffened by elastic ribs, extending back from the pinions. These must be trussed by a thin band of steel, *e e*, fig. 2, for the purpose of diminishing the weight and thickness of the spar. At the front, where the pinions are hinged, there are two levers attached, and drawn together by a spiral spring, *d*, fig. 2, the tension of which is sufficient to balance the weight of the body and machine, and cause the wings to be easily vibrated by the movement of the feet acting on treadles. This spring serves the purpose of the pectoral muscles in birds. But with all such arrangements the apparatus must fail—*length of wing is indispensable!* and a spar thirty feet long must be strong, heavy, and cumbrous; to propel this alone through the air, at a high speed, would require more power than any man could command.

In repudiating all imitations of natural wings, it does not follow that the only channel is closed in which flying mechanism may prove successful. Though birds do fly upon definite mechanical principles, and with a moderate exertion of force, yet the wing must necessarily be a vital organ and member of the living body. It must have a marvellous self-acting principle of repair, in case the feathers are broken or torn; it must also fold up in a small compass, and form a covering for the body. These considerations bear no relation to artificial wings; so in designing a flying-machine, any deviations are admissible, provided the theoretical conditions involved in flight are borne in mind.

Having remarked how thin a stratum of air is displaced beneath the wings of a bird in rapid flight, it follows that in order to obtain the necessary *length* of plane of supporting heavy weights, the surfaces may be superposed, or placed in parallel rows, with an interval between them. A dozen pelicans may fly one above the other without mutual impediment, as if framed together; and it is thus shown how two hundred weight may be supported in a transverse distance of only ten feet.

In order to test this idea, six bands of stiff paper, three feet long and three inches wide, were stretched at a slight upward angle, in a light rectangular frame, with an interval of three inches between them, the arrangement resembling an open Venetian blind. When this was held against a breeze, the lifting power was very great, and even by running with it in a calm it required much force to keep it down. The success of this model led to the construction of one of a sufficient size to carry the weight of a man. Fig. 3 represents the arrangement. *a a* is a thin plank, tapered at the outer ends, and attached at the base to a triangle, *b*, made of similar plank, for the insertion of the body. The boards, *a a*, were trussed with thin bands of iron, *c c*, and at the ends were vertical rods, *d d*. Between these were stretched five bands of holland, fifteen inches broad and sixteen feet long, the total length of the web being eighty feet. This was taken out after dark into a wet piece of meadow land, one November evening, during a strong breeze, wherein it became quite unmanageable. The wind acting upon the already tightly stretched webs, their united pull caused the central boards to bend considerably, with a twisting, vibratory motion. During a lull, the head and shoulders were inserted in the triangle, with the chest resting on the baseboard. A sudden gust caught up the experimenter, who was carried some distance from the ground, and the affair falling over sideways, broke up the right-hand set of webs.

In all new machines we gain experience by repeated failures, which frequently form the stepping-stones to ultimate success. The rude contrivance just described (which was but the work of a few hours) had taught, first, that the webs, or aeroplanes, must not be distended in a frame, as this must of necessity be strong and heavy, to withstand their combined tension; second, that the planes must be made so as either to furl or fold up, for the sake of portability.

In order to meet these conditions, the following arrangement was afterwards tried:—*a a*, figs. 4 and 5, is the main spar, sixteen feet long, half an inch thick at the base, and tapered, both in breadth and thickness, to the end; to this spar was fastened the panels *b b*, having a base-board for the support of the body. Under this, and fastened to the end of the main spar, is a thin steel tie-band, *e e*, with struts starting from the spar. This served as the foundation of the superposed aeroplanes, and, though very light, was found to be exceedingly strong; for when the ends of the spar were placed upon supports, the middle bore the weight of the body without any strain or deflection; and further, by a separation at the baseboard, the spars could be folded back, with a hinge, to half their length. Above this were arranged the aeroplanes, consisting of six webs of thin holland, fifteen inches broad; these were kept in parallel planes, by vertical divisions, two feet wide, of the same fabric, so that when distended by a current of air, each two feet of web pulled in opposition to its neighbour; and finally at the ends (which were each sewn over laths), a pull due to only two feet had to be counteracted, instead of the strain arising from the entire length, as in the former experiment. The end-pull was sustained by vertical rods, sliding through loops on the transverse ones at the ends of the webs, the whole of which could fall flat on the spar, till raised and distended by a breeze. The top was stretched by a lath, *f* and the system kept vertical by staycords, taken from a bowsprit carried out in front, shown in fig. 6. All the front edges of the aeroplanes were stiffened by bands of crinoline steel. This series was for the supporting arrangement, being equivalent to a length of wing of ninety-six feet. Exterior to this, two propellers were to be attached, turning on spindles just above the back. They are kept drawn up by a light spring, and pulled down by cords or chains, running over pulleys in the panels *b b*, and fastened to the end of a swivelling cross-yoke, sliding on the base-board. By working this cross-piece with the feet, motion will be communicated to the propellers, and by giving a longer stroke with one foot than the other, a greater extent of motion will be given to the corresponding propeller, thus enabling the machine to turn, just as oars are worked in a rowing boat. The propellers act on the same principle as the wing of a bird or bat: their ends being made of fabric, stretched by elastic ribs, a simple waving motion up and down will give a strong forward impulse. In order to start, the legs are lowered beneath the baseboard, and the experimenter must run against the wind.

An experiment recently made with this apparatus developed a cause of failure. The angle required for producing the requisite supporting power was found to be so small, that the crinoline steel would not keep the front edges in tension. Some of them were borne downwards and more on one side than the other, by the operation of the wind, and this also produced a strong fluttering motion in the webs, destroying the integrity of their plane surfaces, and fatal to their proper action.

Another arrangement has since been constructed, having laths sewn in both edges of the webs, which are kept permanently distended by cross-stretchers. All these planes are hinged to a vertical central board, so as to fold back when the bottom ties are released, but the system is much heavier than the former one, and no experiments of any consequence have as yet been tried with it.

It may be remarked that although a principle is here defined, yet considerable difficulty is experienced in carrying the theory into practice. When the wind approaches to fifteen or twenty miles per hour, the lifting power of these arrangements is all that is requisite, and, by additional planes, can be increased to any extent; but the capricious nature of the ground-currents is a perpetual source of trouble.

Great weight does not appear to be of much consequence, *if carried in the body*; but the aeroplanes and their attachments seem as if they were required to be very light, otherwise, they are awkward to carry, and impede the movements in running and making a start. In a dead calm, it is almost impracticable to get sufficient horizontal speed, by *mere running* alone, to raise the weight of the body. Once off the ground, the speed must be an increasing one, if continued by suitable propellers. The small amount of experience as yet gained, appears to indicate that if the aeroplanes could be raised in detail, like a superposed series of kites, they would first carry the weight of the machine itself, and next relieve that of the body.

Until the last few months no substantial attempt has been made to construct a flying-machine, in accordance with the principle involved in this paper, which was written seven years ago. The author trusts that he has contributed something towards the elucidation of a new theory, and shown that the flight of a bird in its performance does not require that enormous amount of force usually supposed, and that in fact birds do not exert more power in flying than quadrupeds in running, but considerably less; for the wing movements of a large bird, travelling at a far higher speed in air, are very much slower; and, where weight is concerned, great velocity of action in the locomotive organs is associated with great force.

It is to be hoped that further experiments will confirm the correctness of these observations, and with a sound working theory upon which to base his operations, man may yet command the air with the same facility that birds now do.

The CHAIRMAN: "I think the paper just read is one of great interest and importance, especially as it points out the true mechanical explanation of the curious problem, as to how and why it is that birds of the most powerful flight always have the longest and narrowest wings. I think it quite certain, that if the air is ever to be navigated, it will not be by individual men flying by means of machinery; but that it is quite possible vessels may be invented, which will carry a number of men, and the motive force of which will not be muscular action. We must first ascertain clearly the mechanical principles upon which flight is achieved; and this is a subject which has scarcely ever been investigated in a scientific spirit. In fact, you will see in our best works of science, by the most distinguished men, the account given of the anatomy of birds is, that a bird flies by inflating itself with warm air, by which it becomes buoyant, like a balloon. The fact is, however, that a bird is never buoyant. A bird is immensely heavier than the air. We all know that the moment a bird is shot it falls to the earth; and it must necessarily do so, because one of the essential mechanical principles of flight is weight, without it there can be no momentum, and no motive force capable of moving through atmospheric currents.

"Until I read Mr. Wenham's paper, a few weeks since, I was puzzled by the fact, that birds with long and very narrow wings seem to be not only as efficient fliers, but much more efficient fliers than birds with very large, broad wings. If you observe the flight of the common heron—which is a bird with a very large wing, disposed rather in breadth than in length—you will notice that it is exceedingly slow, and that it has a very heavy, flapping motion. The common swallow, on the other hand, is provided with a long and narrow wing, and I never understood how it was that long-winged birds, such as these, achieved so rapid a flight, until I read Mr. Wenham's paper. Although I do not profess to be able to follow the elaborate calculations which he has laid before us, I think I now understand the explanation he has given. His explanation of the action of narrow wings upon the air is, that it is precisely like the action of the narrow vanes of the ship's screw in water, and that the resisting power of the screw is the same, or nearly the same, whether you have the total area of revolution covered by solid surface, or traversed by long and narrow vanes in rotation.

"If Mr. Wenham's explanation be nearly correct, that supposing this implement (referring to a model) to be carried forward by some propelling power, the sustaining force of the whole area is simply the sustaining force of the narrow band in front. This, however, is a matter which will have to be decided by experiment. It certainly appears to explain the phenomena of the flight of birds. There are one or two observations in the paper I do not quite agree with. Although I have studied the subject for many years, I have not arrived at Mr. Wenbam's conclusion that the upward stroke of a bird's wing has precisely the same effect as a downward stroke in sustaining. An upward stroke has a contrary effect to the downward stroke; it has a propelling power certainly, but I believe that the sustaining power of a bird's flight is due entirely to the downward stroke. I should be glad to hear what Mr. Wenham may have to say upon this. My belief is, that an upward stroke must have, so far as sustaining is concerned, a reverse action to the downward stroke.

"Then with regard to another observation of Mr. Wenham's, that the tails of birds are used as rudders. I believe this to be an entire mistake; for if the tail of a bird could have the slightest effect in guiding, the vane of it must be disposed perpendicularly, and not horizontally, or nearly so, as at present.

"If you cut off the tail of a pigeon, you will find that he can fly and turn perfectly well without it. He may be a little awkward about it at first, but that is because he has lost his balancing power. We all know that it is a common thing to see a sparrow without his tail, therefore, I do not in the least believe that tails have any effect in guiding. They have an important effect in stopping progress, and, undoubtedly, that is one of the necessary elements of turning. If a bird comes close over your head, and is frightened, you will find his claws distended and his tail spread out as a fan, to stop the momentum of his flight. These are the two only observations with which I cannot agree; but as regards the explanation he has given as to the resistance offered by long and narrow wings, he has made an important discovery."

Mr. WENHAM: "With regard to the wing not affording support to the bird during the upward stroke, some of the largest birds move their wings slowly, that is, with a less number than sixty strokes per minute. Now, as a body free to fall must descend fifteen feet in one second, whether in horizontal motion or not, it appears clear to me that there must be some counter-acting effect to prevent this fall. When the wing has reached the limit of the down-stroke, it is inclined upwards in the direction of motion, consequently the rush of air caused by the forward speed, weight, and momentum of the bird against the under surface of the wing, supports the weight, even though the wing is rising in the up-stroke at the time. In corroboration of my theory, I will read an extract from Sir George Cayley, who made a large number of experiments. He says, in page 83, of Vol. xxv., 'Nicholson's Journal':—'The stability in this position, arising from the centre of gravity, being below the point of suspension, is aided by a remarkable circumstance that experiment alone could point out. In very acute angles with the current, it appears that the centre of resistance in the sail does not coincide with the centre of its surface, but is considerably in front of it. As the obliquity of the current decreases, these centres approach and coincide when the current becomes perpendicular to the plane, hence any heel of the machine backwards or forwards removes the centre of support behind or before the point of suspension.'

"From this discovery, it seems remarkable that Sir George Cayley, finding that at high speeds with very oblique incidences the supporting effect became transferred to the front edge, the idea should not have occurred to him that a narrow plane, with its long edge in the direction of motion, would have been equally effective. I may give another illustration. We all know, from our schoolboy experience, that ice which would not be safe to stand upon, is found to be quite strong enough to bear heavy bodies passing over it, so long as rapid motion is kept up, and then it will not even crack. We know, also, that in driving through a marshy part of road, in which you expect the wheels to sink in up to the axles, you may pass over much more easily by increasing the speed. In both these examples there is a greater weight passed over in a given time, and consequently a better support obtained. The ice will not become deflected; neither has the mud time to give way. At a slow speed the same effect may be obtained by extending the breadth of the wheel. Thus, suppose an ordinary wheel to sink ten inches, if you double this width it will sink only five inches; and so on, until by extending the wheel into a long roller you may pass over a quicksand with perfect safety. Now, Nature has carried out this principle in the long wings of birds, and in the albatross it is seen in perfection."

Document 1-3(b), Report on the first wind tunnel built by Francis H. Wenham, 1871.

By the aid of a special subscription the Society has been enabled to present to its members some data with respect to the action of a current of air upon inclined planes of necessarily limited area, but varying angles. When the instrument, designed for this object by Mr. Wenham, was completed, with the aid of Mr. Browning, every facility was afforded for testing its capacities at Messrs. Penns' Engineering Works at Greenwich.

By means of a fan-blower, a current of considerable force was directed through a trunk ten feet long by eighteen inches square. The plane to be acted upon was fixed to the long end of a horizontal arm, which vibrated like the beam of a balance, and bore upon its shorter end a sliding counter weight, so as to balance the weight of any plane which might be fixed at the opposite extremity. The horizontal or direct pressure was read off by a spring steelyard, which was connected to the end of a lever from a vertical spindle close to the base of the machine. The vertical or sliding force due to the various inclinations was read off by an upright spring steelyard.

ANGLES OF INCLINED PLANES.

Forces in lbs.	0°	15°	20°	45°	60°	
						Force of wind $7/10$ in water
Direct	3.24	0.33	0.52	2.4	3.05	Plane 1 square foot.
Vertical	0	1.5	1.8	2.4	1.7	
Direct	3.1	0.43	0.62	2.45	3.05	" 1 circular foot
Vertical	0	1.6	1.75	2.45	1.8	
Direct	2.2	0.23	0.38	1.25	1.62	" 4½in. x 18in. = 81 square inches
Vertical	0	1.05	1.2	1.25	1	
Direct	3.77	0.38	0.57	2.2	2.76	" 18in. x 9in. = 162 square inches.
Vertical	0	1.6	1.85	2.2	1.75	
						Force of wind 1 in water
Direct	4.29	0.62	0.95	3.74	4.75	Plane 1 square foot.
Vertical	0	2.35	2.7	3.47	2.45	
Direct	4.26	0.62	1	3.5	3.24	" 1 circular foot
Vertical	0	2.25	2.85	3.5	1.8	
Direct	2.8	0.3	0.57	1.6	1.52	" 4½in. x 18in.
Vertical	0	1.5	1.8	1.6	1	
						Force of wind $6/10$ in water
Direct	3.24	0.43	0.76	2.7	2.7	Plane 1 square foot.

Vertical	0	1.6	2.05	2.7	1.6		
Direct	3.24	0.43	1.24	2.45	2.96	"	1 circular foot
Vertical	0	1.75	2.5	2.45	1.8		
Direct	2	0.29	0.38	1.3	1.57	"	4½in. x 18in., long edge to wind
Vertical	0	1.25	1.3	1.3	1		
Direct	2.57	0.24	0.43	1.65	2.1	"	4½in. x 18in., end on to wind, 8/10 in water
Vertical	0	0.8	1.3	1.65	1.3		
Direct	*5 mean	1.47	0.76	2.9	3.81		Plane 18in. square = 2¼ square feet
Vertical	0	2	2.4	2.9	2.2		

*Fluctuates 0.57 lbs.

MEAN OF THE ABOVE COLUMNS.

Direct	3.31	0.4	0.68	2.34	2.73	Mean height of water, say 0.73in., which should represent 3.8lbs. per foot; readings are ½th too high for actual force per foot.
Vertical	0	1.6	1.96	2.34	1.26	

These experiments, when all the angles are averaged for errors, seem to indicate the law that the lifting force of inclined planes, carried horizontally through air, is increased in the direct ratio that the sine bears to the length of the plane, or the height to the incline to the base; thus, if instead of stating the angles in degrees, we say "one in ten," or "one in three or four," as the case may be, this will at once express the proportion in which the lifting force exceeds the resistance. The average of all results is very near to this, making a little allowance for the surface friction of the plane through the air. At 45° the two forces are equal; above this the proportions are in the inverse ratio, as the lifting force is then less than the direct.

It has been stated that the resistance of wedges or cones through the air is diminished directly in the ratio that the height or diameter of the base bears to the length of the cone. The experiments do not confirm this, but show that the resistance is less in proportion as the angle becomes more acute.

It will be seen on reference to these tables that as the angles become more acute, the lifting force exceeds the horizontal or power required to propel planes through air in an enormous ratio.

More acute angles than 15° were not experimented upon, but even at this the lift is four times greater than the thrust, and alone serves to abate the mystery relating to the support of weight in flight, which at least in the case of easy flying birds, consists in the action of surfaces at acute angles with the line of motion.

The experiments, though at present somewhat crude and incomplete, show that very oblique incidences or angles, with a small rise, have a remarkably strong lifting force compared with the power required to propel the plane, and that the ratio of the lift to the thrust greatly increases as the angle or rise diminishes, in all probability accounting for the long-sustained flight of birds with motionless wings. It is desirable that these important experiments should be verified and continued; the apparatus will be at the service of any member desirous of repeating them or trying others. For example, up to the present time only flat surfaces have been experimented upon, and as all the sustaining surfaces in the wings of birds are curved or hollow, it remains to be proved what is the relative advantage of this form, for it is upon such data that plans of construction must be based, and on which failure or success must depend, and to determine whether flight is practicable or not. The whole endeavour must be to find a support on the air in such a way that it cannot yield, so that there shall be as little mechanical loss as possible from what is known as "slip."

A most evident example of the enormous loss of power arising from "slip" may be cited in the plans that appear, from time to time, for raising men or machinery by means of vertical screws. Taking 200 lbs. as the lightest total weight of the man and machine, the following result will show how utterly inadequate his power is to raise any but the most trifling weight by such means. Assuming the surface or area of revolution at any number of square feet, the fraction of this weight of 200 lbs. is distributed over each foot of that surface. The force per square foot must therefore represent the wind velocity to produce the reaction or resistance necessary to support the weight. A surface of 25 square feet will then stand thus:—for 200 lbs. there is a resistance of 8 lbs. per foot required. To get this there must be a wind velocity of 3,600 ft. per minute, therefore 3,600 × 200 + 33,000 = 22 horse power. The following table will show the comparative results for four different diameters or areas of screw, making no allowance for the friction of the machine, thus demonstrating the hopelessness of any successful arrangement on this system.

Diam of screw	Area of revolution	Weight to be raised	lbs. per ft.	Velocity in ft. per minute	Horse power
5ft. 8in.	= 25 feet	200	= 8	3,600	= 22
8 0	= 50 "	200	= 4	2,600	= 16
11 3	= 100 "	200	= 2	1,800	= 11
16 0	= 200 "	200	= 1	1,300	= 8

The difference of power required to produce the same effect in the last and first case is nearly threefold; this arises from the increased slip of the smaller area screw.

The theory of the strong lifting power of planes at very oblique incidences, moving rapidly through the air, having now to some extent been practically tested, a few words may not be out of place concerning the position or arrangement of those planes. Mr. Wenham, in a paper read at the first meeting of the Aëronautical Society, brought forward a number of examples in evidence of this great lifting force, but without defining any exact law, merely relying upon it as a fact for the reasons given.

Then came the question, how to obtain the large extent of surface in a compass small enough to secure strength with lightness. In that essay he showed that an apparatus intended to support a weight of even 200 lbs. on one long extended plane, like the wings of a bird, would be an impracticable construction, requiring long and heavy spars. He then proposed to cut the planes into lengths, and superpose them, like a Venetian blind. It has been objected that this is wrong in theory, and that the action of one plane will interfere with the other; but the fact is, that there is no theory in the matter. It is simply a question of construction. The planes are equally effective in detachments, and may evidently be so arranged as not to interfere with each other. Thus, a series of planes made of thin silk, or tissue-paper, may be rigged, one above the other, like kites on a cord, and weighing not more than half an ounce each, and yet have ample strength to sustain one pound.

Though another season has elapsed, the Society has no announcement to make of the existence of a successful machine for aërial locomotion, or for travelling in any desired direction. Yet the retrospect of the past year shows a decided improvement in the form of experiments made to elucidate and define a correct law of action and propulsion of bodies sustained in air. We have no longer the extravagant and impossible theories of gravity and the laws of motion to account for this difficult phenomenon, so many of which were sent to the Society in the early days of its existence, the publication of which would in no way have contributed to the end in view, but, on the contrary, being wholly unsupported by any previous facts or experiments, would have brought discredit and ridicule upon Aëronautics, as the science which it may fairly be considered.

Should the problem be solved, the Aëronautical Society will take a high rank amongst its compeers, and cannot now be denounced as useless till flight for man is proved to be impossible. There are some amongst, not the least earnest of our members, who are quite willing to prove its impossibility, calmly regarding it as a mere scientific question; but, in fact, there is more difficulty in proving an impossibility than in constructing a machine having partial or total success, and we are far from being able to demonstrate the certainty of failure by any known laws, either of principle or construction. However plausible a mere argument unsupported by facts may appear, we have only to raise our eyes to the machinery of nature exemplified in large flying birds, and behold a reality yet incompletely explained, and not scientifically accounted for.

The Society may congratulate itself for having been the means of bringing together and recording a number of facts not lost upon its members, which have produced greater unanimity of purpose, and have, it is presumed, directed the efforts of experimentalists in accordance with a more generally recognized and practicable theory.

Document 1-3(c), Minutes of the Royal Aeronautical Society, 1872.

A General Meeting of the members of this Society was held in the Theatre of the Society of Arts, John Street, Adelphi, on Tuesday evening, the 18th inst. Mr. JAMES GLAISHER, F.R.S., presided.

A new machine, constructed under the direction of the Society, for measuring the relation between the velocity and pressure of the wind, was exhibited.

At the request of the CHAIRMAN,

The minutes of the previous meeting were read by Mr. F. W. BREAREY, the Hon. Secretary.

The Chairman: Ladies and Gentlemen,—the subject which will most naturally attract our attention this evening is that of the experiments which have been made by the apparatus now on the table before us. I had almost forgotten that at our last meeting we spoke of this instrument having been designed. It was not completed so soon as we expected; and, although much time has been occupied in making experiments, the results are not quite so conclusive as could be desired; but so far as they go are important—not only in respect to the problem we now wish to solve, but, as bearing upon the pressure of the wind on the surfaces of planes. I will not now engage your time longer, but I will ask Mr. Wenham, under whose care, in conjunction with Mr. Browning, the experiments were carried out, to give a statement respecting the results. It is an instrument of a kind which I have long desired, and it seems calculated to achieve what we require in this direction with greater accuracy than any other instrument I know. I call upon Mr. Wenham to explain the apparatus.

Mr. WENHAM expressed his regret at the absence of Mr. Browning, who had been associated with them in these experiments. To make this instrument understood, he would explain how it acted as an ordinary anemometer, for ascertaining the direct force of the wind on a plane, when in a vertical direction to its surface. This consists mainly of a vertical steel spindle, supported on a hardened steel centre. Through an eye at the upper end of the spindle, a horizontal arm passes, and is secured by a small cross-pin, which allows the arm to vibrate like the beam of a balance. The long end of the arm carries the planes; and the opposite short one has a sliding counter-weight, which is adjusted so as to exactly balance planes of different sizes at the long end of the arm. Each plane is clamped at the end of a tail rod, which is pivoted through the forked end of the arm, by a vertical steel pin, as close to the plane as possible; the other end of the tail passes loosely through a vertical slot, slightly curved as a radius, from the balance centre of the arm. By this arrangement, the surface of the plane is always kept at right angles to the current, throughout the extent of its horizontal motion. A wooden shield is fixed close before the front of the arm, to protect this and the balance weight from the wind, so that the planes only may be exposed to its force. The action of the instrument, as a single anemometer only, or when the planes are set at right angles to the current of air, is obvious. The direct pressure is read off by the spring steel-yard, which is connected to the end of a lever from the vertical spindle, close to the base of the machine. In order to measure the vertical forces, the planes are set at the requisite angles from a divided sector, whose centre coincides with the clamping screw at the back. The raising force due from the various inclines was read off by the upright spring steel-yard. It was found almost impossible for one observer to read off the horizontal and vertical forces simultaneously during fluctuations, therefore the readings were noted by two persons at a given signal—even this was a matter of some difficulty. The arrangement would be far more useful and perfect as a scientific machine, if fitted with a piece of clockwork, moving a paper cylinder, on which the vertical and direct forces would be simultaneously registered by separate pencils, describing two undulating lines, showing at a glance the relative forces; the experimenter would then have nothing else to attend to, but to see that all other conditions were acting properly.

The CHAIRMAN: I think the remarks by Mr. Wenham are important, especially with regard to the effects produced on the planes at different inclinations. When the plane was placed vertical, the pressure of the blast of air was direct, and tended only to move the plane in a horizontal direction—being that of the direction of the air itself—but when the plane was inclined, a part of the pressure was exerted in raising the plate in a vertical direction, and a part only in exerting a horizontal pressure; so that the latter was less than in the previous case. When the plane was placed at an angle of 45°, the horizontal force and the vertical force were found to be identical, as mentioned in the manner described by Mr. Wenham. It was also found that whether the exposed surface was a circle, a square, or a parallelogram, providing the area was the same, the results were identical to the degree of accuracy to which the readings could be determined. Anyone who had not considered with care the nature of the pressure produced by the flow or rush of a fluid, elastic or incompressible, against a plane surface placed in its course, might imagine that the system of parallel forces was merely equivalent to a single resultant force acting at the centre of pressure, and capable of resolution according to the ordinary parallelogram law. But this of course is not the case, for the particles of the fluid which come in contact with the plane have somehow or other to get out of the way, by gliding along the surface of the plane (as they cannot get through it), and this produces a complication in the neighbourhood of the surface of such a kind as cannot be theoretically predicted. One thing, however, is quite clear, and that is, that the directions of all the small forces acting on the surface certainly are not parallel, and that we must therefore have recourse to experiment. Even the fact that when the inclination of the plane to the current (supposed moving horizontally) is 45°, the vertical and horizontal pressures are equal, is not by any means evident; nor in fact can it be *exactly* true; for supposing (to fix the ideas) that the upper part of the plane is bent over so as to point in a direction opposed to that in which the current is moving, and making an angle of 45° with it, then most of the particles of air in the vicinity of the plane will, in order to get out of the way, be moving downwards along its surface; so that compounding this motion with that of the current, we should expect the horizontal force to be greater than the vertical. The experiments have shown that this difference is not appreciable to the extent to which the instrument can measure it. The same qualification also must be understood to apply to these results, from which it would appear that the pressure was independent of the form of the surface. The velocity of the current in these experiments was measured by a Lind's anemometer, an instrument that has never appeared to me to give very satisfactory results; but still the only one available for the purpose. I regret that the apparatus is considered by Mr. Browning to be too delicate to be used in the open air, but I hope that this will not be always found to be the case. As I have said before, difficulties exist only to be overcome, and some day I trust, we may obtain a series of experiments, in which ordinary wind will replace the use of the artificial current. I see Mr. Brooke present, who helped us with the experiments, and he may be able to say something as to the results gained.

Mr. BROOKE said it was not exactly mentioned, but the fact was notorious to everyone acquainted with mechanics, that in whatever position the plane was placed, the horizontal pressure may be resolved into two—one perpendicular to the plane, the other in the direction of the plane. It was clear that the resolved pressure acting in the direction of the plane was wholly effective in raising the plane. The resolution of the pressure into two was well known to everyone acquainted with the principles of mechanics; but it was to be understood that there were many other facts to be considered. The simple geometrical consideration of the action of the pressure upon the plane, did not involve the necessity for the particles of air which had impinged upon the plane, getting out of the way to enable other particles to impinge upon it. This led, in this experiment, to a result which might have been expected, but which it was important to ascertain. There were two rectangular planes of the same shape and area, and one was capable of being inclined lengthwise, in relation to the wind, and the other crosswise. Supposing the wind to be coming in a given direction (indicated as being towards the speaker) it was quite clear, with the plane inclined lengthwise, there would be less surface of the plane impinged upon than there would be in the transverse direction (indicated on the instrument). The particles which impinged upon the former must move along the plane and had much more difficulty in getting out of the way, than particles which impinged on the plane in the latter position. This would show that the effective pressure of the wind at the same velocity was greater upon the one plane than upon the other. And, conversely, a revolving, or oscillating plane, moving in the former direction (indicated), would move with less force than in the latter direction (indicated). And here was an illustration connected with the wings of birds, particularly of those that had powerful flight—where the wing was exceedingly long and narrow, it struck the wind in that direction (indicated). The experiment showed that from the same amount of surface there would be greater effect upon the air by a long narrow wing, than by a short and broad one of the same area. That was one of the results that had been obtained by these experiments.

Mr. WENHAM: I partly neglected to show how this illustrates the flight of birds. You will find that the lifting power of the smallest angle is nearly five times that of the direct force. We were not able to try less angles. The smaller the angle of inclination, in regard to the current, the less the direct force; and, comparatively, the lifting force is scarcely diminished. At 15 degrees, one force is nearly five times that of the other.

Mr. HARTE asked if, in making those experiments, attempts were made to ascertain any pressure of the wind downwards.

Mr. WENHAM: No! I omitted to mention that. A spirit level was laid across, so as to level the instrument. We had a trunk twelve feet long and eighteen inches square, to direct the current horizontally, and in a parallel course.

THE CHAIRMAN: Certain conditions of current were tried by Lind's Anemometer.

Mr. HARTE: Did you notice, in making these experiments, where the centre of pressure came?

Mr. WENHAM: We were not able to ascertain very accurately. In all cases there was a tendency to lift the front edge.

Mr. HARTE: Did you notice whether, according to the angle, the centre of pressure came forward?

Mr. WENHAM: We found as the angle became more acute, the centre of pressure came nearer to the front edge.

Mr. HALL (of Acton): Was the experiment made with a surface larger than one foot?

The CHAIRMAN: We had one eighteen inches square.

Mr. HALL: A different result would, I think, be attained within two feet from what was attained with one foot.

The CHAIRMAN: We have not spoken of two feet, because the shaft was scarcely large enough to give the even pressure required. We did not feel quite so certain with respect to large planes; and, therefore, the experiments with them are not included in these records; but I am ready to believe that the larger the planes, the larger the results. With areas of six inches, twelve inches, or two feet, the larger the area, the larger are the relative results. I have had three or four anemometers together, and always found this to be the case.

Mr. BROOKE: I rise to make an explanation. The 0 in the return ought to be 90. It ought to be 15, 20, 45, and 90.

Mr. F. W. BREAREY (the Secretary): If there is any gentleman here who could give us any advantage with regard to a fan-blower, we should be glad to avail ourselves of it. The area was so small that we could not expose much surface.

The CHAIRMAN: But we ought to give our thanks to Mr. Penn for the blower he lent to us and for the use of his steam power. The entire work of the shop was stopped during part of the time we occupied it. I should like to ask you to thank Mr. Penn for the facilities he gave us on that occasion for making these experiments. (Applause.)

Thanks were accorded to Mr. Penn by acclamation.

Document 1-4

Samuel P. Langley, *Experiments in Aerodynamics* (Washington, DC: Smithsonian Institution, 1891), excerpt "Langley's Law" reprinted in *The Aeronautical Annual* 1, James Means, ed., (Boston: W.B. Clarke & Co., 1895): 127–128.

Professor Samuel Pierpont Langley's paper, *Experiments in Aerodynamics*, published in 1891 by the Smithsonian Institution, has been called the "first substantive American contribution to aerodynamics." The following passage taken from this paper presents the theory that came to be known as "Langley's Law." This "law" stated that the power required for a vehicle to fly through the air decreased as the velocity increased—or, more simply, the higher the speed, the lower the drag. While this optimistic theory was subsequently realized to be in error, at the time it was published it represented an important new scientific approach to the study of aerodynamics. The problem for aeronautical experimenters and designers during Langley's time was that lift and drag forces were very difficult to measure with any precision, though Lilienthal, the Wrights, and other contemporaries came to suspect that Langley's law was badly mistaken.

Document 1-4, Samuel P. Langley, "Langley's Law," 1891.[28]

"To prevent misapprehension, let me state at the outset that I do not undertake to explain any art of mechanical flight, but to demonstrate experimentally certain propositions in aerodynamics which prove that such flight, under proper direction, is practicable. This being understood, I may state that these researches have led to the result that mechanical sustentation of heavy bodies in the air, combined with very great speeds, is not only possible, but within the reach of mechanical means we actually possess, and that while these researches are, as I have said, not meant to demonstrate the art of guiding such heavy bodies in flight, they do show that we now have the power to sustain and propel them.

Further than this, these new experiments (and theory, also, when reviewed in their light) show that if in such aerial motion, there be given a plane of fixed size and weight, inclined at such an angle, and moved forward at such a speed, that it shall be sustained in horizontal flight, then the more rapid the motion is, the *less*

---

[28] This document does not include six figures that appeared in the original. All six are photographs of Otto Lilienthal making glider flights.

will be the power required to support and advance it. This statement may, I am aware, present an appearance so paradoxical that the reader may ask himself if he has rightly understood it. To make the meaning quite indubitable, let me repeat it in another form, and say that these experiments show that a definite amount of power so expended at any constant rate, will attain more economical results at high speeds than at low ones, *e.g.*, one horse-power thus employed will transport a larger weight at twenty miles an hour than at ten, a still larger at forty miles than at twenty, and so on, with an increasing economy of power with each higher speed, up to some remote limit not yet attained in experiment, but probably represented by higher speeds than have as yet been reached in any other mode of transport—a statement which demands and will receive the amplest confirmation later in these pages."

Document 1-5(a–d)[29]

(a) Otto Lilienthal, "The Problem of Flying," *Annual Report of the Board of Regents of the Smithsonian Institution,* **July 1893 (Washington, DC: Government Printing Office, 1894), pp. 189–194; translated from German article published in** *Prometheus* **4, No. 205 (1893): 769–744.**

(b) Otto Lilienthal, "Practical Experiments in Soaring," *Annual Report of the Board of Regents of the Smithsonian Institution,* **July 1893 (Washington, DC: Government Printing Office, 1894), pp. 195–199; translated from German article published in** *Prometheus* **5, No. 220 (1893).**

(c) Otto Lilienthal, "The Best Shapes for Wings," *Aeronautical Annual* **(Boston: W.B. Clarke & Co., 1897), pp. 35–37; abridged translation from** *Zeitschrift für Luftschiffahrt* **14.**

(d) Vernon, "The Flying Man: Otto Lilienthal's Flying Machine," *McClure's Magazine* **3 (September 1894): 323–31.**

Otto Lilienthal was one of the most inspiring aviation pioneers of the late nineteenth century. His gliding exploits were widely publicized and the articles about him and photographs of his flights captured the imaginations of people in Europe and America. His work galvanized others to action, and his martyrdom in a flying accident in August 1896 motivated others to increase efforts to develop a flying machine. Wilbur and Orville Wright were two men who were so influenced by him; news of Lilienthal's death sparked their initiation of active research into the problem of flight, and their early work was guided by Lilienthal's research and theories.

Representing Otto Lilienthal's evangelical writings on aviation are the first three documents below, translated from German and published in the United States. "The Problem of Flying" and "Experiments in Soaring" are taken from pieces included in 1893 in the German journal *Prometheus*, which subsequently appeared in English in the Smithsonian Institution's 1894 annual report. "The Best

---

[29] All figures for this document are photographss and have been omitted here.

Shapes for Wings" is an abridged translation from the 1897 edition of *Aeronautical Annual*. The final piece, an article on "The Flying Man" that appeared in *McClure's Magazine* in September 1894, captures the technical genius inherent to Lilienthal's designs as well as the widespread popular enthusiasm for his intrepid flying.

Document 1-5(a), Otto Lilienthal, "The Problem of Flying," 1893.

While theoretically no difficulty of any considerable importance precludes flight, the problem can not be considered solved until the act of flying has been accomplished by man. In its application, however, unforeseen difficulties arise of which the theorist can have no conception.

The first obstacle to be overcome by the practical constructor is that of stability. It is an old adage that *"Wasser hat keine Balken"* [Water has no rafters]. What then, shall be said of air?

Leaving out of the question propelling mechanisms which require more than ordinary refinements of construction, theory teaches that a properly constructed flying apparatus may be brought to sail in a sufficiently strong wind; while in still air, such a machine may be made to glide downward upon a slightly inclined path. In the practical application of these two methods, however, it is found that while the apparatus is supported by moving air, it is also subjected to the whims of the wind, which often places it in uncomfortable positions, overturns it, or carries it into higher regions and then precipitates it, headforemost, to the ground. Lowering of the center of gravity is of little avail, nor does the most ingenious change of the wings or the steering surfaces alter the case. There is still no trace of the majestic soaring of the bird, for the wind is a treacherous fellow, who follows his own inclinations and laughs at our art. Therefore let us try the second method, the oblique descent in still air.

According to computation the apparatus should descend at a small angle, reaching the ground at a considerable distance, but this experiment is a success only in short flight. Beyond these the apparatus becomes unmanageable, darts vertically up, turns about, comes to a full stop, stands on its head, and descends with uncomfortable rapidity to the ground, the contact with which will probably have demolished the machine, if it do not turn a lucky somersault and land upon its back. Nor do repeated changes of the center of gravity alter the case beyond making it turn over backward instead of forward, leaving the conditions as unstable as before. Fancy the fate of the man who confides in such an apparatus.

Shall we now give up all hopes of success or shall we try new means to deprive the flying machine of its vicious propensities? This question has been answered in various ways. On the one hand it is thought that it should be possible, by mechanical means, to produce stable flight automatically, and an association of engineers of repute at Augsburg—an excellent proof that investigations of the art of flying have begun to be taken up by willing and self-sacrificing men—has among other things proposed mechanical contrivances for the regulation of soaring.

The apparatus is meant to descend from a captive balloon. By the application of ingenious methods the sailing surfaces (wings) are forced to retain their inclination. According to the report of Engineer M. Von Siegsfeld on the subject, no system has as yet been discovered that would promise sufficient security to any one sailing at a considerable elevation.

As desirable as it is that these investigations should discover safe automatic devices to give stability to soaring, it remains, on the other hand, doubtful whether the dangers attending such flights could even then be obviated. I am of the opinion that the evolution of the flying machine will be similar to that of the bicycle, which was not made in a day, and that this will not be either. Although in soaring the center of gravity may be placed below the center of pressure of the supporting air, it appears that even in this case, on account of the elasticity of the air itself, permanent stability could only be obtained by a constant and arbitrary correction of the position of the center of gravity. This is performed by birds incessantly and it is in virtue of a perfect adaptation of the form of their wings to any aerial motion that their flight appears to us so sure, graceful, and beautiful.

In the same way, a man can move through the air and have the general ability to guide his apparatus by constant shifting of the center of gravity. Descent should not be at first tried from great elevations, for such a feat requires practice. In the beginning, the height should be moderate and the wings not too large, or the wind will soon show that it is not to be trifled with. In fact, under some circumstances, one may be swept off toward still higher regions, the descent front which might well be disastrous. It therefore seems best that the wings should not exceed from 8 to 10 square meters (somewhat over 80 to 100 square feet), or that the experiment should be conducted in any wind blowing more than 5 meters per second (nearly 1,000 feet a minute), which represents a gentle breeze. A good run against the wind, however, and a leap from a safe height of 2 or 3 meters may secure a flight of 15 or 20 meters.

Constant practice will enable the experimenter to withstand a stronger breeze, to increase the surface of the wings to 15 square meters (160 square feet), and to start from a greater elevation, especially if there be a moderate slope beneath him with a soft, yielding surface. After becoming sufficiently expert to deviate from a straight line, the experimenter may enjoy the sensation of flying, but it is always a necessary condition that he should face the wind while descending, as the birds do. If then flight is attempted with the wind, it must be more rapid than the wind, or the result will be very apt to be a dangerous somersault at the time of coming to the ground, so that it is, on the whole, most advisable to follow the lessons of the birds, who ascend and descend against the wind.

I have been experimenting in this way for three years, and the constant progress made in the perfection of my machine, and the increased security it gives has convinced me of the correctness of the plan. At all events, I think it best to perfect the soaring apparatus before attempting flight with moveable wings.

After numerous experiments from low elevations, I gradually ventured to increase the height, and for this purpose I erected a tower-like shed, which, while it gave me room to store my apparatus, enabled me to conduct my experiments from the roof. The illustrations, taken from instantaneous photographs, show one of my securely constructed machines for soaring and the various phases of a soaring experiment.

Figure 1 represents the first leap from the roof, the cut showing the front view of the apparatus, which in some respects resembles the spread wings of a bat, and folds up like those for convenience of storage and transportation. The frame is of willow, covered with sheeting; the entire area contains nearly 150 square feet, and the entire apparatus weighs about 45 pounds. The roof of the tower is rather over 30 feet above the surrounding level, and from this elevation, after sufficient practice, one may glide over a distance of over 50 yards at an angle of descent of from 10 to 15 degrees.

Figures 2, 3, 4 show the porgies of the experiment. While flying freely in the air the proper angle of descent has to be regulated by shifting the center of gravity. Of course, the wind plays a very important part here, and it is only long and constant practice that we can learn to make allowance for its irregularities and to steer the apparatus properly. The capriciousness of the wind may exert unequal pressure on the great expanse of wing, and then it may happen that one wing will be elevated higher than the other.

This is shown in fig. 5. In this case the equilibrium may be restored by a change in the center of gravity, which may be effected by extending the legs as far to the left as possible, and thus adding more weight to the wing on that side. The two steering planes attached to the rear aid in enabling one to keep the face to the wind.

Figure 6 shows the simple manner of grasping the machine. There are no straps or buckles, and yet the connection is perfect. Each arm rests on a cushion attached to the framework, the hands seize a crossbar, and the remainder of the body hangs free.

My recent experiments have been made from hills having an elevation of about 250 feet and sloping uniformly every way at an angle of 10 to 15°. From the lower ridges I have already sailed a distance of over 250 yard. The great difficulty to be encountered in the endeavor to soar comes in learning to guide the flight, rather than in the difficulty of providing power to move the wings.

Progress in the mechanics of flying received at one time a severe check through the utterances of a high authority in physics. Starting with an erroneous hypothesis and putting too high a value on the amount of work required, he claimed that the maximum of possible flight had already been developed in the largest birds, and, as man represented about four times the heaviest of them, human flight was to be discarded as an utter impossibility. Now it must be admitted that the difficulties increase with the size of the flying individual; but flying itself is not the difficulty, for the largest flyers are at the same time the best flyers when once they get going in the air.

The object of this paper is to attempt to dispel old prejudices and to win new adherents for the problem in question. Even considered only as a physical exercise, the sport of flying would create one of the healthiest of all enjoyments and add one of the most effective remedies to the means now adopted for the conquest of those diseases which are now incident to our modern culture.

Document 1-5(b), Otto Lilienthal, "Practical Experiments in Soaring," 1893.

My own experiments in flying were begun with great caution. The first attempts were made from a grass plot in my own garden upon which, at a height of 1 meter from the ground, I had erected a springboard, from which the leap with my sailing apparatus gave me an oblique descent through the air. After several hundred of these leaps I gradually increased the height of my board to 2 1/2 meters, and from that elevation I could safely and without danger cross the entire grass plot. I then went to a hilly section, where leaps from gradually increased elevations added to my skill and suggested many improvements to my apparatus. The readers of *Prometheus* have already been informed of the selection of a piece of ground which enabled me to extend my flights over a distance of several hundred meters. The remainder of the summer since my last publication (in Nos. 204 and 205 of this journal) has sufficed to bring these experiments to a termination and to dispose of some important questions as to the possible results.

Indulging in subtle inquiries and theorizing does not promote our knowledge of flying, nor can the simple observation of natural flight, as useful as it may be, transform men into flying beings, although it may give us hints pointing towards the accomplishment of our purpose. We see buzzards rise skyward without any motion of their wings; we observe how the storks intermingle in the flock with outspread wings and in beautiful spirals; we see, high up in the air, the piratical falcon in quest of booty remain stationary in the wind for minutes at a time. We recognize every spot on his brownish plumage, but we do not perceive the least exertion of his wings to maintain his stationary position, and this small bird of prey is not in the least concerned at our presence. He reciprocates the protection secured for him since Brehm and other naturalists have pointed out his usefulness by undisturbedly precipitating himself to the grass before our eyes, and, seizing a grasshopper, we again see him meters above our heads without having detected the least flapping of his wings during the entire performance.

We notice that constant changes are going on in the force of the wind, but the falcon does not alter his position by a single inch, although having already begun devouring his prey, he can give but divided attention to his flight. Now he bends his head downward and backward, so that the world below must appear to him inverted, and evidently enjoys eating the insect as his talons leisurely pluck it to pieces. In the position in mid-air (which is maintained even during this employment) he appears like an automaton rooted in the wind. Just the faintest balancing motion, apparently serving to compensate for the irregularities of the wind, is perceptible in the extreme points of his wings, which are slightly inclined backwards.

The poise of the falcon in mid-air, which appears to us as a defiance to the law of gravity, may be considered not only the most remarkable, but also the most instructive example of flight.

In observing the majestic, circular soaring of other aërial travellers, one can readily believe that these skillful wing artists understand how to profit by the periodic currents of the air, and in describing spirals instinctively transform the force of the opposing current of air into lifting or suspensive power; but when the bird, without the least movement of his wings, remains stationary in one point of the sky, we are led to infer the existence of a peculiar form of surface which may be held suspended by the application of a uniformly moving wind.

While the existence of this possibility may be demonstrated by elementary experiments, this does not discover the secret of soaring, and though nature conclusively demonstrates that it can not be the want of power that prevents our flying, that knowledge alone does not provide us with wings. Furthermore, while nature points out how it is done, that does not necessarily imply that there may not be found other ways or means of doing it. However we may theorize on the subject, without a practical application of the theory, things will remain unchanged and our flight will only be in imagination or in dreams.

My experiments, then, should form the transition, the first step from theory to practice. Like others, I too have, in the beginning, attempted using machines with moveable wings, but this does not apparently aid in the development of an art of flight. The mark is too high and not immediately attainable, and one's ambition should be fully satisfied by withstanding the wind with wings of the size adapted to flying men. Each flight demands a rising from the ground and a landing; the former is as difficult as the latter is dangerous, and regardless of the most ingeniously constructed apparatus, the art of both will have to be acquired just as the child learns to stand and walk. Anyone desirous of exposing himself unnecessarily to danger and of ruining in a few seconds the carefully constructed apparatus need only expose his machine to the wind without having familiarized himself with its management, and he will soon know what it means to control an apparatus of from 10 to 15 square meters in area, where other people can but with difficulty manage an open umbrella.

To all those who, by their own experience or otherwise, can form a correct idea of the difficulties that present themselves, the instantaneous photographs by Mr. Alexander Krajewsky, accompanying this paper, may be of interest.

In continuation of my formerly published experiments, I endeavor with every new trial to gain more complete control over the wind, and without disregarding any necessary precaution I have already succeeded in at least temporarily retaining a uniform level and even in remaining stationary in the wind for a few seconds. The simplicity of my flying machine, which is controlled by shifting the center of gravity, has compelled me to avoid strong breezes, which however might presumably have aided in securing a stationary position. During my continued flights, however, I have been at times surprised by a sudden increase in the force of the wind which either carried me upward almost perpendicularly or supported me in a stationary position for a few seconds to the great delight of the spectators.

The freedom from accidents in these apparently daring attempts may be considered proof that the apparatus already described offers ample security in carrying out my plan of investigation.

To those who, from a modest beginning and with gradually increased extent and elevation of flight have gained full control over the apparatus, it is not in the least dangerous to cross deep and broad ravines.

It is a difficult task to convey to one who has never enjoyed aerial flight a clear perception of the exhilarating pleasure of this elastic motion. The elevation above the ground loses its terror, because we have learned by experience what sure dependence may be placed upon the buoyancy of the air. Gradual increase of the extent of these lofty leaps accustoms the eye to look unconcernedly upon the landscape below. To the mountain climber the uncomfortable sensation experienced in thrusting his foot into the slippery notch cut in the ice or to a treacherous rubble above deep abysses, with other dangers of the most terrifying nature, may often tend to lessen the enjoyment of the magnificent scenery. The dizziness caused by this, however, has nothing in common with the sensation experienced by him who trusts himself to the air; for the air demonstrates its buoyancy in not only separating him from the depth below, but also in keeping him suspended over it. Resting upon the broad wings of a well-tested flying machine, which, yielding to the least pressure of the body, obeys our directions; surrounded by air and supported only by the wind, a feeling of absolute safety soon overcomes that of danger.

One who has already practiced straight flights for some time will naturally endeavor to next guide his apparatus in a lateral direction, and indeed there is nothing easier than the guiding of the aerodrome, which is accomplished by shifting the center of gravity. The steering blades have nothing to do with this, their function being to keep the machine facing the wind.

Plate XIII, fig. 1 illustrates such a serpentine flight. I started from a hill to the right, the base of which is still visible in the figure, and soared toward the plain below in a somewhat circuitous path. The photograph was taken at the moment when I had almost turned my back to the plain. The view shown in Plate XIII, fig. 2, was taken at a time when I was lifted and carried upward by a suddenly increasing current which impeded progress and rendered me absolutely stationary.

In Plate XIV, flights are represented in geometric perspective. The lowest dotted line, d, e, was described during a calm. Even the expert flyer must descend during a calm at an angle of from 9° to 10°. The run began upon the top of the hill, near a; at b I left firm ground and endeavored to glide along the mountain slope, placing the wings at c at such an angle that the pressure of the wind, L, would not only support the machine, but also carry it forward. This increased the velocity sufficiently to enter at d into the line of stable flight. Such a maneuver is necessary, because a velocity of 9 meters per second is required for a flight in a calm, while but 6 meters were obtained by the run. At e the ground has almost been reached, and by raising the wings slightly in front the momentum is diminished and a landing effected without serious jar.

The second line, $e\,f$, shows a flight in a moderate breeze, in which the proper position with a downward inclination of 6° had been assumed immediately upon starting.

Flight against the wind is slower. The distance to be accomplished may be extended by a carefully determined and properly maintained inclination of the wings; in fact, by careful observation of this the soaring may be extended over a distance equal to ten times the height of the starting point.

During a strong breeze a sinuous line of flight results from the temporary support given by the wind at times. This is shown in the line $b\,g$, though such experiments should be undertaken only by one fully familiar with the management of the apparatus. The indefinable pleasure however experienced in soaring high up in the air, rocking above sunny slopes without jar or noise, accompanied only by the æolian music issuing from the wires of the apparatus, is well worth the labor given to the task of becoming and expert.

It does not seem at all impossible that the continuance of such flights may lead to free, continuous sailing in agitated air.

The results of our present experiments already furnish an indication of the degree of mechanical energy that must be added to that involved in oblique soaring to enable us to gain independent horizontal flight.

The solution of this problem, however, would exceed the purposes of the present article, and I content myself by stating that the conditions of a motor can easily be met, supposing that the propelling mechanism has been properly chosen, and that extraordinary lightness is not even essential.

The interests of the professional flight-essayer demand further experiments in practical flight and the gain of further efficiency. But even to those who only desire to utilize, as a means of sport, the results already obtained, opportunities are offered to promote the interest of the problem of flight and the way for a more ready prosecution of the subject.

The time has passed when every person harboring thoughts of aerial flight can at once be pronounced a charlatan. If we may hope that our aeronautic publications are eventually to be taken seriously by the majority of those skilled in allied subjects, it is important at the outside to awaken the interest of those whose natural concern this great problem should be, but who now shrug their shoulders. We shall then at least be able to show some practical results, and towards these ends we here take the first step.

Document 1-5(c), Otto Lilienthal, "The Best Shapes for Wings," 1897.

The results which we reach by practical flying experiments will depend most of all upon the shapes which we give to the wings used in experimenting.

Therefore there is probably no more important subject in the technics of flying than that which refers to wing formation.

The primitive idea that the desired effects could be produced by means of flat wings has now been abandoned, for we know that the curvature of birds' wings gives extraordinary advantages in flying.

The experiments on the resistance of air to curved surfaces have shown that even very slight curvatures of the wing-profile increase considerably the sustaining power, and thereby diminish the amount of power required in flight.

The wing of a bird is excellent not only because of the curvature of its cross-section, but the rest of its structure and formation also has influence upon the flight. Therefore the outline of the wing is certainly of importance.

It is probable that the form of the cross-section of the wing and flight-feathers *(Schwungfedern)* has a favorable influence upon the flight.

Experiments have not yet been made to show conclusively whether or not the feather structure of a wing endows it with a special quality whereby the sustaining power is increased. With investigators this has been a subject of conjecture. Therefore it is questionable *(auch fraglich)* whether we are wrong if, in constructing flying apparatus, we keep to the bat's wing, which is easier to construct.

Bats fly much better than is generally thought. Two early bats, which I saw flying this summer in broad sunshine and in somewhat windy weather, sailed along so well without flapping their wings that I thought, at first, they were swallows. Of course on evenings when there is no wind, the bat must flutter continually. The early-flying bat is also called evening-sailer *(Abendsegler)* which indicates that its sailing flight has been marked.

The most important point as regards the form of the wing will always be the curvature of its profile. If we examine any bird's wing we find that the enclosed bones cause a decided thickening at the forward edge. The question now is, What part does this thickening play in the action of the curved surface? The thickening is quite considerable, particularly in birds which have long, narrow wings. An albatross in my possession has a breadth of wing 16 centimetres, the thickened part of which measures 2 centimetres; the thickness is therefore 1/8 of the breadth of the wing. As the albatross is one of the best sailers, we can scarcely assume that the comparatively great thickness of the wing at its outer edge has a detrimental effect upon the bird's flight.

For a long time I have assumed that the thickening which all birds' wings have at the front edge produces a favorable effect in sailing flight. By means of free-sailing models I have now learned that nature makes a virtue of necessity, that the thickened front edge is not only harmless, but in sailing flight is helpful *(sondern den Schwebe-effect nicht unerheblich erhöht).*

The experiments are easily tried. It is only necessary to make a number of models of equal size and weight, each one having a different curve in its sustaining surfaces. These models I make of strong drawing paper, the size of the surfaces being about 4 inches in width by 20 inches in length.

The experimenter can let these models sail from any tower or roof in front of which there is an open space. Each model must be made to glide through the air many times until it reaches the ground. Experiments must be made in the stillest possible air.

The lengths of flights are all noted down, and from a long series of experiments the arithmetical mean for each design is computed. The models having the best profiles will make the longest flights. In this way a reliable table can be made which will show the relative merits of the profiles, and will also show quite plainly in which direction the most useful form will have to be developed.

Until now I have endeavored to find out the best proportions for wings by constructing different kinds of sailing apparatus. In this way, of course, many important facts have been ascertained. The construction of full-sized apparatus requires a great deal of time and is expensive, therefore we must welcome a method which permits inquiry into the forms of wings in models which fly automatically. Besides that, it is not every one's business to throw himself into space in a sailing apparatus, although he who would succeed in practical flying can scarcely avoid this way.

Considering the fact that the most important thing is to ascertain what are the best qualities of the natural wing, which is in every respect perfect,—these steadily sailing models offer every one an opportunity of engaging in experiments of this kind. Further, any one who takes up this kind of experiment will find great pleasure in watching the manœuvers of his small flyers, which often vie with the best sailers among birds. I can therefore recommend this occupation not only for the furthering of the science of mechanical flight, but also because it affords a most interesting pastime.

The few measurements made so far by this method are too incomplete to be fit, as yet, for publication. I am preparing, however, a systematic series of experiments, the results of which will be stated when the experiments are finished.

Meanwhile, I cherish the hope that this paper may be an incentive to others to make similar experiments, so that we may sooner reach the desired end.

Document 1-5(d), Vernon, "The Flying Man," 1894.

Herr Otto Lilienthal, of Berlin, who has attained some celebrity as "The Flying Man," was born forty-six years ago in the antiquated little city of Anklarn, near the Baltic coast of Pomerania, about sixty miles to the northwest of Stettin. A residence so near the sea afforded him, in early life, many an opportunity of prosecuting his favorite studies and observations. In later years he migrated with his younger brother, Gustav, his enthusiastic coadjutor in all his researches in the domain of aviatics, to Berlin, where he established, and is now conducting, a large manufactory of small steam engines, whose mechanical appliances furnish him with every facility for the construction of his flying apparatus. He is an accomplished mathematician, and a close observer of nature; and is, besides, endowed in large measure with that poetic instinct which nearly always constitutes one side of even the most practical German character.

THE BIRD'S WING ON LILIENTHAL'S MODEL.

For more than twenty years Herr Lilienthal, with his brother's aid, and in the intervals of more serious occupations, has been studying the subject of aërial navigation. He has taken the flying bird as his teacher. After many experiments with flat wings or plane surfaces, he became convinced that it was the gentle parabolic curve of the wing which enables a bird to sustain itself without apparent effort in the air, and even to soar, without a motion of the wings, against the wind. This he has demonstrated not only by experiment, but by an application of the doctrine of the resolution of forces to the action of the wind upon a concave surface. The circling ascents of the carrier-pigeon, as he rises when released, to gain a general view of the landscape, and to take his bearings before starting on his homeward journey, depend upon this principle. He *flies* with the wind, but he *sails* or *soars* against it. The fins of many fishes and. the web feet of aquatic birds, are strikingly analogous in construction. The sails of a ship assume a similar form. It would be impossible to sail so near the wind if the instrument of propulsion were a rigid flat surface. It is the effort of the sail to get away from the wind, which it gathers in its ample bosom and drives the boat forward, almost in the very teeth of the breeze.

"There are still prominent investigators who *will not see*," said Herr Lilienthal to me, "that the arched or vaulted wing includes the secret of the art of flight. As we came upon the track of this idea, my brother and I, who were then young and wholly without means, used to spare from our breakfasts, penny by penny, the money to prosecute our investigations; and often the 'struggle for life' compelled us to interrupt them indefinitely. While we were devoting every moment of our spare time to the solution of the problem, almost every one in Germany regarded the man who would waste his energies in such unproductive labor as a fool. Years ago the most distinguished professor of mathematics in the Berlin Industrial Academy sent me word that of course it could *do no harm* to amuse myself with such pastimes, but warned me earnestly against putting any money into them. A special commission of experts, organized by the state, had, in fact, laid it down as a fundamental principle, once for all, that it was *impossible* for a man to fly. German societies for the promotion of aëronautics did not then exist, and those subsequently formed were devoted almost entirely to the interests of ballooning.

"I have always regarded the balloon, and the exclusive attention which it so long attracted, as a hindrance rather than a help to the development of the art of flight. If it had never been invented, it is probable that more serious investigations would have been prosecuted towards other solutions of the problem. Since the time of Montgolfier nearly all practical efforts have been directed to the improvement of the balloon. But it has nothing in common with the birds, and it is these that we must take as our model and exemplar. What we are seeking is the means of free motion in the air, in any direction. In this the balloon is of no aid; there is no relation between the two systems."

THE WING OF THE BIRD.

The wing of a bird is divided into three parts, corresponding to shoulder-joint, the forearm, and the hands and fingers of the human frame. The two former, composed largely of bones and muscles and tendons, are comparatively heavy, and their rapid movement demands the expenditure of considerable physical force; the last consists almost entirely of "pen-feathers," or pinions, which move to a certain extent automatically. In the larger birds—the "sailers" or "soarers," which alone are to be considered here—the first two members, with their concave under surfaces, furnish the sustaining power; and the last, being at the greatest distance from the shoulder, or axis of motion, the chief propulsive force. The construction of each member is peculiarly adapted to its special purpose, and it is this which Herr Lilienthal has endeavored to imitate.

An oarsman, on his forward stroke, opposes the blade of his oar almost perpendicularly to the resistance of the water. As he lifts it at the beginning of the backward stroke he "feathers" it, or brings it into a nearly horizontal position, so that its edge cuts the air. The pinions of birds act in precisely the same way. There are other analogies between the wing and the oar. The backstroke of the oar occupies only about half the time of the "pull," and the upstroke of the wing bears about the same relation to the downward beat. Moreover, at certain inclinations of the wing, the upward stroke, while detracting little or nothing from the sustaining power, contributes to the forward movement. As the pinions separate in consequence of the action of the air from above, they present their concave surface obliquely to the resisting medium and act like an oar in "sculling," which, whether moved to the right or the left, impels the boat forward. It is evident that this must greatly lighten the physical exertions of a bird in rapid flight, for in whichever direction he moves his wings he gains propulsive force.

To the conviction that concave or vaulted wings were essential to success, Herr Lilienthal was led not only by the examination of a great variety of natural wings, and by theoretical deduction, but by actual experiment. The means adopted for this purpose were ingenious and simple. He fitted up an apparatus in the form of the fly-fans found above the dining tables of clubs and restaurants, with two long arms revolving horizontally, to the ends of which surfaces of different kinds and degrees of curvature could be affixed in any required position. The motive power was furnished by a weight, and could be exactly measured. There was also an adjustment which enabled the observer to measure the lifting force of various surfaces, moving at different angles of inclination through still air. By this means Herr Lilienthal was enabled to reach conclusions which were of great value to him in the construction of his flying machine; and the most important of them was that the most effective form of wing was that whose convexity, as measured by the versed sine of the arc, should be one-twelfth of the breadth of the wing, or of the length of the chord connecting the opposite edges.

HERR LILIENTHAL'S WINGS.

The flying machine devised and now used by Herr Lilienthal is designed rather for *sailing* than for *flying*, in the proper sense of the term; or, as he says, "for being carried steadily and without danger, under the least possible angle of descent, against a moderate wind, from an elevated point to the plain below." It is made almost entirely of closely woven muslin, washed with collodion to render it impervious to air, and stretched upon a ribbed frame of split willow, which has been found to be the lightest and strongest material for this purpose. Its main elements are the arched wings; a vertical rudder, shaped like a conventional palm-leaf, which acts as a vane in keeping the head always towards the wind; and a flat, horizontal rudder, to prevent sudden changes in the equilibrium.

The operator so adjusts the apparatus to his person that, when in the air, he will be either resting on his elbows or seated upon a narrow support near the front. With the wings folded behind him he makes a short run from some elevated point, always against the wind, and when he has attained sufficient velocity launches himself into the air by a spring or a jump, at the same time spreading the wings which are at once extended to their full breath by atmospheric action; whereupon he sails majestically along like a gigantic seagull. In this way Herr Lilienthal has accomplished flights of nearly three hundred yards from the starting point.

"No one," said Herr Lilienthal to me, "can realize how substantial the air is, until he feels its supporting power beneath him. It inspires confidence at once. With flat wings it would be almost impossible to guard against a fall. With arched wings it is possible to sail against a moderate breeze at an angle of not more than six degrees to the horizon."

The principle is recognized in the umbrella-form universally adopted for the parachute. Try to run with an open umbrella held above the head and slightly inclined backwards, and see what a lifting power it exerts. Mechanical birds have been constructed with flat wings, which, so long as the machinery operated, were able to sustain themselves moderately well and to fly rapidly; but no one has yet succeeded in making any practical use of them. Their course has no intelligent direction; when the motive power gives out, they fall heavily to the earth. *Soaring*, in the sense of rising against the wind as the birds do, is possible only with dome-shaped wings. The aeroplane, or flat wing, when inclined at a certain angle to the breeze, may rise while its momentum continues, but once overcome its power is gone and nothing can restore it.

PROPER DIMENSIONS FOR THE WINGS.

"The curve of a bird's wing," said Herr Lilienthal, "is parabolic ; but the simple parabola differs so little from the arc of a circle that I adopted the latter curve as the more practicable, and the wings which I now use are in the main segments of a spherical surface. They are so constructed that they can be folded together like the wings of a bat, and require very little storage room when not in use.

"It was only gradually that I arrived at the proper dimensions. One does not easily gain an adequate conception of the materiality of the air, and my apprehensions led me at first to make the wings too large. I found that the varying force of the atmospheric currents, modified as they are by the undulations of the earth's surface, endangered my equilibrium in direct proportion to the spread of the wings. Those which I now employ are never more than twenty-three feet from tip to tip, and I am thus enabled, by a simple change of posture, so to alter the position of the centre of gravity as to restore the equilibrium.

"There are limits also to the *breadth* of the wings, or their extension backwards. The operator must be able in a moment to transfer the center gravity so far to the rear as to overcome the action of the air, which might otherwise tend to throw him forward, and precipitate him to the earth. When one feels himself falling, the natural impulse is to stretch out the arms and legs in the direction of the fall; but it is one of the peculiarities of this mode of navigation that the movement must be in the contrary direction, or towards the *upper* side. The centre of gravity is thus shifted to the one side or to the other, forward or backward, and the pressure of the air, acting with greater force on the lighter and broader surface, soon restores the equilibrium. It is not easy to realize in practice at first, but after a short experience the movement becomes almost involuntary.

"When there is no wind, the apparatus acts simply as a parachute. The pressure of the air is directly from beneath, and is equal on all parts of the under surface. I have more than once found myself in this position, when I have utilized the speed attained in a gradual descent, in rising to a greater height in order to soar over some obstacle like a tree or a crowd of people. Under favorable circumstances it is easy to mount to a height even greater than that of the starting point, but the forward motion is thereby partially or wholly neutralized, and it may happen that one comes to a complete standstill in mid-air. In such cases it is only necessary to throw the centre of gravity so far back that the air shall act more powerfully on the forward surface, and the gradual gliding descent is resumed. So in landing I bend backward, exactly as a crow does when alighting in a field, and reach the ground without the slightest shock. The worst that is likely to happen in any case is the breaking of the apparatus; there is little danger to life or limb.

"I am far from supposing that my wings, although they afford the means of *sailing*, and even of *soaring* in the air, possess all the delicate and subtle qualities necessary to the perfection of the art of flight. But my researches show that it is well worthwhile to prosecute the investigations farther."

THE LILIENTHAL MOTOR.

Having demonstrated the practicability of sailing and soaring, Herr Lilienthal has sought, in his recent experiments, to reach a practical solution of the problems of actual flight. The first difficulty to overcome was the discovery of a suitable motor, without which all efforts to fly would be hopeless. If we estimate the ordinary weight of a man at one hundred and sixty pounds, and add to that the weight of the flying apparatus, we have a total burden of at least two hundred pounds to be raised and supported simply by aerial resistance. It is calculated that to overcome the attraction of gravity in such a case requires a force of one and one quarter horse power, which no man is able to exert for more than a very short time.

With such an apparatus as Herr Lilienthal's, steam engines and electric motors are not readily available; but he conceived the ingenious idea of employing, as a motive force, the vapor of liquid carbonic acid, which, under ordinary atmospheric pressure, boils at a temperature far below that at which mercury freezes. His engine requires no fire, nor boiler, nor steam-chest; only a diminutive cylinder with the requisite valve arrangements, which may be readily worked by hand, and a small reservoir of the liquid acid lying close beside it.

The one first constructed was of two horse-power, with a receiver to contain enough carbonic acid to last for two hours, and was attached to the front of the flying apparatus. The whole contrivance, with the necessary machinery to impart motion to the wings, added less than twenty-five pounds to the weight, and this will probably be reduced in the future by the use of some alloy of aluminum, instead of iron, in the manufacture of the heavier portions. The wings were also fitted with rotatory pinions, constructed on the principles already indicated, and capable of automatic action under the pressure of the air.

The first experiments with this apparatus were rather too successful, at least in demonstrating the power of the engine. Unfortunately, the inventor had underestimated the energy of his motor, which acted with such unexpected vigor that the wings were broken, and the modifications thus shown to be necessary will require some time for their completion. Herr Lilienthal confidently expects, however, eventually to solve the problem in this way.

Document 1-6

Hiram Maxim, excerpts from "Natural and Artificial Flight," *Aeronautical Annual* (Boston: W.B. Clarke & Co., 1896), pp. 88–117.

Hiram Maxim was an American engineer who went to Europe in the 1880s to produce and sell small electrical appliances, but he instead gained fame and fortune with his invention in 1884 of a weapon capable of firing bullets in a sustained fully automatic mode, which he called a "machine gun." Working in London in the 1890s he turned his attention to the problem of mechanical flight and designed what was arguably the largest and most expensive attempt at a flying machine in the nineteenth century. His airplane was a failure, and in terms of the advancing science of aerodynamics he contributed little, but the association of his name and the many articles he wrote for popular journals did much to stimulate an increased interest in aerodynamic research. The document included here is one of those articles, published in the *Aeronautical Annual* 1896 edition, and consists of portions of a seven-part "thesis" outlining his own research.

Document 1-6, Hiram Maxim, excerpts from "Natural and Artificial Flight," 1896.

I. INTRODUCTORY.

At the time I commenced my experiments in aeronautics it was not generally believed that it would ever be possible to make a large machine heavier than the air that would lift itself from the earth by dynamic energy generated by the machine itself. It is true that a great number of experiments had been made with balloons, but these are in no sense true flying machines. Everyone who attempted a solution of the question by machines heavier than the air, was looked upon in very much the same light as the man is now who attempts to construct a perpetual motion machine. Up to within a few years, nearly all experiments in aerial navigation by flying machines have been made by men not versed in science, and who for the most part have been ignorant of the most rudimentary laws of dynamics. It is only quite recently that scientific engineers have taken up the question and removed it from the hands of charlatans and mountebanks. A few years ago many engineers would not have dared to face the ridicule which they would be liable to receive if they had asserted that it would be possible to make a machine that would lift itself by mechanical means into the air. However, thanks to the admirable work of Professor Langley, Professor Thurston, Mr. Chanute and others, one may now express his opinion freely on this subject and speculate as to the possibilities of making flying machines, without being relegated to the realm of cranks and fanatics.

During the last five years I have had occasion to write a large number of articles for the public press on this subject, and I have always attempted, as far as it is in my power, to discuss the subject in such a manner as to be easily understood by the unscientific, and I believe that my efforts have done something in the direction of popularizing the idea that it is possible to construct practical flying machines.

In preparing my present work, I have aimed as far as possible to discuss the question in plain and simple language, and to abstain from the use of any formulæ which may not be understood by every one. It has been my experience that if a work abounds in formulæ and tables, even only a few of the scientific will take the trouble to read or understand it. I have therefore confined myself to a plain statement of the actual facts, describing the character of my observations and experiments, and giving the results of the same. All experiments made by others in the same direction have been on a very small scale, and, as a rule, the apparatus employed has been made to travel around a circle, the size of which has not been great enough to prevent the apparatus continually encountering air which had been influenced in some way by the previous revolution.

The first experiments which I conducted were with an apparatus which travelled around a circle 200 feet in circumference, and by mounting some delicate anemometers directly under the path of the apparatus I ascertained that after it had been travelling at a high velocity for a few seconds, there was a well-defined air current blowing downward around the whole circle, so that my planes in passing forward must have been influenced and their lifting effect reduced to some extent by this downward current. My late experiments are the first which have ever been made with an apparatus on a large scale moving in a straight line. In discussing the question of aerial flight with Professor Langley before my large experiments had been made, the Professor suggested that there might be some unknown factor relating to size only which might defeat my experiments, and that none of our experiments had at that time been on a sufficiently large scale to demonstrate what the lifting effect of very large planes would be. A flying machine to be of any value must of necessity be large enough to carry at least one man, and the larger the machine the smaller the factor of the man's weight. Moreover, it is possible to make engines of say from 200 to 400 horse-power, lighter per unit of power than very small engines of from one to two horse-power. On the other hand, it is not advisable to construct a machine on too large a scale, because as the machine becomes larger the relative strength of the material becomes less. In first designing my large machine I intended that it should weigh about 5,000 pounds without men, water, or fuel, that the screw thrust should be 1,500 pounds, and that the total area of the planes should be 5,000 square feet. I expected to lift this machine and drive it through the air at a velocity of 35 miles an hour with an expenditure of about 250 horse-power. However, upon completing the machine I found that many parts were too weak, and these had to be supplanted by thicker and stronger material. This increased the weight of the machine about 2,000 pounds. Upon trying my engines I found that if required they would develop 360 horse-power, and that a screw thrust of over 2,000 pounds could be easily attained, but as an offset against this, the amount of power required for driving the machine through the air was a good deal more than I had anticipated.

II. NATURAL FLIGHT.

During the last 50 years a great deal has been said and written in regard to the flight of birds. Perhaps no other natural phenomenon has excited so much interest and has been so little understood. Learned treatises have been written to prove that a bird is able to develop from 10 to 100 times as much power for its weight as other animals, while other equally learned treatises have shown most conclusively that no greater amount of energy is exerted by a bird in flying than by land animals in running or jumping.

There is no question but what a bird has a higher physical development, as far as the generation of power is concerned, than any other animal we know of. Nevertheless, I think that every one who has made a study of the question will agree that some animals, such as rabbits, exert quite as much power in running in proportion to their weight as a sea-gull or an eagle exerts in flying.

The amount of power which a land animal has to exert is always a fixed and definite quantity. If an animal weighing 100 pounds has to ascend a hill 100 feet high, it always means the development of 10,000 foot-pounds. With a bird, however, there is no such thing as a fixed quantity, because the medium in which the bird is moving is never stationary. If a bird weighing 100 pounds should raise itself into the air 100 feet during a perfect calm, the amount of energy developed would be 10,000 foot-pounds plus the slip of the wings. But, as a matter of fact, the air in which a bird flies is never stationary, as I propose to show; it is always moving either up or down, and soaring birds, by a very delicate sense of feeling, always take advantage of a rising column of air. If a bird finds itself in a column of air which is descending, it is necessary for it to work its wings very rapidly in order to prevent a descent to the earth.

I have often observed the flight of hawks and eagles. They seem to glide through the air with hardly any movement of their wings. Sometimes, however, they stop and hold themselves in a stationary position directly over a certain spot, carefully watching something on the earth immediately below. In such cases they often work their wings with great rapidity, evidently expending an enormous amount of energy. When, however, they cease to hover and commence to move again through the air, they appear to keep themselves at the same height with an almost imperceptible expenditure of force.

Many unscientific observers of the flight of birds have imagined that a wind or a *horizontal* movement of the air is all that is necessary in order to sustain the weight of a bird in the air after the manner of a kite. If, however, the wind, which is only air in motion, should be blowing everywhere at exactly the same speed and in the same direction (horizontally), it would offer no more sustaining power to a bird than a dead calm, because there is nothing to prevent the body of the bird being blown along with the air, and whenever it had attained the same velocity as the air, no possible arrangement of the wings would prevent it from falling to the earth.

The wind, however, seldom or never blows in a horizontal direction. Some experimenters have lately asserted that if it were possible for us to ascend far enough, we should find the temperature constantly falling until at about 20 or 25 miles above the earth's surface the absolute zero might be reached. Now, as the air near the earth never falls in temperature to anything like the absolute zero, it follows that there is a constant change going on, the relatively warm air near the surface of the earth always ascending, and, in some cases, doing sufficient work in expanding to render a portion of the water it contains visible, forming clouds, rain, or snow, while the very cold air is constantly descending to take the place of the rising column of warm air.

On one occasion while crossing the Atlantic in fine weather, I noticed, some miles directly ahead of the ship, a long line of glassy water. Small waves indicated that the wind was blowing in the exact direction in which the ship was moving, and I observed as we approached the glassy line that the waves became smaller and smaller until they completely disappeared in a mirror-like surface which was about 300 or 400 feet wide and extended both to the port and starboard in approximately a straight line as far as the eye could reach. After passing the centre of this zone, I noticed that small waves began to show themselves, but in the exact opposite direction to those through which we had already passed. I observed that these waves became larger and larger for nearly an hour. Then they began to get gradually smaller, when I observed another glassy line directly ahead of the ship. As we approached it the waves completely disappeared, but after passing through it I noticed that the wind was blowing in the opposite direction and that the waves increased in size exactly in the same manner that they had diminished on the opposite side of the glassy zone.

This would seem to indicate that directly over the centre of the first glassy zone, the air was meeting from both sides and ascending, and that at the other glassy zone the air was descending in practically a straight line to the surface of the water where it spread out and set up a light wind in both directions.

I spent the winter of 1890–91 on the Riviera, between Hyères les Palmiers and Monte Carlo. The weather for the most part was very fine, and I often had opportunities of observing the peculiar phenomena which I had already noticed in the Atlantic, only on a much smaller scale. Whereas, in the Atlantic, the glassy zones were from 5 to 20 miles apart, I often found them not more than 500 feet apart in the bays of the Mediterranean.

At Nice and Monte Carlo this phenomenon was also very marked. On one occasion, while making observations from the highest part of the promontory of Monaco on a perfectly calm day, I noticed that the whole of the sea presented this peculiar effect as far as the eye could reach, and that the lines which marked the descending air were never more than a thousand feet from those which marked the centre of the ascending column. At about 3 o'clock in the afternoon, a large black steamer passed along the coast in a perfectly straight line, and I noticed that its wake was at once marked by a glassy line which indicated the centre of an ascending column. This line remained almost straight for two hours, when finally it became crooked and broken. The heat of the steamer had been sufficient to determine this upward current of air.

In 1893, I spent two weeks in the Mediterranean, going by a slow steamer from Marseilles to Constantinople and returning, and I had many opportunities of observing the peculiar phenomenon which I have before referred to. The steamer passed over thousands of square miles of calm sea, the surface being only disturbed by large batches of small ripples separated from each other by glassy streaks, and I found that in no case was the wind blowing in the same direction on both sides of these streaks, every one of them either indicating the centre of an ascending or a descending column of air.

If we should investigate this phenomenon in what might be called a dead calm, we should probably find that the air was rising straight up over the centres of some of these streaks, and descending in a vertical line over the centres of the others. But, as a matter of fact, there is no such thing as a dead calm. The movement of the air is the resultant of more than one force. The air is not only rising in some places and descending in others, but at the same time the whole mass is moving forward with more or less rapidity from one part of the earth to another. So we might consider that, instead of the air ascending directly from the relatively hot surface of the earth and descending vertically in other places, in reality it is moving on an incline.

Suppose that the local influence which causes the up and down motion of the air should be sufficiently great to cause it to rise at the rate of 2 miles an hour, and that the wind at the same time should be blowing at the rate of 10 miles an hour; the motion of the air would then be the resultant of these two velocities. In other words, it would be blowing up an incline of 1 in 5. Suppose now, that a bird should be able to so adjust its wings that it advanced 5 miles in falling 1 mile through a perfectly calm atmosphere; it would be able to sustain itself in an inclined wind, such as I have described, without any movement at all of its wings. If it was able to adjust its wings in such a manner that it could advance 6 miles by falling through 1 mile of air, it would then be able to rise as relates to the earth while in reality falling as relates to the surrounding air.

In conducting a series of experiments with artillery and small guns in a very large and level field just out of Madrid, I often observed the same phenomena as relates to the wind, that I have already spoken of as having observed at sea, except that the lines marking the centre of an ascending or a descending column of air were not so stationary as they were over the water. It was not an uncommon thing when adjusting the sights of a gun to fire at a target at very long range, making due allowances for the wind, to have the wind change and blow in the opposite direction before the word of command was given to fire. While conducting these experiments, I often noticed the flight of eagles. On one occasion a pair of eagles came into sight on one side of the plain, passed directly over our heads and disappeared on the opposite side. They were apparently always at the same height from the earth and soared completely across the plain without once moving their wings. This phenomenon, I think, can only be accounted for on the hypothesis that they were able to feel out with their wings an ascending column of air, that the centre of this column of air was approximately a straight line running completely across the plain, that they found the ascending column to be more than necessary to sustain their weight in the air, and that whereas, as relates to the earth, they were not falling at all, they were really falling some 2 or 3 miles an hour in the air which supported them.

Again, at Cadiz in Spain, when the wind was blowing in very strongly from the sea, I noticed that the sea-gulls always took advantage of an ascending column of air. As the wind blew in from the sea and rose to pass over the fortifications, the seagulls selected a place where they could slide down on the ascending current of air, keeping themselves always approximately in the same place without any apparent exertion. When, however, they left this ascending column, I observed that it was necessary for them to work their wings with great vigor until they again found the proper place to encounter the favorable current.

I have often noticed sea-gulls following a ship. I have observed that they are able to follow the ship without any apparent exertion; they simply balance themselves on an ascending column of air and seem to be quite as much at ease as they would be if they were roosting on a solid support. If, however, they are driven out of this position, I find that they generally have to commence at once to work their passage. If anything is thrown overboard which is too heavy for them to lift, the ship soon leaves them, and in order to catch up with it again, they move their wings very much as other birds do; but when once established in the ascending column of air, they manage to keep up with the ship by doing little or no work. In a head wind we find them directly aft of the ship; if the wind is from the port side, they may always be found on the starboard quarter, and *vice versa.*

Every one who has passed a winter on the northern shores of the Mediterranean must have observed the cold wind which is generally called the *mistral*. One may be out driving, the sun may be shining brightly, and the air be warm and balmy, when, suddenly, without any apparent cause, one finds himself in a cold descending wind. This is the much-dreaded mistral, and if at sea, it would be marked by a glassy line on the surface of the water. On land, however, there is nothing to render its presence visible. I have found that the ascending column of air is always very much warmer than the descending column, and that this action is constantly taking place in a greater or less degree.

From the foregoing deductions I think we may draw the following conclusions:

First, that there is a constant interchange of air taking place, the cold air descending, spreading itself out over the surface of the earth, becoming warm, and ascending in other places.

Second, that the centres of the two columns are generally separated from each other by a distance which may be from 500 feet to 20 miles.

Third, that the centres of greatest action are not in spots, but in lines which may be approximately straight but generally abound in many sinuosities.

Fourth, that this action is constantly taking place over both the sea and the land, that the soaring of birds, a phenomenon which has heretofore been so little understood, may be accounted for on the hypothesis that the bird seeks out an ascending column of air, and that, while sustaining itself at the same height in the air without any muscular exertion, it is in reality falling at a considerable speed through the air that surrounds it.

It has been supposed by some scientists that the birds may take advantage of some vibratory or rolling action of the air. I find, however, from careful observation and experiment, that the motion of the wind is comparatively steady, and that the short vibratory or rolling action is always very near to the earth and is produced by the air flowing over the tops of hills, high buildings, or trees. If a kite is flown only a few feet above the ground, it will be found that the current of air is very unsteady. If it is allowed to mount to 500 feet, the unsteadiness nearly all disappears, while if it is further allowed to mount to a height of 1,500 or 2,000 feet, the pull on the cord is almost constant, and, if the kite is well made, it remains practically stationary in the air.

I have often noticed in high winds, that light and fleecy clouds come into view, say, about 2,000 feet above the surface of the earth, and that they pass rapidly and steadily by, preserving their shape completely. This would certainly indicate that there is no rapid local disturbance in the air in their immediate vicinity, but that the whole mass of air in which these clouds are formed is practically travelling in the same direction and at the same velocity. Numerous aeronauts have also testified that, no matter how hard the wind may be blowing, the balloon is always practically in a dead calm, and if a piece of gold-leaf is thrown overboard even in a gale, the gold-leaf and the balloon never part company in a horizontal direction, though they may in a vertical direction.

Birds may be divided into two classes: first, the soaring birds, which practically live upon the wing, and which, by some very delicate sense of touch, are able to feel the exact condition of the air. Many fish which live near the top of the water are greatly distressed by sinking too deeply, while others which live at great depths are almost instantly killed by being raised to the surface. The swim bladder of a fish is in reality a delicate barometer provided with sensitive nerves which enable the fish to feel whether it is sinking or rising in the water. With the surface fish, if the pressure becomes too great, the fish involuntarily exerts itself to rise nearer the surface and so diminish the pressure, and I have no doubt that the air-cells, which are known to be very numerous and to abound throughout the bodies of birds, are so sensitive as to enable soaring birds to know at once whether they are in an ascending or a descending column of air.

The other class of birds consists of those which only employ their wings for the purpose of taking them rapidly from one place to another. Such birds may be considered not to expend their power so economically as the soaring birds. They do not spend a very large portion of their time in the air, but what time they are on the wing they exert an immense amount of power and fly very rapidly, generally in a straight line, taking no advantage of air currents. Partridges, pheasants, wild ducks, geese, and some birds of passage may be taken as types of this kind. This class of birds has relatively small wings, and carries about 21 times as much weight per square foot of surface as soaring birds do.

[. . .]

IV. THE ADVANTAGES AND DISADVANTAGES OF VERY NARROW PLANES.

My experiments have demonstrated that relatively narrow aeroplanes lift more per square foot than very wide ones, but as an aeroplane, now matter how narrow it may be, must of necessity have some thickness, it is not advantageous to place them too near together. Suppose that aeroplanes should be made ¼ in. thick and be superposed 3 inches apart, that is, at a pitch of 3 inches. One-twelfth part of the whole space through which these planes would have to be driven would be occupied by the planes themselves, and eleven-twelfths would be air space (Fig. 1). If a group of planes thus mounted should be driven through the air at the rate of 36 miles an hour, the air would have to be driven forward at the rate of 3 miles an hour, or else it would have to be compressed, or spun out, and pass between the spaces at a speed of 39 miles an hour. As a matter of fact, however, the difference in pressure is so very small, that practically no atmospheric compression takes place. The air, therefore, is driven forward at the rate of 3 miles an hour, and this consumes a great deal of power, in fact, so much that there is a decided disadvantage in using narrow planes thus arranged.

In regard to the curvature of narrow aeroplanes, I have found that if one only desires to lift a large load in proportion to the area, the planes may be made very hollow on the underneath side; but when one considers the lift in terms of screw thrust, I find it advisable that the planes should be as thin as possible and the underneath side nearly flat. I have also found that it is a great advantage to

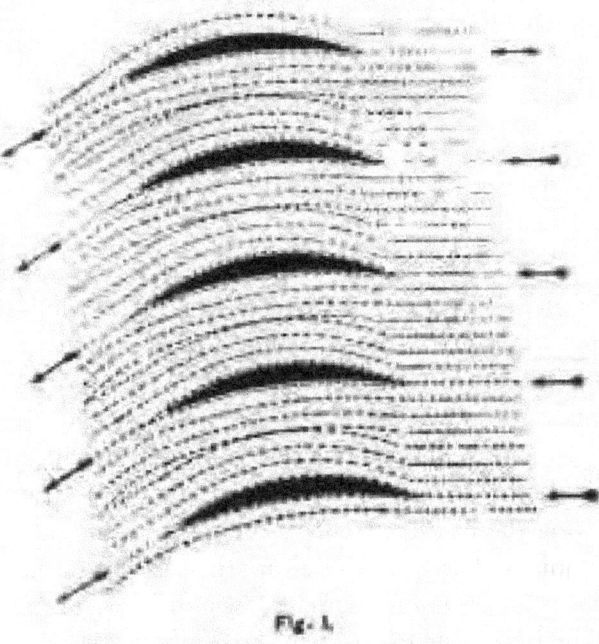

Fig. 1.

arrange the planes after the manner shown in fig. 2. In this manner, the sum of all the spaces between the planes is equal to the whole area occupied by the planes; consequently, the air neither has to be compressed, spun out, or driven forward. I am therefore by this arrangement able to produce a large lifting effect per square foot, and, at the same time, to keep the screw thrust within reasonable limits.

A large number of experiments with very narrow aeroplanes have been conducted by Mr. Horatio Phillips at Harrow, in England. . . . Mr. Phillips is of the opinion that the air in striking the top side of the plane is thrown upward in the manner shown and a partial vacuum is thereby formed over the central part of

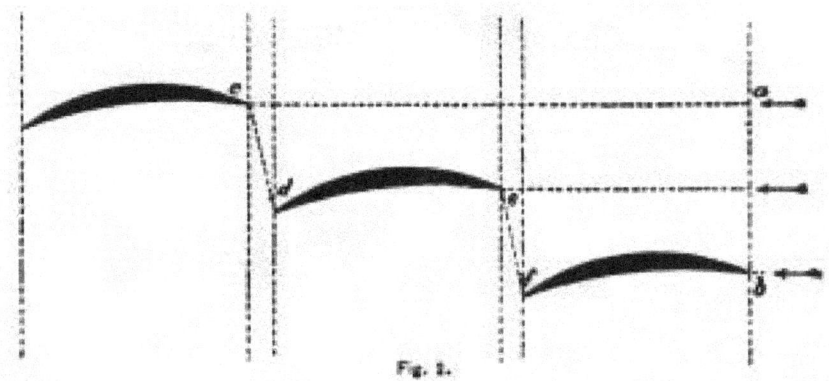

Fig. 2.

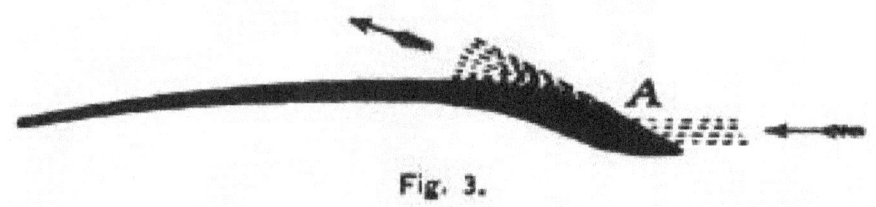

Fig. 3.

the plane, and that the lifting effect of planes made in this form is therefore very much greater than with ordinary narrow planes. I have experimented with these "sustainers" (as Mr. Phillips calls them) myself, and I find it is quite true that they lift in some cases as much as 8 lb. per sq. ft., but the lifting effect is not produced in the exact manner that Mr. Phillips seems to suppose. The air does not glance off in the manner shown. As the "sustainer" strikes the air, two currents are formed, one following the exact contour of the top and the other the bottom. These two currents join and are thrown downward as relates to the "sustainer" at an angle which is the resultant of the angles at which the two currents meet (Fig. 4). These "sustainers" may be made to lift when the front edge is lower than the rear edge because they encounter still air, and leave it with a downward motion.

In my experiments with narrow superposed planes, I have always found that with strips of thin metal made sharp at both edges and only slightly curved, the lifting effect, when considered in terms of screw thrust, was always greater than with any arrangement of the wooden aeroplanes used in Phillips' experiments. It would therefore appear that there is no advantage in the peculiar form of "sustainer" employed by this inventor.

If an aeroplane be made perfectly flat on the bottom side and convex on the top . . . and be mounted in the air so that the bottom-side is exactly horizontal, it pro-

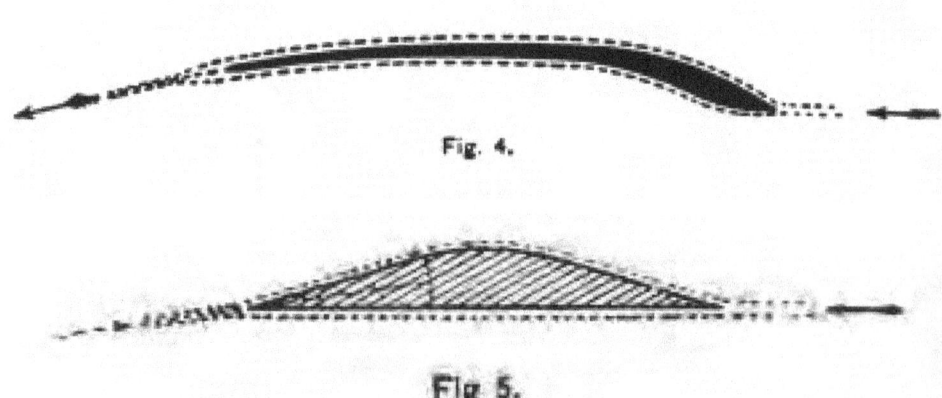

Fig. 4.

Fig 5.

duces a lifting effect no matter in which direction it is run, because as it advances it encounters stationary air which is divided into two streams. The top stream being unable to fly off at a tangent when turning over the top curve, flows down the incline and joins the current which is flowing over the lower horizontal surface. The angle at which the combined stream of air leaves the plane is the resultant of these two angles; consequently, as the plane finds the air in a stationary condition and leaves it with a downward motion, the plane itself must be lifted. It is true that small and narrow aeroplanes may be made to lift considerably more per square foot of surface than very large ones, but they do not offer the same safeguard against a rapid descent to the earth in case of a stoppage or breakdown of the machinery. With a large aeroplane properly adjusted, a rapid and destructive fall to the earth is quite impossible.

In the foregoing experiments with narrow aeroplanes, I employed an apparatus (Fig. 6) [*omitted*] which enabled me to mount my planes at any angle in a powerful blast of air, and to weigh the exact lifting effect and also the tendency to drift with the wind. This apparatus also enables me to determine with a great degree of nicety the best form of an atmospheric condenser to employ.

V. THE EFFICIENCY OF SCREW PROPELLERS.—STEERING, STABILITY, ETC.

Before I commenced my experiments at Baldwyn's Park, I attempted to obtain some information in regard to the action of screw propellers working in the air. I went to Paris and saw the apparatus which the French Government employed for testing the efficiency of screw propellers, but the propellers were so very badly made that the experiments were of no value. Upon consulting an English experimenter who had made a "lifelong study" of the question, he assured me that I should find the screw propeller very inefficient and very wasteful of power. He said that all screw propellers had a powerful fan-blower action, drawing in air at the centre and discharging it with great force at the periphery. I found that no two men were agreed as to the action of screw propellers. All the data or formulæ available were so confusing and contradictory as to be of no value whatsoever. Some experimenters were of the opinion that in computing the thrust of a screw we should only consider the projected area of the blades, and that the thrust would be equal to a wind blowing against a normal plane of equal area at a velocity equal to the slip. Others were of the opinion that the whole screw disk would have to be considered; that is, that the thrust would be equal to a wind blowing against a normal plane equal to the area of the whole disk at the velocity of the slip. The projected area of the two screw blades of my machine is 94 square feet, and the area of the 2 screw disks is 500 square feet. According to the first system of reasoning, therefore, the screw thrust of my large machine, when running at 40 miles an hour with a slip of 18 miles per hour, would have been, according to the well-known formula, $V^2 \times .005 = P$

$18^2 \times .005 \times 94 = 152.28$ pounds.

If, however, we should have considered the whole screw disk, it would have been—
$18^2 \times .005 \times 500 = 810$ pounds.

However, when the machine was run over the track at this rate, the thrust was found to be rather more than 2,000 lbs. When the machine was secured to the track and the screws revolved until the pitch in feet multiplied by the turns per minute was equal to 68 miles an hour, it was found that the screw thrust was 2,164 lbs. In this case it was of course all slip, and when the screws had been making a few turns they had established a well-defined air-current, and the power exerted by the engines was simply to maintain this air-current, and it is interesting to note that if we compute the projected area of these blades by the foregoing formula, the thrust would be—

$68^2 \times .005 \times 94 = 2173.28$ pounds,

which is almost exactly the observed screw thrust. From this, it would appear when the machine is stationary, and all the power is consumed in slip, that only the projected area of the screw blades should be considered. But whenever the machine is allowed to advance, and to encounter new air, the inertia of which has not been disturbed, the efficiency increases in geometrical progression. The exact rate for all speeds I have not yet ascertained. My experiments have, however, shown that with a speed of 40 miles an hour and a screw slip of 18 miles an hour, a well-made screw propeller is 13.1 times as efficient as early experimenters had supposed and attempted to prove by elaborate formulæ.

When I first commenced my experiments with a large machine, I did not know exactly what form of boiler, gas generator, or burner I should finally adopt; I did not know the exact size that it would be necessary to make my engines; I did not know the size, the pitch, or the diameter of the screws which would be the most advantageous. Neither did I know the form of aeroplane which I should finally adopt. It was therefore necessary for me to make the foundation or platform of my machine of such a character that it would allow me to make the modifications necessary to arrive at the best results. The platform of the machine is therefore rather larger than is necessary, and I find if I were to design a completely new machine, that it would be possible to greatly reduce the weight of the framework, and, what is still more, to greatly reduce the force necessary to drive it through the air.

At the present time, the body of my machine is a large platform, about 8 ft. wide and 40 ft. long. Each side is formed of very strong trusses of steel tubes, braced in every direction by strong steel wires. The trusses which give stiffness to this superstructure are all below the platform. In designing a new machine, I should make the trusses much deeper and at the same time very much lighter, and, instead of having them below the platform on which the boiler is situated, I should have them constructed in such a manner as to completely enclose the boiler and the greater part of the machinery. I should make the cross-section of the framework rectangular, and pointed at each end. I should cover the outside very carefully with balloon material, giving it a perfectly smooth and even surface throughout, so that it might be easily driven through the air.

In regard to the screws, I am at the present time able to mount screws 17 ft. 10 in. in diameter. I find, however, that my machine would be much more efficient if the screws were 24 feet in diameter, and I believe with such very large screws, four blades would be much more efficient than two.

My machine may be steered to the right or to the left by running one of the propellers faster than the other. Very convenient throttle valves have been provided to facilitate this system of steering. An ordinary vertical rudder placed just after the screws may, however, prove more convenient, if not more efficient.

The machine is provided with fore and aft horizontal rudders, both of which are connected with the same windlass. If the forward rudder is placed at an angle considerably greater than that of the main aeroplane, and the rear rudder placed flat so as not to lift at all (Fig. 7) and the machine run over the track at a high speed, the front wheels will be lifted from the steel rails, leaving the rear wheels on the rails. If the rudders are placed in the reverse position so that the front rudder is thrown out of action, and the rear rudder lifts to its full extent (Fig. 8), the hind wheels will be lifted from the steel rails, leaving only the forward wheels touching. If both rudders are placed at such an angle that they both lift (Fig. 9), and the machine is run at a very high velocity, all four of the wheels will be lifted from the steel rails. This would seem to show that these rudders are efficient as far as vertical steering is concerned. If the machine should break down in the air it would be necessary to tilt the rudders in the position shown . . . when it would fall to the ground without pitching or diving.

In regard to the stability of the machine, the centre of weight is much below the centre of lifting effect; moreover, the upper wings are set at such an angle that

Fig. 7.—The forward wheels off the track.

Fig. 8. — The rear wheels off the track.

Fig. 9. — All the wheels off the track.

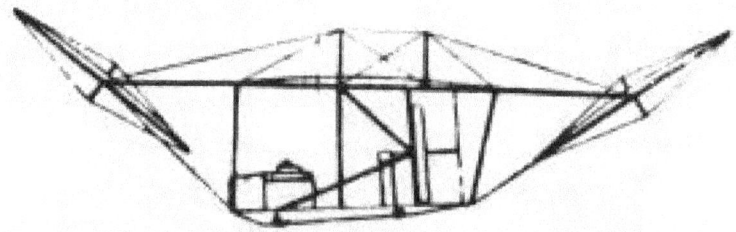

Fig. 10. — Showing the manner of placing the fore and aft rudders in case of a breakage of the machinery.

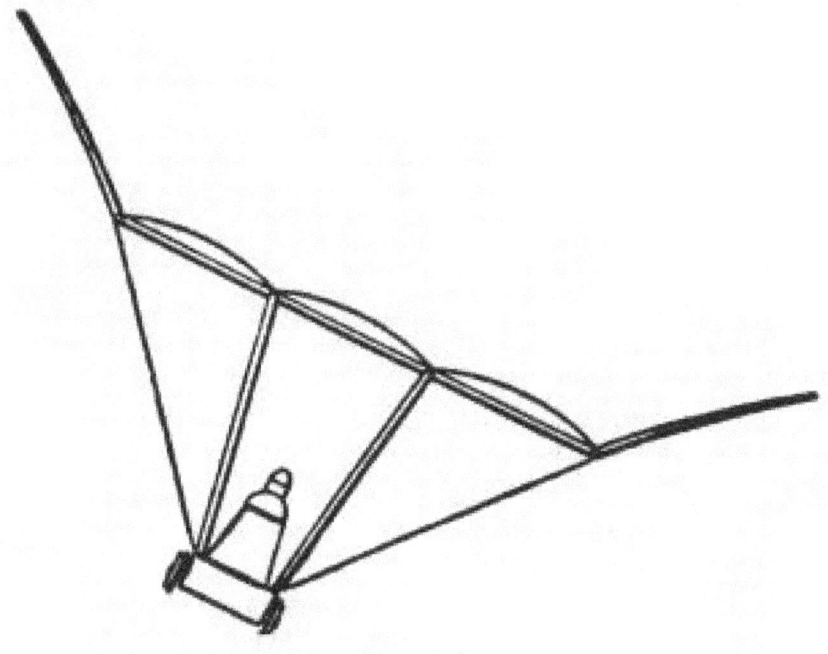

Fig. II.

whenever the machine tilts to the right or to the left, the lifting effect is increased on the lower side and diminished on the higher side (Fig. 11). This simple arrangement makes the machine automatic as far as rolling is concerned. I am of the opinion that whenever flying machines come into use it will be necessary to steer them in a vertical direction by means of an automatic steering gear controlled by a gyroscope. It will certainly not be more difficult to manœuvre and steer such machines than it is to control completely submerged torpedoes.

When the machine is once perfected, it will not require a railway track to enable it to get the necessary velocity to rise. A short run over a moderately level field will suffice. As far as landing is concerned, the aerial navigator will touch the ground while moving forward, and the machine will be brought to a state of rest by sliding on the ground for a short distance. In this manner very little shock will result, whereas if the machine is stopped in the air and allowed to fall directly to the earth without advancing, the shock, although not strong enough to be dangerous to life or limb, might be sufficient to disarrange or injure the machinery.

[. . .]

VII. CONCLUSION.

My large machine, which was injured in my late experiments, has now been repaired and improved, and is quite ready to be used in any other experiments which I may wish to make on the limited area which I now have at my disposal. The railway track on which my experiments have been made is 1,800 feet long and the land on all sides is thickly studded with large trees. When making experiments about 500 feet of the track is used in getting up the necessary speed and 300 feet is utilized in bringing the machine again to a state of rest. My clear run is therefore limited to 1,000 feet, and the time which the machine takes to pass over this length of rail is at the most only a few seconds. It will therefore be seen that it is not an easy matter to conduct experiments in a satisfactory manner. In addition to these experiments with a large machine, I am also conducting a series of experiments in a blast of air issuing from a trunk 3 feet square. The air is set in motion by the action of screw propellers driven by a steam engine of 60 horsepower, and I am able to obtain any atmospheric velocity that I require, from 5 to 90 miles an hour. This apparatus . . . is constructed in such a manner that it enables me to mount in this current of air any object that I wish to experiment with. For instance, a bar of wood 3 inches square is mounted in the blast of air so that one of its sides forms a normal plane perpendicular to the direction of the blast. The engine is then run until the air is passing through the trunk at a velocity of 50 miles an hour. The tendency of this bar of wood to travel in the direction of the air may then be accurately determined, and this is considered as unity. A cylinder exactly 3 inches in diameter may then be mounted and tested in the same manner. The cylinder will of course have less tendency to travel with the air than the square bar of wood, and whatever this tendency is, will be the coefficient of a cylinder. I have provided oval, elliptical, and various other shaped objects to be experimented with, and when the experiments are finished I shall know the exact coefficient of all shapes that it may be practical to use in the framework of a flying machine, and also what effect is produced by placing two or more bodies in close proximity to each other.

In addition to these experiments, I am also able with the same air blast to ascertain the efficiency of various forms of aeroplanes, superposed or otherwise, and placed at all angles, the apparatus being provided with a scale beam which not only enables me to measure the drift, but also to accurately weigh the lifting effect. The aeroplane, or grouping of aeroplanes, in which the drift will go the greatest number of times into the lift will be considered the most satisfactory for the purpose.

Experiments are also being made in the same air blast with a view of ascertaining the condensing and lifting power of various forms of tubes, steam in the condition of exhaust being passed through the tubes while the air is driven between them at any velocity required. The experiments are being made with pure steam and also with steam contaminated with oil, with a view of ascertaining to what extent the efficiency of the condenser is reduced by a film of oil such as may be expected from exhaust steam. These experiments will enable me to ascertain very exactly the weight and the efficiency of atmospheric condensers, the amount that their tubes may be made to lift at various speeds and atmospheric conditions, and will also enable me to select the form which I find most suitable for the purpose.

In navigating a boat, it is only necessary that one should be able to turn it to the right or to the left (port or starboard), but with a flying machine it is not only necessary to steer it to the right or left (horizontally), but also in a vertical direction to prevent it from rearing up forward or pitching, and this, if it is accomplished by hand, will require the constant vigilance of a man at the wheel who can make observations, think, and act instantly. In order to prevent a too rapid up and down deviation of the machine I have constructed it of great length, so that the man at the helm will have more time to think and act. As before stated, however, I am of the opinion that the steering in a vertical direction should be automatically controlled by a gyroscope, and I have made an apparatus which consists of a steam piston acting directly upon the fore and aft rudders, the steam valve being controlled by a gyroscope. As the rudders are moved by the steam, their movement shuts the steam off in exactly the same manner that the moving of a rudder shuts off the steam in the well-known steam-steering apparatus now universally in use on all large steamers.

Now that it is definitely known that it is possible to construct a large machine which is light enough and at the same time powerful enough to raise its own weight and that of its engineers into the air, the next question which presents itself for solution is to ascertain how to steer and control such a machine when actually free from the earth. When it is considered that the machine is of great size and that it is necessary that it should move through the air at a velocity of at least 35 miles an hour in order to leave the ground, it will be obvious that manœuvring experiments cannot be conducted in a circumscribed place such as I now have. It is therefore necessary for me to obtain new and much larger premises where I shall have a very large and level field at my disposal. It is not an easy matter to obtain a field of this character in England, and it is almost impossible to find a suitable place near London. Moreover, experiments of this character, which are of little value unless conducted on a large scale, are exceedingly expensive, in fact, too expensive to be conducted by private individuals. Nevertheless, as my experiments have shown most conclusively that flying machines are not only possible but practicable, I think I am justified in continuing my experiments until a comparatively perfect flying machine has been evolved. When I have obtained possession of a suitable field, I propose to erect a large building which will contain the machine with all its wings in position. The building which I have at present, notwithstanding that it cost $15,000, is not large enough for the purpose, as the wings all have to be taken off before the machine can be housed.

There are so many points that may be improved that I have determined to build a new machine on a somewhat smaller scale, using about 200 or 250 horsepower. I shall make the engines of a longer stroke in proportion to their diameter so as to get a greater piston speed. I shall construct my screw propellers with 4 long and narrow blades, very sharp and thin, and shall make them large enough so that the pressure on the projected area of the blades will be about 10 lbs. per square foot instead of over 20 lbs. as now. This will greatly reduce the waste of power which is now lost in screw slip. As the present boiler has been found larger than is necessary, mv next boiler will be made lighter and smaller, and instead carrying a pressure of 320 lbs. to the square inch, I shall only carry 275 lbs. But the greatest improvement will be made in the framework of the machine, which will be constructed with a view of enabling everything to be driven through the air with the least possible resistance. The main aeroplane will be the same form as now, but placed at an angle of 1 in 113 instead of 1 in 8, and will be used principally for preventing the machine from accidentally falling to the earth. The principal lifting effect will be derived from a considerable number of relatively narrow aeroplanes placed on each side of the machine and mounted in such a position that the air can pass freely between them. The fore and aft rudders will be the same form as those now employed. The condenser will consist of a large number of small hollow aeroplanes about 2 inches wide, made of very thin and light metal and placed immediately behind the screw propellers. They will be placed at such an angle as to lift about 1,000 pounds in addition to their weight and the weight of their contents. Instead of mounting my machine as now on 4 wheels, I propose to mount it on 3, the two hind wheels being about 40 feet apart and the forward wheel placed about 60 feet in front of these. I propose to lay down a track of 3 rails, the sleepers being embedded in the ground so as to produce a comparatively level surface. This railway track should be oval or circular in form so that the machine may be heavily weighed to keep it on the track and be run at a high speed. This will enable me to test the furnace draught, the burner, the steam, the boiler, the engines, the propelling effects of tile screws, and the efficiency of the condenser while the machine is on the ground.

When all the machinery has been made to run smoothly I shall remove all the weight except that directly over the front wheel, and shall place a device between the wheel and the machine that will indicate the lift on the front end of the machine. I shall then run the machine over the track at a velocity which will just barely lift the hind wheels off the track, leaving the front wheel on the track. If the rear end of the machine lifts into the air it will change the angle of the planes and the lifting effect will be correspondingly diminished. This will prevent rising too high. Special wheels with a wide face suitable for running on either the rails or the earth will be provided for the purpose, and when I find that I can keep the hind wheels in the air and produce a varying lifting effect above and below the normal weight resting on the front wheel, I shall remove the weight from the forward wheel and attempt free flight by running the machine as near the ground as possible, making the first attempt by running against the wind, and it will only be after I find that I can steer my machine and manage it within a few feet of the earth, ascend and descend again at will, that I shall attempt high flight.

My experiments have certainly demonstrated that a steam engine and boiler may be made which will generate a horsepower for every six pounds of weight, and that the whole motor, including the gas generator, the water supply, the condenser, and the pumps may be all made to come inside of 11 lbs. to the horsepower. They also show that well made screw propellers working in the air are fairly efficient, and that they obtain a sufficient grip upon the air to drive the machine forward at a high velocity; that very large aeroplanes, if well made and placed at a proper angle, will lift as much as 2 ½ lbs. per square foot at a velocity not greater than 40 miles an hour; also that it is possible for a machine to be made so light and at the same time so powerful that it will lift not only its own weight but a considerable amount besides, with no other energy except that derived from its own engines. Therefore there can be no question but what a flying machine is now possible without the aid of a balloon in any form.

In order to obtain these results it has been necessary for me to make a great number of expensive experiments and to carefully study many of the properties of the air. Both Lord Kelvin and Lord Rayleigh, after witnessing a series of my experiments, expressed themselves as of the opinion that all the mathematical formulæ relating to planes driven through the air at an angle would have to be completely modified. Lord Kelvin himself has written that in some cases my experiments have proved that the conditions were from 20 to 50 times as favorable to the aerial navigator as had heretofore been shown by accepted formulæ, and that the whole mathematical question would require revision.

Experiments of this character unless conducted with great care are exceedingly dangerous. No makeshift or imperfect apparatus should be employed, but the experimenter should have the advantage of the most perfect appliances and apparatus that modern civilization can afford. The necessary plant for conducting experiments in a proper and safe manner is unfortunately much more expensive than the machine itself. If I find that my experiments require more money than I have at my disposal, I feel sure that some future experimenter more fortunate than myself will commence where I leave off, and with the advantages of the knowledge which has been gained by recent experiments will be able to construct a practical flying machine which cannot fail to be a great advantage to mankind.

The numerous and very expensive experiments, conducted on an unprecedented scale, which have made all this possible, and also brought to light new laws relating to the atmosphere, cannot fail to be of the greatest value to mankind, and it is on this basis that I submit the foregoing thesis.

Document 1-7

B. Baden-Powell (Baden Fletcher Smyth), "Present State of Aeronautics," interview in *The Sunday Times* [London], 14 June 1896, reprinted in *The Aeronautical Journal* (January 1897): 4–5.

Major Baden Fletcher Smyth Baden-Powell, a career officer in the British army and brother of the founder of the Boy Scouts (Robert Stephenson Smyth Baden-Powell, first baron Baden-Powell of Gilwell), joined the Aeronautical Society of Great Britain in 1880, fourteen years after it was established. Although his technical abilities were limited, his inventive spirit and enthusiasm for flight were perhaps unmatched. As secretary of the Aeronautical Society (elected in 1897), and later its president (elected while absent on duty in South Africa), B. Baden-Powell closely followed aviation developments internationally, corresponding regularly with Octave Chanute, in America, and many others.

In this document from 1897, Baden-Powell told a reporter (in a 14 June 1896 interview) from the *Sunday Times* of London what he thought about the "present state of aeronautics." In his opinion, though there were many exaggerated reports of aviation firsts, human flight "will come about, and very soon." In the interview, Baden-Powell reported on the work of Langley, Maxim, Lilienthal, as well as that of fellow Englishman Percy Pilcher, and he also described his own work with man-carrying kites. He was also the first in England to call attention to the work of the Wright brothers in his presidential address to the Aeronautical Society of Great Britain in December 1902, one year before the Wright's historic achievement of powered flight. By 1904, Baden-Powell was corresponding with the Wrights. He spent quite a bit of time with Wilbur during the Wright's tour of France in the fall of 1908 and published the article "A Trip with Wilbur Wright" in the December 1908 issue of *Aeronautics* magazine.

Interestingly, in his 1896 interview Baden-Powell saw the predominant value of flying machines resting in military applications: "It is, of course, in war that its great importance will be apparent." This runs contrary to the peace-bringing theme emphasized by the Wrights in the early years after their invention of the airplane.

Document 1-7, B. Baden-Powell, "Present State of Aeronautics," 1896.

The invention of a flying machine, or rather the announcement of a sanguine inventor who believes he has accomplished this, must be to the editor of a paper almost as valuable as the appearance, or reported appearance, of a sea serpent.

Again and again we see paragraphs in the newspapers announcing some marvellous advancement towards the solution of this intricate problem, but as often as not, or a great deal oftener, the invention proves to be merely a design, or an idea, formed by one who has but little real knowledge of the subject. During the last few years, however, several men of undoubted scientific ability and inventive genius have entered the arena, and have proved, both by figures and by experiment, that a flying machine is a possibility, and has come within the range of practical invention. The careful scientist is, however, just the man who does not, as a rule, announce his device and publish full descriptions of it until he is fully convinced that he is on the right track. It is, for this reason, rather difficult to get at the truth as regards the present position of the subject. It will, therefore, be unnecessary for me to refer to the many vague reports one has seen in the papers, except as regards those I know something about. Some rumours have latterly come out about a machine devised by Professor Langley, of the Smithsonian Institute, at Washington. He is, I know, a most capable man, for he has written a book, "Aerodynamics," which, without doubt, is by far the most elaborate scientific study of the question that has been written. He has for years made the most careful experiments, and has deduced many facts with reference to the action of the air on bodies moving through it of the greatest importance to inventors of aerial machines. He seems now to have made some small machine which is reported to be a great success, though I imagine it is nothing more than a model. No full description of it has, I believe, been published, and in all probability the inventor will not hurry to make common property that which has taken him so many years of study and experiment to evolve, at any events, until he has progressed sufficiently to be sure of success.

Then there is the great Maxim machine. Here again we have a very capable and very careful inventor devoting his attention and his money to the subject. The result has been a truly marvellous machine. Doubtless the inventor himself expected better results than it has proved capable of supplying, but nevertheless, though the machine can scarcely be said to have left the earth, yet it has proved a great success as regards the construction of very large and very light apparatus provided with immensely powerful engines. No one, before this was constructed, could have believed in the practicability of making a steam engine of 300 horse power, and placing it in a flying machine presenting some thousands of square feet of surface to the air, the whole apparatus only weighing some three tons.

However, the affair was altogether too large and unwieldy. A slight gust of wind getting under this enormous awning would be apt to upset the whole thing, and do hundreds of pounds' worth of damage. Probably if the machine had really been shot off into midair it would have gone all right, but there was an awkward uncertainty as to how it might come down! I think myself it ought to have been supplied with air bladders, and sent over a cliff out to sea. Then when it fell in the water it would have floated. Now, however, Mr. Maxim talks of constructing a much smatter machine on the same lines, and I shall certainly look forward with the greatest interest for the result, as I believe it will come very near the goal we are searching for.

This machine, and probably Langley's as well, is on what is known as the aeroplane principle; that is, a large fixed plane presenting a small angle to the air is propelled horizontally at a great speed by means of screw propellers, and is forced upwards by the pressure of the air on its under side. It is really simply a large kite, which, instead of being held by a string for the wind to act on it, is driven through the air, and raised much as a kite is when drawn along by its string. This principle has been proved theoretically to give the best results, that is to say, that with a given amount of power a greater weight can be raised by this system of propelling a plane surface presenting a small upward angle to the horizontal. Lord Kelvin and others have, however, pointed out that, notwithstanding this, a machine with vertical screw propellers to lift it straight upwards might have better practical results, even though more power might be required.

Then a good deal of talk is being made about two other inventors, Lilienthal, in Germany, and Pilcher, in England. These machines are very similar in principle, and though of great interest as experimental apparatus, cannot at present be considered as much more than parachutes. They partake of the nature of large wings attached to the human body, and are, therefore, comparatively well under control, but are only capable of "flying" if launched from some high place against, the wind; both inventors however, propose to apply small engines to propel them through the air, and success depends upon whether they can get sufficiently powerful engines without greatly increasing the size of the apparatus, and if they can get sufficiently efficient propellers to drive them at the requisite speed.

You ask also about the Italian machines. All I know is that reports have appeared of the Italian Government being in possession of a number of "air-ships"—and all I can say is that until I hear more of the details I don't believe a word of it.

Navigable balloons I look upon as a different subject. We know, of course, that the French have some, and, as to other nations, my mouth is closed as regards all I know in this line, but I may say at once that though on certain particular occasions, with favourable, i.e., calm weather, they may prove of the very greatest value in war, yet they are so dependent on the absence of wind and other circumstances that I cannot believe they will ever be very much used.

Then as for my own apparatus. Well, that again, is quite another matter. It is not a flying machine in the ordinary acceptance of the term. It is simply a system of large kites only for captive use. That is to say, it is held by a rope to the ground, and is chiefly for use as a lofty observatory for looking over the enemy's lines, and watching the country round about. I believe there are also other useful purposes to which it may be applied, since the length of the tether line is practically unlimited, but I need not speculate on future possibilities. My present object is to get an apparatus to serve instead of a captive balloon without necessitating the transport of cumbersome filling apparatus.

Kite-flying is not so simple a matter as some may suppose. To make an enormous kite like a toy one would be no easy task. Remember, the tail should be as heavy as the kite itself, which would, in my case, amount to some 60 lbs. or more. And then the liability to dive must be done away with, else I am afraid there might be a difficulty in providing observers. My present apparatus, which is now undergoing its trials at Aldershot, folds up so compactly that it call be carried by one man. It is capable of lifting a man to a height of 200 or 300 ft. One thing may be said about it that, besides the balloon, it is the only means known by which a man has been absolutely raised to any height in the air. And it is not entirely dependent on wind, since in calms it may be made to ascend by towing it with horses.

As for the general question of the possibilities of human flight, I for one firmly believe it will come about, and that very soon. It is, of course, in war that its great importance will be apparent. And if it be found necessary to have a large, complicated, and costly machine to effect the purpose, doubtless it will be used for little else. The first nation which can provide a secret flying-machine—and the secret will probably be more in the smaller details of construction than in the general principle—will undoubtedly possess an incalculable advantage in war.

Document 1-8(a–c)

(a) Samuel P. Langley, "Story of Experiments in Mechanical Flight," *Aeronautical Annual* (1897), reprinted in *Researches and Experiments in Aerial Navigation* (Washington, DC: Government Printing Office, 1908), pp. 169–179.

(b) Samuel P. Langley, excerpts from "The New Flying Machine." *McClure's Magazine* 9, No. 2 (June 1897): 647–660.

(c) Alexander Graham Bell, letter in *McClure's Magazine* 9, No. 2, (June 1897): 659; published as "The Aerodromes in Flight," *Aeronautical Annual* (1897), pp. 140–141.

Dr. Samuel P. Langley's work epitomized the nineteenth century "scale model" development approach to solving the problems of flight. As a practical exercise in applied aerodynamics, however, Langley's experiments were plagued with failure. Finally, in 1897, he achieved a signal success with the test flights of his powered model *Aerodrome No. 5*. The following three documents present the story of his experiments with flying scale models. The first is a selection from his 1896 interim report, which he referred to as a "narrative account of my work in aerodromics." A copy of this article was included in the documents sent to Wilbur Wright in 1899 by the Smithsonian Institution.

Regarding this and other reports by Langley, Wright later noted that, "his accounts of the troubles he had encountered and overcome put us on our guard and enabled us to entirely avoid some of the worst of them. He painted so vividly the troubles resulting from excessive lightness that we have been as men vaccinated against that disease."

Langley's success in 1896 with *Aerodrome No. 5* is recounted in an excerpt from an article he published in 1897 on "The New Flying Machine," along with the report written by his good friend and supporter Alexander Graham Bell, who was present to observe the experiments and took photographs of the model in flight.

Following his success with *Aerodrome No. 5* Langley attempted to progress to a full-scale manned model, his *Great Aerodrome*. The widely reported failure of this flying machine in December 1903, mere days in advance of the Wrights success at Kitty Hawk, led to Langley's disgrace and public rejection. Langley died within three years of this final crushing blow, but his few remaining partisans, including Bell, continued for some time thereafter to tout the potential of Langley's Aerodrome.

Document 1-8(a), Samuel P. Langley, "Story of Experiments in Mechanical Flight," 1897.

The subject of flight interested me as long ago as I can remember anything, but it was a communication from Mr. Lancaster, read at the Buffalo meeting of the American Association for the Advancement of Science, in 1886, which aroused my then dormant attention to the subject. What he said contained some remarkable but apparently mainly veracious observations on the soaring bird, and some more or less paradoxical assertions, which caused his communication to be treated with less consideration than it might otherwise have deserved. Among the latter was a statement that a model, somewhat resembling a soaring bird, wholly inert, and without any internal power, could, nevertheless, under some circumstances, advance against the wind without falling; which seemed to me then, as it did to members of the Association, an utter impossibility, but which I have since seen reason to believe is, within limited conditions, theoretically possible.

I was then engaged in the study of astrophysics at the Observatory in Allegheny, Pa. The subject of mechanical flight could not be said at that time to possess any literature, unless it were the publications of the French and English aeronautical societies, but in these, as in everything then accessible, fact had not yet always been discriminated from fancy. Outside of these, almost everything was even less trustworthy; but though, after I had experimentally demonstrated certain facts, anticipations of them were found by others on historical research, and though we can now distinguish in retrospective examination what would have been useful to the investigator if he had known it to be true, there was no test of the kind to apply at the time. I went to work, then, to find out for myself, and in my own way, what amount of mechanical power was requisite to sustain a given weight in the air and make it advance at a given speed, for this seemed to be an inquiry which must necessarily precede any attempt at mechanical flight, which was the very remote aim of my efforts.

The work was commenced in the beginning of 1887 by the construction, at Allegheny, of a turntable of exceptional size, driven by a steam engine, and this was used during three years in making the "Experiments in Aerodynamics," which were published by the Smithsonian Institution under that title in 1891. Nearly all the conclusions reached were the result of direct experiment in an investigation which aimed to take nothing on trust. Few of them were then familiar, though they have since become so, and in this respect knowledge has advanced so rapidly, that statements which were treated as paradoxical on my first enunciation of them are now admitted truisms.

It has taken me, indeed, but a few years to pass through the period when the observer hears that his alleged observation was a mistake; the period when he is told that if it were true, it would be useless; and the period when he is told that it is undoubtedly true, but that it has always been known.

May I quote from the introduction to this book what was said in 1891?

"I have now been engaged since the beginning of the year 1887 in experiments on an extended scale for determining the possibilities of, and the conditions for, transporting in the air a body whose specific gravity is greater than that of the air, and I desire to repeat my conviction that the obstacles in its way are not such as have been thought; that they lie more in such apparently secondary difficulties as those of guiding the body so that it may move in the direction desired and ascend or descend with safety, than in what may appear to be primary difficulties, due to the air itself," and, I added, that in this field of research I thought that we were, at that time (only six years since), "in a relatively less advanced condition than the study of steam was before the time of Newcomen." It was also stated that the most important inference from those experiments as a whole was that mechanical flight was possible with engines we could then build, as one horsepower rightly applied could sustain over 200 pounds in the air at a horizontal velocity of somewhat over 60 feet a second.

As this statement has been misconstrued, let me point out that it refers to surfaces, used without guys, or other adjuncts, which would create friction; that the horsepower in question is that actually expended in the thrust, and that it is predicated only on a rigorously horizontal flight. This implies a large deduction from the power in the actual machine, where the brake horsepower of the engine, after a requisite allowance for loss in transmission to the propellers, and for their slip on the air, will probably be reduced to from one-half to one-quarter of its nominal amount; where there is great friction from the enforced use of guys and other adjuncts; but, above all, where there is no way to insure absolutely horizontal flight in free air. All these things allowed for, however, since it seemed to me possible to provide an engine which should give a horsepower for something like 10 pounds of weight, there was still enough to justify the statement that we possessed in the steam engine, as then constructed or in other heat engines, more than the indispensable power, though it was added that this was not asserting that a system of supporting surfaces could be securely guided through the air or safely brought to the ground, and that these and like considerations were of quite another order, and belonged to some inchoate art which I might provisionally call *aerdromics*.

These important conclusions were reached before the actual publication of the volume, and a little later others on the nature of the movements of air, which were published under the title of "The Internal Work of the Wind" (Smithsonian Contributions to Knowledge, Volume XXVII, 1893, No. 884). The latter were founded on experiments independent of the former, and which led to certain theoretical conclusions unverified in practice. Among the most striking, and perhaps paradoxical of these, was that a suitably disposed free body might, under certain conditions, be sustained in an ordinary wind, and even advance against it without the expenditure of any energy from within.

The first stage of the investigation was now over, so far as that I had satisfied myself that mechanical flight was possible with the power we could hope to command, if only the art of directing that power could be acquired.

The second stage (that of the acquisition of this art) I now decided to take up. It may not be out of place to recall that at this time, only six years ago, a great many scientific men treated the whole subject with entire indifference, as unworthy of attention, or as outside of legitimate research, the proper field for the charlatan, and one on which it was scarcely prudent for a man with a reputation to lose to enter.

The record of my attempts to acquire the art of flight may commence with the year 1889, when I procured a stuffed frigate bird, a California condor, and an albatross, and attempted to move them upon the whirling table at Allegheny. The experiments were very imperfect and the records are unfortunately lost, but the important conclusion to which they led was that a stuffed bird could not be made to soar except at speeds which were unquestionably very much greater than what served to sustain the living one, and the earliest experiments and all subsequent ones with actually flying models have shown that thus far we cannot carry nearly the weights which Nature does to a given sustaining surface without a power much greater than she employs. At the time these experiments were begun, Pénaud['s] ingenious but toy-like model was the only thing which could sustain itself in the air for even a few seconds, and calculations founded upon its performance sustained the conclusion that the amount of power required in actual free flight was far greater than that demanded by the theoretical enunciation. In order to learn under what conditions the aerodrome should be balanced for horizontal flight, I constructed over 30 modifications of the rubber-driven model, and spent many months in endeavoring from these to ascertain the laws of "balancing"; that is, of stability leading to horizontal flight. Most of these models had two propellers, and it was extremely difficult to build them light and strong enough. Some of them had superposed wings; some of them curved and some plane wings; in some the propellers were side by side; in others one propeller was at the front and the other at the rear, and so every variety of treatment was employed, but all were at first too heavy, and only those flew successfully which had from 3 to 4 feet of sustaining surface to a pound of weight, a proportion which is far greater than Nature employs in the soaring bird, where in some cases less than half a foot of sustaining surface is used to a pound. It had been shown in the "Experiments in Aerodynamics" that the center of pressure on an inclined plane advancing was not at the center of the figure, but much in front of it, and this knowledge was at first nearly all I possessed in balancing these early aerodromes. Even in the beginning, also, I met remarkable difficulty in throwing them into the air, and devised numerous forms of launching apparatus which were all failures, and it was necessary to keep the construction on so small a scale that they could be cast from the hand.

The earliest actual flights with these were extremely irregular and brief, lasting only from three to four seconds. They were made at Allegheny in March, 1891, but these and all subsequent ones were so erratic and so short that it was possible to learn very little from them. Pénaud states that he once obtained a flight of 13 seconds. I never got as much as this, but ordinarily little more than half as much, and came to the conclusion that in order to learn the art of mechanical flight it was necessary to have a model which would keep in the air for at any rate a longer period than these, and move more steadily. Rubber twisted in the way that Pénaud used it will practically give about 300 foot-pounds to a pound of weight, and at least as much must be allowed for the weight of the frame on which the rubber is strained. Twenty pounds of rubber and frame, then, would give 3,000 foot-pounds, or 1 horsepower for less than six seconds. A steam engine having apparatus for condensing its steam, weighing in all 10 pounds, and carrying 10 pounds of fuel, would possess in this fuel, supposing that but one-tenth of its theoretical capacity is utilized, many thousand times the power of an equal weight of rubber, or at least 1 horsepower for some hours. Provided the steam could be condensed and the water reused, then the advantage of the steam over the spring motor was enormous, even in a model constructed only for the purpose of study. But the construction of a steam-driven aerodrome was too formidable a task to be undertaken lightly, and I examined the capacities of condensed air, carbonic-acid gas, of various applications of electricity, whether in the primary or storage battery, of hot-water engines, of inertia motors, of the gas engine, and of still other material. The gas engine promised best of all in theory, but it was not yet developed in a suitable form. The steam engine, as being an apparently familiar construction, promised best in practice, but in taking it up, I, to my cost, learned that in the special application to be made of it, little was really familiar and everything had to be learned by experiment. I had myself no previous knowledge of steam engineering, nor any assistants other than the very capable workmen employed. I well remember my difficulties over the first aerodrome (No. 0), when everything, not only the engine, but the boilers which were to supply it, the furnaces which were to heat it, the propellers which were to advance it, the hull which was to hold all these—were all things to be originated, in a construction which, as far as I knew, had never yet been undertaken by anyone.

It was necessary to make a beginning, however, and a compound engine was planned which, when completed, weighed about 4 pounds, and which could develop rather over a horsepower with 60 pounds of steam, which it was expected could be furnished by a series of tubular boilers arranged in "bee-hive" form and the whole was to be contained in a hull about 5 feet in length and 10 inches in diameter. This hull was, as in the construction of a ship, to carry all adjuncts. In front of it projected a steel rod, or bowsprit, about its own length, and one still longer behind. The engines rotated two propellers, each about 30 inches in diameter, which were on the end of long shafts disposed at an acute angle to each other and actuated by a single gear driven from the engine. A single pair of large wings contained about 50 square feet, and a smaller one in the rear about half as much, or in all some 75 feet, of sustaining surface, for a weight which it was expected would not exceed 25 pounds.

Although this aerodrome was in every way a disappointment, its failure taught a great many useful lessons. It had been built on the large scale described, with very little knowledge of how it was to be launched into the air, but the construction developed the fact that it was not likely to be launched at all, since there was a constant gain in weight over the estimate at each step, and when the boilers were completed it was found that they gave less than one-half the necessary steam, owing chiefly to the inability to keep up a proper fire. The wings yielded so as to be entirely deformed under a slight pressure of the air, and it was impossible to make them stronger without making them heavier, where the weight was already prohibitory. The engines could not transmit even what feeble power they furnished, without dangerous tremor in the long shafts, and there were other difficulties. When the whole approached completion, it was found to weigh nearer 50 pounds than 25, to develop only about one-half the estimated horsepower at the brake, to be radically weak in construction, owing to the yielding of the hull, and to be, in short, clearly a hopeless case.

The first steam-driven aerodrome had, then, proved a failure, and I reverted during the remainder of the year to simpler plans, among them one of an elementary gasoline engine.

I may mention that I was favored with an invitation from Mr. Maxim to see his great flying-machine at Bexley, in Kent, where I was greatly impressed with the engineering skill shown in its construction, but I found the general design incompatible with the conclusions that I had reached by experiments with small models, particularly as to what seemed to me advisable in the carrying of the center of gravity as high as was possible with safety.

In 1892 another aerodrome (No. 1), which was to be used with carbonic acid gas, or with compressed air, was commenced. The weight of this aerodrome was a little over $4\frac{1}{2}$ pounds, and the area of the supporting surfaces $6\frac{1}{2}$ square feet. The engines developed but a small fraction of a horsepower, and they were able to give a dead lift of only about one-tenth of the weight of the aerodrome, giving relatively less power to weight than that obtained in the large aerodrome already condemned.

Toward the close of this year was taken up the more careful study of the position of the center of gravity with reference to the line of thrust from the propellers, and to the center of pressure. The center of gravity was carried as high as was consistent with safety, the propellers being placed so high, with reference to the supporting wings, that the intake of air was partly from above and partly from below these latter. The lifting power (i.e., the dead lift) of the aerodromes was determined in the shop by a very useful contrivance which I have called the "pendulum," which consists of a large pendulum which rests on knife edges, but is prolonged above the points of support, and counterbalanced so as to present a condition of indifferent equilibrium. Near the lower end of this pendulum the aerodrome is suspended, and when power is applied to it, the reaction of the propellers lifts the pendulum through a certain angle. If the line of thrust passes through the center of gravity, it will be seen that the sine of this angle will be the fraction of the weight lifted, and thus the dead-lift power of the engines becomes known. Another aerodrome was built, but both, however constructed, were shown by this pendulum test to have insufficient power, and the year closed with disappointment.

Aerodrome No. 3 was of stronger and better construction, and the propellers, which before this had been mounted on shafts inclined to each other in a V-like form, were replaced by parallel ones. Boilers of the Serpolet type (that is, composed of tubes of nearly capillary section) were experimented with at great cost of labor and no results; and they were replaced with coil boilers. For these I introduced, in April, 1893, a modification of the æolipile blast, which enormously increased the heat-giving power of the fuel (which was then still alcohol), and with this blast for the first time the boilers began to give steam enough for the engines. It had been very difficult to introduce force pumps which would work effectively on the small scale involved, and after many attempts to dispense with their use by other devices, the acquisition of a sufficiently strong pump was found to be necessary in spite of its weight, but was only secured after long experiment. It may be added that all the aerodromes from the very nature of their construction were wasteful of heat, the industrial efficiency little exceeding half of 1 per cent, or from one-tenth to one-twentieth that of a stationary engine constructed under favorable conditions. This last aerodrome lifted nearly 30 per cent of its weight upon the pendulum, which implied that it could lift much more than its weight when running on a horizontal track, and its engines were capable of running its 50-centimeter propellers at something over 700 turns per minute. There was, however, so much that was unsatisfactory about it, that it was deemed best to proceed to another construction before an actual trial was made in the field, and a new aerodrome, designated as No. 4, was begun. This last was an attempt, guided by the weary experience of preceding failures, to construct one whose engines should run at a much higher pressure than heretofore, and be much more economical in weight. The experiments with the Serpolet boilers having been discontinued, the boiler was made with a continuous helix of copper tubing, which, as first employed, was about three milli-metres internal diameter; and it may be here observed that a great deal of time was subsequently lost in attempts to construct a more advantageous form of boiler for the actual purposes than this simple one, which, with a larger coil tube, eventually proved to be the best; so that later constructions have gone back to this earlier type. A great deal of time was lost in these experiments from my own unfamiliarity with steam engineering, but it may also be said that there was little help either from books or from counsel, for everything was here sui generis, and had to be worked out from the beginning. In the construction which had been reached by the middle of the third year of experiment, and which has not been greatly differed from since, the boiler was composed of a coil of copper in the shape of a hollow helix, through the center of which the blast from the ælopile was driven, the steam and water passing into a vessel I called the "separator," whence the steam was led into the engines at a pressure of from 70 to 100 pounds (a pressure which has since been considerably exceeded).

From the very commencement of this long investigation the great difficulty was in keeping down the weight, for any of the aerodromes could probably have flown had they been built light enough, and in every case before the construction was completed the weight had so increased beyond the estimate, that the aerodrome was too heavy to fly, and nothing but the most persistent resolution kept me in continuing attempts to reduce it after further reduction seemed impossible. Toward the close of the year (1893) I had, however, finally obtained an aerodrome with mechanical power, as it seemed to me, to fly, and I procured, after much thought as to where this flight should take place, a small house boat, to be moored somewhere in the Potomac; but the vicinity of Washington was out of the question, and no desirable place was found nearer than 30 miles below the city. It was because it was known that the aerodrome might have to be set off in the face of a wind, which might blow in any direction, and because it evidently was at first desirable that it should light in the water rather than on the land, that the house boat was selected as the place for the launch. The aerodrome (No. 4) weighed between 9 and 10 pounds, and lifted 40 per cent of this on the pendulum with 60 pounds of steam pressure, a much more considerable amount than was theoretically necessary for horizontal flight. And now the construction of a launching apparatus, dismissed for some years, was resumed. Nearly every form seemed to have been experimented with unsuccessfully in the smaller aerodromes. Most of the difficulties were connected with the fact that it is necessary for an aerodrome, as it is for a soaring bird, to have a certain considerable initial velocity before it can advantageously use its own mechanism for flight, and the difficulties of imparting this initial velocity with safety are surprisingly great, and in the open air are beyond all anticipation.

Here, then, commences another long story of delay and disappointment in these efforts to obtain a successful launch. To convey to the reader an idea of its difficulties, a few extracts from the diary of the period are given. (It will be remembered that each attempt involved a journey of thirty miles each way.)

November 18, 1893. Having gone down to the house boat, preparatory to the first launch, in which the aerodrome was to be cast from a springing piece beneath, it was found impossible to hold it in place on this before launching without its being prematurely torn from its support, although there was no wind except a moderate breeze; and the party returned after a day's fruitless effort.

Two days later a relative calm occurred in the afternoon of a second visit, when the aerodrome was mounted again, but, though the wind was almost imperceptible, it was sufficient to wrench it about so that at first nothing could be done, and when steam was gotten up the burning alcohol blew about so as to seriously injure the inflammable parts. Finally, the engines being under full steam, the launch was attempted, but, owing to the difficulties alluded to and to a failure in the construction of the launching piece, the aerodrome was thrown down upon the boat, fortunately with little damage.

Whatever form of launch was used, it became evident at this time that the aerodrome must at any rate be firmly held up to the very instant of release, and a device was arranged for clamping it to the launching apparatus.

On November 24 another attempt was made to launch, which was rendered impossible by a very moderate wind indeed.

On November 27 a new apparatus was arranged, to merely drop the aerodrome over the water, with the hope that it would get up sufficient speed before reaching the surface to soar, but it was found that a very gentle intermittent breeze (probably not more than 3 or 4 miles an hour) was sufficient to make it impossible even to prepare to drop the aerodrome toward the water with safety.

It is difficult to give an idea in few words of the nature of the trouble, but unless one stands with the machine in the open air he can form no conception of what the difficulties, are which are peculiar to practice in the open, and which do not present themselves to the constructor in the shop, nor probably to the mind of the reader.

December 1, another failure; December 7, another; December 11, another; December 20, another; December 21, another. These do not all involve a separate journey, but five separate trips were made of a round distance of 60 miles each before the close of the season. It may be remembered that these attempts were in a site far from the conveniences of the workshop and under circumstances which took up a great deal of time, for some hours were spent on mounting the aerodrome on each occasion, and the year closed without a single cast of it into the air. It was not known how it would have behaved there, for there had not been a launch even in nine trials, each one representing an amount of trouble and difficulty which this narrative gives no adequate idea of.

I pass over a long period of subsequent baffled effort, with the statement that numerous devices for launching were tried in vain and that nearly a year passed before one was effected.

Six trips and trials were made in the first six months of 1894 without securing a launch. On the 24th of October a new launching piece was tried for the first time, which embodied all the requisites whose necessity was taught by previous experience, and, saving occasional accidents, the launching was from this time forward accomplished with comparatively little difficulty.

The aerodromes were now for the first time put fairly in the air, and a new class of difficulties arose, due to a cause which was at first obscure—for two successive launches of the same aerodrome, under conditions as near alike as possible, would be followed by entirely different results. For example, in the first case it might be found rushing, not falling, forward and downward into the water under the impulse of its own engines; in the second case, with every condition from observation apparently the same, it might be found soaring upward until its wings made an angle of 60 degrees with the horizon, and, unable to sustain itself at such a slope, sliding backward into the water.

After much embarrassment the trouble was discovered to be due to the fact that the wings, though originally set at precisely the same angle in the two cases, were irregularly deflected by the upward pressure of the air, so that they no longer had the form which they appeared to possess but a moment before they were upborne by it, and so that a very minute difference, too small to be certainly noted, exaggerated by this pressure, might cause the wind of advance to strike either below or above the wing and to produce the salient difference alluded to. When this was noticed all aerodromes were inverted, and sand was dredged uniformly over the wings until its weight represented that of the machine. The flexure of the wings under these circumstances must be nearly that in free air, and it was found to distort them beyond all anticipation. Here commences another series of trials in which the wings were strengthened in various ways, but in none of which, without incurring a prohibitive weight, was it possible to make them strong enough. Various methods of guying them were tried, and they were rebuilt on different designs—a slow and expensive process. Finally, it may be said, in anticipation (and largely through the skill of Mr. Reed, the foreman of the work), the wings were rendered strong enough without excessive weight, but a year or more passed in these and other experiments.

In the latter part of 1894 two steel aerodromes had already been built, which sustained from 40 to 50 per cent of their dead lift weight on the pendulum, and each of which was apparently supplied with much more than sufficient power for horizontal flight (the engine and all the moving parts furnishing over one horsepower at the brake weighed in one of these but 26 ounces); but it may be remarked that the boilers and engines in lifting this per cent of the weight did so only at the best performance in the shop, and that nothing like this could be counted upon for regular performance in the open. Every experiment with the launch, when the aerodrome descended into the water, not gently, but impelled by the misdirected power of its own engines, resulted at this stage in severe strains and local injury, so that repairing, which was almost rebuilding, constantly went on, a hard but necessary condition attendant on the necessity of trial in the free air. It was gradually found that it was indispensable to make the frame stronger than had hitherto been done, though the absolute limit of strength consistent with weight seemed to have been already reached, and the year 1895 was chiefly devoted to the labor on the wings and what seemed at first the hopeless task of improving the construction so that it might be stronger without additional weight, when every gram of weight had already been scrupulously economized. With this went on attempts to carry the effective power of the burners, boilers, and engines further, and modification of the internal arrangement and a general disposition of the parts such that the wings could be placed further forward or backward at pleasure, to more readily meet the conditions necessary for bringing the center of gravity under the center of pressure. So little had even now been learned about the system of balancing in the open air, that at this late day recourse was again had to rubber models, of a different character, however, from those previously used; for in the latter the rubber was strained, not twisted. These experiments took up an inordinate time, though the flight obtained from the models thus made was somewhat longer and much steadier than that obtained with the Pénaud form, and from them a good deal of valuable information was gained as to the number and position of the wings and as to the effectiveness of different forms and dispositions of them. By the middle of the year a launch took place with a brief flight, where the aerodrome shot down into the water after a little over 50 yards. It was immediately followed by one in which the same aerodrome rose at a considerable incline and fell backward with scarcely any advance after sustaining itself rather less than ten seconds, and these and subsequent attempts showed that the problem of disposing of the wings so that they would not yield and of obtaining a proper "balance" was not yet solved.

Briefly it may be said that the year 1895 gave small results for the labor with which it was filled, and that at its close the outlook for further substantial improvement seemed to be almost hopeless, but it was at this time that final success was drawing near. Shortly after its close I became convinced that substantial rigidity had been secured for the wings; that the frame had been made stronger without prohibitive weight, and that a degree of accuracy in the balance had been obtained which had not been hoped for. Still there had been such a long succession of disasters and accidents in the launching that hope was low when success finally came.

I have not spoken here of the aid which I received from others, and particularly from Dr. Carl Barus and Mr. J. E. Watkins, who have been at different times associated with me in the work. Mr. R. L. Reed's mechanical skill has helped me everywhere, and the lightness and efficiency of the engines are in a large part due to Mr. L. C. Maltby.

Document 1-8(b), Samuel P. Langley, excerpts from "The New Flying Machine," 1897.

Has the reader had enough of this tale of disaster? If so, he may be spared the account of what went the same way. Launch after launch was successively made. The wings were finally, after infinite patience and labor, made at once light enough and strong enough to do the work, and now in the long struggle the way had been fought up to the face of the final difficulty, in which nearly a year more passed, for the all-important difficulty of balancing the aerodrome was now reached, where it could be discriminated from other preliminary ones, which have been alluded to, and which at first obscured it. If the reader will look at the hawk or any soaring bird, he will see that as it sails through the air without flapping the wing, there are hardly two consecutive seconds of its flight in which it is not swaying a little from side to side, lifting one wing or the other, or turning in a way that suggests an acrobat on a tight-rope, only that the bird uses its widely outstretched wings in place of the pole.

There is something, then, which is difficult even for the bird, in this act of balancing. In fact, he is sailing so close to the wind in order to fly at all, that if he dips his head but the least he will catch the wind on the top of his wing and fall, as I have seen gulls do, when they have literally tumbled toward the water before they could recover themselves.

Beside this, there must be some provision for guarding against the incessant, irregular currents of the wind, for the wind as a whole—and this is a point of prime importance—is not a thing moving along all-of-a-piece, like water in the Gulf Stream. Far from it. The wind, when we come to study it, as we have to do here, is found to be made of innumerable currents and counter-currents which exist altogether and simultaneously in the gentlest breeze, which is in reality going fifty ways at once, although, as a whole, it may come from the east or the west; and if we could see it, it would be something like seeing the rapids below Niagara, where there is an infinite variety of motion in the parts, although there is a common movement of the stream as a whole.

All this has to be provided for in our mechanical bird, which has neither intelligence nor instinct, without which, although there be all the power of the engines requisite, all the rigidity of wing, all the requisite initial velocity, it still cannot fly. This is what is meant by balancing, or the disposal of the parts, so that the airship will have a position of equilibrium into which it tends to fall when it is disturbed, and which will enable it to move of its own volition, as it were, in a horizontal course.

Now the reader may be prepared to look at the apparatus which finally has flown. (See diagram above.) [*Diagram on p. 143*] In the completed form we see two pairs of wings, each slightly curved, each attached to a long steel rod which supports them both, and from which depends the body of the machine, in which are the boilers, the engines, the machinery, and the propeller wheels, these latter being not in the position of those of an ocean steamer, but more nearly amidships. They are made sometimes of wood, sometimes of steel and canvas, and are between three and four feet in diameter.

The hull itself is formed of steel tubing; the front portion is closed by a sheathing of metal which hides from view the fire-grate and apparatus for heating, but allows us to see a little of the coils of the boiler and all of the relatively large smokestack in which it ends. The conical vessel in front is an empty float, whose use is to keep the whole from sinking if it should fall in the water.

This boiler supplies steam for an engine of between one and one and one-half horsepower, and, with its fire-grate, weighs a little over five pounds. This weight is exclusive of that of the engine, which weighs, with all its moving parts, but twenty-six ounces. Its duty is to drive the propeller wheels, which it does at rates varying from 800 to 1,200, or even more, turns a minute, the highest number being reached when the whole is speeding freely ahead.

The rudder, it will be noticed, is of a shape very unlike that of a ship, for it is adapted both for vertical and horizontal steering. It is impossible within the limits of such an article as this, however, to give an intelligible account of the manner in which it performs its automatic function. Sufficient it is to say that it does perform it.

The width of the wings from tip to tip is between twelve and thirteen feet, and the length of the whole about sixteen feet. The weight is nearly thirty pounds, of which about one-fourth is contained in the machinery. The engine and boilers are constructed with an almost single eye to economy of weight, not of force, and are very wasteful of steam, of which they spend their own weight in five minutes. This steam might all be recondensed and the water re-used by proper condensing apparatus, but this cannot be easily introduced in so small a scale of construction. With it the time of flight might be hours instead of minutes, but without it the flight (of the present aërodrome) is limited to about five minutes, though in that time, as will be seen presently, it can go some miles; but owing to the danger of its leaving the surface of the water for that of the land, and wrecking itself on shore, the time of flight is limited designedly to less than two minutes.

I have spared the reader an account of numberless delays, from continuous accidents and from failures in attempted flights, which prevented a single entirely satisfactory one during nearly three years after a machine with power to fly had been attained. It is true that the aërodrome maintained itself in the air at many times, but some disaster had so often intervened to prevent a complete flight that the most persistent hope must at some time have yielded. On the 6th of May of last year I had journeyed, perhaps for the twentieth time, to the distant river station,

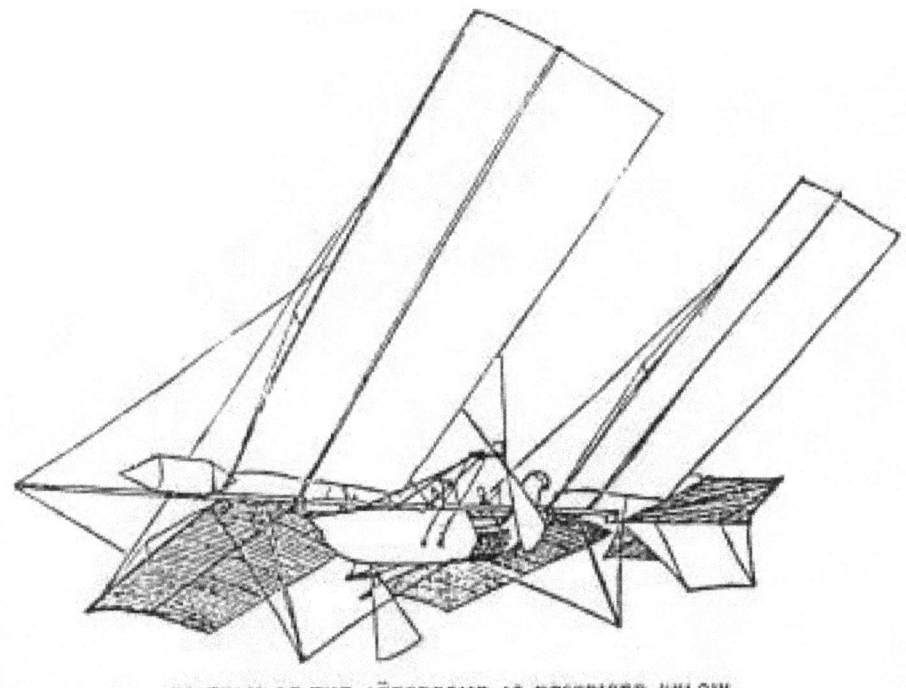

DIAGRAM OF THE AËRODROME AS DESCRIBED BELOW.

and recommenced the weary routine of another launch, with very moderate expectation indeed; and when, on that, to me, memorable afternoon the signal was given and the aërodrome sprang into the air, I watched it from the shore with hardly a hope that the long series of accidents had come to a close. And yet it had, and for the first time the aërodrome swept continuously through the air like a living thing, and as second after second passed on the face of the stop-watch, until a minute had gone by, and it still flew on, and as I heard the cheering of the few spectators, I felt that something had been accomplished at last, for never in any part of the world, or in any period, had any machine of man's construction sustained itself in the air before for even half of this brief time. Still the aërodrome went on in a rising course until, at the end of a minute and a half (for which time only it was provided with fuel and water), it had accomplished a little over half a mile, and now it settled rather than fell into the river with a gentle descent. It was immediately taken out and flown again with equal success, nor was there anything to indicate that it might not have flown indefinitely except for the limit put upon it. . . .

On November 28th I witnessed, with another aërodrome of somewhat similar construction, a rather longer flight, in which it traversed about three-quarters of a mile, and descended with equal safety. In this the speed was greater, or about thirty miles an hour. . . . We may live to see airships a common sight, but habit

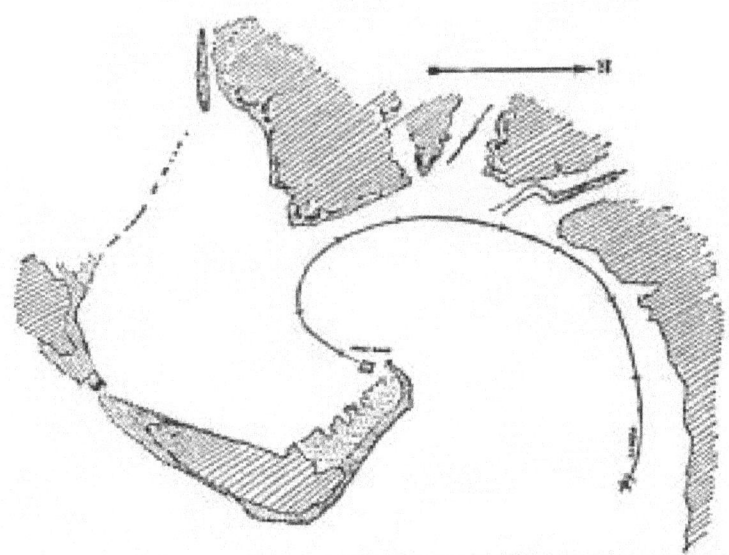

DIAGRAM SHOWING THE COURSE OF THE AERODROME IN ITS FLIGHT ON THE POTOMAC RIVER AT QUANTICO. SEE PAGE 659.

has not dulled the edge of wonder, and I wish that the reader could have witnessed the actual spectacle. "It looked like a miracle," said one who saw it, and the photograph, though taken from the original conveys but imperfectly the impression given by the flight itself.

And now, it may be asked, what has been done? This has been done: a "flying-machine," so long a type for ridicule, has really flown; it has demonstrated its practicability in the only satisfactory way—by actually flying, and by doing this again and again, under conditions which leave no doubt.

There is no room here to enter on the consideration of the construction of larger machines, or to offer the reasons for believing that they may be built to remain for days in the air, or to travel at speeds higher than any with which we are familiar; neither is there room to enter on a consideration of their commercial value, or of those applications which will probably first come in the arts of war rather than those of peace; but we may at least see that these may be such as to change the whole conditions of warfare, when each of two opposing hosts will have its every movement known to the other, when no lines of fortification will keep out the foe, and when the difficulties of defending a country against an attacking enemy in the air will be such that we may hope that this will hasten rather than retard the coming of the day when war shall cease.

I have thus far had only a purely scientific interest in the results of these labors. Perhaps if it could have been foreseen at the outset how much labor there was to be, how much of life would be given to it, and how much care, I might have hesitated to enter upon it at all. And now reward must be looked for, if reward there be, in the knowledge that l have done the best I could in a difficult task, with results which it may be hoped will be useful to others. I have brought to a close the portion of the work which seemed to be specially mine—the demonstration of the practicability of mechanical flight—and for the next stage, which is the commercial and practical development of the idea, it is probable that the world may look to others. The world, indeed, will be supine if it do not realize that a new possibility has come to it, and that the great universal highway overhead is now soon to be opened.

Document 1-8(c), Alexander Graham Bell, "The Aerodromes in Flight," 1897.

Through the courtesy of Dr. S. P. Langley, Secretary of the Smithsonian Institution, I have had, on various occasions, the privilege of witnessing his experiments with aerodromes, and especially the remarkable success attained by him in experiments made upon the Potomac river on Wednesday, May 6, 1896, which led me to urge him to make public some of these results.

I had the pleasure of witnessing the successful flight of some of these aerodromes more than a year ago, but Dr. Langley's reluctance to make the results public at that time prevented me from asking him, as I have done since, to let me give an account of what I saw.

On the date named two ascensions were made by the aerodrome, or so-called "flying-machine," which I will not describe here further than to say that it appeared to me to be built almost entirely of metal, and driven by a steam-engine which I have understood was carrying fuel and a water supply for a very brief period, and which was of extraordinary lightness.

The absolute weight of the aerodrome, including that of the engine and all appurtenances, was, as I was told, about 25 pounds, and the distance from tip to tip of the supporting surfaces was, as I observed, about 12 or 14 feet. The method of propulsion was by aerial screw-propellers, and there was no gas or other aid for lifting it in the air except its own internal energy.

On the occasion referred to, the aerodrome, at a given signal, started from a platform about 20 feet above the water, and rose at first directly in the face of the wind, moving at all times with remarkable steadiness, and subsequently swinging around in large curves of, perhaps, a hundred yards in diameter, and continually ascending until its steam was exhausted, when, at a lapse of about a minute and a half, and at a height which I judged to be between 80 and 100 feet in the air, the wheels ceased turning, and the machine, deprived of the aid of its propellers, to my surprise did not fall, but settled down so softly and gently that it touched the water without the least shock, and was in fact immediately ready for another trial.

In the second trial, which followed directly, it repeated in nearly every respect the actions of the first, except that the direction of its course was different. It ascended again in the face of the wind, afterwards moving steadily and continually in large curves accompanied with a rising motion and a lateral advance. Its motion was, in fact, so steady, that I think a glass of water on its surface would have remained unspilled. When the steam gave out again, it repeated for a second time the experience of the first trial when the steam had ceased, and settled gently and easily down. What height it reached at this trial I cannot say, as I was not so favorably placed as in the first; but I had occasion to notice that this time its course took it over a wooded promontory, and I was relieved of some apprehension in seeing that it was already so high as to pass the tree-tops by 20 or 30 feet. It reached the water 1 minute and 31 seconds from the time it started, at a measured distance of over 900 feet from the point at which it rose.

This, however, was by no means the length of its flight. I estimated from the diameter of the curve described, from the number of turns of the propellers as given by the automatic counter, after due allowance for slip, and from other measures, that the actual length of flight on each occasion was slightly over 3,000 feet. It is at least safe to say that each exceeded half an English mile.

From the time and distance it will be noticed that the velocity was between 20 and 25 miles an hour, in a course which was taking it constantly "up hill." I may add that on a previous occasion I have seen a far higher velocity attained by the same aerodrome when its course was horizontal.

I have no desire to enter into detail further than I have done, but I cannot but add that it seems to me that no one who was present on this interesting occasion could have failed to recognize that the practicability of mechanical flight had been demonstrated.

ALEXANDER GRAHAM BELL.

Document 1-9(a–d)

(a) Octave Chanute, "Conditions of Success in the Design of Flying Machines," (1898) Bulletin No. XXIII, appeared in *American Magazine of Aeronautics* 1 (July 1907).

(b) Augustus M. Herring, "A Solution to the Problem of the Century," *The Aeronautical Annual* 3 (1897): 54–75.

(c) Octave Chanute, "Experiments in Flying: An Account of the Author's Own Inventions and Adventures," *McClure's Magazine* 15 (June 1900): 127–133.

(d) Octave Chanute, excerpt from "Aerial Navigation," *Cassier's Magazine* 20 (May 1901): 111–123.

Octave Chanute was the central figure in the exchange of aeronautical information worldwide at the end of the nineteenth century, and, more than anyone else, he brought the various pioneering researchers, experimenters, promoters, and inventors together into a trans-Atlantic community of aviation enthusiasts. He corresponded extensively and wrote widely in his efforts to advance the state of aeronautics.

Reflecting some of Chanute's own experimental work to perfect a flying machine, the documents presented here include his "Experiments in Flying" from the June 1900 issue of *McClure's Magazine*, and selections from Chanute's report "Aerial Navigation: Balloons and Flying Machines from an Engineering Standpoint." This latter article was printed in the May 1901 issue of *Cassier's Magazine*; Chanute read a revised and updated version with the same title before the American Association for the Advancement of Science on 30 December 1903. The Smithsonian Institution published this paper in its Annual Report of 1904.

Chanute's associate, Augustus Moore Herring is, without doubt, one of the more enigmatic actors in the drama associated with the invention of the airplane. Besides serving as pilot for Chanute's experimental glider flights in 1896, the young Herring made important contributions to Chanute's *Katydid* glider design. Notably this involved Herring's idea for a moveable—or "regulating"—tail that reacted positively to gusts.

After Herring parted company with Chanute and then spent the summer gliding at the Indiana Dunes in 1896, he continued his own aeronautical experiments in southern Michigan. On 11 October 1898, Herring made a powered hop in a machine of his own design. This took place at Silver Beach on Lake Michigan, near the town of St. Joseph. On October 27, the *Benton Harbor News* reported on the event: "During the flight, which lasted some 8 or 10 seconds, Prof. Herring's feet seemed almost to graze the ground while the machine skimmed along on a level path over the beach. The landing was characterized by a slight turning to the left and slowing of the engine, when the machine and operator came as gently to rest on the sand as a bird instinct with life." Matthias Arnot, Herring's financial backer, was present that day on Silver Beach, and he carried a camera. Unfortunately, Arnot was allegedly so surprised by what he saw that he did not take a picture until Herring was in the process of landing. The resulting photo showed the propeller of Herring's airplane revolving but the skids of his machine were not airborne but plowing up the sand, clearly in contact with the ground.

Even if Herring's eighty-eight-pound machine of 1898 did get slightly airborne, it lacked a control system and could only be controlled by the pilot shifting his weight. John Meiler, Herring's great-grandson, who has researched his ancestor's aeronautical career, concedes that the 1898 machine lacked such a system, but he claims that his great-grandfather had equipped a different flying machine with controls as early as 1894. "He realized the potential need for an auxiliary kind of control," Meiler has stated. "But he considered [his 1898 aircraft] an experimental flying machine," one that represented "just one more step in a lifetime of work," not its definitive culmination.

To this day, there are Herring family members and others in the St. Joseph area who believe that Augustus Herring made a powered flight in an airplane before the Wrights and that he deserves far greater recognition. Unfortunately for his reputation, there will always be the Arnot photograph as well as a negative report from his former associate Octave Chanute. After his "successful experiment" at Silver Beach on 11 October, Herring invited Chanute to travel to St. Joseph to witness his next flight attempt, on 16 October. Chanute came, but the air-compressed engine meant to drive the airplane failed to cooperate. The next day, when Herring damaged his plane trying to get it airborne, Chanute left town, thinking Herring a fraud. If he had waited a few more days, his opinion might have changed. According to eyewitness reports, Herring flew his machine successfully on 22 October 1898. Sixty-three years after the event, in 1961, a Mr. Sam Lessing, a hot dog vender at the time in 1898, told the *St. Joseph Herald-Press* that he had been among the witnesses to the event: "I saw it. . . . It couldn't turn, just flew straight ahead."

In his book *A Dream of Wings: Americans and the Airplane (1981)*, historian Tom D. Crouch reviews the fate of Herring's aeronautical career: "Augustus Moore Herring, badly shaken by the news of the Wright success, was unable to believe that these two newcomers to the field had so quickly solved those problems he had been unable to overcome. In 1908 he was a competitor with the Wrights for the first U.S. Army airplane contract, but withdrew his bid without producing a finished machine. The following year he entered a short-lived partnership with Glenn H. Curtiss that ended in a long-running lawsuit with Curtiss and an out-of-court settlement of $500,000 to Herring's heirs. He died in 1926" (p. 307).

Document 1-9(a), Octave Chanute, "Conditions of Success in the Design of Flying Machines," 1898.

After many centuries of failure, it is believed that we are at last within measurable distance of success in Aerial Navigation; that there will be two solutions; one with a dirigible balloon, which will chiefly be used in war, and the other with dynamic, bird-like machines which will possess so much greater speed and usefulness that they should preferably engage the attention of searchers.

I have, of late years, experimented with six full-sized gliding machines carrying a man, comprising three different types, and having reached some definite opinions as to the conditions of eventual success with power driven machines, it is ventured to state them briefly for the benefit of other experimenters; for final success will probably come through a process of evolution, and the last successful man will need to add but little to the progress made by his predecessors.

It is true that the most important component of the future flying machine will be the very light motor. It is the lack of this which has hitherto forbidden dynamic flight and restricted dirigible balloons to inefficient speeds, but it is also true that dynamic flight is impossible unless the stability is adequate. The progress made in light motors within the last ten years has been very great; Maxim, Langley, and Hargrave have produced steam engines weighing but about 5 kilogrammes to the horsepower, and hundreds of ingenious men are now improving the gas engine so rapidly that there is good hope that we shall soon be in possession of a prime mover which shall approximate in lightness the motor muscles of birds, which are believed to weigh but 3 to 9 kilogrammes per horsepower developed.

But even with a very light motor, success cannot be attained until we have thoroughly mastered the problem of equilibrium in the air. This fluid is so evasive, the wind so constantly puts it into irregular motion, that it imposes great difficulties even upon a bird, endowed as he is both with an exquisite organization, with life-instinct and with hereditary skill. It is to this one problem of equilibrium that I have devoted all my attention, in the belief that an inanimate artificial machine must be endowed with automatic stability in the air, and that experiments indicate that this can be achieved.

The wind is constantly in turmoil; it strikes the apparatus at different points and angles, and this changes the position of the center of pressure, thus compromising the equilibrium. To re-establish the latter requires either that the center of gravity, (or weight) shall be shifted to correspond, or that the supporting surfaces themselves shall be shifted, thus bringing back the center of pressure over the center of gravity. Birds employ both methods; they shift the weight of parts of their bodies, or they shift either the position or angle of their wings. It is believed that only the shifting of the wings is open for use for an artificial apparatus.

General Conditions.

It is inferred, therefore, that inventors who begin by working upon an artificial motor, and who endeavor to evolve a complete flying machine at once, are beginning at the wrong end, and are leaving behind them two very important prerequisites.

1st. That the apparatus shall possess automatic stability and safety under all circumstances.

2nd. That the apparatus shall be so light and small as to be easily controlled in the wind by the personal strength of the operator.

The general stability in the line of flight, the steering, can be obtained by a rudder, but the automatic equilibrium must be secured in two directions; first transversely to the apparatus, and secondly fore and aft. Very good results have been automatically obtained to the transverse stability by imitating the attitude of the soaring birds, the underlying principle of which consists in a slight dihedral angle of the wings with each other, either upward or downward, but the very best application of this principle is not yet evolved, and it requires more experimenting. Experimenters have found but little difficulty in securing stability in this transverse direction, but it must be worked out thoroughly.

The longitudinal equilibrium is, however, the most precarious and important. I have tested three methods of securing it automatically.

First, by setting the tail at a slight upward angle with the supporting surfaces, so as to change the angle of incidence of the latter through the action of the "relative wind" on the upper or lower surface of the tail. This is known as the "Penaud" tail; it is susceptible to great improvement in details of construction, as has been abundantly proved, but it is not yet certain that it will counteract all movements of the center of gravity in meeting sudden wind gusts.

Second, by pivoting the wings at their roots, so that they may swing backward and forward horizontally, thus bringing back automatically the center of pressure over the center of gravity, whenever a change occurs in the "relative wind." The so-called "multiple-wing" gliding machine was of this type, and it reduced the movement of the aviator required to meet wind gusts to about 25 millimeters. It cannot, however, be said its construction is perfected.

Third, by hinging vertically the supporting surfaces to the mainframe of the apparatus, so that these surfaces shall change their angle of incidence automatically when required. This last method has only been tested in models, other engagements having prevented experiments this year (1898). The other two methods have been applied to full-sized machines carrying a man. They have given such satisfactory results that not the slightest accident has occurred in two years of experimenting, but their adjustment has not yet reached the consummation originally aimed at; i.e., that the aviator on the gliding machine shall not need to move at all, and that the apparatus shall automatically take care of itself under all circumstances except in landing.

I shall be glad to furnish more minute descriptions to those who may want to repeat these experiments, or to apply the principles to machines of their own. The stability of an apparatus is the very first thing to work out before it is attempted to apply an artificial motor. This cannot be too strongly insisted upon, and the best way of accomplishing this prerequisite is to experiment with a full-sized gliding machine carrying a man. This utilizes the reliable force of gravity until such time as the automatic equilibrium is fully attained. Then, and not until then, it becomes safe to apply a motor.

When artificial power comes to be applied, it is probable that the best motor to use at the beginning will be found in a compressed air engine, supplied from a reservoir upon the apparatus. This is not a prime mover, but it is reliable and easily applied. It will probably afford a flight for but a few seconds, but this will enable the aviator to study the effects of the motor and propeller on the equilibrium of his machine. When this is thoroughly ascertained another motor may be substituted, such as a steam or a gasoline engine, which will produce longer flights, but this will require long and costly experimenting to obtain a light and reliable engine.

Another most important requisite is that the first apparatus with a motor shall be of the smallest dimensions which it is possible to design, and shall therefore carry only one man. This requisite is for four reasons: 1st. In order to keep down the relative weight, which increases as the cube of the dimensions, while the supporting surfaces increase approximately as the square; 2nd. In order to secure adequate control of the apparatus in the wind; 3rd. To diminish the power required for the motor; and 4th . To have as little inertia as possible to overcome in landing. The whole apparatus should be so light and small that the aviator shall carry it about on his shoulders and control it in the wind. This can only be accomplished with a gliding machine. My double-decked machine was of ample strength, with 12.5 square meters of supporting surface, weighing 11 kilograms, and carried a man perfectly on a relative wind of 10 meters per second. It showed an expenditure of 2 horsepower obtained from gravity. It is believed that a power machine can be built with 16 square meters of carrying surface, and a weight of 41 kilograms which will carry a man and a motor of 5 horsepower, if the latter with its propellers and shafts does not weigh more than 5 or 6 kilograms per horsepower. In fact, this has been done with a compressed air motor machine, but the apparatus thus far has produced doubtful results, in consequence of the defects of the motor. It is firmly believed that it will be a great mistake to experiment with a large and heavy machine, for it would probably be smashed upon its first landing, before its possibilities could be ascertained.

The speed first aimed at should be about 10 meters per second, and to achieve this the following are good proportions:

Sustaining surfaces	0.15 square meters per kilogram.
Sustaining surfaces	3.00 square meters per horsepower.
Equivalent head surface	0.25 square meters per horsepower.
Weight sustained	20.00 kilograms per horsepower.

Details of Construction.

The general arrangement and detail of the construction will conform, of course, to the particular design to be tested by the experimenter, but some useful hints may be given. There need be no hesitation as to the materials to employ. The frame should be or wood, which although weaker than bamboo is more reliable and permits the shaping of the spars so as to diminish the head resistance. It has been found by experiment that the best cross-section resembles that of a fish, with the greatest thickness about one-third of the distance from the front edge; this reduces the resistance to coefficients of one-sixth to one-tenth that of a plane of equal area, while a round section, such as that of bamboo, gives a coefficient of about one-half. The spars of the frame can best be joined together with lashing of glued twine or with very thin steel tubing, preferably silver or nickel-plated. The stays or tension members should be of the best steel wire, also nickel-plated and oiled to prevent rust. A very important detail, not yet worked out, consists in connecting the wires to the framework so that they shall pull alike. The supporting surfaces should preferably be of balloon cloth or Japanese silk, varnished with two or three coats of Pyroxelene (collodion) varnish which possesses the property of shrinking the fabric upon drying, so as to make it drum-like.

(A good recipe for this varnish is as follows:—Take 60 grams of gun cotton no.1, dampen it with alcohol to make it safe to handle, and dissolve it in a bottle containing a mixture of 1 liter of alcohol and 3 liters of sulphuric ether. When well dissolved, add 20 grams of castor oil and 10 grams of Canadian Balsam. This is to be kept in a corked can, and poured in small quantities into a saucer, whence it is applied thinly with a flat brush. Two coats will generally be sufficient. It dries very quickly, glues together all the laps in the fabric, and shrinks it in drying.)

An expeditious way of fastening the surfaces to the frame consists in stretching them as tight as possible and then doubling them back around the spar, the flap so made to then fasten temporarily with pins; the first coat of varnish will glue the surfaces together, and the pins may be withdrawn if desired.

Although it is preferable that some of the rear portions shall be flexible, the supporting surfaces and the framework must be sufficiently stiff not to change their general shape when under motion. This indicates bridge construction for the framework and therefore the super-imposing of surfaces. Very little supporting or parachute action will be lost by this, for even when struck at right angles by the wind, Thibaut found that a square plane placed behind another of equal size, and spaced at a distance equal to the length of its side, still experienced a pressure of 0.7 that on the front plane. The supporting surfaces will of course be arched in the direction of flight in accordance with the practice inaugurated by Lilienthal, who showed that they possessed angles of incidence of 3 degrees, five times the lifting power of the planes. It is not probable that success will be achieved in Aerial Navigation with flat sustaining surfaces.

Proportion of Parts.

In proportioning the parts, the factor of safety for static loads should generally be 3, never less than 2, and preferably 5 for the parts subject to the more important strains. These are computed in the same way as they are for bridges, with the difference, however, that the support (on the air), is to be considered as uniformly distributed, and the load is to be assumed as concentrated at the center. It is not believed that it is practicable to calculate the strains due to possible shocks upon landing. They must be taken into consideration in a general way, but the utmost efforts will be made to avoid them.

The sustaining power will be calculated in the manner given by Lilienthal in Mocdcbeek's "Taschenbuch für Flugtechniker und Luftschiffer." He does not, however, fully explain how to calculate the resistance; this consists of the "drift" or horizontal component of normal pressure, plus or minus the tangential pressure, and of the "head resistance" of the framework, or the motor if any, and the body of the operator. As an example how to compute this I give the calculations for the "multiple wing" gliding machine of 1896, which was constructed before experiments, that shows how the head resistance could be further reduced by adopting better cross-sections for the framework.

Area Head Resistance, Multiple Wing Machine.

Description	No.	Dimensions Millimeters	Square Meters	Co-efficient Resistance	Equivalent sq. Meters
Front edge of wings	10	2225 x 12.70	.28257	–	.14128
Main wing arms	10	1956 x 12.70	.24841	⅓	.08280
Ribs of top Aeroplane	3	1346 x 6.35	.02564	1	.02564
Posts of top Aeroplane	4	1829 x 12.70	.09291	⅓	.03097
Posts connecting front wings	8	1280 x 12.70	.12995	⅓	.04332
Posts carrying pivots	2	823 x 19.05	.03135	⅓	.01045
Curved prow pieces	3	914 x 24.50	.06717	1	.06717
Front bow braces	2	731 x 12.70	.01857	⅓	.00619
Rear bow braces	2	841 x 12.70	.02136	⅓	.00712
Cross struts bow & frame	2	670 x 12.70	.01702	⅓	.03613
Rear wing braces	4	2134 x 12.70	.10840	⅓	.03613
Rudder braces	2	1219 x 12.70	.03096	⅓	.01032
Rudder struts	2	548 x 12.70	.01392	⅓	.00464
Wire stays 61 meters		61000 x 1.27	.07747	1-½	.11620
Spring wire stays 8 meters		8000 x 1.27	.01016	1-½	.01524
Rubber springs	6	1300 x 1.00	.00780	1	.00780
Sundry projecting parts		Say	.01198	1	.01198
Aviator's body		Say	.46450	1	.46450
			1.66014		1.08742

In order to calculate the resistance, we must first ascertain the requisite speed for support and the consequent "drift." The front wings measure 13.34 square meters and carry all the weight, they are set at a positive angle of 3 degrees, for which the Lilienthal normal co-efficient η is 0.546. Using the well known formula $W = k s v^2 \eta \cos \alpha$ in which W is the weight, k is the air co-efficient, s the surface, v the velocity, η the Lilienthal co-efficient (0.11) and α the angle of incidence, and calling W=86 kilos, we have for the support:

86 = 0.11 x 13.34 x v2 x 0.546 x cos 3°;

and as cos 3° = 0.9986, we have for the speed:

v = √ 86 / 0.11 x 13.34 x 0.546 x 0.9986 = 10.37 meters.

Whence we have for the front wings:

    Rectangular pressure   0.11 x 10.372 = 11.829 kilos. per square meter.
    Normal pressure at 3°   11.829 x 13.34 x 0.546 = 86.16 kilograms.
    Lift at 3°   86.16 x 0.9986 = 86 kilograms
    Drift at 3°   86.16 x Sine 3° = 4.51 kilograms

The Tangential pressure upon the font wings is zero at 3°. The "drift" on the rear wings, which measures 2.74 sq. meters, and were set at a negative angle of 3°, consists in the product of their surface by the rectangular pressure, (Lilienthal's θ) which at this angle is positive, and the horizontal component of the normal (Lilienthal's η) which is negative at 3°, the latter being obtained by multiplying h by the sine of 3°. We have therefore:

Drift rear wings = 11.829 x 2.74 (0.013 = 0.212 x 0.05233) = 0.98k.

The head resistance is the important factor, and depends upon the shapes which are adopted for the framing to evade air resistance and to secure low co-efficients. It has to be calculated in detail, and the table herewith given recapitulates the various elements of the area of head resistance of the multiple wing machine, reduced by co-efficients to an equivalent area for further calculations.

The rectangular pressure of a speed of 10.37 meters per second being 11.829 kilos per square meter, we have therefore for the whole resistance:

Drift front wings	11.829 x 13.34 x 0.546 [x] 0.52333	= 4.51 kilos
Drift rear wings	11.829 x 2.74 (0.043 – 0.0126)	= 0.98 "
Tangential component at 3°		= 0.00 "
Head resistance	11.829 x 1.087	= 12.86 "
	Total resistance	= 18.35 "

As the speed is 10.37 meters per second, the power required to overcome this total resistance is:

Power 18.35 x 10.37 = 190.28 kilogrammeters or 2.53 horsepower, and as the weight is 86 kilos the angle of descent as a gliding machine ought to be:

Angle 18.35/86 = 0.2134 or tangent of 12°

In point of fact, the apparatus glides generally at this angle and frequently at angles of descent of 10 or 11 degrees, this being probably due to an ascending wind along the hillsides, and fully verifying this mode of calculating the resistance.

In the "double-decker" gliding machine, in which the framing was better designed, the resistance was calculated at 14.46 kilos, and it absorbed a horsepower in gliding in still air. By employing still better cross sections of framework, and especially by placing the aviator in a horizontal position, the head resistance could be reduced by at least one-third, but this particular attitude of the man would involve some risk of accident in landing, and is considered to be too dangerous to be employed in the preliminary experiments. It will be noticed in the table that the resistance of the wire stays is given a co-efficient of 1-½, while theoretically, being cylindrical, their co-efficient should be about ½. This allowance is based upon experience. Wire stays produce undue resistance, and this is probably due to the fact that they vibrate like violin strings when the apparatus is under rapid motion, and thus produce a greater resistance than that due to their rounded cross-section.

The power required will be seen to differ very materially from that indicated by the formula recently proposed in France, which is based on the assumption that the total wing surface, in square meters, multiplied by the co-efficient of air resistance (i.e., the number of kilogrammes carried by a square meter, at a speed of one meter per second) must at least be equal to the cube of the weight of the apparatus in kilogrammes; divided by the square of the power exerted by the motor in kilogrammes, or, K S T² = P³ from which in our own case we would draw:

0.11 x 13.34 x T² = 86³ , or

T = √ 86³ / 0.11 x 13.34 = 658.4 klgm

or 8.78 horsepower, which is more than three times the power calculated by the method here given and tested by actual experiment and measuring.

It must be remembered, however, that the 2.53 and the 2 horsepower, which have been found sufficient to sustain 86 kilogrammes in the air, are the *net* horsepower absorbed by the gliding machines. When a propeller and a motor are added, it will be necessary to allow for the losses in efficiency incident to those adjuncts, and so provide about twice the power to the engine which is indicated by the resistance multiplied by the speed. A safe rule of approximation will be to allow that each nominal horsepower at the engine will sustain 20 kilogrammes, and that each kilogramme of the total weight of the apparatus will require 0.15 square meters of surface to sustain it at speeds of about 10 meters per second. When greater speeds become practicable and safe, the surfaces may be reduced below this so that at 20 meters per second they may be but about 0.05 square meters per kilo, instead of 0.15 square meters per kilo indicated above, and this would permit reducing the head area of the framing, but unless the co-efficient for the aviator's body was in some way reduced the resistance and power required would be greater, because of the higher speed.

These are the conditions and considerations which experiments with full-sized gliding machines, carrying a man, have thus far indicated as necessary to observe in order to achieve success with a dynamic flying machine provided with a motor. The most important of them are:

FIRST, that the automatic equilibrium and safety shall first be secured before an attempt is made to apply a motor, and SECOND, that the apparatus shall be made as small and light as possible, so that the aviator may sustain its weight before taking his flights.

Document 1-9(b), Augustus M. Herring, "A Solution to the Problem of the Century," 1897.

Perhaps no subject offers more scope to the imagination than the benefits and changes for mankind which would result from a practical solution of the problem of manflight. At the same time there is probably no problem which the inventive skill of man has ever attacked which apparently offers, at first sight, more numerous and easy ways of unravelment, and yet which, on careful investigation, develops greater or more unexpected difficulties. In beginning the experiments the methods apparently open might be roughly divided into four classes.

The first of these would comprise all those machines in which the whole or part of the weight was lifted by a balloon or gas-bag; the second, all those forms of apparatus which were intended to sustain or lift their weight with screw propellers revolving on vertical axes; the third, those machines which were intended to sustain their weight (and that of the operator) on flapping or beating wings; the fourth, and last, class would contain the aeroplane, or more properly the aerocurve machines; for the aeroplane may now safely be said to have disappeared from competition with the more efficient form of surface.

The limitations of the navigable balloon are now pretty well recognized. To obtain a speed of even 20 miles per hour, a spindle-shaped envelope of very large size is necessary, and the result at its best is an exceedingly frail and bulky machine, whose ultimate speed capacity is insufficient for wind velocities which frequently occur even near the ground. Its chief defects are great bulk and extreme frailty; for the envelope, in proportion to its relative size, is not many times stronger than a soap bubble. An instance may be cited in support of this in the large navigable balloon built for the Antwerp exposition, which became tilted up during a trial when the rush of gas to the higher end burst the balloon.

The future utility of the navigable balloon is still the subject of differences of opinion; it is, however, certain that whatever may be its ultimate practical advantages as a flying machine, the drawbacks of enormous size and frailty are sure to offer a considerable offset to them.

The vertical screw machines have much to recommend them, but there are far greater difficulties offered to their production than would be supposed. The ability to rise directly into the air from any given spot would be an exceedingly desirable quality. And hence we find that the great majority of experimenters who attack the problem of dynamic flight begin here, starting with a plan of some modification of this type of machine. The stumbling-blocks, however, are soon met. Not least among them is the fact that when the surfaces which form the blades of the screws are revolved over one spot (as they must be to rise directly into the air) they do not give any considerable lifting effect in proportion to the power consumed; for where one might from the theory even of the aeroplane expect a lift of possibly 100 pounds per horsepower, the best result the inventor can produce on a practical scale is pretty sure to be less than one-seventh of that figure. In fact, the lift with the lightest engines we can build is likely to be but little, if any, more than the weight of the machine itself. With engines weighing much more than 4 or 5 pounds per horsepower (250 times as powerful weight for weight as a man), practical success with this type of apparatus is not possible.

The third class, or the beating wing machines, are subject to the same disadvantages in regard to the enormous power required as those of the vertical screw type. In addition to this, the question of maintaining a stable equilibrium in windy weather still further greatly complicates them, so much so, in fact, that there is but small hope of practical machines operated on this principle ever being produced.

It is unnecessary to point out that any combination in a machine of the principles involved in either of the above three classes would still subject it to the fundamental objections of at least one of the classes. These objections are so formidable that, to the great majority of the foremost workers in this field, there now appears but one main principle left, and upon this there is an ever-increasing hope, if not certainty, that flight will be accomplished. This principle is the one which underlies the aeroplane and aerocurve; namely, that when a thin surface is driven rapidly through the air, and is slightly inclined to its path, the equivalent of a pressure is developed on the side which is exposed to the air current—i.e., the under side—which is much greater than the driving force necessary to produce it. If an arched surface (arched in the line of motion) with the hollow side undermost be substituted for a plane, we have an aerocurve. Its chief advantage is that it possesses a higher efficiency. Another, but minor, difference is that it is not necessary to incline an aerocurve in order to develop a pressure on the hollow side when it is moved through the air.

The one advantage which the dynamic or power machine of the aerocurve type has over the vertical screw is the fact that it can, through the agency of the surfaces, convert the relatively small push of the screw propellers into a much larger lifting effect.

It is interesting to note that the first approach to human flight of modern times was attained only by the use of the aerocurve, when early in 1894 the late Otto Lilienthal, of Berlin, Germany, built a huge bat-like machine, with curved rigid wings, on which he was able to "slide" downhill on the air, 150 feet or so at a time.

Practice with this machine soon enabled him to start from very high places, and his flights became correspondingly longer. Early in the beginning of these trials, he became aware, as his writings show, of the enormous power and disturbing effect which those ever-present irregularities in the wind produce, and which, in a large measure, were the cause of his losing his life—a sad accident which has taken from the field of aerodynamics one of, if not, the ablest of its workers; for both the practical and theoretical work of Lilienthal in the new science is of the greatest value, and will be so recognized when more generally understood.

In his first articles Lilienthal repeatedly cautioned others against attempting to glide in winds which exceeded 7 meters a second (about 15 $\frac{1}{2}$ miles an hour), as being excessively dangerous. However, when he made the improvement on his machine of superimposing two smaller surfaces and thereby reduced the "tip to tip" measurement from about 24 feet to 18 feet, the diminished leverage upon which the gusts could act enabled him to sail in stronger winds, so that he even experimented in winds of 22 miles an hour. This, without further improvement in the automatic stability of his machine, was an unwise thing to do, and the accident which occasioned his death, on the 9th of last August, is, more or less correctly, attributed to it; nevertheless, the immediate cause was undoubtedly the result of defects in the machine itself, which had been allowed to deteriorate and get out of repair. In his double-deck machine the upper surface was joined to the lower one by two or three small vertical posts and numerous wires. Probably some of these wires had become rusted or so weakened that they broke when the machine was struck by a heavy puff, and so allowed the upper surface to tilt back and suddenly stop the headway of the machine, but not that of the unfortunate operator, who swung round and round over the apparatus as it pitched to the ground. This tendency to revolve over backward is frequently set up by a very strong sudden gust striking the machine squarely in front; it can, however, be counteracted by a quick movement of the operator's body and legs toward the front. A serious defect in the design of the Lilienthal apparatus is here seen, for on it the operator's position is somewhat strained, and his movement very limited, owing to the fact that he is obliged to hold to a small bar with both hands while his weight is carried on his elbows, which rest, a little farther back, on a portion of the main frame. (See Plate IX, Fig. 2.)

These defects suggested themselves to the writer when, in the summer of 1894, he built a machine similar in many respects to that of Lilienthal. It differed from his in two important particulars: first, the upward movement of the horizontal tail was limited; second, the range through which the operator could shift his weight was nearly three feet instead of about eight inches. To obtain this range of movement the weight of the body, when in flight, rested upon two horizontal bars fitting under the armpits. (See Plate IX, Fig. 3.)

No very startling results, however, were obtained with this machine or with the three subsequent ones, the longest flight attained being only 187 feet in length. Experiment with these machines, nevertheless, furnished a great deal of valuable information. No one who has not experimented with a machine of the Lilienthal type can form any accurate conception of the tremendous power and lifting effect which 130 to 150 square feet of concave surface can exert. It is with an apparatus of this kind that a novice first becomes fully aware that no wind is anything like constant, and that the power of those much-talked-of "gusts" is real, and not imaginary.

Anyone wishing to begin experiment with a gliding machine cannot be too cautious in the selection of an experimental station. Nothing could be more dangerous than to start from a flat roof or a precipitous cliff, or to begin experiment in a locality where surrounding objects, such as hills, buildings, or even large neighboring trees, are likely to break up the wind into swirls and eddies. What is most desirable—in the beginning, at least—is a hill surrounded by country that is as level as possible. Both the starting and landing points should be on comparatively soft earth, free from stones, bushes, and snags. Perhaps the best station of all is to be had where there are high, bare sandhills or dunes, facing a large body of water. The slope of such a hill (to a beginner) is of as much importance as anything else; it must be steep at the top and run off gradually as it nears the bottom so that when he has gained proficiency enough he may start from near the top in calm weather and yet have his flights always close to the hill-side. If this be so and the soil be comparatively soft, the operator can easily save himself from a dangerous fall which might result from a poor start or a breakage of the machine.

In the first experiment with a full-sized gliding machine, a man's natural instincts irresistibly impel him to move in the wrong direction when the balance of the apparatus is disturbed. It is, therefore, at first, impossible to distinguish between effects produced by one's own errors and those produced by wind changes, but in time three separate causes of unsteady flying become easily distinguishable from each other; namely, improper adjustment of the machine, errors of the operator, and changes in the trend, velocity, or direction of the wind.

When the mastery of the machine becomes about as perfect as possible very much of this unsteadiness disappears. Nevertheless, with a wind as steady as winds ever are,—even after having blown for hundreds of miles over absolutely level prairie, or, as in the case of our later experiments, having come in an unobstructed path for several hundred miles over the waters of Lake Michigan,—we found that the effect was by no means a steady one, but was such as to indicate that they were broken up into an inconceivable number of irregularities in pressure, velocity, and direction, in spite of the fact that a light anemometer showed fluctuations in velocity of seldom more than 10 or 12 per cent in readings of 5 seconds duration, taken 10 to 20 seconds apart.

In a wind of 9 to 10 miles per hour with a simple machine of the Lilienthal type the disturbances of the wind are barely noticeable, but at 12 miles they are quite apparent, at 14 miles they require considerable practice to combat, at 16 $\frac{1}{2}$ miles, even with best skill at command, a flight is more or less risky; and when the wind blows above 18 miles per hour it is dangerous, even with a total load of 182 pounds (machine, 27 pounds, operator, 155 pounds) on 130 square feet of surface.

Nothing, perhaps, is more surprising than the power which a gust in even a 14 or 15 mile wind will occasionally exhibit, such, for instance, as sometimes happens to an inexperienced or careless operator, who, in facing the wind with the machine preparatory to making a start, suddenly finds himself lifted anywhere from 2 to 10 feet above his starting place. These flights are invariably backward, and are due to mismanagement in allowing the wind to catch under the surface of the machine while the operator is too far back on it to exert a proper control. In the hands of a skilled person, the flights, in mild winds, generally appear to an on-looker as remarkably smooth, and even in spite of the fact that the operator is seen to frequently shift his position on the machine with considerable rapidity; yet in slightly stronger winds—those of 15 to 16 ½ miles per hour (mean velocity)— the irregularities become very perceptible to the spectator, who may sometimes see the apparatus rock and toss not unlike a ship in a rough sea.

To appreciate the causes which render a gliding machine, or in fact any machine of the aerocurve type, unstable, it is necessary to understand, in a measure, both the peculiarities of the wind and the effect they have on the position of the center of pressure of the surfaces.

In order for any apparatus in free air to be in equilibrium it is necessary, of course, that the center of pressure should be in the same vertical line as the center of gravity. It is not, as many believe, absolutely necessary that the center of weight should be beneath the sustaining surfaces; this may be demonstrated by trying the small paper model shown in Plate XII, Fig. 1, which, if not weighted too heavily, will always fly with the "fin" side up, even if dropped with the weight and fin side undermost.

When a surface is inclined to the air through which it is moving, the lifting pressure is not uniform, but is very much greater toward that edge which is first struck by the current. On a square *plane* 100 inches on a side, the center of all the lifting pressures may be anywhere between the center of the figure and as far forward (apparently) as 14 inches from the front edge, according to the angle and speed at which the surface is presented to the air. The travel sidewise might be even more, granting that gusts may come from either side. With an aerocurve the travel is probably seldom more than $\frac{1}{5}$ as much as it would be with a plane. In practice with any ordinary gliding machine of large surface it is found that the gusts come from any quarter: in front of the apparatus, from the extreme left to the extreme right, and in a wind of over 22 miles per hour they follow each other with such suddenness and with such extreme changes that it is absolutely impossible to shift one's weight in time to counteract them. These conditions, which had to be met, made it imperative to seek for automatic stability along very different lines from any that had heretofore been tried.

The changes or gusts which have the most influence in disturbing the machine are seldom of more than half a second duration, and oftentimes they last less than half that time; yet in so short an interval, it has frequently happened to me, in my experiments with my first three gliding machines, that in less than half a second the lateral equilibrium was so far disturbed that the lateral axis of the machine would make an angle of 35 to 40 degrees with the horizontal. At other times, the angle of advance (the angle at which the surfaces are presented to the air) was so much increased that nearly every bit of the headway was destroyed. On two occasions, the change in direction, both vertically and to the side, was so violent and sudden as to shake my hold loose of the machine.

During my last flight on a Lilienthal type of machine, while experimenting in a wind of about 18 miles an hour, the machine was struck twice in quick succession by a gust from the right. The first impulse raised that side until the apparatus stood at an angle of about 40 degrees; the second impulse, which came between ¼ and ¾ of a second later, increased the inclination to nearly a vertical one, so that one wing pointed to the ground and the other to the zenith. Anticipating a complete overturning of the machine (as did happen) I let go my hold and dropped to the sand below, a distance of not more than 12 or 14 feet, where I landed on my feet, but on the left wing of the overturned machine, which had drifted under me as I fell. This accident damaged the machine so much that it was not rebuilt. We recognized from these experiments that the disturbances increased much more rapidly than the mean velocity of the wind; also that in winds of 18 miles or over, it was impossible for a man to shift his weight far enough and rapidly enough on a single surface machine to keep it in proper equilibrium under all circumstances.

In addition, the conclusions were reached, first, that the angle of advance *must* be automatically maintained, with almost absolute certainty, at a very small angle; second, that the lateral equilibrium should also be largely, if not wholly, automatic, and be maintained by some more effective method than a dihedral angle between the surfaces, or by placing the operator far beneath the apparatus. The first of these is considerably the most important, as disturbances of the angle of advance generally entail disturbances of the lateral equilibrium as well.

In the beginning of experiment to obtain longitudinal stability, three clearly defined methods (each with its own limitations developed by experiment) appeared open.

The first and simplest method is to find such a surface, or grouping of surfaces, that the displacement of the center of pressure is very great for very small changes of the angle of incidence; the second method is to find such a form of surface that its center of pressure remains in one spot, no matter from what angle the relative wind may come; the third method is to provide a separate mechanism to either take up or counteract the disturbing effects of the wind changes.

The first method may be said to have been practically attained (as far as it is attainable by such an arrangement) in the various modifications of the Hargrave kite, and also in its predecessor, the Brown biplane, or, a little better still, in what might be called a "bicurve,"—that is, a biplane machine on which slightly arched surfaces have been substituted for planes. Such a model is shown in Plate IX, Fig. 4, B being the front aerocurve. This model will maintain a very good equilibrium for gliding flight so long as the center of gravity is anywhere between C and the rear edge of B; but each change in the position of the center of gravity corresponds, of course, to a different angle of flight, *i.e.*, to an angle at which the center of pressure of the combined surfaces coincides with the vertical line through the center of gravity of the whole apparatus.

The center of pressure travels toward B as the angle of flight is diminished, and in the reverse direction when it is increased. This great range in the possible position of the center of pressure is a measure of the corresponding change in efficiency which the rear surface undergoes at various angles of flight. This efficiency diminishes very rapidly as the angle of flight is diminished; at very flat angles its useful effect disappears altogether. Not only this, but at small angles of incidence, those under .8 degrees, the longitudinal stability of the arrangement disappears as well. From many experiments with gliding models of this type I found it impossible to obtain glides which represented a travel of over 4 lineal feet for each foot of height lost, unless the model was so weighted that no part of the weight rested on the rear surface. Also, that the power required to support any given weight at an angle of 15 degrees or less is about twice as much as would be needed on superimposed surfaces of the same size held at the same angles of inclination.

When the model shown in Plate IX, Fig. 4 (or any similar one) is so loaded that each surface must carry about half the total weight, the apparatus will take up an angle of about 26 degrees with the relative wind. Under this condition a dynamic model would require between 40 and 46 per cent of its weight in thrust to keep it "afloat," or if liberated as a gliding model it will travel forward a little over *twice* as far as it descends vertically. At the flattest angle, about 15 degrees, at which it maintains a good equilibrium, it will glide only 3 to 3 ¼ times as far as it falls.

As the rear surfaces in an apparatus with following surfaces come more and more in the "wake" of the front elements, the relative supporting effect of the rear becomes less and less as the angle of flight is diminished. Thus, through the phenomena of "interference," the relative efficiency (as a lifting factor) of the rear surfaces becomes greater or less (as the angle of flight is increased or diminished)—a corresponding travel of the center of pressure results which maintains the longitudinal equilibrium, but at the expense of extra weight of the apparatus and considerable additional power.

This fundamental principle—that of interference—underlies the stability of the Malay, Eddy, Bazin, Lamson, Chanute, and Hargrave kites. It is still further applied in the three last named, in which the vertical keels form pairs of Brown's "biplanes," which maintain the lateral equilibrium as well.

It would seem that one might be able to avail of the wonderful stability which a system of following surfaces exhibits, by so grouping the surfaces *vertically* as well as horizontally that they could not interfere. It is comparatively easy to so space them that interference is practically avoided, but from several hundred experiments in this direction I have invariably found *that the automatic equilibrium is always impaired in direct proportion as the front and rear surfaces cease to interfere*. The further conclusion arrived at from these experiments was that in following surface machines, a low efficiency is essential to insure safe equilibrium. Quantitively this efficiency is so low that probably less than 30 pounds can be carried per horsepower when the surfaces are loaded to a greater extent than one pound per square foot of area. Consequently, though dynamic models might be made that would work satisfactorily on this plan, a full-sized machine to carry even one man would offer no such encouragement to its projector, chiefly because the weight of a machine increases much more rapidly than its surface or supporting power. Following surfaces therefore are not available.

Just as experiment and careful measurement made this fact clear, a new prospective method of obtaining automatic equilibrium began to open up. A study of the peculiarities in the travel of the center of pressure of variously arched surfaces indicated the possibility of evolving such a form that the center of the lifting pressures would remain in the same spot for all angles of inclination. The result of much investigation in this line is shown in Plate XI, Fig. 1. Strictly speaking, this piece of apparatus is a gliding model, and as such in a wind of 30 miles an hour or less it possesses a perfect equilibrium, owing to the fact that the position of its center of pressure is almost absolutely constant for all angles of incidence between plus 90 degrees and minus 20 degrees; the same is also approximately true whether the wind strikes from in front or "abeam." It will also fly as a tailless kite, but as such is somewhat inferior to the Hargrave, both in steadiness and in the "angle of the string," yet the lifting effect is probably four times as great per square foot of surface. The projected area of the dome is not quite 6 square feet, yet in some experiments it has registered a pull of over 40 pounds on a spring-balance, and on one occasion repeatedly broke a cord tested to 69 pounds.

As a flying machine it would have the advantages of being able to sustain great weight on a very small surface, and would require but a slow speed to do so. A dome machine of only 9 $\frac{1}{2}$ feet in diameter would be of sufficient size to carry a man in gliding flight at a speed (relative to the air) of only 20 to 21 miles per hour. But its drift is so great that it would not carry the operator horizontally more than 2 $\frac{1}{4}$ to 2 $\frac{3}{4}$ times the height from which it started. As a great thrust of the screw is far more costly in power than great speed, such a machine could hardly be a practical success. However, owing to the very low sailing speed, a dynamic machine to carry one man might be built which would fly. It would be of little practical value, owing to the excessive power required, and its limited speed capacity. This line of experiment was therefore laid aside in the spring of 1896.

By way of explanation I may here add that from a great number of previous experiments with various devices—such as modifications of the drag rudder, gyrostat, and pendulum regulators—I had come to the belief (which all my more recent experiments have only served to strengthen) that *the action of any device to maintain a machine in safe equilibrium must be such that it prepares the machine for each impending wind change before that change actually occurs*, and that any device which tends to forcibly right the flying machine *after* it has departed from an even keel more often produces (in the open air) greater *unsteadiness* than the reverse. The reason for this is not far to seek. It lies chiefly in the fact that the most formidable, as well as the most frequent, disturbances met within the natural wind are cycloid gusts or rotating masses of air which frequently give a machine (or model) powerful double impulses. These impulses are generally opposite in their effect, and succeed each other by irregular intervals varying from about one-fourth to one second apart; and, therefore, a regulator, such as a pendulum or gyrostat mechanism, which begins to act on the machine after it is disturbed, and continues to do so until it regains an even keel, is often the means of greatly augmenting the second impulse of the pair, or the first of a new gust which may strike the apparatus from a different quarter. I do not mean to say that such devices will not work at all; on the contrary, they can be made to give very good results in fairly mild weather, but they all fail in winds of much less velocity than those which any practical machine must be able to contend with.

This conclusion reached, it would appear that the methods left would be found only through a careful study of the wind changes themselves. As before stated, I had become aware almost from the beginning of my gliding experiments of several distinct kinds of disturbances, the most formidable being very sharp, well-defined changes in the velocity and direction of the wind, which last but a fraction of a second and appear to come in pairs. Their distinguishing characteristic (besides their much greater suddenness and power) is, that in practically all cases they are preceded by a perceptible warning which generally consists of a slight strengthening of the wind followed by a momentary calm, which in turn is followed immediately by the "gust" in its fullest force. During the momentary freshening, the wind either comes from or veers in the direction from which the gust proper will strike. These changes can easily be verified by an observer in a strong wind by noting the effects as he feels them on his face.

There are many observations which might be given to corroborate the theory that practically all the gusts which have any material effect in disturbing the angle of advance or the lateral equilibrium of an apparatus are of a rotary character. They are, in fact, nothing more or less than diminutive tornadoes which travel, however, much more rarely on vertical axes than on diagonal or horizontal ones. In a few cases the axis of a gust is found to be horizontal and parallel with the wind. In the majority they are nearly horizontal, but across the direction of the mean wind. The direction of rotation is usually backward, i.e., in the reverse direction that a wheel would have in rolling over the ground. In what may be called steady winds the swirls are of much greater diameter, and the out-flowing eddies, or the momentary freshening of the wind, precede them by a longer interval.

The observations which led to a recognition of the rotary character of the wind changes also were the means which furnished an explanation of the action of certain simple devices, previously found to work with success as far back as 1890 on a small dynamic model which has been illustrated in a previous issue of this Annual. I was, therefore, not wholly in the dark in commencing experiments to produce a regulator which should prepare the apparatus to meet each particular "gust" before it arrived.

The horizontal regulator of the dynamic model was slightly modified to adapt it more closely to the new theory, and in May, 1896, applied to the kite shown in Plate XI, Fig. 2. This kite, which is here shown in a 28-mile wind, possessed such perfect power in maintaining the surfaces at a small angle with the wind, through changes which would otherwise prevent it from flying at all, that in momentary freshening or changes, it would rise until the strings passed the zenith and made an angle of 6 to 8 degrees beyond the vertical. The average angle maintained by the surfaces with the horizontal varied between such narrow limits that it could not be easily detected by the eye. From a number of observations it was found possible to set the regulator to maintain an angle of between 2 and 3 degrees (above the horizontal), and calculations from the weight and surface of the kite, the pull on the string, and its angle above the horizontal show that the lift and drift of the kite correspond very closely indeed with the theoretical ones computed from the annexed tables.

Later in the summer this regulating device was improved and its use extended so as to counteract the rotating columns whose axes were more or less vertical, and thus preserve the lateral equilibrium of the apparatus. With this change it was applied to the gliding machine shown in Plate XIII, Fig. 1. By its use the safe limit of wind in which experiments could be carried on was raised from 16 ½ miles per hour (with the simple Lilienthal, or 20 miles with the Lilienthal double deck) to over 30 miles, and with it the maximum length of flight was increased from 187 feet to 359 feet; at the same time the rocking and tossing of the apparatus was reduced to such an extent that an on-looker could not in any of the 150 to 200 flights detect that the apparatus in flight ever departed from an even keel, either laterally or longitudinally,—i.e., the angle of advance was maintained perfectly at the very flattest angle. It is evident from repeated measurements that this angle never exceeded 4 degrees with the relative wind.

The difference in the amount of ascending trend of the wind at different times and at various points in front of the hill made great differences in the length of flights. The results of an average flight in calm air are here given:

Net projected area of 2 supporting surfaces, 134 square feet; size, 16 feet 2 inches x 4 feet 4 inches.

Net area of horizontal tail (which receives a pressure on its upper side), 19 square feet.

Weight of machine, 23 pounds.

Weight of operator, 155 pounds.

Press upper side of tail (acting as weight), about 7 pounds.

Total weight carried by 134 square feet, 185 pounds.

Total weight carried per square foot area, 1.37 pounds.

At the time of the following experiment the air was nearly calm, the only trace of wind was from the northeast; the flight was made by running downhill toward the north. Length of flight, 242 feet from last footprint to first at landing; time of flight, 7.4 seconds (in the air); difference in level between points was 42 ½ feet. Speed of machine was therefore practically 22 miles an hour. The wind pressure is 22 x 22 x .005 = 2.42 pounds per square foot. The proportion of this as a sustaining factor was 1.37/2.42, or 57 per cent. By referring to the tables hereto appended it will appear that this amount of lift (57 per cent of the normal pressure) corresponds to a positive angle of the surfaces of between 3 and 4 degrees, and by referring to the third column we find that the drift of the surfaces is (.0525 for 3 degrees and .0582 for 4 degrees) about .056 times the total weight of machine and operator and negative pressure on the tail, or .056 x 185 = 10.36 pounds, which is drift of the surfaces alone; to it we must add the head resistance offered by the framing of the machine and that offered by the operator's body. The framing consists of 64 lineal feet of timber which forms the main arms of the wings; this has a thickness of an inch across the wind, and therefore exposes a cross-section surface of about 3.3 square feet. The upright posts are 64 feet in collective length, being on an average 6/10 of an inch in width (across the wind). Their area is therefore practically 3.2 square feet. But as they are sharpened more or less to lessen the resistance they offer to the wind, the total area offered by the woodwork, instead of being 3.3 + 3.2 = 6 ½ square feet, is equivalent to only 2 ½ square feet. Besides this the framings of the tail and vertical rudder expose an equivalent of half a square foot of surface. The regulator and its cords, bands, etc. expose .52 of a square foot, and 160 lineal feet of wire .05 of an inch in diameter, expose the equivalent of .5 square foot more, making the total equivalent area exposed equal to 4.02 square feet. This at 22 miles an hour would offer a resistance of 4.02 x 2.42 = 9.73 pounds; to this must be added the resistance offered by the 5 square feet of the operator's body, arms, and legs. This brings the total resistance to: Resistance of surfaces, or drift, 10.36 pounds; head resistance of machine, 9.73 pounds; and resistance offered by operator, 12.1 pounds; total = 32.19 pounds. This moved over a distance of 242 feet would consume 242 x 32.19 = 7,790 foot-pounds, which would be furnished by the weight of the machine and operator (178 pounds), descending through a vertical distance of 7,790 ÷ 178 = 43 feet 9 inches, against an actual measured height of only 42 ½ feet. The difference in energy can easily be accounted for in either of three ways: First, a slight overestimate of the resistance offered by the operator's body; second, the presence of a slight ascending current of air; or, lastly, that the speed gained in running down the hill at the start was greater than 22 miles an hour. (It is possible to gain a speed of 26 miles if the weight is about half supported on the machine.) It may be interesting to point out in passing that the energy (7,790 foot-pounds) absorbed in keeping the machine and operator afloat during 7.4 seconds represents barely 2-horsepower, but less than one-third of this is drift of the surfaces. It is, however, now pretty well known that it would take at least 3-horsepower to produce a thrust of 32 pounds, even with as large screws as could be conveniently carried on a machine of this size.

The details of the regulating mechanism of neither this nor the "three-deck" machine have been here given, as they are now the subject of applications for patents; nevertheless, to anyone wishing to repeat the experiments I shall be pleased to give all the information necessary.

During October I constructed a new machine of the same general design as the "double deck" but provided with three superimposed surfaces instead of two. In this a considerable change was made in the mechanism which governed the lateral equilibrium. Instead of depending upon the power in the small eddies which precede a rotating gust to operate the machine, their power was used only to work the valves of a mechanism operated by compressed air; in this way the regulation (which is accomplished through a reflex action) became much more powerful and prompt. The tests of the new gliding machine showed that a considerable advance had been made, in that the limit of wind velocity in which flights were safe was raised from 31 ½ miles an hour to over 48, and the maximum length of flight increased from 359 feet to 927 feet (best) and 893 feet (second best); at the same time it was found quite safe to turn the apparatus and fly at a considerable angle with the wind. It was by this means chiefly that the length of flight was increased, as the longer flights were made while "quartering" on the wind; that is, the apparatus after starting (the start must always be made dead against the wind) was kept pointing at an angle from 15 to 35 degrees with the wind, according to the strength of the latter, while the apparatus itself moved along a course nearly, but not quite, at right angles to the wind. This enabled me to keep close to the hillside and take advantage of the rising current of air flowing over the slope. In a few of the flights it would have been possible to have landed on a higher point than the starting one, owing to irregularity of the wind which occasionally raised me, after having gone several hundred feet, to a level above my starting place; these rises were only momentary, and all the flights as a whole were on a descending grade. As the slope both to the right and left had several clumps of small trees which it was necessary to steer over or around (according to the height at which the machine happened to be while it passed near them), these "quartering" flights were not made to any great extent.

With a machine on which the angle of advance is automatically controlled with a fair degree of accuracy, the steering requires but little more effort than a bicycle, and at the same speed, *i.e.*, above 20 miles an hour, I have much doubt in my mind whether a bicycle (on a level road) could be turned on a much shorter radius than a flying machine. It is possible to land within less than 5 feet of any predetermined spot if it be selected well within the range of flight of the starting place.

In a few of the glides with the last machine I attempted to carry an additional weight, in the shape of a bag partially filled with sand. This bag was fastened between the middle and bottom surfaces and, beginning with about 12 pounds, the weight was gradually increased until 41 lbs. were carried without materially shortening the length of flight. The heavier weight considerably increased the difficulty in landing in light wind, owing to the greater speed relative to the ground. In high winds it was of very little hindrance either in starting or landing. The object in view in experimenting with the weight was to ascertain the power required on a dynamic machine and to test the manageability of the apparatus with a weight equivalent to the necessary engines and supplies. The result was so very encouraging that I have since then commenced constructing the engines. One of the pair is shown in Plate XIII, Fig. 2. It develops (alone) about two-thirds of the total thrust-power needed; its weight is only 12 pounds; its action is, however, a little irregular, and on that account is still the subject of experiment; it is a gasolene engine of the Otto cycle type.

Not least among the interesting results brought out by these gliding experiments is the fact, which becomes more and more evident from repeated experiment, that there is a very great difference in the supporting power of the air, whether one faces the natural wind or advances through still air; for while a natural wind of 18 to 19 miles per hour is sufficient (over level ground) to support the double-deck machine and operator, and will even momentarily raise them directly in the air for a foot or two, the same machine requires a minimum speed of 22 miles to support the same weight at the same angle in still air.

On Plate XIV are given drawings of the three-deck machine; another drawing will be found on Plate XII, Fig. 2, giving exact sizes of struts, etc., in cross-section. If built to scale the machine will have 227 square feet (net) surface. This is, however, 40 per cent more than a man of average weight ever requires except in a calm or in winds of less than 10 miles an hour. In winds over 12 miles an hour a beginner will get along better with the upper surface removed. If this machine is built on such a scale that the dimensions are only two-thirds of those given, it will be of ample size (103 square feet surface) for the average operator in any wind of over 25 miles an hour. It is better not to reduce the size of main spars and struts at all from the sizes given for the larger machine when constructing the surfaces on a smaller scale. The sizes given are for the best grade of black or silk spruce only; this wood will stand at least 16,000 pounds to the square inch; it must be straight grain and entirely free from flaws.

There is, perhaps, no better sport imaginable than coasting through the air, especially so where the flights are comparatively long. In moderate winds, of 18 to 25 miles per hour, the path of the machine is often quite horizontal for a hundred feet or so after leaving the hillside, until, in fact, the rising current which flows over the hill has been cleared. If during the first part of the flight an operator wishes to keep near the ground he may do so by moving an inch or two forward on the machine; he will find, however, that in thus sailing downward through an ascending wind the speed increases at a tremendous rate.

Perhaps the most trying ordeal is experienced when the machine unexpectedly encounters a strongly ascending current of air which may raise the operator, in some instances, 40 or 50 feet above his line of flight. Such occurrences are comparatively frequent in a wind of 30 miles or over, but are not dangerous so long as the regulating mechanism remains in working order, as the machine then retains an absolutely level keel. I have twice been raised as much as 40 feet above my starting point without either myself or those who were on the ground being able to detect any change whatever in the inclination of either axis of the machine. Considering the fact that the rise through even such a distance seldom takes more than $1\frac{1}{4}$ to $1\frac{1}{2}$ seconds, the automatic stability of the machine would seem to be well attained.

After having adjusted the regulators, and repeatedly tested them in a number of short flights, and at the same time having found the correct position for his weight, all the beginner need do, after starting, is to keep as still as possible and he will make a very creditable flight. If it be necessary to steer to the right or left, moving the body over to that side and a little forward will accomplish the result. For ordinary steering it is seldom necessary to do more than stick out one leg toward the side to which you wish to turn. If you meet a very strongly ascending trend of wind it is manifested by an increase in the weight which appears to rest on your arms; in such a case the vertical rise may be greatly diminished by moving 2 or 3 inches forward of the normal position as long as the rise continues; it is better, though, to simply stick the legs out in front.

Descending currents of air diminish the weight on the arms and give one the sensation experienced in a quick-starting elevator on a down trip. So far, out of, possibly, over 300 trials with the regulated machines, a descending current has never brought the machine quite down but once, but even then the dropping speed was not too great to make a comparatively easy landing possible. On the other hand, the machines have been momentarily raised above their line of flight in probably 2 flights out of every 5. And in winds above 25 miles an hour the machines have risen above the starting point in as many as 75 per cent. of the flights. The highest rise was probably little short of 60 feet. The most difficult thing a beginner has to learn is how to land, *i.e.*, when to move back on the machine in order to check its headway; this knowledge can only be gained by actual experiment.

Document 1-9(c), Octave Chanute, "Experiments in Flying," 1900.

It is considerably over forty years since I first became interested in the problem of flight. This presented the attraction of an unsolved problem which did not seem as visionary as that of perpetual motion. Birds gave daily proof that flying could be done, and the reasons advanced by scientists why the performance was inaccessible to man did not seem to be entirely conclusive, if sufficiently light motors were eventually to be obtained. There was, to be sure, a record of several thousand years of constant failures, often resulting in personal injuries; but it did not seem useless for engineers to investigate the causes of such failures, with a view to a remedy. I, therefore, gathered from time to time such information as was to be found on the subject, and added thereto such speculations as suggested themselves. After a while this grew absorbing, and interfered with regular duties, so that in 1874 all the accumulated material was rolled up into a bundle and red tape tied around it, a resolution being taken that it should not be undone until the subject could be taken up again without detriment to any duty. It was fourteen years before the knot was untied.

Meantime a considerable change had taken place in the public attitude on the question. It was no longer considered proof of lunacy to investigate it, and great progress had been made in producing artificial motors approximating those of the birds in relative lightness. The problem was, therefore, taken up again under more favorable circumstances. A study was begun of the history of past failures, and the endeavor was made to account for them. In point of fact, this produced a series of technical articles which swelled into a book, and also led to the conclusion that, when a sufficiently light motor was evolved, the principal cause of failure would be that lack of stability in the air which rendered all man-ridden flying machines most hazardous; but that, if this difficulty were overcome, further progress would be rapid.

Experiments were, therefore, begun to investigate this question of stability and safety, and, if possible, to render the former automatic. These experiments were hundreds in number, and were, at first, very modest. They consisted in liberating weighted paper models of various shapes, either ancient or new, with gravity as a motive power, and observing their glides downward. This was done in still air. After a while, resort was had to larger models, with muslin wings and wooden frameworks, carrying bricks as passengers; and these were dropped from the house-top in the early morning when only the milkman was about. Very much was learned as to the effect of the wind; and then tailless kites of all sorts of shapes were flown, to the great admiration of small boys. During the seven or eight years within which this work was carried on, some glimmerings were obtained of the principles involved, and some definite conclusions were reached. But it was only after Lilienthal had shown that such an adventure was feasible that courage was gathered to experiment with full-sized machines carrying a man through the air.

Otto Lilienthal was a very able German engineer and physicist. He demonstrated that concave wings afforded, at very acute angles, from three to seven times as much support as flat wings in the air. He made, from 1891 to 1896, more than 2,000 successful glides, the longest being about 1,200 feet, upon machines of his own design, launching himself into the air from a hilltop and gliding down against the wind. In 1895, he endeavored to add a motor, but found that this complicated the handling so much that he went back to his gliding-device. It was while experimenting with a double-decked machine of this character, which probably was in bad order, that he fell and was killed, in August, 1896. Thus perished the man who will probably be credited by posterity with having pointed out the best way to preliminary experiments in human flight through the air.

Just before this dismal accident, I had been testing a full-sized Lilienthal machine. I discarded it as hazardous, and then tested the value of an idea of my own. This was to follow the same general method, but to reverse the principle upon which Lilienthal had depended for maintaining his equilibrium in the air. He shifted the weight of his body, under immovable wings, as fast and as far as the sustaining pressure varied under his surfaces. This shifting was mainly done by moving the feet, as the actions required were small except when alighting. My notion was to have the operator remain seated in the machine in the air, and to intervene only to steer or to alight; moving mechanism being provided to shift the wings automatically, so as to restore the balance when endangered. There are several ways in which this can be done. Two of them have been worked out to a probable success in my experiments, and there is still a third which I intend to test in due course.

To make such experiments truly instructive, they should be made with a full-sized machine and with an operator riding therein. Models seldom fly twice alike in the open air (where there is almost always some wind), and they cannot relate the vicissitudes which they have encountered. A flying-machine would be of little future use if it could not operate in a moderate wind; hence the necessity for an operator to report upon what occurs in flight, and to acquire the art of the birds. My own operations were conducted from that point of view, with the great disadvantage, however, that being over threescore years of age, I was no longer sufficiently young and active to perform any but short and insignificant glides in such tentative experiments; the latter being directed solely to evolving the conditions of stability, and without any expectation of advancing to the invention of a commercial flying-machine. I simply tested various automatic devices to secure equilibrium, and, with great anxiety, employed young and active assistants. The best way to carry on such adventures is first to select a soft place on which to alight. This is well secured on a dry and loose sand-hill, and there ought to be no bushes or trees to run into. Our party found such sand-hills, almost a desert, in which we pitched our tent, on the shore of Lake Michigan, about thirty miles east of Chicago. The main hill selected was ninety-five feet high; but the highest point started from was sixty-one feet above the beach, as the best instruction was to be obtained from short glides at low speeds.

With parties of from four to six persons, five full-sized gliding-machines (one rebuilt) were experimented with in 1896, and one in 1897. Out of these, two types were evolved, the "Multiple-Wing" and the "Two-Surfaced," which are believed to be safer than any heretofore produced, and to work out fairly well the problem of automatic equilibrium. The photographs herewith reproduced, many of them heretofore unpublished, are from snapshots taken of these two types. In 1896, very few photographs were taken, all the attention being devoted to studying the action of the machines, and the one picture shown is the sixth permutation of the "Multiple-Wing" machine, so-called. In 1897, there was more leisure to take snap-shots, as the machine used was a duplication of the "Two-Surfaced" of 1896, supplied with a regulating mechanism designed by Mr. A. M. Herring, my assistant. Each photograph was taken from a different experiment (there were about 1,000 glides); but the point of view was varied, so as to exhibit the consecutive phases of a single flight. The frog-like appearance of some of the legs is due to the speed.

The first thing which we discovered practically was that the wind flowing up a hillside is not a steadily flowing current like that of a river. It comes as a rolling mass, full of tumultuous whirls and eddies, like those issuing from a chimney; and they strike the apparatus with constantly varying force and direction, sometimes withdrawing support when most needed. It has long been known, through instrumental observations, that the wind is constantly changing in force and direction; but it needed the experience of an operator afloat on a gliding-machine to realize that this all proceeded from cyclonic action; so that more was learned in this respect in a week than had previously been acquired by several years of experiments with models. There was a pair of eagles, living in the top of a dead tree about two miles from our tent, that came almost daily to show us how such wind effects are overcome and utilized. The birds swept in circles overhead on pulseless wings, and rose high up in air. Occasionally there was a side-rocking motion, as of a ship rolling at sea, and then the birds rocked back to an even keel; but although we thought the action was clearly automatic, and were willing to learn, our teachers were too far off to show us just how it was done, and we had to experiment for ourselves.

The operator stands on a hill-side. He raises up the apparatus, which is steadied by a companion, and quickly slips under and within the machine. He faces the wind. This wind buffets the wings from side to side, and up or down, so that he has much difficulty in obtaining a poise. This is finally accomplished by bracing the cross-piece of the machine's frame against his back, and depressing the front edge of the wings so that they will be struck from above by the wind. His arm-pits rest on a pair of horizontal bars, and he grasps a pair of vertical bars with his hands. He is in no way attached to the machine, so that he may disengage himself instantly should anything go wrong. Then, still facing dead into the wind, he takes one or two, never more than four, running steps forward, raising up the front edge of the apparatus at the last moment, and the air claims him. Then he sails forward into the wind on a generally descending course. The "Multiple-Wing" machine was provided with a seat, but, goodness ! there was no time to sit down, as each glide of two to three hundred feet took but eight to twelve seconds, and then it was time to alight. The latter phase of the problem had been the subject of meditation for months, and the conclusion had been reached to imitate the sparrow. When the latter approaches the street, he throws his body back, tilts his outspread wings nearly square to the course, and on the cushion of air thus encountered he stops his speed and drops lightly to the ground. So do all birds. We tried it with misgivings, but found it perfectly effective. The soft sand was a great advantage, and even when the experts were racing there was not a single sprained ankle.

The rebuilt "Multiple-wings" were pivoted at their roots, and vibrated backward and forward on ball-bearings, restrained by rubber springs. As the wind varied, they adjusted themselves thereto, and brought back the supporting air pressure over the operator, thus reestablishing the threatened balance. This was done automatically. But in consequence of various defects in construction and adjustment, the operator still had to move one or two inches, as against the from seven to fifteen inches of movement required by the Lilienthal apparatus. Some two or three hundred glides were made with the "Multiple-wing" without any accident to man or machine, and the action was deemed so effective, the principle so sound, that the full plans were published in the *Aëronautical Annual* for 1897, for the benefit of experimenters desiring to improve on this apparatus.

There is no more delightful sensation than that of gliding through the air. All the faculties are on the alert, and the motion is astonishingly smooth and elastic. The machine responds instantly to the slightest movement of the operator; the air rushes by one's ears; the trees and bushes flit away underneath, and the landing comes all too quickly. Skating, sliding, and bicycling are not to be compared for a moment to aërial conveyance, in which, perhaps, zest is added by the spice of danger. For it must be distinctly understood that there is constant danger in such preliminary experiments. When this hazard has been eliminated by further evolution, gliding will become a most popular sport.

The "Two-surfaced" machine, so-called, produced longer and more numerous glides. There were perhaps 700 or 800, at a rate of descent of about one foot in six; so that while the longest distance traversed was 360 feet, we could have sailed 1,200 feet, had we started from a hill 200 feet high. In consequence of the speed gained by running, the initial stage of the flight is nearly horizontal, and it is thrilling to see the operator pass from thirty to forty feet overhead, steering his machine, undulating his course, and struggling with the wind gusts which whistle through the guy wires. The automatic mechanism restores the angle of advance when compromised by variations of the breeze; but when these come from one side and tilt the apparatus, the weight has to be shifted to right up the machine. This is generally done by thrusting out the feet toward the side which has been raised, a movement which is just the reverse of what would be instinctively made on the ground, but which becomes second nature to an expert. These gusts sometimes raise the machine from ten to twenty feet vertically, and sometimes they strike the apparatus from above, causing it to descend suddenly. When sailing near the ground, these vicissitudes can be counteracted by movements of the body of three or four inches; but this has to be done instantly, for neither wind nor gravity will wait on meditation. At a height of 300 or 400 feet the regulating mechanism would probably take care of these wind gusts, as it does, in fact, for their minor variations. The speed of the machine is generally about seventeen miles an hour over the ground, and from twenty-two to thirty miles an hour relative to the air. Constant effort was directed to keep down the velocity, which was at times fifty-two miles an hour. This is the purpose of the starting and gliding against the wind, which thus furnishes an initial velocity without there being undue speed at the landing. The highest wind we dared to experiment in blew at thirty-one miles an hour; when the wind was stronger, we waited and watched the birds.

There was a gull came fishing over the lake, and took up his station over its very edge, about 100 feet high in air. The wind was blowing a steady gale from the north at sixty-one measured miles an hour. The bird breasted it squarely, and without beat of wing maintained for five minutes his position of observation. Occasionally there was a short rocking motion fore and aft, or from side to side. At times he was raised several feet and drifted backward; at others he drooped down; but he never flapped once. It is evident that he derived from the wind alone all the power required to remain afloat and to perforate the blast without drifting back. Whether man will ever be able to perform this feat, which has been termed "aspiration," is perhaps doubtful, but there is no mistake about the observation.

The only thing we could not ascertain was whether our practice hill, 350 feet to his leeward, produced an ascending trend in the wind about the bird, who was level with its summit.

Another day a curious thing occurred. We had taken one of the machines to the top of the hill, and loaded its lower wings with sand to hold it while we went to lunch. A gull came strolling inland, and flapped full-winged to inspect. He swept several circles above the machine, stretched his neck, gave a squawk, and went off. Presently he returned with eleven other gulls, and they seemed to hold a conclave, about 100 feet above the big new white bird which they had discovered on the sand. They circled round after round, and once in a while there was a series of loud peeps, like those of a rusty gate, as if in conference, with sudden flutterings, as if a terrifying suggestion had been made. The bolder birds occasionally swooped downward to inspect the monster more closely; they twisted their heads around to bring first one eye and then the other to bear, and then they rose again. After some seven or eight minutes of this performance, they evidently concluded either that the stranger was too formidable to tackle, if alive, or that he was not good to eat, if dead, and they flew off to resume fishing, for the weak point about a bird is his stomach.

We did not have the slightest accident to lament during all our experiments. These were chiefly performed by two young, active men, who took turns, and who became expert in a week; but then, we attempted no feats and took no chances. Toward the last, we gained such confidence in the machines that we allowed amateurs to try them under guidance. Half a dozen performed fairly well, but awkwardly of course. One of them was our cook, who was by profession a surgeon, and one was a newspaper reporter who had succeeded in finding his way to the camp. Another was a novice; he was picked up by a wind gust, raised forty feet vertically, and gently set down again. Any young, quick, and handy man can master a gliding-machine almost as soon as a bicycle, but the penalties for mistakes are much more severe. After all, it will be by the cautious, observant man—the man who accepts no risks which he can avoid, perhaps the ultra-timid man—that this hazardous investigation of an art now known only to the birds will be most advanced. Not even the birds could have operated more safely than we; but they would have made longer and flatter glides, and they would have soared up into the blue.

In my judgment, neither of the machines above described is as yet perfected, and I believe it is still premature to apply an artificial motor. This is sure to bring about complications which it is preferable to avoid until the equilibrium has been thoroughly evolved. I, therefore, advise that every plausible method of securing stability and safety shall be tested, that many such experiments shall be made, first with models, and then with full-sized machines, and that their designers shall practice, practice, practice; to make sure of the action, to proportion and adjust the parts, and to eliminate hidden defects. If any feat is attempted, it should be over water, in order to break the fall, should any occur. All this once accomplished, it will be time enough to apply a motor; and it seems not improbable that the gliding-machine will furnish the prototype. This step-by-step process is doubtless slow and costly, but it greatly diminishes the chance of those accidents which bring a whole line of investigation into contempt. We have no reason to believe that, contrary to past experience, a practical flying-machine will be the result of the happy thought of one or of two persons. It will come rather by a process of evolution: one man accomplishing some promising results, but stopping short of success; the next carrying the investigation somewhat further, and thus on, until a machine is produced which will be as practical as the "safety" bicycle, which took some eighty years for its development from the original despised velocipede.

Since the above described experiments were tried, another deplorable accident has come to re-inculcate the necessity for extreme caution. Mr. Percy S. Pilcher, a young, accomplished, and enthusiastic English engineer, lost his life September 30, 1899, while making experiments in soaring with a machine of his own design upon Lilienthal principle. He had already formed hundreds of glides since 1894, and had introduced a method of towing the machine with horses, by means of a long cord with multiplying tackle, so that he could rise from level ground. On this occasion, a first successful flight was made; but on the second trial, after a height of some thirty feet had been gained, a snap was heard, the tail was seen to collapse, and the apparatus dived forward, and fell to the ground, Mr. Pilcher receiving injuries from which he died two days later. He doubtless was the victim of his own amiability, for his apparatus had been wet by a shower, so that the canvass of the tail had shrunk, thus producing undue strains upon the bamboo stretcher, the wind was gusty, and the weather very unfavorable; but as many persons had come from a distance to witness the experiments, Mr. Pilcher did not like to disappoint them, and accepted the undue risks which cost him his life. He was less than thirty-four years of age, a skilful and earnest mechanician, who had already built the oil-engine and screw which he meant to apply to his machine.

Notably enough, he had written to me some eighteen months before for leave to copy and test one of my machines, which leave, with instructions, had, of course, been gladly given. The machine had been built, and was to have been tried on the following day. It is a curious coincidence that Lilienthal is said to have also built a machine, quite original with him, upon the same principle as that above alluded to, and that this also was to have been tested within a day or two of the owner's death. It is idle to speculate on what would have been the result; but then accidents might have happened in my own work, and I am profoundly thankful that we were spared such anguish.

Having been compelled, for the last two years, to give all my time and attention to a practical business, I have been unable to experiment; but I have had an expert testing models of a third method of securing automatic stability, which I hope to experiment full-sized.

Aside from the more imaginative and eccentric inventors, there are now a number of scientific investigators who are working to bring about the solution of this difficult problem; and it is not at all improbable that some experimenter will succeed, within a year or so, in making a flight of something like a mile with a motor. This is now fairly feasible, and there are several inventors who are preparing to attempt it. But between this achievement and its extension to a journey, or even to its indefinite repetition, there will intervene many accidents. Nor is there a fortune to be made by the first successful man. Experimenters who wish to advance the final solution of the quest surely and safely must work without expectation of other reward than that of being remembered hereafter; for, in the usual course of such things, it will be the manufacturers who will reap the pecuniary benefits when commercial flying-machines are finally evolved. There will probably be two types of these, one of them a machine for sport, with a very light and simple motor, if any, carrying but a single operator, and deriving most of its power from wind and gravity, as do the soaring birds. This will be used in competitions of skill and speed, and there will be no finer or more exciting sport. The other future machine will probably be of a journeying type. It will be provided with a powerful, but light, motor and with fuel for one or two days' travel. It will preferably carry but a single man, and will be utilized in exploration and in war. Its speed will be from thirty to sixty miles an hour at the beginning, and eventually much greater, for it is a singular fact that the higher speeds require less power in the air, within certain limits, than low speeds. At high velocities, the surfaces may be smaller, lie at flatter angles, and offer less resistance, but the pressure then increases on the framework, and the ultimate speed may not be more than 80 or 100 miles an hour.

Neither of these machines seems likely to compete with existing modes of transportation. But be this as it may, every improvement in transportation, whether in cheapness, in comfort, or in speed, soon develops new and sometimes unexpected uses of its own; so, even with sober anticipation of the benefits to be realized, investigators and public spirited men may well afford to advance the solution of a problem which has so warmly appealed to the imagination of men for the past forty or fifty centuries.

Document 1-9(d), Octave Chanute, excerpt from "Aerial Navigation," 1901.

FLYING MACHINES.

Some imaginative investigators have, therefore, resumed the search, antedating by far the invention of the balloon, for the conditions to be observed in devising a practical flying machine in imitation of the birds. Much of past work has been fanciful and crude, but there are now aeronautical societies and technical publications in most countries which promote sound research for both balloons and flying machines, and an international congress on these subjects was held at Paris in September, 1900. This congress had an elaborate programme for papers and discussions. Only abstracts of the proceedings have been published at this time of writing, but as no great advance or discovery was announced, this paper will not be incomplete if it merely recapitulates what was already known.

Attempts at artificial flight date back to the very dawn of history, but such attempts have been impeded by two main obstacles so prohibitive as at times to cause the investigation to be classed with that into the possibility of perpetual motion. These main obstacles are, first, the extreme danger in man's attempts to acquire the art of the birds, or to fly with an apparatus as yet imperfect; and second, the lack of an artificial motor as light, in proportion the power developed, as bird machinery. We know, approximately, that the motor muscles of birds develop such an output of energy that a full horse-power, if such were produced, would weigh but from six to twenty pounds, while our most powerful locomotives, with their tenders, weigh about two hundred pounds to the horse-power. This great gap has been partly closed within the last ten years, and investigators into flying devices are no longer regarded as visionaries.

It was about 1889 that a number of competent men simultaneously, but independently, took up the problem. Apparently the times were ripe, and already the advance since then has been greater than during the preceding three centuries. Langley, in the United States, and Maxim, in Great Britain, had become convinced that the current coefficients used for oblique air reactions were incorrect, and began experiments of their own to ascertain the facts. They both selected plane surfaces for trial, and reached much the same results as to the lifting power and resistance which are to be obtained from the air at various angles of inclination and at various speeds. These proved to be many times greater than those given by the ancient formulæ, and these experimenters determined upon building actual flying apparatus which should be driven by power.

Aside from the general design and the form and arrangement of the supporting surfaces, the important elements to consider are the ratio of those surfaces to the weight and the comparative power and speed required. There has been considerable range in these elements in the various experiments, and these will be described so as to bring out the important points. Langley began producing working models of flying machines in 1891. He tested various arrangements of wings and of surfaces, and successively applied steam, gasoline, and carbonic acid gas motors, and then steam again, struggling also with various methods of launching the apparatus into the air. At last, in May, 1896, one of his machines made two flights of about half a mile each in about one minute and a half, and later, in November, another machine flew once, more than three quarters of a mile. The whole apparatus weighed 30 pounds and spread 70 square feet of supporting surface, or in the ratio of $2\frac{1}{3}$ square feet to the pound. The steam-engine was of 1 horse-power and weighed 7 pounds, thus sustaining in flight 30 pounds per horse-power, but running down very soon because the water in the boiler was exhausted.

Maxim began on a much larger scale. He undertook the construction of a full-sized flying machine to carry three men, spreading 4000 square feet of supporting surface, and weighing 8000 pounds, thus affording only half a square foot to the pound. It was provided with a compound steam-engine and a boiler of 363 H. P., a marvel of ingenuity and mechanical skill. This apparatus was provided with wheels, and was placed upon a railway of 8 feet gauge, being restrained from premature flight by a pair of outside wooden rails placed above the wheels. Many experiments were made to test the speed required and the lifting effect, during which various mishaps were encountered and repaired. After several years of this study, the machine, in 1894, unexpectedly undertook free flight, by bursting through the upper rails during one of the tests. It flew, perhaps, 300 feet, but steam was at once shut off, and the apparatus alighted and was broken. The damage was repaired, but as the machine sustained only about 28 pounds per horse power, requiring a speed of 36 miles an hour, and as it had been very costly and other business pressed, Mr. Maxim did not resume his experiments. He is understood to be engrossed by his great manufacturing interests in gun and shipbuilding, some of the profits from which may hereafter be invested in another flying machine. This will probably be provided with a petroleum motor, now being experimented with, from which he expects even better results than from the marvelous steam-engine previously built by him, which latter weighed, with its boiler and a condenser, less than 10 pounds per horse-power.

About contemporaneously with Langley and Maxim, Hargrave, in Australia, Phillips, in Great Britain, and Tatin and Ader, in France, besides many others, experimented with flying devices. Hargrave began in 1885, and produced about twenty working models, propelled by clockwork, by rubber, by compressed air, and by steam. His last steam engine weighed about 10 pounds per H. P., but he hopes to improve upon this with a gasoline engine. He employs for his models comparatively very large surfaces, as much as 5 square feet to the pound to be lifted, a proportion which may not be realised in full-sized machines, and hence requires speeds of only 10 miles per hour, with which he has succeeded sustaining 79 pounds to the horse-power. He has invented the new form of kite which bears his name, and designs to suspend a motor and propeller, as well as himself, below a team of these novel cellular arrangements, and to fly through the air by towing the kites.

Phillips has been experimenting a long while, and has reached the conclusion that very narrow wings, somewhat like slats, are the most effective. He produced, in 1893, a machine looking like a Venetian blind on wheels, driven by a steam-engine. With this he is said to have lifted about 72 pounds to the horse-power, at speeds of 28 miles an hour, with surfaces in the proportion of one-third of a square foot to the pound of weight. The stability was, however, so defective that the apparatus, which weighed 402 pounds in all, could not be trusted in free flight, and made only brief skims. Tatin is a veteran experimenter. In 1879 he produced a flying model, driven by compressed air, with which he made many flights and had some breakages. In 1897 he produced, in connection with Dr. Richet, a model weighing 72 pounds, driven by a steam-engine. He obtained a lift of 55 pounds to the horsepower, with speeds of about 40 miles an hour, and surfaces of about 1 square foot to the pound; but the maximum flight was only 460 feet, much inferior to Langley's, and the equilibrium was defective. He says that he believes that he can overcome this defect, which has hitherto brought to grief every power-driven machine which has flown more than thrice, but this remains to be seen.

Ader is a French electrical engineer who has intermittently been engaged in aerial investigations for thirty years. He has built three full-sized machines, one, in 1872, to be driven by manpower, which, of course, was found inadequate, one steam-driven in 1891, at the expense of a banker, which produced indifferent results; and a third at the expense of the French Government, in 1897, which cost half a million francs. This was tested on the Satory field of manœuvers, with the most rigid precautions to guard the secrets of this war engine. The construction reproduced almost servilely the anatomical structure of birds. The surfaces were in the proportion of one-quarter of a square foot to the pound, and the whole apparatus weighed 1100 pounds. It was driven by a steam-engine of 40 H. P., weighing about 7 pounds per H. P., and was provided with screw propellers. The speed required for support was about 50 miles an hour, and the apparatus sustained 27 pounds per H. P., or somewhat less than Maxim's or Langley's.

No data have been published as to the tests, but it is said that wind squalls produced a quick descent, and that further experiments were abandoned. The equilibrium was probably so defective that it was deemed wise by the French Government to spend no more money on the machine. It was shown at the Paris Exhibition last year. All experts who have seen it agree that it is a wonderful piece of mechanical workmanship, that the motor is adequate, and that the wings are capable of sustaining all the weight, notwithstanding all their comparative exiguity.

The men above mentioned are but a tithe, perhaps a hundredth, of those who have been planning and experimenting with power-driven flying machines, but they are here picked out as the men who have accomplished the more notable successes. We now come upon a small group of investigators who believed that it was premature to apply artificial power to a flying machine until the proper arrangements and shape of the supporting surfaces were evolved, and their management in the air worked out by long practice.

First and chief among these was Lilienthal, a German engineer and physicist. During ten or fifteen years he made a series of elaborate experiments upon the best shapes for artificial wings, published a book on the subject, showing the superiority of arched forms, and, about 1891, brought out his first form of gliding machine, with which, after careful training, he was enabled to make many personal flights from hillsides, using gravity as a motive power. He gradually improved upon this with different machines, the last being a double-decker, weighing about 50 pounds, and carrying his own weight of 170 pounds in addition. The surface was 151 feet square, being thus in the proportion of about three-quarters of a square foot to the pound, and with this he made many glides, at angles of descent of about one in six, the maximum distance being 1200 feet, and depending, of course, upon the height from which he started. The timing of the flights showed that about 110 pounds were sustained per horse-power, at speeds of 23 miles an hour. In 1895 Lilienthal applied to his apparatus a carbonic acid gas motor of $2\frac{1}{2}$ H. P.; which was found, however, to affect the equilibrium so seriously that it was given up. He resumed gliding, and had, altogether, made about 2000 flights, with only trifling accidents, when, in August, 1896, he was upset in the air by a wind gust, fell, and was killed, to the great loss of aviation, which he would, doubtless, have advanced further.

Lilienthal was imitated by Pilcher, an English engineer, who modified the apparatus and made hundreds of glides between 1895 and 1899. His machine spread eighty-five hundredths of a square foot to the pound, and showed 100 pounds to be supported to the horse-power, obtained from gravity, at speeds of 25 miles an hour. He provided himself with a gasoline motor, but did not get far enough along to apply it. Towards the last he devised a method of starting up from level ground by towing the apparatus with horses, and in one of these experiments, in September, 1899, taking undue risks in order not to disappoint visitors, he was upset in the air and killed.

The writer of this has emulated Lilienthal and Pilcher, and thus far without disaster. He has confined his endeavours wholly to the evolution of automatic stability, making the supporting surfaces movable instead of the man, an arrangement the reverse of that of his predecessors. He has had about 1000 glides made by assistants, with two different types (five machines) without the slightest accident.

The proportion of surfaces was three-quarters of a square foot to the pound, the speed, 22 miles per hour, and the weight sustained, 89 pounds per horse-power. He has been experimenting by proxy with a third type of movable surfaces, which be intends to test full-sized, and he holds that it is entirely premature to introduce an artificial motor.

Full particulars concerning the last three experimenters will be found in the "Aeronautical Annual" for 1896 and 1897.

Since these experiments a further advance has been achieved by Messrs. Wilbur and Orville Wright, who have produced a double decked gliding machine in which the operator is placed in a horizontal position, thus opposing to forward motion 1 square foot, instead of 5 square feet, when he is upright, and they have further reduced the resistance of the framing by adopting improved shapes, so that the aggregate head resistances are reduced to about one-half of those which previously obtained. The experiments were made on the North Carolina coast, in the United States, in October, 1900.

The above is the record of what has been accomplished within the last decade in an investigation heretofore relegated to what had been termed "cranks." While a journey of 1200 miles has been made with a globular balloon, at the will of the wind, trips of only 5 or 6 miles have been made at the will of the operator, in calm weather, with fusiform balloons. With flying machines a maximum flight of three-quarters of a mile has been made by a model provided with a motor, and thousands of glides, up to 1200 feet, have been made by men-ridden machines with the aid of gravity, which latter power imposes no extra weight upon the apparatus and is always in good order.

The maximum speed of the dirigible balloon thus far is about 18 miles an hour, with eventual possibilities up to 44 miles per hour, while the speed of flying machines has already been about 50 miles an hour, with possibilities to 60 or 100 miles an hour, which speeds are attained by some swift birds. While fusiform balloons will, therefore, constitute one solution, and while they will be gradually improved and will serve in war, and perhaps in exploration, it now seems probable that future developments will chiefly appertain to flying machines. To make these a success the two main problems must be worked out; first, the motor, which must be very light, and, second, the stability, which is even more important, and which should be automatic.

To appreciate the difficulties appertaining to the motor, we may consider the difference in amount of power required for land and for aerial transportation. An American "Consolidation" locomotive will develop about 1000 H. P. and haul 2000 tons upon a level railway. Hence, it will haul 4000 pounds per horse-power. But this 1000 H. P., weighing, say, 100 tons with its tender, could impart a speed of only 57 miles an hour to Count Zeppelin's air ship, were this vessel (a manifest absurdity) able to lift the engine and to bear the resulting air pressure due to the speed. Neither could the locomotive sustain itself in the air if attached to a weightless flying machine.

We have seen that, with motor-driven apparatus, the best that has been positively done thus far has been to sustain from 27 to 55 pounds per horse-power by impact upon the air. Gliding machines, it is true, using gravity as a motive power, show 89 to 110 pounds sustained per horse-power, but these figures must be considerably reduced when an artificial motor is substituted, in order to cover the inevitable mechanical losses in the machinery and in the propeller. Much has been done within the past decade towards reducing the weight of motors. Steam-engines have been produced weighing but 10 pounds per horse-power, and the latest gasoline motor, that of Buchet, is said to weigh only 12 ½ per H. P.; but much remains to be done to render machines working so nearly up to the limit of endurance absolutely reliable and safe in the air. Numerous and costly experiments are required to accomplish this. It now seems probable that the successful aerial motor will be some form of gasoline engine, using air instead of water as a working fluid, and thus saving weight. But those who know how tedious and slow has been the development of the steam-engine will have no very sanguine expectations of the early attaining of perfection in the gasoline motor.

It is still more imperative that the whole apparatus shall not fall by losing its balance in the air. It must maintain its equilibrium and be reasonable safe under all the vicissitudes of flight,—in starting, in sailing, in alighting, and in wind gusts. The bare statement of this requirement meets with ready assent, and yet how few of the investigators have the will or the patience to spend the time, and to take the risk, to learn the art of the birds by personal experiments with gliding machines. The writer has been advocating this method for some years, he has confined his researches to its advance, and he sees no reason to change his views.

The underlying principle of maintaining equilibrium in the air is that the centre of pressure upon the sustaining surfaces shall at all times be upon the same vertical line as the centre of gravity due to the weight of the apparatus. In calm air this is fairly secured, but in a wind the centre of pressure is constantly shifted by the turmoils of the air, for it advances or recedes with the diminution or increase of the angle of incidence. There are several ways of counteracting this difficulty. The centre of gravity may be shifted back or forward to coincide again with the vertical line passing through the new centre of pressure; this is the method employed by Lilienthal and by Pilcher, which they applied by shifting the position of their personal weight. Or the centre of pressure may be brought back into a vertical line with a fixed centre of gravity, either by changing the angle of incidence, or by shifting the surfaces themselves. These latter are the methods which have been experimented upon by the writer in three different ways:—

1st. Affixing a horizontal tail (the Pénaud tail) at an angle to the supporting surfaces. This catches the air on its upper or lower surface, and shifts the angle of incidence of the wings, and, consequently, the centre of pressure.

2d. Pivoting the wings at their roots, so as to move horizontally. These are arranged so that the impact of the air shall bring them back into the proper position.

3d. Pivoting the surfaces so as to rock vertically. This is arranged so that the impinging air shall automatically shift the angle of incidence, and, therefore, the centre of pressure.

The third arrangement is believed to be the best, but one cannot be sure, inasmuch as all the adjustments above indicated are most delicate. Simple as the principles seem to be, it requires years of experiment to apply them properly. The positions of the pivots, the strength and adjustments of restraining springs, and the best position for the centre of gravity involve thousands of cut-and-try experiments, first with models, and then with full-sized gliding machines carrying a man. The important feature is that the man shall remain stationary.

In regard to the shape of surfaces to be employed, Lilienthal demonstrated that concavo-convex wings, like those of birds, are far superior in supporting power to planes, and the latter have now been practically abandoned by aviation experts. The amount of sustaining surfaces experimented with has varied, as above mentioned, from 1/4 square foot to 5 square feet to the pound of weight, this corresponding to speeds of from 50 miles down to 10 miles an hour, to obtain support. The amount required evidently depends upon the speed, but the larger the surfaces the greater is the weight. In the remote future it is probable that small surfaces will obtain, thus reducing the amount of required framework and the consequent head resistance, but until the problem of equilibrium has been fully solved it will be preferable to employ surfaces of about 1 square foot to the pound, involving speeds of 20 to 25 miles an hour in order to promote safety in alighting. With smaller surfaces we may hope to sustain eventually as much as 80 pounds per indicated horse-power, but something will depend upon the efficiency of the propeller.

The propeller is the next thing to be considered after the equilibrium has been secured and a reliable motor worked out. Both Hargrave and Lilienthal gave preference to flapping vanes over the screw propeller, but other experimenters prefer screws. It is yet too soon to draw definite conclusions on this question, and it opens a field for further experimenting.

We can, however, already calculate approximately the proportions, the strength and weight, the supporting efficiency, the speed, and the power required for a projected flying machine, so as to judge of the practicability of a design. Indeed, the mathematics of the subject have been so far evolved that engineering computations may eventually replace vague speculation in the domain of aerial navigation.

But after the problem has been worked out to a mechanical success, the commercial uses of aerial apparatus will be small. The limitations of the balloon have already been mentioned; such craft will be slow, frail, and very costly. We are now sufficiently advanced in the design of flying machines to perceive some of their limitations. They will be comparatively small and cranky, require much power, carry little extra weight, and depend for their effective speed, on each journey, whether they go against the wind or with it, so that they cannot compete with existing modes of transportation in cheapness or in carrying capacity. It is true that high speeds may be attained, and this may serve in war, in exploration, perhaps in mail transportation, and in sport; but the loads will be very small, and the expenses will be great. But flying machines will develop new uses of their own; and as mankind has always been benefited by the introduction of new and faster modes of transportation, we may hope that successful aerial navigation will spread civilization, knit the nations closer together, make all regions accessible, and perhaps so equalize the hazards of war as to abolish it altogether, thus bringing about the predicted era of universal peace and good-will.

Document 1-10(a–c)

(a) Orville Wright, excerpt from deposition of 13 January 1920, *The Papers of Wilbur and Orville Wright*, Marvin W. McFarland, ed. (New York, NY: McGraw-Hill Book Company, 1953), p. 3.

(b) Wilbur Wright, letter to the Smithsonian Institution, 30 May 1899, *The Papers of Wilbur and Orville Wright*, pp. 4–5.

(c) Wilbur Wright, letter to Octave Chanute, 13 May 1900, *The Papers of Wilbur and Orville Wright*, pp. 15–19.

The first document includes a response by Orville Wright, on 13 January 1920, to the question, "When and under what circumstances did you and Wilbur Wright first become interested in the problem of flight?" His brief statement sets up the two letters that follow with comments on how the Wrights were interested in flight even as children and how as young adults they began to tackle the problems of heavier-than-air flight, by first writing to the Smithsonian Institution for information.

The second and third documents are two letters written by Wilbur Wright that express his and his brother's early interest and commitment to flight research. The first of these letters was written in 1899 to the Smithsonian Institution, where Professor Langley was working, to request articles on aeronautics. In response to it, Richard Rathbun, the Assistant Secretary of the Smithsonian Institute, sent the Wrights four pamphlets: *Empire of the Air* (1893) by Louis-Pierre Mouillard; *The Problem of Flying and Practical Experiments in Soaring* (1893) by Otto Lilienthal; *Story of Experiments in Mechanical Flight* (1897) by Samuel P. Langley; and *On Soaring Flight* (1897) by E.C. Huffaker. Rathbun's letter of response also referred the Wrights to a list of other works, including *Progress in Flying Machines* (1894) by Octave Chanute and *Experiments in Aerodynamics* (1891) by Samuel P. Langley, along with the 1895, 1896, and 1897 issues of *The Aeronautical Annual*. The second was a 1900 letter to Octave Chanute. The search for an appropriate location to test their designs mentioned in the letter to Chanute led the Wrights to the Outer Banks of North Carolina and the winds of Kitty Hawk.

Document 1-10(a), Orville Wright on the brothers' interest in flight, 1920.

Our first interest began when we were children. Father brought home to us a small toy actuated by a rubber spring which would lift itself into the air. We built a number of copies of this toy, which flew successfully. By "we" I refer to my brother Wilbur and myself. But when we undertook to build the toy on a much larger scale it failed to work so well. The reason for this was not understood by us at the time, so we finally abandoned the experiments. In 1896 we read in the daily papers, or in some of the magazines, of the experiments of Otto Lilienthal, who was making some gliding flights from the top of a small hill in Germany. His death a few months later while making a glide off the hill increased our interest in the subject, and we began looking for books pertaining to flight. We found a work written by Professor Marey on animal mechanism which treated of the bird mechanism as applied to flight, but other than this, so far as I can remember, we found little.

In the spring of the year 1899 our interest in the subject was again aroused through the reading of a book on ornithology. We could not understand that there was anything about a bird that would enable it to fly that could not be built on a larger scale and used by man. At this time our thought pertained more particularly to gliding flight and soaring. If the bird's wings would sustain it in the air without the use of any muscular effort, we did not see why man could not be sustained by the same means. We knew that the Smithsonian Institution had been interested in some work on the problem of flight, and, accordingly, on the 30th of May 1899, my brother Wilbur wrote a letter to the Smithsonian inquiring about publications on the subject. . . .

Document 1-10(b), letter by Wilbur Wright to the Smithsonian Institution from Dayton, Ohio, 30 May 1899.

I have been interested in the problem of mechanical and human flight ever since as a boy I constructed a number of bats of various sizes after the style of Cayley's and Pénaud's machines. My observations since have only convinced me more firmly that human flight is possible and practicable. It is only a question of knowledge and skill just as in all acrobatic feats. Birds are the most perfectly trained gymnasts in the world and are specially well fitted for their work, and it may be that man will never equal them, but no one who has watched a bird chasing an insect or another bird can doubt that feats are performed which require three or four times the effort required in ordinary flight. I believe that simple flight at least is possible to man and that the experiments and investigations of a large number of independent workers will result in the accumulation of information and knowledge and skill which will finally lead to accomplished flight.

The works on the subject to which I have had access are Marey's and Jamieson's books published by Appleton's and various magazine and cyclopaedic articles. I am about to begin a systematic study of the subject in preparation for practical work to which I expect to devote what time I can spare from my regular business. I wish to obtain such papers as the Smithsonian Institution has published on this subject, and if possible a list of other works in print in the English language. I am an enthusiast, but not a crank in the sense that I have some pet theories as to the proper construction of a flying machine. I wish to avail myself of all that is already known and then if possible add my mite to help on the future worker who will attain final success. I do not know the terms on which you send out your publications but if you will inform me of the cost I will remit the price.

Document 1-10(c), letter from Wilbur Wright to Octave Chanute, 13 May 1900.

For some years I have been afflicted with the belief that flight is possible to man. My disease has increased in severity and I feel that it will soon cost me an increased amount of money if not my life. I have been trying to arrange my affairs in such a way that I can devote my entire time for a few months to experiment in this field.

My general ideas of the subject are similar to those held by most practical experimenters, to wit: that what is chiefly needed is skill rather than machinery. The flight of the buzzard and similar sailers is a convincing demonstration of the value of skill, and the partial needlessness of motors. It is possible to fly without motors, but not without knowledge & skill. This I conceive to be fortunate, for man, by reason of his greater intellect, can more reasonably hope to equal birds in knowledge, than to equal nature in the perfection of her machinery.

Assuming then that Lilienthal was correct in his ideas of the principles on which man should proceed, I conceive that his failure was due chiefly to the inadequacy of his method, and of his apparatus. As to his method, the fact that in five years time he spent only about five hours, altogether, in actual flight is sufficient to show that his method was inadequate. Even the simplest intellectual or acrobatic feats could never be learned with so short practice, and even Methuselah could never have become an expert stenographer with one hour per year for practice. I also conceive Lilienthal's apparatus to be inadequate not only from the fact that he failed, but my observations of the flight of birds convince me that birds use more positive and energetic methods of regaining equilibrium than that of shifting the center of gravity.

With this general statement of my principles and belief I will proceed to describe the plan and apparatus it is my intention to test. In explaining these, my object is to learn to what extent similar plans have been tested and found to be failures, and also to obtain such suggestions as your great knowledge and experience might enable you to give me. I make no secret of my plans for the reason that I believe no financial profit will accrue to the inventor of the first flying machine, and that only those who are willing to give as well as to receive suggestions can hope to link their names with the honor of its discovery. The problem is too great for one man alone and unaided to solve in secret.

My plan then is this. I shall in a suitable locality erect a light tower about one hundred and fifty feet high. A rope passing over a pulley at the top will serve as a sort of kite string. It will be so counterbalanced that when the rope is drawn out one hundred & fifty feet it will sustain a pull equal to the weight of the operator and apparatus or nearly so. The wind will blow the machine out from the base of the tower and the weight will be sustained partly by the upward pull of the rope and partly by the lift of the wind. The counterbalance will be so arranged that the pull decreases as the line becomes shorter and ceases entirely when its length has been decreased to one hundred feet. The aim will be to eventually practice in a wind capable of sustaining the operator at a height equal to the top of the tower. The pull of the rope will take the place of a motor in counteracting drift. I see, of course, that the pull of the rope will introduce complications which are not met in free flight, but if the plan will only enable me to remain in the air for practice by the hour instead of by the second, I hope to acquire skill sufficient to overcome both these difficulties and those inherent to flight. Knowledge and skill in handling the machine are absolute essentials to flight and it is impossible to obtain them without extensive practice. The method employed by Mr. Pilcher of towing with horses in many respects is better than that I propose to employ, but offers no guarantee that the experimenter will escape accident long enough to acquire skill sufficient to prevent accident. In my plan I rely on the rope and counterbalance to at least break the force of a fall. My observation of the flight of buzzards leads me to believe that they regain their lateral balance, when partly overturned by a gust of wind, by a torsion of the tips of the wings. If the rear edge of the right wing tip is twisted upward and the left downward the bird becomes an animated windmill and instantly begins to turn, a line from its head to its tail being the axis. It thus regains its level even if thrown on its beam ends, so to speak, as I have frequently seen them. I think the bird also in general retains its lateral equilibrium, partly by presenting its two wings at different angles to the wind, and partly by drawing in one wing, thus reducing its area. I incline to the belief that the first is the more important and usual method. In the apparatus I intend to employ I make use of the torsion principle. In appearance it is very similar to the "double-deck" machine with which the experiments of yourself and Mr. Herring were conducted in 1896-7. The point on which it differs in principle is that the cross-stays which prevent the upper plane from moving forward and backward are removed, and each end of the upper plane is independently moved forward or backward with respect to the lower plane by a suitable lever or other arrangement. By this plan the whole upper plane may be moved forward or backward, to attain longitudinal equilibrium, by moving both hands forward or backward together. Lateral equilibrium is gained by moving one end more than the other or by moving them in opposite directions. If you will make a square cardboard tube two inches in diameter and eight or ten long and choose two sides for your planes you will at once see the torsional effect of moving one end of the upper plane forward and the other backward, and how this effect is attained without sacrificing lateral stiffness. My plan is to attach the tail rigidly to the rear upright stays which connect the planes, the effect of which will be that when the upper plane is thrown forward the end of the tail is elevated, so that the tail assists gravity in restoring longitudinal balance. My experiments hitherto with this apparatus have been confined to machines spreading about fifteen square feet of surface, and have been sufficiently encouraging to induce me to lay plans for a trial with [a] full-sized machine.

My business requires that my experimental work be confined to the months between September and January and I would be particularly thankful for advice as to a suitable locality where I could depend on winds of about fifteen miles per hour without rain or too inclement weather. I am certain that such localities are rare.

I have your *Progress in Flying Machines* and your articles in the *Annuals* of '95, '96, & '97, as also your recent articles in the *Independent*. If you can give me information as to where an account of Pilcher's experiments can be obtained I would greatly appreciate your kindness.

Document 1-11(a–b)

(a) Wilbur Wright, letter to Bishop Milton Wright from Kitty Hawk, North Carolina, 23 September 1900, *The Papers of Wilbur and Orville Wright*, pp. 25–27.

(b) Wilbur Wright, "Some Aeronautical Experiments," lecture delivered at the meeting of the Western Society of Engineers, Chicago, Illinois, 18 September 1901, *The Papers of Wilbur and Orville Wright*, pp. 99–118.

The following two documents by Wilbur Wright present different perspectives on the Wright brothers' first gliding experiments at Kitty Hawk, North Carolina. The first is an excerpt from a letter written to his father at the beginning of their first season of flight tests there in 1900. The second is a lecture given in Chicago before the Western Society of Engineers after the conclusion of their second season in 1901. Preceding Wilbur Wright's lecture was an introduction by Octave Chanute, who hosted Wilbur during his stay in Chicago.

Chanute played a critical role in boosting the morale of the Wrights, especially Wilbur's declining confidence, after the many frustrations and disappointments of their 1900 and 1901 gliding seasons at Kitty Hawk. While a guest in Chanute's home (which was "cluttered up" with "models of flying machines suspended from the ceiling so thick that you could not see any ceiling at all"), Wilbur had long talks with the veteran engineer, many of them about the merits of Lilienthal's airfoil data. In his Chicago speech, Wilbur had not questioned the validity of any of Lilienthal's data, but he would start doing so soon thereafter, a change of mind reflected in how he edited his speech for subsequent publication. When he returned to Dayton, Wilbur poured over all available data with his brother Orville, including the poor performance of their own wings. They determined that they would need to start over from scratch, building from an experimental research program of their own. This decision led them to construct a makeshift wind tunnel in which they systematically tested a whole new range of airfoil shapes, some 200 of them. From this test program they attained the aerodynamic data they needed to design their highly successful glider of 1902 and the historic 1903 *Flyer* that resulted from it.

Document 1-11(a), letter from Wilbur Wright to Milton Wright, 23 September 1900.

I have my machine nearly finished. It is not to have a motor and is not expected to fly in any true sense of the word. My idea is merely to experiment and practice with a view to solving the problem of equilibrium. I have plans which I hope to find much in advance of the methods tried by previous experimenters. When once a machine is under proper control under all conditions, the motor problem will be quickly solved. A failure of motor will then mean simply a slow descent & safe landing instead of a disastrous fall. In my experiments I do not expect to rise many feet from the ground, and in case I am upset there is nothing but soft sand to strike on. I do not intend to take dangerous chances, both because I have no wish to get hurt and because a fall would stop my experimenting, which I would not like at all. The man who wishes to keep at the problem long enough to really learn anything positively must not take dangerous risks. Carelessness and overconfidence are usually more dangerous than deliberately accepted risks. I am constructing my machine to sustain about five times my weight and am testing every piece. I think there is no possible chance of its breaking while in the air. If it is broken it will be by awkward landing. My machine will be trussed like a bridge and will be much stronger than that of Lilienthal, which, by the way, was upset through the failure of a movable tail and not by breakage of the machine. The tail of my machine is fixed, and even if my steering arrangement should fail, it would still leave me with the same control that Lilienthal had at the best. The safe and secure construction & management are my main improvements. My machine is more simple in construction and at the same time capable of greater adjustment and control than previous machines.

I have not taken up the problem with the expectation of financial profit. Neither do I have any strong expectation of achieving the solution at the present time or possibly any time. My trip would be no great disappointment if I accomplish practically nothing. I look upon it as a pleasure trip pure and simple, and I know of no trip from which I could expect greater pleasure at the same cost. I am watching my health very closely and expect to return home heavier and stronger than I left. I am taking every precaution about my drinking water.

**Document 1-11(b), Wilbur Wright, "Some Aeronautical Experiments,"
Chicago, 18 September 1901.**

The difficulties which obstruct the pathway to success in flying machine construction are of three general classes: (1) Those which relate to the construction of the sustaining wings. (2) Those which relate to the generation and application of the power required to drive the machine through the air. (3) Those relating to the balancing and steering of the machine after it is actually in flight. Of these difficulties two are already to a certain extent solved. Men already know how to construct wings or aeroplanes, which when driven through the air at sufficient speed, will not only sustain the weight of the wings themselves, but also that of the engine, and of the engineer as well. Men also know how to build engines and screws of sufficient lightness and power to drive these planes at sustaining speed. As long ago as 1893 a machine weighing 8,000 lbs. demonstrated its power both to lift itself from the ground and to maintain a speed of from thirty to forty miles per hour; but it came to grief in an accidental free flight, owing to the inability of the operators to balance and steer it properly. This inability to balance and steer still confronts students of the flying problem, although nearly ten years have passed. When this one feature has been worked out the age of flying machines will have arrived, for all other difficulties are of minor importance.

The person who merely watches the flight of a bird gathers the impression that the bird has nothing to think of but the flapping of its wings. As a matter of fact this is a very small part of its mental labor. To even mention all the things the bird must constantly keep in mind in order to fly securely through the air would take a considerable part of the evening. If I take this piece of paper, and after placing it parallel with the ground, quickly let it fall, it will not settle steadily down as a staid, sensible piece of paper ought to do, but it insists on contravening every recognized rule of decorum, turning over and darting hither and thither in the most erratic manner, much after the style of an untrained horse. Yet this is the style of steed that men must learn to manage before flying can become an everyday sport. The bird has learned this art of equilibrium, and learned it so thoroughly that its skill is not apparent to our sight. We only learn to appreciate it when we try to imitate it. Now, there are two ways of learning how to ride a fractious horse: one is to get on him and learn by actual practice how each motion and trick may be best met; the other is to sit on a fence and watch the beast a while, and then retire to the house and at leisure figure out the best way of overcoming his jumps and kicks. The latter system is the safest; but the former, on the whole, turns out the larger proportion of good riders. It is very much the same in learning to ride a flying machine; if you are looking for perfect safety, you will do well to sit on a fence and watch the birds; but if you really wish to learn, you must mount a machine and become acquainted with its tricks by actual trial.

Herr Otto Lilienthal seems to have been the first man who really comprehended that balancing was the *first* instead of the *last* of the great problems in connection with human flight. He began where others left off, and thus saved the many thousands of dollars that it had theretofore been customary to spend in building and fitting expensive engines to machines which were uncontrollable when tried. He built a pair of wings of a size suitable to sustain his own weight, and made use of gravity as his motor. This motor not only cost him nothing to begin with, but it required no expensive fuel while in operation, and never had to be sent to the shop for repairs. It had one serious drawback, however, in that it always insisted on fixing the conditions under which it would work. These were that the man should first betake himself and machine to the top of a hill and fly with a downward as well as a forward motion. Unless these conditions were complied with, gravity served no better than a balky horse—it would not work at all. Although Lilienthal must have thought the conditions were rather hard, he nevertheless accepted them till something better should turn up; and in this manner he made some two thousand flights, in a few cases landing at a point more than a thousand feet distant from his place of starting. Other men, no doubt, long before had thought of trying such a plan. Lilienthal not only thought, but acted; and in so doing probably made the greatest contribution to the solution of the flying problem that has ever been made by any one man. He demonstrated the feasibility of actual practice in the air, without which success is impossible. Herr Lilienthal was followed by Mr. Pilcher, a young English engineer, and by Mr. Chanute, a distinguished member of the society I now address. A few others have built machines, but nearly all that is of real value is due to the experiments conducted under the direction of the three men just mentioned.

The balancing of a gliding or flying machine is very simple in theory. It merely consists in causing the center of pressure to coincide with the center of gravity. But in actual practice there seems to be an almost boundless incompatibility of temper which prevents their remaining peaceably together for a single instant, so that the operator, who in this case acts as peacemaker, often suffers injury to himself while attempting to bring them together. If a wind strikes a vertical plane, the pressure on that part to one side of the center will exactly balance that on the other side, and the part above the center will balance that below. This point we call the center of pressure. But if the plane be slightly inclined, the pressure on the part nearest the wind is increased, and the pressure on the other part decreased, so that the center of pressure is now located, not in the center of the surface, but a little toward the side which is in advance. If the plane be still further inclined the center of pressure will move still farther forward. And if the wind blow a little to one side, it will also move over as if to meet it. Now, since neither the wind nor the machine for even an instant maintains exactly the same direction and velocity, it is evident that the man who would trace the course of the center of pressure must be very quick of mind; and he who would attempt to move his body to that spot at every change must be very active indeed. Yet this is what Herr Lilienthal attempted to do, and did do with most remarkable skill, as his two thousand glides sufficiently attest. However he did not escape being overturned by wind gusts several times, and finally lost his life through a breakage of his machine, due to defective construction.

The Pilcher machine was similar to that of Lilienthal, and like it, seems to have been structurally weak; for on one occasion, while exhibiting the flight of his machine to several members of the Aeronautical Society of Great Britain, it suddenly collapsed and fell to the ground, causing injuries to the operator which proved sadly fatal. The method of management of this machine differed in no important respect from that of Lilienthal, the operator shifting his body to make the centers of pressure and gravity coincide. Although the fatalities which befell the designers of these machines were due to the lack of structural strength, rather than to lack of control, nevertheless it had become clear to the students of the problem that a more perfect method of control must be evolved. The Chanute machines marked a great advance in both respects. In the multiple-wing machine, the tips folded slightly backward under the pressure of wind gusts, so that the travel of the center of pressure was thus largely counterbalanced. The guiding of the machine was done by a slight movement of the operator's body toward the direction in which it was desired that the machine should go. The double-deck machine built and tried at the same time marked a very great structural advance, as it was the first in which the principles of the modern truss bridges were fully applied to flying machine construction. This machine in addition to its greatly improved construction and general design of parts also differed from the machine of Lilienthal in the operation of its tail. In the Lilienthal machine the tail, instead of being fixed in one position, was prevented by a stop from folding downward beyond a certain point, but was free to fold upward without any hindrance. In the Chanute machine the tail was at first rigid, but afterward, at the suggestion of Mr. Herring, it was held in place by a spring that allowed it to move slightly either upward or downward with reference to its normal position, thus modifying the action of the wind gusts upon it, very much to its advantage. The guiding of the machine was effected by slight movements of the operator's body, as in the multiple-wing machines. Both these machines were much more manageable than the Lilienthal type, and their structural strength, notwithstanding their extreme lightness, was such that no fatalities, or even accidents, marked the glides made with them, although winds were successfully encountered much greater in violence than any which previous experimenters had dared to attempt.

My own active interest in aeronautical problems dates back to the death of Lilienthal in 1896. The brief notice of his death which appeared in the telegraphic news at that time aroused a passive interest which had existed from my childhood, and led me to take down from the shelves of our home library a book on *Animal Mechanism* by Prof. Marey, which I had already read several times. From this I was led to read more modern works, and as my brother soon became equally interested with myself, we soon passed from the reading to the thinking, and finally to the working stage. It seemed to us that the main reason why the problem had remained so long unsolved was that no one had been able to obtain any adequate practice. We figured that Lilienthal in five years of time had spent only about five hours in actual gliding through the air. The wonder was not that he had done so little, but that he had accomplished so much. It would not be considered at all safe for a bicycle rider to attempt to ride through a crowded city street after only five hours' practice, spread out in bits of ten seconds each over a period of five years; yet Lilienthal with this brief practice was remarkably successful in meeting the fluctuations and eddies of wind gusts. We thought that if some method could be found by which it would be possible to practice by the hour instead of by the second, there would be hope of advancing the solution of a very difficult problem. It seemed feasible to do this by building a machine which would be sustained at a speed of 18 miles per hour, and then finding a locality where winds of this velocity were common. With these conditions, a rope attached to the machine to keep it from floating backward would answer very nearly the same purpose as a propeller driven by a motor, and it would be possible to practice by the hour, and without any serious danger, as it would not be necessary to rise far from the ground, and the machine would not have any forward motion at all. We found, according to the accepted tables of air pressures on curved surfaces that a machine spreading 200 square feet of wing surface would be sufficient for our purpose, and that places could easily be found along the Atlantic coast where winds of 16 to 25 miles were not at all uncommon. When the winds were low, it was our plan to glide from the tops of sand hills, and when they were sufficiently strong, to use a rope for our motor and fly over one spot. Our next work was to draw up the plans for a suitable machine. After much study we finally concluded that tails were a source of trouble rather than of assistance; and therefore we decided to dispense with them altogether. It seemed reasonable that if the body of the operator could be placed in a horizontal position instead of the upright, as in the machines of Lilienthal, Pilcher and Chanute, the wind resistance could be very materially reduced since only one square foot instead of five would be exposed. As a full half horsepower could be saved by this change, we arranged to try at least the horizontal position. Then the method of control used by Lilienthal, which consisted in shifting the body, did not seem quite as quick or effective as the case required; so, after long study, we contrived a system consisting of two large surfaces on the Chanute double-deck plan, and a smaller surface placed a short distance in front of the main surfaces in such a position that the action of the wind upon it would counterbalance the effect of the travel of the center of pressure on the main surfaces. Thus changes in the direction and velocity of the wind would have little disturbing effect, and the operator would be required to attend only to the steering of the machine, which was to be affected by curving the forward surface up or down. The lateral equilibrium and the steering to right or left was to be attained by a peculiar torsion of the main surfaces, which was equivalent to presenting one end of the wings at a greater angle than the other. In the main frame a few changes were also made in the details of construction and trussing employed by Mr. Chanute. The most important of these were: (1) the moving of the forward main cross-piece of the frame to the extreme front edge; (2) the encasing in the cloth of all cross-pieces and ribs of the surfaces; (3) a rearrangement of the wires used in trussing the two surfaces together, which rendered it possible to tighten all the wires by simply shortening two of them.

With these plans we proceeded in the summer of 1900 to Kitty Hawk, North Carolina, a little settlement located on the strip of land that separates Albemarle Sound from the Atlantic Ocean. Owing to the impossibility of obtaining suitable material for a 200 square-foot machine, we were compelled to make it only 165 square feet in area, which, according to the Lilienthal tables, would be supported at an angle of three degrees in a wind of about 21 miles per hour. On the very day that the machine was completed the wind blew from 25 to 30 miles per hour, and we took it out for trial as a kite. We found that while it was supported with a man on it in a wind of about 25 miles, its angle was much nearer twenty degrees than three degrees. Even in gusts of 30 miles the angles of incidence did not get as low as three degrees, although the wind at this speed has more than twice the lifting power of a 21-mile wind. As winds of 30 miles per hour are not plentiful on clear days, it was at once evident that our plan of practicing by the hour, day after day, would have to be postponed. Our system of twisting the surfaces to regulate the lateral balance was tried and found to be much more effective than shifting the operator's body. On subsequent days, when the wind was too light to support the machine with a man on it, we tested it as a kite, working the rudders by cords reaching to the ground. The results were very satisfactory, yet we were well aware that this method of testing is never wholly convincing until the results are confirmed by actual gliding experience.

We then turned our attention to making a series of actual measurements of the lift and drift of the machine under various loads. So far as we were aware this had never previously been done with any full-size machine. The results obtained were most astonishing, for it appeared that the total horizontal pull of the machine, while sustaining a weight of 52 pounds, was only 8.5 lbs., which was less than had previously been estimated for head resistance of the framing alone. Making allowance for the weight carried, it appeared that the head resistance of the framing was but little more than 50 percent of the amount which Mr. Chanute had estimated as the head resistance of the framing of his machine. On the other hand it appeared sadly deficient in lifting power as compared with the calculated lift of curved surfaces of its size. This deficiency we supposed might be due to one or more of the following causes: (1) That the depth of the curvature of our surfaces was insufficient, being only about 1 in 22, instead of 1 in 12. (2) That the cloth used in our wings was not sufficiently airtight. (3) That the Lilienthal tables might themselves be somewhat in error. We decided to arrange our machine for the following year so that the depth of curvature of its surfaces could be varied at will, and its covering airproofed.

Our attention was next turned to gliding, but no hill suitable for the purpose could be found near our camp at Kitty Hawk. This compelled us to take the machine to a point 4 miles south, where the Kill Devil sand hill rises from the flat sand to a height of more than 100 feet. Its main slope is toward the northeast, and has an inclination of 10 degrees. On the day of our arrival the wind blew about 25 miles an hour, and as we had had no experience at all in gliding, we deemed it unsafe to attempt to leave the ground. But on the day following, the wind having subsided to 14 miles per hour, we made about a dozen glides. It had been the original intention that the operator should run with the machine to obtain initial velocity, and assume the horizontal position only after the machine was in free flight. When it came time to land he was to resume the upright position and light on his feet, after the style of previous gliding experimenters. But on actual trial we found it much better to employ the help of two assistants in starting, which the peculiar form of our machine enabled us readily to do; and in landing we found that it was entirely practicable to land while still reclining in a horizontal position upon the machine. Although the landings were made while moving at speeds of more than 20 miles an hour, neither machine nor operator suffered any injury. The slope of the hill was 9.5 deg., or a drop of 1 foot in 6. We found that after attaining a speed of about 25 or 30 miles with reference to the wind, or 10 to 15 miles over the ground, the machine not only glided parallel to the slope of the hill, but greatly increased its speed, thus indicating its ability to glide on a somewhat less angle than 9.5 deg., when we should feel it safe to rise higher from the surface. The control of the machine proved even better than we had dared to expect, responding quickly to the slightest motion of the rudder. With these glides our experiments for the year 1900 closed. Although the hours and hours of practice we had hoped to obtain finally dwindled down to about two minutes, we were very much pleased with the general results of the trip, for setting out as we did, with almost revolutionary theories on many points, and an entirely untried form of machine, we considered it quite a point to be able to return without having our pet theories completely knocked in the head by the hard logic of experience, and our own brains dashed out in the bargain. Everything seemed to us to confirm the correctness of our original opinions: (1) that practice is the key to the secret of flying; (2) that it is practicable to assume the horizontal position; (3) that a smaller surface set at a negative angle in front of the main bearing surfaces, or wings, will largely counteract the effect of the fore and aft travel of the center of pressure; (4)

that steering up and down can be attained with a rudder, without moving the position of the operator's body; (5) that twisting the wings so as to present their ends to the wind at different angles is a more prompt and efficient way of maintaining lateral equilibrium than shifting the body of the operator.

When the time came to design our new machine for 1901, we decided to make it exactly like the previous machine in theory and method of operation. But as the former machine was not able to support the weight of the operator when flown as a kite, except in very high winds and at very large angles of incidence, we decided to increase its lifting power. Accordingly, the curvature of the surfaces was increased to 1 in 12, to conform to the shape on which Lilienthal's table was based, and to be on the safe side, we decided also to increase the area of the machine from 165 square feet to 308 square feet, although so large a machine had never before been deemed controllable. The Lilienthal machine had an area of 151 square feet; that of Pilcher, 165 square feet; and the Chanute double-decker, 134 square feet. As our system of control consisted in a manipulation of the surfaces themselves instead of shifting the operator's body, we hoped that the new machine would be controllable, notwithstanding its great size. According to calculations it would obtain support in a wind of 17 miles per hour with an angle of incidence of only 3 degrees.

Our experience of the previous year having shown the necessity of a suitable building for housing the machine, we erected a cheap frame building, 16 feet wide, 25 feet long, and 7 feet high at the eaves. As our machine was 22 feet wide, 14 feet long (including the rudder) and about 6 feet high, it was not necessary to take the machine apart in any way in order to house it. Both ends of the building, except the gable parts, were made into doors which hinged above, so that when opened they formed an awning at each end, and left an entrance the full width of the building. We went into camp about the middle of July, and were soon joined by Mr. E. C. Huffaker, of Tennessee, an experienced aeronautical investigator in the employ of Mr. Chanute, by whom his services were kindly loaned, and by Dr. G. A. Spratt, of Pennsylvania, a young man who has made some valuable investigations of the properties of variously curved surfaces and the travel of the center of pressure thereon. Early in August, Mr. Chanute came down from Chicago to witness our experiments, and spent a week in camp with us. These gentlemen, with my brother and myself, formed our camping party, but in addition we had in many of our experiments the valuable assistance of Mr. W. J. Tate and Mr. Dan Tate, of Kitty Hawk.

The machine was completed and tried for the first time on the 27th of July in a wind blowing about 13 miles an hour. The operator having taken a position where the center of pressure was supposed to be, an attempt at gliding was made; but the machine turned downward and landed after going only a few yards. This indicated that the center of gravity was too far in front of the center of pressure.

In the second attempt the operator took a position several inches further back but the result was much the same. He kept moving further and further back with each trial, till finally he occupied a position nearly a foot back of that at which we had expected to find the center of pressure. The machine then sailed off and made an undulating flight of a little more than 300 feet. To the onlookers this flight seemed very successful, but to the operator it was known that the full power of the rudder had been required to keep the machine from either running into the ground or rising so high as to lose all headway. In the 1900 machine one fourth as much rudder action had been sufficient to give much better control. It was apparent that something was radically wrong, though we were for some time unable to locate the trouble. In one glide the machine rose higher and higher till it lost all headway. This was the position from which Lilienthal had always found difficulty to extricate himself, as his machine then, in spite of his greatest exertions, manifested a tendency to dive downward almost vertically and strike the ground head on with frightful velocity. In this case a warning cry from the ground caused the operator to turn the rudder to its full extent and also to move his body slightly forward. The machine then settled slowly to the ground, maintaining its horizontal position almost perfectly, and landed without any injury at all. This was very encouraging, as it showed that one of the very greatest dangers in machines with horizontal tails had been overcome by the use of a front rudder. Several glides later the same experience was repeated with the same result. In the latter case the machine had even commenced to move backward, but was nevertheless brought safely to the ground in a horizontal position. On the whole, this day's experiments were encouraging, for while the action of the rudder did not seem at all like that of our 1900 machine, yet we had escaped without difficulty from positions which had proved very dangerous to preceding experimenters, and after less than one minute's actual practice had made a glide of more than 300 feet, at an angle of descent of 10 degrees, and with a machine nearly twice as large as had previously been considered safe. The trouble with its control, which has been mentioned, we believed could be corrected when we should have located its cause. Several possible explanations occurred to us, but we finally concluded that the trouble was due to a reversal of the direction of the travel of the center of pressure at small angles. In deeply curved surfaces the center of pressure at 90 degrees is near the center of the surface, but moves forward as the angle becomes less, till a certain point is reached, varying with the depth of curvature. After this point is passed, the center of pressure, instead of continuing to move forward, with the decreasing angle, turns and moves rapidly toward the rear. The phenomena are due to the fact that at small angles the wind strikes the forward part of the surface on the upper side instead of the lower, and thus this part altogether ceases to lift, instead of being the most effective part of all, as in the case of the plane.

Lilienthal had called attention to the danger of using surfaces with a curvature as great as one in eight, on account of this action on the upper side; but he seems never to have investigated the curvature and angle at which the phenomena entirely cease. My brother and I had never made any original investigation of the matter, but assumed that a curvature of one in twelve would be safe, as this was the curvature on which Lilienthal based his tables. However, to be on the safe side, instead of using the arc of a circle, we had made the curve of our machine very abrupt at the front, so as to expose the least possible area to this downward pressure. While the machine was building, Messrs. Huffaker and Spratt had suggested that we would find this reversal of the center of pressure, but we believed it sufficiently guarded against. Accordingly, we were not at first disposed to believe that this reversal actually existed in our machine, although it offered a perfect explanation of the action we had noticed in gliding. Our peculiar plan of control by forward surfaces, instead of tails, was based on the assumption that the center of pressure would continue to move farther and farther forward, as the angle of incidence became less, and it will be readily perceived that it would make quite a difference if the front surface instead of counteracting this assumed forward travel, should in reality be expediting an actual backward movement. For several days we were in a state of indecision, but were finally convinced by observing the following phenomena: (Figure 1) We had removed the upper surface from the machine and were flying it in a wind to see at what angles it would be supported in winds of different strengths. We noticed that in light winds it flew in the upper position shown in the figure, with a strong upward pull on the cord c. As the wind became stronger, the angle of incidence became less, and the surface flew in the position shown in the middle of the figure, with a slight horizontal pull. But when the wind became still stronger, it took the lower position shown in the figure, with a strong downward pull. It at once occurred to me that here was the answer to our problem, for it is evident that in the first case the center of pressure was in front of the center of gravity and thus pushed up the front edge; in the second case, they were in coincidence, and the surface in equilibrium; while in the third case the center of pressure had reached a point even behind the center of gravity, and there was therefore a downward pull on the cord. This point having been definitely settled, we proceeded to truss down the ribs of the whole machine, so as to reduce the depth of curvature. In Figure 2, line 1, shows the original curvature; line 2, the curvature when supporting the operator's weight; and line 3, the curvature after trussing.

On resuming our gliding, we found that the old conditions of the preceding year had returned; and after a few trials, made a glide of 366 feet and soon after one of 389 feet. The machine with its new curvature never failed to respond promptly to even small movements of the rudder. The operator could cause it to almost skim the ground, following the undulations of its surface, or he could cause it to sail out almost on a level with the starting point, and passing high above the foot of the hill, gradually settle down to the ground. The wind on this day was blowing 11 to 14 miles per hour. The next day, the conditions being favorable, the machine was again taken out for trial. This time the velocity of the wind was 18 to 22 miles per hour. At first we felt some doubt as to the safety of attempting free flight in so strong a wind, with a machine of over 300 square feet, and a practice of less than five minutes spent in actual flight. But after several preliminary experiments we decided to try a glide. The control of the machine seemed so good that we then felt no apprehension in sailing boldly forth. And thereafter we made glide after glide, sometimes following the ground closely, and sometimes sailing high in the air. Chanute had his camera with him, and took pictures of some of these glides, several of which are among those shown.

We made glides on subsequent days, whenever the conditions were favorable. The highest wind thus experimented in was a little over 12 meters per second—nearly 27 miles per hour.

It had been our intention when building the machine to do the larger part of the experimenting in the following manner: When the wind blew 17 miles an hour, or more, we would attach a rope to the machine and let it rise as a kite with the operator upon it. When it should reach a proper height the operator would cast off the rope and glide down to the ground just as from the top of a hill. In this way we would be saved the trouble of carrying the machine uphill after each glide, and could make at least 10 glides in the time required for 1 in the other way. But when we came to try it we found that a wind of 17 miles, as measured by Richard's anemometer, instead of sustaining the machine with its operator, a total weight of 240 lbs., at an angle of incidence of 3 degrees, in reality would not sustain the machine alone—100 pounds—at this angle. Its lifting capacity seemed scarcely one third of the calculated amount. In order to make sure that this was not due to the porosity of the cloth, we constructed two small experimental surfaces of equal size, one of which was airproofed and the other left in its natural state; but we could detect no difference in their lifting powers. For a time we were led to suspect that the lift of curved surfaces little exceeded that of planes of the same size, but further investigation and experiment led to the opinion that (1) the anemometer used by us over recorded the true velocity of the wind by nearly 15 percent; (2) that the well known Smeaton coefficient of $.005V^2$ for the wind pressure at 90 degrees is probably too great by at least 20 percent; (3) that Lilienthal's estimate that the pressure on a curved surface having an angle of incidence of 3 degrees equals .545 of the pressure at 90 degrees is too large, being nearly 50 percent greater than very recent experiments of our own with a special pressure testing machine indicate; (4) that the superposition of the surfaces somewhat reduced the lift per square foot, as compared with a single surface of equal area.

In gliding experiments, however, the amount of lift is of less relative importance than the ratio of lift to drift, as this alone decides the angle of gliding descent. In a plane the pressure is always perpendicular to the surface, and the ratio of lift to drift is therefore the same as that of the cosine to the sine of the angle of incidence. But in curved surfaces a very remarkable situation is found. The pressure instead of being uniformly normal to the chord of the arc, is usually inclined considerably in front of the perpendicular. The result is that the lift is greater and the drift less than if the pressure were normal. Lilienthal was the first to discover this exceedingly important fact, which is fully set forth in his book, *Bird Flight the Basis of the Flying Art*, but owing to some errors in the methods he used in making measurements, question was raised by other investigators not only as to the accuracy of his figures, but even as to the existence of any tangential force at all. Our experiments confirm the existence of this force, though our measurements differ considerably from those of Lilienthal. While at Kitty Hawk we spent much time in measuring the horizontal pressure on our unloaded machine at various angles of incidence. We found that at 13 degrees the horizontal pressure was about 23 lbs. This included not only the drift proper, or horizontal component of the pressure on the side of the surface, but also the head resistance of the framing as well. The weight of the machine at the time of this test was about 108 lbs. Now, if the pressure had been normal to the chord of the surface, the drift proper would have been to the lift (108 lbs.) as the sine of 13 degrees is to the cosine of 13 degrees, or .22 x 108 / .97 = 24+ lbs.; but this slightly exceeds the total pull of 23 lbs. on our scales. Therefore, it is evident that the average pressure on the surface instead of being normal to the chord was so far inclined toward the front that all the head resistance of framing and wires used in the construction was more than overcome. In a wind of 14 miles per hour, resistance is by no means a negligible factor, so that tangential is evidently a force of considerable value. In a higher wind which sustained the machine at an angle of 10 degrees, the pull on the scales was 18 lbs. With the pressure normal to the chord, the drift proper would have been .17 x 98 / .98 = 17 lbs., so that, although the higher wind velocity must have caused an increase in the head resistance, the tangential force still came within one pound of overcoming it. After our return from Kitty Hawk we began a series of experiments to accurately determine the amount and direction of the pressure produced on curved surfaces when acted upon by winds at the various angles from zero to 90 degrees. These experiments are not yet concluded, but in general they support Lilienthal in the claim that the curves give pressures more favorable in amount and direction than planes; but we find marked differences in the exact values, especially at angles below 10 degrees. We were unable to obtain direct measurements of the horizontal pressures of the machine with the operator on board, but by comparing the distance traveled in gliding with the vertical fall, it was easily calculated that at a speed of 24 miles per hour the total horizontal resistances of our machine, when bearing the operator, amounted to 40 pounds which is equivalent to about 2 ½ horsepower. It must not be supposed, however, that a motor developing this power would be sufficient to drive a man-bearing machine. The extra weight of the motor would require either a larger machine, higher speed, or a greater angle of incidence, in order to support it, and therefore more power. It is probable, however, that an engine of 6 horsepower, weighing 200 pounds, would answer the purpose. Such an engine is entirely practicable. Indeed, working motors of one half this weight per horsepower (9 pounds per horsepower) have been constructed by several different builders. Increasing the speed of our machine from 24 to 33 miles per hour reduced the total horizontal pressure from 40 to about 35 pounds. This was quite an advantage in gliding as it made it possible to sail about 15 percent further with a given drop. However, it would be of little or no advantage in reducing the size of the motor in a power-driven machine, because the lessened thrust would be counterbalanced by the increased speed per minute. Some years ago Prof. Langley called attention to the great economy of thrust which might be obtained by using very high speeds, and from this many were led to suppose that high speed was essential to success in a motor-driven machine. But the economy to which Prof. Langley called attention was in foot pounds per mile of travel, not in foot pounds per minute. It is the foot pounds per minute that fixes the size of a relatively low speed, perhaps not much exceeding 20 miles per hour, but the problem of increasing the speed will be much simpler in some respects than that of increasing the speed of a steamboat; for, whereas in the latter case the size of the engine must increase as the cube of the speed, in the flying machine, until extremely high speeds are reached, the capacity of the motor increases in less than simple ratio; and there is even a decrease in the fuel consumption per mile of travel. In other words to double the speed of a steamship (and the same is true of the balloon type of airship) eight times the engine and boiler capacity would be required, and four times the fuel consumption per mile of travel; while a flying machine would require engines of less than double the size, and there would be an actual decrease in the fuel consumption per mile of travel. But looking at the matter conversely, the great disadvantage of the flying machine is apparent, for in the latter no flight at all is possible unless the proportion of horsepower to flying capacity is very high; but on the other hand a steamship is a mechanical success if its ratio of horsepower to tonnage is insignificant. A flying machine that would fly at a speed of 50 miles an hour with engines of 1,000 horsepower, would not be upheld by its wings at all at a speed of less than 25 miles an hour, and nothing less than 500 horsepower could drive it at this speed. But a boat which could make 40 miles per hour with engines of 1,000 horsepower, would still move 4 miles an hour even if the engines were reduced to 1 horsepower. The problems of land and water travel were solved in the 19th century because it was possible to begin with small achievements and gradually work up to our present success. The flying problem was left over to the 20th century, because in this case the art must be highly developed before any flight of any considerable duration at all can be obtained.

However, there is another way of flying which requires no artificial motor, and many workers believe that success will first come by this road. I refer to the soaring flight, by which the machine is permanently sustained in the air by the same means that are employed by soaring birds. They spread their wings to the wind, and sail by the hour, with no perceptible exertion beyond that required to balance and steer themselves. What sustains them is not definitely known, though it is almost certain that it is a rising current of air. But whether it be a rising current or something else, it is as well able to support a flying machine as a bird, if man once learns the art of utilizing it. In gliding experiments it has long been known that the rate of vertical descent is very much retarded and the duration of the flight greatly prolonged, if a strong wind blows *up* the face of the hill parallel to its surface. Our machine, when gliding in still air, has a rate of vertical descent of nearly 6 feet per second, while in a wind blowing 26 miles per hour up a steep hill, we made glides in which the rate of descent was less than 2 feet per second. And during the larger part of this time, while the machine remained exactly in the rising current, *there was no descent at all, but even a slight rise*. If the operator had had sufficient skill to keep himself from passing beyond the rising current, he would have been sustained indefinitely at a higher point than that from which he started. The illustration shows one of these very slow glides at a time when the machine was practically at a standstill. The failure to advance more rapidly caused the photographer some trouble in aiming, as you will perceive. In looking at this picture you will readily understand that the excitement of gliding experiments does not entirely cease with the breaking up of camp. In the photographic darkroom at home we pass moments of as thrilling interest as any in the field, when the image begins to appear on the plate and it is yet an open question whether we have a picture of a flying machine, or merely a patch of open sky. These slow glides in rising currents probably hold out greater hope of extensive practice than any other method within man's reach, but they have the disadvantage of requiring rather strong winds or very large supporting surfaces. However, when gliding operators have attained greater skill, they can, with comparative safety, maintain themselves in the air for hours at a time in this way, and thus by constant practice so increase their knowledge and skill that they can rise into the higher air and search out the currents which enable the soaring birds to transport themselves to any desired point by first rising in a circle and then sailing off at a descending angle. This illustration shows the machine, alone, flying in a wind of 35 miles per hour on the face of a steep hill, 100 feet high. It will be seen that the machine not only pulls upward, but also pulls forward in the direction from which the wind blows, thus overcoming both gravity and the speed of the wind. We tried the same experiment with a man on it, but found danger that the forward pull would become so strong that the men holding the ropes would be dragged from their insecure foothold on the slope of the hill. So this form of experimenting was discontinued after four or five minute trials.

In looking over our experiments of the past two years, with models and full-sized machines, the following points stand out with clearness:

1. That the lifting power of a large machine, held stationary in a wind at a small distance from the earth, is much less than the Lilienthal table and our own laboratory experiments would lead us to expect. When the machine is moved through the air, as in gliding, the discrepancy seems much less marked.

2. That the ratio of drift to lift in well-shaped surfaces is less at angles of incidence of five degrees to 12 degrees than at an angle of three degrees.

3. That in arched surfaces the center of pressure at 90 degrees is near the center of the surface, but moves slowly forward as the angle becomes less, till a critical angle varying with the shape and depth of the curve is reached, after which it moves rapidly toward the rear till the angle of no lift is found.

4. That with similar conditions, large surfaces may be controlled with not much greater difficulty than small ones, if the control is effected by manipulation of the surfaces themselves, rather than by a movement of the body of the operator.

5. That the head resistances of the framing can be brought to a point much below that usually estimated as necessary.

6. That tails, both vertical and horizontal, may with safety be eliminated in gliding and other flying experiments.

7. That a horizontal position of the operator's body may be assumed without excessive danger, and thus the head resistance reduced to about one fifth that of the upright position.

8. That a pair of superposed, or tandem surfaces, has less lift in proportion to drift than either surface separately, even after making allowance for weight and head resistance of the connections.

Document 1-12(a–b)

(a) Octave Chanute, excerpt from remarks at a meeting of the Aéro-Club, Paris, 2 April 1903, translated from Ernest Archdeacon, *La Locomotion*, 11 April 1903, pp. 225–227, in *The Papers of Wilbur and Orville Wright*, pp. 654–673.

(b) Wilbur Wright, excerpts from "Experiments and Observations in Soaring Flight," address at the meeting of the Western Society of Engineers, Chicago, 24 June 1903, published in *Journal of the Western Society of Engineers* (December 1903), in *The Papers of Wilbur and Orville Wright*, pp. 318–330.

One of the most influential reports on the glider experiments done in America, and especially of the recent work of Wilbur and Orville Wright, was an address given in Paris by Octave Chanute in the spring of 1903. At the time Chanute made these remarks, the Wrights were applying what they had learned from their gliding experiments to the design of a powered airplane.

The Wright brothers' third season of flight tests at Kitty Hawk in 1902 crossed a threshold in the development of flight. With their newly designed glider they had, for the first time, a machine in which they could really learn how to fly. In the following document, Wilbur Wright describes this breakthrough season of flight experiments. This passage is taken from an address given at the 1903 meeting of the Western Society of Engineers, as a follow-up to his earlier address presented in September 1901. These remarks were made on the eve of the Wrights' departure for a fourth season of tests at Kitty Hawk, where they planned to fly their new powered airplane based on the glider described here.

Document 1-12(a), Octave Chanute, excerpt from remarks at a meeting of the Aéro-Club, Paris, 2 April 1903.

The strains arise from the irregularities and turmoils of the wind. The current does not come as an evenly flowing stream, but as a series of swirling waves, as shown in the smoke issuing from a chimney. These waves strike the apparatus either on one side or on the other, from below or from above, and constantly tend to upset it. The velocity and force with which the machine is struck by the wave depend both upon the distance from the center of rotation of the latter and upon the speed of the current, and this is probably the reason why anemometers show such varying pressures. A flying machine must meet and overcome all these vicissitudes and this must be done instantly.

In 1896 and 1897, being impressed that the equilibrium of the bird was partly automatic, and that the problem of stability in the wind was the first which must be solved, I undertook some experiments near Chicago, Illinois, with full-sized gliding machines carrying a man, in order to study equilibrium, and that alone. I caused to be built five machines of four different types. The first was a Lilienthal apparatus, in order to start from the known before passing to the unknown, and three of the other machines were based upon the reverse of the Lilienthal type, that is to say that instead of re-establishing the equilibrium (when compromised by the variations of the wind) by displacing the body of the operator, and therefore that of the center of gravity, as did Lilienthal and Pilcher, the new machines were based upon the theory that it was possible so to arrange the carrying surfaces that they should move automatically under the action of the wind and bring back the center of pressure vertically over the center of gravity—a condition absolutely necessary in order to maintain equilibrium.

One of these machines, called the "multiple-wing," is shown in figures 1 to 5. The supporting wings are at the front and are superposed and braced together. They turn upon ball bearings marked B, and are restrained in front by rubber springs which allows a certain amount of horizontal movement. If the *relative wind* increases, thus requiring a lesser angle of incidence to sustain the weight, the wings are blown backward, the apparatus oscillates towards the front, and the angle becomes smaller. Once the squall passed, the springs bring the wings back to their normal position. The rear wings are flexible and merely aid in balancing, and an aeroplane, which might be replaced by an additional pair of wings, surmounts the whole. The whole apparatus weighed 15.2 kilograms, the supporting area of the front wings being 13.33 square meters and that of the rear wings 2.74 m^2. This machine proved very nearly automatic, the operator having to move but 25 millimeters, in ordinary glides, as against 127 mm. with the Lilienthal and 63 mm. with the "two-surface" machine, which is next to be described.

The "two-surface" machine gave the best results and the longest glides. It is shown in figures 6 to 10. The supporting surfaces are of varnished silk and are affixed to a framework similar to a bridge truss, and the operator is below in an upright position, being sustained under the armpits by two horizontal bars. Equilibrium is obtained through a horizontal tail, similar to that invented by Pénaud, but with elastic fastenings devised by Mr. Herring. This tail generally makes an angle of 7 to 8 degrees with the carrying surfaces, and by receiving the relative wind below or above automatically changes the angle of attack according to the exigencies of the moment. The frame was of wood and the weight was 11 kg.

and the surface 12.45 square meters, which easily sustained an operator weighing 71 kg. More than one thousand glides were made with these machines in 1896 and 1897 without accident, the following tables giving an idea of the results with each of the machines:

SOME GLIDES OF MULTIPLE-WING MACHINE—WEIGHT 86 KG., MOUNTED

Operator	Length (meters)	Time (Seconds)	Speed M. per sec.	Remarks
Herring	45.11	7.0	6.45	
Avery	53.04	7.6	7.00	
Herring	50.60	7.5	6.72	
Avery	55.77	7.9	7.00	Angle of descent not measured but about 10° to 11°
Herring	52.43	7.8	6.70	

SOME GLIDES OF TWO-SURFACE—WEIGHT 82 KG., MOUNTED

Operator	Length (meters)	Time (seconds)	Angle descent	Height fallen (meters)	Speed M. per sec	Rate of descent	Kg.	Remarks
Avery	60.64	8	10°	10.5	7.62	1 in 5.75	106.	
Herring	71.32	8.7	7 ½°	9.3	8.23	1 in 7.69	87.	
Avery	77.11	....	10 ½°	14	....	1 in 5.50		Time not taken
Herring	72.84	....	11°	14.1	....	1 in 5.24		Time not taken
"	67.06	9	....	....	7.42	....		Angle not taken
"	71.62	10.3	....	....	7.	....		Angle not taken
Avery	78.03	10.2	8°	10.8	7.6	1 in 7.18	86.	
Herring	109.42	14.	10°	18.9	7.8	1 in 5.75	110.	

The slope of the hill being steep and the wind ascending, it was estimated that the resistance required 2 horsepower, instead of the 1 ⅛ H.P. shown by the work done as calculated. The drawings of these various machines were published in the *Aeronautical Annual* for 1897, edited by Jas. Means, Boston, U.S.A., and amateurs were invited to repeat and to improve upon the performances.

In 1902 I caused to be built an apparatus to obtain automatic stability by a third method. This consists in pivoting the sustaining surfaces about 4/10 of their width from the front, and restraining them by springs. If the relative wind increases, the center of pressure moves backward and tends to give the surfaces a smaller

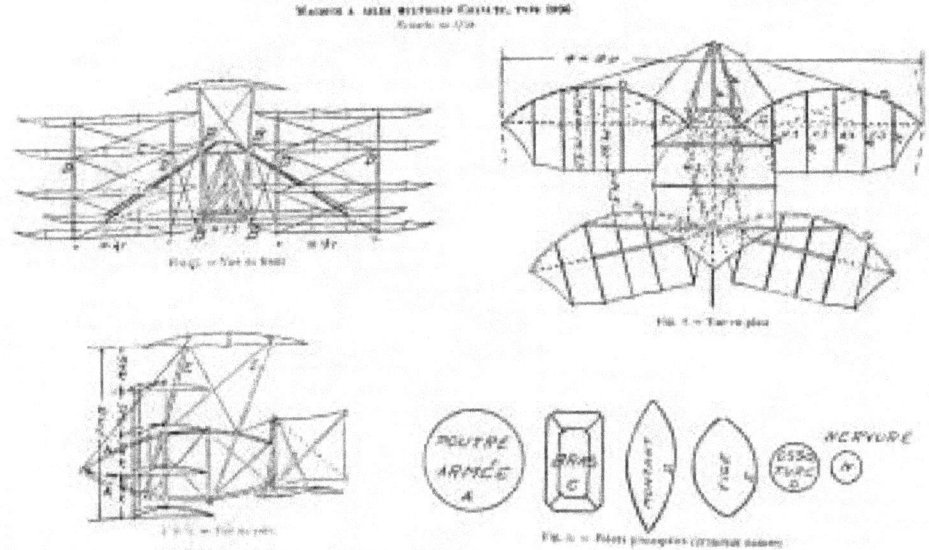

angle of attack. Only a few tests were made in 1902, but more are to be carried on in 1903. The apparatus is shown in figures 11 to 15 for the benefit of amateurs and a photograph of one of the glides is herewith reproduced.

All these experiments were made without accident. It is true that in the beginning all possible precautions were taken. The practice ground selected was on the soft sand hills which border the southern edge of Lake Michigan, some 50 kilometers from Chicago, where there are but few trees or bushes. No glides were made in very gusty winds, or those exceeding 50 kilometers to the hour, and Messrs. Herring and Avery, my assistants, were alone allowed to experiment. Later on, more confidence was gained and visitors were allowed to make short glides under instruction from the experts. All succeeded well, and even the cook became almost an expert in a short time. It must, however, be recognized that this sport is dangerous and that all who would engage in it must take all possible precautions before venturing upon it.

The invitation to amateurs to repeat these experiments remained unacted upon till 1900, when Messrs. Wilbur & Orville Wright of Dayton (Ohio), took up the question. They have accomplished such advance upon all previous practice that the rest of this paper will be devoted to giving an account thereof. The improvements which they have introduced are the following:

1st. Placing the horizontal rudder or tail at the front, a position which proves more efficient in acting upon the air.

2nd. Placing the operator prone on the machine, thus diminishing by $\frac{4}{5}$ the resistance due to his body.

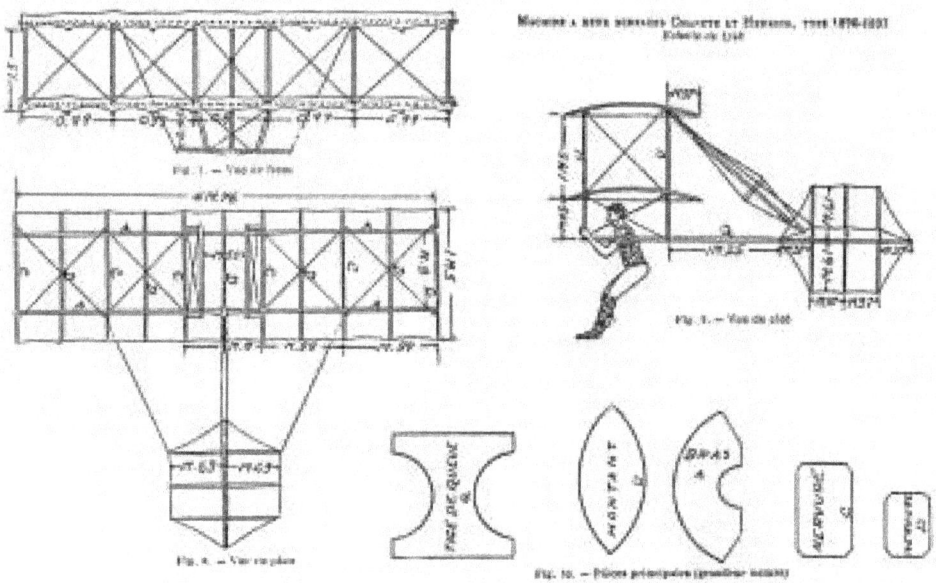

3rd. Warping the wings to steer to right or left.

The practice ground chosen by the Wright brothers, after an inquiry made of the National Signal Bureau of the United States, is far superior to the hills near Chicago. It is situated at Kitty Hawk near the ocean in North Carolina, and consists of a soft sand hill about 30 meters high, on a tongue of land, without a single tree, bush or grass. At its foot a sand beach one kilometer wide extends to the sea, and regular sea breezes blow daily from the ocean. Moreover the spot is a desert and quite inaccessible to the curious, who are always in the way when one is seeking for unknown phenomena.

The Wright apparatus of 1900 measured 5.64 meters across by a width of 1.52 m., there being two surfaces aggregating 15.6 square meters, and weighing 21.8 kg. It chiefly served to study the result of placing the horizontal rudder at the front, the apparatus being loaded with sandbags or weights and flown as a kite. Having ascertained that the machine worked well, the Wrights ventured to make a few glides with good success, and postponed to the next season learning the difficult art of the birds.

The method of conducting the experiments necessarily differed from those adopted by Lilienthal and his imitators. In those machines the operator, being on his feet, carries the apparatus, poises it in the wind, and runs forward until he leaves the ground. With the two-surface machine, for instance, which weighed but 11 kg. (for 12.45 m. surface) the aviator could walk about with his wings on his shoulders. The Wrights, desiring to build a larger and heavier machine and to place the aviator horizontally, had to devise a new method of getting under way.

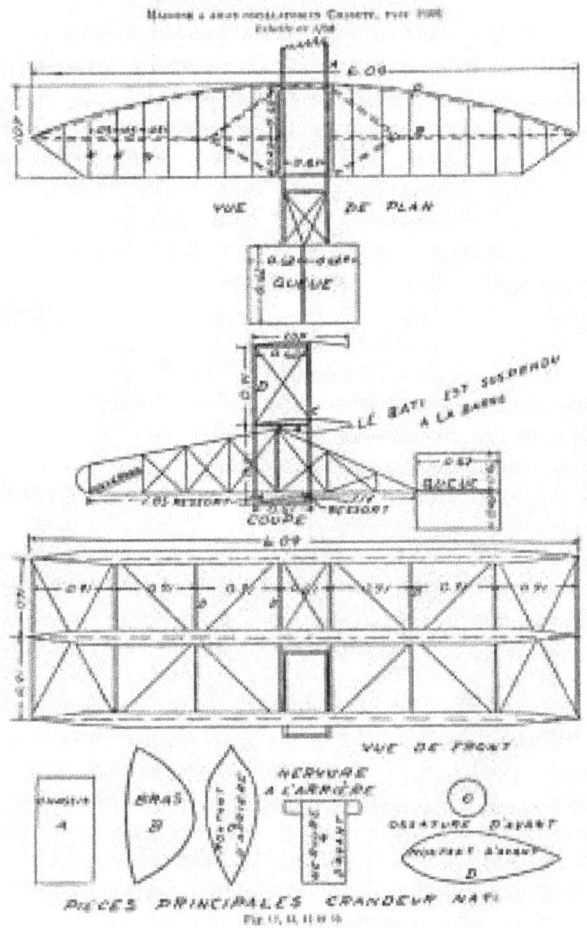

This consisted in employing two assistants each grasping one extremity of the machine, and running forward against the wind. The aviator, placed in the center, first remains on his feet, runs a few steps and, as speed and supporting power increase, tips forward prone on the framing, hooking his feet on the rear transverse bar, and as soon as he feels himself well carried by the air, he cries, "Let go." From this on gravity serves as a motor; he glides along down the slope of the hill, keeping near the ground, meeting the irregularities of the wind by the action of the front rudder and steering to the right or left, by the torsion of the wings, which are framed loosely, and by the vertical rudder behind. When the foot of the hill is reached, a vigorous action of the front rudder causes the machine to shoot upward a little, thus increasing the resistance and diminishing the speed. The machine then alights upon the sand and stops after a short slide upon its shoes.

The 1901 machine was 6.7 meters across by a width of 2.13 m., the two surfaces being spaced 1.42 m. vertically and giving a surface of 27.1 m^2, with a weight of 45.4 kg. This apparatus performed several hundred glides without accident, but although the head resistance was less, by reason of the horizontal position of the aviator, the angles of descent were nearly the same (8° to 10°) as with the machines in which the aviator stood upright. Now the important thing for a gliding apparatus is that the fall through the air shall be the least possible. We can always lengthen the glides by starting from a higher point, but the angle of descent shows at once the ratio between the propelling force (the weight) and the horizontal resistance, and the latter should be reduced to a minimum before proceeding to other details.

The horizontal resistance consists of three factors:

1st. The *drift*, depending upon the angle of incidence.

2nd. The *head resistance* of the various thicknesses.

3rd. The *tangential* force, recognized by Lilienthal.

The *drift* is the horizontal resultant of the normal pressure and has been thoroughly written upon heretofore, so that it need not be here discussed. Recent experiments seem to show, however, that the friction of the air upon the surfaces is not negligible, and should be separately added.

The *head* resistance is the most important, and is the one in which improvements may be achieved, by shaping the framing so as to give the lowest possible coefficients, by diminishing the number of parts, and by reducing all guy wires and similar vibrating parts to a minimum.

The tangential force only applies to arched surfaces, and presents the curious property that at certain angles of incidence it acts as a propulsive force. Those who may wish to learn more about this are referred to Lilienthal's book, *Der Vogelflug als Grundlage der Fliegekunst*, and to his chapter in Moedebeck's *Taschenbuch für Flugtechniker und Luftschiffer*.

We arrive at the aggregate horizontal resistance by three different processes, as follows:

1st. The speed required for support and the angle of incidence with the relative wind, being first ascertained as accurately as possible, the normal pressure is computed by the aid of Lilienthal's table for arched surfaces, which will be found in Moedebeck's *Taschenbuch*, and this normal is multiplied by the sine of the angle of application to obtain the *drift*. To this is added the *head resistance* due to the framing, the aviator and adjuncts, which is obtained by measuring accurately all the thicknesses and applying to each a coefficient due to its form. It is interesting in this connection to know that beams fishlike in cross section offer but one sixth the resistance on a plane of the same area as their "master section" while vibrating wires offer about twice the resistance due to their "master section." All this is best obtained by drawing up a tabular schedule of all the parts and deducing there from the equivalent or fictive area of resistance of the apparatus, which, when multiplied by the air pressure due to the speed, gives the aggregate *head resistance*; to the sum of the latter with the *drift* is added or deducted the *tangential* force, as obtained from Lilienthal's table, according to whether it is positive or negative. In other words:

Resistance = drift + head resistance ± tangential. This process is tedious and slow, but it is well worth the trouble, for it indicates those parts of the apparatus in which resistance can be reduced.

2nd. The second process furnishes a check upon the first and consists in floating the mounted machine in a wind of sufficient intensity or towing it by a cord at sufficient speed to obtain support and measuring the actual pull with a hand scale. This cannot be done very often as it involves considerable preparation, but it is a check not to be disputed.

3rd. The third process consists in simply multiplying the weight by the sine of the angle of descent. In fact, as the speed of the glides is almost always uniform it is evident that the forces balance, and the resistance along the path is to the weight as the sine of the angle to the radius. It is found as the result of a great many such computations that the three processes agree very closely.

The Messrs. Wright concluded from their experiments in 1901 that there were a number of defects in the apparatus used that year. That the shape of the surfaces was not the best possible and that they became deformed under the pressure of the relative wind. During the succeeding winter they made a whole series of laboratory experiments upon various surfaces, and built a third machine for experiment in 1902.

This last apparatus was 9.75 meters across, by a width of 1.52 m., the two surfaces being spaced 1.42 m. apart, and giving 28.4 m2. of supporting surface. The front rudder was 1.4 m². in area, and the weight was 53 kg. Fig. 16 to 20 show this apparatus. The main pieces are shown full size. They were all of wood, the front spar and the main arm were of an American wood for which pine would be a substitute; these parts were either imbedded in the cloth, or in a sheath or strip glued on afterwards. The ribs were of ash, steamed and bent, with a curvature of 1/20 one third of the chord from the front. All fastenings were made by lashing the sticks together, and it is well to dampen the twine with thin glue when wrapping it around. The surfaces consisted of tightly woven cotton cloth such as that used for balloons. It was not varnished but would have been the better for it. The braces were of piano wire and it was generally aimed to have a factor of safety of 10 times the breaking weight. A light pair of shoes or skis was attached to the framing and spar at the front.

The machine was tested on the practice ground already described, in September and October, 1902, and it gave very superior results to those obtained with preceding machines. The angles of descent were flatter and the weight sustained per horsepower was greater. The two brothers glided alternately and they soon attained almost complete mastery over the inconstancies of the wind. They met the wind gusts and steered as they willed. They did not venture to sweep much more than one quarter circle, so as not to lose the advantage of a head wind, but

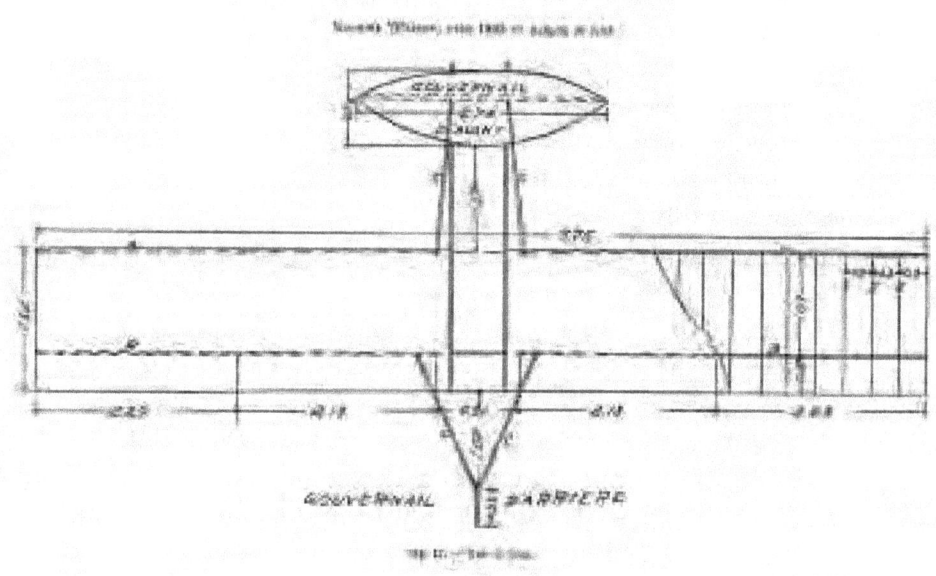

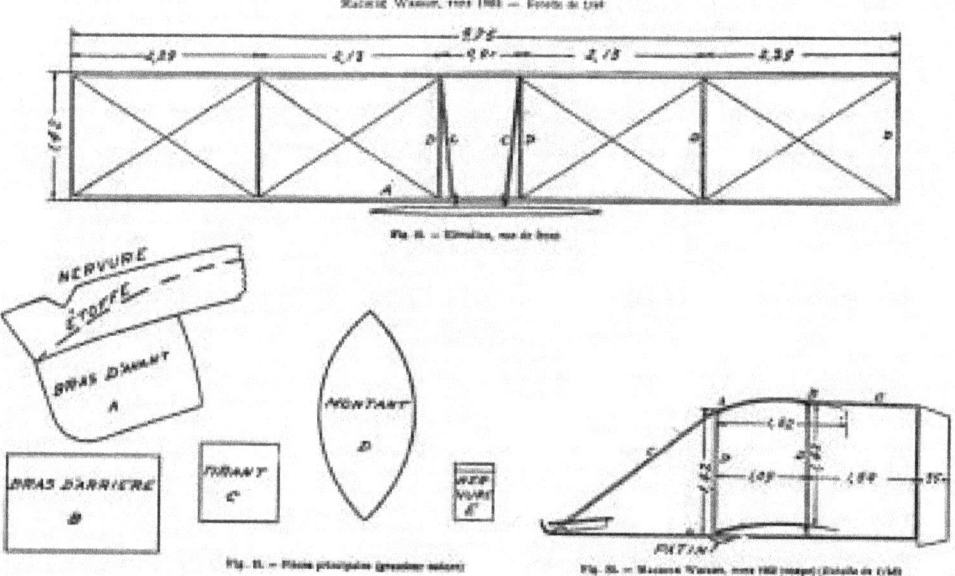

they constantly improved in the control of the machine and in learning the art of the birds. Some 800 glides were made of which the following are representative:

The variations mainly result from more or less skill in each glide and consequent slight swerving.

The last column in the table is that which best indicates the advance made over previous practice. The machine of Mr. Maxim supported 12.7 kg., per horsepower, that of Professor Langley 13.6 kg., while previous glides from Lilienthal down showed 45 kg. per horsepower. It is quite true that when a motor is added the weight sustained per horsepower is diminished about one half by reason of the inevitable losses due to the motor, the transmission and the propeller, but Messrs. Wright have so far diminished the resistances that they now sustain 62 kg. per horsepower, and the time is evidently approaching when, the problems of equilibrium and control having been solved, it will be safe to apply a motor and a propeller.

In point of fact the Messrs. Wright are now gliding very nearly as well as the vulture, which generally descends one meter in ten (5°45') in calm air. It is therefore not impossible that man shall eventually master sailing flight and learn, when once well in the air and at speed, to imitate the soaring birds which circle and rise through the force of the wind alone when circumstances are favorable; which conditions chiefly consist in an ascending trend of the wind, such as is frequently found in hot countries, those being the regions which are frequented by sailing birds.

This does not mean that the future flying machine, if such is developed, can entirely do without a motor. This adjunct will be indispensable, not only to fly in those regions where the wind seldom has an ascending trend, but also to restore the speed or the equilibrium very quickly under many circumstances, such as a very sudden wind gust, or a descent too near to the ground. It is easily perceived, however, that when the motor need only be used occasionally the length of possible journeys of flying apparatus will be greatly increased in the countries where circumstances favor sailing flight.

It is thus seen that there has been a gradual evolution from Lilienthal onward. That gliding machines have been made much safer and are now fairly under control. There is doubtless a chance for further improvement, and this can only be accomplished by experiment. The present indications are that such experiments will be extensively undertaken, and a word of warning may well be given in concluding this article.

Let all exert the utmost possible prudence in conducting gliding experiments. First select suitable ground, a soft sand hill is the best, so isolated from other hills as to have no cross currents or whirling winds. Begin practice very gradually, starting from a small elevation and learning by degrees the control of the machine and the best way of alighting safely. Follow the slope of the hill closely, so as to have but a short distance to fall if a false maneuver is made. Examine the apparatus after each glide, and if anything has gotten out of order, repair it at once before making another glide. Try no feats which involve being high in air except over an appropriate sheet of water, but do not resort to that method of experimenting upon ordinary occasions. It involves too much loss of time in drying oneself and the machine, and the latter is likely to become distorted.

More important than all, do not try to beat previous records. This leads to taking risks and to producing accidents. It is well to have competitions at which several amateurs practice together, because they learn from each other and because they are then more disposed to have a surgeon present, a precaution which I have always taken, but we must always remember that the important things to be secured are control of the machine and safe landings, without regard to the distance glided over, the latter mainly depending upon the elevation started from.

Progress and safety will be greatly promoted by beginning with small machines just as the Wrights have done. An apparatus 6 meters by 1.3 meters, giving with two surfaces arched to $1/25$ of their width some 15 square meters of supporting area, is ample to carry a man and will be found far more manageable than the last machine of Messrs. Wright of which the drawings are here given. As the cost is not great, being from 500 to 1500 francs, according to the amount of care bestowed on the apparatus, it is feasible to build new machines from time to time in order to introduce the improvements which constantly suggest themselves. It is hoped that when many experimenters get at work such progress shall be made as materially to advance the time when aviation shall become practical.

Document 1-12(b), Wilbur Wright, excerpts from "Experiments and Observations in Soaring Flight," Chicago, 24 June 1903.

The prime object in these experiments was to obtain practice in the management of a man-carrying machine, but an object of scarcely less importance was to obtain data for the study of the scientific problems involved in flight. Observations were almost constantly being made for the purpose of determining the amount and direction of the pressures upon the sustaining wings; the minimum speed required for support; the speed and angle of incidence at which the horizontal resistance became least; and the minimum angle of descent at which it was possible to glide. To determine any of these points with exactness was found to be very difficult indeed, but by careful observations under test conditions it was possible to obtain reasonably close approximations. . . .

In addition to the work with the machine we also made many observations on the flight of soaring birds, which were very abundant in the vicinity of our camp. Bald eagles, ospreys, hawks, and buzzards gave us daily exhibitions of their powers. The buzzards were the most numerous and were the most persistent soarers. They apparently never flapped except when it was absolutely necessary, while the eagles and hawks usually soared only when they were at leisure. Two methods of soaring were employed. When the weather was cold and damp and the wind strong, the buzzards would be seen soaring back and forth along the hills or at the edge of a clump of trees. They were evidently taking advantage of the current of air flowing upward over these obstructions. On such days they were often utterly unable to soar except in these special places. But on warm clear days when the wind was light they would be seen high in the air soaring in great circles. Usually however it seemed to be necessary to reach a height of several hundred feet by flapping before this style of soaring became possible. Frequently a great number of them would begin circling in one spot, rising together higher and higher till finally they would disperse, each gliding off in whatever direction it wished to go. At such times other buzzards only a short distance away found it necessary to flap frequently in order to maintain themselves. But when they reached a point beneath the circling flock they too began to rise on motionless wings. This seemed to indicate that rising columns of air do not exist everywhere, but that the birds must find them. They evidently watch each other and when one finds a rising current the others quickly make their way to it. One day when scarce a breath of wind was stirring on the ground, we noticed two bald eagles sailing in circling sweeps at a height of probably 500 feet. After a time our attention was attracted to the flashing of some object considerably lower down. Examination with a field glass proved it to be a feather which one of the birds had evidently cast. As it seemed apparent that it would come to earth only a short distance away some of our party started to get it. But in a little while it was noticed that the feather was no longer falling but on the contrary was rising rapidly. It finally went out of sight upward. It apparently was drawn into the same rising current in which the eagles were soaring, and was carried up like the birds.

The days when the wind blew horizontally gave us the most satisfactory observations, as then the birds were compelled to make use of the currents flowing up the sides of the hills and it was possible for us to measure the velocity and trend of the wind in which the soaring was performed. One day four buzzards began soaring on the northeast slope of the Big Hill at a height of only ten or twelve feet from the surface. We took a position to windward and about 1,200 feet distant. The clinometer showed that they were $4\frac{1}{2}$ to $5\frac{1}{2}$ degrees above our horizon. We could see them very distinctly with a field glass. When facing us the under side of their wings made a broad band on the sky, but when in circling they faced from us we could no longer see the underside of their wings. Though the wings then made little more than a line on the sky the glass showed clearly that it was not the under side that we saw. It was evident that the buzzards were soaring with their wings constantly inclined above five degrees above the horizon. They were attempting to gain sufficient altitude to enable them to glide to the ocean beach three fourths of a mile distant, but after reaching a height of about 75 feet above the top of the hill they seemed to be unable to rise higher, though they tried a long time. At last they started to glide toward the ocean but were compelled to begin flapping almost immediately. We at once measured the slope and the wind. The former was 12 ½ degrees; the latter was six to eight meters per second. Since the wings were inclined 5 degrees above the horizon and the wind had a rising trend of fully 12 degrees, the angle of incidence was about 17 degrees. The wind did not average more than seven meters, 15 miles an hour. For the most part the birds faced the wind steadily, but in the lulls they were compelled to circle or glide back and forth in order to obtain speed sufficient to provide support. As the buzzard weighs about .8 pounds per square foot of wing area, the lifting power of the wind at 17 degrees angle of incidence was apparently as great as it would have been had it been blowing straight upward with equal velocity. The pressure was inclined 5 degrees in front of the normal, and the angle descent was 12 ½ degrees.

On another day I stood on top of the West Hill directly behind a buzzard which was soaring on the steep southern slope. It was just on a level with my eye and not more than 75 feet distant. For some time it remained almost motionless. Although the wings were inclined about five degrees above the horizon, it was not driven backward by the wind. This bird is specially adapted to soaring at large angles of incidence in strongly rising currents. Its wings are deeply curved. Unless the upward trend amounts to at least eight degrees it seems to be unable to maintain itself. One day we watched a flock attempting to soar on the west slope of the Big Hill, which has a descent of nearly nine degrees. The birds would start near the top and glide down along the slope very much as we did with the machine but we noticed that whenever they glided parallel with the slope their speed diminished, and when their speed was maintained the angle of descent was greater than that of the hill. In every case they found it necessary to flap before they had gone two hundred feet. They tried time and again but always with the same results. Finally they resorted to hard flapping till a height of about 150 feet above the top of the hill was reached, after which they were able to soar in circles without difficulty. On another day they finally succeeded in rising on almost the same slope, from which it was concluded that the buzzard's best angle of descent could not be far from eight degrees. There is no question in my mind that men can build wings having as little or less relative resistance than that of the best soaring birds. The bird's wings are undoubtedly very well designed indeed, but it is not any extraordinary efficiency that strikes with astonishment but rather the marvelous skill with which they are used. It is true that I have seen birds perform soaring feats of almost incredible nature in positions where it was not possible to measure the speed and trend of the wind, but whenever it was possible to determine by actual measurement the conditions under which the soaring was performed, it was easy to account for it on the basis of the results obtained with artificial wings. The soaring problem is apparently not so much one of better wings as of better operators.

Document 1-13(a–b)

(a) Orville Wright, diary entry for 17 December 1903, published as "Diary of the First Flight," *Collier's* 25 (December 1948): 33.

(b) Wilbur and Orville Wright, statement to the Associated Press, Dayton, Ohio, 5 January 1904.

The following documents are companions to the Wright brothers' telegram, sent to their father in Dayton, announcing the historic achievement of powered flight on 17 December 1903. The first is the diary entry for that historic day, composed by Orville, and the second is the public statement of their achievement, subsequently prepared by the Wrights for release to the press. The latter was given to the local representative of the Associated Press in Dayton (a man by the name of Frank Tunison), but he was not interested. Articles about the Wrights containing part of this statement did appear in miscellaneous newspapers, beginning on 6 January 1904. But only the *Dayton Press* carried the full statement contained here. Instead of the carefully crafted, low-key statements deliberately thought out by the Wrights, what appeared in newspapers about their flying—when anything appeared at all—were imaginative, inaccurate accounts that disturbed the Wrights and misled the public into thinking that this story was no different than all the others that came before—that was, mostly fanciful and not worthy of serious attention.

It is truly astonishing how the press bungled the story of the Wright's achievement. Not only did the press botch the story in 1903, but it continued to neglect or completely distort it for the next five years, well into 1908. One aviation historian, the distinguished Richard K. Smith, has gone so far as to say that "the relationship between the Wrights and the news medium of their day is one of the most grotesque stories of the 20th century and it was by no means the fault of the Wrights" ["Not a Success—But a Triumph: 80 Years Since Kitty Hawk," *Naval War College Review* 36 (November–December 1983): 13]. As indicated in the historic telegram itself, the Wrights chose to inform the press with carefully worded statements and continued to do so for months, but the press paid no attention. December 1903 and early 1904 were not good times for the press to give much credence to anyone's claims of a successful flying machine. Just nine days before the alleged achievement at Kitty Hawk, the press had seen, with its own eyes, the debacle in the Potomac River involving Dr. Samuel P. Langley's full-scale aerodrome. If the great Professor Langley, one of America's greatest scientists, could not solve the problems of flight, then how could a couple of unknown bicycle mechanics from Dayton, Ohio?

Document 1-13(a), Orville Wright, diary entry for 17 December 1903.

Thursday, December 17, 1903

When we got up a wind of between 20 and 25 miles was blowing from the north. We got the machine out early and put out the signal for the men at the station. Before we were quite ready, John T. Daniels, W. S. Dough, A. D. Etheridge, W. C. Brinkley of Manteo, and Johnny Moore of Nags Head arrived. After running the engine and propellors a few minutes to get them in working order, I got on the machine at 10:35 for the first trial. The wind, according to our anemometers at this time, was blowing a little over 20 miles (corrected) 27 miles according to the Government anemometer at Kitty Hawk. On slipping the rope the machine started off increasing in speed to probably 7 or 8 miles. The machine lifted from the track just as it was entering on the fourth rail.

Mr. Daniels took a picture just as it left the tracks. I found the control of the front rudder quite difficult on account of its being balanced too near the center and thus had a tendency to turn itself when started so that the rudder was turned too far on one side and then too far on the other. As a result the machine would rise suddenly to about 10 ft. and then as suddenly, on turning the rudder, dart for the ground. A sudden dart when out about 100 feet from the end of the tracks ended the flight. Time about 12 seconds (not known exactly as watch was not promptly stopped). The lever for throwing off the engine was broken, and the skid under the rudder cracked. After repairs, at 20 min. after 11 o'clock Will made the second trial. The course was about like mine, up and down but a little longer over the ground though about the same in time. Dist. not measured but about 175 ft. Wind speed not quite so strong. With the aid of the station men present, we picked the machine up and carried it back to the starting ways. At about 20 minutes till 12 o'clock I made the third trial. When out about the same distance as Will's, I met with a strong gust from the left which raised the left wing and sidled the machine off to the right in a lively manner. I immediately turned the rudder to bring the machine down and then worked the end control. Much to our surprise, on reaching the ground the left wing struck first showing the lateral control of this machine much more effective than on any of our former ones. At the time of its sidling it had raised to a height of probably 12 to 14 feet. At just 12 o'clock Will started on the fourth and last trip. The machine started off with its ups and downs as it had before, but by the time he had gone three or four hundred feet he had it under much better control, and was traveling on a fairly even course. It proceeded in this manner till it reached a small hummock out about 800 feet from the starting ways, when it began its pitching again and suddenly darted into the ground. The front rudder frame was badly broken up, but the main frame suffered none at all. The distance over the ground was 852 feet in 59 seconds. The engine turns was 1071, but this included several seconds while on the starting ways and probably about a half second after landing. The jar of landing had set the watch on machine back so that we have no exact record for the 1071 turns. Will took a picture of my third flight just before the gust struck the machine. The machine left the ways successfully at every trial, and the tail was never caught by the truck as we had feared.

After removing the front rudder, we carried the machine back to camp. We set the machine down a few feet west of the building, and while standing about discussing the last flight, a sudden gust of wind struck the machine and started to turn it over. All rushed to stop it. Will who was near the end ran to the front, but too late to do any good. Mr. Daniels and myself seized spars at the rear, but to no purpose. The machine gradually turned over on us. Mr. Daniels, having had no experience in handling a machine of this kind, hung on to it from the inside, and as a result was knocked down and turned over and over with it as it went. His escape was miraculous, as he was in with the engine and chains. The engine legs were all broken off, the chain guides badly bent, a number of uprights, and nearly all the rear ends of the ribs were broken. One spar only was broken.

After dinner we went to Kitty Hawk to send off telegram to M. W. While there we called on Capt. and Mrs. Hobbs, Dr. Cogswell and the station men.

Document 1-13(b), statement by the Wright brothers to the Associated Press, 5 January 1904.

It had not been our intention to make any detailed public statement concerning the private trials of our power "Flyer" on the 17th of December last; but since the contents of a private telegram, announcing to our folks at home the success of our trials, was dishonestly communicated to the newspapermen at the Norfolk office, and led to the imposition upon the public, by persons who never saw the "Flyer" or its flights, of a fictitious story incorrect in almost every detail; and since this story together with several pretended interviews or statements, which were fakes pure and simple, have been very widely disseminated we feel impelled to make correction. The real facts were as follows:

On the morning of December 17th, between the hours of 10:30 o'clock and noon, four flights were made, two by Orville Wright and two by Wilbur Wright. The starts were all made from a point on the level sand about two hundred feet west of our camp, which is located a quarter of a mile north of the Kill Devil sand hill, in Dare County, North Carolina. The wind at the time of the flights had a velocity of 27 miles an hour at ten o'clock, and 24 miles an hour at noon, as recorded by the anemometer at the Kitty Hawk Weather Bureau Station. This anemometer is thirty feet from the ground. Our own measurements, made with a hand anemometer at a height of four feet from the ground, showed a velocity of about 22 miles when the first flight was made, and 20 1/2 miles at the time of the last one. The flights were directly against the wind. Each time the machine started from the level ground by its own power alone no assistance from gravity, or any other source whatever. After a run of about 40 feet along a monorail track, which held the machine eight inches from the ground, it rose from the track and under the direction of the operator climbed upward on an inclined course till a height of eight or ten feet from the ground was reached, after which the course was kept as near horizontal as the wind gusts and the limited skill of the operator would permit. Into the teeth of a December gale the "Flyer" made its way forward with a speed of ten miles an hour over the ground and thirty to thirty-five miles an hour through the air. It had previously been decided that for reasons of personal safety these first trials should be made as close to the ground as possible. The height chosen was scarcely sufficient for maneuvering in so gusty a wind and with no previous acquaintance with the conduct of the machine and its controlling mechanisms. Consequently the first flight was short. The succeeding flights rapidly increased in length and at the fourth trial a flight of fifty-nine seconds was made, in which time the machine flew a little more than a half mile through the air, and a distance of 852 feet over the ground. The landing was due to a slight error of judgment on the part of the aviator. After passing over a little hummock of sand, in attempting to bring the machine down to the desired height, the operator turned the rudder too far; and the machine turned downward more quickly than had been expected. The reverse movement of the rudder was a fraction of a second too late to prevent the machine from touching the ground and thus ending the flight. The whole occurrence occupied little, if any, more than one second of time.

Only those who are acquainted with practical aeronautics can appreciate the difficulties of attempting the first trials of a flying machine in a twenty-five mile gale. As winter was already well set in, we should have postponed our trials to a more favorable season, but for the fact that we were determined, before returning home, to know whether the machine possessed sufficient power to fly, sufficient strength to withstand the shocks of landings, and sufficient capacity of control to make flight safe in boisterous winds, as well as in calm air. When these points had been definitely established, we at once packed our goods and returned home, knowing that the age the flying machine had come at last.

From the beginning we have employed entirely new principles of control; and as all the experiments have been conducted at our own expense without assistance from any individual or institution, we do not feel ready at present to give out any pictures or detailed description of the machine.

Document 1-14(a–d)

(a) Wilbur and Orville Wright, letter to the Secretary of War, Dayton, Ohio, 9 October 1905, *The Papers of Wilbur and Orville Wright*, **pp. 514–515.**

(b) U.S. Army Signal Corps, "Specification No. 486, Advertisement and specification for a heavier-than-air flying machine," issued 23 December 1907, copy in Milton Ames Collection, Historical Archives, NASA Langley Research Center, Hampton, Virginia.

(c) Wilbur and Orville Wright, letter to General James Allen, U.S. Army Signal Corps, 27 January 1908, *The Papers of Wilbur and Orville Wright*, **p. 856.**

(d) Octave Chanute, excerpts from "Recent Aeronautical Progress in the United States," *The Aeronautical Journal* **[London] (12 July 1908): 52–55.**

While the anniversary of the invention of the airplane is marked from the Wright brothers' successful test flights of 1903, the inaugural of the airplane's development into a practical, useable device was marked by a subsequent improved Wright airplane's successful completion of government trials, which led to the Wright's first sale of a flying machine on 2 August 1909.

This significant step in the development of aviation is marked through the following four documents, and conclude this chapter. In 1905, Wilbur and Orville Wright initiated a correspondence with the Secretary of War offering their aircraft for sale to the government after a public demonstration of its capabilities. After a further bit of prodding, in 1907 the United States Army issued the second document, an official statement of specifications regarding the aerodynamic performance of a flying machine to be used by the Signal Corps as an observation aircraft. The third document is the Wright brothers' responding bid. Rounding out this group of documents is a summary by Octave Chanute published in Great Britain reviewing the "Recent Aeronautical Progress in the United States."

With Wilbur away in Europe demonstrating one of their airplanes to astonished crowds, Orville was left to pilot the airplane for the Army. The military tests were held at Fort Myer, Virginia, just outside of Washington, D.C., beginning on 3 September 1908. A week of impressive demonstrations, in which the aircraft met or exceeded requirements, dramatically opened the eyes of America to the amazing technological achievement of the Wright brothers and the promise of aviation. Thereafter, not even the unfortunate crash on 17 September that injured Orville and killed his official passenger, Lieutenant Thomas E. Selfridge, could stifle the growing appreciation that the day of the airplane was finally at hand.

Orville recovered from his injuries, and with Wilbur's help the tests resumed in July 1909. In the government contract awarded to the Wrights following the acceptance of their flying machine by the U.S. Army, the brothers received $30,000 for the delivery of one aircraft and the training of two pilots. The price included a $5,000 bonus for exceeding the stipulated speed requirement by a recorded average speed of 42.583 miles per hour over the 10-mile test course.

Document 1-14(a), letter from the Wright brothers to the Secretary of War, 9 October 1905.

Some months ago we made an informal offer to furnish to the War Department practical flying machines suitable for scouting purposes. The matter was referred to the Board of Ordnance and Fortification, which seems to have given it scant consideration. We do not wish to take this invention abroad, unless we find it necessary to do so, and therefore write again, renewing the offer.

We are prepared to furnish a machine on contract, to be accepted only after trial trips in which the conditions of the contract have been fulfilled; the machine to carry an operator and supplies of fuel, etc., sufficient for a flight of one hundred miles; the price of the machine to be regulated according to a sliding scale based on the performance of the machine in the trial trips; the minimum performance to be a flight of at least twenty-five miles at a speed of not less than thirty miles an hour.

We are also willing to take contracts to build machines carrying more than one man.

Document 1-14(b), U.S. Army Signal Corps, "Specification No. 486, Advertisement and specification for a heavier-than-air flying machine," 23 December 1907.

Signal Corps Specification, No. 486

Advertisement and specification for a heavier-than-air flying machine.

To the Public:

Sealed proposals, in duplicate, will be received at this office until 12 o'clock noon on February 1, 1908, on behalf of the Board of Ordnance and Fortification for furnishing the Signal Corps with a heavier-than-air flying machine. All proposals received will be turned over to the Board of Ordnance and Fortification at its first meeting after February 1 for its official action.

Persons wishing to submit proposals under this specification can obtain the necessary forms and envelopes by application to the Chief Signal Officer, United States Army, War Department, Washington, D.C. The United States reserves the right to reject any and all proposals.

Unless the bidders are also the manufacturers of the flying machine they must state the name and place of the maker.

Preliminary.—This specification covers the construction of a flying machine supported entirely by the dynamic reaction of the atmosphere and having no gas bag.

Acceptance.—The flying machine will be accepted only after a successful trial flight, during which it will comply with all requirements of this specification. No payments on account will be made until after the trial flight and acceptance.

Inspection.—The Government reserves the right to inspect any and all processes of manufacture.

General Requirements

The general requirements of the flying machine will be determined by the manufacturer, subject to the following conditions:

1. Builders must submit with the proposals the following:

(a) Drawings to scale showing the general dimensions and shape of the flying machine which they propose to build under this specification.

(b) Statement of the speed for which it is designed

(c) Statement of the total surface area of the supporting planes

(d) Statement of the total weight.

(e) Description of the engine which will be used for motive power.

(f) The material of which the frame, planes, and propellers will be constructed. Plans received will not be shown to other bidders.

2. It is desirable that the flying machine should be designed so that it may be quickly and easily assembled and taken apart and packed for transportation in army wagons. It should be capable of being assembled and put in operation in about one hour.

3. The flying machine must be designed to carry two persons having a combined weight of about 350 pounds, also sufficient fuel for a flight of 125 miles.

4. The flying machine should be designed to have a speed of at least forty miles per hour in still air, but bidders must submit quotations in their proposals for cost depending upon the speed attained during the trial flight, according to the following scale:

40 miles per hour, 100 per cent.
39 miles per hour, 90 per cent.
38 miles per hour, 80 per cent.

37 miles per hour, 70 per cent.
36 miles per hour, 60 per cent.
Less than 36 miles per hour rejected
41 miles per hour, 110 per cent.
42 miles per hour, 120 per cent.
43 miles per hour, 130 per cent.
44 miles per hour, 140 per cent.

5. The speed accomplished during the trial flight will be determined by taking an average of the time over a measured course of more than five miles, against and with the wind. The time will be taken by a flying start, passing the starting point at full speed at both ends of the course. This test subject to such additional details as the Chief Signal Officer of the Army may prescribe at the time.

6. Before acceptance a trial endurance flight will be required of at least one hour during which time the flying machine must remain continuously in the air without landing. It shall return to the starting point and land without any damage that would prevent it immediately starting upon another flight. During this trial flight of one hour it must be steered in all directions without difficulty and at all times under perfect control and equilibrium.

7. Three trials will be allowed for speed as provided for in paragraphs 4 and 5. Three trials for endurance as provided for in paragraph 6, and both tests must be completed within a period of thirty days from date of delivery. The expense of the tests to be borne by the manufacturer. The place of delivery to the Government and trial flights will be at Fort Myer, Virginia.

8. It should be so designed as to ascend in any country which may be encountered in field service. The starting device must be simple and transportable. It should also land in a field without requiring a specially prepared spot and without damaging its structure.

9. It should be provided with some device to permit of a safe descent in case of an accident to the propelling machinery.

10. It should be sufficiently simple in its construction and operation to permit an intelligent man to become proficient in its use within a reasonable length of time.

11. Bidders must furnish evidence that the Government of the United States has the lawful right to use all patented devices or appurtenances which may be a part of the flying machine, and that the manufacturers of the flying machine are authorized to convey the same to the Government. This refers to the exclusive purchase of patent rights for duplicating the flying machine.

12. Bidders will be required to furnish with their proposal a certified check amounting to ten per cent of the price stated for the 40-mile speed. Upon making the award for this flying machine these certified checks will be returned to the bidders, and the successful bidder will be required to furnish a bond, according to Army

Regulations, of the amount equal to the price stated for the 40-mile speed.

13. The price quoted in proposals must be understood to include the instruction of two men in the handling and operation of the flying machine. No extra charge for this service will be allowed.

14. Bidders must state the time which will be required for delivery after receipt of order.

James Allen
Brigadier General, Chief Signal Officer of the Army.
Signal Office,
Washington, D. C., *December 23, 1907.*

Document 1-14(c), letter from Wilbur and Orville Wright to General James Allen, 27 January 1908.

We herewith inclose a bid for furnishing the Signal Corps with a heavier-than-air flying machine, in accordance with Specification No. 486, of December 23, 1907, together with a certified check for two thousand five hundred dollars ($2,500.00).

The machine we propose to deliver is designed to weigh between 1,100 and 1,250 lbs. with two men on board, and for a speed of forty miles an hour. It will have an area of 500 square feet in the supporting planes; and will be propelled by a four-cycle, water-cooled gasoline motor. The frames of the planes will be constructed of spruce and ash covered with cotton muslin; the propellers of spruce and linen.

We have made the date of delivery of the machine 200 days, in order to provide sufficient time for increasing the speed of the machine now under construction, in case Requirement No. 5 is to be interpreted literally. If, however, Requirement No. 5 is interpreted to mean an average of the speeds with and against the wind over a measured course, which is the correct method to give and average corresponding to flight made in still air, as specified in Requirement No. 4, we would be able to make delivery at a much earlier date.

We inclose a photograph of our machine of 1905, which was similar to the one we now propose to furnish. We would request that this, as well as the drawings, be kept confidential.

Document 1-14(d), Octave Chanute, excerpts from "Recent Aeronautical Progress in the United States," London, 1908.

The public attitude in the United States in 1906 and 1907 concerning aërial navigation has been one of expectancy and apathy. The announcements of the marvellous success achieved by Wright Brothers, which every investigation seemed to confirm, must have deterred many searchers from experimenting at all, until they know how much remained to be accomplished in aviation. . . .

Meanwhile Aëro Clubs have sprung up like mushrooms all over the country. The leading one still is the Aëro Club of America in New York, which was organized in 1905. It has held several exhibitions, has promoted the publication of an interesting book, "Navigating the Air," the establishment of a correspondence School of Aëronautics by Mr. Triaca, and an effort is now being made in connection with the Club to raise a fund of £5,000 to be offered in prizes for Aviation.

In St. Louis and Chicago local aëro clubs propose to organize balloon races to be held in 1908. In these and other cities, members are encouraged to engage in the sport by owning and riding balloons themselves, and it remains to be seen how long the enthusiasm will last.

Two monthly Aëronautical Magazines have been started, one in New York and one in St. Louis, but it is yet to be ascertained how well they will be supported.

The Jamestown Exposition of 1907 organized an aëronautical exhibit which amounted to but little, as well as an Aëronautical Congress which brought out few papers, but searchers have been building apparatus to be experimented with in the summer of 1908.

The most distinguished of these is Dr. Alexander Graham Bell, the inventor of the Telephone, who has been experimenting with his tetrahedral kite. He tested on December 6th, 1907, his gigantic man-lifting kite "Cygnet," consisting of 3,393 wing cells, presenting 1,966 square feet of oblique surfaces, and weighing with the floats and passenger an aggregate of 600 pounds. This was towed into the middle of a lake and raised against a wind of 21 miles an hour by a tug-boat. It exhibited that perfect stability which all previous experiments indicated, and upon the wind's dying away it descended gently from a height of 168 feet, but was broken on striking the water. It is to be tested again during the summer of 1908 with a view to eventually apply a motor. There is no question as to the automatic equilibrium of this form of apparatus, but it is possible that the inferior lifting power of the oblique surfaces and the resistance of so many front edges will make the design less favorable for a flying machine than other forms.

Dr. Bell then generously provided the means (he gives credit to his wife) and organized the so-called "Aërial Experiment Association" with its headquarters at Hammondsport, New York, to give his assistants a chance to experiment their own ideas. The first result was the construction of the motor-driven aëroplane "Red Wing," chiefly designed by Lieutenant T. Selfridge, which made its trial trip on sleigh runners on the ice of Lake Leuka on March 9th, 1908. It is a double-decked apparatus 43 feet across, the surfaces being arched both fore and aft and from tip to tip; the upper aëroplane being bowed downward somewhat in the attitude of the gull and the lower aëroplane, which is 6 feet shorter, being bowed upward somewhat to the attitude of the vulture when soaring. The total surface is 386 square feet and the total weight, including the aviator, is 570 pounds. It is driven by a Curtiss motor of 40-horse power, actuating a screw propeller.

At the very first attempt the apparatus left the ice after travelling only 200 feet and flew a total distance of 319 feet from the point where it left the ice to the point of descending. It alighted somewhat clumsily and broke one strut, this being the first public exhibition of the flight of a heavier-than-air machine in America. . . .

The main interest, however, attaches to the pending United States Government tests of flying machines which are under contract for delivery next August. On December 23rd, 1907, the United States Army Signal Corps issued invitations to tenders, which produced much amazement. European and American journals have said that these specifications assume that flying machines are almost a usual method of transportation, and that the terms are so exacting as to seem unreasonable. The Signal Services officers answer that the specifications were drawn up after interviews with some of the inventors and merely cover what they said they could perform, while some clauses were added to prevent the Government's being trifled with, and that the tests will be conducted with judicious reason and liberality. More especially does this apply to the granting but three trials each for the speed test and the endurance test of one hour, which might be defeated on each occasion by some fortuitous and trifling circumstance. . . .

The performances of the Wright Brothers have been viewed with incredulity because of the mystery with which they have been surrounded in the hope of a rich money reward, yet it is now generally conceded that they have accomplished all that they have claimed, i.e., to have made a first dynamic flight in 1903, to have mastered circular courses in 1904, making 105 flights, the longest of which was three miles, and to have obtained thorough control over their apparatus in 1905, making 49 flights, the longest of which was 24 miles, consisting of 30 sweeps over a circular course at an average speed of 38 miles an hour. Since then they have made no flights, having been engaged in negotiations with a view to marketing their invention.

Now they have made a contract with the United States Government to furnish a flying machine under those formidable specifications. They set to work at once. They have built parts of more than one machine, so as to guard against bad breakages, and have returned to their old experimental grounds near Kitty Hawk, North Carolina. This is situated on a long sand spit, two or three miles wide, between the waters of Pamlico Sound and the Atlantic Ocean. It is about as inaccessible a spot near civilisation [sic] as can well be, being almost a desert, occupied by a few fishermen and a Government lifesaving station. Near the camp is "Kill Devil Hill," a cone of drifted sand about 100 feet high, on which former gliding experiments were made.

Here the Wrights have established themselves and begun their practice, for it is only by strenuous practice that the mastery of the air is to be obtained. They are said to be proceeding with great caution, testing every part and peculiarity of the machine, the longest flight yet reported (May 14th, 1908) being eight miles, followed, however, by a serious breakage on landing, said to be due to a false maneuvre, in consequence of some change in the location of the levers which control the rudders. It is stated that the wreck was so complete that the parts will be shipped back to Dayton, Ohio, where the craft will be rebuilt.

The Wrights are understood to have until August 27th to deliver their machine to the Signal Corps for testing, so that there will be sufficient time to resume practice after the machine is repaired. Whether this practice will take place on the same ground or elsewhere is not known. The spot is very secluded, but the ubiquitous reporter has found the camp and is sending "news" both true and untrue, to the great annoyance of [the] Wright Brothers. . . .

An amusing struggle has resulted. The reporters are frantic for information, and the Wrights most determined that no description be given of their apparatus. It is probable that many contradictory cablegrams will have been received in Great Britain when the present paper reaches the Hon. Secretary.

The Wright brothers stand a fair chance of passing the tests and having their machine accepted. They may be defeated by some accident during the preliminary trials or the formal tests, but the present writer is sure that all the members of the Aëronautical Society of Great Britain will join him in the hope that the best of luck will attend the demonstration.

Document 1-15

Simon Newcomb, "Aviation Declared a Failure," *Literary Digest* 37, 17 October 1908, 549.

The press was not the only group to bungle the Wright brothers' story and to minimize the importance of what they had achieved and what the flying machine signified for society. One of the most vocal "bah-humbugs" in the early 1900s happened to be one of the greatest scientists of his day, Professor Simon Newcomb (1835–1909).

Newcomb was a giant in the field of celestial mechanics. His work in the late nineteenth century on the orbital motion of the planets of the Solar System provided the cornerstone of the nautical and astronomical almanacs of the United States and Great Britain, not just in the early 1900s but as recently as 1984. Albert Einstein acknowledged the importance of Newcomb's work in the development of his own theory of relativity. Born in England, he immigrated in 1854 at age 19 to Maryland to join his father. Newcomb taught himself most of the mathematics and astronomy he knew, taking a job in 1857 in the American Nautical Almanac Office, which was located in Cambridge, Massachusetts. Admitted to Harvard, he graduated in 1858, and three years later he took an appointment to the Naval Observatory at Washington, D.C. The next ten years he spent determining the positions of celestial objects using various telescopes, some of his own design. In 1877, at age forty-two, he became director of the American Nautical Almanac Office, now located in Washington, D.C., and initiated the work on celestial motion for which he would become most famous. In 1884, Newcomb became professor of mathematics and astronomy at The Johns Hopkins University, staying there until 1893. He served as editor of the *American Journal of Mathematics* for many years and was a founding member and the first president (1899–1905) of the American Astronomical Society. He also served as president of the American Mathematical Society from 1897 to 1898. By the time the Wright brothers made their historic first flight in 1903, Newcomb had risen to the top of the astronomical community in the United States, the recipient of many of the highest national and international awards given to scientists, including Fellow of the British Royal Society.

Newcomb loved to travel. He spoke French and German fluently and enough Italian and Swedish to travel easily in those countries. An avid hiker, at age seventy (in 1905) he climbed to the chalet high up the side of the Matterhorn in the Swiss Alps, a feat almost unprecedented for a man of his age. One may not expect that such a brilliant and adventurous man would be so pessimistic about the future of aviation, but no one was more outspoken or pungent in his pessimism about the usefulness of flying than Newcomb.

What follows is an article published in the *Literary Digest* in October 1908 revealing Newcomb's ongoing pessimism. Newcomb in fact had been writing about aviation in condescending terms since the 1890s. In one article published in *McClure's Magazine* around the turn of the century, Newcomb had written: "Man's desire to fly like a bird is inborn in our race, and we can no more be expected to abandon the idea than the ancient mathematician could have been expected to give up the problems of squaring the circle." In other words, he believed the problems of flight could never be solved. In the article below, Newcomb's skepticism about the future of aviation is quoted at length. The caption under the picture of Newcomb that ran with the story read "The eminent American astronomer. He doubts if aviation will ever be of much practical value."

Newcomb's was not the only voice still expressing such negativism. In 1910, the *New York Telegraph* called a proposal from a Texas congressman for a study of the possibility of U.S. airmail operations "ludicrous." The paper mocked: "Love letters will be carried in rose-pink aeroplane, steered with Cupid's wings and operated by perfumed gasoline" [quoted in Roger E. Bilstein, *Flight in America: From the Wrights to the Astronauts* (Baltimore, MD: Johns Hopkins University Press, 1984), pp. 16–17].

Though Newcomb died in 1909, his attitudes and others like them did not fade as quickly as one might think. But as aviation caught Americans' attention in the early 1910s, the public's misgivings greatly diminished. When the airplane proved a dynamic instrument in the Great War (1914–1918), most doubt about the practicality of aviation disappeared. Still today, though, there are millions of people who share the basis of Newcomb's resistance to flying. As he is quoted as saying in the article below: "Is it not evident, on careful consideration, that the ground affords a much better bade than air ever can? Resting upon it we feel safe and we know where we are. In the air we are carried about by every wind that blows."

Though the world's emerging aerodynamic specialists might surely have argued with Newcomb's point of view, not even they could undervalue the many dangers, uncertainties, and inefficiencies of machines flying through the air. In an important sense, aerodynamicists would be spending the next 100 years answering Newcomb's complaint.

Document 1-15, Simon Newcomb, "Aviation Declared a Failure," 1908.

A pessimist has been defined as a man who, when offered the choice of two evils, takes both. Prof. Simon Newcomb in his recent article on "The Problem of Aerial Navigation" (*The Nineteenth Century*, London, September) varies the formula somewhat. When offered two perfectly good methods of navigating the air, he rejects both. He concludes that the disadvantages of both dirigible balloons and aeroplanes outweigh all existing and possible advantages, and that the solid earth is good enough for him, anyhow. We have not space for the professor's extended bill of particulars, but will proceed at once to his general statement of the case against aviation. First summing up the advantages in one paragraph, only to overwhelm them in the next, be says:

"Let us . . . in fairness see what is to be placed on the credit side. First and almost alone among these is the fact that steam transportation on land requires the building of railways, which are so expensive that the capital involved in them probably exceeds that invested in all other forms of transportation. Moreover, there are large areas of the earth's surface not yet accessible by rail, among which are the poles and the higher mountains. All such regions, the mountains excepted, we may suppose to be attainable by the perfected air-ship of the future. The more carefully we analyze these possible advantages, the more we shall find them to diminish in importance. Every part of the earth's surface on which men now live in large numbers, and in which important industries are prosecuted, can now be reached by railways, or will be so reached in time. True, this will involve a constantly increasing investment of capital. But the interest on this investment will be a trifle in comparison with the cost and drawbacks incident to the general introduction of the best system of aerial transportation that is even ideally possible in the present state of our knowledge. . . ."

"May we not say . . . that the efforts at aerial navigation now being made are simply most ingenious attempts to substitute, as a support of moving bodies, the thin air for the solid ground? And is it not evident, on careful consideration, that the ground affords a much better base than air ever can? Resting upon it we feel safe and know where we are. In the air we are carried about by every wind that blows. Any use that we can make of the air for the purpose of transportation, even when our machinery attains ideal perfection, will be uncertain, dangerous, expensive, and inefficient, as compared with transportation on the earth and ocean. The glamour which surrounds the idea of flying through the air is the result of ancestral notions, implanted in the minds of our race before steam transportation had attained is present development. Exceptional cases there may be in which the air-ship will serve a purpose, but they are few and unimportant."

Professor Newcomb admits that in certain special cases flyers or balloons may accomplish what could be done in no other way. For instance, he thinks it not unlikely that the pole may be first reached by a dirigible balloon. The balloon as an engine of war, however, he regards as an impossibility, and he reassures the Englishmen who have been looking forward apprehensively to a vertical bombardment of British towns by a fleet of German dirigibles. He says in conclusion:

"In presenting the views set forth in the present article the writer is conscious that they diverge from the general trend, not only of public opinion, but of the ideas of some able and distinguished authorities in technical science, who have given encouragement to the idea of aerial navigation. Were it a simple question of weight of opinion he would frankly admit the unwisdom of engaging in so unequal a contest. But questions of what can be done through the application of mechanical power to bodies in motion have no relation to opinion. They can be determined only by calculations made by experts and based upon the data and principles of mechanics."

"If any calculations of the kind exist, the writer has never met with them, nor has he ever seen them either quoted or used by any author engaged in discussing the subject. So far as his observation has extended, the problem has been everywhere looked upon as merely one of experiments ingeniously conducted with all the aid afforded by modern apparatus. He has seen no evidence that any writer or projector has ever weighed the considerations here adduced, which seem to him to bring out the insuperable difficulties of the system he has been discussing, and the small utility to be expected from it even if the difficulties were surmounted."

If he is wrong in any point—and he makes no claim to infallibility—it must be easy to point out in what his error consists. He therefore concludes with the hope that if his conclusions are ill-founded their fallacy will be shown, and that if well-founded they may not be entirely useless in affording food for thought to those interested in the subject.

Experts in aviation and the members of aero-clubs will not be apt to agree with all this, but it will certainly be useful in counteracting the ardor of those enthusiasts who think that we are all going to fly to Europe before the end of 1909.

Chapter 2

Building a Research Establishment

The American Way

> *The only people so far who have been able to get at something like accurate results from wind-tunnel experiments are the workers at the Experimental Station at Langley Field, which is run by the National Advisory Committee for Aeronautics of the United States of America. Thanks to the wealth of the United States and the high intelligence of those who are charged with the task of aeronautical experiments, workers in the American research establishments have acquired knowledge that is in many ways far ahead of anything we have in this country. And they have it very largely by what is called "ad hoc research,"—that is to say, going and looking for the solution of one particular problem, instead of experimenting around blindly in the hope that something may turn up, after the fashion which is known as "basic research."*

The above editorial comment by C. G. Grey, a prominent British aeronautical engineer and editor of the aviation journal *The Aeroplane*, appeared in the 6 February 1929 issue of that journal.

From the Wright Bicycle Shop to the Langley Full-Scale Tunnel

For those who followed the ingenious Wright brothers into the air after 1903, the maturation of the airplane depended on a growing and increasingly sophisticated understanding of aerodynamics, for there was still much about flight that was unknown. The route to greater aeronautical knowledge in the post-Wright era was not a straight highway. To the extent a map even existed, it offered a maze of twisting roads involving trial-and-error design of new flying machines; dogged pragmatic testing; deeper scientific inquiry; and, perhaps most importantly, a shrewd combination of the best that both theory and experiment had to offer.

Yet for the aeronautical scientists and engineers in pursuit of aviation progress in the early twentieth century, there was perhaps no surer course to progress than the one laid out by the Wright brothers themselves. Unfortunately, not everyone in the brave new world of aeronautics understood the Wrights' path to success. Many wrongly interpreted their invention of the airplane either as the heroic act of ingenious mechanical tinkerers or as a basic scientific discovery, rather than as a solid technological program of engineering research and development.

In fact, the Wrights' first successful flying machines, starting with their 1902 glider and the 1903 airplane that followed, not only represented key breakthroughs in their efforts to master heavier-than-air flight, but also bore testimony to the irreplaceable value of combining careful laboratory experiments with actual flight testing.

In 1915, twelve long years after the epochal Wright flight, the U.S. Congress established the National Advisory Committee for Aeronautics, the predecessor of present-day NASA, with the mission "to supervise and direct the scientific study of the problems of flight with a view to their practical solution." Eventually this agency recreated the Wrights' formula for success in a series of federal research laboratories.[1] The NACA accomplished this, originally, by building an extraordinary community of aeronautical engineers, scientists, technicians, and test pilots at the Langley Memorial Aeronautical Laboratory (LMAL) in Tidewater, Virginia—the NACA's first and, until 1941, only research facility.[2] At NACA Langley, the laboratory staff faced, and eventually resolved, many fundamental questions about how to plan and conduct institutional aerodynamic research. These questions dealt with issues of how research should be focused, what sort of facilities should be built, what type of experimental investigations should be carried out, and how experimentalists and theoreticians could work together fruitfully. The early NACA community also dealt with the issue of whether science or engineering should be in control of the research program, and especially whether the German academic laboratory model could be transferred successfully to America—or whether an American lab would have to find its own way. Through the resolution of these issues, in conjunction with a growing list of important achievements in aeronautics, a sustainable organizational identity for Langley and for the subsequent NACA laboratories began to evolve. The result was a dynamic and highly creative organization that, although its original facility was named after scientist Samuel P. Langley, actually reflected more the systematic engineering approach of Wilbur and Orville Wright, with an emphasis on a search for practical solutions. Throughout this sometimes torturous process, the NACA helped a fledgling American aircraft industry advance the airplane in a few short decades to an astoundingly high level of technological performance and corresponding importance in modern society.

---

[1] On the history of the National Advisory Committee for Aeronautics, the place to start is Alex Roland's *Model Research: The National Advisory Committee for Aeronautics* (Washington, DC: NASA SP-4103, 1985), two volumes.

[2] NACA Langley's history is told in James R. Hansen's *Engineer in Charge: A History of the Langley Aeronautical Laboratory* (Washington, DC: NASA SP-4305, 1987).

In building a test device like the wind tunnel and then using it systematically to generate reliable data, the Wright brothers forged a solid link between aeronautical research and the design of successful aircraft. SI Negative No. A-2708-G

Center stage in NACA research, from the beginning, was a unique device designed for basic aerodynamic investigation: the "wind tunnel." Following the initial achievement of flight, wind tunnels quickly became the central research facilities at aeronautical laboratories all over the world. This, too, was a legacy of the Wright brothers' approach, for the aerodynamic shape of their landmark aircraft had evolved directly out of testing conducted in a simple little wind tunnel they built in their Dayton bicycle shop. From these wind tunnel experiments in 1901, they garnered the empirical insights needed to design the first truly effective flying machine, their breakthrough glider of 1902.

The Wright brothers did not invent the wind tunnel, and the basic concept behind it predated their work by roughly 400 years. During the Renaissance, the brilliant technological dreamer Leonardo da Vinci recognized that air blown past a stationary object produced the same effect as the object itself moving at the same relative speed through the air, the fundamental concept upon which wind tunnels operate. Da Vinci expressed this idea in his statement from the *Codex Atlanticus*, "As it is to move the object against the motionless air so it is to move the air against the motionless object." Sir Isaac Newton later recognized this same principle.

From these humble beginnings, the first person to apply the concept to practical aerodynamic research seems to have been Benjamin Robbins (1707–1751), a brilliant English mathematician who conducted a series of fundamental experiments on the ballistic properties of artillery projectiles. To evaluate the air resistance of different shapes, Robbins constructed a device known as a "whirling arm." This apparatus consisted of a four-foot horizontal arm attached to a vertical spindle that was rotated by the force of a falling weight. Robbins mounted test shapes with similar cross-section areas on the end of his whirling arm, and found that the speed of the arm's rotation varied considerably with objects of different profiles. From this he concluded that different shapes produced different amounts of air resistance, identifying the factor subsequently understood as aerodynamic "drag."[3]

Following Robbins came John Smeaton (1724–1792), an English civil engineer interested in sources of power for practical applications. Smeaton used a whirling arm device of his own making to investigate the function of windmill sails. In 1759, he described his experiments and the elaborate apparatus used in his tests in a seminal paper presented to the Royal Society of London. In this report, he outlined the need for such equipment by noting a basic problem in aerodynamic research: relying solely on natural wind. "In trying experiments on windmill sails, the wind itself is too uncertain to answer the purpose; we must therefore have recourse to an artificial wind," wrote Smeaton, adding that tests "may be done two ways: either by causing the air to move against the machine, or the machine to move against the air." This principle of relative motion, the same observed earlier by da Vinci, proved to be the key to the future development of the wind tunnel. However, at the time, and indeed for the next hundred years, relatively few of Smeaton's colleagues or successors either understood or accepted it.[4]

Nevertheless, there were some provident individuals intrigued with the investigation of flight in the nineteenth century who utilized Smeaton's whirling arm apparatus to understand aerodynamic principles. Sir George Cayley, for one, whose work outlined the basic shape of the airplane (see Chapter 1 of this volume), used a five-foot whirling arm that achieved tip speeds of up to twenty feet per second

---

[3] John D. Anderson, Jr., provides a concise technical summary of the contributions of Leonardo da Vinci, Benjamin Robbins, and other pioneers to the genesis of early aerodynamic concepts in *A History of Aerodynamics and Its Impact on Flying Machines* (Cambridge, England: Cambridge University Press, 1997), pp. 14–27 and 55–57.

[4] On John Smeaton, see Anderson, *A History of Aerodynamics*, pp. 58–61 and 76–79.

to measure the lift and drag properties of different wing shapes.[5] Later in the century, some of the most important whirling arm experiments of the pre-flight era were conducted by Professor Samuel P. Langley, who built a large testing apparatus with a thirty-foot arm on the roof of Pittsburgh's Allegheny Observatory in 1886. Powered by a ten-horsepower steam engine, this whirling arm produced tip speeds of nearly 150 feet per second, the equivalent of about 100 miles per hour. But while Langley's research produced much valuable information, his frustrations with the whirling arm also epitomized the limitations of this type of aerodynamic testing machine. Because its sixty-foot diameter made the device too large to be used indoors, Langley was forced to conduct his tests outside where the results were influenced by the vagaries of natural wind and atmospheric conditions. Even worse, the rotation of the arm itself created currents that disrupted the air and compromised the test results.[6] Clearly, researchers needed a superior apparatus if they were to obtain reliable results in a controlled laboratory setting.

A completely new type of device invented by Francis H. Wenham (1824–1908) offered the solution. In 1867 Wenham, a British marine engineer, submitted a proposal to the Aeronautical Society of Great Britain to build a novel machine that applied the principles garnered from earlier studies of hydrodynamic water channels to the new field of aerodynamic research. The Aeronautical Society sponsored Wenham's project, and in 1871 he completed an apparatus equipped with a steam-powered fan that blew air through "a trunk 12 feet long and 18 inches square, to direct the current horizontally, and in parallel course." It was, in essence, the world's first "wind tunnel"—though this term would not be coined until forty-two years later.[7]

From the very beginning, wind tunnels proved a boon to aerodynamic discovery, and the technical data they generated established a new and much more viable basis of knowledge about virtually all phenomena of flight. From Wenham's inaugural wind tunnel investigations came the realization that the aerodynamic lifting forces working on wing surfaces were much greater than Newtonian theory predicted. This early result bred real confidence that powered flight was possible, and it bequeathed much-needed credibility to the serious scientific study of aeronautics. Following Wenham's lead, more students of flight adopted the wind tunnel, start-

---

[5] Sir George Cayley's significance in aerodynamic research is analyzed in Tom D. Crouch, *A Dream of Wings: Americans and the Airplane, 1875–1905* (Washington, DC: Smithsonian Institution Press, 1981, 1989), pp. 27–29, 33–35, 47–48, and 63–64.

[6] See Crouch, *A Dream of Wings*, pp. 48–52, for insights into Samuel P. Langley's research equipment.

[7] N. H. Randers-Pherson, "Pioneer Wind Tunnels," *Smithsonian Miscellaneous Collections* 93 (19 January 1935): 1–2. See also F. H. Wenham, "On Forms of Surfaces Impelled Through the Air and Their Effects on Sustaining Weights." On Wenham's contributions to aerodynamics, see Anderson, *A History of Aerodynamics*, pp. 116–117 and 119–126.

ing with fellow Englishman Horatio Phillips, who in 1884 used a wind tunnel of his own design to establish that cambered airfoils developed significantly more lift than flat planes. By 1896, when Hiram Maxim constructed a wind tunnel to advance his aerodynamic research (outlined in the Maxim document in Chapter 1), the use of wind tunnels had spread to France and other European nations. In that year, wind tunnel technology also crossed the Atlantic to the Massachusetts Institute of Technology (MIT), where an inventive graduate student in engineering by the name of A. J. Wells built the first American wind tunnel, a makeshift arrangement that diverted air from a building ventilation duct for simple aerodynamic tests of flat planes.[8]

By the dawn of the twentieth century, though wind tunnel development was still in its infancy, awareness of these valuable tools had spread through the small community of aviation advocates working to achieve mechanical flight. What was most significant for the subsequent history of aviation was that Wilbur and Orville Wright knew enough about wind tunnels to turn to one when they were confronted with anomalous data about the lifting power of wings. As outlined in the previous chapter, the Wrights reached this point after the frustrating 1901 season of flight tests, when they were forced to face the fact that their gliders, based on Otto Lilienthal's coefficients, simply did not achieve the predicted amount of lift. From this they reasoned that Lilienthal's tables might possibly contain errors, which led them to conduct tests of previous assumptions and then move on to develop their own experimental data.

Before building a wind tunnel, the Wrights utilized a clever apparatus of their own design, consisting of a freely rotating horizontal wheel mounted on the front of a bicycle. An airfoil model and a control shape were mounted on opposite sides of this wheel, and the forward movement of the bicycle was used to generate a flow of air past these test surfaces. Results obtained from tests with this device confirmed their suspicion that the Lilienthal tables were indeed faulty, but the method was neither precise nor consistent enough to develop data for a new set of tables. For this purpose, the Wrights required a wind tunnel. In their bicycle workshop they constructed a six-foot-long square trough sixteen inches wide, through which they channeled a twenty-seven-mile-per-hour airflow produced by a two-blade fan driven by an electrical shop motor. As part of their tunnel they also designed a set of balances to measure the relative lift and drag on test airfoil sections. With their "laboratory" thus equipped, sometime around 22 November

---

[8] Albert J. Wells describes his wind tunnel device in his bachelor's thesis, "An Investigation of Wind Pressure upon Surfaces," (pp. 4–23), completed at MIT in 1896. For descriptions of the tunnels devised by Horatio Phillips and Hiram Maxim, see Randers-Pherson, "Pioneer Wind Tunnels," pp. 3–4.

Reproduction of the 1901 Wright drag balance. Orville actually built two balances, one for lift measurements and the other for drag. Model airfoils could be mounted on both balances and easily changed. Lift and drag forces calculated from these measurements clearly indicated that the previously accepted aerodynamic coefficients on which the design of airfoils were being based were grossly in error. SI Negative No. A-41899-B

1901 they began a short but intensive two-week sequence of tests, through which they systematically and precisely evaluated over 150 different airfoil shapes. By the conclusion of these tests the Wrights had amassed the greatest body of aerodynamic data in the world, and from this pinnacle of understanding they were armed with the knowledge to build a true working flying machine. Noted historian and Wright biographer Tom Crouch observed of this period of wind tunnel testing that "[b]oth brothers would look back on these few weeks in November and December 1901 as the psychological peak of their joint career in aeronautics."[9] In retrospect, this short stretch of time also proved to be a turning point in the

---

[9] Tom D. Crouch, *The Bishop's Boys: A Life of Wilbur and Orville Wright* (New York, NY: W.W. Norton & Co., 1989), pp. 227–228. See also Crouch, A Dream of Wings, pp. 246–248.

entire history of technology, because it prepared the way, only two years later, for the successful flight of a powered airplane.

On 7 December 1901 the Wright's sister Katherine noted in a message to their father that "[t]he boys have finished their tables of the action of the wind on various surfaces, or rather they have finished their experiments." The important work of analyzing and applying the results lay ahead, and it is at this point that the documentary trail of Chapter 2 begins, with a series of letters written by Wilbur Wright to Octave Chanute and George Spratt, a Chanute disciple who, with his mentor, would subsequently join the Wrights at Kitty Hawk for the momentous 1902 flying season.

Like the airplane itself, the use of wind tunnels quickly advanced beyond the work of the Wright brothers. In unrelated work in 1901, Professor Albert F. Zahm (1862–1954) began operating an impressively sized forty-foot-long tunnel with a six-foot-square cross section at Catholic University of America in Washington, D.C. This large test apparatus was entirely enclosed in a building specifically designed for its operation, making it the world's first true wind tunnel laboratory. Zahm used this facility to conduct pioneering tests on skin friction, and went on to become an important figure in the institutional foundation of American aerodynamic research.[10]

Although the Wrights, Zahm, and other Americans continued to pursue aerodynamic discoveries, it was the Europeans who quickly came to dominate the field in the first quarter of the twentieth century. Between 1903 and the start of World War I in 1914, no less than ten wind tunnels began operation in Europe, as bellicose governments concerned with the military applications of aircraft invested in new aeronautical laboratories. This government interest produced such important facilities as the British National Physical Laboratory outside London and the influential German laboratory at the University of Göttingen. The latter facility featured an innovative closed-circuit wind tunnel designed by Professor Ludwig Prandtl, who was at the time perhaps the greatest scientific mind delving into aerodynamic phenomena. Its signature design feature was the incorporation of a return passage, which kept all the air moving through the tunnel instead of allowing the air to circulate uncontrolled through the building. Although Prandtl's tunnel was simple, with a constant two- by two-meter cross section throughout, his design was vastly more efficient and eventually spawned a whole new generation of wind tunnels. With a lower volume of air in motion, the closed tunnel required

---

[10] A. F. Zahm, "New Methods of Experimentation in Aerodynamics" [paper presented at the meeting of the American Association for the Advancement of Science, Pittsburgh, PA, 20 June 1902], in *Aeronautical Papers of Albert F. Zahm, Ph.D., 1885–1945* (Notre Dame, IN: University of Notre Dame, 1950).

less power than comparably sized open tunnels; furthermore, the quality of the air and its flow could be more precisely controlled. The majority of modern wind tunnels utilize this closed-circuit concept, and their lineage can be traced to the original tunnels of Ludwig Prandtl.[11]

In neighboring France, renowned structures engineer Gustav Alexandre Eiffel (1832–1923) invested not only his legendary energy but also his considerable fortune in aeronautical research, establishing two wind tunnel facilities in and around Paris. In 1909, he built a 1.5 meter diameter tunnel at Champs de Mars, near the famous tower that bears his name. Two years later, after conducting over 4,000 tests in his first tunnel, Eiffel constructed a new and much larger wind tunnel laboratory at Auteuil. (Eiffel published a book entitled *La Resistance de l'Air* to report on his initial research, and a 1913 translation of this volume by American naval observer Jerome C. Hunsaker was the first to use the term "wind tunnel" to describe these facilities.) An important focus of Eiffel's tests involved pressure distribution on airfoil surfaces, and he used his wind tunnels to reveal that the reduction in surface pressure on the top surface of a wing was more significant in generating aerodynamic lift than the pressure increase on its lower surface. Eiffel also conducted important research on propeller aerodynamics, and he was the first to test models of complete airplanes in a wind tunnel. This work helped to establish a clearer understanding of the correlation between test results and the actual performance of full-size aircraft. Eiffel was also able to establish empirically the validity of the relative motion principle, providing scientific proof for the fundamental theory underlying wind tunnel simulation that began back with da Vinci's idea.[12]

While the epicenter of aerodynamic research and development shifted to Europe by the start of the Great War, small pockets of aeronautical enthusiasts existed in the United States, many of whom were not at all happy that the homeland of the Wright brothers had given away its early advantage to the French, Germans, and British. American research lacked a strong organizational backing, and a few farsighted individuals who appreciated the long-term importance of aviation to the United States lobbied for a governmental commitment to aeronautical

---

[11] For the history of the aerodynamics research organized under Ludwig Prandtl's leadership at the University of Göttingen, see Paul A. Hanle, *Bringing Aerodynamics to America* (Cambridge, MA: MIT Press, 1982). Those who read German should consult Julius C. Rotta, *Die Aerodynamische Versuchsanstalt in Göttingen* (Göttingen, Germany: Vanderhoeck & Ruprecht, 1990), especially pp. 45–47. For a detailed analysis of Prandtl's contribution to aerodynamics, see Anderson, *A History of Aerodynamics*, pp. 251–260.

[12] For an English translation of Eiffel's own description of his wind tunnels, see G. Eiffel, *The Resistance of Air and Aviation: Experiments Conducted at the Champs-de-Mars Laboratory*, trans. Jerome C. Hunsaker (London: Constable & Co.; and Boston: Houghton Mifflin & Co., 1913). For an assessment of Eiffel's tunnels and their contributions to aerodynamics, see Anderson, *A History of Aerodynamics*, pp. 267–282.

research that could compete with the European establishments. Some of the most dedicated enthusiasts were naval officers Jerome C. Hunsaker, David W. Taylor, and Washington Irving Chambers. In 1912, Chambers produced a report on aviation for the U.S. Navy (part of which is included in this chapter), which included a detailed proposal for a national aerodynamic laboratory. Following this, David Taylor enlisted the technical assistance of Albert Zahm to design a closed-circuit wind tunnel with an eight-foot-square test section, completed in 1913 at the Washington Navy Yard.

On the academic side, MIT, where wind tunnel pioneer A. J. Wells held a position on the faculty, initiated efforts to develop a program in aeronautical research. MIT president Richard Maclaurin personally investigated Britain's National Physical Laboratory in 1910, and shortly thereafter he approached the U.S. Navy about a cooperative effort to establish a course of study in aeronautical engineering. When Jerome Hunsaker finished a master's degree at MIT in 1912, Maclaurin asked the Navy to allow him to remain for three years as an instructor. Before taking up this assignment, Hunsaker accompanied Zahm on part of the latter's six-month inspection of European aeronautical research facilities, sponsored by the Smithsonian Institution. An excerpt from Zahm's report of this trip is included in the following documents. After his return to the United States, Hunsaker used plans of the British National Physical Laboratory's (NPL) four- by four-foot open-circuit wind tunnel to construct a duplicate facility on the MIT campus.[13] The Smithsonian published papers by Hunsaker and five other MIT professors in 1916 as *Reports on Wind Tunnel Experiments in Aerodynamics*, one of the earliest comprehensive reports on American aeronautical research. Hunsaker's description of wind tunnel testing at MIT is excerpted from these papers and included in the document section of this chapter.

Finally the United States government took action and, with a two-paragraph rider attached to a 3 March 1915 naval appropriations bill, Congress established a new agency, the National Advisory Committee for Aeronautics (NACA). This charter—with its specific charge "to supervise and direct the scientific study of the problems of flight with a view to their practical solution"—guided the NACA for the next forty-three years, until on 1 October 1958 it became the nucleus of the new National Aeronautics and Space Administration. In the beginning, however, the NACA was just a single panel of advisers with a modest $5,000 annual budget to tackle the ambitious task of coordinating both civilian and military aeronautical research and development programs. Initially, the NACA sponsored research in

---

[13] For an excellent treatment of Jerome C. Hunsaker's role in the progress of American aeronautical research, see William F. Trimble, *Jerome C. Hunsaker and the Rise of American Aeronautics* (Washington, DC: Smithsonian Institution Press, 2002).

No organization would ever do more to foster the development of wind tunnel technology than the National Advisory Committee for Aeronautics. This photograph shows a meeting of the NACA in January 1921. Around the table, from left to right, sat Professor Charles F. Marvin, chief of the U.S. Weather Bureau; Dr. John F. Hayford of Northwestern University; Orville Wright; Major Thurman H. Bane, chief of the engineering division of the U.S. Army; Paul Henderson, second assistant to the postmaster general; Rear Adm. William A. Moffett, chief of the Navy Bureau of Aeronautics; Dr. Michael I. Pupin of Columbia University; Rear Adm. D. W. Taylor, chief of the U.S. Navy's bureau of construction and repair; Dr. Charles D. Walcott, secretary of the Smithsonian Institution and chairman of the NACA; and Dr. Joseph S. Ames of The Johns Hopkins University, chairman of the NACA's executive committee. NASA Image #NACA-1921

the Washington Navy Yard tunnel and at Stanford University, where Dr. William F. Durand pursued a systematic delineation of the best propeller shapes.[14] It was not long, however, until the Committee realized that it needed a fully staffed laboratory of its own that could investigate all aspects of aeronautical research. With two special appropriations in 1916 and 1917 of nearly $190,000 to establish an aeronautical laboratory, the NACA sought a site where they could share a flying field with the military. After consideration they settled on a location near Hampton, Virginia, where the U.S. Army was setting up an air base christened Langley Field. Construction at Langley began in April 1917, just as the United States entered World War I.

America's entry into the war injected a sense of urgency into the nation's aviation research, but the timing could not have been worse for the concept of a joint

---

[14] On the Washington Navy Yard tunnel and its aeronautical research, see J. Norman Fresh, "The Aerodynamics Laboratory—The First 50 Years," *Aero Report* 1070 (Washington, DC: Department of the Navy, 1964), pp. 7–14.

The U.S. Army's first-ever wind tunnel, built at McCook Field in Ohio in 1918, could reach extremely high speeds for its day. It had a twenty-four-blade fan that spanned five feet in diameter, which could push the airspeed (within its small fourteen-inch diameter closed-throat test area) to a little over 450 miles per hour. The tunnel is now on display in the Air Force Museum in Dayton. SI Negative No. A-1855

civilian-military laboratory. Both the army and the navy felt that the civilian laboratory at Langley Field could not be adequately equipped and staffed in time to solve wartime problems. Thus, using the rationale of wartime expediency, both services put their efforts into their own resources. The navy expanded its Washington Navy Yard laboratory, while the army established a facility of its own at McCook Field near Dayton.

To handle the increasing work load at the Washington Navy Yard, David Taylor persuaded Albert Zahm in 1917 to leave Catholic University and take a new position as head of the navy aeronautical laboratory. There at the Navy Yard in 1918, a second wind tunnel began operation. This was an unsophisticated copy of a 1912 British NPL open-circuit, forty-five-MPH design with a four- by four-foot test section. The navy used the new wind tunnel mainly for tests on airfoils and for instrument development and calibration, freeing the older but larger eight-foot tunnel for studies involving scale models of complete aircraft. Zahm supervised the navy's aeronautical laboratory until 1930, and during his thirteen-year career at the Washington Navy Yard he conducted important research concerning skin friction, component drag, instrumentation, and wind tunnel design, with results presented in more than thirty-five published papers.

On the army side, the wartime Air Production Board established by Congress authorized the establishment of a "temporary" aviation engineering and experimental facility at McCook Field in September 1917. To staff the facility, the army pulled its research people from the still-incomplete Langley Field. Initially, work at McCook Field focused on production matters, but the need for basic research rapidly became apparent, resulting in the construction of a small wind tunnel in 1918. The army's wartime research concentrated on the refinement of engines, propellers, and a variety of aircraft-operation issues, but also saw the beginnings of an aerodynamic research program utilizing both wind tunnels and in-flight testing. An excerpt of a July 1918 report from the Airplane Engineering Department at McCook, "Full Flight Performance Testing," is included in this chapter's documents.

By the end of the war, Langley Field was essentially out of the army's aviation research efforts, leaving the NACA alone in Virginia to develop its own independent program. Although the NACA began its own flight testing at Langley Field in 1919, it was not until the following year that its first wind tunnel was ready for operation. This "NACA Wind Tunnel No. 1," an open-circuit design with a five-foot test section, was nothing more than another American copy of an outdated British NPL tunnel. While limited, this first tunnel did allow NACA engineers to practice the application of airflow theory and the design of wind tunnel equipment.[15] This critical initial experience is reflected in early NACA technical reports on the "Design of Wind Tunnels" from 1919 and 1920. Other documents provided in this chapter also bear witness to the immature state of the American aeronautical engineering community around 1920 and signal how far that community had to advance to catch up with the Europeans.

The formative nature of the entire field of aerodynamic research in the 1920s is captured in an interesting sequence of documents beginning with a 23 August 1920 call from Dr. Joseph Ames, physics professor and then vice-chairman of the NACA Committee on Aerodynamics (Ames would later serve as the chairman of the NACA's Main Committee, from 1927 to 1939), for the "standardization of wind tunnels"—or as Ames described it, "information which would enable one to connect the data published" from all the various wind tunnels around the world. In answer to Ames's appeal for standardization came letters from many of the greatest names in aerodynamic research, including Dr. Prandtl at the University

---

[15] The design of the NACA's first wind tunnel is examined in Donald D. Baals and William R. Corliss, *Wind Tunnels of NASA* (Washington, DC: NASA SP-440, 1981), pp. 2–3.

Two wind tunnel researchers pose near the entrance end of Langley's five-foot Atmospheric Wind Tunnel (AWT), where air was pulled into the test section through a honeycomb arrangement meant to smoothen the flow. NASA Image #L-1990-04342 (LaRC)

of Göttingen, Dr. Zahm at the Washington Navy Yard, Dr. Durand at Stanford, and many representatives of the nascent American aircraft industry. A selection of these responses to Ames are included in the documentary collection, and they provide a snapshot into what the leading aerodynamic researchers thought needed to be done in order to improve the reliability of wind tunnel data around the world and to build an international network that would allow aerodynamic experts to better learn from each other. The NACA eventually drew up specifications for comparative tests in wind tunnels followed by wind tunnel personnel at Langley. While the overall objective of worldwide standardization was never fully realized, this early attempt toward it brought the minds of international experts together on basic issues of aerodynamic research, and it raised fundamental questions about the problems of testing scale models in wind tunnels.

A comparable call for more exactitude in aerodynamic testing came a few months later in 1920 when Commander Jerome Hunsaker, at that time with the U.S. Navy Bureau of Construction and Repair, recommended that the NACA pursue a systematic program comparing wing characteristics as ascertained in wind tunnel tests with those learned in free flight. (Hunsaker became a committee member of the NACA in 1922 and later served as NACA chairman from 1941 to 1956.) The proposal, which the NACA embraced and carried out at Langley, led to the establishment of new research methods that produced important results. For the first flight tests the NACA obtained a Curtiss JN4H "Jenny" biplane, and Langley model makers built two models of this aircraft for wind tunnel testing. The engineers used one model for tests measuring lift, drag, and moments, while the other was outfitted for pressure-distribution tests. The wind tunnel data from the models was then compared to measurements obtained during actual flight tests of the full-size Jenny. Such comparative tests were first tried a decade earlier by Gustav Eiffel, but the Langley program sparked by Hunsaker's proposal proved far more extensive and was accomplished with considerably greater precision. Learning a great deal from research programs that combined wind tunnel experiments with free flight tests, the NACA began to stake out a crucial role for itself as a research establishment. Unlike the Army and Navy programs, which tended to be more development-oriented than research-oriented, the NACA focused on basic aerodynamic problems that affected civilian and military aviation alike.

In a roundabout way that was nevertheless typical of the serendipitous path of basic research, Langley's JN4H comparison studies also helped lead the NACA to an altogether unexpected destination, the design of a revolutionary new type of wind tunnel. The test program resulting from Hunsaker's proposal confirmed that wind tunnel testing was a valid predictor of aircraft performance and behavior, but it also pointed NACA researchers toward another well-known, but poorly understood, problem concerning scale effects. Prior to 1920, no wind tunnel ever built had been able to address scale effects with any success. Yet by that time people working in aeronautical research realized that the forces generated by a scale model were not, in fact, proportional to the model's scale. A novice to the field might assume that a $\frac{1}{20}$-scale model of a airplane placed into an air stream moving $\frac{1}{20}$ as fast as the actual airplane flew would generate roughly the same forces of lift and drag as a full-scale machine in flight, but that was far from the truth; the scale model actually generated considerably less. Early aerodynamicists had developed empirical coefficients to "scale-up" the data, but the NACA comparison tests showed just how unreliable these coefficients were. The air itself could not be "scaled-down" to model size, and air properties, such as density and temperature, manifested themselves the same way in the wind tunnel as they did around a full-size airplane. To gain the maximum value from wind tunnel testing, an answer to the scaling problem had to be found.

The conceptual tools needed were found in the hydrodynamic work of Osborne Reynolds (1842–1912). Experimenting with the characteristics of fluid flows at the

University of Manchester in England, in 1883 Reynolds established experimentally that the transition from laminar or smooth flow to turbulent flow always occurred when certain variables exceeded a critical value defined by a specified flow parameter. The dimensionless number that resulted came to be known as the "Reynolds number," and it became the key to understanding scale factors. This fundamental finding, acknowledged as "a stunning discovery," eventually enabled researchers to establish a direct quantitative link between experiments with scale models used in wind tunnels and the airflow patterns of full-scale designs. In essence, it meant that the testing of scale models in wind tunnels could lead not only to meaningful theoretical results, but also to practical design applications.[16] It took nearly forty years for Reynolds's breakthrough to make the transfer from scientific esoterica into a practical factor for aircraft engineers, but the tremendous expansion of wind tunnel work in the United States in the 1920s was in part driven by a growing appreciation of the significance of Reynolds number effects.

In demonstrating that a fluid flow suddenly changed from laminar to turbulent as speed increased, Reynolds showed that the forces a moving fluid exerted on a body depended on the fluid's velocity, density, and viscosity, and on key dimensions of the body itself, such as length or diameter. To achieve the "Reynolds number," he combined these parameters into a mathematical expression where all of the dimensions cancelled one another out. Because it was dimensionless, the Reynolds number could be used to compare fluid-flow forces around similarly shaped, but differently sized, objects. One could achieve what came to be called "dynamical similarity" by varying different parameters, such as decreasing the velocity or increasing the density, to produce the same Reynolds number for different tests.

In theory, then, it was possible for aerodynamicists to make an excellent correlation between model tests and aircraft performance, but this was easier said than done. When a $\frac{1}{20}$-scale model was tested in the NACA's original atmospheric wind tunnel in 1920, the Reynolds number for the test represented no better than $\frac{1}{10}$ that of the corresponding full-scale flight. In principle, larger models could have been used, but wingspans greater than about 3 $\frac{1}{2}$ feet could not be used in the five-foot-diameter test section of the first NACA wind tunnel due to aerodynamic interference from the tunnel walls. In addition this atmospheric wind tunnel also lacked the power to run at the high speed necessary to generate the required Reynolds number, as such a velocity was simply not practical in an open wind tunnel.

---

[16] See Osborne Reynolds, "An Experimental Investigation of the Circumstances Which Determine Whether the Motion of Water in Parallel Channels Shall Be Direct or Sinuous, and the Law of Resistance in Parallel Channels," *Proceedings of the Royal Society* (London, 1883), pp. 84–89. On Reynolds's contribution to fluid mechanics, particularly the concept of the "Reynolds number," see Anderson, *A History of Aerodynamics*, pp. 109–114.

Dr. Joseph S. Ames at his desk at NACA headquarters in the early 1920s. Ames was a founding member of the NACA, appointed by President Woodrow Wilson in 1915. He served as chairman of the NACA's Main Committee from 1927 to 1939. NASA Image #LAL90-3738 (Ames)

Two possible solutions to the problem emerged almost simultaneously in the year 1920. Both originated in Europe, and both were offered for "sale" to the NACA in America. In each case, the technical solution was to increase the density/viscosity factors in the Reynolds number calculation. The first concept came from Wladimir Margoulis, a Russian-born aerodynamicist and protégé of Nikolai Joukowski and Gustav Eiffel, who had just started work for the NACA's Office of Aeronautical Intelligence in Paris as technical consultant and translator. Margoulis's proposal was to replace air with carbon dioxide and completely seal this atmosphere in a fully enclosed wind tunnel. Because carbon dioxide's density was over 1 ½ times greater than that of air, the Reynolds number of any test in such a chamber would be correspondingly higher.

The second concept (though it is impossible to resolve whose idea actually developed first) came from Dr. Max Munk, a brilliant star student of Ludwig Prandtl, whose laboratory at Göttingen was the leading aerodynamic research facility in the world. In 1916, Prandtl had designed a major new wind tunnel, the virtues of which American aeronautical observer and NACA consultant Edward P. Warner extolled in his 1920 "Report on German Wind Tunnels and Apparatus." Unlike his earlier 1908 tunnel, which had a constant cross section, Prandtl's second-generation apparatus merged the tapered diffuser of an open tunnel with a closed-circuit design, resulting in much-improved air management. Prandtl based the design on the principle that the potential energy (static pressure) and kinetic energy (velocity) of air could be interchanged in different parts of a wind tunnel to enhance performance. The open test section of the new Göttingen wind tunnel measured two meters in diameter, but by enlarging the return duct and thereby reducing the velocity of the returning air, friction and the associated power requirements were reduced—yet because slower-moving air exerted a higher static pressure than high-speed air, the air's momentum was retained. The return duct, cast in concrete (and in this particular tunnel uniquely placed in a vertical orientation underneath the floor) possessed guide vanes to turn the air flow around the corners. Two other particularly significant features of this design were the incorporation of a stilling chamber and a contraction cone ahead of the testing area that acted to reduce turbulence and then accelerate the air passing into the test section. The design was so efficient that a 300-horsepower motor rotating a four-bladed fan was sufficient to produce test section wind velocities of 170 feet per second. While historians sometimes call Prandtl's earlier 1908 design the first "modern" wind tunnel, it is actually his second Göttingen tunnel that truly

The young NACA first built its reputation as an outstanding aeronautical research institution on the strength of its Variable Density Tunnel (VDT). The tank for the VDT arrived at Langley by rail from its manufacturer, the Newport News (Va.) Shipbuilding & Dry Dock Company in February 1922. It was an eighty-five-ton pressure shell with walls made from steel plate lapped and riveted according to a practice standard in steam-boiler construction. NASA Image #L-1990-04352 (LaRC)

deserves the accolade, for this revolutionary design established a new standard and became the model for subsonic wind tunnels around the world.

It was the environment and technical culture of Prandtl's laboratory that structured the thinking and nurtured the remarkable talent of Max Munk, who in 1917 earned not one but two doctorates at Göttingen (in both engineering and physics). With the exception of Prandtl, no one knew the Göttingen wind tunnels better than Munk—and as subsequent achievements in wind tunnel design proved, no one, not even Prandtl, better understood the principles and potential of wind tunnel technology.

In early 1920, Munk proposed the idea of a wind tunnel built inside a pressure vessel so that tests could be run under high pressure, thus increasing the density of the air as much as twentyfold. The thirty-year-old temperamental genius tempted the American research establishment with his concept in a personal letter sent from Germany to the navy's Hunsaker, who knew of Munk's work at Göttingen from his continuous review of German aeronautical activities. Hunsaker informed NACA Chairman Dr. Joseph Ames of Munk's idea, and Ames persuaded the rest of the Committee, which was hard-pressed for talented aerodynamicists, to offer Munk a position as a technical consultant. To employ the German aerodynamicist in the United States required two special orders from President Woodrow Wilson, one to allow a recent enemy into the country, and the other to authorize him to hold a government job. When these were secured, Munk arrived in Washington, D.C., in late 1920 and began seven turbulent years of NACA employment, first as a technical assistant in the NACA's Washington office and later as chief of aerodynamics at Langley.[17]

The acceptance of Munk's idea and the design of such a bold new type of wind tunnel turned the NACA from a second-rate player into a world leader in aerodynamic research. This tunnel, known as the Variable Density Tunnel, or VDT, went into service at Langley as "NACA Wind Tunnel No. 2" in 1922, and its results were vastly superior to those obtained with any previous tunnel design.

Two NACA technical papers by Munk excerpted as documents for this chapter provide insights into the revolutionary nature of the VDT's design. Externally, it appeared to be little more than a large cylindrical tank with hemispherical ends. But, in fact, a five-foot-diameter wind tunnel was mounted inside such that air flowed through a central test section, then past a fan that blew the air back around via an annular return passage. The entire tank could be pressurized to 300 pounds per square inch (20 atmospheres), sufficient to produce Reynolds numbers for tests of $1/20$-scale models that were equivalent to full-scale flight. An externally

---

[17] On Munk's coming to the NACA, see Hansen, *Engineer in Charge*, pp. 72–78 and 84–95 and Roland, *Model Research*, pp. 87–98.

mounted 250-horsepower synchronous motor turned a seven-foot-diameter propeller to produce test speeds of seventy-five feet per second. Small windows in the tank permitted technicians to view the test section during operation, and a hatch at one end provided access for mounting test specimens and for tunnel maintenance.

Langley engineers quickly ran a wide variety of tests in the VDT, including studies of several model airplanes, to validate the high-pressure concept. But the most long-lasting and significant investigations involved airfoils. Through extensive use of the VDT, the NACA drew accurate performance curves for the commonly used airfoils of the era, and then extended the investigation to develop entire families of airfoils with similar characteristics. This iterative process eventually led to the development of the highly refined airfoils prominent in both the aircraft design revolution of the 1930s and the laminar flow developments of the 1940s, areas that will be addressed in subsequent chapters. Even now, it is difficult to overestimate the effect of these airfoil test programs on aeronautical development. The unique capabilities of the VDT and the valuable library of airfoil data it generated greatly advanced the state of aerodynamics not only in the United States but also around the world.

NACA Langley's chief of aerodynamics, Elton W. Miller, inspects his researcher's installation of a Sperry M-1 Messenger airplane into the lab's new Propeller Research Tunnel (PRT) in early January 1927. This was the first complete, full-scale airplane ever to be tested in a wind tunnel in the United States. NASA Image #L-01892 (LaRC)

As a civilian organization, the NACA was relatively unencumbered with military security requirements during this period, and it made the airfoil information readily available to airplane designers through a formal series of "Technical Reports." These "TRs" were carefully prepared publications that not only helped the American aircraft industry select its wing shapes but also served as basic textbook material for an entire generation of up-and-coming aeronautical engineers. A number of documents in this chapter and in the chapters that follow are excerpts from classic NACA Technical Reports.

The VDT provided for unprecedented airfoil research at high Reynolds numbers, but it was not well suited for certain other investigations, such as propeller testing. Model propellers were unsuitable because they did not deflect during operation the same way full-size propellers did, a source of such significant error as to invalidate model tests. Testing of full-size propellers, on the other hand, required an evaluation of actual performance on an airplane in expensive and often dangerous flight tests, and even then accurate measurements were difficult to obtain. Though the NACA sponsored propeller research in Stanford University's Eiffel-type tunnel started in 1917, until the mid-1920s no valid laboratory method of studying the entire propulsion system and its relationship with the body of the aircraft existed. The NACA's response to this situation is shown in documents starting with a series of memos from Dr. Munk and others at the NACA that trace the genesis of the world's first wind tunnel of considerable size, the Propeller Research Tunnel (PRT) of 1926–1927.[18] These contemporary documents are complemented by a retrospective account from the autobiography of Langley engineer Fred E. Weick, who built and first operated the NACA Propeller Research Tunnel (PRT) at the Langley laboratory.

The PRT featured a Prandtl-style tunnel with a huge twenty-foot open test section. The enormous power required for a tunnel of this size was provided by two navy surplus 1,000-horsepower submarine engines, which produced wind velocities of over 160 feet per second, or the equivalent of 110 miles per hour. The balance supported a full-size airplane body or a substitute "test fuselage" equipped with an onboard dynamometer to measure engine torque directly. Special methods were devised to measure blade deflection optically during tests. When it went into operation in 1927, the PRT quickly proved its worth for propeller testing in conditions that were close to those in actual flight, but because it was the first tunnel large enough to accommodate full-size airplane fuselages its use soon expanded to

---

[18] On the design, construction, and early operation of the NACA's Propeller Research Tunnel, see Fred E. Weick and James R. Hansen, *From the Ground Up: The Autobiography of an Aeronautical Engineer* (Washington, DC: Smithsonian Institution Press, 1988), pp. 49–59, as well as Hansen, *Engineer in Charge*, pp. 87–90.

Four hundred thirty-four feet long, 222 feet wide, and ninety feet high, the building housing the thirty- by sixty-foot Full-Scale Tunnel dominated the Langley scene. Its location along the Little Back River, a tidal river off the nearby Chesapeake Bay, occasionally caused flooding problems for the tunnel during hurricanes and nor'easters. NASA Image #EL-1999-00405 (LaRC)

include drag studies of other aircraft components, such as landing gears, tail planes, and cooling systems. The latter tests in the PRT led to the development of the celebrated NACA cowling for radial aircraft engines and Langley's first Collier Trophy Award. (These events are detailed in the next chapter, found in the forthcoming Volume 2, which deals with the design revolution in aircraft aerodynamics.) These crucial results from the PRT, following closely on the heels of the contributions of the VDT, catapulted the NACA Langley research laboratory to a position of unparalleled importance in American aviation.[19]

The PRT also inspired the NACA to build an even larger wind tunnel. The close connection that existed between the engineering design of the PRT and the conception of Langley's next mammoth facility, the historic thirty- by sixty-foot Full-Scale Tunnel completed in 1931, has not been fully appreciated by historians. But thanks to the discovery of a series of neglected NACA memos from 1925,

---

[19] For a historical analysis of the NACA cowling program, see James R. Hansen, "Engineering Science and the Development of the NACA Cowling," in *From Engineering Science to Big Science: The NACA and NASA Collier Trophy Research Project Winners*, ed. Pamela E. Mack (Washington, DC: NASA SP-4219, 1998), pp. 1–28. The latter is an expanded version of Hansen's chapter on the cowling in *Engineer in Charge*.

The first tests of an aircraft in the FST involved a Vought O3U-1 "Corsair." In the summer of 1931, the NACA used the navy airplane for some preliminary tests to check out the FST and as the subject of the first publicity photographs taken of FST operations. NASA Image #EL-1999-00425 (LaRC)

reproduced in the document section of this chapter, one can now see how NACA engineers actually anticipated the FST in the building of the PRT. On 7 April 1925, after reviewing a seven-page memo on a "Proposed Giant New Wind Tunnel" from an enthusiastic assistant aeronautical engineer named Elliott G. Reid, Langley engineer-in-charge Leigh M. Griffith asked the NACA Washington office, Why not, after finishing the PRT, plunge ahead with an even larger tunnel capable of testing a full-size aircraft? "If we could actually fly the same model that we test in the tunnel," Griffith wrote, "we would have the unquestioned means of investigating airplane performance and characteristics in the most direct, accurate, convenient, and conclusive manner."

In other words, the NACA would have a means of eliminating the scale effect factor altogether. The VDT showed that scale-model tests at high Reynolds numbers produced accurate data about airfoils, but was not effective for testing fuselages. The PRT, on the other hand, demonstrated the value of testing the synergistic component characteristics of full-size aircraft, but were not large enough to include

The cavernous test section of the FST also came in handy at NACA conferences. In this picture from May 1934, attendees at the NACA's annual aircraft engineering conference posed beneath a Boeing P-26A "Peashooter." Present in this photo, among other notables, were Orville Wright, Charles Lindbergh, and Howard Hughes. NASA Image #EL-1996-00157 (LaRC)

the full span of the wings. So while both tunnels produced valuable breakthroughs, neither was capable of totally reproducing the conditions experienced by an entire airplane in flight. It was the desire to cross this new threshold in aerodynamic research, prompted by the work of the VDT and PRT, that inspired NACA engineers to design and build a wind tunnel large enough for full-scale testing.

The NACA Full-Scale Tunnel (FST), completed in 1931, proved gigantic in every respect and loomed over every other structure at Langley. A closed-circuit wind tunnel with an unprecedented thirty- by sixty-foot open test section, the FST was large enough to handle airplanes or large-scale models with wingspans of up to forty-five feet. Two monstrous thirty-five-foot-diameter propellers in the dual return ducts, each driven by a 4,000 horsepower electric motor, circulated almost 160 tons of air through the 838-foot-long circuit and produced wind velocities of nearly 120 miles per hour. These speeds were sufficient to enable measurements that could be confidently extrapolated to cover the aircraft's entire speed range because scale factors were minimized or eliminated entirely. The remarkable FST and the aerodynamic research it enabled was a crowning achievement for the

In 1928, the NACA replaced its original Atmospheric Wind Tunnel ("Wind Tunnel No. 1") with two tunnels—a five-foot vertical tunnel and a seven- by ten-foot Atmospheric Wind Tunnel. An NACA engineer sets up a test in Langley's seven- by ten-foot AWT. Though an all-purpose facility, its main purpose was to study stability and control problems. NASA Image #EL-1999-00418 (LaRC)

NACA, and it marked how far American aeronautic research institutions had come since the end of World War I.[20]

The NACA also built other wind tunnels along with the highlighted VDT, PRT, and FST. The first pioneering high-speed research was begun in 1927 with a small eleven-inch tunnel that used the exhaust air released when the VDT was depressurized to produce brief flows approaching the speed of sound. The concept proved successful and led to the construction of two improved high-speed tunnels in the next decade. In 1929 Langley's obsolete "Wind Tunnel No. 1" was dismantled and replaced with two new tunnels, a seven- by ten-foot Atmospheric Wind Tunnel (AWT) and a five-foot Vertical Wind Tunnel. Completed in 1930, the new

---

[20] On the design and history of the NACA's Full-Scale Tunnel, see Baals and Corliss, *Wind Tunnels of NASA*, pp. 22–23, plus Hansen, *Engineer in Charge*, pp. 101–105, 194–202, and 447–449.

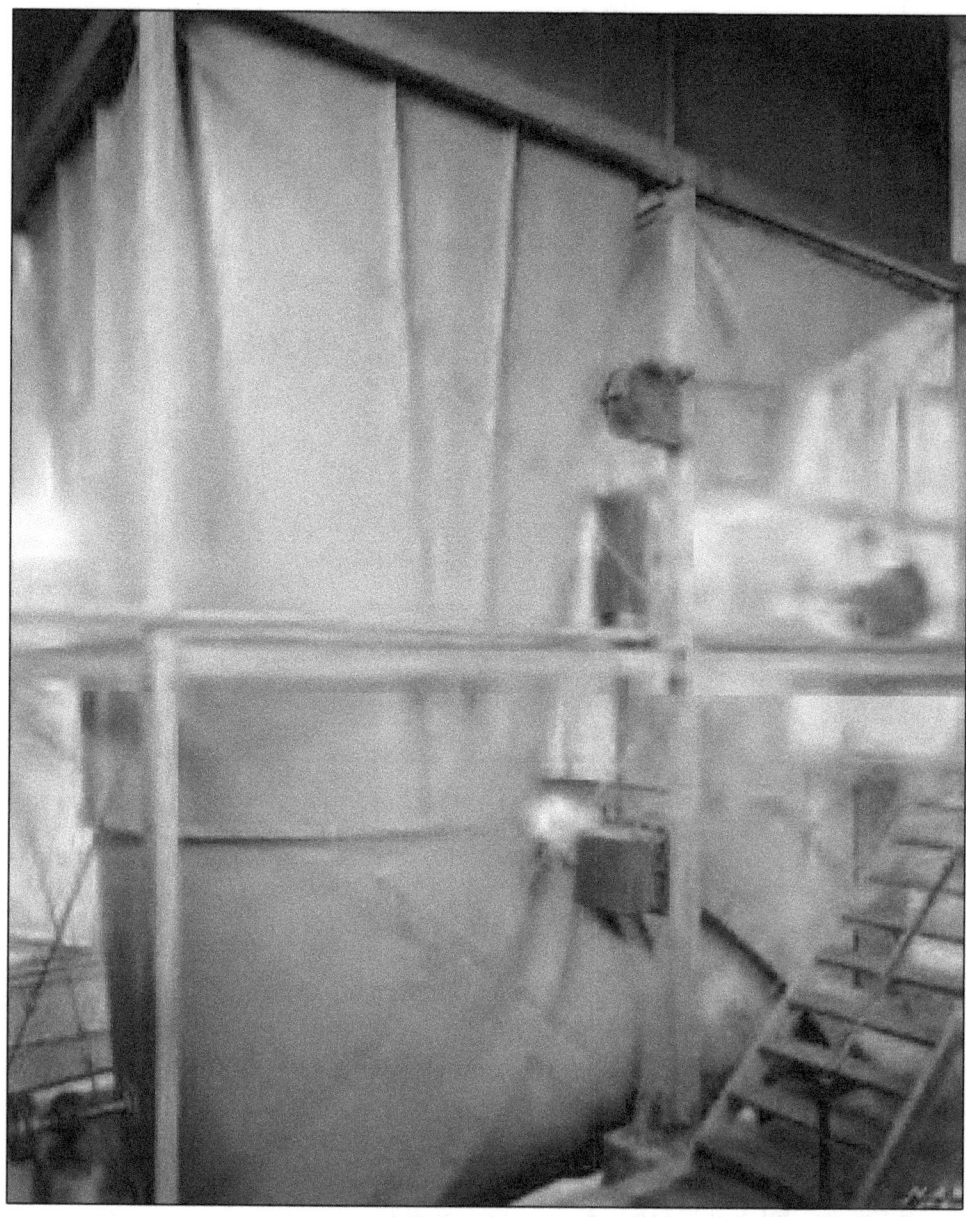

Housed in the same building as the seven- by ten-foot AWT, Langley researchers used the five-foot Vertical Wind Tunnel mainly for spin tests. An engineer kneels on a platform next to the test chamber. In the foreground is the device's (closed) return passage. Like the AWT, the tunnel had an open throat. The vertical tunnel stood thirty-one feet tall and was twenty feet long and ten feet wide. It was not ready for operation until early 1931. NASA Image #EL-1999-00410 (LaRC)

Another major center of U.S. aerodynamic research—this one sponsored by the U.S. Army—flourished at McCook Field near Dayton, Ohio. During World War I, McCook engineers made significant contributions to airplane and engine development, but even greater contributions came in the decade after the war. Like NACA Langley, the aircraft engineering division at McCook combined systematic wind tunnel testing with a full program of flight research. But beyond that, the McCook operation also engaged more directly in the actual design, construction, and operational problems of aircraft, which were beyond the mandate of NACA research. In 1927, the army closed McCook and moved its flying field and associated units to a major new parcel of land dubbed Wright Field. Eventually the site became Wright-Patterson AFB. SI Negative No. A-1848

AWT was a marked improvement over the 1920 tunnel it replaced, and it remained a workhorse facility at Langley for many years. The novel Vertical Tunnel, with its upward flowing air stream, was an innovative concept that offered an unprecedented opportunity to investigate aircraft spins, a leading cause of disastrous crashes. An excellent example of adapting an existing technology to a new use, the Vertical Tunnel of 1930 was the prototype for additional spin tunnels at Langley, as well as for the concept of free-flight wind tunnel testing, which the NACA also pioneered.[21]

By the beginning of the 1930s, the NACA had assembled a collection of wind tunnels at Langley whose collective capabilities surpassed those of any other

---

[21] All of the early NACA tunnels are discussed in Baals and Corliss, *Wind Tunnels of NASA*. Appendix D of Hansen's *Engineer in Charge* provides a comprehensive catalog of all the facilities that were developed at NACA Langley from 1917 to 1958.

Between 1926 and 1930, the Daniel Guggenheim Fund for the Promotion of Aeronautics disbursed over $3 million for the creation of aeronautical engineering programs at several American universities. In association with these programs, a number of new wind tunnel facilities came to life. Trustees of the Guggenheim fund included, left to right standing, J. W. Miller, secretary; F. Trubee Davison; Elihu Root, Jr.; Hutchinson Cone; Charles Lindbergh; Harry Guggenheim, the fund's president; Dr. Robert Millikan; and, left to right seated, John D. Ryan; Daniel Guggenheim, the fund's creator; Orville Wright; and Dr. William F. Durand. SI Negative No. A-3519

aeronautical laboratory in the world, an achievement highlighted in the C. G. Grey editorial reproduced at the beginning of this chapter. That a European authority acknowledged America's leading position in aerodynamic research was particularly significant, considering how far behind the United States had been just fifteen years earlier.

The NACA was not alone in the establishment of American aerodynamic research institutions during this period. The United States Army Air Service expanded its aerodynamic work at McCook Field after World War I and constructed a five-foot-diameter open wind tunnel there in 1922. An important area of aerodynamic research in this facility became the search for solutions to aircraft "flutter," the often catastrophic uncontrolled oscillations of wings and control surfaces in flight, a subject that NACA researchers also pursued vigorously beginning in the late 1920s. When McCook Field closed in 1927, the Army moved this tunnel to the new Wright Field (now Wright-Patterson Air Force Base) near Dayton, where the five-foot tunnel remained in service into the 1990s as an educational learning tool for the Air Force Institute of Technology (AFIT). And as already noted in the work of Albert Zahm, during this period the U.S. Navy also sponsored important research in the wind tunnels of the Washington Navy Yard.

Along with the government investment, there was also a growing private interest in advancing aerodynamic research in the 1920s. A leading figure in this was multimillionaire philanthropist Daniel Guggenheim, who devoted part of his mining fortune to foster the development of academic programs in aeronautics at American universities. In 1925, when only the MIT and the University of Michigan offered degrees in aeronautical engineering, Guggenheim donated half a million dollars to New York University to establish a School of Aeronautics. The next year he founded the Daniel Guggenheim Fund for the Promotion of Aeronautics with an endowment of $2.5 million. Between 1926 and 1930, the Guggenheim Fund awarded major grants to seven prestigious engineering schools, expanding the programs at MIT and the University of Michigan and starting new programs at Stanford, the California Institute of Technology (Caltech), the University of Washington, the Georgia Institute of Technology (Georgia Tech), and the University of Akron. A large portion of these grants went into construction of wind tunnel laboratories, but, significantly, some funds were earmarked for the hiring of exceptional professors and researchers. For example, a Guggenheim grant allowed Clark and Robert Millikan to recruit Theodore von Kármán, one of Europe's most outstanding theoretical aerodynamicists, to come to the United States as the new director of Caltech's Guggenheim Aeronautical Laboratory (GALCIT), where, for the next thirty years, he would play a dominant role in shaping the growth of theoretical aeronautics in the United States. By mid-century, over 90 percent of the nation's leading aeronautical engineers were graduates of Guggenheim-funded colleges.[22]

Another significant factor in the formation of the American aeronautical engineering community in the 1920s and 1930s was the diaspora of experienced research engineers from NACA Langley into the larger world of aeronautics. Dozens of early NACA employees moved on to accept important positions in the aircraft industry or at universities. For example, Montgomery Knight and Elliot G. Reid

---

[22] On the Guggenheim connection with aviation, see Richard P. Hallion, *Legacy of Flight: The Guggenheim Contribution to American Aviation* (Seattle, WA: University of Washington Press, 1977). See also Hallion's chapter, "Daniel and Harry Guggenheim and the Philanthropy of Aviation," in *Aviation's Golden Age: Portraits from the 1920s and 1930s*, ed. William M. Leary (Iowa City: University of Iowa Press, 1989), pp. 18–34.

left Langley to help start the Guggenheim programs in aerodynamics at Georgia Tech and Stanford, respectively. These programs in turn produced a new crop of aeronautical engineers for the future expansion of the NACA and the civilian aircraft manufacturing industry. This process of networking and cross-fertilization with the broader American aeronautical community proved to be one of the NACA's most important contributions to aerodynamics and American aeronautics generally.

To be sure, it was people—primarily engineers and scientists—who perceived the needs of aeronautics, envisioned the flying machines and wind tunnels, brought them to life, and thought up and performed the aerodynamic test programs. With such a complex and exciting technology as aviation, it should not be surprising that the field attracted some of the finest minds available. But with brilliance frequently comes ego and a high degree of individualism, a combination that can make it difficult to establish and maintain a sense of community among professionals with different ideas and opposing modes of operation and cultural norms.

Over time, in America as elsewhere, an international aeronautical research community formed as practitioners in various countries began to realize that significant progress resulted at least as much from cooperation and the free exchange of ideas as from secrecy and cut-throat competition. Indeed, corporate and national competitions remained intense in the inter-war period. Americans for their part wanted to catch up with and surpass the Europeans in the field of aeronautics. Documents in this chapter involving the NACA's Office of Aeronautical Intelligence—primarily John Jay Ide's reports from Paris back to the NACA in Washington, D.C., on what was developing at Europe's many aeronautical centers in the early 1920s—certainly need to be evaluated with the American military and commercial goals of the early inter-war period in mind. For the United States to be on the cutting edge of aeronautical science and technology, its aeronautical specialists had to know what its European rivals were up to in their laboratories, aircraft industries, and military installations. Although complete histories of NACA and U.S. military intelligence in the field of aeronautics have not yet been written, it is clear from what is known, and from the progressive evolution of the American propeller-driven airplane into the World War II era, that the intelligence mission succeeded in major respects.

Unsavory manifestations of national loyalties surfaced from time to time just about everywhere. Without dwelling on them, this chapter offers one arresting insight into the chauvinism of an early NACA researcher, Frederick H. Norton, Langley's chief physicist from 1920 to 1923. In 1921 Norton wrote to NACA Headquarters complaining that Dr. Max Munk, only recently arrived to the NACA from Germany, had the audacity to propose the use of a German airfoil section for a small helicopter he was considering. An MIT graduate, who one might think possessed a less parochial view, Norton could not understand how NACA leadership would allow Munk to use any foreign airfoil, especially one from Germany, the wartime

Dr. Max M. Munk, the Variable Density Tunnel's creator, inspecting the machine not long after its installation. One peered into the machine through two small portals on the side. NASA Image #EL-1999-00258 (LaRC)

enemy, when there were plenty of good American airfoils from which to choose.

Munk himself continued to be embroiled in conflict; it was not easy for anyone to work with this temperamental genius, either during the early years at NACA Langley or in subsequent jobs. Whether the problems with Munk at Langley (which eventually resulted in Munk's dismissal from the NACA in 1927) represented simply a clash of personalities or something deeper like a culture clash has been a matter of some interesting historical analysis and interpretation.

One way to understand the Munk affair is to consider that two basic approaches to aerodynamic research exist: the empirical approach, where experimentation and practical assumptions seek solutions to problems; and the theoretical approach, where a mathematical analogy is created to foster an understanding that can be used to solve problems. Such a dichotomy oversimplifies the research process, unquestionably, but it serves to illustrate the mindsets of two emphases in research that definitely exist—and that sometimes even break into camps. In the era following the Wrights' first flight covered by this chapter, both empiricists and theorists contributed important elements to the overall picture, and what began in many places as a mutual lack of understanding and distrust of each other's methods slowly merged into a mature discipline with room, and a need,

Von Kármán (black coat and tie) sketches out a plan on the wing of an airplane as members of his Jet-Assisted Takeoff (JATO) engineering team looks on. Clark Millikan stands to von Kármán's far right with Martin Summerfield in between. To von Kármán's immediate right is Caltech rocket pioneer Frank J. Molina. The man in uniform is Capt. Homer A. Boushey, who later that day (23 August 1941) became the first American to pilot an airplane that used JATO solid propellant rockets. NASA Image #JATO-VONKARMAN (Ames)

for both. Yet the maturing did not take place without occasional trouble.

The roots of aeronautical empiricism penetrated deep, and dated at least as far back as Cayley. Generally speaking, all of the aeronautical pioneers from Cayley to the Wrights were experimenters who learned through careful observation of their successes and failures. By the 1920s, this long tradition had led not only to successful flight and a proliferation of practical flying machines, but also to the establishment of aeronautical laboratories dedicated to expanding the knowledge of flight through experimental means. The empirical investigations performed by several of these researchers, including Francis Wenham, Horatio Phillips, Gustav Eiffel, Albert Zahm, and the Wrights, created a firm base for later theoretical work. Faced with a dearth of knowledge about aerodynamics, they built machines with which reliable measurements could be made. In so doing, they not only learned how various objects interacted with an airstream, but also furnished themselves and others with the fundamental facts needed to construct theoretical models that could describe and predict those interactions with increasing accuracy.

In the United States during the first decades of the twentieth century, empirical methods and straightforward approaches to solving practical problems suited the mood of the engineering culture and the abilities of its practitioners well; this certainly proved to be the case at the NACA laboratories, where the engineer—not the scientist—gained the upper hand. Nonetheless, the contributions of science and, perhaps even more importantly, the public perception of science greatly influenced the course of NACA research. American scientists sought a fundamental understanding of natural phenomena, as European scientists did, but science and engineering were often seen as one and the same by the American public; typically, what engineers did—at least what they did successfully—was credited as "science." Engineering accomplishments, solutions to practical problems built out of concrete or steel rather than new knowledge for its own sake, were commonly celebrated as "scientific" achievements. In America, "science" solved problems and built things, and the field was largely open to anyone who could prove himself. Like the rest of American society, a sense of democracy permeated the enterprise. This is not to imply there was no management hierarchy in American technology—far from it—but ideas could originate from lower strata and receive consideration. Subordinates in many organizations, including the NACA, felt free to debate and test notions coming to them from the top down. Even more often, ideas percolated from the bottom up.

The European model, and especially the German, differed substantially. In the physical sciences, mathematics was seen as the route to true understanding, and the objective of an investigation was the development of a mathematical formula that described the phenomenon.[23] Although an admirable objective, this could lead to formulae that were extremely difficult, if not impossible, to solve when applied to practical engineering problems. At the very least, such a technical culture required its engineers and scientists to be good mathematicians and to think abstractly. The national technical cultures of most European nations remained hierarchical and, unless one had the benefits of aristocracy, scientists and engineers earned their positions in the upper echelons of a profession through many years of formal study and subservient work. In such cultures, the leader of an organization held a commanding position over the ideas and methods that the entire organization would use.

Ludwig Prandtl, Max Munk, and Theodore von Kármán were outstanding products of the German system and, not coincidentally, they proved to be among

---

[23] See Paul A. Hanle, *Bringing Aerodynamics to America* (Cambridge, MA: MIT Press, 1982), for an insightful analysis of the "Prandtl School" and the German technological culture of aerodynamic research in the early 1900s.

the first and finest theoretical aerodynamicists. Prandtl led the way for his two brilliant students. In demonstrating his ability to tackle practical problems in innovative ways by building the first closed wind tunnels, he built, not successful airplanes, but an edifice of fundamental aerodynamic understanding that was unexcelled anywhere. Prandtl wanted not to observe and measure lift so much as he wanted a mathematical explanation of it; this ambition underlay most of his published papers, including his classic statement of 1921, "Applications of Hydrodynamics to Aeronautics," written expressly for NACA publication (excerpts of which are published in this chapter's documents). In this quest for theoretical enrichment, Prandtl was not alone. Other contemporary Europeans, notably Nikolai Joukowski in Russia and F. W. Lanchester in England, also devoted most of their careers to a search for satisfying theoretical explanations of this complex phenomenon. None of them completely solved the problem, but all of them contributed vital pieces to the puzzle. Munk and von Kármán acquired their mentor's appreciation of the value of theoretical aerodynamics, and they both brought it, along with their considerable intelligence and talent, to the United States in the 1920s. Both men were well known to prominent American aeronautical figures, notably to Hunsaker and Zahm, who recommended them highly. Once in America, however, their careers took decidedly different directions.

Von Kármán, who arrived later (in 1929), came to the U.S. after being actively recruited by Caltech's Robert Millikan. He moved into a university research environment that shared characteristics with his former school, the Technische Hochschule in Aachen, Germany. At Caltech, von Kármán directed GALCIT, but he also continued to work with graduate students and pursue his own research. He adapted well to academic life in America, and he quickly gained the respect of the American aeronautical community.[24]

Munk's American career started out perhaps even more meteorically but later came crashing down. Munk started his NACA employment in the early 1920s, working almost exclusively by himself on theoretical problems in the Washington, D.C., office. Besides refining and carrying out his idea for the VDT, he explored a number of critical aerodynamic problems and published an important series of reports. (The NACA would eventually publish over forty of Munk's papers.) But Munk wanted to be in charge of aeronautical research at Langley.

Only a few of the many documents that exist in the NACA archives related to problems with Munk need to be seen to appreciate the clash of ideas and attitudes that ultimately led to a revolt against Munk at Langley—and to his forced departure from the NACA. One of them published in this chapter involved Munk's

---

[24] Theodore von Kármán's life is covered in fascinating detail in Michael H. Gorn's *The Universal Man: Theodore von Kármán's Life in Aeronautics* (Washington, DC: Smithsonian Institution Press, 1992).

supervision of the construction of the VDT. In a 6 October 1921 memo to Washington, Langley Chief Physicist Frederick Norton expressed resentment at Munk's overbearing manner and vague directions during construction of the tunnel, complaining that "Dr. Munk does not seem to have any clear idea as to what he wishes in the engineering design, excepting that he is sure that he does not want anything that [I or my men] suggest." The best story about the difficulties of working with Munk comes from Langley engineer Fred E. Weick's autobiography *From the Ground Up* (Washington, DC: Smithsonian Institution Press, 1999), an excerpt from which is included in this chapter. Moving from the Navy Bureau of Aeronautics, where he worked as a civilian engineer, to the NACA in 1925, Weick took on the responsibility of actually building the Propeller Research Tunnel, which was also a Munk concept. Unfortunately for Weick, Munk's incredibly demanding supervision of the PRT work nearly drove Weick crazy trying to find ways to please the man and still do the job correctly. To Weick's credit, he found a way of working with—and mostly around—Munk, and the facility was successfully built.

Other documents in this chapter reveal different perspectives on the Munk affair. A 16 November 1926 memo from Munk to George Lewis, the NACA's director for research in Washington, D.C., indicated that Langley was "at present pretty well filled up with problems; we are really overstocked." On the surface, Munk's memo simply expressed concern over the need to give "the fullest amount of thought and interest" to the research problems his laboratory staff was already busy exploring. Reading between the lines, however, one senses that Munk may actually have been more concerned about his own personal control of the research program at Langley and wanted to prevent new ideas from reaching NACA Headquarters that he had not evaluated first.

Relations with Munk were difficult from the start, even when he spent most of his time in Washington; the problems definitely intensified when George Lewis sent him to Langley for extended periods. The Langley engineers considered Munk an arrogant outsider, and Munk's efforts to fit in, to the extent he made them, failed miserably. Perhaps it was because he saw himself as intellectually and technically superior to those around him—and made that belief clear to others, insisting they conform to his own ideas and conclusions. While they respected Munk's abilities, the engineers at Langley, many of them capable people in their own right, balked at his autocratic rule. Resignations of key people, like Norton, began in 1923; the rate increased after Munk was assigned to Langley as chief of aerodynamics in 1926, culminating with the mass resignation of all the section heads less than a year later. Lewis still hoped to retain Munk in Washington, but Munk, not yet grasping the "culture-shock" nature of the problem, chose to leave the NACA. When he left, the section heads returned.

Even though Munk's personality doomed him at Langley, he succeeded in introducing theoretical aerodynamics to the laboratory, and he helped to usher in an important change in the way the LMAL worked. Munk, like Prandtl, saw the connection between empirical and theoretical aeronautics and how each could help advance the other. One of his seminal papers, "General Theory of Thin Wing Sections," published by the NACA as Technical Report 142 in 1922, resulted from his conviction that theoretical aerodynamicists would not be able to generate improved airfoils from scratch using only the mathematical methods suggested by Kutta and Joukowski. Instead, Munk reversed the process. He decided to start with an empirically proven airfoil and then fit an analysis to it. Starting from a known design allowed him to validate the theory, something that had eluded many theorists. His theory was not perfect, but it allowed for an easier and more accurate prediction of wing performance. Over the next few years, a small group of scientists and engineers with better analytical capabilities and backgrounds joined the LMAL staff. Although conflicts between the empiricists and the theorists never disappeared totally, empirical and theoretical results came increasingly into agreement, as both the experimental tools and the analytical methods improved. Inexorably, the two camps pulled together into a mature discipline.

By the early 1930s, aeronautical research and development had come of age in the United States. Vigorous debates about the quality of the direction being followed by the NACA still took place, notably the fiery public back and forth in 1930 and 1931 between *Aero Digest*'s opinionated editor Frank Tichenor, the NACA's main detractor, and *Aviation*'s Dr. Edward P. Warner, its principal defender. (Their exchange of editorials is included in this chapter's documents.) But, in the opinion of most observers and colleagues in the aeronautics community, the program of research being pursued by the NACA by the 1930s, in cooperation with the military air services, the aircraft industry, and the universities, seemed to be exactly what the country needed not just for better airplanes but for global aeronautical hegemony.

In two decades' time, American aeronautical research had grown from having virtually nothing in the way of laboratory facilities to a position of preeminence in the world, with more than twenty major wind tunnels in operation for universities, aircraft manufacturers, the military, and the NACA. Equally important were the bright, trained, and experienced researchers who could effectively use these experimental tools and increasingly powerful analytical methods—combined with their own innate creativity and vision—to generate new knowledge and solve the fundamental problems facing a rapidly growing technology. The stage was set for an honest-to-goodness revolution in aerodynamics and aircraft design.

The Documents

Document 2-1(a–d)

(a) **Wilbur Wright, letter to George A. Sprat, 15 December 1901.**

(b) **Wilbur Wright, letter to Octave Chanute, 15 December 1901.**

(c) **Wilbur Wright, letter to Octave Chanute, 23 December 1901.**

(d) **Wilbur Wright, letter to Octave Chanute, 19 January 1902.**

All of the above documents are found in
The Papers of Wilbur and Orville Wright, 1899–1948,
Marvin W. McFarland, ed. (New York, NY: McGraw-Hill), 1953.
The originals are in the Library of Congress, Washington, D.C.

The successful flights of Wilbur and Orville Wright on 17 December 1903 were due in no small part to the brothers' systematic pursuit of aeronautical knowledge. Recognizing that tables developed by John Smeaton and Otto Lilienthal—tables the brothers had used in designing their disappointing 1900 and 1901 gliders—included significant errors, the Wrights designed and built novel research devices and used them systematically to generate reliable data that pointed to a successful design. As he did throughout the project, Wilbur Wright frequently communicated their progress and questions with several prominent figures in early aviation. These letters, a sample of many written during the Wrights' wind tunnel, or "trough," experiment period, discuss their test results, how these results compared to Lilienthal's data, and their efforts to obtain "a perfectly straight current of wind" for the tests. While the work was distinctly empirical in nature, the details of measurement and experiment design, especially those noted in Wright's 15 December 1901 letter to Octave Chanute, clearly show the brothers' attention to accuracy and repeatability—vital elements of any empirical program. Also of interest is Wilbur's 23 December 1901 rejection of Chanute's offer to try and interest Andrew Carnegie in supporting the Wright's work (in a 19 October 1901 letter not included herein) and his pragmatic analysis of an aviation competition in a letter to Chanute dated 19 January 1902.

Document 2-1(a), letter from Wilbur Wright to George A. Sprat, 15 December 1901.

We were pleased to receive your letter and the photograph of your new testing machine. It seems quite ingeniously designed and I think should give good results. As you say, the greatest trouble will probably be with the changeableness of the wind. If I understand you properly, the machine is intended for locating the center of pressure at any angle (or rather locating the angle for any center of pressure), and for finding the direction of the resultant pressure as measured in degrees from the wind direction, so that the ratio of lift to drift is easily obtained, the lift being the cotangent and the drift being the tangent of the angle at which the arm stands. Does the machine also measure the lift, in terms of per cent of the pressure at 90°? so that you can make tables like that of Lilienthal?

I think I told you in my last that we had been experimenting with a lift measuring machine. We have carried our experiments further and have made a measurement of the lifts of about 30 surfaces at angles of 0°, 2 ½°, 5°, 7 ½°, 10°, 12 ½°, 15°, 17 ½°, 20°, 25°, 30°, 35°, 40°, & 45°. The results have rather surprised us as we find at angles of 7° to 15° *with some surfaces* a greater lift than Lilienthal gives in his table. Our #7 surface, which is a rectangle 1:6 with a depth of curve of ¹⁄₁₂ the chord, has a *lift* of one hundred and nineteen per cent at 17 ½ degrees. Lilienthal only claims about 80 per cent. But at 3° our measurement is way below him. I will try to send you a blueprint showing the lifts of some of the surfaces we have tested. Some surfaces which lift big at very small angles are no good at large angles & *vice versa*. We have not attempted to trace the travel of the center of pressure except that by holding some of the surfaces between the tips of our fingers we were able to roughly determine which ones tended to reverse and which did not. It seems that surfaces with rather flat upper sides and thickened front edges lift more *at small angles* than plain curves and have little reversal of the travel of the center of pressure. Thickening the front edge does not seem to add near as much to the drift as I expected though it adds some.

We have found less drift with surfaces ¹⁄₂₀ deep than with curves ¹⁄₁₂ deep. What is your experience?

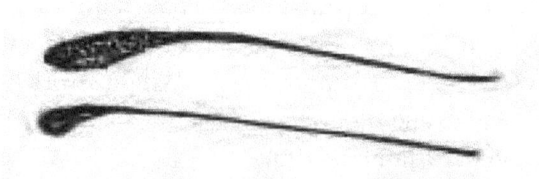

Document 2-1(b), letter from Wilbur Wright to Octave Chanute, 15 December 1901.

I have your letter of 11th with enclosures and have read all with much interest. Moedebeck evidently is a balloonist rather than an aviator. Mr. Lilienthal says that the results obtained by his brother with the apparatus *exactly coincided* with their joint measurements. My study of Otto Lilienthal's writings leads me to question whether this is not too strong a term. After Lilienthal began gliding I do not recall that he ever recommended 3° as being a specially desirable angle of incidence. In his calculations of glides he invariably uses angles of from 9° to 12° incidence. If he had found the great advantage in the smaller angles which the tables pronounced best, he would surely have mentioned it. From this I am led to think that Lilienthal himself had noticed that there was a discrepancy between his glides and his tables, at small angles especially. There is also some question whether a surface about 8 ft. x 20 ft., that is, 1:2.5, would give the exact result obtained with the surface of very different aspect which was used in the experiments on which the tables were based.

Prof. Marvin's article has evidently been prepared with some care and is well worth close study. I quite agree with him that data of a glide in which the forces are not in equilibrium throughout the glide or the portion of it on which the data are based must be worthless for purposes of calculation. You will remember that at Kitty Hawk we said that the #3 glide of Aug. 8th, A.M., was really the only one of much value for purposes of calculation because in it, alone, the speed from the instant of starting to the instant of landing was very nearly uniform. The angle of incidence varied scarcely at all; the line of motion was exactly opposite to the direction of the wind so that the conditions were equivalent to a glide of about double the speed in still air. Owing to the practical impossibility of obtaining correct data I have never considered glides of very great value for purposes of calculation.

I notice that Prof. Marvin holds about the same view on the tangential that I held last July, *viz.*, that the front edge must be lower than the rear edge or the bird or machine can not go forward. But the sight of a buzzard which maintained its speed with its wings constantly pointed *above* the horizon, and the fact that our machine pulled less than the weight X tang. of the angle of incidence, in spite of the head resistance of the framing, led me to suspect that Lilienthal might be right about the tangential. Our recent experiments are so clear on this point that I can no longer doubt that a suitably arched surface can glide forward in a descending course of ten degrees with the front edge of the surface pointed as much as *five degrees above the horizon.*

I send you blueprints of some of the preliminary charts of our recent measurements of lifts. They were made on common paper so that the blueprints from them scarcely show the small squares, and this makes it hard to read the exact values, but you can see the general result. The lines run *exactly according to our observations* at the points at which measurements were made, *viz.*, 0°, 2 ½°, 5°, 7 ½°, 10°, 12 ½°, 15°, 17 ½°, 20°, 25°, 30°, 35°, 40°, & 45°. In connecting these points we have used our best judgment as to the course in the intervening spaces. I am not sure but that there is room for some improvement as to this. I send three of our series charts and one large one with a selection of surfaces of various types.

The aspects and curvatures are shown on the back of the large sheet. On the whole the finished charts give me a better impression of the accuracy of the

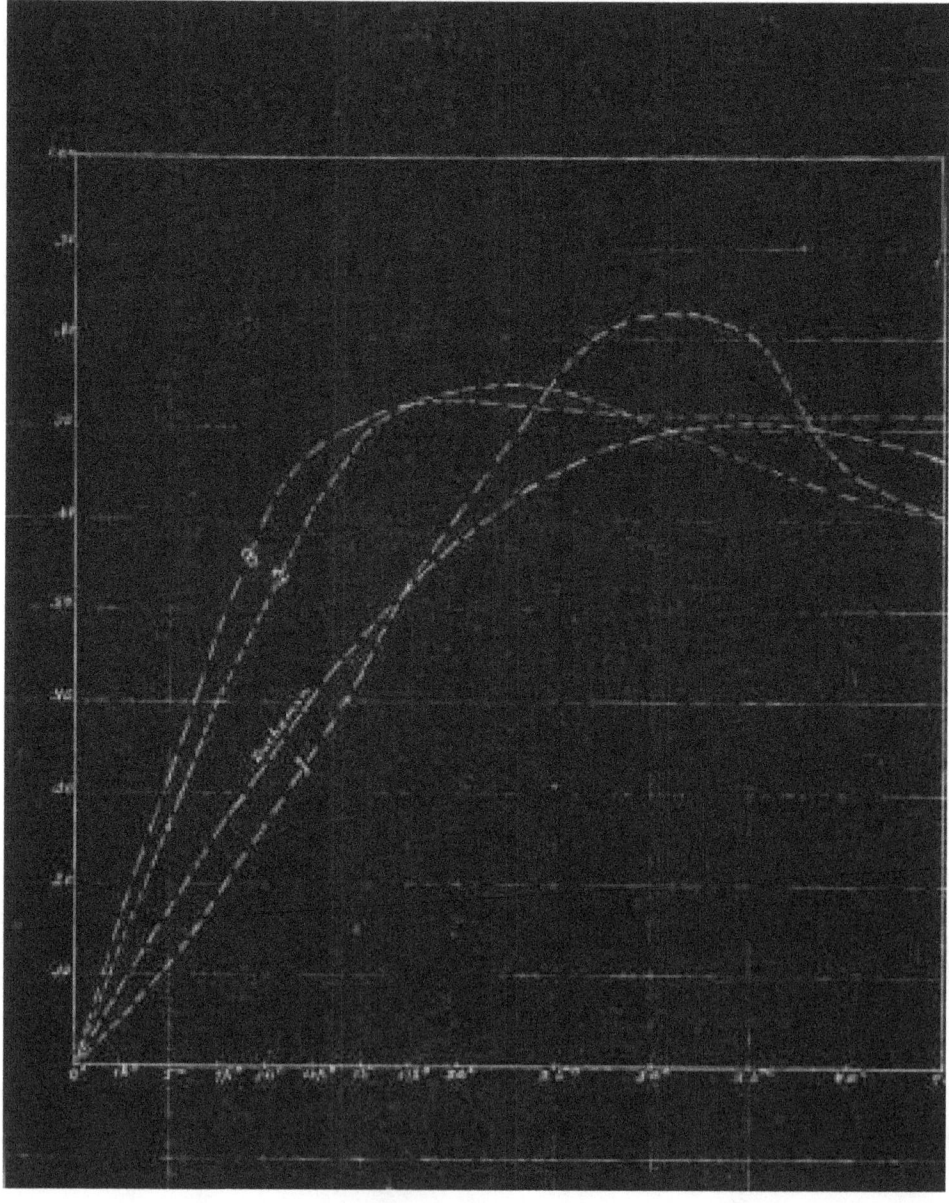

results than I had after plotting the first few lines, as I find that many kinks in the lines, which at first seemed to be due to errors of the observations or imperfections in the machine, are now seen to be due to the surfaces themselves. Thus when #7 was plotted we found an uncalled-for depression at 10° and a hump at 12 ½°, and though we verified the observations we were not entirely easy in our minds till we plotted #8 and #9 and found that they had exactly corresponding depressions and humps. The fact that the hump on #9 comes exactly at the depression on #7 makes #9 the greater lifter at 10°.

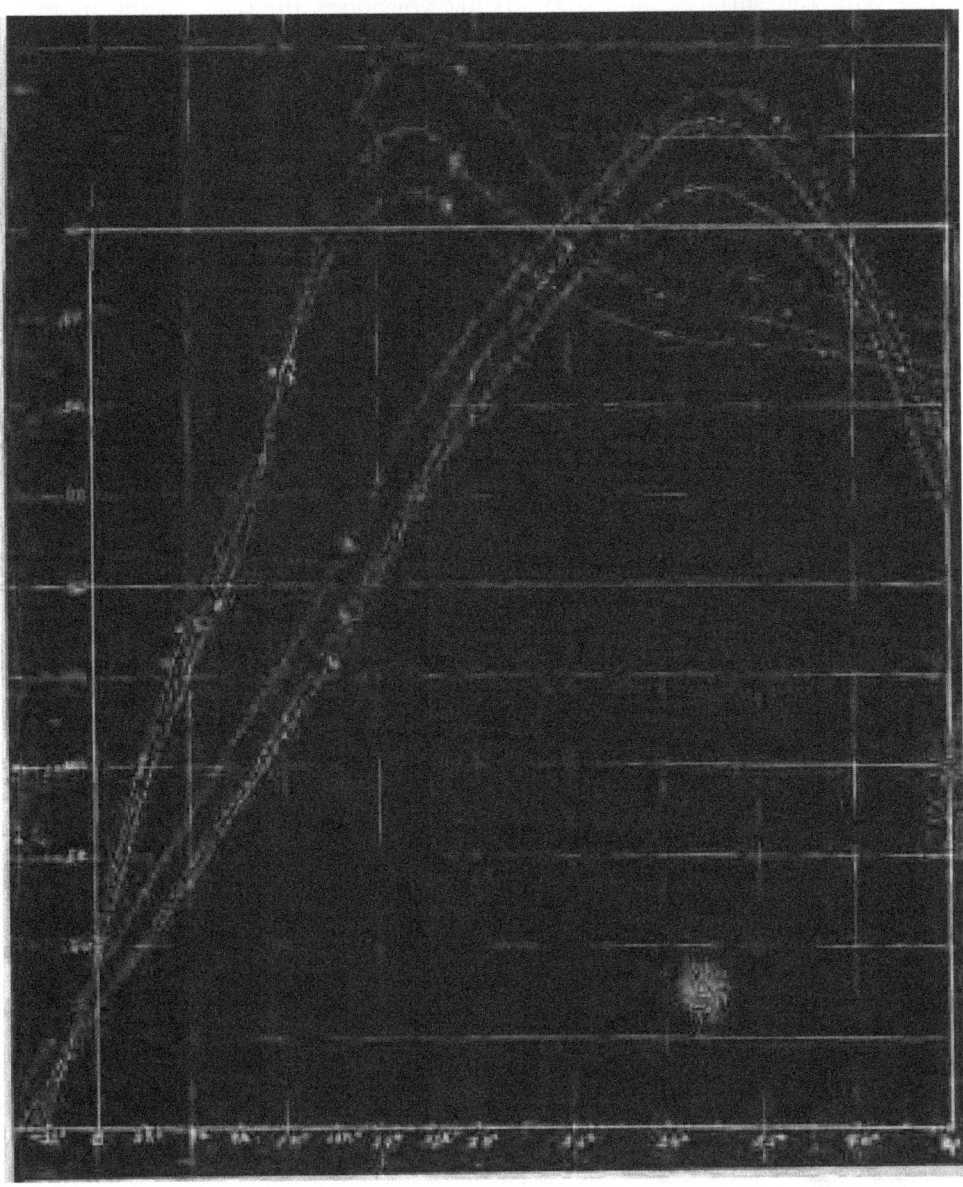

You will also note that #4, #5, & #6 have corresponding depressions at 10°–12° and humps at 17°–20°. When we first measured #10 we struck a snag at 12 ½°, for the machine, which at the lower angles had been recording a reasonably regular increase in lift with each successive observation, suddenly refused to indicate any increase and, though we examined the machine carefully and verified the angle of incidence, we could not believe that the observation was correct till we had remeasured 10° also. Afterward we found that this was a characteristic of all surfaces having the curvature well to the front. Arcs rise in regular peaks while parabolic surfaces show summits like volcanoes, the crater being more or less marked according to the depth of curvature.

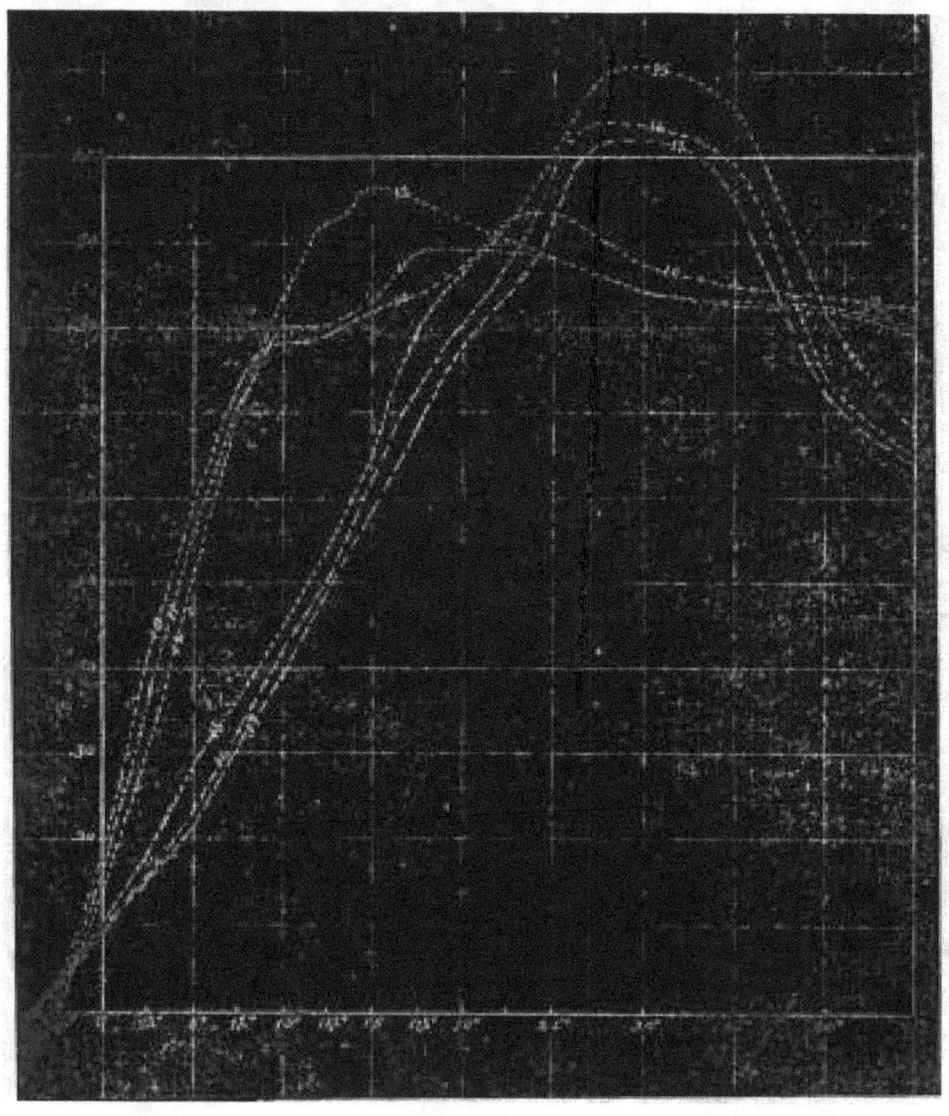

It has been a great advantage we think to make a systematic measurement of several typical series of surfaces rather than to work blindly on all sorts of shapes, as a study of the series plates quickly discloses the general principles which govern lift and tangential and thus renders the search for the *best* shapes much easier. A mere glance shows that while increasing the ratio of breadth to length does not increase the maximum lift to any great extent it does cause the maximum to be reached at a smaller angle; and the wider the spread from tip to tip the smaller the angle at which large lifts can be obtained.

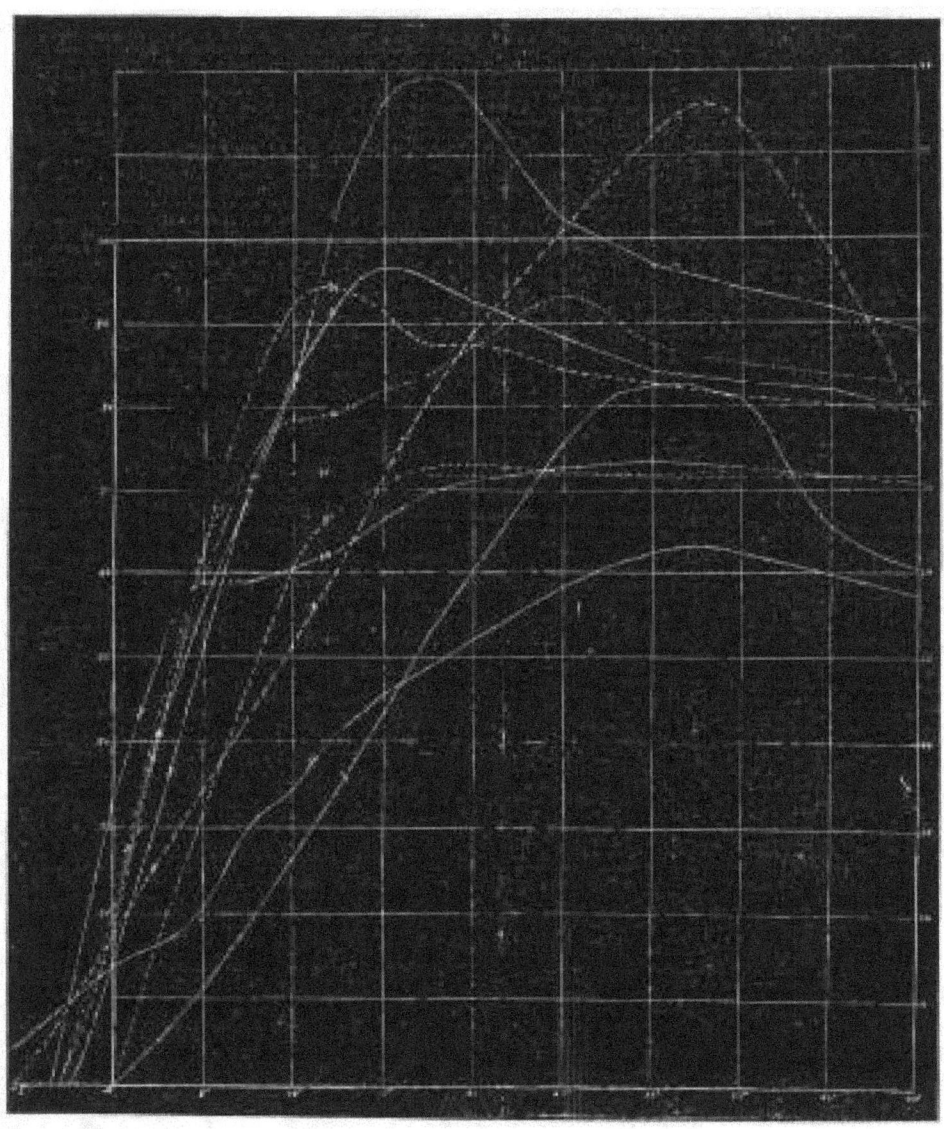

By comparing #1, #4, #5, & #6 it is seen that the effect of curving the surface is to give a steady increase in the lift of all angles without affecting the angle of maximum pressure. Varying the depth of curvature has a less marked effect than the experiments of Lilienthal would indicate in the matter of lift, but when we come to consider tangential the difference is very marked. The great advantage of moving the maximum curvature well forward is in the matter of center of pressure, though it seems also to cause an increase in lift at smaller angles, and in general gives a slightly more favorable tangential at angles of 4°–10°. Thickening the front edge has a marked tendency to give big lifts at small angles—*vide* #20, #25, & #35; and is even better than moving the curvature forward in its effect on center of pressure. It somewhat increases the head resistance, however, though not near so much as the great increase in thickness might be expected to cause.

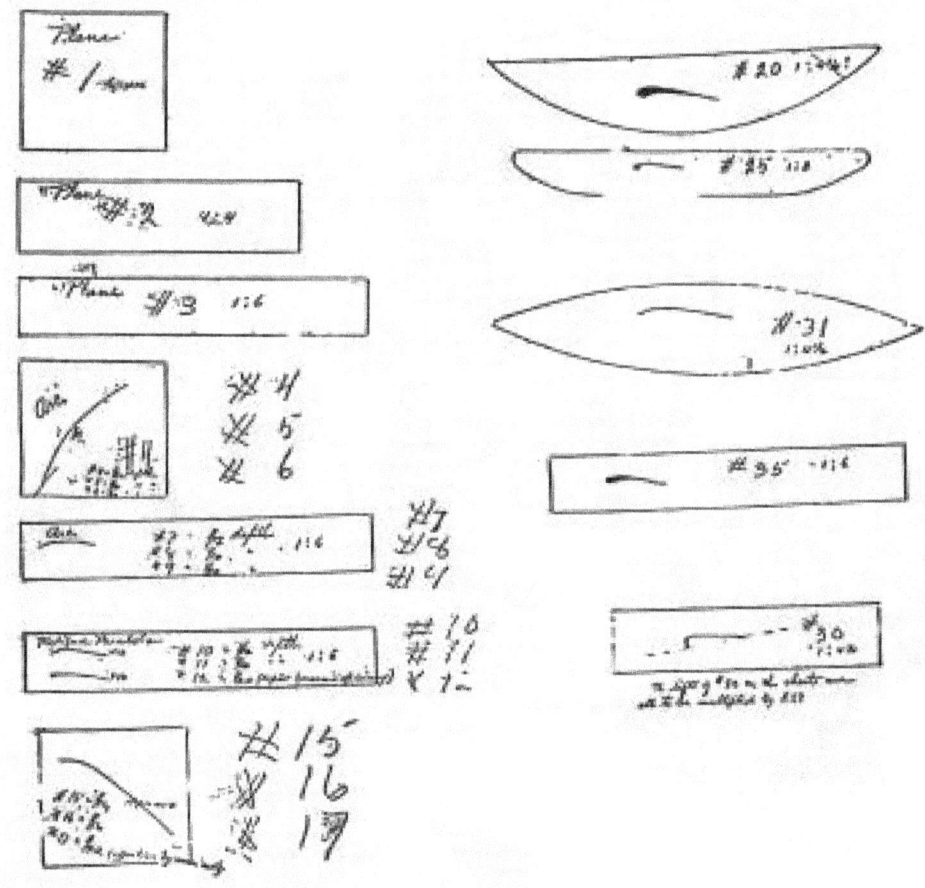

The charts showing the effect of superposing are not completed, but there seems to be some indication that they will tend to establish a general law that (eliminating connections) the lift and tangentials of a set of superposed or following surfaces spaced about their length apart [are] approximately equal to [those] of a single surface of similar profile or curvature having a breadth equal to that of one surface and a length equal to the sum of their lengths. That is, two 1:6 surfaces would give the same results as one 2:6 surface; and four of them that of a 4:6 surface. From this it would appear that superposing reduces the efficiency of the individual surfaces. In considering a double-deck or triple-deck machine, as compared with a single surface of equal area, it would seem the single surface can be cut up as in Fig. 1 and the parts superposed without loss, but if cut as in Fig. 2 there is a loss. Superposing may be used to reduce the fore-and-aft dimension but not the lateral.

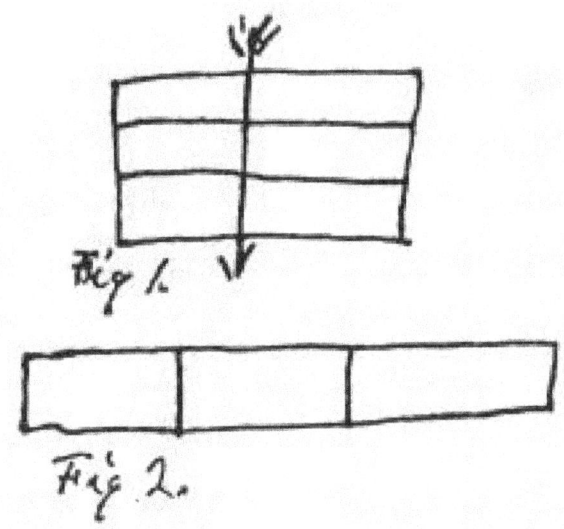

A somewhat similar law seems to hold when comparing rectangles with segments or triangles. Surface may be cut off at one point and added at another without affecting the efficiency of the surface. Thus if the corners be cut off at aa and added at $a'a'$ thus forming a triangle, the effect remains about the same, so long as the spread from tip to tip remains the same. It is at least roughly true that both in superposing and in reshaping a single surface the efficiency depends on the ratio of maximum breadth to area, rather than on the ratio of breadth to length.

I regret that we did not have time to carry some of these experiments further, but having set a time for the experiments to cease, we stopped when the time was up. At least two thirds of my time in the past six months has been devoted to aeronautical matters. Unless I decide to devote myself to something other than a business

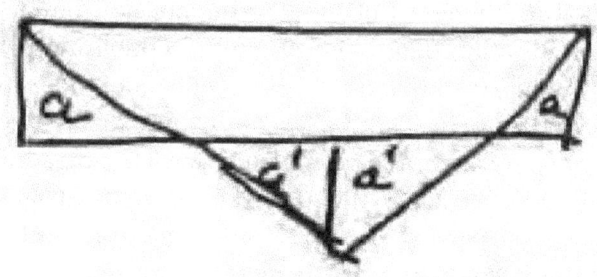

career I must give closer attention to my regular work for a while. I hope at some later time to resume these investigations and also to carry out a plan I have considered of obtaining an accurate measurement of the value of P90.

You will note that #7 reaches a lift of 119 per cent at 17 $\frac{1}{2}$°, a greater amount than any preceding investigation has found, so far as I am aware. And that #4 reaches almost as high at 30°. By means of an entirely different instrument I have confirmed the fact that the *normal pressure* of #4 at 20° (twenty) is slightly more than equal size at 90°. At 30° the normal pressure of #4 is much greater than the normal plane. At about 45° the normal pressure of #4 is equal to the normal pressure of a plane at about 70°. If these high values of #7 and #4 at 17 $\frac{1}{2}$° and 30° are too high, then the lift of #1 at 17 $\frac{1}{2}$° is also too high in the same ratio. In fact all the measurements of all the surfaces were made under conditions which make all the measurements true if any of them are true.

I will return the Marvin document in a few days.

[p.s.] #12 has the highest dynamic efficiency of all the surfaces shown.

Document 2-1(c), letter from Wilbur Wright to Octave Chanute, 23 December 1901.

I am returning the Marvin papers herewith. I should have sent them much sooner but that I hoped to be able to complete some notes to send with them. I find however that it is more difficult than I expected and though I have made several attempts I have not yet been able to get everything shaped up to suit my ideas of the real operation of the forces which arise in glides. I think that the *real* angle of incidence should be not the angle bounded by the chord of the surface and the line of the path, but an angle bounded by the line which marks the negative angle at which lift begins and the line of the *relative wind*. The true normal should be perpendicular to the line of no lift instead of to the chord, and the tangential is the deflection of the direction of the *resultant pressure* from this true normal. You will see that all this greatly increases the complication of calculations but I see no other way of arriving at theoretically perfect results. For rough work much of this refinement would be unnecessary, but in *defining the meaning of terms* we should aim to get them *exactly* as they really are. I will send you my ideas on some of these points as soon as I can get them properly straightened out. I get lost now and then.

I return the Langley letter which I have read with regret. I was not unprepared for his decision as some remarks Mr. Huffaker made last summer gave the impression that Lilienthal was not in high favor among the Washington group of workers for some reason. Mr. Huffaker seemed to think that Lilienthal had been overestimated. Since seeing his book I can not help thinking that he is underestimated, and that he will stand even higher when the doubts with which some of his most important discoveries have been accepted are finally cleared away. It seemed to me that the publication of his book in English would not only be of very great value to all aeronautical workers but it would be a well deserved tribute to the memory of a man who spent much money, an immense amount of his time, and finally his life in carrying out investigations which he gave freely to the world.

Your offer to assist in figuring out the results of our recent experiments is thankfully received. The labor itself is not so tremendously great, though there are several days' work required; but I have felt the need of a verification of our calculations to guard against blunders. If you desire I will send you our data as read from the machine, with directions for translating them into per cent of the pressure on a square plane of equal area at 90°. I will also send you photographs of our instruments and try to make it clear just how they operated. As to the accuracy of the results I think I am very safe in saying that the possible error is less than one twentieth. I think the average is much closer than this. We spent nearly a month getting a *straight* wind, but finally were able to get a current whose direction did not vary one eighth of one degree. A possible error of one twentieth, or 5%, is by no means insignificant but it does not greatly reduce the practical value of the tables. For purposes of comparing the lifts of different surfaces at the same angle or different angles, or of comparing the relative lifts of the same surface at different angles, the tables will be more accurate still, as some of the most serious sources of error are eliminated when the exact value of the common resistance against which the surfaces are weighed is not required to be fixed. I think about 2% would easily cover the errors which could arise from errors in mounting the surface at the exact angle desired, variations in wind direction, and errors in reading. This of course does not include mere blunders.

Our measurements of the tangential include the edge resistance of the surfaces and I can devise no way of eliminating this from the measurements. On the whole I am inclined to think that it is possible with large surfaces to get rather more favorable tangentials than the small surfaces show. The thickness of our regular surfaces was about three per cent of the fore-and-aft dimension. This you will notice is more than the relative thickness of the largest spars of our gliding machines. We did not use thinner material because it was deemed to be of the utmost importance that the surface should be sufficiently strong and stiff to prevent any twisting or warping under pressure of the wind. A distortion too small to be noticed and almost too small to be measured might cause quite serious errors in measurements where great exactness was desired. I am inclined to think that this has been a more serious matter than most investigators have supposed.

It was with very great reluctance that we discontinued our experiments at the time we did, but there were so many things yet to be investigated that it was very evident that we would be unable to spare the time to carry them through at the present. We got all that we originally set out for, so we thought it a favorable time to take a recess. Then, too, we saw that any further time consumed now would seriously impair our chance of a trip to Kitty Hawk next fall.

As to your suggestion in regard to Mr. Carnegie, of course nothing would give me greater pleasure than to devote my entire time to scientific investigations; and a salary of ten or twenty thousand a year would be no insuperable objection, but I think it possible that Andrew is too hardheaded a Scotchman to become interested in such a visionary pursuit as flying. But to discuss the matter more seriously, I will say that several times in the years that are past I have had thoughts of a scientific career, but the lack of a suitable opening, and the knowledge that I had no special preparation in any particular line, kept me from entertaining the idea very seriously. I do not think it would be wise for me to accept help in carrying our present investigations further, unless it was with the intention of cutting loose from business entirely and taking up a different line of lifework. There are limits to the neglect that business will endure, and a little pay for the time spent in neglecting it would only increase the neglect, without bringing in enough to offset the damage resulting from a wrecked business. So, while I would give serious consideration to a chance to enter upon a new line of work, I would not think it wise to make outside work too pronounced a feature of a business life. Pay for such outside work would tend to increase the danger. The kindness of your offers to assist, however, is very much appreciated by us.

Document 2-1(d), letter from Wilbur Wright to Octave Chanute, 19 January 1902.

I am sending you herewith photo and description of our pressure-testing machine. It is our belief that the method and construction employed entirely avoid errors from the following sources: (1) Variation in wind velocity; (2) Variation in temperature and density of the atmosphere; (3) Travel of center of pressure; (4) Variation in angle of incidence owing to movements of the mounting arms. The first two causes gave Mr. Langley trouble; while the 3rd & 4th vitiate somewhat the *natural wind* experiments of Lilienthal. Gravity and centrifugal force are also rendered nugatory.

Our greatest trouble was in obtaining a perfectly straight current of wind, but finally, by using a wind straightener, and changing the resistance plane to a position where its ill influence was much reduced, and also by breaking it up into a number of narrow vertical surfaces instead of a single square, we obtained a current very nearly constant in direction. The instrument itself was mounted in a long square tube or trough having a glass cover. After we began to make our record measurements we allowed no large object in the room to be moved and no one except the observer was allowed to come near the apparatus, and he occupied exactly the same position beside the trough at each observation. We had found by previous experience that these precautions were necessary, as very little is required to deflect a current a tenth of a degree, which is enough to very seriously affect the results. I will send another batch of data in a few days.

Your letter from St. Louis of course interested us very much. The newspapers of yesterday announce that the fair will be held in 1903 as originally planned. If this be final there will be little time for designing and building a power machine which is, I suppose, the only kind that could hope to be awarded a prize of any size. Whether we shall compete will depend much on the conditions tinder which the prizes are offered. I have little of the gambling instinct, and unless there is reasonable hope of getting at least the amount expended in competing I would enter only after very careful consideration. Mathematically it would be foolish to spend two or three thousand dollars competing for a hundred thousand dollar prize if the chance of winning be only one in a hundred. However we shall see about the matter later.

Meanwhile it will be just as well for me to postpone the paper on our late experiments on pressures & tangentials till we have decided whether or not we shall compete, as it would be hardly advisable to make public information which might assist others to carry off the prize from us. If the exposition authorities should deem it advisable to offer some preliminary prizes for papers on such subjects with a view to getting into the hands of all the competitors the best possible information and thus rendering the final contest of machines more exciting, I would place our tables in competition, but otherwise we ought to delay publication for a short time at least. This injection of the mercenary idea into the flying problem is really a nuisance in some respects.

Document 2-2(a–c)

(a) Washington Irving Chambers, excerpt from "Report on Aviation," 21 September 1912, *Annual Report of the Secretary of the Navy for 1912* (Washington, DC, 1912).

(b) Jerome C. Hunsaker, Assistant Naval Constructor, *Report on Facilities for Aeronautical Research in England, France, and Germany, Part III—Germany*, undated (ca. November 1913), NASA Record Group 255, Hunsaker Biography File, Entry 3, Box 12, National Archives, Washington, D.C.

(c) Albert F. Zahm, excerpt from *Report on European Aerodynamical Laboratories* (Washington, DC: Smithsonian Institution, 1914).

A few months after the untimely death of Wilbur Wright in 1912, Captain Washington Irving Chambers of the United States Navy submitted "Report on Aviation" to the Navy Department's Bureau of Navigation. Chambers began his report by outlining possible uses of aircraft in naval warfare. Concerned about the lagging state of aviation research in the United States, especially when compared to what was being carried out in Europe, Chambers concluded his report with a detailed proposal for the establishment of "a national aerodynamical laboratory" in the United States. It is this part of the document that is reproduced here. While Chambers's specific recommendations were not acted on at the time, his report stands as a farsighted statement of the need for a national research establishment.

Following the success of the Wright Brothers, European interest in aeronautical research intensified, but, incredibly, it languished for a decade in the United States. Between 1903 and the start of World War I, no less than ten wind tunnels began operation in Europe, while only two were built in America. Much of the European work was due to a realization, particularly after the Wrights' 1908 demonstration flights in France, that airplanes could provide a military advantage, something of great value in an increasingly antagonistic Europe. In England, Italy, and Germany, governments invested in both airplanes and new aeronautical laboratories, such as the one at the National Physical Laboratory in London and a laboratory at Göttingen, Germany, that featured an innovative, closed-circuit wind tunnel. Private funds underwrote new facilities in Russia and France, including two laboratories built by Gustav Eiffel in and near Paris. The United States, on the other hand, cared little about the looming catastrophe in Europe, preferring to remain neutral and, thus, seeing little need for an investment in military aviation. The long Wright-Curtiss patent fight over ownership of rights to the airplane further dampened America's enthusiasm for airplanes and aeronautical development.

Within the military, a few farsighted naval officers, including Washington Irving Chambers, David W. Taylor, and Jerome C. Hunsaker, realized the long-term importance of aviation to the military and lobbied for an aeronautical laboratory. An investigative board chaired by Chambers recommended a national aerodynamic laboratory in its 1912 "Report on Aviation." This report includes considerable detail about the nature of the work to be done in such a laboratory and the necessary facilities, and it goes so far as to suggest a joint military-civilian agency "advisory committee" to direct the operation. Although Chambers' recommendation for joint military-civilian management angered his superior officers—unlike Taylor, Hunsaker, and other pioneers of naval aviation, Chambers was never promoted to rear admiral—his report described a research organization very close to the one adopted for the National Advisory Committee on Aeronautics three years later.

The Chambers report also inspired the navy to take action on its own. Taylor, assisted by two lieutenants and Zahm, was developing a design for a closed-circuit wind tunnel. With encouragement from the Chambers Report, Taylor managed to obtain the funds to build it, and the eight- by eight-foot Washington Navy Yard tunnel, the first wind tunnel built by the United States Government, began operation in March 1914.

While America did not invest in aeronautical laboratories to the extent that European nations did before World War I, a few interested parties stayed abreast of the European developments and pressed for the establishment of such facilities in the United States. Supported by the Smithsonian Institution, Catholic University Professor Albert F. Zahm and Assistant Naval Constructor Jerome C. Hunsaker, two of the leading figures in American aviation, visited the major European research facilities in 1913 to evaluate their capabilities and progress in aeronautics. Hunsaker first submitted his findings to his superiors in the navy in an extensive three-part report. Part III of his *Report on Facilities for Aeronautical Research in England, France, and Germany* covers six facilities and gives a good picture of the state of the German aeronautical art. The work of Ludwig Prandtl at the University of Göttingen particularly impressed the perceptive Hunsaker. While he found the Göttingen work to be "characterized by poor equipment made the most of by men of extraordinary ability," he noted that the Kaiser Foundation was underwriting a new wind tunnel there. Hunsaker's report also mentions Prandtl's theoretical work in computing dirigible resistance using hydrodynamic theory, work that would be a major foundation for the development of theoretical aerodynamics.

Zahm's *Report on European Aeronautical Laboratories*, a public report of the same inspection trip published by the Smithsonian Institution in 1914, described the European research laboratories and apparatus in considerable detail, but this report did not contain the kind of analysis and insightful evaluation of the facilities and work that can be found in Hunsaker's report. Nevertheless, Zahm's report showed that America was considerably behind the Europeans. This report, being a published document, was widely circulated in aviation and political circles, and it proved to be the catalyst that spurred the creation of the National Advisory Committee for Aeronautics.

Document 2-2(a), Washington Irving Chambers, excerpt from "Report on Aviation," 21 September 1912.

INFLUENCE OF FOREIGN LABORATORIES

Little more than a year ago our knowledge of the effect of air currents upon aeroplane surfaces was almost entirely a matter of theory. The exact information available was so meager that aeroplanes were built either as copies, slightly modified, of other machines, or else by way of haphazard experiment. This state of affairs pertains to some extent in the United States today, although in Europe aeroplane construction is now largely based on scientific data obtained at notable aerodynamic laboratories.

The intuitive, hasty, and crude methods of the pioneer can not succeed in competition with the accurate and systematic methods of the scientific engineer, and it is beginning to dawn upon our perceptions that through lack of preparation for the work of the scientific engineer, i.e., through delay in establishing an aerodynamic laboratory, a waste of time and money, a decline of prestige, and an unnecessary sacrifice of human life has already resulted.

Students of aviation do not need to be informed of the practical necessity for aerodynamic laboratories. They have repeatedly pointed out, in aeronautical publications, the immense commercial advantages to be anticipated from the establishment of at least one in this country, and they have naturally expected that some philanthropic patriot of wealth and scientific interest would come to the rescue with a suitable endowment fund that would enable such work to be started in short order without Government aid. The fact that no patriot has responded is disappointing, in view of the large private donations that have done so much for aviation in France, but in my opinion, it simply indicates something lacking in the manner of disseminating information concerning the importance of the subject. I am not willing to believe that our people will refuse to establish one when they are fully acquainted with the advantages to humanity and to sane industrial progress, and when a reasonable concrete proposition is advanced for their consideration.

It is now my purpose to submit such a proposition, and, in doing so, I will follow briefly, in general outline, the ideas advanced in an address to the Fifth International Aeronautic Congress by one of the greatest authorities in the world, the Commandant Paul Renard, president of the International Aeronautic Commission.

A NATIONAL AERODYNAMIC LABORATORY

Before considering the character of the work to be done and some details of the needed plant, it will facilitate matters to show what should not be done at such a laboratory.

There are those who dream of supplying the laboratory with all the instruments known to mechanics, to physics, and even to chemistry, in order to have a creditable and complete national institution. They would concentrate in one locality all the scientific instruments and acumen available, with the false idea that economy would result. This would be a grave error.

The financial resources, however great, are sure to be limited, and a too ambitious or a superfluous installation would squander the sources of power and indirectly menace the initiative of other industries. The character of the new work to be done demands that everything should be rejected that can be dispensed with readily in order that appliances specially needed in the new work may be provided and that these appliances be of the latest and most efficient types.

For the sake of economy, not only of money but of time and intellectual energy, tests and experiments that can be executed as well or better elsewhere by existing establishments should be avoided. For example, it is unnecessary to install a complete set of instruments and implements for testing the tensile strength of materials or their bending and crushing strength. Many other establishments permit of such work. If the laboratory is to be located in Washington, where certain advantages exist, such work could be readily done at the navy yard, where other facilities exist such, for instance, as the testing of models for hydroaeroplanes and flying boats. The Bureau of Standards and Measures and other Government branches in Washington also offer facilities which it would not be wise to duplicate in such a laboratory.

I do not think that such an institution should be burdened with measuring the power of motors or preoccupied with the details of their performances. This may be done at various other Government establishments, and it is understood that the Automobile Club of America is also equipped for this work.

Nor is it necessary to have a complete chemical laboratory under the pretext of studying questions relating to the chemistry of fuel or the permeability of balloon envelopes.

I do not wish to convey the idea that an aerodynamic laboratory should be deprived entirely of such facilities and that it should be obliged to seek minor information from other establishments when that information may be more economically obtained by a duplicate plant on a small scale. Such duplicate conveniences, however, should be regarded as strictly accessory; but it should be well understood that whenever important researches can be prosecuted as well or better elsewhere, dependence should be placed on those other establishments where such work is a specialty.

TWO DISTINCT CLASSES OF WORK

An aerodynamic laboratory should be devoted to (1) experimental verification, (2) experimental research. The first is concerned with testing the qualities of existing appliances, propellers, sustaining surfaces, control mechanism, etc. Usually these tests are made at the request of interested parties (as is now the case with water models at the navy-yard model basin). A constructor or a designer will bring, for example, a propeller and will wish to know its power or thrust at a given speed on the block or on a moving appliance under the conditions of flight, or he may bring several propellers to compare their performances and to ascertain what power they absorb at different speeds.

One of the very successful appliances devoted to this work at St. Cyr is a movable car, in which an aeroplane may be mounted and tested at speeds in perfect safety as to its strength, its efficiency, and the suitability of its control mechanism. This device is specially adapted to make actual service tests of sustaining surfaces, in other words, to try out in perfect safety the relative efficiencies of finished aeroplanes. It is a most important adjunct, as it supplements and rounds out the important research work on models in the closed laboratory.

Tests of this character, i.e., verification tests, constitute, so to speak, standard work. They are performed at the request of manufacturers, clubs, independent investigators, and other interested parties on condition of payment for the actual cost of the work. They therefore contribute to the support of the establishment.

The tests of verification, however, notwithstanding their great utility, do not constitute either the most important or the most interesting work of the laboratory. The research work, which prosecutes continuously and patiently systematic, thorough, and precise investigation of new ideas, or of old ideas with new applications, with the specific intention of discovering laws and formulas for advancing the progress of aerial navigation, is of greater importance, because it is the short cut to substantial efficiency, economy, improvement, and prestige.

This work is concerned with developing adequate methods of research in all branches of aerial navigation and in furnishing reliable information to all students, engineers, inventors, manufacturers, pilots, navigators, strategists, and statesmen. The knowledge thus gained should be disseminated regularly through publications, lectures, open-air demonstrations, and by exhibitions of apparatus, instruments, materials, and models—in fact, by all the facilities of the aerodrome, the showroom, the library, and the lecture room.

An exact knowledge of aerodynamics can best be acquired in such a laboratory by experimentation with standard scale models in air tunnels such as those used by M. Eiffel and others. In this way reliable data is obtained of the air resistance to be encountered and the efficiency at various velocities, the amount of lift, the effect of varying impact at different angles of attack on the stability—in fact, all the exact data which, reduced to curves and diagrams, enables the engineer to design a machine in a scientific manner. From such data the performance of a new machine can be closely predicted. The performance of the finished product can be verified later as before described.

Much of the research work will be prosecuted at the request of technical men outside of the institution, to whom the laboratory should offer, gratuitously as far as possible, its material and personal resources.

THE COUNCIL AND ORGANIZATION

To obtain benefit from these researches it will be necessary to know that they are worth the time and expense, and a body of men—a council or a board of governors—should be authorized to accept or reject requests for this work. This will be a delicate task, but the principal duty of the council should be to establish and to correct from time to time a program of the research work to be executed by the director and his staff and to coordinate the work to the best advantages within the limits of the money available. The disbursement of the Government funds, however, and the responsibility therefore should be entirely under the director.

With the actual state of aerial navigation and its deficiencies as a guide it will be the policy of the council to concentrate effort upon such points as seem most important, promising, and interesting for the time being.

I do not think there would be any doubt, if we had the laboratory in working order now, but that all questions relating to improvement in stability, automatic control, and safety in general would have the right of way.

The council or board, which in England is called the "advisory committee," should be representative of other Government departments than that employing the director, and should be independent of the director and his administrative staff. It might be possible for the director to act as a member of the council and, if so, it would conduce to harmony and expedition.

The council should not be a large body, but should be composed mostly of specialists of unquestioned ability, men interested in the development of aerial navigation in various branches of the Government and in its useful and safe adaptation to commerce and sport.

Whatever the ability of this council it should not be allowed to pretend that it has a monopoly of aeronautic acumen. Many brilliant and worthy ideas may originate outside of the establishment which it will be wise to investigate. And to avoid any possibility of the council being charged with narrow prejudice, it is indispensable that it be not composed entirely of specialists. In a few words, it should comprise representative men who are also learned and technical men, with broad vision and reputation, whose presence will guarantee to industrial investigators that their ideas will be treated in an unpartisan or unbiased spirit. I will not attempt to suggest the composition of this council or board, but it is evident that the Army and Navy should each be adequately represented on it.

ENDOWMENTS, PRIZES, AND REWARDS

If the laboratory should obtain, in addition to the funds required for prosecuting researches by its staff, any endowments of financial aid in excess of immediate needs (and I am confident it will eventually), it would accomplish useful work by offering prizes and granting rewards for important results achieved outside of the institution. The division of rewards would be one of the functions of the council, and it is possible that this would be one of the best uses of such resources, after the success of the laboratory is assured.

The complete role of an ideal aerodynamic laboratory can be summed up now in a few words in the natural order of establishment: (1) Execution of verification tests by means of nominal fees; (2) facilities to technical men for prosecuting original research; (3) execution of researches in accordance with a program arranged by the council, and (4) reward of commendable results accomplished outside of the laboratory.

NATURE OF THE PLANT

Researches and tests can be made on either a large or a small scale, preferably on both.

The use of small models can be made prolific in results because of the comparatively small cost, provided we understand the laws governing transformation into the full sized products. For model work a large plant is unnecessary. M. Eiffel has done very valuable work in a very small establishment.

Certain classes of tests with large models, such, for example, as the block test of propellers, do not require much space. But the conditions are altered when such tests are made on a machine in motion. These more difficult tests are absolutely indispensable and very important to the usefulness of an official laboratory.

Experiments and tests with small models being comparatively inexpensive, private establishments often undertake their execution, but when we attempt to draw conclusions from their results we are obliged to admit that the laws of comparison with full-sized machines are debatable the world over. Comparisons are sensibly true between small surfaces and larger surfaces that have been extended proportionately to the square of the linear dimensions, even to surfaces five or ten times larger, but when we pass to much larger surfaces, as we are obliged to, we are forced to adopt formulas with empirical coefficients, about which there is indefinite dispute.

The difficulty can be overcome only by precise experiments upon large surfaces, and such experiments, whatever the manner in which they are performed, will be costly. If privately executed, the financial returns would not cover the cost.

The laboratory should comprise, therefore, two distinct parts, one devoted to experiments on small-scale models and the other to experiments on surfaces of large dimensions. But in both parts precise and thorough work is necessary.

When we have studied separately each element of an aeroplane, for example, it will be necessary to test the complete apparatus. An aerodrome annex is therefore necessary, or, at least, the laboratory should be located in proximity to an aerodrome of which it can make use. In order that the observations may not only be qualitative but quantitative, it will be necessary to follow all the movements of the complete machine to know at each instant the speed, the inclination, the thrust of the propellers, the effective horsepower, and, in fact, to conduct a true open-air laboratory for aircraft after the manner of certain tests that have been prolific of results in France.

The English have established close relations between the royal aircraft factory and their laboratory, the function of the former being the reconstruction and repair of aeroplanes, the test of motors, and the instruction of mechanics.

LOCATION OF THE LABORATORY

The location of the model-testing plant, the headquarters of the administration staff, requires comparatively small space, and there is no reason why it should be remote from a city or from intellectual and material resources. It is advantageous to have it easy of access to many interested people who are not attached to it.

The location of the open-air laboratory should obviously be at an aerodrome as near as may be convenient to the model-testing plant or headquarters. Close proximity of the two parts is desirable, but not necessary. The high price of land near a large city obliges the aerodrome annex of foreign plants to be located at a distance, but we are fortunate in having here at Washington ideal conditions for the location of both parts. The model laboratory should obviously be located on the site of Langley's notable work at the Smithsonian Institution, where the nucleus, an extensive library of records, and a certain collection of instruments, are still available. The National Museum is also an ideal location for the historical collection of models that will result.

No more ideal location for the annex, the open-air laboratory, or aerodrome exists in all the world than that afforded by the as yet undeveloped extension of Potomac Park. This is Government property which is of doubtful utility as a park only, but which would be of immense utility and interest as a park combined with a scientific plant of the character under consideration.

There is no reason why the public should be excluded from such a practice field, but there is much to recommend that it be open to the public under proper regulations as to the traffic, especially on occasion of certain tests or flights of an educational value. It is of sufficient area, about 1 square mile. It is about 2 miles long, is almost entirely surrounded by broad expanses of water, and, while convenient of access, is so situated that the public may be readily excluded when tests of a dangerous character are in process of execution. The fine driveways that will be required as a park will offer excellent facilities for the practice work of the aerodrome and for the moving test cars that should be supplied.

One of the most attractive features of this location is the advantage it offers as an ideal aerodrome for both the Army and the Navy, for both land and water flying and the opportunity it affords for cooperation in all branches of the work of instruction and experimentation. Furthermore, it is near to the shop facilities of the navy yard, the accommodations of the Washington Barracks, the conveniences of various Government hospitals, and it would doubtless add to the information and interest of the near-by War College Staff and the General Board of the Navy. Its location would enable our statesmen in Congress and a great number of officials in all departments to keep in touch at first hand with the progress of aeronautics, with the quality of the work done, and with the manner in which the money appropriated was being expended. The educational facilities afforded by the work and by the lectures would be invaluable to the course of instruction for Army, Navy, and civil students of aeronautics.

As Washington is a mecca for business people of all parts of the country, a laboratory located here would be convenient in a commercial sense, especially in view of its southerly location, which renders the open aerodrome available for use throughout the greater part of the year. The only objection that I can see to the Potomac Park extension is that the ground will require a considerable clearing, but the trees on the harbor side of the location would not necessarily require removal.

THE APPARATUS NEEDED

It is useless to discuss here the various instruments and methods that have been a source of some dispute abroad. All have some good feature, but time has shown where some of the cumbersome and unnecessary installations may be eliminated to advantage and where others may be improved. The new plant of M. Eiffel, at Auteuil, may be regarded as a model for the wind tunnel and the aerodynamic balance. A duplicate of that plant alone would be of inestimable value. The last volume published by M. Eiffel is a forcible example of the value of his discoveries by this method with respect to the angle of incidence and the displacements of the center of pressure. It seems to merit the utmost confidence, although the details of his installation differ from those at Chalais, at Koutchino, at the Italian laboratory, and others. This method permits of testing the resistance of body structures, the sustaining power of surfaces, the tractive power of propellers, and the influence of transverse or oblique currents. If "free drop" apparatus at uniform speed be regarded as indispensable to obtaining the coefficients of air resistance to solid bodies of different shapes, it is possible that the interior of the Washington Monument could be used to advantage, as was the Eiffel Tower, without disturbance of the main function of that noble structure. This would be an excellent place from which to observe the stability or action of falling models cast adrift at an altitude of 500 feet under varying atmospheric conditions. The free drop of full-sized models would of course require the use of kites or captive balloons.

The moving car previously referred to for tests of verification would be the most useful open-air plant and would soon repay the outlay required by the value of the information obtained from its use. A miniature duplicate of this method for preliminary tests on models with a wire trolley would be of value in a hall of large dimensions. It would be useful in winter work but not invaluable.

The track of the open-air vehicle at St. Cyr is too restricted to give the best results. The car can not circulate continuously at high speed and maintain the speed for a sufficient length of time. An ideal endless track may readily be arranged at the Potomac Park extension, preferably of rectangular form with rounded corners. A railway track would be preferable, but excellent results could be obtained from auto trucks run on macadamized roadbeds. Good results could be obtained by the use of suitable hydroaeroplanes or flying boats suitably equipped with instruments.

At the aerodrome annex ample facilities should be provided for measuring the wind velocity at various heights and at different points. The convenient installation of recording anemometers and the employment of kites or captive balloons should be considered.

A branch of the United States Weather Bureau could readily be established at the aerodrome here in connection with the investigation of meteorological phenomena affecting the movements of aeroplanes in flight and as an adjunct to the national laboratory.

Exactly measured bases and posts of observation are also required, as well as instruments of vision or photographic apparatus, to permit of following machines in their flights and of preserving the records for study.

One of the most useful installations for recording advanced information is an actual aeroplane itself equipped with instruments adapted to record, while in flight, much of the information that is desired. Such machines are already in use in France and in England.

It will be in perfect harmony and convenient to the laboratory to obtain all the services of an aircraft factory from the Washington Navy Yard, where facilities already exist for the reconstruction and repair of aeroplanes, the test of motors, and the instruction of mechanics. But this should not be allowed to interfere with our policy of relying upon private industry for the purchase of new machines, for the sake of encouraging the art among private builders.

It will suffice to merely mention the hangars or sheds required of the local accessories, such as drafting room, office, and minor repair shops. The character and location of these present no difficulties, but they should not be made the principal part of the institution as they are in several elaborately equipped foreign laboratories. The power plant, however, is a subject for careful consideration and the economy effected by M. Eiffel in his new installation at Auteuil is worthy of study.

COST

I have seen estimates varying from $250,000 to $500,000 for such a plant, but inasmuch as $100,000, with an annuity of $3,000 donated by M. Henry Deutsch de la Meurthe to the University of Paris for the establishment of the aeronautical laboratory at St. Cyr, seems to have been sufficient for a very creditable though somewhat deficient plant, I will venture an opinion that $200,000 would be sufficient in our case. Although the same plant would cost more in this country, I assume that some of the buildings required are already available at the Smithsonian Institution. If located elsewhere the cost would be considerably more than the sum named.

A COMMISSION RECOMMENDED

Inasmuch as more definite information regarding the actual cost of a dignified and creditable but modest and sufficient installation should be obtained, and as the details of the plan, the scope, the organization, and the location of such an important undertaking should not be left to the recommendations of one man, I respectfully recommend that a commission or board be appointed to consider and report to the President, for recommendation to Congress, on the necessity or desirability for the establishment of a national aerodynamic laboratory, and on its scope, its organization, the most suitable location for it, and the cost of its installation.

W. Irving Chambers

Document 2-2(b), J. C. Hunsaker, Report on Facilities for Aeronautical Research in England, France, and Germany, Part III—Germany, *undated (ca. November 1913).*

Part III—Germany
(a) Modellversuchsoustalt fur Luftschiffart und Flugtechnik and der Universitat, Göttingen,
Director: Dr. Prandtl.

In 1907, the Motorluftschiff-Studien-Gesellschaft was founded to promote the German air craft industry. Various banks, commercial houses, and industrial works contributed heavily. The parent society later floated stock companies to build the German Wright aeroplane, and the Parseval air ship, and founded aero clubs in the principal cities. Money was allotted to the University of Gottingen for experimental research, and in 1908-9 a laboratory was built there. The director of the laboratory was Dr. Prandtl, head of the department of applied mechanics, at the University. He has given a great part of his time to aeronautical research since that date while still holding his position at the University and delivering lectures there on mathematics and mechanics. He has had only the assistance of his students and a few mechanics. There have usually been two students in the laboratory working for a doctorate. These men remained only a year or two and were replaced by new men. In view of the changing personnel, the excellence and continuity of the research is most remarkable. However, it is not likely that men of the ability of Fuhrmann, Fokke, and others could have been hired as assistants for a reasonable sum.

The Gottingen work is characterized by poor equipment made the most of by men of extraordinary ability.

In June, 1913, the Moroluftschiff-Studien-Gesellschaft had expended its capital and had accomplished its purpose.

In the first place, the German Wright Company has been distanced by its competitors in the aeroplane field, which is proof that the society's object of creating the industry has been attained. In the second place, the Parseval air ship, made by the Luftfahrzeug Gesellschaft, has been successfully developed and is financially profitable. Germany leads the world in air ship building. Accordingly the M-S-G has gone into liquidation and the Gottingen laboratory grant is stopped.

However, model research will be continued by a large grant from the Kaiser Foundation. The old wind tunnel will be abandoned, and a new and more powerful one built. It is expected that new buildings will be erected in Gottingen and that Dr. Prandtl will leave the University to be the director. Dr. Prandtl is now inspecting the aeroplane laboratories of Europe. I met him in Paris at the Eiffel laboratory, and in England at the National Physical Laboratory. Both wind tunnels he considers good. He is not yet ready to announce the type of his new tunnel, but he will not reproduce his old one.

It seems that with a closed circuit great difficulty is ahead in securing a uniform stream of air if high speeds are used.

The closed circuit tunnel at Gottingen gives a uniform stream of air of velocity 10 meters per second but nearly two years have been spent adjusting baffles, screens, and honey combs to attain a good result.

The tunnel and its work are well known to the readers of aeronautic literature. Descriptions will be found in:—Jahrbuch der Motorluftschiff-Studien-Gesellschaft for the years 1908-1910-1911-1912, published by Gustav Braunbeck, Berlin.

Also in—Jahrbuch der Luftfahrt by A. Vorreiter, 1910-11-12, published by J.F. Schmanns, Munich.

Also in—Zeitschrift—fur Flugtechnik und Motorluftschaffahrt, a periodical published by Oldenburg, Berlin.

The balance used by Dr. Prandtl is very delicate, and also very complicated. The best description is given by Vorreiter. Its precision is less than 3/4%.

A special device is used for model propeller testing in connection with the regular balance.

A great deal of work has been done on balloon shapes, aeroplane wings and model propellers.

A most interesting piece of work has been the computation of the head of resistance of a model dirigible by hydrodynamic theory and its verification by experiment. The discrepancy found is considered to be skin friction.

A research on propellers has been made to determine the distribution of pressure over the blade in motion. A model in wax was covered with copper by electro deposit and the wax melted out. The hollow copper model then had holes bored at intervals over the blades. One hole at a time was opened and the internal pressure transmitted through a hollow shaft and slip joint to a manometer.

The form of a Parseval air ship and the arrangement of fins and rudders is the result of model research at Gottingen. As Count Zeppelin is an officer of the M-S-G, considerable work has been done on Zeppelin models but none of it is published.

As the wind tunnel is not to be duplicated in the new laboratory no detail criticism of it is given here. The general statement can be made that an irregular flow is created which must be smoothed out by a multiplicity of obstructions in the channel. That irregularity has been removed is more to the credit of the personnel of the laboratory than to the design of the wind tunnel.

It appears that there is no relation between model research at Gottingen and full scale experiment. The laboratory is devoted primarily to scientific research and at odd times undertakes industrial testing for private firms.

(b) Zeppelin Versuchsanstalt, Friederichshafen.

The Zeppelin company's works are not open to visitors, and I was unable to secure admission. It is generally known that the company has a very complete engine testing plant and is equipped with a complete meteorological station including an outfit of pilot balloons for sounding the upper air levels.

In addition there is machinery for testing the strength of materials of construction.

A large and rough wind tunnel of rather blast of wind is reported to be used to study the stability of route of air ship models.

(c) Technische Hochschule Aachen, Aachen.

Director of laboratory—Prof. Reisswer.

Under Prof. Reisswer at Aachen courses in aeronautics and aerodynamics have been given to students who elected to attend. No design work was given. A large wind tunnel of square section 2 m. by 2 m. has been built. The tunnel is open to the outdoor air at one end and at the other is a powerful exhauster. A velocity of 30 meters per second can be maintained and provided there is no wind outside, the current is very uniform. There are no baffles or honey combs in the channel. A 75 H.P. motor is used.

It is noted that Eiffel obtains 30 meters per second in a 2 meter channel with less than 50 H.P.

When there is any wind outside the air stream is not uniform. Much time is lost waiting for a calm day and an advantage must be taken of still times at dawn and evening.

No expensive balance has been designed, and work is largely confined to propeller testing. The model propeller attached to the axis of a small electric motor is suspended by two wires in a canal. The torque is measured by the difference in pull on the two wires. The thrust is measured by the pull on a third wire led up stream.

There is no permanent staff working with the tunnel and no great amount of work is done at present. This winter one of the assistants from Gottingen will take charge.

(d) Versuchsanstalt für Flugwesen der Kgl. Technischen Hochschule zu Berlin, Spandauer Weg, Reinickendorf-West, Berlin.
Director:—Major Von Parseval.
Asst. Dr. Quittner.

The laboratory is in process of construction under the direction of Maj. Von Parseval, the inventor of the Parseval dirigible.

An office building of 4 rooms has been erected beside an old wooden air ship shed of the Parseval Company.

This shed is 25 by 70 by 22 meters high.

Down the center is built a trestle 5m high and 70m long carrying a track on which a dynamometer carriage runs.

The carriage is drawn by a cable leading over suitable blocks to two sets of sand bags. The first set fall and accelerate the car in a few meters while the second set continue to fall and maintain the car in uniform motion. The weight of sand will be adjusted after trials. Automatic brakes are provided.

The car is some 2m high and is designed to measure lift, drift, and center of pressure of aeroplane wings of span not greater than 10 meters. Forces are measured against hydraulic pressure boxes and recorded graphically on ordinary steam engine indicators from which the springs are removed.

Great precision of measurement is not attempted. It is hoped to study the effect on aeroplane coefficients of change in area.

It is not likely that a high speed can be reached.

A small whirling arm, radius 5.5 meters, is completed. This arm is turned by falling weights. The trouble from a following wind is avoided by giving only two complete turns for a test. Only head resistance forces can be measured. A steam engine indicator drum records the force on the model. A velocity of 12 meters per second is used.

It is expected to study small surfaces to compare their coefficients with those for large surfaces tested on the railway.

The staff consists of the director Major v. Parseval, and his assistant Dr. Quittner, with 3 mechanics.

The courses in aeronautics at the Hochschule include lectures on aeroplane and dirigible construction by Major v. Parseval, with the mathematics and mechanics of flight given by other members of the faculty. No practical design is attempted. Aeronautical studies are optional.

(e) Prof. Dr. Fr. Ahlborn, 23 Ufer Strasse, Hamburg.

Prof. Ahlborn is well known to naval architects for his contributions to "Schiffbau," on stream line flow in water. He first called attention to the inherent stability of the East Indian Zanonia leaf, and was the means of causing Igo Etricn of Vienna to develop the Zanonia form aeroplane.

All of his investigations have been conducted in a tank of water some 3 x 3 x 15 feet long. Models at very low speed were drawn through the water by an electric carriage running above the tank.

He found that fine beech sawdust boiled until it reached the density of water could be used to show stream line phenomena. Various photographic methods were developed which have been described in his papers. At the present time, the Hamburg American Line and the City of Hamburg are to build him an experimental model tank for the further study of fluid motion.

All of Prof. Ahlborn's results are qualitative and serve to illustrate fluid flow at very low speeds. Their application to aeronautics is not apparent.

The Massachusetts Institute of Technology has purchased some 30 slides from Prof. Ahlborn which represent his most important work. Permission is given to copy them provided they are not published.

It seems that work such as Prof. Ahlborn has done should go hand in hand with wind tunnel research. No one has yet been able to delineate the flow of air in a high speed wind tunnel. At the National Physical Laboratory the whole stream becomes cloudy if tobacco smoke is used and the camera only shows a fog. Mr. Eiffel has tried salammoniac vapor without success. At Gottingen cork dust has been tried and abandoned.

(f) Deutschen Versuchsanstalt für Flugwesen, Adlershof, bei Berlin.
Director:—Dr. Bendemann.

Dr. Bendemann conducted the test for the Kaiser Prize for the best German aeroplane motor. This competition showed the necessity for having in Germany a motor testing establishment when impartial comparisons could be made.

The German Society of Mechanical Engineers has now endowed a laboratory of which Dr. Bendemann is to be director.

Five motor testing beds in separate buildings are completed. Five motors can thus be tested at the same time without interference due to noise and flying oil. The main building contains offices and an aeroplane hanger. A tower 100 feet high surmounts it. It is expected to suspend an aeroplane from cables reaching up inside this tower to the top in such a manner that by the method of oscillations the moment of inertia about the various axes may be measured. Work of this nature is not yet commenced.

No wind tunnel will be built.

Experiments will be conducted on full-scale aeroplanes in flight, carrying recording instruments. Work will be started as soon as the proper instruments are completed.

The first experiments will be to send up aeroplanes with a form of recording dynamometer attached to the controls. The force exerted by the pilot and the rapidity of his movements will thus be studied.

Designs are being made for a dynamometer car to run on the railroad and to carry a full size aeroplane. The car will be pushed by a locomotive and high speeds are expected.

There seem to be unlimited funds.

The staff already consists of the Director and 10 assistants.

The laboratory is next to the Johannistahl flying grounds.

[signed] J. C. Hunsaker

Document 2-2(c), Albert F. Zahm, excerpt from
Report on European Aerodynamical Laboratories, 1914.

GENERALITIES

Places visited.—During August and September, 1913, in company with Jerome C. Hunsaker, Assistant Naval Constructor, U. S. N., I visited the principal aeronautical laboratories near London, Paris, and Göttingen, to study, in the interest of the Smithsonian Institution, the latest developments in instruments, methods, and resources used and contemplated for the prosecution of scientific aeronautical investigations. Incidentally we visited many of the best aerodromes (flying fields) and air craft factories in the neighborhood of those cities, and took copious notes of our observations. We also visited many aeronautical libraries, book stores, and aero clubs, in order to prepare a comprehensive list of the best and latest publications on aerial navigation and its immediately kindred subjects. In each of the countries, England, France and Germany, we spent about two weeks. We were made welcome at all the places visited, and thus established personal relations which should be valuable in future negotiations with the aeronautical constructors and investigators in those countries. But these incidental visits and studies, though they may prove serviceable, do not seem germane to the present report. Neither does it seem advisable to take more than passing notice of the aeronautical laboratories themselves in those manifold details which have been already published in large and comprehensive reports now accessible in the Smithsonian Library.

Organization, resources, and scope of the laboratories.—The laboratories examined by us are in particular (1) the aeronautical research and test establishments of the British government near London; (2) the Institut Aerotechnique de St. Cyr, and the Laboratoire Aerodynamique Eiffel, both near Paris; (3) the Göttingen Modelversuchsanstalt at Göttingen; and (4) the newly organized laboratory adjoining the flying field at Johannisthal near Berlin, known as the "Deutsche Versuchsanstalt für Luftfahrt zu Adlershof."

These establishments resemble each other in some important features, but differ in others. All are devoted to both academic and engineering investigations. All are directed by highly trained scientific and technical men. The directors are not merely executives; they are the technical heads—scientists or engineers specifically qualified by superior training in aeronautical engineering and its immediately cognate branches—who initiate the researches, and assist their technical staffs in devising apparatus, interpreting results, and making systematic reports.

The establishments differ in their organization, resources, and equipments, and, to a considerable extent, in the scope and character of their investigations. Of the five institutions mentioned, the one in England and the one at Göttingen are now supported largely by governmental appropriations; and the other three are maintained by private capital, allotted as required, or accruing from fees or endowment funds. Again, the laboratories near London, at St. Cyr, and at Adlershof are practically unlimited in the scope of their researches, while Eiffel's and the Göttingen laboratory have confined their activities substantially to wind-tunnel experiments.

The aeronautical researches of the British Government are in charge of the British Advisory Committee for Aeronautics, a self-governing civilian organization which was appointed by the Prime Minister of England to work out theoretical and experimental problems in aeronautics for the army and navy, and comprises twelve to fourteen expert men, under the presidency of Lord Rayleigh. This committee initiates and directs investigations and tests at the Royal Air Craft Factory, at the National Physical Laboratory, at the Meteorological Office, at Vickers Sons, and Maxim's etc. It expends, in performing its regular functions, a sum exceeding the income of any private aeronautical laboratory and received directly from the government treasury.

The committee is primarily occupied with work for the government, but also performs researches and tests for private individuals, for suitable fees, but without guaranteeing secrecy as to the results. The work of the committee is manifold and comprehensive. Whirling-table measurements, wind-tunnel measurements, testing of engines, propellers, woods, metals, fabrics, varnishes, hydromechanic studies, meteorological observations, mathematical investigations in fluid dynamics, the theory of gyroscopes, aeroplane and dirigible design—whatever studies will promote the art of aircraft construction and navigation may be prosecuted by this committee. A detailed program and the results of actual investigations have been published in the annual report of this committee.

M. Eiffel has paid from his personal fortune all the expenses of his plant and elaborate researches, though it is understood that he may sometimes charge nominal fees for investigations made for private individuals who wish exclusive rights to the data and results obtained. The general director of the laboratory is Eiffel himself—who initiates the researches and publishes the results. He has in immediate charge two able engineers, MM. Rith and Lapresle, aided by three trained observers who are skilled draughtsmen. Two mechanics and one janitor complete the personnel. The work of the laboratory is all indoors, and is confined to researches in aerodynamics alone, or more specifically to wind-tunnel measurements and reports thereon.

The institute at St. Cyr was founded by Deutsch de la Meurthe, who gave $100,000 for the original plant and has provided $3000 per year, during his life, for maintenance. It was presented by him to the University of Paris, and is now under the general direction of the professor of physics, M. Maurain, aided by a technical staff and a large advisory council of eminent engineers, scientists, and officers of the university, officers of the French government, and members of various clubs and aeronautical organizations. The staff comprises the director in charge and his assistant, together with such students, two or three at a time, as may come as temporary volunteers from the University of Paris.

The institute conducts large-scale experiments in the open field as well as indoor researches, makes investigations for general publication or for private interests, on payment of suitable fees, and permits private persons to conduct researches in the laboratory. The work is practically unlimited, as is the case in the English aeronautical laboratories. A special feature of the institute is its three-quarter mile long track with electric cars for tests on large screws, large models, and full-size aeroplanes.

The Göttingen aerodynamical laboratory was begun as a private enterprise, but is now to be enlarged and maintained in part by financial aid of the Kaiser Foundation. The original building, with its wind tunnel, was erected in 1908 after the plans of its director, Prof. Prandtl of the University of Göttingen, at a cost of 20,000 marks supplied by the Motorluftschiff-Studien-Gesellschaft. Its available income is said to be $7,000 a year. The enterprise was inaugurated on a small scale because of the uncertainty, at that date, as to the practical value of such an establishment. The work of this laboratory, as in Eiffel's, has been practically limited to wind-tunnel experiments, though Prof. Prandtl has written some valuable theoretical investigations, and is reported to be undertaking large-scale experiments in the open air by use of a car on a level track, as at St. Cyr.

The Deutsche Versuchsanstalt für Lutftfahrt zu Adlershof has been recently founded by the Verein Deutscher Ingeniure. The laboratory adjoins the great Flugplatz, with its two square kilometer flying field surrounded by numerous air craft factories, scores of hangars, an aero club house, and a grand stand. Major Von Tschudi, a retired German officer, is general manager of the organization which operates the flying field in the interest of all aero manufacturers and experimentalists, whether civilian or governmental. Dr. Eng. F. Bendeman is director of the laboratory and has ten assistants, comprising, among others, Dr. Fuhrman, who was former assistant in the Göttingen laboratory. I have not ascertained the financial resources of the laboratory, but a prelude to its present operations was a competition involving some three score German aeronautical motors, for the Kaiser Prize, and additional contributions from the country at large, aggregating in all 125,000 marks. It is understood that the laboratory is liberally supported, is unlimited in the scope of its work, and will conduct both indoor researches and field experiments similar to those at St. Cyr.

After this general view, a technical account of the foregoing aeronautical establishments may be useful.

Document 2-3(a–c)

(a) U.S. Congress, "An Act Establishing an Advisory Committee for Aeronautics," Public Law 271, 6 3rd Congress, 3rd Session, passed 3 March 1915.

(b) George P. Scriven, "Letter of Submittal" (Chairman's Letter for the first *Annual Report*, 9 December 1915), *NACA Annual Report for 1915* (Washington, DC).

(c) George P. Scriven, "Existing Facilities for Aeronautic Investigation in Government Departments," *NACA Annual Report for 1915* (Washington, DC).

Zahm's *Report on European Aeronautical Laboratories*, published by the Smithsonian Institution in 1914, revealed just how far the United States trailed Europe in aeronautical research. Armed with this new information, the Smithsonian's regents recommended that Congress establish a national advisory committee for aeronautics, based largely on the British Advisory Committee for Aeronautics organization, in February 1915. The navy, content with its own aerodynamical laboratory at the Washington Navy Yard, did little to support the recommendation, but it ended up in an ironic supporting role nevertheless. The text that established the National Advisory Committee for Aeronautics comprised only two paragraphs, and Congress included them in a Naval Appropriations Act that it passed on 3 March 1915. While only $5,000 per year for five years was appropriated, this brief rider set the tone for the organization, stating, "That it shall be the duty of the Advisory Committee for Aeronautics to supervise and direct the scientific study of the problems of flight, with a view to their practical solution, and to determine the problems which should be experimentally attacked, and to discuss their solution and their application to practical questions." Over the next two decades, that small investment in the NACA would grow into several million dollars, and the NACA would build a superb aeronautical laboratory and research program.

Congress defined the NACA to include members from the army, navy, Smithsonian Institution, weather bureau, and bureau of standards, as well as members from the broader aeronautical community. In so doing, it intended to coordinate both military and civilian aeronautical research and development through one agency that was charged with the authority to conduct research in any laboratory that might be assigned to it. During its earliest years, the NACA sponsored research in the Washington Navy Yard tunnel and elsewhere, such as William F. Durand's propeller research at Stanford University, but the committee members soon realized that they needed a comprehensive laboratory and staff that could support all facets of aeronautical research. Indeed, one of the NACA's first tasks was to examine the available facilities in the United States. The 1915 annual report outlined the capabilities and work loads of five government agencies in a section entitled "Existing Facilities for Aeronautic Investigation in Government Departments," and concluded "that utilizing all facilities at present available, the progress that can be made will be fragmentary and at best lack the coordination that is necessary to accomplish . . . the important work now in sight." Armed with this conclusion, the NACA's first chairman, Army Brigadier General George P. Scriven, proposed a joint military-civilian facility, but Navy Secretary Josephus Daniels objected, fearing it could jeopardize the navy's research plans. With its Washington Navy Yard wind tunnel in operation and Jerome Hunsaker involved in research at MIT, the navy was already actively engaged in aerodynamical research in 1915, and Daniels saw no need to change course. Nevertheless, he allowed an $82,516 NACA appropriation, which included $53,580 for construction of a new laboratory, to be attached to a 1916 Naval Appropriations Act. (The following year's Naval Appropriations Act provided the NACA with an additional $107,000, but subsequent NACA funding came from civil appropriations acts.) Given the navy's reticence, the NACA worked with the army to find a site where they could build and share a flying field, and where the NACA could build its laboratory. They settled on a location near Hampton, Virginia, and construction at Langley Field began in April 1917, just as the United States entered World War I.

Document 2-3(a), U.S. Congress, "An Act Establishing an Advisory Committee for Aeronautics," 1915.

Public Law 271, 63d Cong., 3d sess., passed 3 March 1915 (38 Stat. 930).

An Act making appropriations for the naval service for the fiscal year ending June thirtieth, nineteen hundred and sixteen, and for other purposes.

An Advisory Committee for Aeronautics is hereby established, and the President is authorized to appoint not to exceed twelve members, to consist of two members from the War Department, from the office in charge of military aeronautics; two members from the Navy Department, from the office in charge of naval aeronautics; a representative each of the Smithsonian Institution, of the United States Weather Bureau, and of the United States Bureau of Standards; together with not more than five additional persons who shall be acquainted with the needs of aeronautical science, either civil or military, or skilled in aeronautical engineering or its allied sciences: Provided, That the members of the Advisory

Committee for Aeronautics, as such, shall serve without compensation: Provided further, That it shall be the duty of the Advisory Committee for Aeronautics to supervise and direct the scientific study of the problems of flight, with a view to their practical solution, and to determine the problems which should be experimentally attacked, and to discuss their solution and their application to practical questions. In the event of a laboratory or laboratories, either in whole or in part, being placed under the direction of the committee, the committee may direct and conduct research and experiment in aeronautics in such laboratory or laboratories: And provided further, That rules and regulations for the conduct of the work of the committee shall be formulated by the committee and approved by the President.

That the sum of $5,000 a year, or so much thereof as may be necessary, for five years is hereby appropriated, out of any money in the Treasury not otherwise appropriated, to be immediately available, for experimental work and investigations undertaken by the Committee, clerical expenses and supplies, and necessary expenses of members of the committee in going to, returning from, and while attending meetings of the committee: Provided, That an annual report to the Congress shall be submitted through the President, including an itemized statement of expenditures.

Document 2-3(b), George P. Scriven, Chairman's Letter for the first NACA Annual Report, 1915.

LETTER OF SUBMITTAL.
NATIONAL ADVISORY COMMITTEE FOR AERONAUTICS,
STATE, WAR, AND NAVY BUILDING
Washington, D. C., December 9, 1915.

The PRESIDENT:

In compliance with the provisions of the act of Congress approved March 3, 1915 (naval appropriation act, Public, No. 273, 63d Cong.), the National Advisory Committee for Aeronautics has the honor to submit herewith its annual report for the period from March 3, 1915, to June 30, 1915, including certain recommendations for future work and a statement of expenditures to June 30, 1915.

The committee was appointed by the President on April 2, 1915, and held its first meeting for organization on April 23, 1915. On June 14 the President approved rules and regulations which had been formulated by the committee for the conduct of its operations.

By the act establishing the committee, an appropriation of $5,000 a year for five years was made immediately available. Of the appropriation for the first year, ending June 30, 1915, there was expended a total of $3,938.94, as shown by the itemized statement in the accompanying report, and the unobligated balance of

$1,061.06 was covered into the Treasury as required by law.

In order to carry out its purposes and objectives, as defined in the act of March 3, 1915, the committee submits herewith certain recommendations and an estimate of expenses for the fiscal year ending June 30, 1917. The estimates in detail were submitted through the Secretary of the Navy.

Attention is invited to the appendixes of the committee's report, and it is requested that they be published with the report of the committee as a public document.

It is apparent to the committee that there is a large amount of important work to be done to place aeronautics on a satisfactory foundation in this country. Competent engineers and limited facilities are already available and can be employed by the committee to advantage, provided sufficient funds be placed at its disposal, as estimated for the fiscal year 1917.

What has been already accomplished by the committee has shown that although its members have devoted as much personal attention as practicable to its operations, yet in order to do all that should be done, technical assistance should be provided which can be continuously employed. There are many practical problems in aeronautics now in too indefinite a form to enable their solution to be undertaken. The committee is of the opinion that one of the first and most important steps to be taken in connection with the committee's work is the provision and equipment of a flying field together with aeroplanes and suitable testing gear for determining the forces acting on full-sized machines in constrained and in free flight, and to this end the estimates submitted contemplate the development of such a technical and operating staff, with the proper equipment for the conduct of full-sized experiments.

It is evident that there will ultimately be required a well-equipped laboratory specially suited to the solving of those problems which are sure to develop, but since the equipment of such a laboratory as could be laid down at this time might well prove unsuited to the needs of the early future, it is believed that such provision should be the result of gradual development.

The investigations which the committee proposes in its program for the coming year can only be carried out to a satisfactory degree, with the limited facilities already existing, provided sufficient funds are made available. The estimates of the committee are based on such line of action, and on the assumption that a flying field can be placed at its disposal on Government land. If, however, such facilities be not practicable at this time, some progress may still be made by the utilization of the facilities of the Government aeronautic stations at Pensacola and San Diego.

The estimate of expenses for the fiscal year ending June 30, 1917, is as follows:

For carrying into effect the provisions of the act approved March third, nineteen hundred and fifteen, establishing a National Advisory Committee for Aeronautics, there is hereby appropriated, out of any money in the Treasury not otherwise appropriated, for experimental work undertaken by the committee, assistants and the necessary unskilled labor, equipment, supplies, office rent, and the necessary traveling expenses of the members and employees of the committee, personal services in the field, and in the District of Columbia: Provided, That an annual report to the Congress shall be submitted through the President, including an itemized statement of expenditures, $85,000.

The committee, therefore, submits its report, recommendations, and estimates to your favorable consideration.

Very respectfully,
George P. Scriven,
Brigadier General, Chief Signal Officer of the Army,
Chairman.

Document 2-3(c), George P. Scriven, "Existing Facilities for Aeronautic Investigation in Government Departments," NACA Annual Report for 1915.

For the conduct of the work outlined, limited facilities already exist in different Government departments about as described in general terms in the following. These facilities can be augmented by the facilities described as existing the different technical institutions, etc., previously referred to:

A. The Bureau of Standards is well equipped for carrying on all investigations involving the determination of the physical factors entering into aeronautic design, and is prepared to take up such matters as are of sufficient general interest to warrant same.

B. The Navy Department is equipped with a model basin and wind tunnel at the Washington Navy Yard, with adequate shop facilities for carrying on the work in a limited way, and is also constructing at the Washington Navy Yard a plant for the testing of aeronautic motors and devices involved in their operation, which will be in commission at an early date. Also, under the Navy Department steady progress is being made in attacking practical problems involved in the development of the Navy aeronautic service at its station at Pensacola, and theoretical and practical designs are in hand in the Bureaus of Construction and Repair and Steam Engineering.

C. The War Department has limited facilities at the flying school at San Diego, for investigations of interest to that branch of the service, and is able to carry out in a limited way experiments of interest to the service on full-sized machines, for which work it has the assistance of technical experts.

D. The Weather Bureau is well equipped for the determination of the problems of the atmosphere in relation to aeronautics, and Prof. Marvin, a member of the advisory committee, is the chairman of a subcommittee engaged on this problem.

The work, however, will necessarily be limited until the necessary funds for more extensive work become available. There is already available in the records of the bureau much information of value which requires compilation in a form suited to aeronautic requirements, and this work is the subject of a preliminary report included in the annual report of the committee.

E. The Smithsonian Institution has been engaged for a number of years on the compilation of the bibliography of aeronautics, and is prepared to continue this work for at least two years more with the funds at its disposal. The institution has also contributed funds toward the development of the work of the subcommittee of the Weather Bureau in its investigation of the problem of the atmosphere in relation to aeronautics.

CONCLUSIONS.

From the above, it will be apparent that utilizing all facilities at present available, the progress that can be made will be fragmentary and at best lack that coordination which is necessary to accomplish in a direct, continuous, and efficient manner, and as rapidly as practicable, the important work now in sight. If the committee is to be prepared to keep pace with the increasing needs of the very rapid development already under way, stimulated by the unusual conditions existing in Europe, the facilities and technical assistance recommended are essential. While the needs at present are principally those which have an important bearing on military preparedness, the committee is of the opinion that aeronautics has made such rapid strides that when the war is over there will be found available classes of aircraft and a trained personnel for their operation, which will rapidly force aeronautics into commercial fields, involving developments of which today we barely dream.

Respectfully submitted.

George P. Scriven,

Brigadier General, Chief Signal Officer of the Army,
Chairman.

Document 2-4

Jerome C. Hunsaker, excerpt from "The Wind Tunnel of the Massachusetts Institute of Technology," *Reports on Wind Tunnel Experiments in Aerodynamics* (Washington, DC: Smithsonian Institution, 1916).

The Massachusetts Institute of Technology became interested in aeronautical research and the idea of creating a course of study in aeronautical engineering in 1913—about the same time the navy began to pursue aeronautical research. MIT President Richard Maclaurin approached the navy about a cooperative effort and asked that Jerome Hunsaker, who had taken a masters degree at MIT in 1909, be assigned to the school for three years. The navy agreed, and Hunsaker reported to the school early in 1914, but only after his return from a previously scheduled European inspection trip with Albert Zahm.

While visiting the National Physical Laboratory in England during that trip, Hunsaker obtained plans for the NPL's four- by four-foot open tunnel and its latest balance instrument. Using the NPL plans, Hunsaker and MIT erected a duplicate tunnel on campus in 1914 and began experimental work. Two years later, the Smithsonian Institution published *Reports on Wind Tunnel Experiments in Aerodynamics*, written by Hunsaker and five other MIT professors. One of the earliest comprehensive reports on organized American aeronautical research, it describes the tunnel and some fundamental research being conducted. More importantly, it shows a degree of cooperation on the working level between several government agencies—cooperation that was not readily apparent at the executive level as the debate over a national aeronautical laboratory peaked.

Document 2-4, Jerome C. Hunsaker, excerpt from
"The Wind Tunnel of the Massachusetts Institute of Technology," 1916.

An aeroplane or airship in flight has six degrees of freedom, three of translation and three of rotation, and any study of its behavior must be based on the determination of three forces—vertical, transverse, and longitudinal—as well as couples about the three axes in space. Full-scale experiments to investigate the aerodynamical characteristics of a proposed design naturally become mechanically difficult to arrange. The experimental work is much simplified if tests be made on small models as in naval architecture, and a further simplification is made by holding the model stationary in an artificial current of air instead of towing the model at high speed through still air to simulate actual flying conditions.

The use of a wind tunnel depends on the assumption that it is immaterial whether the model be moved through still air or held stationary in a current of air of the same velocity. The principle of relative velocity is fundamental, and the experimental discrepancies between the results of tests conducted by the two methods may be ascribed on the one hand to the effect of the moving carriage on the flow of air about the model and to the effect of gusty air, and on the other hand to unsteadiness of flow in some wind tunnels.

The wind tunnel method requires primarily a current of air which is steady in velocity both in time and across a section of the tunnel. The production of a steady flow of air at high velocity is a delicate problem, and can only be obtained by a long process of experimentation. A study was made of the principal aerodynamical laboratories of Europe from which these conclusions were reached: (1) That the wind tunnel method permits a leisurely study of the forces and couples produced by the wind on a model; (2) that the staff of the National Physical Laboratory, Teddington, England, have developed a wind tunnel of remarkable steadiness of flow and an aerodynamical balance well adapted to measure with precision the forces and couples on a model in any position; and (3) that the results of model tests made at the above laboratory are applicable to full scale aircraft.

Consequently it was decided to reproduce in Boston the four-foot wind tunnel of the National Physical Laboratory, together with the aerodynamical balance and instruments for velocity measurement. Dr. R. T. Glazebrook, F. R. S., director of the National Physical Laboratory, most generously presented us with detail plans of the complete installation, including the patterns from which the aerodynamical balance was made. Due to this encouragement and assistance we have been able to set up an aerodynamical laboratory with confidence in obtaining a steady flow of air of known velocity. The time saved us by Dr. Glazebrook, which must have been spent in original development, is difficult to estimate.

The staff of the National Physical Laboratory have developed several forms of wind tunnel in the past few years. In 1912–13 Mr. Bairstow and his assistants conducted an elaborate investigation into the steadiness of wind channels as affected by the design both of the channel and the building by which it is enclosed. The conclusions reached may be summarized as follows:

(1) The suction side of a fan is fairly free from turbulence.

(2) A fan made by a low pitch four-bladed propeller gives a steadier flow than the ordinary propeller fan used in ventilation, and a much steadier flow than fans of the Sirocco or centrifugal type.

(3) A wind tunnel should be completely housed to avoid effect of outside wind gusts.

(4) Air from the propeller should be discharged into a large perforated box or diffuser to damp out the turbulent wake and return the air at low velocity to the room.

(5) The room through which air is returned from the diffuser to the suction end of the tunnel should be at least 20 times the sectional area of the tunnel.

(6) The room should be clear of large objects.

The wind tunnel of the Institute of Technology was built in accordance with the English plans, with the exception of several changes of an engineering nature introduced with a view to a more economical use of power and an increase of the maximum wind speed from 34 to 40 miles per hour.

Upon completion of the tunnel an investigation of the steadiness of flow and the precision of measurements was made in which it appeared that the equipment had lost none of its excellence in reproduction in the United States.

As will be shown below, the current is steady both in time and across a cross-section within about 1 per cent in velocity. Measurements of velocity by means of the calibrated Pitot tube presented by the National Physical Laboratory are precise to one-half of 1 per cent. Force and couple measurements on the balance are precise to one-half of 1 per cent for ordinary magnitudes. Calculated coefficients which involve several measurements of force, moment, velocity, angle, area, and distance, as well as one or more assumptions, can be considered as precise to within 2 per cent. It is believed that it is not practicable to increase the precision of the observations to such an extent that the possible cumulative error shall be materially less than the above.

DESCRIPTION OF WIND TUNNEL

A shed 20 by 25 by 66 feet houses the wind tunnel proper, 16 square feet in section, and some 53 feet in length (pl. 1). Air is drawn through an entrance nozzle and through the square tunnel by a four-bladed propeller, driven by a 10 H. P. motor. Models under test are mounted in the center of the square trunk on the vertical arm of the balance to be described later.

The air entering the mouth passes through a honeycomb made up of a nest of 3-inch metal conduit pipes 2 feet 6 inches in length. This honeycomb has an important effect in straightening the flow and preventing swirl.

Passing through the square trunk and past the model, the air is drawn past a star-shaped longitudinal baffle into an expanding cone. In this the plans of the National Physical Laboratory were departed from by expanding in a length of 11 feet to a cylinder of 7 feet diameter. This cone expands to 6 feet in the English tunnel. M. Eiffel affirms that the working of a fan is much improved by expanding the suction pipe in such a manner as to reduce the velocity and so raise the static pressure of the air. Since the fan must discharge into the room, the pressure difference that the fan must maintain is thus reduced. Also with a larger fan the velocity of discharge is reduced, and the turbulence of the wake kept down.

The propeller works in a sheet metal cylinder 7 feet in diameter, and discharges into the large perforated diffuser. The panels of the latter are gratings and may be interchanged fore and aft. The gratings are made of 1 ½-inch stock with holes 1 ½ by 1 ½ inches. Each hole is then a square nozzle one diameter long. The end of the diffuser is formed by a blank wall. The race from the propeller is stopped by this wall and the air forced out through the holes of the diffuser. Its velocity is then turned through 90 degrees. The area of the diffuser holes is several times the sectional area of the tunnel, and the holes are so distributed that the outflow of air is fairly uniform and of low velocity (pl. 2, fig. 1).

A four-bladed black walnut propeller (pl. 2, fig. 2) was designed on the Drzwiecki system and has proved very satisfactory. In order to keep down turbulence a very low pitch with broad blades had to be used. To gain efficiency such blades must be made thin. It then became of considerable difficulty to insure proper strength for 900 R. P. M. as well as freedom from oscillation.

The blade sections were considered as model aeroplane wings and their effect integrated graphically over the blade. The blade was given an angle of incidence of 3 degrees to the relative wind at every point for 600 R. P. M. and 25 miles per hour. The pitch is thus variable radially.

To prevent torsional oscillations, the blade sections were arranged so that the centers of pressure all lie on a straight line, drawn radially on the face of the blade.

Fig. 2. ENTRANCE NOZZLE, SHOWING END OF HONEYCOMB

This artifice seems to have prevented the howling at high speeds commonly found with thin blades. The propeller has a clearance of ½ inch in the metal cylinder.

The propeller is driven by a "silent" chain from a 10 H. P. interpole motor beneath it. The propeller and motor are mounted on a bracket fixed to a concrete block and are independent of the alignment of the tunnel. Vibration of the motor and propeller can not be transmitted to the tunnel as there is no connection.

The English plans for power contemplate a steady, direct current voltage. Such is not available here. A 15 H. P. induction motor is connected to the mains of the Cambridge Electric Light Company. This motor then turns at a speed proportional to the frequency of the supply current for a given load. Fluctuations of voltage are without sensible effect, and the frequency may be taken as practically constant.

The induction motor is directly connected to a 12 H. P. direct current generator, which is turned at constant speed and which generates, therefore, a constant direct current voltage for given load.

By change of the generator field rheostat and motor field rheostat the propeller speed can be regulated to hold any wind velocity from 4 to 40 miles per hour. The control is very sensitive. Left to itself, the speed of the wind in the tunnel will vary by 2 per cent in 2 or 3 minutes. This variation is so slow that by manipulation of the rheostats the flow can be kept constant within ½ per cent. The cause of the surging of the air is not understood, but is probably due to hunting of the governor of the prime mover in the Cambridge power house causing changes in frequency too small to be apparent. The gustiness of outdoor winds seems to have no effect, although the building is not airtight.

AERODYNAMICAL BALANCE

The aerodynamical balance (pl. 3) was constructed by the Cambridge Scientific Instrument Company, England, to the plans and patterns of the National Physical Laboratory. The balance is described in detail by Mr. L. Bairstow in the Technical Report of the Advisory Committee for Aeronautics, London, 1912–13. For details of operation and the precision of measurements reference may be made to the original article.

In general, the balance consists of three arms mutually at right angles representing the axes of coordinates in space about and along which couples and forces are to be measured. The model is mounted on the upper end of the vertical arm which projects through an oil seal in the bottom of the tunnel.

The entire balance rests on a steel point, bearing in a steel cone. The point is supported on a cast-iron standard secured to a concrete pillar, which in turn rests on a large concrete slab. The balance is then quite free from vibration of the floor, building, or tunnel.

The balance is normally free to rock about its pivot in any direction. When wind blows against the model, the components of the force exerted are measured by determining what weights must be hung on the two horizontal arms to hold the model in position. Likewise the balance is free to rotate about a vertical axis through the pivot. The moment producing this rotation is balanced by a calibrated wire with graduated torsion head.

Force in the vertical axis is measured by means of a fourth arm. The model for this measurement is mounted on a vertical rod which slides freely on rollers inside the main vertical arm of the balance. The lower end of this rod rests on one end of a horizontal arm having a knife edge and sliding weight.

For special work on moments, the interior vertical rod is replaced by another having a small bell crank device on its head which converts a moment about the center of the model into a vertical force to be measured as above (pl. 4).

In this way provision is made for the precise measurement of the three forces and the three couples which the wind may impress on any model held in any unsymmetrical position to the wind.

The balance is fitted with suitable oil dash pots to damp oscillations, and devices for limiting the degrees of freedom to simplify tests in which only one or two quantities are to be measured. The balance can be adjusted to tilt for 1/10,000 pound force on the model. In general, the precision of measurements is not so good as the sensitivity, and in the end is limited by the steadiness of the wind and the skill of the observer.

The weights and dimensions of the balance were verified by the National Physical Laboratory, where also the torsion wires were calibrated.

For ordinary forces, weighings may be considered correct to 0.5 per cent. Naturally for very small forces, such as the rolling moment caused by a small angle of yaw, the measurements can not be so precise.

[*Not included here are the final sections of Hunsaker's report, concerning the alignment of the wind tunnel and the means by which tunnel instrumentation measured air velocity.*]

Document 2-5

G. I. Taylor, "Pressure Distribution Over the Wing of an Aeroplane in Flight," British Advisory Committee for Aeronautics, *Reports and Memoranda*, No. 287 (London, 1916).

Early airplane designers found themselves confronting a paradox. They could obtain repeatable data on new designs using models in wind tunnels—something almost impossible to achieve in flight—but the forces and moments measured on wind tunnel models did not agree with what an actual airplane experienced in flight. While many recognized the paradox, only a few researchers even approached a possible solution to it prior to the introduction of dynamical similarity concepts in the 1920s. Gustav Eiffel grappled with this problem around 1910, using what he called "augments," developed from his experiments, to adjust model data and predict full-size performance. Although Eiffel concluded that "the calculations are in each case in complete accord with the actual conditions observed in flight," his augments were not sufficiently accurate to be universally applicable. With other wind tunnel work, however, Eiffel showed how measurements of the air pressure distribution across the surface of a wing could be correlated to the lift and drag generated by the wing. This work pointed the way toward a useful design technique, but much more work was needed to correlate performance data of models and full-size airplanes before the technique could be fully exploited.

A major problem in determining this correlation was obtaining useful data from airplanes in flight. No one knew how to measure propeller performance and efficiency accurately, so drag could not be directly determined from engine performance. G. I. Taylor reasoned that Eiffel's pressure-distribution techniques would work on an airplane in flight. He devised a rib with pressure taps that could be mounted in a wing along with a clever means to measure and record pressure-distribution data in flight. Taylor still found that a correction, which he believed was due to skin friction, was necessary, but his device worked, and it produced lift and drag coefficient curves of a similar shape to those obtained from wind tunnel models at all but the lowest lift coefficients.

Document 2-5, G. I. Taylor, "Pressure Distribution Over the Wing of an Aeroplane in Flight," 1916.

SUMMARY.—The comparison of the results of experiments performed with model aerofoils in wind channels with the actual performance of an aeroplane in flight has proved extremely difficult. It has been possible to find the relationship between angle of incidence and lift coefficient for a full-scale machine, but when attempts are made to find the resistance of the wing of an aeroplane in flight, various causes combine to diminish the accuracy of the result. In the first place the thrust of the airscrew is the subject of considerable uncertainty, for it depends on the power of the engine and on the efficiency of the airscrew, neither of which are known accurately. If the thrust of the airscrew were known, the total resistance of the aeroplane would be known, but considerable uncertainty would still exist as to what proportion of the total resistance is due to the wings.

In spite of these various sources of error, there is strong evidence to show that the resistance of a full-scale aeroplane differs from the predictions made as a result of model experiments.

The measurements here described were undertaken with a view to getting further and more direct information as to the relationship between model and full-scale aerofoils. Simultaneous measurements of the pressures at various points round a certain section of the lower wing of a B.E.2C aeroplane were made by means of an apparatus which registered photographically the heights of the liquid in a number of manometer tubes. Observations were taken at various speeds ranging from 50 to 95 miles per hour.

The pressures were integrated, and the normal and longitudinal forces and moments, which act on the wing in the experimental section, were found.

The inclinometer measurements of the angle of incidence corresponding with various lift coefficients were then used to find the drag coefficient of the section.

The results are shown in Fig. 12. For the purpose of comparison the drag curves obtained from model experiments and also from the performance tests of full-scale aeroplanes, are shown in the same figure. It will be seen that the present measurements agree with the performance tests over the greater part of the curves, but that for small lift coefficients the pressure-integration curve indicates a higher drag than the performance tests. The drag curve for the model lies below the pressure-integration curve over its whole range, but on the other hand, the general shapes of the two curves are very similar.

It will be possible to say more about the relationship between the present measurements and the model results when the pressure distribution around a similar section of a model has been measured. This work is now being carried out at the National Physical Laboratory.

The pressure distribution has also been used to obtain the moment coefficient. A curve showing the moment coefficients for various lift coefficients is given. (Fig. 11)

1. The section chosen for these experiments was the old B.E.2C section with a hollow undersurface. A brass rib was made to drawing, and eighteen small copper tubes were fixed to it in the positions shown in Fig. 1. These were filled at the

ends with brass plugs, through which holes 1/32nd inch diameter were drilled. These holes were flush with the surface of the wing.

2. The holes were numbered α, β, $1a$, 2, 3, 4, 5, 6, 7, 8 on the upper surface, and 9, 10, 11, 12, 13, 14, 15 and $1b$ on the lower surface. There was also a hole numbered (1) in the leading edge of the wing. The positions of the holes are given in terms of co-ordinates in Table (1), the axis of x being along the chord and the axis of y the perpendicular through the leading edge. x and y are expressed as fractions of the chord, so that x varies from 0 to 1.

3. The pressures at the holes were measured simultaneously by means of an apparatus shown in Fig. 2, which consisted of twenty glass manometer tubes, connected at their lower ends to a reservoir which was full of alcohol. Their upper ends were connected by means of rubber tubes with the pipes which led through the interior of the wing to the pressure holes. When the aeroplane was at rest the alcohol stood about half-way up the tubes.

4. When the aeroplane was in flight the height of the liquid in the tubes connected with the various holes registered the pressure at the holes. These heights were recorded photographically. A small electric light, some ten inches from the tubes, was used to cast shadows of the columns of liquid on to some bromide paper, which could be wound between two light-tight boxes past the tubes. The bromide paper was pressed up against the tubes by means of a back (shown leaning up against the apparatus in Fig. 2), and the whole apparatus was shut in by a hinged light-tight door (shown open in Fig. 2). To make an exposure, the small electric light was switched on for half a second.

5. The shadow of the meniscus at the top of the alcohol was very sharp, and it was possible to read the height of the liquid in the tube to $\frac{1}{10}$th of a millimetre. A specimen of the records obtained is shown in Fig. 3.

6. On testing the apparatus, it was found that the maximum error in measuring pressure by means of the photographic records was 0.4 mm. of alcohol, and the probable error was a little more than 0.1 mm. It was found, however, that the error for a given position of the liquid in a tube was always the same. A table of errors was therefore constructed, and by using this table the pressures could be obtained correct to 0.1 mm. of alcohol.

7. The whole apparatus was hung in the aeroplane by rubber suspensions, so that it might not be affected by the vibration of the aeroplane.

8. As a further preventive to possible rapid movements of the alcohol in the tubes the rubber pipes, which served to connect them with the alcohol reservoir, were constricted by passing them through a series of holes in a brass rod. The sizes of these holes were adjusted by means of screws, so that the time taken by the alcohol to come to rest after a sudden change in speed of the aeroplane was about fifteen seconds.

9. In order to measure the height of the meniscus at the top of the fluid in each tube, it was necessary to know what direction in the plane of the photograph was horizontal. For this purpose the two outside tubes in the apparatus were both connected with the static pressure side of the pitot tube. A straight line was drawn on each photograph touching the two outside meniscuses, and this was used as a base line from which the pressures in the other tubes were measured. This line is shown in Fig. 3.

10. The aeroplane had a constant weight of 2,020 lbs. during the experiments; the lift coefficient and angle of incidence should therefore depend only on the reading of the pitot tube. For a given angle of incidence the pressure at each of the holes should be a given fraction of ρV^2, r being the density of the air and V the velocity of the aeroplane.

11. Hence, to each reading of the pitot tube there should correspond a definite pressure at each hole, whatever the density of the air might be. This was specially convenient because observations could be taken at whatever height the air happened to be calmest, without the necessity of considering its density.

12. One of the central tubes was connected with the high pressure side of the pitot tube, and the reading of the liquid in this was used instead of an air speed indicator to obtain an accurate measure of the airspeed of the machine.

13. In order to find what value of ρV^2 corresponded with a given pitot reading, it was necessary to fly the machine at known speeds through air of known density, and to take the corresponding readings of the liquid in the pitot tube. This was done on a speed course of known length, and it was found that at 60 and at 80 miles per hour the pressure difference between the static pressure tube and the pitot tube was $0.475 \rho V^2$. When a pressure head is placed facing the wind in a wind channel, this pressure difference is $\frac{1}{2} \rho V^2$. The discrepancy between the two seems to be due to the disturbing effect of the wings of the machine, which probably increases the pressure in the static pressure tube without affecting the pressure in the pitot head.

14. The pressure at each hole is expressed in all cases as a fraction of ρV^2. This fraction is found by multiplying the ratio of the head of liquid in the tube connected with the hole to the head in the tube connected with the pitot tube by 0.475. The readings for different air speeds of the aeroplane are shown graphically in Figs. 4, 5, and 6. In these diagrams each hole, as numbered on the right-hand side of each curve, is represented by a special mark, and smooth curves have been drawn through the points in order to eliminate accidental errors as far as possible. The magnitude of the accidental errors can be estimated by noticing the distances of the observed points from the smooth curves. In the case of hole 2, for instance, the accidental error which may be expected is about $0.003 \rho V^2$, while in the case of holes 10, 1b, and 15 it is much greater, being about $0.01 \rho V^2$.

15. In the case of hole 10 the source of error was detected. It was found that the fabric was inclined to come away from the rib in the immediate neighbourhood of this hole, so that the pressure recorded was probably intermediate between the pressure outside and the pressure inside the wing. In the cases of holes 1b and 15 it seems probable that the errors are due to the peculiar conditions of air flow, which seem to exist immediately underneath a turned-down leading edge.

16. *Pressure integration.*—Having determined the pressure at each hole as accurately as seems possible from present observations, it remains to estimate by integration the forces which act on the aerofoil in the neighbourhood of the section chosen for the pressure distribution experiments.

17. The integrations were effected graphically by the method used by Jones and Patterson in the case of a model aerofoil. Curves were plotted for various speeds of the aeroplane, ranging from 50 to 95 miles per hour by intervals of 5 miles per hour, showing the pressure round the aerofoil as a function of x and as a function of y. These curves, when integrated by means of a plantimeter give the normal and longitudinal force coefficients.

18. Curves were also plotted showing the relationship between (pressure multiplied by x) and x. These curves give the part of the moment coefficient due to the component of pressure perpendicular to the chord. The part of the moment coefficient due to the component parallel to the chord is so small as to be negligible.

19. The points on the pressure curves corresponding to the various holes are taken from the curves of Figs. 4, 5, and 6. In drawing the pressure curves shown in Figs. 7, 8, and 9, it will be seen that it is possible to vary their forms to a certain extent, while at the same time keeping them to the determined points. When the first set of experiments was made there were no pressure holes between 1 and 1a. Under these circumstances it was found that very considerable variations in the areas of the curves in Fig. 8 were possible; accordingly the holes a and b were constructed. By making the curves pass through the points corresponding with these two new holes, and bearing in mind certain geometrical limitations, which the curves must comply with, it was found that the differences in their area, due to various possible ways of drawing the curves were not large enough to affect the results.

20. The geometrical limitations alluded to above are obvious. In the case of the normal force and moment curves of Figs. 7 and 9, the curves must lie between $x = 0$ and $x = 1$, and the values of x must increase steadily in going along the curve from the point corresponding with the nose to the point corresponding with the trailing edge. In the case of the longitudinal force curves of Fig. 8, the values of y must increase steadily along the curves from $y = 0$ to $y = 0.0793$.

21. Besides these geometrical limitations there is a hydro-dynamical one. The pressure can not be greater, at any point on the wing, than $\frac{1}{2} \rho V^2$ above the pressure in the surrounding undisturbed air; and, moreover, it appears certain that, at

some point near the nose of the wing, the pressure will attain this value. It seems probable that the pressure in the pitot tube is $\frac{1}{2}\rho V^2$. In this case the pressure in the static pressure tube would be $\frac{1}{2}\rho V^2 - 0.475\rho V^2$, or $0.025\rho V^2$, above the pressure in the surrounding air; and since the pressures measured by the instrument are the differences between the pressures in the holes and the pressure in the static pressure tube, the maximum possible pressure on the diagrams should be $0.475\rho V^2$. On looking at Figs. 4 and 6 it will be seen that the point of maximum pressure, which presumably corresponds with the point where the air divides to pass over and under the wing, crosses hole 1, in the nose, when the aeroplane is flying at 53 miles per hour, while at about 80 miles an hour it crosses hole a. At the highest speed attained—95 miles per hour—it had not crossed the next hole, β.

22. The normal force, longitudinal force, and moment coefficients derived by integrating the curves are given in the 6th, 7th and 8th columns of Table (2). They are denoted by the, symbols $[k_z]_s$, $[k_x]_s$, and $[k_m]_s$, the "s" being used to show that the coefficients apply to the experimental section only. In order to get the lift and drag coefficients of the section explored, it is necessary to find the angle of incidence of the section. The aeroplane was measured and it was found that the angle of incidence of the wing in the neighbourhood of the experimental rib was 18´, or 0.3° less than the mean angle of incidence. The mean angle of incidence has been measured for different mean lift coefficients by means of an inclinometer, and the relationship between them is shown graphically in Fig. 10.

23. The mean lift coefficients for various speeds were obtained by dividing the weight of the aeroplane by ρV^2 and by the area of the wings. The weight of the aeroplane was 2,020 lbs. The area of its wings was 384 square feet. Hence if V be expressed in miles per hour, the mean lift coefficient

The mean angle of incidence, θ, was found from the curve in Fig. 10. The angle of incidence of the experimental rib was 0.3° less than this. These angles are tabulated for various speeds in column 5, Table (2).

24. The lift and drag coefficients of the section, $[k_l]_s$ and $[k_d]_s$, are related to $[k_z]_s$ and $[k_x]_s$ by the relations

$[k_l]_s = [k_z]_s \cos(\theta - 0.3°) - [k_x]_s \sin(\theta - 0.3°)$,

$[k_d]_s = [k_x]_s \cos(\theta - 0.3°) - [k_z]_s \sin(\theta - 0.3°)$,

They are tabulated in columns 9 and 10 of Table (2).

25. In order to find the "scale effect" between the forces on the experimental section and those on the same section of a model, curves of the lift and drag coefficients of the section should be drawn for various angles of incidence. On the other hand, in order to compare the drag of the section with the drag of the whole wing,

as calculated from the performance of an aeroplane, it is in some ways more convenient to draw curves representing the drag of the section for different values of the lift coefficient of the whole machine. For this purpose it is necessary to correct the lift coefficient of the whole machine to what it would be if the mean angle of incidence were the same as that of the experimental section. This is done by taking the lift coefficient corresponding with the angle of incidence of the experimental section from the curve in Fig. 10 instead of the actual mean lift coefficient given in column 2 of Table (2). These corrected lift coefficients are given in column 4.

26. *Results.*—The results have been plotted in the form of curves representing the lift, drag, and moment coefficients of the experimental section for various values of the lift coefficient of the whole aeroplane. These curves are shown in Figs. 11 and 12. In the case of the drag curve it must be remembered that the drag obtained from pressure integration, shown as the full curve in Fig. 12, is less than the actual drag because no account is of tangential forces (*i.e.*, skin friction) in pressure integration. The effect of skin friction may be allowed for roughly by adding 0.0035 to the drag coefficient. After making this allowance the final drag curve resulting from these experiments is shown as curve B, Mg. 12.

27. For the purpose of comparison the drag curves obtained from model experiments (curve D) and also from the performance tests of full-scale aeroplanes (curve C) are shown in the same figure. It will be seen that pressure distribution experiments indicate a higher drag than that obtained by either of the other methods, but it must be remembered that curve B represents the drag on a certain section of the lower plane of a biplane, while curves C and D apply to the mean drag of both wings. It is quite possible also that the drag at the ribs is greater than the drag at the intermediate points of the wing where the fabric sags below the true wing section; this is not probable, however, in view of the fact that model tests show that the drag of a scalloped wing is the same as that of a smooth wing.

28. It is not possible to compare the moment coefficient of a model with $[k_m]_s$ because no measurements of the forces acting on a biplane fitted with B.E.2C section have been made. On the other hand it has been found that the moment curve for a single B.E.2C section wing is practically identical with the moment curve for R.A.F. 6. The moment coefficients of the lower wing of a biplane of R.A.F. 6 section have been measured, and it seems probable that they will be nearly the same as the moment coefficients we require.

29. The moment curve for the lower wing of a biplane of R.A.F. 6 section is shown as a dotted curve in Fig. 11. It will be seen that it lies very near the moment curve for the experimental section.

TABLE 1.

TABLE SHOWING THE x AND y CO-ORDINATES OF THE EXPERIMENTAL TUBES.

Tube.	x	y
1	0·0015	0·0052
1a	0·013	0·0230
2	0·040	0·0408
3	0·086	0·0592
4	0·185	0·0748
5	0·385	0·0772
6	0·584	0·0684
7	0·783	0·0478
8	0·930	0·0221
9	0·931	0·0025
10	0·785	0·0077
11	0·583	0·0151
12	0·387	0·0212
13	0·193	0·0215
14	0·099	0·0135
15	0·051	0·0073
1b	0·026	0·0034
α	0·0016	0·0105
β	0·0077	0·0172

TABLE 2.

	Complete Aeroplane.				Experimental Portion.				
1.	2.	3.	4.	5.	6.	7.	8.	9.	10.
Air Speed m.p.h.	k_f	Angle of incidence. θ	k_f at $\theta = 0.3°$.	Angle of incidence. $y = 0.3$.	$[k_z]_e$	$[k_x]_e$	$[k_m]_e$	$[k_f]_e$	$[k_d]_e$
50	0·407	8·65	0·397	8·35	0·322	−0·0126	−0·1096	0·320	0·0343
55	0·337	6·40	0·327	6·10	0·275	−0·0031	−0·1026	0·272	0·0249
60	0·283	4·47	0·272	4·17	0·230	+0·0024	−0·0917	0·230	0·0191
65	0·241	3·56	0·230	3·26	0·195	0·0060	−0·0826	0·194	0·0171
70	0·208	2·68	0·197	2·38	0·164	0·0083	−0·0736	0·163	0·0147
75	0·181	1·93	0·170	1·63	0·139	0·0102	−0·0689	0·139	0·0142
80	0·159	1·37	0·147	1·07	0·119	0·0115	−0·0624	0·119	0·0138
85	0·141	0·91	0·129	0·61	0·103	0·0123	−0·0603	0·103	0·0134
90	0·126	0·53	0·114	+0·23	0·187	0·0132	−0·0556	0·087	0·0136
95	0·113	0·19	0·101	−0·11	0·075	0·0149	−0·0563	0·075	0·0148

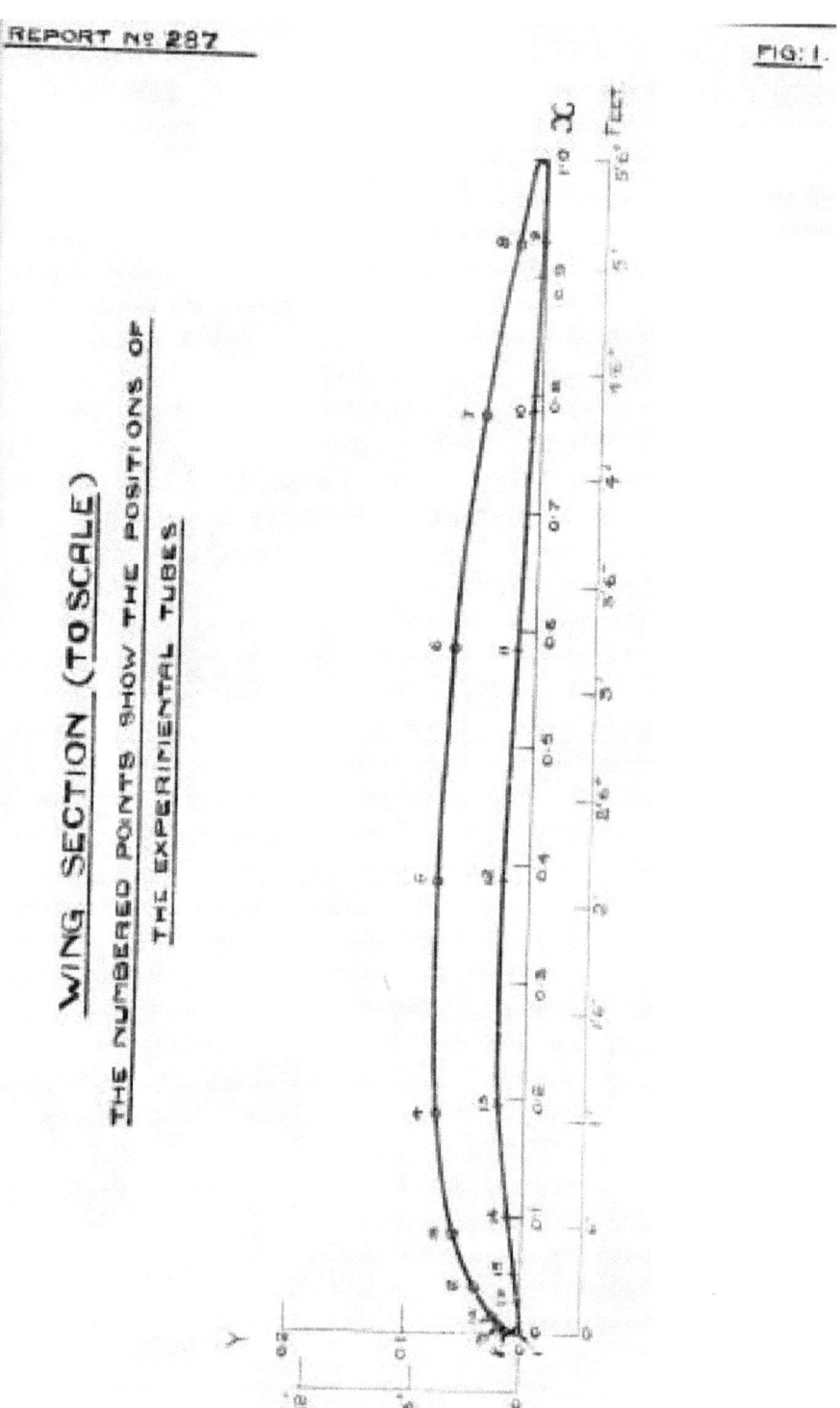

Fig. 2.

Fig. 3.

Document 2-5

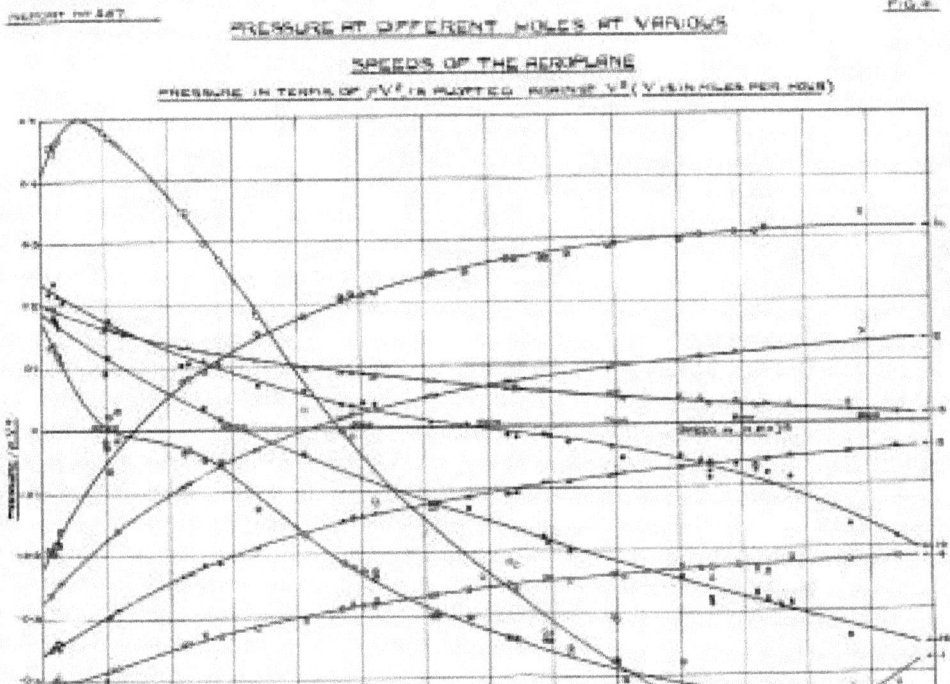

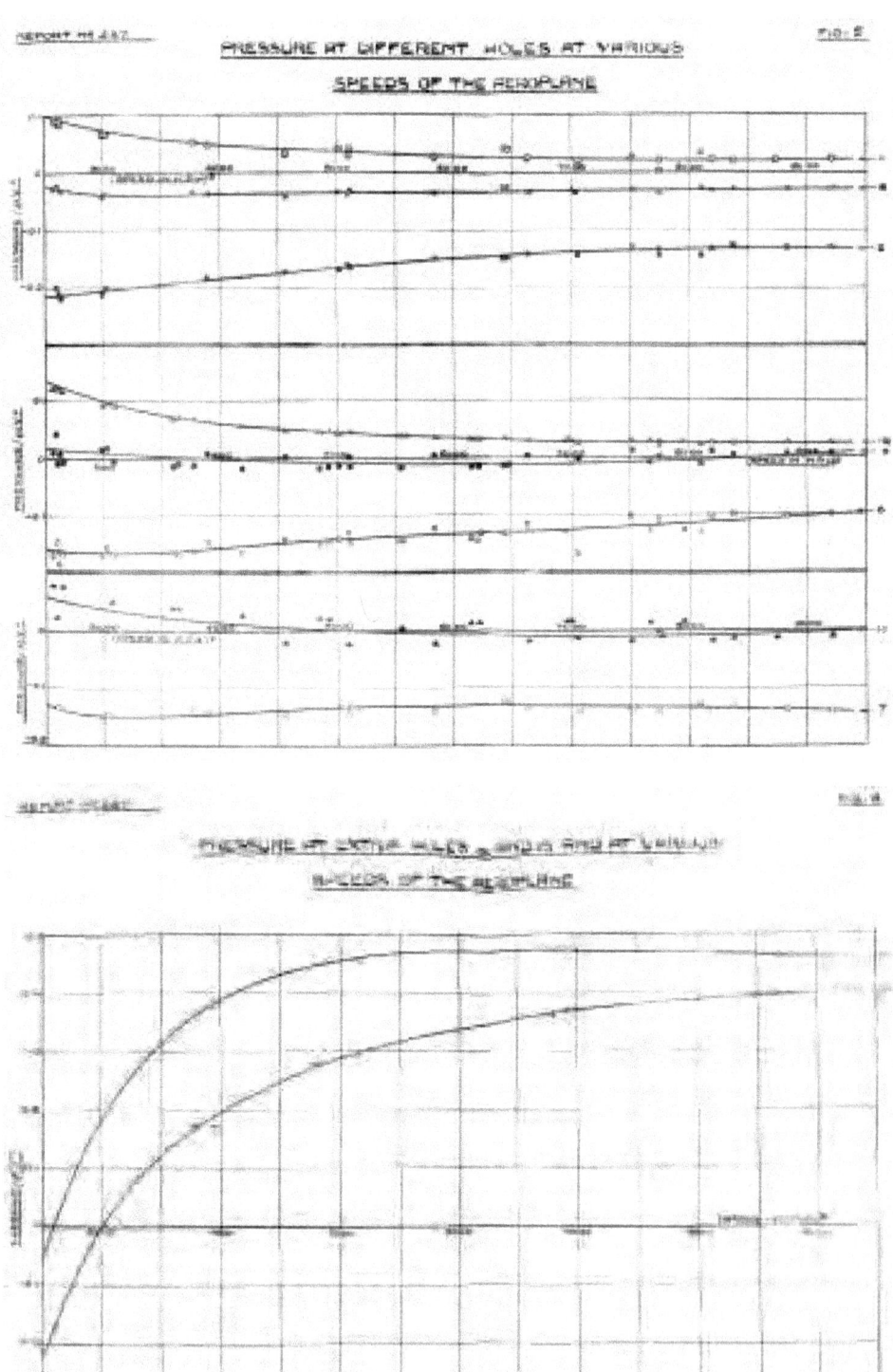

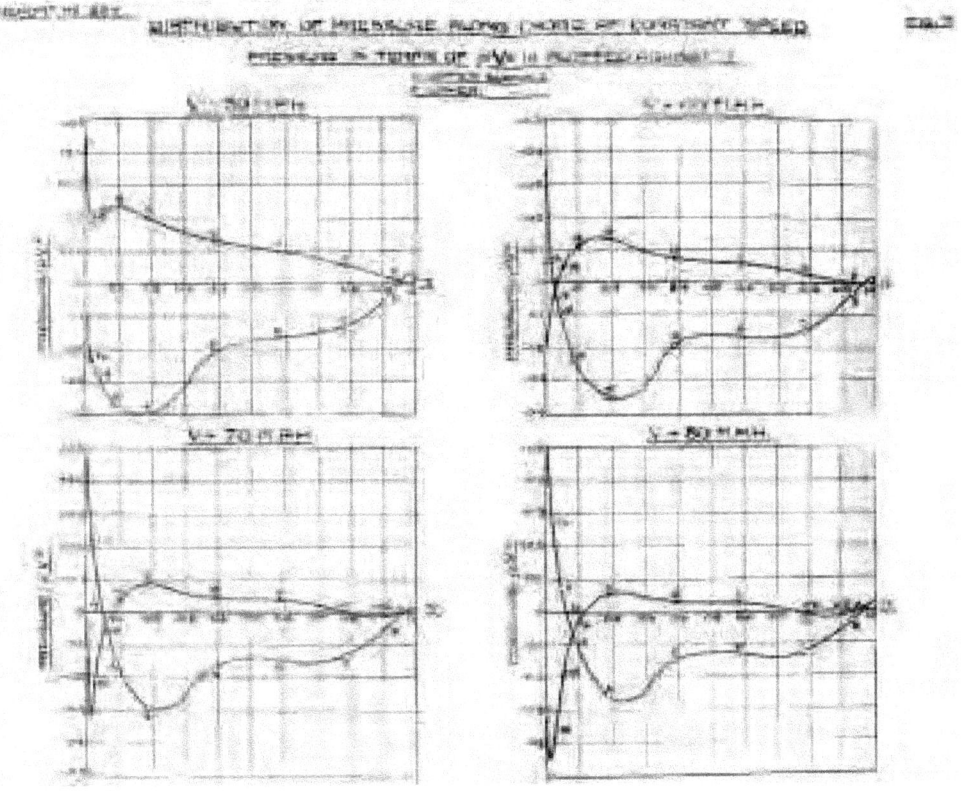

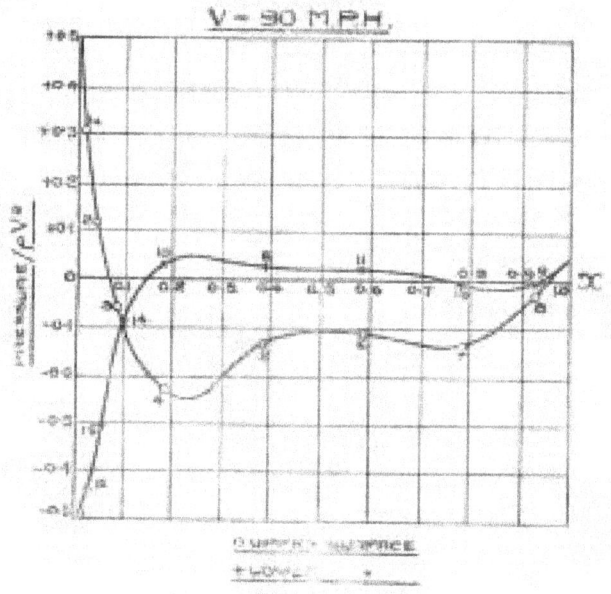

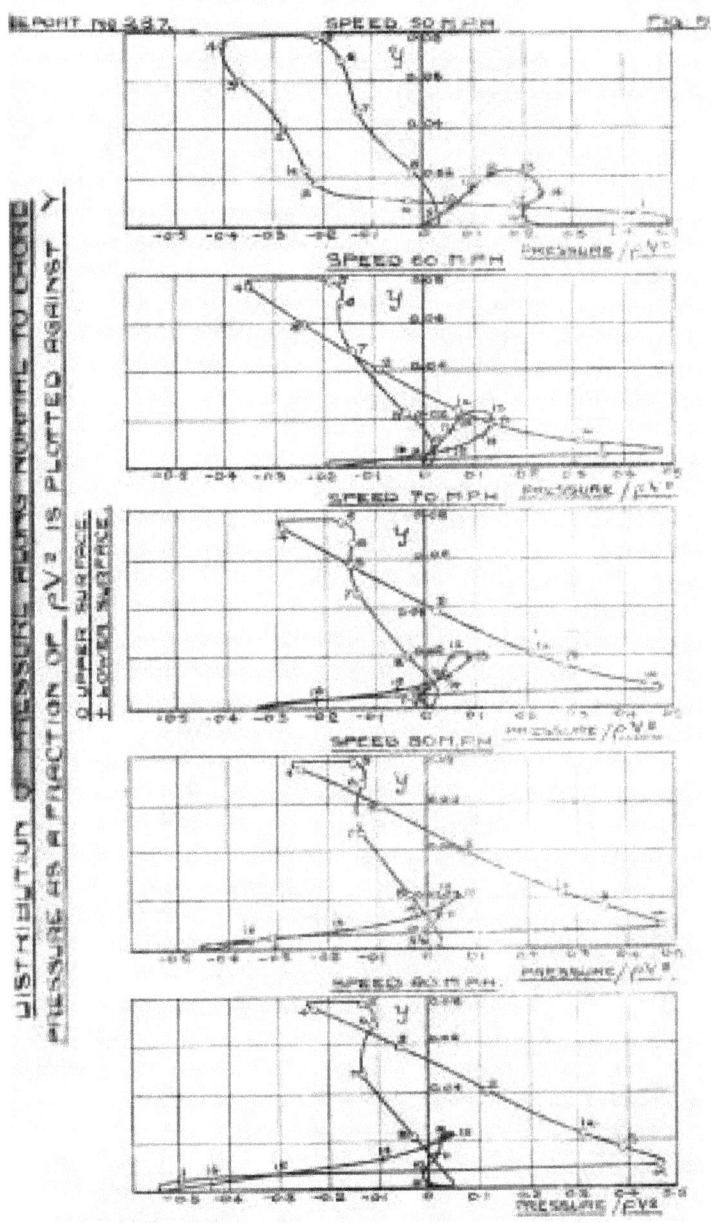

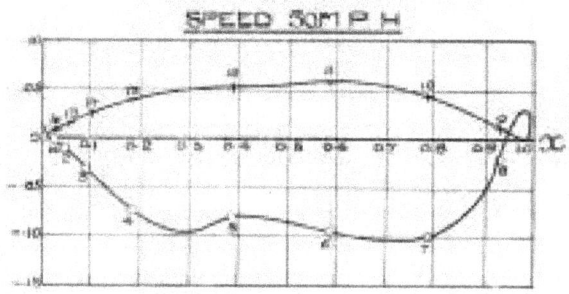
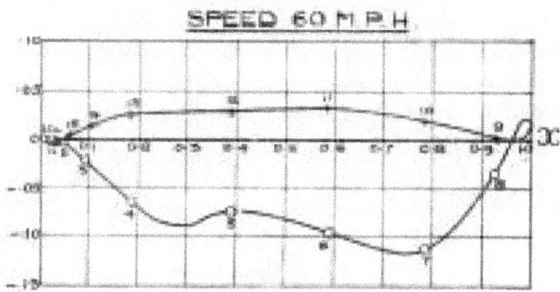
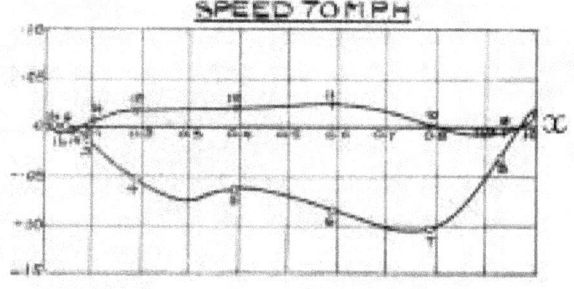

REPORT No 287   FIG: 9.B

MOMENT COEFFICIENT DIAGRAMS

ABSCISSAE ARE VALUES OF x
ORDINATES ARE PRODUCTS OF (PRESSURE/ρV^2)
AND x

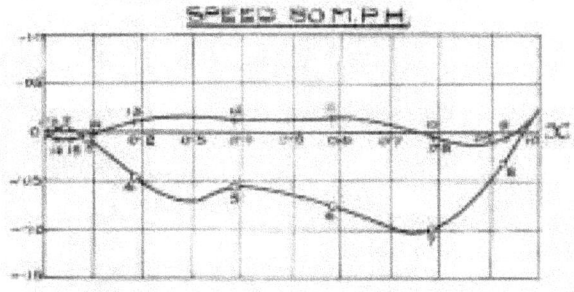

SPEED 80 M.P.H.

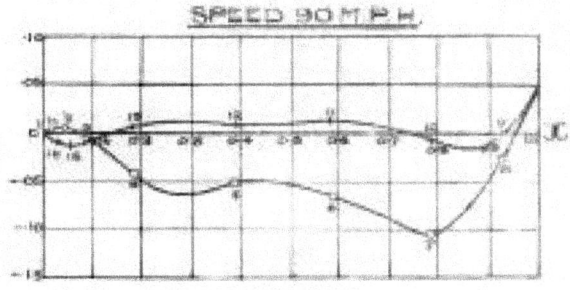

SPEED 90 M.P.H.

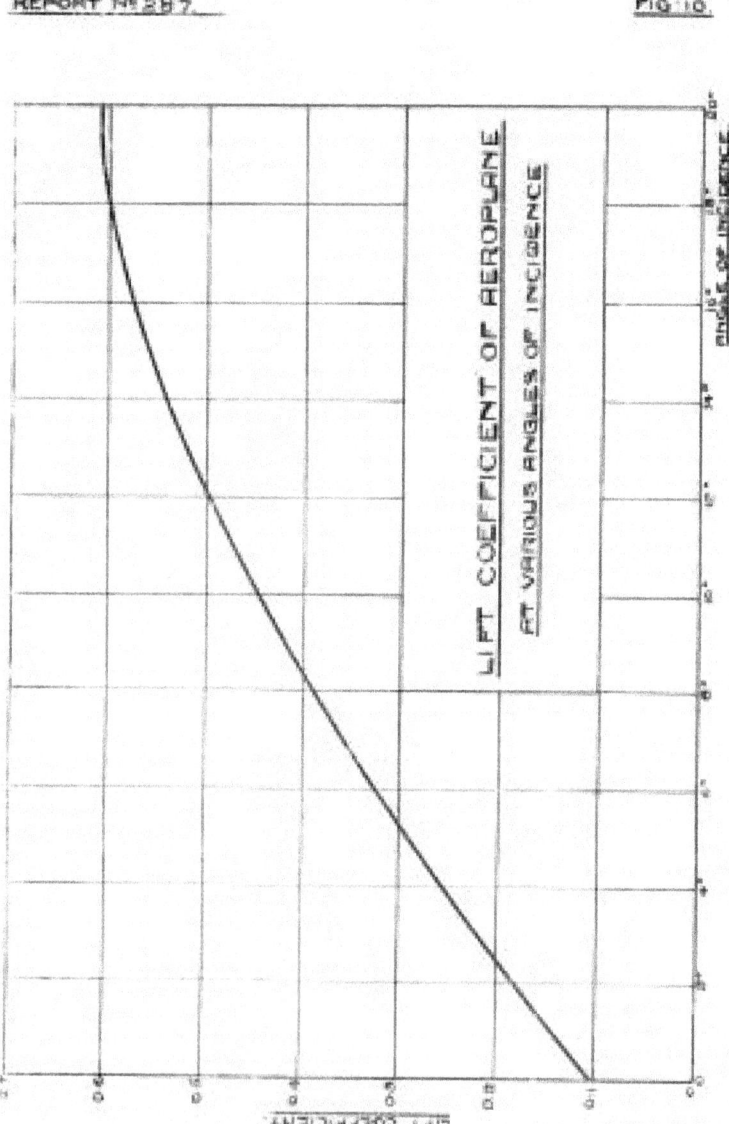

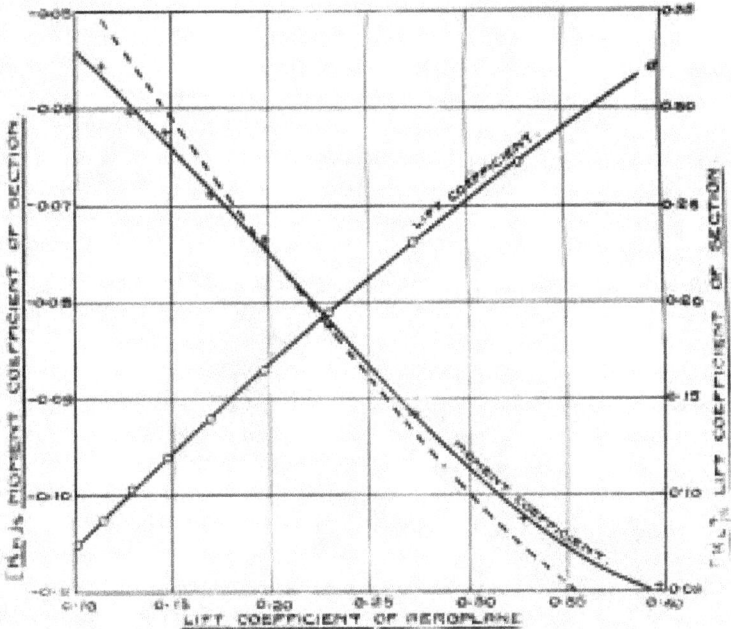

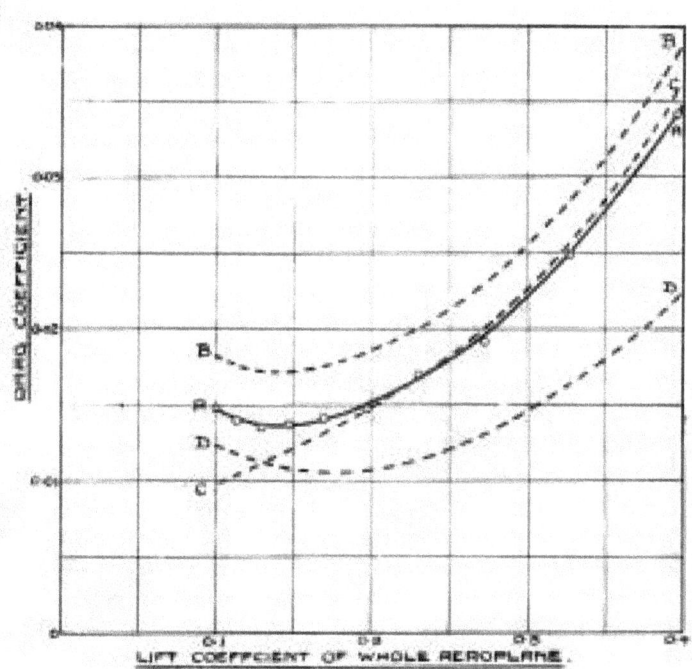

Document 2-6

Henry T. Tizard, "Methods of Measuring Aircraft Performances," Aeronautical Society of Great Britain, *Aeronautical Journal*, No. 82 (April–June 1917).

To be useful, aircraft flight testing demands accuracy, repeatability, and continuity in two crucial areas: flight instruments and test flight methods. As World War I progressed in Europe, the fledgling air forces and airplane manufacturers increasingly realized that no standards or consistency existed for such basic performance parameters as rate of climb and altitude. Even the idea of "level flight" was an uncertain concept due to the variations experienced with aneroid altimeters. If, however, airplanes were to be thoroughly evaluated and their performance envelopes determined for the pilots who would fly them, some standards were essential. England's Royal Aircraft Factory led the effort in establishing what H. T. Tizard called "the general principles of . . . scientific testing of airplanes," principles quickly adopted by the Royal Flying Corps, in which Tizard served as a pilot with the rank of captain.

Tizard's "Methods of Measuring Aircraft Performances" was notable in two areas. First, he explained the effects that normal variations in air temperature, pressure, and density had on the accuracy of contemporary aircraft instruments, particularly altimeters and airspeed indicators, and he discussed methods to recognize and minimize errors. Perhaps of greater interest are his comments regarding the flyers— those now known as test pilots. Here we see possibly the earliest published recognition that test pilots play a unique role in aircraft development that requires special training. Tizard mentioned this repeatedly in his paper, noting that, "it is . . . the flyer on whom the accuracy of the tests depends. I feel that too great stress can not be laid on this; he is the man who does most of the experiments, and . . . he requires training and a great deal of practice."

Tizard himself went on to a distinguished career in aeronautical research. Knighted for his contributions, in 1935 Sir Henry Tizard was appointed the chair of the RAF's newly established Committee for the Scientific Survey of Air Defense. In that role he sponsored R. A. Watson-Watt's development of Radio Direction Finding, or "radar," and helped to develop the vital air defense system that defeated the World War II German aerial assault in the strategically crucial 1940 Battle of Britain.

Document 2-6, Henry T. Tizard, "Methods of Measuring Aircraft Performances," 1916.

AEROPLANE TESTING.

The accurate testing of aeroplanes is one of the many branches of aeronautics which have been greatly developed during the war, and especially during the last year. For some months after the war began a climb to 3,000 to 5,000 feet by aneroid and a run over a speed course was considered quite a sufficient test of a new aeroplane; now we all realise that for military reasons certainly, and probably for commercial reasons in the future, it is the performance of a machine at far greater heights with which we are mainly concerned. In this paper I propose to give a short general account of some of the methods of testing now in use at the Testing Squadron of the Royal Flying Corps, and to indicate the way in which results of actual tests may be reduced, so as to represent as accurately as possible the performance of a machine independently of abnormal weather conditions, and of the time of the year. For obvious reasons full details of the tests and methods employed can not yet be given. So far as England is concerned, I believe that the general principles of what may be called the scientific testing of aeroplanes were first laid down at the Royal Aircraft Factory. Our methods of reduction were based on theirs to a considerable extent, with modifications that were agreed upon between us; they have been still further modified since, and recently a joint discussion of the points at issue has led to the naval and military tests being coordinated, so that all official tests are now reduced to the same standard. It should be emphasised that once the methods are thought out scientific testing does not really demand any high degree of scientific knowledge; in the end the accuracy of the results really depends upon the flyer, who must be prepared to exercise a care and patience unnecessary in ordinary flying. Get careful flyers whose judgment and reliability you can trust and your task is comparatively easy; get careless flyers and it is impossible.

At the outset it may be useful to point out by an example the nature of the problems that arise in aeroplane testing. Suppose that it is desired to find out which of two wing sections is most suitable for a given aeroplane. The aeroplane is tested with one set of wings, which are then replaced by the other set, and the tests repeated some days later. The results might be expressed thus:—

	A Wings.	B Wings.
Speed at 10,000 ft.	90 m.p.h.	93 m.p.h.
Rate of climb at 10,000 ft.	25ft. a minute.	300ft. a minute.

Now the intelligent designer knows, or soon will know, that firstly an aneroid may indicate extremely misleading "heights," and secondly, that even if the actual height above the ground is the same in the two tests the actual conditions of atmospheric pressure and temperature may have been very different on the two days. He will therefore say, what does that 10,000 mean? Do you mean that your aneroid read 10,000 feet, or do you mean 10,000 feet above the spot you started from, or 10,000 feet above sea level? If he proceeds to think a trifle further he will say—what was the *density* of the atmosphere at your 10,000 feet; was it the same in the two tests? If not, the results do not convey much. There he will touch the keynote of the whole problem, for it is on the density of the atmosphere that the whole performance of an aeroplane depends; the power of the engine and the efficiency of the machine depend essentially on the density, the resistance to the motion of the machine through the air is proportional to the density, and so finally is the lift on the wings. None of these properties are proportional solely to the pressure of the atmosphere, but to the density, that is the weight of air actually present in unit volume. It follows that it is essential when comparing the performances of machines to compare them as far as possible under the same conditions of atmospheric density, not as is loosely done at the same height above the earth, since the density of the atmosphere at the same height above the earth may vary considerably on different days, and on the same day at different places.

At the same time, in expressing the final results, this principle may be carried too far. Thus, if the speed of a machine were expressed as 40 metres a second at a density of 0.8 kilogrammes per cubic metre, the statement, though it may be strictly and scientifically accurate, will convey nothing to 99 per cent of those directly concerned with the results of the test. The result is rendered intelligible and indeed useful by the form "90 m.p.h. at 10,000 feet," or whatever it is. With this form of statement, in order that all the statements of results may be consistent and comparative, we must be careful to mean by "10,000 feet" a certain definite density, in fact the average density of the atmosphere at a height of 10,000 feet above mean sea level. This is what the problem of "reduction" of tests boils down to; what is the relation between atmospheric density and height above sea level? This knowledge is obtained from meteorological observations. We have collected all the available data—mostly unpublished—with results shown in the following table:—

Table I.

MEAN ATMOSPHERIC PRESSURE, TEMPERATURE AND DENSITY AT VARIOUS HEIGHTS ABOVE SEA LEVEL.

Height in kilometers	Height in equivalent feet.	Mean pressure in millibars.	Mean temp. in absolute degrees centigrades.	Mean density in kgm. per cubic meter.
0	0	1,014	282	1.253

1	3,280	900	278	1.128
2	6,560	795	273	1.014
3	9,840	699	268	0.909
4	13,120	615	262	0.818
5	16,400	568	255	0.735
6	19,680	469	248	0.658
7	22,960	407	241	0.589

These are the mean results of a long series of actual observations made by Mr. W.H. Dines, F.R.S. It is convenient to choose some density as standard, call it unity, and refer all other densities as fractions or per centages of this "standard density." We have taken, in conformity with the R.A.F., the density of dry air at 76m.m. pressure and 16 degrees centigrade as our standard density; it is 1.221 kgm. per cubic metre. The reason this standard has been taken is that the air speed indicators in use are so constructed as to read correctly at this density, assuring the law:

$$p = \tfrac{1}{2}\rho V^2$$

where V is the air speed, p the pressure obtained, ρ the standard density.

In some ways it would doubtless be more convenient to take the average density at sea level as the standard density, but it does not really matter what you take so long as you make your units quite clear. Translated into feet, and fraction of standard density, the above table becomes:—

Table II.

Height in feet.	Per centage of standard density.
0	102.6
1,000	99.4
2,000	96.3
3,000	93.2
4,000	90.3
5,000	87.4
6,000	84.6
*6,500	83.3
7,000	81.9
8,000	79.2
9,000	76.5
*10,000	74.0
11,000	71.7
12,000	69.5
*13,000	67.3

14,000	65.2
15,000	63.0
16,000	61.1
*16,500	60.1
17,000	59.1
18,000	57.1
19,000	55.2
20,000	53.3

*6,500 feet is introduced as corresponding roughly to the French test height of 2,000 metres. 10,000 feet similarly corresponds roughly to comparing aeroplane test performances to the French standard of 3,000 metres, and similarly for 13,000 and 16,500 feet.

Let us briefly consider what these figures mean. For example, we say that the density at 10,000 feet is 74 per cent of our standard density, but it is not meant that at 10,000 feet above mean sea level the atmospheric density will always be 74 per cent of the standard density. Unfortunately for aeroplane tests this is far from true. The atmospheric density at any particular height may vary considerably from season to season, from day to day, and even from hour to hour; what we do mean is that if the density at 10,000 feet could be measured every day, then the average of the results would be, as closely as we can tell at present, 74 per cent of the standard density.

The above table may therefore be taken to represent the conditions prevailing in a "normal" or "standard" atmosphere, and we endeavour, in order to obtain a strict basis of comparison, to reduce all observed aeroplane performances to this standard atmosphere, i.e., to express the final results as the performance which may be expected of the aeroplane on a day on which the atmospheric density at every point is equal to the average density at the point. Some days the aeroplane may put up a better performance, some days a worse, but on the average, if the engine power and other characteristics of the aeroplane remain the same, its performance will be that given.

It must be remembered that a standard atmosphere is a very abnormal occurrence; besides changes in density there may occur up and down air currents which exaggerate or diminish the performance of an aeroplane, and which must be taken carefully into account. They show themselves in an otherwise unaccountable increase or decrease in rate of climb or in full speed flying level at a particular height.

We now pass to the actual tests, beginning with a description of the observations which have to be made and thereafter to the instruments necessary. The tests resolve themselves mainly into

(a) A climbing test at the maximum rate of climb for the machine.

(b) Speed tests at various heights from the "ground" or some other agreed low level upwards.

Experience agrees with theory in showing that the best climb is obtained by keeping that which is frequently called the air speed of an aeroplane, namely, the indications of the ordinary air speed indicator, nearly constant whatever the height. [In other words ρV^2 is kept constant.] We can look at this in this way. There is a limiting height for every aeroplane above which it can not climb; at this limiting height, called the ceiling of the machine, there is only one speed at which the aeroplane will fly level, at any other air speed higher or lower it will descend. Suppose this speed is 55 m.p.h. on the air speed indicator. Then the best rate of climb from the ground is obtained by keeping the speed of the machine to a steady indicated 55 m.p.h. Fortunately a variation in the speed does not make very much difference to the rate of climb; for instance, a B.E.2C with a maximum rate of climb at 53 m.p.h. climbs just as fast up, say, to 5,000 feet at about 58 m.p.h. This is fortunate as it requires considerable concentration to keep climbing at a steady air speed, especially with a light scout machine; if the air is at all "bumpy" it is impossible. At great heights the air is usually very steady, and it is much easier to keep to one air speed. It is often difficult to judge the best climbing speed of a new machine; flyers differ very much on this point, as on most. The Testing Squadron, therefore, introduced some time ago a rate of climb indicator intended to show the pilot when he is climbing at the maximum rate. It consists of a thermos flask, communicating with the outer air through a thermometer tube leak. A liquid pressure gauge of small bore indicates the difference of pressure between the inside and outside of the vessel. Now, when climbing, the atmospheric pressure is diminishing steadily; the pressure inside the thermos flask tends therefore to become greater than the outside atmospheric pressure. It goes on increasing until air is being forced out through the thermometer tubing at such a rate that the rate of change of pressure inside the flask is equal to the rate of change of atmospheric pressure due to climbing. When climbing at a maximum rate, therefore, the pressure inside the thermos flask is a maximum. The pilot therefore varies his air speed until the liquid in the gauge is as high as possible, and this is the best climbing speed for the machine.

What observations during the test are necessary in order that the results may be reduced to the standard atmosphere? Firstly, we want the time from the start read at intervals, and the height reached noted at the same time. Here we encounter a difficulty at once, for there is no instrument which records height with accuracy. The aneroid is an old friend now of aeronauts as well as of mountaineers, but although it has often been tentatively exposed, it is doubtful whether 1 per cent of those who use it daily realise how extraordinarily rare it is that it ever does what it is supposed to do, that is, indicate the correct height above the ground, or starting place. The faults of the aeroplane aneroid are partly unavoidable and partly due to those who first laid down the conditions of its manufacture. An aneroid is an instrument which in the first place measures only the pressure of the surrounding air. Now if p_1 and p_2 are the pressures at two points in the atmosphere, the difference of height between these points is given very closely by the relation,

$$h = k\theta \log_e \frac{p_1}{p_2}$$

where θ is the average temperature, expressed in "absolute" degrees, of the air between the two points and k is a constant. It is obvious that if we wish to graduate an aneroid in feet we must choose arbitrarily some value for θ. The temperature that was originally chosen for aeroplane aneroids was 50 degrees Fahrenheit, or 10 degrees centigrade. An aneroid, as now graduated, will therefore only read the correct height in feet if the atmosphere has a uniform temperature of 50 degrees Fahrenheit from the ground upwards, and it will be the more inaccurate the greater the average temperature between the ground and the height reached differs from 50 degrees Fahr. Unfortunately 50 degrees F. is much too high an average temperature; to take an extreme example it is only on the hottest days in summer, and even then very rarely, that the average temperature between the ground and 20,000 feet will be as high as 50 degrees F. On these very rare occasions an aneroid will read approximately correctly at high altitudes; otherwise it will always read too high. In winter it may read on cold days 2,000 feet too high at 16,000 feet, i.e., it will indicate a height of 18,000 feet when the real height is only 16,000. It is always necessary therefore to "correct" the aneroid readings for temperature. The equation

$$H = \frac{273 + t}{283} \cdot h$$

gives us the necessary correction. Here H is the true difference in height between any two points, t the average temperature in degrees centigrade between the points, and h the difference in height indicated by aneroid. It is convenient to draw a curve showing the necessary factors at different temperatures, some of which are given in the following table:—

Table III.
Aneroid Correction Factors

Temperature.	Correction factor.
70° Fahr.	1.040
50°	1.000
30°	0.961
10°	0.922
-10°	0.883

For example, if a climb is made through 1,000 feet by aneroid and the average temperature is 10 degrees Fahr., the actual distance in feet is only 1,000 x 0.922 = 922 feet. The above equation is probably quite accurate enough for small differences of height—up to 1,000 feet say—and approximately so for bigger differences.

The magnitude of the correction which may be necessary shows how important it is that observations of temperature should be made during every test. For this purpose a special thermometer is attached to a strut of the machine, well away from the fuselage, and so clear of any warm air which may come from the engine. The French, I believe, do not measure temperature, but note the ground temperature at the start of a test, and assume a uniform fall of temperature with height. This, undoubtedly, may lead to serious errors. The change of temperature with height is usually very irregular, and only becomes fairly regular at heights well above 10,000. [*Tizard illustrates this with a chart entitled "Variations of Temperature with Height," omitted here, which displays the data with a series of curves.*]

The aneroid being what it is, one soon comes to the conclusion that the only way to make use of it in aeroplane tests is to treat it purely as a pressure instrument. For this reason it is best to do away with the zero adjustment for all test purposes and lock the instrument so that the zero point on the height scale corresponds to the standard atmospheric pressure of 29.9 inches or 760mm. of mercury. Every other height then corresponds to a definite pressure; for instance, the locked aneroid reads 5,000 feet when the atmospheric pressure is 24.8ins., and 10,000 feet when it is 20.7ins., and so on. If the temperature is noted at the same time as the aneroid reading, we then know both the atmospheric pressure and temperature at the point, and hence the density can be calculated, or, more conveniently, read off curves drawn for the purpose. The observations necessary (after noting the gross aeroplane weight, and nett or useful weight carried) are therefore, (i.) aneroid height every 1,000 feet, (ii.) time which has elapsed from the start of the climb, and (iii.) temperature; to these should be added also (iv.) the air speed, and (v.) engine revolutions at frequent intervals. The observed times are then plotted on squared paper against the aneroid heights and a curve drawn through them. From this curve the rate of climb at any part (also in aneroid feet) can be obtained by measuring the tangent to the curve at the point. This is done for every 1,000 feet by aneroid. The true rate of climb is then obtained by multiplying the aneroid rate by the correction factor corresponding to the observed temperature. These true rates are then plotted afresh against standard heights and from this curve we can obtain the rates of climb corresponding to the standard heights 1,000, 2,000, 3,000, etc. Knowing the change of rate of climb with height, the time to any required height is best obtained by graphical integration. The following table gives the results of an actual test:—

Table IV.

Machine   Engine
Date 27/12/16.

From Curve

Height in Aneroid ft.	Observed temp Fahr.	Per centage of standard density	Observed time	Rate of climb in Aneroid ft.	Real rate of climb (corrected for temp.)	Standard height	% of standard density	Time	Rate of climb
0	36°	0.00							
1,000	37°	101.0	1.00	835	814	1,000	99.40	1.20	775
2,000	38°	97.2	2.10	735	718	2,000	96.30	2.56	685
3,000	36°	94.0	3.70	640	655	3,000	93.26	4.11	610
4,000	36°	90.7	5.40	560	544	4,000	90.25	5.85	545
5,000	36°	87.4	7.25	540	495	5,000	87.32	7.80	490
6,000	33°	84.7	9.40	450	435	6,000	84.50	9.96	435
7,000	30°	82.1	11.90	405	389	7,000	81.80	12.40	385
8,000	26°	79.9	14.25	365	347	8,000	79.16	15.14	345
9,000	22°	77.6	17.00	330	312	9,000	76.55	18.20	310
10,000	23°	74.7	20.25	310	294	10,000	74.00	21.61	280
11,000	21°	72.2	23.60	280	264	11,000	71.70	25.41	245
12,000	20°	69.8	27.40	230	216	12,000	69.50	29.81	210
13,000	17°	67.7	31.90	195	182	13,000	67.32	35.13	170
14,000	12°	65.9	37.90	150	139	14,000	65.17	41.33	130
15,000	8°	64.1	45.25	110	101	14,500	64.11	46.23	105

The corrections are often much greater than those necessary in the above case.

It will be noticed that the rate of climb of this machine is approximately halved for a difference in height of 5,000 feet. Now it is possible to get a difference in density near the ground of as much as 15 per cent between a hot day in summer and a cold day in winter. This corresponds to a difference in height of 5,000 feet, so that this machine would climb off the ground on a hot day at only half the rate that it would on a very cold day. Variation in atmospheric density, combined with the errors of an aneroid, fully account for the observed difference between a "good climbing day" and a "bad climbing day."

At least two climbing tests of every new machine are carried out up to 16,000 feet or over by aneroid. If time permits three or more tests are made. The final results given are the average of the tests and represent as closely as possible the performance on a standard day, with temperature effects, up and down currents, and other errors eliminated.

If we produce the rate of climb curve upwards it cuts the height axis at a point at which the rate of climb would be zero, and therefore the limit of climb reached. This is the "ceiling" of the machine.

SPEEDS.

His 16,000 feet, or whatever it is, reached, the flyer's next duty is to measure the speed flying level by air speed indicator at regular intervals of height (generally every 2,000 feet) from the highest point downwards. To do this he requires a sensitive instrument which will tell him when he is flying level. The aneroid is quite useless for this purpose, and a "statoscope" is used. The principle of this instrument is really the same as that of a climbmeter. It consists of a thermos flask connected to a small glass gauge, slightly curved, but placed about horizontally. In this gauge is a small drop of liquid, and at either end are two glass traps which prevent the liquid from escaping either into the outside air or into the thermos flask. As the machine ascends the atmospheric pressure becomes smaller, and the pressure in the flask being then higher than the external pressure, the liquid is pushed up to the right hand trap, where it breaks, allowing the air to escape. On descending the reverse happens; the liquid travels to the left, breaks, and air enters the flask. When flying truly level the drop remains stationary, moving neither up nor down. The instrument is made by the British Wright Co.

The flyer or the observer notes the maximum speed by the air speed indicator, i.e., the speed at full engine throttle. At one or more heights also, he observes the speeds at various positions of the throttle down to the minimum speed which will keep the machine flying at the height in question. The petrol consumption and the engine revolutions are noted at the same time, as well, of course, as the aneroid height and temperature. Accurate observation of speeds needs very careful flying—in fact much more so than in climbing tests. If the air is at an all bumpy observations are necessarily subject to much greater error, since the machine is always accelerating and decelerating. The best way to carry out the test seems to be as follows. The machine is flown first just down hill and then just up hill and the air speeds noted. This will give a small range between which the real level speed must lie. The flyer must then keep the speed as steadily as possible on a reading midway between these limits, and watch the statoscope with his other eye. If it shows steady movement, one way or the other, the air speed must be altered accordingly by 1 m.p.h. In this way it is always possible at heights where the air is steady to obtain the reading correct at any rate to 1 m.p.h., even with light machines, provided always sufficient patience is exercised. The r.p.m. at this speed are then noted.

One difficulty, however, can not be avoided. If at any height there is a steady up or down air current, then though the air may appear calm, i.e., there may be no "bumps," the air speed indicator reading may be wrong, since to keep the machine *level* in an up current it is necessary to fly slightly down hill relatively to the air. Such unavoidable errors are, however, eliminated to a large extent by the method of taking speeds every 2,000 feet, and finally averaging the results.

We must now consider how the true speed of the aeroplane is deduced from the reading of the air speed indicator. It is well known that an air speed indicator reads too low at great heights—for example, if it reads 70 m.p.h. at 8,000 feet the real speed of the machine through the air is nearer 80 m.p.h. The reason for this is that the indicator, like the aneroid, is only a pressure gauge—a sensitive pressure gauge, in fact, which registers the difference of pressure between the air in a tube with its open end pointing forward along the lines of flight of the machine, and the real pressure (the static pressure) of the external air. This difference of pressure is as nearly as we can judge by experiment $= \frac{1}{2} pV^2$ (where p is the density of the air and V the speed of the machine), provided that the open end of the tube is well clear of wings, struts, fuselage, etc., and so is not affected by eddies and other disturbances. Now assuming this law, air speed indicators are graduated to read correctly, as I have said above, at a density of 1.221 kgm. per cubic metre, which we have taken as our standard density and called "unity." It corresponds on an average to a height of about 800 feet above sea level.

Then suppose the real air speed of an aeroplane at a height of "h" feet is V m.p.h., and the indicated air speed is 70 m.p.h., this means that the excess pressure in the tube due to the speed is proportional to 1×70^2,

$$\text{or } \rho \times V^2 = 1 \times 70^2,$$

where ρ is the density at the height in question, expressed as a fraction of the standard density. To correct the observed speed, we therefore divide the reading by the square root of the density. Thus, observation of the maximum speed of an aeroplane at a height of 8,000 feet by the locked aneroid gave 80 m.p.h. on the indicator, the temperature being 31 degrees Fahr. From the curve we find that the density corresponding to 8,000 feet and 31 degrees is 0.85 of standard density. The corrected airspeed is therefore:—

$$\frac{80}{\sqrt{.85}} = 86.7 \text{ M.P.H.}$$

This "corrected" air speed will only be true if the above law holds, that is to say, if there are no disturbances due to the pressure head being in close proximity to struts or wings. It is always necessary to find out the magnitude of this possible error, that is, to calibrate the air speed indicator, and the only way to do this is to measure a real air speed of the aeroplane at some reasonable altitude for easy observation by actual timed observations from the ground, and from these timed results check those deduced from the air speed indicator readings. This calibration is the most important and difficult test of all, since on the accuracy of the results depends the accuracy of all the other speed measurements. It can either be done by speed trials over a speed course close to the ground, or when the aeroplane is flying at a considerable height above the ground. In the Testing Squadron we have till lately attached more importance to the latter method, mainly because the conditions approximate more to the conditions of the ordinary air speed measurements at different heights, and because the weather conditions are much steadier and the flyer can devote more attention to flying the machine at a constant air speed than he can when very close to the ground.

One method is to use two camera obscuras, one of which points vertically upwards and the other is set up sloping towards the vertical camera. At one important testing centre the cameras are about ¾ mile apart, and the angle of the sloping camera is 45°. By this arrangement, if an aeroplane is directly over the vertical camera it will be seen in the field of the sloping camera if its height is anywhere between 1,000 and 16,000 feet, although at very great heights it would be too indistinct for measurements except on a very clear day. The height the tests are usually carried out is 4,000 feet to 6,000 feet.

The aeroplane is flown as nearly as possible directly over the vertical camera, and in a direction approximately at right angles to the line joining the two cameras. The pilot flies in as straight a line and at as constant an airspeed as he can. Observers in the two cameras dot in the position of the aeroplane every second. A line is drawn on the tables of each camera pointing directly towards the other camera, so that if the image of the aeroplane is seen to cross the lines in the one camera it crosses the line in the other simultaneously. From these observations it is possible to calculate the height of the aeroplane with considerable accuracy; the error can be brought down to less than 1 part in 1,000 with care. Knowing the height, we can then calculate the speed over the ground of the aeroplane by measuring the average distance on the paper passed over per second by the image in the vertical camera. If x inches is this distance, and f the focal length of the lens, the ground speed is $x \times h/f$ feet per second.

It is necessary to know also the speed and direction of the wind at the height of the test. For this purpose the pilot or his observer fires a smoke puff slightly upwards when over the cameras, and the observer in the vertical camera dots in its trail every second. The height of the smoke puff is assumed to be the same as that of the aeroplane—it probably does not differ from this enough to introduce any appreciable error in the results. The true speed through the air is then found.

[*Tizard then shows this graphically in the form of a simple ABC triangle in which length AB represents the ground speed of the aeroplane as measured in the camera, CB represents the velocity and direction of the wind, and the length AC represents the true air speed of the machine.*]

The tests are done in any direction relative to the wind, and generally at three air speeds, four runs being made at each air speed.

The advantages of this method are:—

(1) Being well above the earth the pilot can devote his whole attention to the test.
(2) Within reasonable limits any height can be chosen, so that it is generally possible to find a height at which the wind is steady.
(3) It does not matter if the pilot does not fly along a level path so long as he does so approximately. What is more important is that he should fly at a constant air speed.
(4) It is not necessary that there should be any communication between the two cameras, although it is convenient. The two tracks are made quite independently, and synchronised afterwards from the knowledge that the image must have passed over the centre line simultaneously in the two cameras.

The main disadvantage is that somewhat elaborate apparatus is necessary, but this is of not much importance in a permanent testing station.

There are often periods in war time, however, when an aeroplane has to be tested quickly, and low cloud layers and other causes prevent the camera test from being carried out. It is then necessary to rely on measurements of speeds near the ground for the calibration of the air speed indicator. In this method the aeroplane is flown about 50 feet off the ground, and is timed over a measured run. There are two observers, one at each end of the course; when the aeroplane passes the starting point the observer sends a signal and starts his stop-watch simultaneously; the second observer starts his stop-watch when he hears the signal, and in his turn sends a signal and stops his watch when the aeroplane passes the finishing point. By this double timing, errors due to the so-called "reaction time" of the observers are practically eliminated, for the observer at the end of the course tends to *start* his watch late, while the first observer *stops* his late. The mean of the two observations gives the real time. Four runs, two each up and down the course, are done at each air speed, the pilot or his observer noting carefully the average air speed during the run. Observations of the atmospheric pressure and temperature from which the density can be obtained are also taken. The average strength and direction of the wind during each trial are noted from a small direct reading (or recording) anemometer and the speed corrected in the same way as in the camera tests. If there is a strong cross wind the aeroplane may have to be pointed at a considerable angle to the course, and this makes the test a very difficult one to carry out well. Generally speaking, it is only reliable when the wind is quite light, not more, at any rate, than 10 m.p.h. Even this is too strong if it is a cross wind.

A further difficulty is that at high speeds, over 100 m.p.h., an aeroplane may take quite a considerable time to accelerate up to a steady speed, and so it must fly level for a long distance each end before reaching the actual course. At the testing station previously alluded to the course is a mile long, and there is a clear half mile or more at each end, but it is doubtful whether even this distance is enough for the machine to attain steady speed before the starting point. Finally, the flyer of a single-seater is generally too busy watching the ground to do more than glance at his air speed indicator more than a few times during the run. Doubtless it would be better in such a case to use some form of recording air speed instrument, although then other difficulties would arise.

Having gotten the true air speed from camera or speed course tests, and knowing the density at the height at which the test was carried out, we obtain what the air speed indicator should have read by multiplying the measured air speed by the square root of the density. By comparing this with the actual reading of the indicator we obtain the necessary correction. The whole procedure may be shown best by a table giving part of the results of a camera test made at the beginning of the year.

A summary of the complete speed tests may now be given. Firstly, the air speed and engine revolutions are noted flying level at full throttle every 2,000 feet approximately, by aneroid. From the aneroid reading and the temperature observations at each height the density is obtained. The reading of the air speed indicator is then first corrected for instrumental errors by adding or subtracting the correction found by calibration tests over the cameras or speed course. This number is then again corrected for height by dividing by the square root of the density. The result should give the true air speed, subject, of course, to errors of observation. The numbers so obtained are plotted against the "standard" heights, i.e., the average height in feet corresponding to the density during the test. A smooth curve is then drawn through the points and the air speeds at standard heights of 3,000, 6,500, 10,000, 13,000, and 16,500 read off the curve. These heights are chosen because they correspond closely with 1, 2, 3, etc., kilometres. The indicated engine revolutions are also plotted against the standard heights, because these observations form a check on the reliability of the results; also the ratio of speed to engine revolutions at different heights may give valuable information with regard to the propeller.

[At this point Tizard includes a set of tables that present data from tests of air speed at height, showing the need for the outlined adjustments in order to obtain reliable and accurate results. Tizard also presents another set of figures showing curves drawn from the calculated data. In it the air speeds lie very closely on a smooth curve except at one point—about 10,000 feet—where the author believes they were probably affected by a downward current of air.]

In a brief paper it is impossible to do more than explain the more important of the "performance" tests of aeroplanes, considered solely as flying machines. For military purposes a number of tests are necessary, some of which can not easily be reduced to figures. Nor can it be supposed for an instant that the methods outlined here are final; aeroplane testing, like all other work connected with aeroplanes, is only in its infancy; and as time goes on, and knowledge accumulates, better methods and instruments will evolve. There are some who lay considerable emphasis on the necessity of every test instrument being self-recording, and although this scheme appears at first sight Utopian and would relieve the pilot of a single-seater of considerable trouble, there are many objections to it when considered in detail, not the least of which is the difficulty of getting new and elaborate instruments made at a time when all manufacturers are fully engaged on other important work. When an observer can be taken I would personally place much more reliance on direct observations at the present time, and one great advantage of direct observation is that the results are there, and no time is lost through the failure of a recording instrument to record, a circumstance which is not unknown in practice. So far as we use recording instruments, we use them only as a check on direct observations, although we may probably adopt recording air speed indicators for the calibration tests of single seaters. But whether recording or direct reading instruments are used, it is as I said before, the flyer on whom the accuracy of the tests depends. I feel that too great stress can not be laid on this; he is the man who does most of the experiments, and like all experimenters in every branch of science, he requires training and a great deal of practice. Although the methods themselves may be greatly changed, this much may perhaps be claimed, that the general principles on which they are founded are sound, and will only be altered in detail. The importance of the work can hardly be exaggerated; model experiments are notoriously subject to scale and other corrections, which if not carefully scrutinised may be very misleading, and it is only by accurate full-scale work in addition that we can hope to maintain a steady improvement in the efficiency of aeroplanes.

[*The published paper includes the following transcript of the discussion after its presentation by Captain Tizard.*]

FIFTH MEETING, 52nd SESSION.

An ordinary general meeting of the Society was held in the Theatre of Royal Society of Arts, London, on Wednesday, March 7th, 1917, at 8:00 p.m. There was a large attendance of members and guests. The chair was to be the Right Hon. Lord Sydenham, G.C.I.E., F.R.S.

Captain H. J. TIZARD, of the R.F.C., Associate Fellow, read a paper, illustrated by slides, on "Methods of Measuring Aircraft Performances."

On the conclusion of the lecture a discussion followed.

Squadron-Commander BUSTEED: I regret that the Naval Testing Department is not as far advanced as one would like it to be. A good deal of useful work has been done, but the R.N.A.S. and R.F.C. Department had adopted different density standards, though these were now the same.

I appreciate the necessity for instruments, but my experience goes to show that the machine instruments were most required for single-seaters, and unfortunately, after the pilot had managed to get in, there was very little room for them. They were also a source of trouble in getting tests through quickly; readings taken by pilots had proved very fair.

Lieutenant G. H. MILLAR, R.N.V.R., said that in his opinion it was a pity that the standard atmosphere which had been adopted was a purely empirical one; he would have preferred one based on a given temperature and pressure at sea-level and a uniform rate of fall of temperature. Some months previously he had calculated such a standard atmosphere, taking as the condition at sea-level a pressure of 760mm. and a temperature of 15 deg. C., with a fall of 1.5 deg. C. per 1,000 ft. rise, and the curve of density against height thus obtained did not differ greatly from that given in this paper, the difference varying between 400 and 700 ft. for height from 0 to 20,000 ft. By reducing the assumed ground temperature the curves could be brought nearly into coincidence. The advantage of such a standard over the empirical one was that it could be calculated at any time by remembering two constants. He also thought that the unit of density should cetainly be the density at zero height for the standard atmosphere adopted. No advantage was gained by using the density for which the speed indicators were initially calibrated, since the instrument had to be calibrated in the machine in any case, and in practice instruments were found to be anything up to 20 per cent out. With regard to calibration in the machine, calibration at height had the disadvantage that the speed range of the machine was reduced, unless the machine was flown slightly downhill for the higher speeds. He had found that it was best to take four pairs of runs at different speeds over as wide a range of speed as possible, and even with quite rough methods of timing, the four spots usually came very nearly on a straight line. He was inclined to doubt Captain Tizard's statement that the best climbing speed was the same at all heights, although probably little was sacrificed by climbing throughout at the "ceiling" speed. He stated that terms were badly needed for the quantities v/ρ and n/ρ where "v" was the speed, "n" the r.p.m., and "ρ" the density. These quantities were of great importance in considering the aerodynamic properties of the machine and propeller (apart from the engine) and the relations between them, and the angle of incidence and angle of ascent or descent were independent of the height or density. He wished to express his admiration for the thoroughness with which the R.F.C. tests were carried out.

Captain GRINSTED: The principle of the methods of testing of aeroplanes and of the reduction of the results to a standard basis as now established and improvements in accuracy of testing can now be made only by improving the instruments by which measurements are made.

Captain Tizard objected to the use of an aneroid as a height-measuring instrument, and preferred to use it simply as a pressure-measuring instrument.

Even as such the aneroid is not perfect, and in saying that it measures pressure it is given too good a character. Owing to its lag it does not give a correct measurement of pressure when the rate of change of pressure is at all rapid. I should like to know if Captain Tizard has found difficulty in obtaining instruments sufficiently free from lag for the purpose of accurate aeroplane testing.

The measurement of performance is now confined to tests of speed and climb. There are other things of importance, such as the rate at which the aeroplane can be brought on to a given bank or its direction of flight turned through a given angle which should be measured when comparing performances of aeroplanes. I should like to know if Captain Tizard has considered methods of making such measurements.

Mr. BERTRAM COOPER: I should like to ask Captain Tizard if he could tell us something about the lag of the climbmeter. We have had several "lags" mentioned tonight, but not this "lag," which seems to me to be a pretty serious one.

The action of the instrument depends on the accumulation of pressure in the bottle, which is relieved by the leak. It will be clear, therefore, that the reading will always lag behind the real state of affairs at the spot where it is made, the exact amount depending on how fast the upward or downward journey to that spot was made. For instance, a pilot could stall his machine when seeking his best "climb" and the instrument would still tell him he was climbing when he was, in fact, falling owing to stalling. Moreover, the error here would be aggravated by the "gravity error" on the liquid. This liquid would be relatively lighter owing to the falling, and would consequently tend to show a rate of climb in excess of that actually appropriate to the pressure difference that existed. And this leads me to ask Captain Tizard what is the most serious error in practice, the "lag" error or the "gravity" error? I notice he said that the instrument was not satisfactory near the ground. I take it that is chiefly because of the gravity error and "bumpy" flying, but is not the "lag" error serious at all heights?

Captain FARREN: The methods of measuring aeroplane performance described by Captain Tizard are, as he said, only different in some minor points from those in use at the R.A.F., where they serve both for the testing of new types and for reducing the full-scale experiments on aeroplane resistance, etc., which have been going on there for some time. The methods were, in fact, arrived at to a great extent by discussion between the R.A.F. and the Testing Squadron. The same standards of density are used, and we agree with him generally on the superiority of ordinary instruments and good observers over automatic recording instruments. We have not had so much experience as he has had with single-seaters, which are rather a different problem from two-seaters, demanding much more skill from the pilot, but it seems that even here automatic recorders have disadvantages.

With regard to measuring speeds at heights, Captain Tizard is of the opinion that it is not possible to fly level except by using a statoscope. My experience is that in the case of certain very expert pilots a flight taking as long as ten minutes can sometimes be made, during which the aneroid shows no appreciable movement. (The aneroids used are very high-class instruments, with 20 ft. divisions.) I realise that this does not really mean that no height is gained or lost in the test, because every aneroid is known to possess lag. But under these circumstances on the average the height difference between the beginning and end of the run can not be more than about 100 ft. in a length of about 12 miles—corresponding to a slope of 1 in 600 or so, which represents a correction to the speed of well under the error of observation. But undoubtedly the statoscope gives generally much better results. The instrument in use at the R.A.F. is similar in principle to the one shown, but very much smaller, occupying a space 1 in. by 1 in by 6 ins, approximately. This gives very satisfactory results in use.

With regard to the rate-of-climbmeter—which, it may be interesting to know, was christened the "coffeeometer" by the pilots at the R.A.F., on account of the thermos flask used on the first instrument!—this was first shown me by Captain Dobson (then of the Testing Squadron) in July, 1916. A search in the Instrument Stores at the R.A.F. brought to light an exactly similar instrument—made in Germany! It is apparently a standard balloon instrument, but the credit of introducing it into aeroplane testing is due to the Testing Squadron. This instrument again has been much reduced in size, and occupies about the same space as the statoscope, referred to above. In use it suffers from one disadvantage—any vertical acceleration, such as that which occurs as the result of a change in speed, causes the indicating column to move on account of the change in effective gravity. As a result only very gradual changes in speed must be made in searching for the best climbing speed. An attempt has been made to develop a dial indicator (in which the defect would not appear), but without success, on account of the large volume of air enclosed in the diaphragm.

Captain Tizard laid stress on the necessity for very careful work on the part of the pilot. I think too much emphasis can not be put on this point. We are now emerging from the middle age of aeronautics—when flying was, to the ordinary man, a kind of magic, practised by a sort of superman who daily carried his life in his hands, but nevertheless continued to survive in spite of the apparently rash things he habitually did. To some extent the fear of the passing generation of flyers that flying would become cheapened and commonplace helped to keep alive this idea. They saw their living vanishing. It must be admitted that their fears were justified. It is difficult to estimate aright the value of their work. We are too near to see it in its true place. I think we can be sure that history will not be unjust to them or stinting in its acknowledgments. But it is evident that nowadays it is becoming easier and easier to fly—also less risky. The "magic" has gone. In its place we find a new branch of engineering—a new science. For accurate and useful work nowadays skill and nerve are still essential, but to these care, thoroughness, and training in making accurate observations must be added. Everyone who has had anything to do with aeroplane testing knows how it was common talk that A. always got a better climb out of a machine than anyone else. Perhaps he did it by willpower or some other occult practice; anyhow, it was beyond us. Naturally A. gained in many ways, and it can not be reckoned against him that he did not make any special efforts to dispel the idea. Pilots are human. But the real truth is that A. was possessed of a power of accurate and thorough workmanship, which always, in any kind of work, brings the best results.

Aeroplane testing, as a part of aeroplane designing, demands for satisfactory results the highest training. It occupies no special place by virtue of this—it merely comes into line with the rest of engineering. Now, one can learn to fly in a month—even in England in war time—but an engineer's training requires years. It is evidently necessary, therefore, that engineers—men with scientific training and trained to observe accurately, to criticise fairly, to think logically—should become pilots, in order that the development of aeroplanes may proceed at the rate at which it must proceed if we are to hold that place in the air to which we lay claim—the highest.

I wish to add the following remarks:—

In the years immediately preceding the war aeronautics suffered very much from a lack of full-scale experiments. Money was but grudgingly given, and the foresight of the Government in this matter was not conspicuous. As a nation we are remarkable for our inertia. After the outbreak of war for some time little improvement was evident, but gradually the state of affairs became better. At the present moment we are in a fairly good position, but it is necessary to make provision for "after the war." At the moment experiments are not killed—as they used to be—for lack of money. But after the war the inevitable reaction will almost certainly mean that a partial slump will occur. Money will be scarce and aeronautics will suffer in company with other activities. It is here that the trade must help. They must realise that if they are to build up aeronautics as a branch of engineering they must be prepared to experiment thoroughly. They must provide money and manufacturing facilities for testing and for full-scale experiments. Men will not be lacking. In no branch of engineering have we ever had to want for men—aeronautics has special attractions which will ensure a steady supply of the best. But only if the prospects are sufficiently attractive. A stinting policy here will only result in other countries beating us. It has been our unhappy experience in the past in more than one science to see our brains and our energies wasted owing to lack of encouragement from those who could and should have given it. We have seen other countries gifted with more foresight take our ideas—and our men—and forge ahead. Eventually we have generally managed to regain some of our losses.

But in the keener struggle which is to come in every trade we must not go back to the old tactics, or we shall not find Fate so kind to us. It is to be hoped that Captain Tizard's lecture will cause aeroplane manufacturers to see that if they are prepared to treat their productions as other engineers do, to provide for testing and experiment on a liberal scale as is done in every other kind of profession, then they will reap their reward.

Captain TIZARD replied.

Lord SYDENHAM expressed on behalf of those present their indebtedness to Captain Tizard for his interesting and valuable paper.

A vote of thanks was then offered to Lord Sydenham for presiding, and the meeting terminated.

Document 2-7(a–b)

(a) Jerome C. Hunsaker, Assistant Naval Constructor, letter to H. M. Williams, Managing Editor, Aviation and Aeronautical Engineering, 11 March 1918, Hunsaker Collection, Box 1, Folder 1, File A.

(b) "Education in Advanced Aeronautical Engineering," *NACA Annual Report for 1920* (Washington, DC), p. 20.

The following two documents illustrate the nascent state of American aerodynamic education at the end of the second decade of the twentieth century. In the spring of 1918, Jerome Hunsaker was busy overseeing the design of airplanes, airships, catapults, and aircraft engines for the U.S. Navy. At that time he wrote to H. M. Williams, the managing editor of *Aviation* magazine, to praise their decision to publish a much-needed textbook on aerodynamics and aeronautical engineering. Two years later in its sixth annual report, the NACA published a resolution adopted in April 1920 calling for both the military services and American universities to establish courses in "advanced aeronautical engineering."

Document 2-7(a), letter from Jerome C. Hunsaker to H. M. Williams, 1918.

March 11, 1918.

Dear Mr. Williams:

I am very glad indeed to learn that you are to bring out in book form the "Course in Aerodynamics and Aeronautical Engineering" by Klemin and Huff, as published serially in your paper.

I know there is a real need for a thorough treatment of the subject such as this course presents. A large part of the work is fundamental and hence will not quickly pass out of date as is unfortunately the case with a great deal of technical literature.

Let me add to my good wishes for the success of the book, my suggestion that you expedite the printing.

Very truly yours,
Jerome C. Hunsaker
Asst. Naval Constructor, U.S.N.

Document 2-7(b), "Education in Advanced Engineering," NACA Annual Report for 1920.

At the semiannual meeting of the full committee in April, 1920, consideration was given to the question of education in advanced aeronautical engineering. This meeting was attended by all the members of the committee connected with universities: Drs. Ames, Durand, Hayford, and Pupin, and it is deemed worthy of special notice that each of these members individually expressed his approval of the resolution which was adopted at that meeting in the following terms:

Whereas it is deemed essential to the development of aviation in America for military and naval purposes that advanced instruction in aeronautical engineering be given to military and naval officers at a competent educational Institution; and

Whereas the public demand for such instruction will in all probability not be sufficient to justify or permit the offering of such advanced courses in more than one institution at the present time; and

Whereas such an advanced course is now being given at the Massachusetts Institute of Technology; and

Whereas it is deemed further essential that actual experience with aerodynamic research should form a part of such advanced Instruction: Therefore be it

Resolved, That the National Advisory Committee for Aeronautics hereby recommends to the Secretary of War and to the Secretary of the Navy the adoption of a continuing policy for the instruction of officers in advanced aeronautical engineering, and that for the next three years classes of 15 Army officers and 15 Navy officers be detailed annually to take such instruction in advanced aeronautical engineering at the Massachusetts Institute of Technology at the expense of the War and Navy Departments, respectively.

Resolved further, That, in connection with the course in advanced aeronautical engineering, the National Advisory Committee for Aeronautics cooperate in every way with the Massachusetts Institute of Technology by offering to its faculty and students the facilities for investigations in aerodynamics and experimental work on actual airplanes at the committee's research laboratory, Langley Field, Va.

Resolved further, That the National Advisory Committee for Aeronautics offer to give at various engineering universities courses of lectures in advanced aeronautical engineering by members of its engineering staff.

Resolved further, That the National Advisory Committee for Aeronautics recommend that educational institutions generally not consider the establishment of courses in aeronautical engineering at the present time, as it is the opinion of the committee that the demand for such instruction outside of the Government service is not sufficient, and competent instructors for such courses are not available.

This resolution was transmitted to the Secretary of War and to the Secretary of the Navy. The War Department, acting on the committee's recommendation, secured the necessary authority from Congress to detail 25 officers for special instruction at the Massachusetts Institute of Technology. It is understood that the Navy has not secured similar authority. The committee therefore strongly recommends to Congress that similar authority be given for the detail of naval officers for such special training. At the present time both services are weak in respect to the number of officers sufficiently educated in aeronautical engineering. The committee considers that the diligent prosecution of a continuing program of education will be of great value within a few years in the development of military and naval aviation.

Document 2-8(a–b)

(a) United States Army, excerpts from "Full Flight Performance Testing," *Bulletin of the Airplane Engineering Department, U.S.A.* 1, No. 2 (July 1918): 20–49.

(b) "Special Aerodynamic Investigations," *NACA Annual Report for 1919* (Washington, DC), pp. 27–28.

Early in World War I, military officials saw that the NACA's Langley Laboratory would not be ready in time to meet its research needs. Accordingly, the U.S. Army built a temporary facility at McCook Field, in Dayton, Ohio, and moved its Airplane Engineering Department personnel from Langley. Because the army was interested in practical research to quickly identify and solve problems that directly affected aircraft production, performance, and reliability, much of McCook's work involved flight testing. Recognizing that "the possibilities of error in full flight testing are very great," one of McCook's first investigations examined methods to standardize test methods and ascertain the accuracy of instruments such as altimeters and airspeed indicators.

"Full Flight Performance Testing" was one of McCook's earliest publications, and it shows the army's practical approach to flight research. The following excerpts include the report's introduction and the second section that describes the basic types of instruments recommended by the McCook Field engineers for recording aerodynamic data from flight tests.

Noting the success of McCook's approach, as well as the slow progress of construction at Langley, the NACA sought to work with the army at McCook. The committee, which included the commanding officer at McCook Field, Colonel Thurman H. Bane, approved a broad program of "scientific work" at McCook. Outlined under "Special Aerodynamic Investigations" in the *NACA Annual Report for 1919*, this work marked the beginning of a fruitful cooperation between the Army and the NACA.

Document 2-8(a), United States Army, excerpts from
"Full Flight Performance Testing," 1918.

AERONAUTICAL RESEARCH DEPARTMENT REPORT

The possibilities of error in full flight testing are very great, both as regards the use of instruments, the methods of observation, and the corrections applicable for varying atmospheric conditions. This article has been written as a definite summary of the subject to facilitate standardization of the methods of testing and recording of results. It deals solely with standard performance tests and climb and speeds at varying altitude, with no consideration of stability, controllability or radiator and engine performance.

In Section I is included a review of such physical data as is required for a complete understanding of the subject. The points considered in this section are: formulae for density of air; density in grammes per cubic meter and pounds per cubic foot; reduction to per centage of standard density; standard atmosphere; density values at various relationship between pressure, temperature and altitude at constant temperature; Halley's formula for temperature, and corrections; Bureau of Standards altitude pressure curve; calibration chart for altimeter, and an alignment chart for altitude.

In Section 2 are described the commonest and most useful forms of instruments employed in performance testing, with their principles and calibration. These include air speed indicators whose utility is obvious; barographs and altimeters for measuring pressures and allowing altitudes to be deduced therefrom; strut thermometers so that necessary temperature corrections may be made; tachometers whereby the r.p.m. of the motor may be obtained; statoscopes to enable the pilot to fly level at altitudes; recording drums used on various types of measuring instruments, and anemometers and wind vanes.

Finally, in Section 3 are described the methods employed in the calibration of the air-speed meter, in measuring climb and speed at altitude, and the various methods of correcting and recording results.

In the Appendix a standard form for recording results is submitted. This form has been adopted for use at McCook Field.

Simplicity of presentation rather than an exhaustive, scientific treatment has been sought.

[. . .]

Section 2
AIR SPEED INDICATORS

Several types of air speed indicators, based on a number of principles, have been employed in the past, such as pressure instruments with a plate balanced by a spring, rotating vane anemometers, and hot wire anemometers. For a number of practical and theoretical reasons, instruments of this type have been discarded, and attention is now concentrated on instruments measuring differences of pressure transmitted from two Pitot tubes or Pitot and Venturi tubes.

Pitot and Venturi Tubes

Since all air speed indicators based on pressure differences are of the Pitot, or Pitot and Venturi type, a simple explanation of the principles involved will be included.

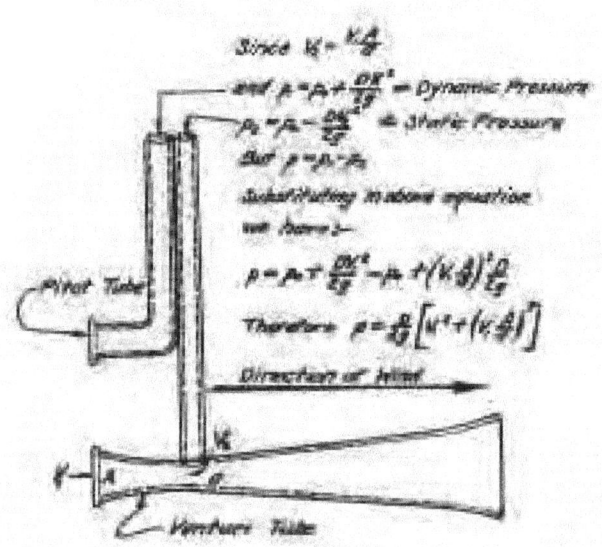

Fig. 9—Combination of Pitot and Venturi tubes

(a) *Pitot tube.*—In Fig. 10 is shown a Pitot tube in diagrammatic form. It consists of two concentric tubes, the inner open to the wind, the outer closed and communicating with the current of air only by a series of fine holes. The tubes are connected to the two arms of a pressure gauge, which measures the difference in pressure between them. The inner tube, open to the wind, brings the air impinging on it to rest, and the pressure on it is, therefore, a measure of both the static pressure in stream and of the kinetic energy head of the stream. If p is the static pressure of the stream, V the velocity, the total pressure on the inner tube will be given by

$$p + \frac{DV^2}{2g}$$

The outer tube, on the other hand, being closed to the wind, will, if the holes are small enough, read the static pressure of the air flow p.
Hence the differences in pressures read on gauge will be

$$\frac{DV^2}{2g}$$

and the gauge reading will be a measure of the velocity.
Pitot tubes with suitable gauges are widely used in laboratory practice, but owing to the small difference in head $DV^2/2g$, the forces acting on the gauge are very small and hard to record.

(b) *Combination of Pitot and Venturi.*—To increase the pressure differences, and thus get practicable forces on the gauges, the Venturi tube is coupled with the pressure part of the Pitot. Such a combination is shown diagrammatically in Fig. 9. Here the velocity at the throat will be considerably greater than that acting on the suction side of a Pitot, and therefore has a considerably greater effect. The mathematical theory of the Venturi is a little more complicated than that of the Pitot and the theoretical suction heads are not always in accord with practical results. From the simple formula of Fig. 9 it can be seen that the gauge readings will be proportional to $DV^2/2g$, hence are a measure of the velocity.

Correction for density and reduction to standard density in air speed meters.—From the preceding considerations of the Venturi and Pitot tubes, it is seen that the forces on the gauge are proportional to $A D V^2$, where D is the density, V the velocity and A some constant depending on the instrument. The airspeed reading equation is therefore

$$R = ADV^2$$

Airspeed meters being calibrated at 16 deg. C. and 760 mm., they will only be correct at the standard density D corresponding to this condition.

If the instrument gives a certain reading V_1 at density D_1 then true reading V_t

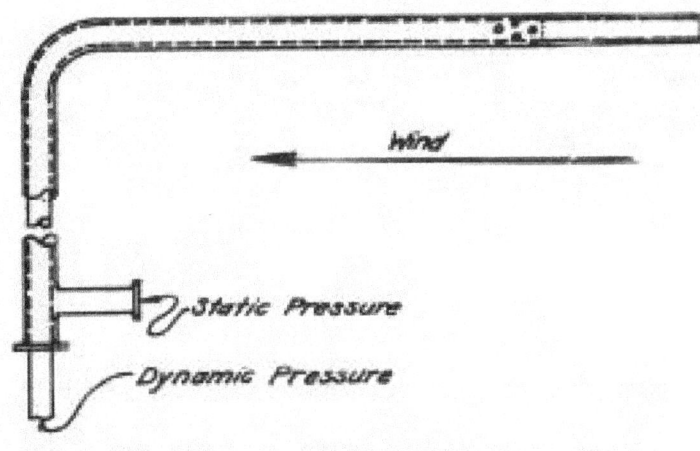

Fig. 10—Standard form of Pitot tube

will be given by equation: $AD_1V_t^2 = AD_0V_1^2$

and $$V_t = V_1 \sqrt{\frac{D}{D_1}}$$

Since $$\frac{D}{D_1} = \frac{273-t}{27+t+h} \cdot \frac{760}{P}$$

$$V = v \sqrt{\frac{273 + t}{273 - t_1} \cdot \frac{760}{P}}$$

The correction can be applied either by computation or from the curves of Figs. 2 and 3 where densities for varying temperatures and pressures are given as per centages of the standard density, as well as the values of the ratios ÷D0/D1. On certain occasions it may be quicker to use table 3.

Approximate air speed correction at heights.—A very useful table furnished by the Technical Department, British Aircraft Production, allows air speed corrections at heights to be made with fair accuracy, on the assumption of certain standard conditions. Its use is not recommended for the computation of performance results, but may be very handy as a check. The table employs mean value of the density at various altimeter heights. Ground temperature of 16 deg. C. and pressure of 760 mm. are assumed, and a lapse rate of 1.75 deg. C. per 1000 ft. ascent.

Table 3
Multiplying Factors to Reduce Speed Readings at Varying Pressure and Temperatures to Standard Density Pressure in Millimeters of Mercury

Temperature, degrees centigrade	200	250	300	350	400	450	500	550
-40∞	1.750	1.566	1.429	1.324	1.236	1.168	1.109	1.057
-30	1.790	1.599	1.459	1.351	1.265	1.191	1.130	1.080
-20	1.830	1.631	1.492	1.380	1.290	1.218	1.155	1.100
-10	1.852	1.661	1.521	1.407	1.318	1.239	1.175	1.121
0	1.899	1.673	1.550	1.430	1.340	1.264	1.120	1.142
10	1.924	1.729	1.574	1.459	1.365	1.286	1.220	1.162
20	1.961	1.757	1.602	1.483	1.388	1.308	1.241	1.182
30	1.990	1.787	1.630	1.508	1.410	1.330	1.261	1.204
40		1.815	1.661	1.533	1.437	1.354	1.281	1.224
Temperature, degrees centigrade	600	650	700	750	800	850	900	950
-40∞	1.010	.944	.936	.904				
-30	1.031	.982	.956	.925	.894			
-20	1.051	1.012	.975	.942	.911			
-10	1.075	1.031	.995	.960	.930	.902		
0	1.094	1.051	1.012	.978	.949	.920	.893	
10	1.115	1.070	1.030	.996	.965	.937	.910	
20	1.132	1.089	1.050	1.012	.982	.954	.926	
30	1.151	1.098	1.068	1.030	.997	.979	.941	
40	1.170	1.120	1.078	1.050	1.013	.984	.958	

Typical Recording Air Speed Meter

In the Toussaint-Lepère air speed meter the dynamic pressure of the wind is measured by a combination of Pitot tube and Venturi meter. This pressure is transmitted to a clock work recording device by a gauge consisting of bellows and a tension spring.

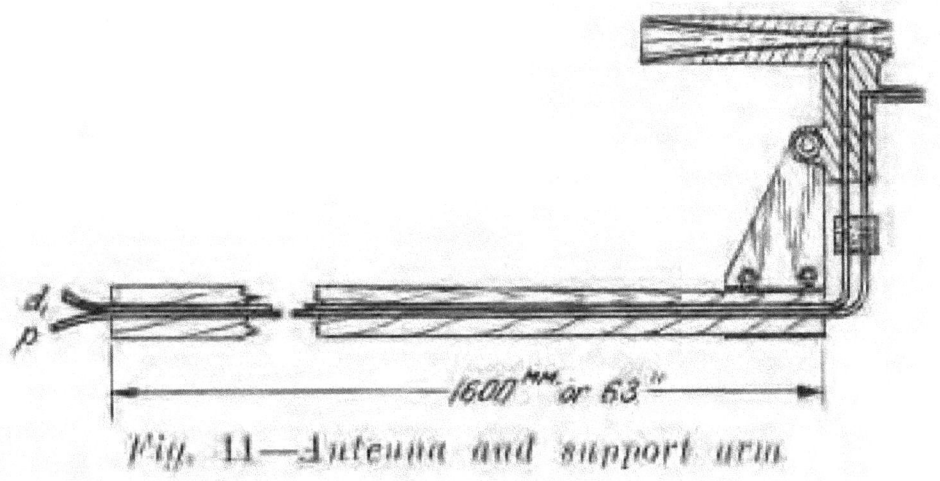

Fig. 11—Antenna and support arm.

The Pitot tube and Venturi meter are combined in a small casting conveniently called by the French *antenna* and similar to that of many other speed indicators. This antenna is shown in Fig. 11. The Venturi is carefully proportioned to give the maximum possible suction with a given air speed. The antenna is supported by a long, slender, hollow arm of light wood which contains the tubes transmitting the pressure to the recording device as shown in Fig. 12. It is fastened to this arm by a light, adjustable clip, in order that the antenna may be turned directly into the wind.

The recording device is shown in Fig. 14 and diagrammatically in Fig. 13. It has the ordinary clockwork drum and pen. These are described elsewhere. The gauge consists of two movable circular plates S1 and S2, rigidly connected by a rod ab. The plates form the tops of the bellows f1 and f2. The sides of these bellows are made of thin rubber that is very flexible, the bottoms are formed by the fixed plates m and n. The suction from the Venturi is led to the airtight chamber c-c, and so acts on top of the plate S1. The pressure from the Pitot is led to the under side of the plate S2. The top of S2 and the bottom of S1 are open to the air inside of the box. Thus a variation of that pressure causes no motion of the rod ab which is moved only by the difference of the pressure transmitted from the antenna. The rod ab is constrained to move vertically by the form bar linkage a-d-c-b. The link bc carries on one end the marking pen g; on the other a counter weight for the movable parts of the instrument. At the end of this link is fastened the spring R, whose tension balances the pressure of the pen. This spring is so placed that the displacement of the pen is nearly proportional to the wind speed. The recording apparatus is enclosed in a box about 9 in. x 6 in. x 5 in., total weight about 4 ¼ lbs. The apparatus slides out of this box to facilitate adjustment of paper on the drum.

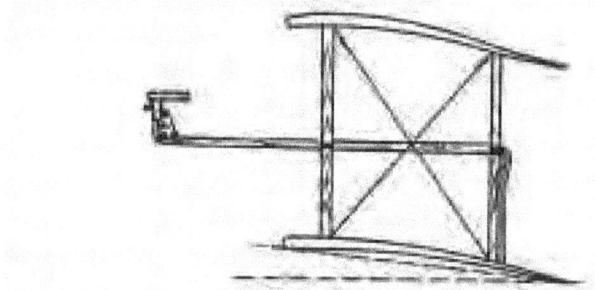

Fig. 12—Installation of Toussaint-Lepère air speed meter

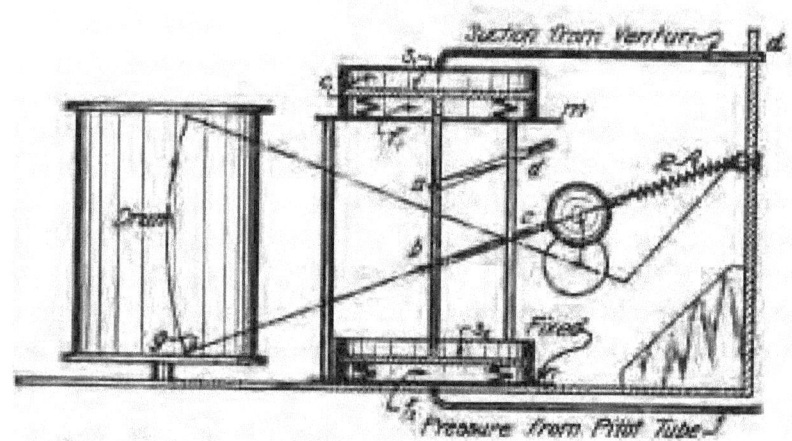

Fig. 13—Recording apparatus of Toussaint-Lepère air speed meter

The complete speed indicator must be calibrated and a chart or table made for converting the readings on this drum into true wind speeds. This chart of course is only correct for readings in air of standard density.

The recording apparatus is suspended in the air by elastic cords or may be held by the passenger in a two-seater. The antenna must not be placed near any obstructions or disturbance including the slip-stream, body, etc. The supporting arm is fastened to any convenient part of the airplane such as a strut by tape or a fitting (see Fig. 21). The antenna is then adjusted to point directly into the wind. With this instrument as with all air speed meters, a test run in flight must be made over a measured course to determine the effect of interference of the plane upon the air flow to the antenna, and to find the correction due to this interference.

The Foxboro-Zahm Direct Reading Air Speed Meter

In Figs. 15 and 16 are shown views of a very widely used combination of the Foxboro indicating box and the Zahm Pitot-Venturi tube (now adopted as standard by the Signal Corps). The pressure lead of the Pitot enters the small cylinders located in the indicating case which in itself is made air-tight by a gasket under the cover. The, suction of the Venturi is transmitted to the case itself. When a difference of pressure exists between the inside and outside of the two cylinders, they elongate or contract. The motion is transmitted to the pointer by means of links to a circular rack which engages a pinion on the spindle.

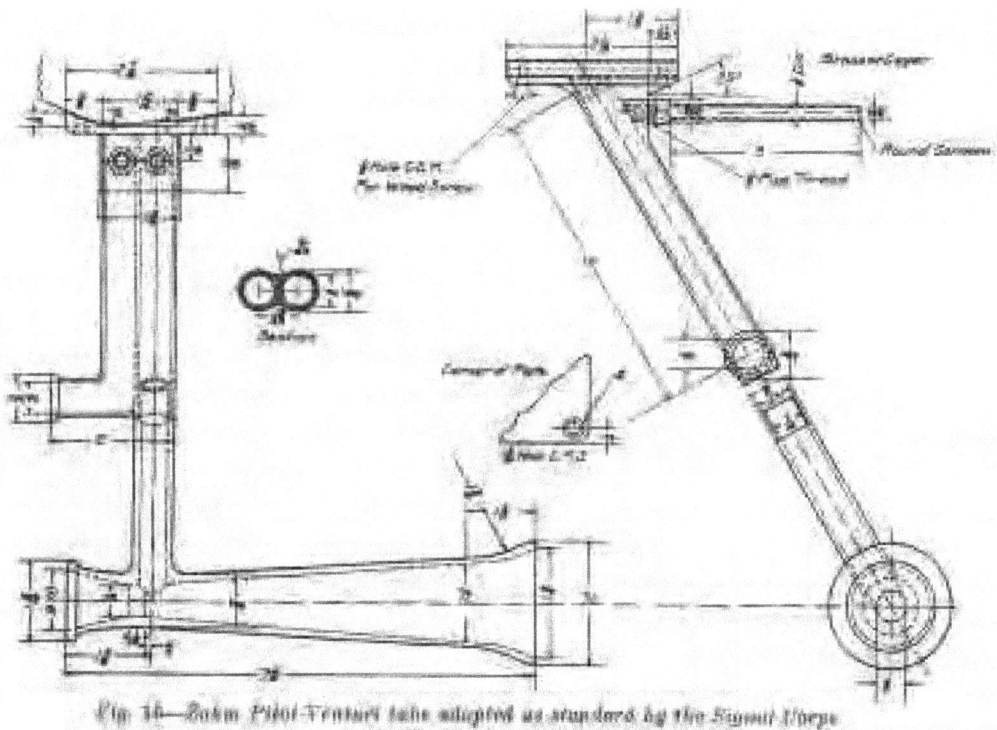

Fig. 16—Zahm Pitot-Venturi tube adopted as standard by the Signal Corps

It may be useful to include also the old type Foxboro head which is widely used. This is shown in Fig. 17. The tube in front presents a large opening to the wind. In this opening is fitted a conical guard pointing into the wind, behind which is located the opening to a small pipe. This pipe is the only outlet from the cup-shaped opening, and it transmits the dynamic pressure to the gauge. The wind passing the tube creates a suction in the space inclosed by the frustrum of the cone. A small pipe, seen in the photograph at the base of the cone to the left, transmits this suction to the gauge.

In airspeed meters, since the movement of the aneroid boxes is proportional to the square of velocity, the scale on the dial is not uniformly graduated; and were it not for a compensating device, the divisions of the scale for the higher velocities would increase rapidly as the velocity increased. The small springs fastened to the aneroid boxes shown in Fig. 15 restrict the movement of the boxes and shorten the scale divisions.

Table 4
Air Speed Corrections at Heights

Apparent speed instrument reading m.p.h.	Corrected Speeds at Heights (m.p.h.)			
	6500 ft.	10000 ft.	15000 ft.	20000 ft.
40	44	46	50	54
45	49	52	56	60
50	65	58	62	67
55	60	63	68	74
60	66	69	75	80
65	71	75	81	87
70	76	81	87	94
75	82	86	93	100
80	87	92	99	107
85	93	98	106	114
90	98	104	112	120
95	104	109	118	127
100	109	115	124	131
105	115	121	130	140
110	120	127	137	147
115	120	132	143	154
120	131	138	140	161
125	137	144	155	[170]
130	142	150	161	171
135	147	156	168	171
140	153	161	174	187
145	158	167	180	194
150	161	173	186	201
155	169	179	193	207

Miscellaneous Air Speed Meters

The British R.A.F. IV-A air speed indicator head is of the Pitot type modified in construction. The dynamic pressure tube and the static pressure tube are entirely separate, held parallel about 2 in. apart by a small fitting. The dynamic tube is just a plain tube open to the wind, but the static pressure tube is closed by a small streamlined cap. The holes are drilled well back along the cylindrical part of the tube. The leads are separate and the head is very easy to make. The gauge, shown diagrammatically in Fig. 18, is typical of most of those used for this type of air speed indicator. The whole gauge is made airtight by rubber gaskets. Inside there are two diaphragms A and B made of thin flexible metal. The top of B and the bottom of A are fixed to the case. The movable ends push through small rods to the cross-arm C on the spindle D. At the end of this spindle is an arm E which engages a quadrant suitably geared to the pointer. The motion of the pointer is opposed by a light hairspring. The dynamic pressure is led to the inside of the diaphragms, the static merely inside of the case. The diaphragms therefore tend to expand and so the gauge is sensibly independent of gravity and centrifugal force, and entirely free of the pressure in the cockpit. Other makers of gauges have different managements of diaphragms and different mechanism, but the principle is the same.

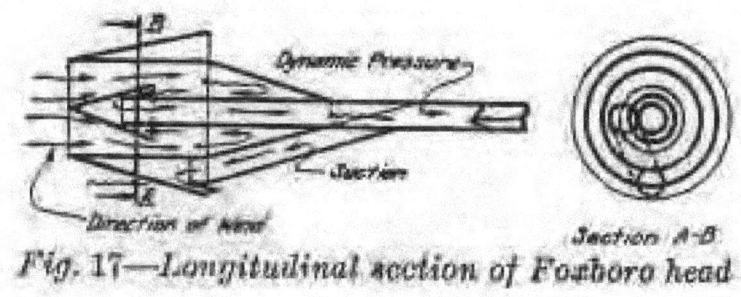

Fig. 17—Longitudinal section of Foxboro head

Another British instrument, the Ogilvie indicator-head is merely a Pitot tube. The gauge is different from the usual type. It has a single airtight chamber divided into two parts by a flexible rubber diaphragm. The static pressure is on one side, the dynamic on the other. A light silk thread is attached to the center of this diaphragm. The thread is kept taut at all times by a very light hairspring. The whole mechanism is very delicate, almost too fragile for rugged work. Later types of Ogilvie indicators have a gauge made similar to the R.A.F. IV.-A.

Badin Double Venturi Head

The Badin type of head is a double Venturi meter as in Fig. 19. The small inner meter has its exit at the throat of the outer meter. This greatly increases the suction at a given wind speed; a very desirable quality, especially on slow speed machines or dirigibles. The Badin system appears to have only the suction lead from the head to the gauge. This is not good practice as the total pressure in the cockpit may be quite different from the static pressure at the head.

The Sperry Venturi speed meter is also of the double Venturi type. There are

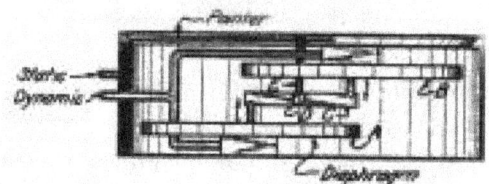

Fig. 18—Section of British R. A. F. Mark IV. air speed indicator gauge

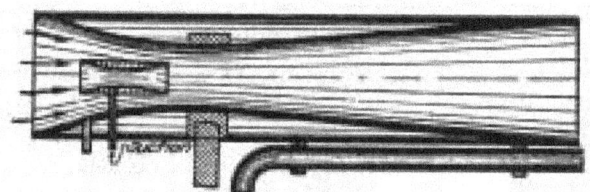

Fig. 19—Badin double throat Venturi meter

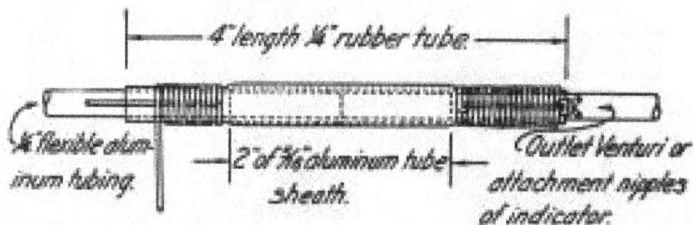

Fig. 20—Showing method of connecting aluminum tubing to Venturi meter and air speed meter

three leads for pressure difference. One, the suction, from the throat of the small Venturi, one from an open tube pointing back along the outside of the outer Venturi, and the third from the front edge of the case of the outer Venturi. Readings may be taken between the first and either of the others. There seems to be a distinct disadvantage in using the tube pointing back for this gives a reading less than the static pressure and so reduces the available pressure difference. For use on high speed airplanes there is a Sperry head of the Pitot tube type.

Important Practical Points

Attachment of Pitot tube heads.—A simple attachment as shown in Fig. 21 is customary, but this is open to the objection that the air stream is interfered with by the strut. It is much better to offset the instrument from the strut.

Connecting up Venturi-Pitot tube with airspeed indicating instrument.—The Venturi-Pitot tube is connected with the airspeed indicator by flexible aluminum tubing. In using the aluminum tubing sharp bends and kinks must be avoided. It is absolutely necessary that the connections at all joints be airtight. For this reason the following method is recommended for making connections between the aluminum tubing and the outlets of the Venturi, or the nipples of the indicators, or between sections of the tubing. (See Fig. 20.)

(1) Slip a 4 in. length of standard rubber tubing, $\frac{1}{4}$ in. bore over the 2 in. length of the $\frac{5}{16}$ dia. aluminum sheath, so that the ends of the rubber tube extend 1 in. beyond the extremities of the sheath.

(2) Butt the ends of the aluminum tube and the connection, and slide the sheath in the rubber tube over the joint so that the joint comes at the middle of the sheath.

(3) Bind the two ends of rubber tubing with wire. First tie the wire near the sheath with a simple knot, leaving one short end free, which is pressed down along the tube and bound under. The wire is wrapped around the tube and when the wrapping is finished the two ends of the wire are twisted together and cut off, leaving a $\frac{1}{4}$ in. stub to prevent slipping. In binding the rubber care should be taken not to cut it.

Document 2-8(b), "Special Aerodynamical Investigations," NACA Annual Report for 1919.

SPECIAL AERODYNAMIC INVESTIGATIONS.

In the summer of 1919 the executive committee approved a program of scientific work to be carried out at McCook Field, Dayton, Ohio, under the supervision of its aerodynamical expert, Dr. George de Bothezat. This work involved: First, the theoretical analysis of the full performance of an airplane in steady flight; second, the development of new instruments and methods in order to measure in a single test flight the full performance characteristics of an airplane; third, the analysis of the full performance record of an airplane and deductions there from as to how the efficiency of an airplane can be increased by minor changes; fourth, the making of such minor changes in a given type of airplane to be followed by a second full performance test; fifth, the checking of the results against the original theory and the necessary modifications of the theory to permit in the future the determination of all the performance characteristics of an airplane in steady flight by mathematical calculation.

Dr. de Bothezat has been stationed at Dayton since August, 1919, and a small staff has been selected from members of the engineering division of the Air Service and assigned by the commanding officer at McCook Field to work with him in the prosecution of this work. The present commanding officer at McCook Field, Col. Thurman H. Bane, is also a member of the National Advisory Committee for Aeronautics. This work has been successfully inaugurated with the hearty cooperation of the officials and civilian engineers of the Air Service.

Document 2-9(a–c)

(a) "Office of Aeronautical Intelligence," *NACA Annual Report for 1918* (Washington, DC), pp. 24–25.

(b) John J. Ide, Technical Assistant in Europe, to the NACA, excerpts from "Report on Visit to England, July 1–22, 1921," 4 August 1921, Ide Collection, Box 8, National Air and Space Museum, Washington, D.C.

(c) John J. Ide, Technical Assistant in Europe, to the NACA, "Wind Tunnel at Issy-les-Moulineaux," 8 December 1921, Ide Collection, Box 8, National Air and Space Museum, Washington, D.C.

While the young NACA actively pursued the construction of a laboratory to seek new aeronautical knowledge as America fought in a world war, the committee recognized that considerable progress continued to be made in other countries and that some means of managing the growing body of "scientific and technical data relating to aeronautics" was essential. Thus, the committee established the Office of Aeronautical Intelligence in 1918. As outlined in the *NACA Annual Report for 1918*, the Office of Aeronautical Intelligence worked closely with the military and naval intelligence offices, including "special committees stationed at London, Paris, and Rome to collect information regarding all phases of the scientific and technical study of war problems."

In December 1918, a naval reserve ensign by the name of John J. Ide wrote the chief of naval operations (aviation) requesting that he be assigned to the naval attaché at the American Embassy in Paris to follow the progress of European aviation. Although the navy originally denied his request, Ide persisted and ultimately received a Paris posting in 1921 as the NACA's technical assistant in Europe. In this role, Ide repeatedly visited manufacturers and aeronautical laboratories throughout Europe, and he submitted many detailed reports on airplanes, wind tunnels, instruments, engines, and other aviation-related matters to the committee on a regular basis. The two documents reproduced herein are typical of Ide's reports. The reader will note that Ide took great care not to judge the merits or applicability of what he observed, but rather to report what he had learned as factually as possible.

The NACA treated these reports as confidential, but the information proved useful to the American military, manufacturers, and the staff at Langley. For an example, note Ide's description of an optical method to observe propeller blade bending under the "Aerodynamical Department" section of his 4 August 1921 report. While not totally successful in England, there can be little doubt that this idea influenced the Langley researchers who developed a successful technique to optically measure blade bending in the Propeller Research Tunnel a few years later.

Document 2-9(a), "Office of Aeronautical Intelligence," NACA Annual Report for 1918.

OFFICE OF AERONAUTICAL INTELLIGENCE.

In January, 1918, the need for a central governmental depository in Washington for scientific and technical data relating to aeronautics was recognized, and the Aircraft Board suggested that the National Advisory Committee for Aeronautics was the logical governmental agency for the collection and classification of such data to be made available to the military and naval air services in this country. This committee, accordingly, established an Office of Aeronautical Intelligence and adopted rules and regulations for the handling of its work.

The committee has made the necessary arrangements at home and abroad for the collection of such data. There are many sources of obtaining such information, the chief at the present time being the research information committee, organized under the National Research Council in January 1918, by funds provided by the Council of National Defense. It consists of the Director of Military Intelligence, Director of Naval Intelligence, and Dr. S. W. Stratton as chairman.

The purpose of the research information committee is to serve as a collector and distributor of scientific and technical information regarding all war problems. Special committees stationed at London, Paris, and Rome collect information regarding all phases of the scientific and technical study of war problems and transmit the same to the central committee in Washington for distribution to the interested services. Similarly, these special committees receive information from Washington and transmit the same to the interested services abroad.

Since February, 1918, Dr. William F. Durand, chairman of the National Advisory Committee for Aeronautics, has served as scientific attaché to the American Embassy in Paris, representing the National Research Council on this research information service, and has, in addition, acted as special representative of the Aircraft Board at the International Aircraft Standardization Conferences in London in February and in October, 1918, besides serving as a special liaison officer in aeronautical matters between France and the United States.

In September, 1918, Dr. W.C. Sabine, head of the department of technical information of the Bureau of Aircraft Production and a member of the National Advisory Committee for Aeronautics, was placed in charge of the Office of

Aeronautical Intelligence of the National Advisory Committee for Aeronautics, with the title of director of scientific and technical data.

Many valuable documents dealing with important research problems in aeronautics have been secured by the Office of Aeronautical Intelligence, and copies have been distributed to those concerned with the problems involved.

The committee has established in connection with its Office of Aeronautical Intelligence, and particularly for the use of its engineering staff, a small selected library, containing the most useful and valuable aeronautical and technical books and publications.

Document 2-9(b), John J. Ide, excerpts from
"Report on Visit to England, July 1–22, 1921."

American Embassy,
7 Rue de Chaillot, Paris, XVIe
August 4, 1921.

CONFIDENTIAL

From: Technical Assistant in Europe, U.S.N.A.C.A..
To: National Advisory Committee for Aeronautics, Washington, D.C.
Subject: Report on Visit to England, July 1–22, 1921.

On July 1st I went from Paris to London.

R.A.F. Pageant.

On July 2nd I witnessed the R.A.F. Pageant at Hendon. The Pageant was a remarkable display of the proficiency attained by the Royal Air Force in formation flying, fighting, stunting, and bombing. The airplanes used, with one exception, were standard service types developed during or shortly after the war. The exception was the Siddeley "Siskin," which replaced the Westland "Wagtail" in the mock duel with the Nieuport "Nighthawk."

Siddeley "Siskin".

The Siddeley "Siskin" (Figs. 1 and 2) has been built by the Armstrong-Siddeley Company of Coventry. It is a single-seater fighter, originally designed to take a 320 HP A.B.C. "Dragonfly" radial engine. After the failure of this engine to fulfill expectations, the design of the "Siskin" was slightly changed to accommodate the Siddeley 300 HP 14 cylinder radial engine. This engine, although considerably heavier than the "Dragonfly," is very reliable and develops its rated power. The "Siskin" clearly outmaneuvered the "Nighthawk" in the mock fight.

As seen from the illustrations, the "Siskin" has one pair of inclined struts on each side of the fuselage. Ailerons are fitted only to the upper plane which has a certain amount of overhang. Two sets of struts run from the fuselage to the upper plane, forming two W's.

Tests have been discontinued for the present with a single cylinder water-cooled engine of 8 x 11 in. bore and stroke. It has developed 120 HP and there are rumors that Beardmore is going to construct a slow speed ungeared airship engine of six of the cylinders, the total weight to be 1600 lbs. Tests are proceeding with direct fuel injection with a monosoupape air-cooled cylinder. The fuel is forced into the cylinder under pressure, mixing with air sucked in through the valve.

I saw a Siddeley 150 HP 7 cylinder radial engine being tested. The Siddeley radials, designed by Capt. Green, are very highly considered at the R.A.E. While fairly heavy, the 150 HP model weighing 400 lbs. complete, they are very reliable.

The R.A.E. is still occupied with redesigning the A.B.C. "Dragonfly" engine. As redesigned, the aluminum heads overlap the steel cylinders by 1 inch. The heads have fins to assist cooling. The crankshaft and master rod have been made heavier, and the induction system is quite new. Six of these engines have been finished to be fitted to Nieuport "Nighthawks."

A universal engine test bench, similar to the standard type but lightened, has been constructed for the purpose of being installed in a Handley Page, the fuselage of which has been fitted with an aluminum lined chamber to take the bench. A small propeller, placed in the nose of the fuselage, is to be connected with the engine under test. Mr. Smith stated that it was possible that tests with this apparatus might not be carried out as one of the engine testing rooms is to be converted into an altitude chamber, Squadron Leader Norman having studied the installation at the Bureau of Standards while in America.

<u>Aerodynamical Department.</u>

I was taken through the 7 x 7 ft. and the 4 x 4 ft. wind tunnels. The screen fitted in the newer 7 x 7 ft. tunnel, which has an engine of 200 HP, effects a saving of power in the order of 15% by smoothing out the air flow. This screen consists merely of a brick wall in which there is an air space between each brick.

Considerable work has been done in propeller design for wind tunnels. Contrary to expectations, a four-bladed propeller gave more even flow than one with six blades.

Observations have been made of the bending of propeller blades. Small white squares have been marked along the otherwise black blades of a propeller. The bending effect at various speeds has been observed by means of reflected light. These experiments have come to a standstill owing to inability to express mathematically the curve of the blade.

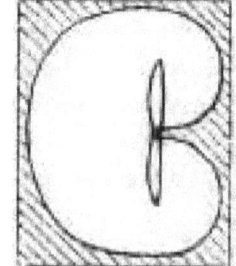

A small propeller has been enclosed in a box suitably shaped on the inside so that the propeller makes its own air. The section of the box is shown in the sketch given herewith.

Through windows on the side of the box the behavior of the air stream is observed by the movements of threads suspended from another thread.

The experiments in recording the pressures along propeller blades have been interrupted by an accident, the plate holding the manometers having broken loose and smashed them. Contrary to previous reports, no difficulty was experienced with fastening the tubes to the propeller blades.

[. . .]

<u>National Physical Laboratory.</u>

On July 19th I visited the National Physical Laboratory at Teddington, where I was taken through the wind tunnels by Mr. Nayler.

The 7 x 14 ft. wind tunnel has been completed and is in operation. The power is supplied by two 200 HP engines with a synchronized gear which has been developed by Vickers. A perforated brick wall, similar to that used at the R.A.E., divides the room into two sections. The principal advantage of this wall according to Mr. Nayler, is the fact that the wind tunnel can be made very much shorter with it than without it, thus saving a considerable amount of space. No balance has yet been installed in this tunnel. At present the 7 x 14 ft. tunnel is being used for the measurement of rotary derivatives, such as the rolling due to rolling (Lp), the yawing due to rolling (Np). and the rolling due to yawing (Lr). A model of an S.E.5 suspended on wires as shown in the accompanying sketch, is used for these measurements.

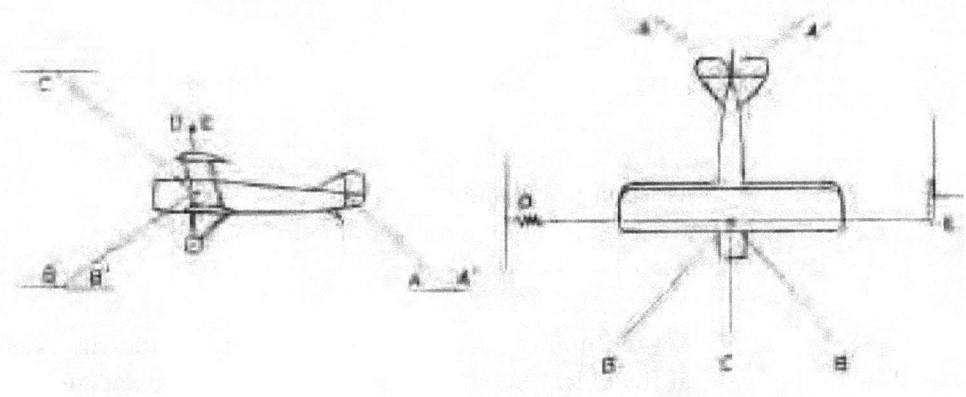

The amplitude of yaw or roll is ascertained by measuring the movement of a spot of light reflected from a lamp by a mirror on the side of the fuselage to a strip of thin paper placed at the observer's station at the side of the tunnel. The model is placed in motion by moving the arm (E), to which one end of the wire running thwartships above the model is secured, there being a spiral spring at the other end of this wire.

One of the 7 x 7 ft. wind tunnels is being used for pressure plotting over a model rigid airship hull of the R 33 type. Mr. Nayler stated that this test is a result of the fact that the bow girders of several Zeppelins collapsed when making turns at high speeds. Although the British had as yet experienced no trouble from this source, it was desired to ascertain the pressures to which the airship was subjected.

In another 7 x 7 ft. tunnel there is being conducted a series of tests for thrust and torque of a family of air screws. There are six air screws in the family, all having the same section but with various pitch diameter ratios. The diameter of all the air screws is 3 ft. 6 in.

Mr. Nayler also conducted me through the material testing laboratory. A new method of fatigue testing of metals has been developed by Mr. Gough of the N.P.L. It is expected that a report of this method will shortly be published in "The Engineer," as the manuscript has been accepted.

By Mr. Gough's method, the ultimate strength of the material under test can be determined in about 15 minutes instead of the considerable number of hours necessary by the present methods. The new method is based upon the principle that a change in molecular construction resulting in an increase in the amplitude of vibration occurs after a short period of test. Light is employed for the measurement of the vibrations.

On July 22d I returned to Paris from London.

Respectfully,

[signed] John Jay Ide.

Document 2-9(c), John J. Ide, "Report on Wind Tunnel at Issy-les-Moulineaux," 1921.

December 8, 1921

From: Technical Assistant in Europe, U.S.N.A.C.A.
To: National Advisory Committee for Aeronautics, Washington, D.C.
Subject: Wind Tunnel at Issy-les-Moulineaux.

I recently visited the wind tunnel under construction at the headquarters of the French Aeronautical Technical Service at Issly-les-Moulineaux outside the gates of Paris. This tunnel is housed in a building of brick, steel and glass about 100 ft. wide and 210 ft. long which was almost completed at the time it was decided to use it for a wind tunnel. The building is not particularly well adapted for the purpose having numerous columns and trusses which project into the interior and impede the smooth flow of air around the tunnel. Also, not much advantage can be taken of good atmospheric conditions by admitting outside air as there is only one large door which is at the entrance cone end.

The tunnel itself, of the Eiffel type, is constructed entirely of reinforced concrete and it was made to the designs of the Aerodynamical Section of the Technical Service by the company of which M. Caquot, formerly director of the Technical Section, is the head. The tunnel is in a large cement pit about eight feed deep and taking up the entire floor space of the building with the exception of gangways along the sides and ends. The tunnel is supported by two concrete walls running longitudinally and about eight feet high leaving a free air passage under the tunnel. The entrance nozzle is set back very far from the front end of the building. The section of the outer walls of the collector has been made square instead of circular and of a width equal to the diameter of the front end of the collector. This permits the outer walls of the collector to be parallel to the walls and floor of the room and also eliminates any break in contour caused by the experimental chamber. Aft of the latter, the outer walls are gradually faired into the diffuser.

The diameter of the tunnel at the experimental section is 3 meters (9.84 ft.) and at the propeller end 7 meters (22.96 ft.). The propeller itself is to have 6 blades with square tips and each blade is to be an arc of a circle. The pitch of the blades will be variable, though not while in motion. It was stated that the efficiency of the propeller was in the neighborhood of 75 per cent. The propeller will be driven by an electric motor of 1000 H.P. which it is expected will enable a speed of 80 meters (262 ft.) per second to be realized.

Behind the diffusor is the concrete stand for the motor. In order to change the direction of the air leaving the diffusor the two forward faces of the stand are curved thus in plan:

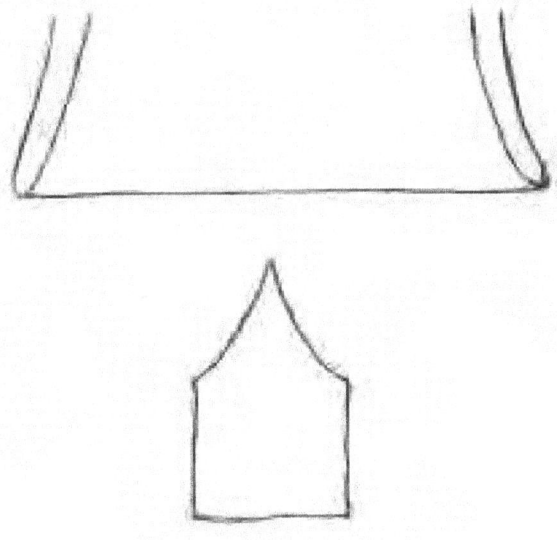

It is proposed to continue these curves upward and outward by wooded partitions.

Entrance to the experimental chamber is by means of a passage with a staircase arranged in the thickness of one of the supporting walls of the tunnel. There are two doors to insure an air lock. The chamber itself is very large being roughly a cube of about 20 ft., a side into which project the ends of the diffusor and collector. The only natural light is that which comes through the latter. In the centre of the floor is a square trap closed by a platform. When it is desired to have large models, etc. brought into the experimental chamber, movement of a lever lowers the platform sufficiently for it to be rolled clear of the opening on rails attached to the inner faces of the two longitudinal walls below the tunnel.

Arrangements will be made so that the tunnel can be partially or completely closed at the experimental section if desired. There are a number of hooks flush with the concrete surface of the diffusor and collector cones near the experimental section. These hooks can be pulled outward and to them can be attached the supports of the cones of a small tunnel of 80 cm. (2.62 ft.) diameter placed concentrically with the large tunnel. By using the same propeller and the power available for the large tunnel it is expected that a speed of 400 meters (1312 ft.) per second will be reached by the 80 cm. tunnel.

The model suspension will be by wires instead of by a spindle. It is understood that the balance will be of a dynamometric type having various gauges to which the wires supporting the models will be attached, according to the strength of the forces to be measured.

It is hoped to have the wind tunnel completed by July 1922.

[signed] John Jay Ide.

Document 2-10(a–b)

(a) D. W. Taylor, letter to Frederick C. Hicks, 9 December 1919, Hunsaker Collection, Box 3, File H.

(b) Edward P. Warner, excerpts from "Report on German Wind Tunnels and Apparatus," October 1920, Box 18, Folder 24, McDermott Library Archives, University of Texas–Dallas. [Also published, with the same title, in *Aerial Age Weekly* (15 November 1920): 275–277.]

While the NACA worked to gather information through its Office of Aeronautical Intelligence, other Americans with a particular interest in foreign developments did all they could to stay abreast of these activities as well. U.S. Navy officer Jerome Hunsaker was among the first to cultivate friendships with aeronautical experts outside the United States. After a tour of the primary European laboratories in 1913, he maintained these relationships during and after World War I. As the 9 December 1919 response—from his superior Rear Admiral D.W. Taylor to an inquiry from Congressman Frederick C. Hicks—shows, the expert intelligence provided by sources like Hunsaker proved useful to politicians and policy makers as well as researchers. Of particular interest are the comments on Japan's purchase of French technology to aid its own fledgling aviation industry. Although relative newcomers to aeronautics in 1919, Japanese progress over the next two decades would provide the island nation with formidable aerial weapons for World War II.

Edward P. Warner, an aviation consultant who had been the NACA's first chief scientist, prepared an independent report to the NACA covering his 1920 inspection of German laboratories at Göttingen, Aachen, and Dessau that he also published in *Aerial Age Weekly* later that year. Coverage of the Aachen and Dessau installations was brief, but Warner went into considerable detail concerning Ludwig Prandtl's second wind tunnel at Göttingen. The detail was warranted, for this was the first truly modern, closed-circuit wind tunnel, with efficient turning vanes in the corners and a variable cross-sectional area for improved air management, lower power consumption, and reduced turbulence. Prandtl's balance apparatus for this tunnel also impressed Warner with its accuracy and simplicity of operation.

Document 2-10(a), letter from D. W. Taylor to Frederick C. Hicks, 1919.

December 9, 1919

My dear Mr. Hicks:

Referring to your letter of the 4th inst., relative to the Aviation Program of Foreign Countries, my information in regard to Aviation in Japan is not very full, but it is possible that Captain Craven can give you something additional. The last we have heard about Japan was that they had a mission of some 24 officers of the army and navy, together with civilian professors representing the engineering colleges in Japan, which has made an inspection of aviation in England, France, Italy, and the United States and has now returned to Japan and, presumably, will make definite recommendations.

There is a rumor that the Japanese have purchased the rights to manufacture one or two French aviation engines developed during the War and are equipping one of their arsenals to turn out this engine. I understand that French experts have gone, or will go, to Japan to assist in this development. The Japanese are also supposed to be purchasing in France planes left over as excess War material. I have no information that they have purchased anything in the line of large flying boats. I know that they made no attempt to do so while in the United States.

With regard to Italy, I don't believe that the Government has under way any military or naval program, but the Caproni Company is attempting to develop their large bomber for passenger carrying with the idea of running a commercial business for the expected tourists trade. The Italian Government is also rebuilding one or more of their semi-rigid airships so as to fit them to carry a large number of passengers. This appears to be an attempt to make some use of the excess stocks of airships left over from the War.

In France such development as is proceeding appears to be entirely directed toward commerce, and the French have arranged for a subsidy to concerns which will maintain machines and aviators available for military purposes in time of war. The French also are commencing the construction of rigid airships primarily for commercial travel between France and Africa, but I understand the Government's connection is very close as the ships are being designed by the Technical Section of the Army, but are to be operated for commercial purposes.

In England all seems to be in turmoil. As you know, General Seely under Secretary of State for Air has resigned, giving as his reason the impossibility of the existing arrangement which places the Royal Air Force under Mr. Churchill, the War Minister. The opposition press intimates that Mr. Churchill having absorbed the Royal Air Force is attempting by political maneuver to obtain the post of Minister of Defense, and absorb the Admiralty in addition, thus carrying consolidation to the limit. The Royal Air Force started off the beginning of the year with an ambitious program for the development of civil aviation, but due to the economic crisis their funds have been cut to about one half the original credit, which appears to be no more than enough to wipe out outstanding war obligations and close out their various contracts. The result is that building has stopped on practically all types, in particular the great program of rigid airships is suspended for lack of funds. The only new airship building in England is R-38 which is being completed for the United States Navy.

The Royal Air Force is under fire from three sides; from the public for gross extravagance and bad administration and waste of public funds, because they did not cancel war-time contracts with a firm hand, but have permitted certain favored concerns to continue the production of aircraft contracted during the War with a result that now the cut has come, the layoff of men is demoralizing, and the amount of money already spent is wasted. The second attack comes from the Aeronautical trade or industry, which objects to military control in the development of commercial aviation and complains of the lack of results and positive action in stimulating this development as promised. The third attack comes from those interested in the efficiency of the Fleet with a charge that since the Air Ministry has taken control no progress has been made in Naval Aviation, but on the contrary conditions have become progressively worse. The development of large flying boats has been allowed to come to a stop because of lack of sympathy with the type in the Royal Air Force. The flying boat type and the construction of airships of the Zeppelin type for use of the Fleet has been stopped for lack of funds because the Air Ministry has frittered away its fund on other things which are of no benefit to the Navy, so it is alleged.

The net result in England is a very pretty row and it is difficult to see exactly what is going on because of the smoke. I enclose, herewith, extracts from the British aeronautical press whose tone will give you an idea of conditions.

With regard to your request for reports of the National Committee for Aeronautics, I forward, herewith, the reports mentioned. The correct title of this Board is "National Advisory Committee for Aeronautics," as given by the Sundry Civil Act for 1920. This Committee was originally called "Advisory Committee for Aeronautics" in the Naval Bill of 1916 which carried funds for the first year. This title was changed because of confusion with a British Committee of identical title.

I am not in very close touch with the Helium Plant which comes under Admiral Griffin, but I understand that a Helium Board which coordinates the interests of the Army and the Navy in this matter is directing the development of this project, and that things are going along in accordance with the agreed plan, and so far as I know without any friction or difficulty. You know there have been some difficulties in the past with the Bureau of Mines of the Interior Department which was urging a technical process for the separation of Helium which other experts did not favor in war time as it was experimental. Possibly, what you have heard is a result of this.

Very respectfully,

[signed] D. W. Taylor.

Document 2-10(b), Edward P. Warner, excerpts from "Report on German Wind Tunnels and Apparatus," 1920.

REPORT ON GERMAN WIND TUNNELS AND APPARATUS
By
Edward P. Warner

The National Advisory Committee for Aeronautics, in view of the important research work that has been conducted in the German wind tunnels at Göttingen, Aachen, Dessau, and Friedrichshafen, requested Professor Edward P. Warner, the Committee's Acting Technical Assistant in Europe, to submit the following report, descriptive of the above mentioned wind tunnels, together with methods of operation and details of the apparatus used.

It is appropriate that any discussion of aerodynamical work in Germany should begin with Göttingen and with Prof. Prandtl, where the first serious work of the kind was undertaken, before the war, and where the most extensive and interesting results have been obtained both in respect of wind tunnel testing and of purely mathematical investigations.

There are, at the present time, two wind tunnels at the Göttingen laboratory, one being the original 1-meter tunnel of pre-war days, but moved into a new building; the other a newer installation 2 meters in diameter. During the war both tunnels were kept in service, but that is impossible with the present shortage of funds and of employees, and the small tunnel is now seldom used. The small tunnel was described by Dr. Zahm in his report to the Smithsonian Institution in 1913, and need not here be gone into in detail. The newer and larger one is built on essentially the same principle, but with substantial modification in detail. In the first place, the plane of the closed circuit which the air follows has been turned from the horizontal to the vertical, the return passage being underground. The whole tunnel, except the portion immediately around and adjacent to the test section, is made of concrete, so there is no question of air-tightness.

The section, 2 meters in diameter at the throat, expands in each direction to a diameter of 4.5 meters before the first turns are reached, and this size is maintained all the way around the return, making an area ratio of 5 to 1. The air is still guided around the turns by means of vanes, but these vanes are no longer adjustable in position or inclination to secure regularity of flow, such adjustment having been found unnecessary, nor are they now made in honey comb form.

They are cast in the concrete, are of crescent form with a maximum thickness of about one-fifteenth of the chord, which is about 18 inches, and are spaced approximately 8 inches apart. The vanes are not arranged to cause all the particles of air to swing about a common center, as might perhaps be expected, but are all of the same size and all curved to the same radius, so that the outside boundary of the tunnel comes to a corner and the section is increased at the turns.

The stream is not enclosed at its throat, although a tube is available which can be put into place to partially restrain the flow if desired. The stream being enclosed at every other point of its travel, the pressure at the throat is the atmospheric pressure in the free air, while that in the return passage is raised above atmospheric, and no experimental chamber or air-lock is necessary.

The only means provided for regularizing the flow is a honeycomb placed in the large section of the tunnel, before the contraction to the throat begins. This honeycomb is made up of metal plates separated by corrugated strips, so that the cells have a rather eccentric form, approximately semi-circular. The mean effective radius of these cells is about $\frac{3}{8}$", the length 8 inches. Prof. Prandtl lays great stress on the value of placing the honeycomb where he has it, before the entrance cone, in order to avoid turbulence and minute eddies which he believes are produced by a honeycomb of the ordinary type. Admittedly there is some loss in regularity of flow as ordinarily judged when the honeycomb is moved farther away from the throat, yet the regularity at Göttingen seems to be very good. Incidentally, I found Capt. Toussaint quite convinced of the merits of Prandtl's placing of the honeycomb and preparing to adopt a similar disposition at St. Cyr, and the matter was also receiving serious consideration at the N.P.L.

The current is produced by a four-bladed wooden propeller of ordinary type, driven by a 300 H.P. electric motor. The maximum wind speed obtainable is 50 m. per sec., which, if the motor is working at rated power without overload, corresponds to an energy factor of 1.36 (a result not in any way remarkable). The motor is outside the tunnel and drives the propeller by a shaft passing through the wall at the turn.

The forces and moments on the models are ultimately weighed on ordinary platform balances, the only special apparatus being that which transmits the forces from the roof of the tunnel, or, more properly, from the platform erected above the airstream, down to the balances which are placed conveniently for the observers standing on the floor. The apparatus is primarily of interest in that the support in all cases is solely by wires, no spindles being used under any conditions. Furthermore the measurements of lift and drag are direct and entirely separate, no moments entering in until the balances are reached.

In the case of an airship model or other similar streamline body, the support is by five wires. Four of these are arranged in two pairs the two members of a pair attaching to the model at the same point and then diverging, their plane being perpendicular to the direction of the wind. One pair is attached about a quarter of the way back from the nose of the model, the other about three-fifths of the way back. The fifth wire is attached at the same point as the forward pair, and runs forward exactly parallel to the wind direction. At a point about two feet forward of the nose of the airship this wire terminates in a small ring to which are also fixed two other wires, one running vertically upwards and the other obliquely downwards and forwards to the floor at the end of the entrance cone. This diagonal wire is simply fixed to a screw-eye in the floor, while the vertical wire is attached to a crank on a rocking beam, another offset crank on which bears against a vertical rod running down to the pan of the platform balance. The pull in the vertical wire can thus be weighed directly, and, knowing the directions of the other two wires attached to the ring, the pull in the horizontal wire, which is equal to the drag of the model, can be calculated. As a matter of fact, I believe, although I am not certain, that the oblique wire runs off at just 45°, so that the balance reads the drag directly with ordinary weights. The total pull in each pair of vertical wires is similarly measured, the two members of a pair being attached to two cranks on a single beam so that the total is obtained directly. The lift on the model is then equal to the sum of the two readings (correction having been made for the tensions due to the weight of the model), and the pitching moment can be directly calculated from their difference. The angle of attack is adjusted by raising and lowering the rear beam and wires. Of course, as the angle is changed the horizontal distance between the points of attachment of the two sets of wires changes, and, if the rear beam were raised or lowered vertically, it would be impossible for both pairs of wires to remain truly vertical. The two beams are therefore connected together by a link forming, with the two sets of wires and the line connecting their points of attachment in the model, a parallelogram. The five wires suffice to restrain the model from all motions except those in roll, and these are resisted by its own weight. As an additional safeguard against rolling, a wire may be attached to the lower surface of the model and run downwards and backwards over a pulley, a heavy weight being hung at its free end, thus introducing a constant correction to both lift and drag. Despite this precaution, the model of an airship which was being tested while I was at the laboratory several times started oscillating badly and it was necessary to reach into the current with a stick and steady the model just before the final balance readings were taken.

In the case of wings, the same balances are used and the method is essentially the same, but there are differences in details of attachment. The model is made with four hooks cut from thin sheet metal mounted on it, three of these hooks being distributed along the leading edge and the fourth carried by a rod which projects about one and a half chord lengths to the rear of the trailing edge. The two hooks mounted near the tips on the leading edge carry wires running vertically to the forward lift beam, the center hook carries the wire which runs forward to measure the drag, and the rear hook, on the rod, forms the point of attachment for two wires running off in a V to the rear lift beam. The model is then definitely fixed except in regard to yaw. Should any particular model show a tendency to oscillate in that respect it can easily be checked by running two wires from each corner of the leading edge to the forward lift beam, the whole set of four then appearing from the front as a W. This is to be done as a regular practice at the Zeppelin laboratory.

The spindle correction for wings is obtained by substituting for the model a T made of two pieces of stream-line wire about $\frac{5}{8}$" x $\frac{3}{16}$" in section. The cross-arm of the T has a length equal to the span of the aerofoil, the shank a length equal to the distance between the leading edge of the aerofoil and the rear point of attachment used for the determination of pitching moments. This T is hung up in exactly the same manner as the model, and the drag measured. The resistance of the stream-line wires being known with fair accuracy by computation and by previous experiment, the effect of the supporting wires can be determined at once by subtraction. In the case of an airship model the method is the same, but a single pointed rod is used in place of the T, the rod being held parallel to the wind direction. The method is not absolutely satisfactory in this case, as no allowance is made for the interference between the supporting wires and the model, which is quite different from the interference between the supporting wires and the rod, but the per centage error from this cause is undoubtedly very small.

The wing models in use at Göttingen are all made of plaster, scraped to form while still soft. One or, more usually, two aluminum plates are used as the core. The finish is excellent, as good as I have ever seen on plaster models, although I think it is no better than on the best of the models made for the Garden City laboratory. The standard size of wing model at Göttingen is 1 m. by 20 cm., an aspect ratio of 5.

The method of regulating the wind speed is of particular interest, as it is entirely automatic and seems to work with absolute perfection. The speed is fundamentally measured, in accordance with the usual practice by measuring the pressure difference between the large and the small sections of the tunnel. The pressure difference is measured by leading the pressure which differs from atmospheric in under a cup the rim of which dips in oil, and by weighing the "lift" on this cup with a balance, just as is done at the Leland Stanford tunnel. In this case, since the pressure under the cup is greater than atmospheric, the weights have to be hung on the same side of the balance axis with the cup. On the other arm of the balance is an oil dash-pot, while a long pointer runs downwards from the balance axis and moves between two contacts. As the pointer makes contact on one side or the other a servo motor is started in one direction or the other, steadily moving a rheostat and increasing or decreasing the motor speed, in such a way as to bring the wind speed back to normal, until the contact is broken and the pointer once more floats freely. In order to secure a greater range and a more sensitive maintenance of speed, the pointer is made with a subsidiary contact in the form of a light flat spring on each side of the main contact, and these subsidiary contacts actuate a fine rheostat. If the deviation from normal speed is really large, the unbalancing of the beam overcomes the resistance of the light contact spring and the beam moves, far enough to bring the main pointer into contact with its stops and actuate the coarse rheostat. In order to prevent the fine rheostat being brought up against its stops, a contact is automatically made when the handle draws near to the limit of its travel and this moves the coarse rheostat one step. The whole adjustment is now automatic. There is a push-button starter, and, the motor once started, the operator has only to hang the proper weight on the balance (the weights are marked directly in meters per sec.) and go away and leave it.

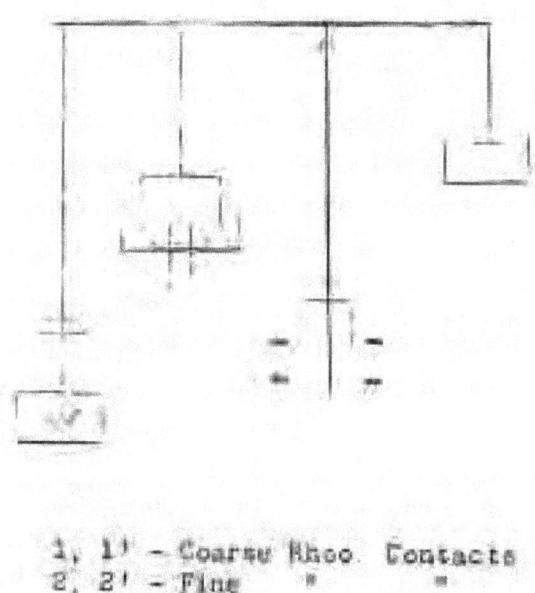

1, 1' – Coarse Rheo. Contacts
2, 2' – Fine   "   "

Although it appears that not much stability work has been done at Göttingen, a balance for measuring all forces and moments simultaneously is now under construction and is nearly completed. For this purpose, the model is hung from a platform, the model being heavily enough ballasted to keep all the suspension wires taut. These suspension wires are rigidly attached to the platform. The platform is held in position by six wires, two running vertically, two parallel to the wind stream, and the remaining pair horizontally and perpendicular to the tunnel axis, and the simultaneous measurement of the tensions in all these wires permits calculation of the six forces and moments acting. This is much the same in principle as the roof balance which has long been used for certain special experiments at the N.P.L. and to the all-wire-support roof balance which is now being designed at the R.A.E. The most interesting and original feature of the Göttingen balance is the automatic adjustment of the six balances, all of which are arranged in a row before the operator.

[*At this point the author goes into great detail about the design of the wind tunnel balance, which is omitted here.*]

No propeller testing has been done at Göttingen recently, but a propeller balance of unique type has been designed and partially constructed. The propeller under test will be driven through bevel gears and a vertical shaft from a motor outside the wind-stream, and the whole apparatus, motor, transmission, and propeller, will be carried on floats in a tank of water or oil. The thrust is to be measured by the combined pull in two wires running parallel to the wind direction, the torque by the pull which has to be applied to a vertical wire attached to one side of the floating platform in order to keep that platform horizontal. The cross-wind force and pitching and yawing moments can also be obtained with ease if desired by running other wires in suitable directions. In order that the measurements of torque may be sensitive, the inherent stability of flotation must be small, so that the angle of tilt due to the torque would be large were it not for the restraining wire. In order to accomplish this the metacentric height must be reduced or, since metacentric height is equal to I/V and V is fixed by the weight of the apparatus, the moment of inertia of the water-line section about a longitudinal axis must be made as small as possible without actually bringing the metacenter below the center of gravity. This is done in the Göttingen apparatus by making the floats proper of such a size that they will be completely submerged and the surface of the water will only be broken by the tubes which connect the floats to the platform. The propellers tested are to be one meter in diameter.

Great confidence is felt at Göttingen that the Prandtl theory of wing action, together with the work done along the same lines by Munk and Betz, now furnishes a practical tool for the engineer. There is no question of the consistency of the results obtained within a certain field, and notably in predicting the effect of changes in aspect ratio. A series of tests on wings of the same section but of different aspect ratios was recently made, and when the results obtained were reduced to aspect ratio 5 by Prandtl's formulae all the points lay, well within the experimental error, on a single curve. In view of this, Prandtl considers further testing for aspect ratio effect on new sections as quite unnecessary.

Aerodynamic Work at Aachen.

A regular course in aerodynamics is given by Prof. Karman at the Technische Hochschule at Aachen, and there are a number of students undertaking investigations in that field. The laboratory equipment, however, is very limited. The only wind tunnel available is on the roof of the building and takes its air from the free atmosphere, with no protection from gusts. It is also very short, and the flow is so irregular that it would be impossible to work with wings or, indeed, to do any work at all with a balance. The tunnel is two meters in diameter and has a 100 H.P. motor, realizing a speed of 33 meters per sec. There is no expansion in the exit cone, the sirocco fan which produces the current being of the same diameter as the throat of the tunnel. The experimental chamber is enclosed as in the Eiffel type, and the irregularity of flow in the stream is attested by a pronounced circulation of air all around the chamber. The honeycomb cells appear too large for best results, being about eight inches in diameter and two feet long. The most interesting thing about the Aachen tunnel itself is its noise-making characteristic. Up to about 25 meters per sec., while not by any means silent, the amount of noise is not unusual for the corresponding size and speed. At that speed there is a sudden change, and above 25 m. per sec. there is an ear-splitting shriek totally unlike anything that I ever heard from a wind tunnel before. The most peculiar thing about it, however, is that if one thrusts an arm into the stream until the hand is near its middle the high-pitched noise instantly stops, recommencing as soon as the arm is withdrawn. The question of noise in wind tunnels is at present considered as a vital one all over Europe, having come up for discussion without any effort on my part at Aachen, St. Cyr, Teddington, and Farnborough, and a multitude of different causes are ascribed. The problem is a very interesting one allied to, but more complex than, the general question of propeller noise, on which Bryan and Lanchester have already done some work and on which the former presented a paper to the British Association last year. Apparently no one has even attempted to explain completely the strange behavior of the Karman tunnel, but it is evidently due to the formation and dissipation of eddies with an accompanying strong reduction of pressure and density. The sound is not unlike the whine of a high-velocity bullet as it passes overhead.

At the present time the Aachen tunnel is working chiefly on the heat dissipation from and the flow of air through radiators. In carrying out these experiments a section of the radiator is closed off from the remainder, so that no water can flow through it, and minute holes are made in the sides of these isolated passages. Connecting the holes thus made to a gauge, the static air pressure is found at several points between those where the air enters and where it leaves the radiator, and the resistance can be studied and computed from the pressure gradient thus found, in the manner commonly employed in examining the flow of fluids through pipes. Water is pumped through the radiator at variable speeds and variable inlet temperatures and the heat dissipation is computed in the usual way.

Prof. Karman anticipates doing some work at very high velocities with the aid of a 100 H.P. centrifugal compressor delivering air at a pressure of one-half atmosphere, but no start has been made as yet.

Karman is now working on the theory of turning wings, or wings which rotate simultaneously, but not necessarily with constant or equal angular velocities, about an axis in the plane of the chords and about another axis parallel to the first but at a considerable distance from the plane of the chords. If the two rotations are properly adjusted to each other there is a lift and also a driving force, the action being something similar to that of flapping flight and eliminating the need for a propeller. The investigations now being made are directed towards finding the most efficient type of rotation, and will be followed by actual experiments with such wings. Another professor at Aachen is specializing in the study of the equations of viscous flow and attempting to apply them to aircraft parts, apparently proceeding somewhat along the lines on which Bairstow is now working.

A small water-channel is being used by one of Karman's research students for photographing the flow through diverging and converging passages of various sorts with a view to analyzing the flow in the exit and entrance cones of wind tunnels. Aluminum powder is used for making the flow visible. The most striking thing about the results so far obtained is the abrupt change in type of flow which appears about ten seconds after the motion has commenced, the flow initially "filling" a diverging cone and then breaking away from the walls, passing straight down the center as a stream of approximately constant cross-section with a region of dead-water on each side. Unfortunately, no moving pictures have as yet been taken to show the exact mechanism of the change. The alteration of the flow probably depends at least in part on the cessation of acceleration and the establishment of steady conditions.

During the war Prof. Karman was in Austria and had there a very large wind tunnel, which, however, does not appear to have been run for very long or in a very scientific manner. The tunnel was 3.2 meters in diameter, and propellers up to a diameter of 2.8 meters were tested, the ratio between these figures being far larger than has ever been considered safe elsewhere. The speed attained was 40 meters per second. The propeller thrust and torque were measured by the pendulum method, the whole apparatus being carried by four parallel wires. The tunnel was apparently rather short in proportion to its diameter, and the motor driving the propeller under test and the supporting wires were placed outside the mouth, a long shaft projecting along the entrance cone to drive the propeller. The whole tunnel was on wheels so that it could be moved to one side when the pendulum apparatus and driving motor were to be used for making static tests. The cost of the tunnel was one million kroner, or $200,000 at normal rates of exchange.

The Junkers Laboratory at Dessau.

This wind tunnel is hardly to be compared with Prandtl's, as the aim in its construction was entirely different. The Junkers laboratory is admittedly solely commercial in its aims, its only object being to furnish data for improving the Junkers airplanes. None of the work done there has been published, and there is no intention of publishing any. The experiments at Dessau have been chiefly concerned with the development of improved sections for cantilever wings, although a little stability testing of an elementary sort has been undertaken, and the constant attempt has been to speed up and systematize the making of routine tests with a moderate degree of accuracy, rather than to seek increased refinement in measurement.

The Junkers tunnel is octagonal in section, but the octagon is not a regular one, the breadth of the stream being 1.2 meters, the height only 0.9. A fan is used instead of a propeller for producing the air current. The tunnel is of true Eiffel type as regards the experimental chamber, and a couple of inches of water is left standing on the floor at all times to minimize leakage and to suppress the dust. The drive is by an electric motor of 100 H.P., and the maximum wind-speed is 39 meters per second, giving an energy ration of .497/.

The steadiness of flow in the Dessau laboratory seems distinctly inferior to that at Göttingen, and there is no question that there is more disturbance of the air in the room and spreading of the stream. The honeycomb is similar to that used by Prof. Prandtl as far as the form and arrangement of the cells are concerned, but it is placed at the beginning of the throat instead of at an enlarged section. The cells are ½" in mean diameter and 2" long.

The balance at the Junkers laboratory is of interest chiefly because it is completely autographic. The measurements depend entirely on springs, no weights being used. The balance carriage slides longitudinally and is carried on two steel rods running parallel to the wind direction. The weight of the carriage is considerable, and the friction in sliding on fixed rods would be intolerably large, but this friction is practically eliminated by rotating the rods at high speed through a belt drive—a very ingenious and successful method. The longitudinal motion of the carriage is proportional to the drag, and is transmitted through a linkage to a pencil which moves vertically over a record-sheet. The lift is measured by the tilt of a beam which carries the model and which is also restrained by springs, and is also transmitted through a linkage to a pencil making its record on the same sheet as the first one. The angle of attack is changed steadily and progressively by a motor which winds up a wire attached to the trailing edge of the model, and which, at the same time, rotates the drum carrying the record-sheet at a constant speed. The making of a complete test requires only about twenty minutes. No attempt is made to keep the speed constant automatically, the rheostats being set at a beginning of a run and then left alone. It is assumed that the changes in speed due to voltage changes are negligible.

The wing models tested are made of wood, and painted so that I could not examine the method of construction in detail. A very large number of thick wings have been tested, at least a hundred models being stored in their cabinets at the time of my visit, and work is still continuing in the effort to improve the present Junkers wing. The best results obtained to date with thick wings at Dessau are: max. L/D 17.0, max. Lc .65, at 35 meters per second. The maximum lift coefficient is not in any way remarkable, and the high maximum L/D is largely accounted for by the way in which the model was supported. All tests are made with the aerofoil carried by a stream-line spindle about $3/16"$ wide passing into the model at the center of the span and near the leading edge, and it is well known that such a support, when its resistance is corrected for in the usual approximate fashion, leads to values of the drag which are far below the correct figures.

The Zeppelin Wind Tunnel at Friedrichshafen.

The only remaining wind tunnel of any importance is that which is under construction, but not yet completed, for the Zeppelin Airship Company. To be sure, there is a tunnel at Adlershof, but this is old and little used, and was fully described in the report submitted by Dr. Zahm after his tour in 1913 and published by the Smithsonian Institution.

The Zeppelin tunnel has been designed by, and is being built under the supervision of Dr. Max Munk, and is naturally similar in many respects to the Göttingen installation, since it was at Göttingen that Dr. Munk received all his aerodynamical training and much of the apparatus there was designed by him. The Zeppelin tunnel, however, is considerably larger than either of those at Göttingen, and will be the largest tunnel in the world from the time of its opening until the new 7 x 14-foot channel at the N.P.L. goes into action. The diameter of the wind-stream at Friedrichshafen is 3 meters. The drive will be by two Maybach engines coupled together in tandem, delivering a total of 500 H.P., and a speed of 30 meters per sec. is anticipated.

The most original feature of the tunnel design lies in the provision of air-tight gangways, of much larger section than the tunnel itself or even than the large end of the exit cone, for the return of the air. The principle is the same as a Göttingen, but the mechanical execution is somewhat different, largely because of the great size of the laboratory. There are two return passages, one on each side of the tunnel, instead of a single one, and these passages take the form of gangways each about three meters wide and six meters high, with air-tight windows and air-locks and without any guide vanes at the corners. The passages are shaped at the ends so that there will be as gradual a change of section and of direction as possible, the whole, including the entrance cone, being made in concrete. The exit cone, however, is of wood. Since the gangways are air-tight, the throat section can be open to the atmosphere, just as at Göttingen, and admission to the test room is perfectly free at all times, no air-lock or other control being necessary. In order to permit access to the test chamber there is a subway under each of the gangways.

The slope of the exit cone is small, the vertex angle being 7-$\frac{1}{2}$°, and the length is not sufficient to give as large an area ratio as is the usual custom. The length of the cone is to be 15 meters and the diameter of the propeller 5 meters, the propeller being driven with the same gear reduction as is employed in the Maybach engines on Zeppelin airships.

No honeycomb has been fitted as yet, and Dr. Munk hopes to avoid the use of any straightening device, but his hopes apparently have no sound basis.

The measuring instruments will be of the same general type as those at Göttingen, but naturally much stronger. The whole weighing apparatus is to be installed on an overhead platform where the observers will work, the support being by steel rails carried by concrete piers. The most striking feature about the balance is its construction, which follows Zeppelin airship lines faithfully, the cross-beams used for weighing the lift being built-up duralumin lattice girders. The total weight of the moving parts will certainly be less than in any other balance ever built for so large a VL. Of the two bridges, the forward one will be fixed rigidly to the rails, but the other will be mounted on wheels and will be allowed to move longitudinally as the angle of attack is changed so that the horizontal distance between the two bridges will always remain equal to the horizontal distance between the lower ends of the two sets of wires. The adjustment of the angle of attack is by raising one of the bridges, the supports for which slide on vertical rods. The bridge is raised with a cable which is wound up on a drum by turning a graduated handwheel.

The speed is to be controlled automatically by the same method as that employed at Göttingen, the servo motor operating the engine throttles instead of a rheostat.

The Friedrichshafen laboratory will work for the parent airship company and for all its subsidiaries, of which there are a considerable number, including the airplane firm at Seemoos (formerly located at Lindau) where the Dornier flying boats are built, and the aircraft company at Staaken, for which Rohrbach is the chief engineer.

September, 1920.

Document 2-11

"Free Flight Tests," *NACA Annual Report for 1918* (Washington, DC) pp. 20–21.

Like the British Royal Aircraft Factory, the NACA sought to develop "as complete tests as possible of the performance, in all respects, of airplanes while in the air under normal conditions," and it established a subcommittee on free flight tests to pursue that end. There was, however, a larger goal in the committee's mind. As this portion of the *Annual Report for 1918* indicates, the committee specifically wanted to develop the means to correlate the "information gained from all sorts of tests on the ground" with flight-test data to obtain not only a validation of ground-based methods, but a synergy of laboratory and flight testing as well.

Document 2-11, "Free Flight Tests," **NACA Annual Report for 1918.**

Free Flight Tests

The general purpose of the work of the subcommittee on free flight tests is to obtain as complete tests as possible of the performance, in all respects, of airplanes while in the air under normal conditions. The general purpose of these tests is to supplement and make more valuable the information gained from all sorts of tests on the ground, including tests of engines and tests of airplane parts and airplane models in wind tunnels. It is obvious that the actual performance in the air, when it becomes known, is the best possible basis for future progress.

The committee now has, in a late stage of development, instruments for recording in the air the torque and revolutions per minute of the engine, the thrust of the propeller, the air speed, the angle of attack, and the inclination of the wing chord to the true horizon. It is proposed to complete this development as promptly as possible, and to get these instruments in action in the air, presumably on a D.H.4 airplane, to determine the power-plant performance and the relations in the air between the lift, drag, air speed, and angle of attack.

When such tests have been successfully demonstrated as possible, by making them, the next steps on the program of the committee are to analyze the results and show what conclusions can be drawn from them.

The committee then proposes, in due time, to extend the free-flight tests to such quantities as will help to develop the stability characteristics of airplanes, possibly to furnish some information as to the stresses in various parts of an airplane in operation.

To secure the necessary degree of accuracy and reliability in free-flight observations, the new instruments have in each case been so designed as to give a continuous autographic record.

Document 2-12(a–b)

(a) Edward P. Warner, F. H. Norton, and C. M. Hebbert, *The Design of Wind Tunnels and Wind Tunnel Propellers*, NACA Technical Report No. 73 (Washington, DC, 1919).

(b) F. H. Norton and Edward P. Warner, *The Design of Wind Tunnels and Wind Tunnel Propellers, II*, NACA Technical Report No. 98 (Washington, DC, 1920).

Although American aeronautical researchers usually chose to pursue a more empirically based path in designing wind tunnels than did their German counterparts, they were nevertheless interested in a thorough understanding of these machines and the aerodynamic phenomena they could produce. During the NACA's early years, Frederick Norton and Edward Warner led an investigation at Langley to determine the key aspects of wind tunnel design that engineers could use with confidence. Their report, *The Design of Wind Tunnels and Wind Tunnel Propellers*, published in two parts, clearly reveals the empirical approach to design optimization employed by the Americans for many years, and it shows the limited state of theoretical aerodynamics in the United States at the time. The publication of these two technical reports also demonstrates the NACA's commitment to making useful information about experimental tools and methods as well as developments in aircraft design available to the aviation community.

Document 2-12(a), Edward P. Warner, F. H. Norton, and C. M. Hebbert, **The Design of Wind Tunnels and Wind Tunnel Propellers,** *NACA Technical Report No. 73, 1919.*

THE ELEMENTARY THEORY OF THE FLOW OF AIR THROUGH WIND TUNNELS.

If the air flowing through a wind tunnel and back through the room from the exit to the entrance of the tunnel followed Bernouilli's theorem with exactness, there would be no change in the energy, possessed by a given particle of air, except for the loss due to friction as the kinetic energy lost on issuing from the tunnel would be restored in the form of pressure energy. The power required to maintain the flow would then be

$$P = m \times h_f$$

where h_f is the head (in feet of air) lost by friction and m is the mass of air flowing per second. As the same mass of air must pass every point in the tunnel, the product of mean air speed by cross-section area must be a constant for its whole length, neglecting compressibility and changes in temperature during the passage. Since the major part of the frictional losses occur in the reduced section of the tunnel (provided that it is not very short and that the diffuser is not so constructed as excessively to hamper the travel of the air from the tunnel back into the room), h_f would be practically independent of the size and angle of the exit cone, and the power consumed would also be independent of these factors.

As a matter of fact, the conditions of flow are not simple enough to permit the direct application of Bernouilli's theorem. Borda has shown that the loss of energy when fluid moving at high velocity in a pipe is discharged abruptly into a large room or reservoir is equal to the kinetic energy initially possessed. The kinetic energy is not converted into pressure energy as the theory indicates that it should be, and it is therefore profitable to use an exit cone of considerable length, in order that part of the kinetic energy may be saved by conversion into the potential form before the sudden discharge into the room. The length to which it is desirable to prolong the cone is limited by the growing loss by friction within the exit cone itself. A more rapid conversion of the kinetic energy by increasing the vertex angle of the exit cone is forbidden by the unwillingness of the air to change its course suddenly and follow the walls of the exit cone. If the vertex angle be made too large the effect is almost the same as that of an abrupt increase in cross section. Eiffel, as the result of an elaborate theoretical and experimental research on tunnels having exit cones generated by straight lines, has come to the conclusion that the vertex angle of the exit cone should be not more than 7°, and that the diameter at the large end of the exit cone should be three times that at the small end. It is necessary to base the dimensions of a tunnel on a compromise, as the arrangement which would give the absolute maximum of efficiency would have to be housed in a building of prohibitive size. The overall length can be materially reduced at the cost of a slight increase in power, and the first cost of the building, depending on its dimensions, must be balanced against the cost of operation, which varies with the power of the motor and so with the efficiency of the tunnel. The relations to be observed among the various dimensions of the tunnel and the angles of the cones will be discussed more fully elsewhere. Knowing the power consumed by a tunnel, its diameter, and the speed of the air, the total losses can easily be computed for that particular speed, and the magnitude of the figure thus obtained will serve as a measure of the efficiency of operation of the tunnel. Since, however, the losses vary with the speed, they can not be compared directly for two tunnels unless they are run at the same speed. The factor most commonly used for comparisons between tunnels is the ratio of the kinetic energy possessed by the air passing through the tunnel in unit time to the work done by the motor in unit time. This is sometimes called the "over-all efficiency," but is herein alluded to as the "energy ratio." The term efficiency in this connection is misleading, as the two quantities introduced into the ratio are not directly connected, but merely happen to have the same dimensions and so to be convenient for the purpose. Furthermore, the value of the ratio is very commonly more than 1, and is sometimes very much more.

To determine the manner in which the power consumed varies with speed, and so determine the validity or otherwise of the above relation, as well as to find the relation which must be preserved among the various factors in order that geometrically similar tunnels may be strictly comparable, the Theory of Dimensions may be used. The method pursued need not be gone into in detail, as it has been described many times before, and it will suffice to summarize the results. It appears that, if the compressibility of the air and the action of gravity on it be assumed to be of negligible importance at the speeds employed, the power consumed is proportional, for geometrically similar tunnels, to the cross-section area and to the cube of the speed, provided that VD/v, where V is the air speed, D the tunnel diameter, and v the coefficient of kinematic viscosity, is maintained constant. Experiments conducted with a model tunnel at Langley Field and fully described elsewhere in this report, as well as those carried on by Durand, Castellazzi, and others, show that the "energy ratio" varies but little with changes of VD/v and it is therefore safe to apply the results of model experiments to full-sized tunnels, even though the speeds may not be strictly in inverse ratio to the diameters. In general, the "energy ratio" increases as VD/v increases, and it therefore requires less power to drive a tunnel than would be predicted from a direct application of the results of tests on a model of the tunnel and propeller.

The useful work done by a propeller is equal to the product of the thrust by the speed of flow of the fluid through the propeller disk. The thrust of a wind tunnel propeller is then

$$\frac{m \times h_f}{V'}$$

where V' is the speed of the air past the propeller, and this equation holds good whether Bernouilli's theorem is followed or not, so long as h_f is the total loss of head from all causes.

$$m = w/g \times A' \times V'$$

A' being the cross-section area at the propeller, and the propeller thrust is therefore equal to the weight of a column of air having a height equal to the total loss of head and a cross-sectional area equal to the disc area of the propeller. Since the power is proportional to the cube of the speed, the thrust varies as its square.

If the factors causing departures from Bernouilli's theorem are neglected, the useful work done in moving the air against friction will be, as already mentioned, independent of the degree of expansion of area in the exit cone, and so of the diameter of the propeller. Under these conditions, in fact, the advantage in respect of power consumed would rest with the short exit cone and small propeller, as the propeller efficiency is highest for a large value of the "slip function" and this is obtained by making the speed of the air through the propeller high and keeping down the diameter of the propeller. Assuming that the output of work is the same in all cases, the thrust will be inversely proportional to the speed of air through the propeller, or directly proportional to the disk area.

LAWS OF SIMILITUDE FOR WIND TUNNEL PROPELLERS.

It is obvious from a study of the Drzewiecki theory of propeller action that a series of propellers of similar blade form and width-diameter ratio, all working at the same true angle of attack, will give thrusts approximately proportional to $N^2 D^4$, where N is the engine speed in revolutions per unit time and D the propeller diameter. This proportion can be demonstrated by the Theory of Dimensions to hold exactly true for geometrically similar propellers of perfect rigidity, but it is very nearly correct even where propellers of different pitches are concerned. It has been shown that the thrusts of a series of propellers designed to drive the same wind tunnel or geometrically similar tunnels is proportional to $N^2 D^4$, and also to the cross-section area, which, in turn, varies as D^2. It follows from these two relations that $N^2 D^2$ must be a constant, and the peripheral speed of the propeller required to draw air through a wind tunnel at any particular speed will therefore be quite independent of the diameter of the propeller if the power required is independent of that diameter. It follows as an obvious corollary that, if the power required is not independent of the degree of expansion in the exit cone, the peripheral speed of the propeller will be least under the same conditions as those for which the power required has its minimum value.

It is easily demonstrable that the stresses, both those due to centrifugal force and those due to bending by the air pressure, in a series of geometrically similar propellers depend only on the peripheral speed, and that they vary as the square of that quantity. There is therefore a limiting peripheral speed which can not be exceeded with safety. For wooden propellers, it is unsafe to run the peripheral speed much beyond 60,000 feet per minute, or 305 meters per second, and it is better to stay well inside this figure. In the case of an airplane or airship where large power must be taken on a single propeller the peripheral speed can be reduced by gearing down, as the engine speed decreases more rapidly than the propeller diameter increases. In the wind tunnel, it has just been shown that this is not the case, and that the peripheral speed, and so the stress, actually increases if the propeller diameter is enlarged beyond a certain point. There is then a clearly defined upper limit to the power which it is safe to apply to driving the propeller in any given wind tunnel, and therefore a limit to the maximum speed attainable. This maximum can only be raised by reducing the losses and so improving the over-all efficiency of the plant.

Since the power required to secure a given speed with a given "energy ratio" is proportional to the cross-sectional area of the tunnel, and is also proportional to VN^2D^4, the propellers in a series of tunnels of different diameters operating, at the same speed and having the same "energy ratio," all work at the same value of N^2D^2, and so of the peripheral speed. This leads to the rather astonishing conclusion that the peripheral speed necessary to produce a given air speed depends only on that air speed and on the energy ratio, and is not at all affected by the size of the tunnel or of the propeller (except indirectly, in so far as these factors have an effect on the energy ratio). For any value of the energy ratio, then, there is a limiting air speed which can not be exceeded without running the peripheral speed up beyond the limits of safety, and this speed is the same for large tunnels as for small, although the actual power consumed of course varies with the tunnel diameter. In order to realize the highest possible wind speed the power coefficient of the propeller must be made as large as possible. This can be done by using many blades and by making them of high sections set at relatively large angles of attack. If the velocities desired are too high to be obtained in this way, it will be necessary to use two or more propellers arranged in tandem, acting like a multi-stage compressor.

It has been shown that
$$P = K_1 D_1^2 V_1^3$$
and also that
$$P = K_2 V_2 N^2 D_2^4$$

where the subscripts 1 and 2 denote, respectively, the conditions existing in the experimental chamber and at the propeller, and K_1 and K_2 are experimental constants depending on the type of tunnel and propeller. Since $D_1^2 V_1 = D_2^2 V_2$, if the velocity across the exit cone at the propeller is uniform, the first of these relations may be written
$$P = K_1 D_2^2 V_2 V_1^2$$

Dividing this by the second of the relations above,
$$K_2 N^2 D_2^2 = K_1 V_1^2$$
and
$$\frac{V_1}{ND} = \sqrt{\frac{K_2}{K_1}}$$

The ratio of the air speed to the peripheral speed is thus a constant for a given tunnel and its value for any particular tunnel depends only on the type of installation—not at all on its size. Values of V_1/ND_2 for a few tunnels are tabulated herewith:

Name.	V_1(m./sec.).	N(r.p.s.).	D_2(m).	V_1/ND_2.
Eiffel, Auteuil[1] 31.8	3.83	3.80	2.18	
Leland Stanford, Jr.	24.0	6.77	3.35	1.06
Langley Field, model	41.5	68.3	0.610	1.00
N. P. L., 4-foot	15.24	22.5	1.6761	0.40
Curtiss, 4-foot	34.5	22.92	2.44	0.62
Curtiss, 7-foot	42.8	20.00	3.66	0.58
McCook Field	221.0	29.50	1.52	4.92

[1]This tunnel was square and the ratio of V1 to V2 is therefore equal to the ratio of the cross-section areas and not to that of the squares of diameters at the minimum section and at the propeller.

Castellazzi's experiments.

Number of blades.	Blade width, diameter.	V_1(m./sec.).	N (r. p. s.).	D_2(m).	$V1/ND_2$.
24	0.0435	25.0	17.25	0.600	2.41
24	.0300	25.0	19.17	.600	2.17
16	.0650	25.0	16.67	.600	2.50
12	.0435	25.0	20.50	.600	2.03
8	.0650	25.0	19.33	.600	2.15
6	.0650	25.0	22.17	.600	1.88

It will be noted that the highest value of V_1/ND_2 in this table, with one exception, is 2.50, and this value was obtained in a tunnel of very efficient type in combination with a propeller having a total blade width equal to one-third of its circumference. Analysis by the Drzewiecki method leads to the belief that it will be possible to raise V_1/ND_2 to 3, but that this figure can hardly be exceeded with propellers resembling those now in use. The exception mentioned above, the small tunnel at McCook Field, has a fan of special type and will be discussed later.

If the allowable peripheral speed be taken as 285 meters per second, ND^2 is 90.6 meters per second. If V_1/ND_2 be assumed to be 3, the limiting value for V is 271.8 meters per second, or 607 miles an hour. This is a considerably higher speed than has yet been attained, or than is ever likely to be desired in connection with the study of aircraft. If higher speeds should be needed they can be secured either by the use of a multiplicity of propellers in series or, up to a certain point, by the use of a fan with an abnormally large hub and short blades entirely filling the periphery of the hub, as in the McCook Field tunnel, where the hub diameter is two-thirds of the total diameter. If V_1/ND_2 is raised to 5, a value only a little higher than that in the McCook Field tunnel, the limiting air speed for the peripheral speed given above is increased to 453 meters per second, or 1,012 mile an hour. ([*Footnote:*] In this analysis the change of density of the air, due to decrease of static pressure with increasing speed, is neglected. This does not lead to a very large error, as both the propeller thrust and the frictional resistance to the passage of the air increase with the air density, the former varying more rapidly than the latter.)

The assumption has so far been made that the air has a free passage across the whole area swept by the propeller. Of course the hub always blocks off a part of this area, but it has usually been an insignificant fraction. If the propeller diameter is n times the hub diameter, the proportion of the area blocked off is $1/n^2$, and the speed of the air across the propeller blades, assuming a uniform distribution everywhere outside the hub, is increased in the ratio $\dfrac{1}{1-1/n^2}$.

If the propeller be made, as is the common practice, with a constant blade width, and if the lift coefficient be assumed constant all along the blade, the portion of the total thrust given by the part of the blade inside of any given point is very nearly proportional to the cube of the radius at that point. For example, one eighth of the thrust would be given by the inner half of the blades if they extended clear to the center, with no hub at all. The use of a hub, or the covering up of part of the blades with a "spinner," therefore decreases the thrust in the ratio $1-1/n^3$. Since useful power is equal to this product of the thrust by the speed across the propeller disk, the net change in power, due to hub or spinner, is

$$\dfrac{1-\dfrac{1}{n^3}}{1-\dfrac{1}{n^2}} = \dfrac{n^3-1}{n^3-n}$$

The increase in power coefficient by the use of a spinner, the propeller pitch being adjusted to give the same angle of attack of the blades with or without the spinner, is 5 per cent for a spinner of hub one-quarter the diameter of the propeller, 17 per cent when the ratio is one-half, and 27 per cent when, as in the McCook Field tunnel, it is two-thirds. Furthermore, the use of a very large hub makes it possible to use more blades and make their total width a larger fraction of the circumference of the circle swept by the blades. In the McCook field fan there are 24 blades, and their total width is approximately equal to the circumference of the hub.

Where very high speeds are desired, as in the calibration of air-speed meters, a throttling insert has sometimes been used to reduce the section of a large tunnel. The effect is to increase the speed, but usually much less than is expected. If the "energy ratio" remained constant, halving the diameter of the tunnel would increase the speed available with a given expenditure of power by 59 per cent. A change of this sort usually, however, diminishes the energy ratio unless the tunnel is of the type combining a long straight portion with conical ends, and permitting the extension of the cones back into the straight cylindrical part. The use of a throttling insert in a tunnel with a short experimental chamber, like those used by Eiffel and Crocco, is almost certain to lead to a large drop in energy ratio, and the increase of speed by halving the diameter in such a laboratory would probably be less [than] 50 per cent. Furthermore, it is necessary for best results that the propeller ordinarily used be replaced by one especially designed for use in conjunction with the throttling insert. If the diameter of the tunnel be halved the area at the smallest section is divided by four, and, even with an increase of 59 per cent in speed at the throat or in the experimental chamber, the speed of the air past the propeller is reduced by 60 per cent. Since the propeller diameter and its normal rotational speed to develop the rated power are unchanged, the propeller for use with the throttling insert must have a much smaller effective pitch than that employed with the full section, if the maximum of efficiency is to be obtained.

RELATIVE ADVANTAGE OF SMALL AND LARGE TUNNELS.

It has just been shown that the gain in speed by reducing the diameters by the use of throttling insert is disappointingly small. This leads naturally to a study of the best size of wind tunnel to be employed, and of the relation between speed and size which should be sought.

In the construction of aerodynamical laboratories, as the attempt has been made to approach ever more nearly to full-flight conditions, two divergent schools of practice have grown up. The first, best represented by the National Physical Laboratory in England, has constantly increased the diameter of the wind stream, and so increased the size of model which may be tested, but has remained content with relatively moderate wind speeds. The second, on the other hand, has concentrated its efforts on the pumping of the air across a small section at enormous velocity.

In comparing the merits of the high speed and the large diameter tunnels, there are three points which must be borne in mind. In the first place, the highest possible value of LV (LV being the criterion of dynamic similarity) is to be obtained with a minimum expenditure of power. Secondly, the interference between the model and its support is to be reduced to a minimum, and, finally, that disposition should be favored which enables us to secure the greatest accuracy in the construction of the models.

It has been shown that

$$P = KAV^3 + K_1 D^2 V^3$$

where D is the diameter of the tunnel and K_1 is a constant.

In order to avoid interference between the model and the walls of the tunnel, the ratio of maximum span to tunnel diameter must not exceed a certain value (usually about 0.4). Setting L, the span of the model, proportional to D, we can then modify the above equation:

$$P = K_1 D^2 V^3 = K_2 \frac{L^2 V^3}{f_3}$$

The power required to drive the fan will therefore be least, for any given value of LV, in that tunnel where the diameter is largest and the speed is smallest.

The relative magnitude of the interference between the model and its support, the so-called "spindle effect," depends on the ratio of the spindle diameter to the linear dimension of the model. Its reduction is a matter of very vital importance, the spindle correction undoubtedly being the largest single source of error in most wind-tunnel tests.

The bending moment in the spindle at any point (say one chord length from the wing tip) is proportional to the product of the span by the force acting on the model.

$$M = C_1 L F = C_2 L(L^2 V^2) = C_2 L^3 V^2$$

If d is the diameter of the spindle, the relations between the bending moment, fiber stress, and deflection may be written:

$$f = \frac{Md}{I} = \frac{C L^3 V^2}{d^3}$$

$$\delta = \frac{Ml^2}{E \cdot I} = \frac{C L^5 V^2}{d^4}$$

if the material of the spindle be the same in all cases.

If the maximum fiber stress be limited to a definite value,

$$\frac{d^3}{L^3} = C_3 V^2$$

$$\frac{d}{L} = \left[\frac{L^2}{d^2}\right] \times V^{2/3}$$

The ratio of spindle diameter to model size, and consequently the spindle interference, will therefore be greatest in the high-speed, small diameter tunnel.

If, as is usually the case, it is stiffness and not strength which prescribes the diameter of the spindle, and if the deflection be limited to a determined value, the required spindle size is given by the equation:

$$\left[\frac{d}{l}\right] = \frac{f \cdot L \cdot V}{S}$$

$$\frac{d}{a} = f \cdot \sqrt{[L \cdot V]} / V$$

For a given value of LV, then, d/l will be least when the speed is low and the tunnel diameter large. The advantage of the large tunnel on this score is even greater than appears at first, as a larger spindle deflection is permissible with a large tunnel than with a small one. In fact, the permissible deflection increases nearly as rapidly as does the tunnel diameter.

In respect of the third consideration, accuracy of construction of the model, the superiority of the large tunnel, permitting the use of a large model, is so manifest as hardly to call for discussion. A model of 3-foot span can include many parts, such as fittings and wires, which it is quite hopeless to put on one of half that size.

So far, the advantage has rested with the large diameter in every particular. It has one disadvantage in that the size and weight of the balance are much increased, longer weighing arms, heavier counterweights, and a general strengthening up of the apparatus are necessitated. Furthermore, the initial cost of the building to house a large tunnel is very high. In the writer's opinion, however, the advantages far outweigh the drawbacks, and any future development of wind tunnels for model testing should proceed along the lines of increasing the diameter rather than the speed.

All that has been said against high speeds applies, of course, only to tunnels for the testing of models. Speeds equal to the speeds of flight of airplanes are essential for the calibration of instruments.

DESIGN OF WIND TUNNEL PROPELLERS BY THE DRZEWIECKI THEORY.

It is possible, if the rate of flow of the air through a wind-tunnel propeller be known, to predict the performance of the propeller by the Drzewiecki theory. Indeed, the application of that theory to wind-tunnel propellers is rather simpler than its application to the airplane, as there is no in-draught correction to contend with. If the velocity at the minimum section of the tunnel is given, the velocity through the propeller can be computed with absolute accuracy on the assumption that the distribution across the exit cone is uniform. This assumption can only justify itself in the results of the analysis derived from it as a basis.

The best way of checking the accuracy of the analytical method of design is to apply it to a propeller already working satisfactorily. This has been done with the propeller used in the model wind-tunnel experiments described in a later section of the report. The angle of the relative wind to the plane of the propeller can be computed from the wind speed, and it is then possible, knowing the angles of blade setting, to work and find the angle of attack of each blade element. Having this, the power consumed by the propeller and its efficiency can be found in the usual way. This was done for two cases. In the first case the tunnel was of the Eiffel type, with an enlarged experimental chamber, and the calculated power checked the actual consumption within the experimental error (about 2 per cent, owing to uncertainty as to motor losses). In the second case the air stream was inclosed throughout, a cylindrical tube being carried across the experimental chamber, and the power consumed was about 15 per cent more than that calculated. It is considered that both of these tests showed a very fair check and that the use of the Drzewiecki theory for design is amply justified. The average error, both in these and in other cases which have been tried, is in the direction of underestimation of the power consumption.

In designing a propeller for a new tunnel it is necessary to make an estimate of the energy ratio, and so of the speed for a given power. If the estimate is too low, the propeller pitch will be made too low, and the propeller will work at an inefficiently small angle of attack. The speed will be higher than that estimated, but still not so high as it would be with a proper propeller. If the propeller blades are made too narrow, or if too few blades are used, the full power of the motor will not be absorbed at the rated revolutions per minute. The speed will then fail to reach the value expected for the rotational speed realized, the angle of the relative wind to the plane of the propeller will fall below the estimated value, and the angle of attack of the blade elements will become inefficiently large. Any change of this sort from the designed conditions of operation tends to correct itself, as the larger angle of attack increases the power consumed and the thrust given by the propeller. This in turn speeds up the air and brings the angle of attack to a lower value. It is for this reason that fairly satisfactory results have so frequently been secured with propellers chosen almost at random, but the best efficiencies can only be obtained with a propeller designed especially for the conditions under which it is to operate. The commonest faults in the design of wind-tunnel propellers have been either to overestimate the energy ratio for a projected tunnel or to underestimate the total blade width required for the absorption of the given power at the most efficient angle of attack. The result in both eases is to cause the blades to work at too large an angle of attack.

There is some doubt as to the manner in which the angle of attack should vary along the blades. Most wind-tunnel propellers in which the Drzewiecki sytem was used at all have been designed for a constant angle of attack, but since, as was just noted, the propellers have usually been made too small to absorb the full power of the motor, they actually work at an angle of attack larger than that desired and increasing from the tip to the root of the blade. In the design of a propeller for the Langley Field wind tunnel, the opposite disposition has been

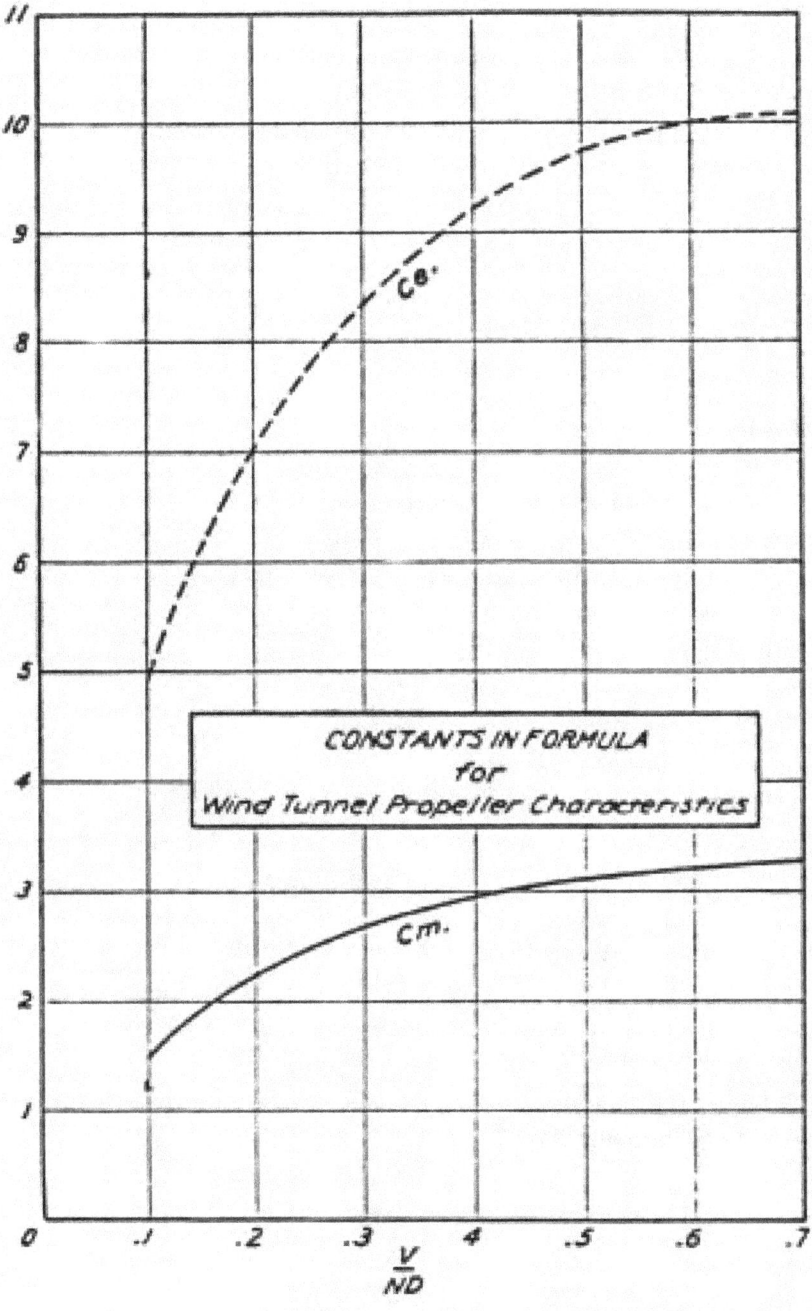

Fig. 1.

deliberately chosen, the angle of attack being made largest near the tips and decreased toward the hub in order that the air may be drawn out along the sides of the exit cone and in order that the larger part of the thrust may come on the most efficient portion of the blades. No experimental data on the effect of this arrangement of the blade sections are available as yet.

In order to make it easy to estimate the number of blades and the blade width required in a propeller for a tunnel, assuming that the wind speed, power consumption, and revolutions per minute are known, a number of propellers have been computed for a variety of conditions and the results expressed by a formula and a curve. The power is given by the formula

$$P = \frac{V^3 \cdot b \cdot n \cdot D^2 \cdot N^2 \cdot h}{C \times 10^8}$$

where P is the horsepower input of the motor, V the air-speed through the propeller in meters per second, b the blade width in centimeters, N the revolutions per minute, n the number of blades, D the propeller diameter in meters, and C a constant, the magnitude of which depends on the pitch of the propeller. C is plotted against V/ND_2 in figure 1. If English units be used, V being given in miles an hour, D in feet, and b in inches, a factor 10^9 replaces 10^8 in the denominator of the power formula given above, and C is given by the dotted curve in figure 1. The theoretical basis for the derivation of this formula is the same as that for a formula derived by the writer, and previously published, for the power consumption of airplane and airship propellers.

The efficiency of wind-tunnel propellers is usually very low, and the maximum attainable depends largely on the magnitude of the pitch ratio. In the propeller designed for the Langley Field tunnel the calculated efficiency is 58 per cent. In figure 2, probable propeller efficiencies have been plotted against V/ND_2. The efficiencies there predicted may be exceeded when the peripheral speed is low, so that thin sections can be used over the whole length of the blade, or when a very large hub or spinner is used to cover up the less efficient parts. In order to give an idea of the range of values of V/ND_2 employed in successful tunnels, a few are tabulated below, the data being taken from the table under "Laws of Similitude for Wind Tunnel Propellers".

	V_2/ND_2
Leland Stanford, Jr.	.0265
N. P. L., 4-foot	.27
Curtiss, 4-foot	.20
Curtiss, 7-foot	.20
Langley Field model	.25
McCook Field	.48

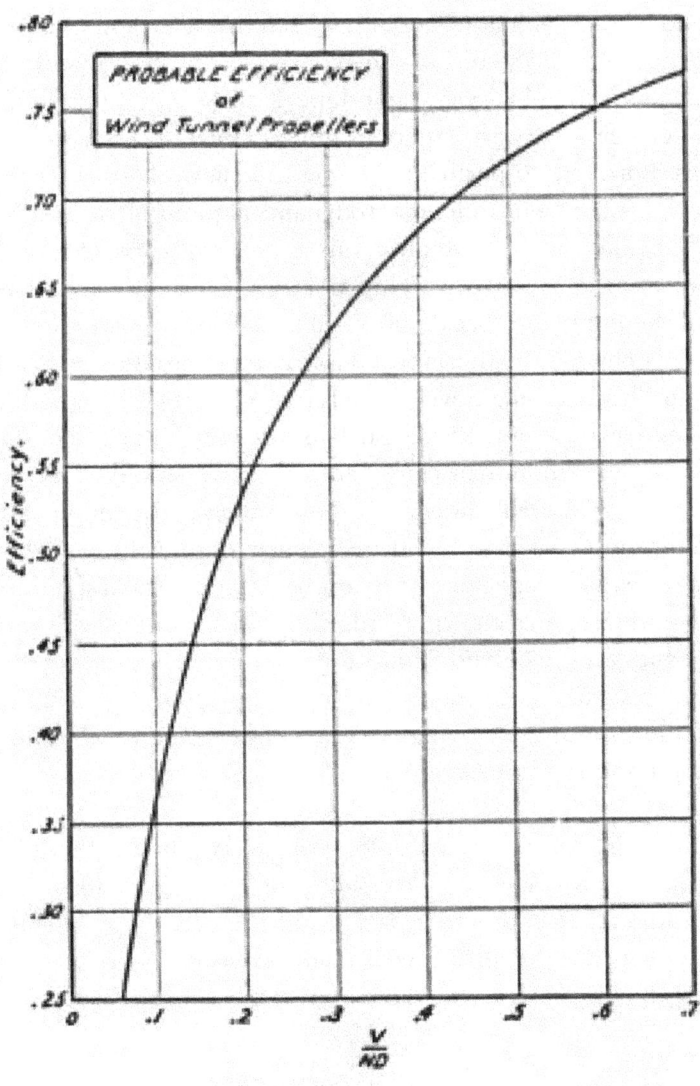

Fig. 2.

THE FORMS OF ENTRANCE AND EXIT CONES.

There has been a great deal of discussion and dispute as to the best form for the cones in which the air acquires and loses its speed, and further experiment is desirable. The effect which changes in the form of these cones have on the efficiency is, however, much less than has commonly been supposed, judging from experiments recently performed at Langley Field and reported in another part of this paper.

In the absence of data to indicate the best form, most of the wind tunnels which have been constructed have used, at least on the exit side of the experimental chamber, the frustrum of a right cone generated by a straight line. This was true of the N.P.L. and all their imitators, and it has been true also of most of the tunnels designed with an eye to the results of the experiments of Crocco and Castellazzi, and using long exit cones of very gradual slope. A surface of this type has at least the advantage of being easy to generate and to fabricate from wood or sheet metal. There is, however, no particular reason to believe that it is the most efficient that can be constructed from an aerodynamical point of view. Eiffel and his followers, on the other hand, have always used cones of curving form. It seems fair to assume that the loss in diverging nozzle is partially dependent on the deceleration of the fluid, and that the loss will usually be least where the deceleration is least. It is obvious, furthermore, that the flow through the exit cone will be smoothest and least turbulent when the form of the cone is smooth, and that any abrupt change of slope of the walls, such as that at the juncture of the parallel portion of the tunnel with an exit cone generated by a straight line, is liable to cause the lines of flow to break away from the contour of the tunnel wall, and to establish a region of "dead-water" and turbulence around the periphery of the exit cone. The smoothness of a curve can best be judged by taking differences, or, if the equation of the curve is known, by plotting the derivative. This was done in designing the cones for the Leland Stanford, Jr., tunnel. The plotting of the curve of acceleration for a tunnel will then serve the double purpose of indicating the smoothness of the curve and of giving the maximum rate at which the velocity of the air is changing, and so the maximum force necessary for accelerating the moving stream.

A curve of velocity against distance along the axis of the tunnel can be drawn on the assumption that velocity is inversely proportional to the square of the diameter of the tunnel. This, of course, is true only for velocity parallel to the axis, and entirely neglects the radial component. In order to obtain the acceleration from this curve, the derivative giving acceleration is written

$$\frac{dv}{dt} = \frac{dv}{dx} \times \frac{dx}{dt} = v \times \frac{dv}{dx}$$

The acceleration at any point along the tunnel is therefore equal to the product of the ordinate of the curve just described by its slope at that point. These factors can be found graphically or, in the case of a curve for which the equation is know, analytically.

In the case of a straight cone, for example, the formula for diameter at any point is

$$D = D_1 + (D_2 - D_1) \times \frac{x}{l}$$

where D_1 and D_2 are the diameters at the small and large ends of the cone, respectively, l the length of the cone, and x the distance from the small end. Then

and

The acceleration is equal to the product of these expressions, or

If the exit cone is generated by rotating about the axis of the tunnel a parabola having its vertex at the junction of the exit cone with the straight portion the formula for diameter of the cone at any point becomes

The acceleration may then be obtained by the same steps just employed for the straight cone.

In figure 3, the velocity and acceleration, as well as the cone diameter, are plotted against x for cones of these two forms. The units are meters and seconds, and the curves relate to a tunnel having an exit cone tapering in diameter from 1.5 meters to 3 meters in a length of 6 meters, and a wind speed of 50 meters per second. It appears that the straight cone is far inferior, judged by the criteria laid down above, to that of parabolic form. The maximum acceleration for the first is more than two and a half time that for the second, and there is a large discontinuity in the acceleration curve for the straight cone, as might be anticipated from the discontinuity in the slope of the side of the tunnel. The parabolic form gives zero acceleration at the point where the air emerges from the exit cone. There is some question as to the desirability of using a reverse curve which will have tangents parallel to the axis of the tunnel at both its ends, and so securing zero acceleration at both ends of the exit cone. The air has to be slowed down some time, and there would seem to be little advantage in bringing it to a constant velocity as it leaves the retaining walls of the exit cone if it is to be decelerated again the instant that it is free from those walls. Also, the current of air, since it is to be turned through an angle of 180° and travel back through the room to the entrance of the tunnel, must acquire a radial velocity either inside the exit cone or immediately after it has left it. No gain is apparent from a construction which permits the air to acquire a certain amount of radial velocity and then straightens it out again, only to force it to turn outward once more a few feet father along its path. The effect of a reversal in the curve of the walls near the large end of the exit cone is certainly slight, as very good results have been obtained both with and without such a reversal.

The form of the entrance cone appears to have but little effect on the "energy ration," and this is in accord with the results of hydraulic experiments, where it is always found that the loss in a converging nozzle is much less than that in a

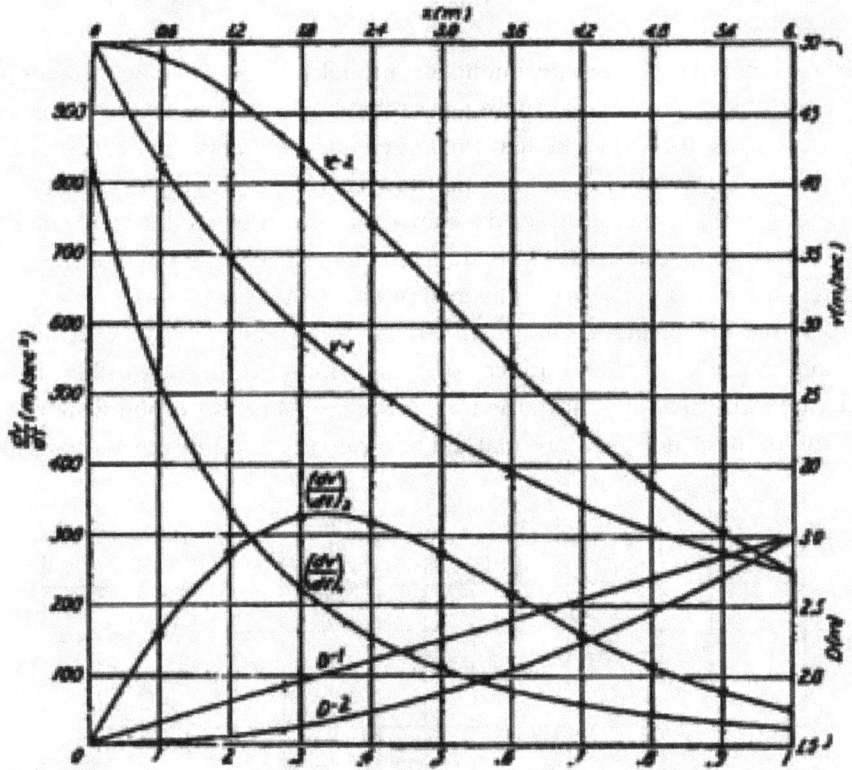

Fig. 3.—Velocities and accelerations of fluid in exit cones.

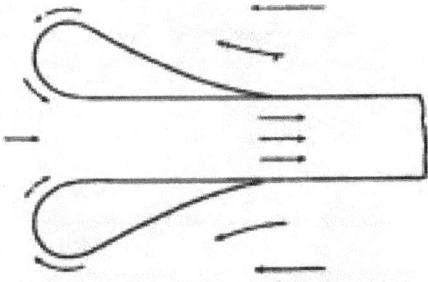

Fig. 4.—Fairing of entrance to N. P. L. tunnel.

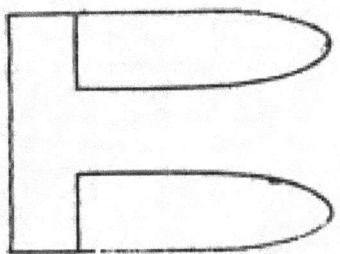

Fig. 5.—Proposed fairing of entrance cone.

diverging one, and that the nozzle can converge very abruptly without seriously increasing the loss.

Most of the European experiments on model tunnels have been made with straight entrance cones. While these are probably as efficient as any other type, they must have avena contracta near the large end, causing turbulence which persists into the experimental chamber, and there is further eddying and disturbance due to the turning of the air around a sharp corner at the small end of the cone. To avoid these difficulties and to secure as steady a flow as possible in the experimental chamber it is the almost universal practice, in actual tunnels, to make the entrance cone of curing form. It has been found at the National Physical Laboratory that even if the entrance cone, or bell-mouth, as it is called there, is curved around until a tangent to the wall at the large end is perpendicular to the axis of the tunnel, there still are marked and persistent eddies in the neighbor-

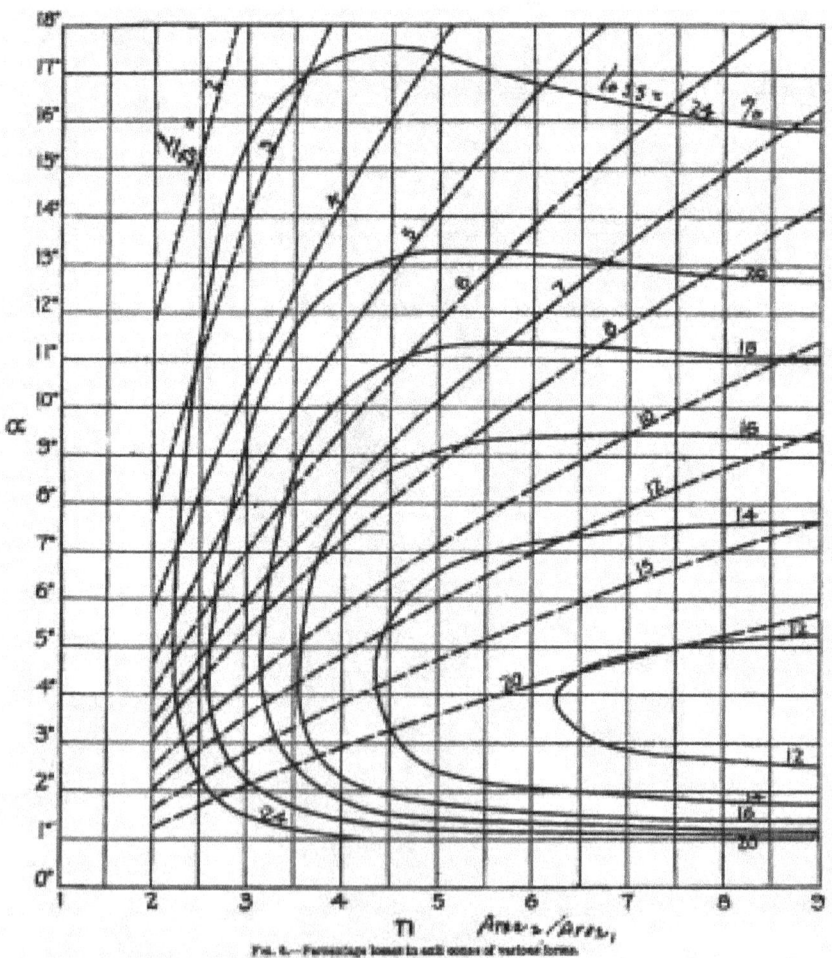

hood of the sharp edge. To entirely eliminate this edge it is not the practice at the N.P.L. to carry the bell-mouth around, as shown in figure 4, until it meets the straight portion of the tunnel. This method has not been adopted at Langley Field, as it is desired to make some experiments on the full-sized tunnel with the normal entrance cone, but provision has been made for building a fairing to extend clear around to the experimental chamber, as shown by the dotted lines in figure 5, so giving the air a perfectly smooth passage.

THE THEORY OF LOSSES IN THE EXIT CONE.

The losses in the exit cone of a wind tunnel arise from three sources. The first is the friction against the walls, and is best determined by Fritzsche's formula for fluid friction. The second is the diverging angle of the cone, which, as already noted, always leads to a loss of energy as compared with the ideal conditions expressed by Bernouilli's theorem. The magnitude of this loss is determined with satisfactory accuracy by a formula devised by Fliegner. Finally, there is a loss due to the sudden release of the air from the exit cone and its passage into the room, where its velocity drops almost to zero. This loss was shown by Borda to be equal to the kinetic energy possessed by the air at the large end of the cone. These losses, and their relation to the factors entering into wind tunnel design, together with all the losses in other parts of the tunnel, have been fully addressed by Eiffel, and it is not necessary to repeat his work here. For the benefit of those designing tunnels, however, a set of curves has been plotted which make it possible to read off at once the loss in a straight conical exit cone of any type, and to determine, given the limiting conditions, such as the size of a building to house the tunnel, the characteristics of the best exit cone of that particular case. Since from 80 per cent to 90 per cent of the total losses in a tunnel (not including those in the propeller) occur in the exit cone, the problem of designing a tunnel with a high energy ratio is essentially a problem of reducing the losses in the exit cone.

In figure 6 the ordinates are the vertex angles of exit cones, the abscissae, the ratio of the cross-section area at the large end of the cone to the cross-section area where models are tested, at the throat or in the experimental chamber. The family of curves drawn in full lines are curves of equal loss, and the number which one bears expresses the loss in the exit cone as a per centage of the kinetic energy possessed by the air at the smallest section of the tunnel. For example, if there were no losses except those in the exit cone, a tunnel having an exit cone of form corresponding to any point on the curve marked 20 would have an energy ratio of 5. The nearly straight dotted lines running across the sheet diagonally correspond to various constant lengths of exit cone, and they are marked with the ratio of length to diameter at the small end.

To illustrate the use of this chart in choosing an exit cone a few illustrative examples will be given.

1. A tunnel is to be 2 meters in diameter. In order to keep the size and cost of the building within reasonable limits, it is desired that the length of the exit cone shall not exceed 20 meters. Subject to this limitation, the cone is to be chosen for maximum efficiency.

The ratio of length to diameter here is 10. Passing along the dotted line bearing that number, it is seen that it cuts the curve of 16 per cent loss at two points and that it does not cut the 14 per cent curve at all, but that it approaches nearest to the latter at the point (α = 6.8 degrees, n = 4.8). It is usually best to make n a little smaller than the value for the minimum loss in the exit cone, as a reduction in n in the diameter at the large end of the cone and so in the propeller diameter, and it has already been shown that this is favorable to propeller efficiency. It would probably be best, in this case, to take n = 4.3, α = 6.1 degrees, or some other combination in that immediate neighborhood.

2. A very large wind tunnel is to be built, and, in order that the propeller diameter may not be unreasonably large, as well as to keep down the height of the building, the propeller diameter is limited to twice the diameter of the tunnel at the minimum section.

If the ratio of diameters at the ends of the exit cone is 2, n = 4. Drawing a vertical from the scale of abscissae at this point, it is seen that it approaches nearest to the 14 per cent curve at (α = 4.5 degrees). The length of the exit cone for this angle is 13 times the minimum diameter. It would not be advisable, under these conditions, to choose the cone for the absolute maximum efficiency, as the length could be decreased 4 ½ diameters at a cost of only 5 per cent increase in total power by increasing α to 6.7 degrees. Since the curvature of the constant power curves is not abrupt, the conditions can be changed considerably from those for minimum loss without very much affecting the efficiency, and it is almost always worthwhile to make some concession of efficiency in order to reduce the dimensions of the building and of the tunnel itself.

EXPERIMENTS ON MODEL WIND TUNNELS.

The first set of experiments conducted dealt with a model of the wind tunnel for Langley Field, as it was originally planned. The tunnel was of the Eiffel type, with a large experimental chamber, and this chamber was reproduced to the proper scale in the model. All of the models used were one-fifth the size of the large tunnel, the experimental chamber being 30.5 cm. in diameter in the models. The entrance and exit cones were made of plaster over a base of wall board, and were shellacked, so that a very smooth surface was secured. The plaster was scraped to form, as soon as it is set, with a steel template rotated about a shaft running along the axis of the tunnel. An exit cone is shown, with the template in place and ready to apply the plaster, in figure 7. The drive was by belt from a 2-horsepower induction motor, and the propeller was four-bladed. The blade had

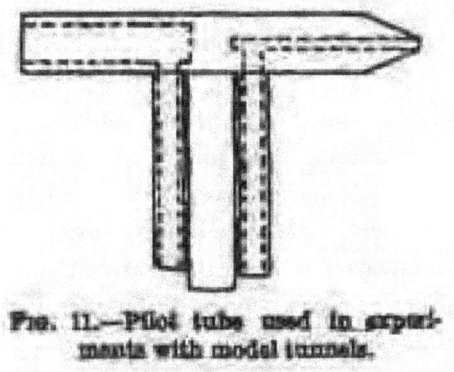

Fig. 11.—Pitot tube used in experiments with model tunnels.

a constant width of 4.5 cm. The speed of the propeller was measured with a Veeder liquid tachometer, and the power consumption with a polyphase watt-meter. The tunnel and instruments ready for use are shown from two points of view in figures 8 and 9, the propeller in figure 10. Figure 11 illustrates the Pitot tube used for measuring the wind speed. A hole 1 mm. in diameter is bored in the tapering end, and communicates with one of the two hypodermic tubes passing down the shank. The static pressure is secured inside a hole 2.5 mm. in diameter drilled from the other end of the tube, and this hole communicates with the other hypodermic tube. The piece between the two hypodermic tubes is a solid rod to provide stiffness. Since the static pressure points to the rear, the pressure in that side of the gauge is less than the true static, and the readings are higher than they theoretically should be. The tube was calibrated against a standard Pitot in the wind tunnel of the Bureau of Standards, and was found to have a constant of 1.167 (i.e., the readings of the small Pitot tube were 16.7 per cent higher than they theoretically should have been). This Pitot tube was very insensitive to rotations in all planes, as it could be turned 20 degrees without affecting the reading more than 8 per cent. This is a great advantage where, as in traversing the cones, the direction of flow of air is uncertain. The dynamic head on the Pitot tube was measured by an alcohol gauge, shown in figure 12. Only one side of the gauge is ordinarily used. Since the glass tube is raised and lowered by a micrometer screw so that the meniscus of the alcohol stands opposite the same mark on the tube for each reading there are no corrections, such as are required in the ordinary Krell manometer, for varying diameter of the glass tube or for changing level of the fluid in the reservoir.

The mode of procedure in each complete test was to make traverses of the entrance and exit cones and the experimental chamber at several points, measuring the wind speed at several radii, and then to make runs at a number of different speeds, measuring the wind speed at the point where a model would be placed for test and the power consumption. The energy ratio and the manner of its variation with speed was determined from this set of runs.

The traverses for the original model are plotted in figure 13. The points A and B were in the entrance cone, A being at the large end of the cone, B midway between the ends. C was in the experimental chamber, 5 cm. from the entrance cone side. D was 20 cm. downstream from the entrance cone and E was 46 cm. from the entrance cone, 15 cm. from the exit. F, G, and H were in the exit cone, and were equally spaced along its length, H being in front of the propeller and as close as it could be placed without danger of having the tube struck by the blades. The exit cone in this model was parabolic in form. The location of the point D corresponded to that at which the model is to be placed in the full-sized tunnel.

The speed in the entrance cone had a maximum at the center and one near the wall, the one near the wall being higher than that in the center. The maximum occurred with 2 cm. of the wall. On going still farther out the speed dropped rapidly, due to friction. The velocity in the experimental chamber near the entrance cone was constant, as nearly as could be detected, over 90 per cent of the diameter of the stream. On going farther downstream the velocity distribution became more irregular, the speed being a maximum at the center and dropping off steadily toward the edges of the stream. The ratio of the velocity 75 per cent of the way out to the edge of the stream to that at the center was 1.00 at C, 0.97 at D, and

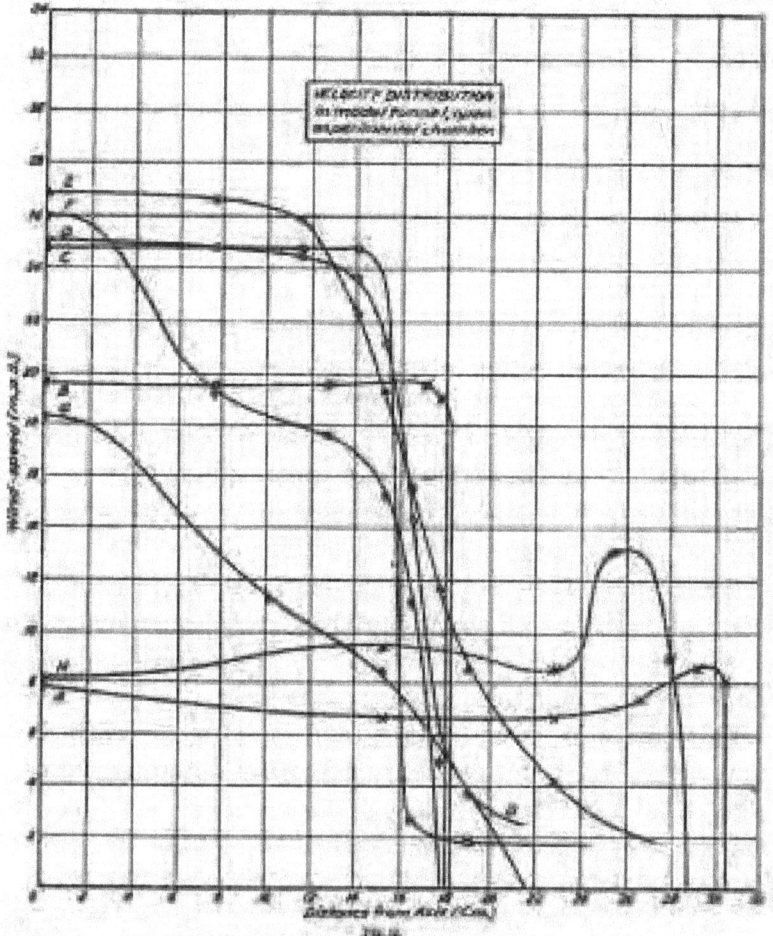

0.96 at E. The edge of the stream was not sharply defined, even very near to the point of issuance from the entrance cone, and at E, three-quarters of the way across the chamber, the velocity dropped off in a smooth curve from very near the center of the stream out to far beyond its normal boundaries.

The velocity distribution in the exit cone was exceedingly strange. The velocity dropped off rapidly from a maximum at the center, so that the stream appears actually to contract rather than to expand in the exit cone. As in the entrance cone, there was another maximum near the wall, but it was farther from the wall than was that at the entrance, and the velocity was much lower than at the center. Directly in front of the propeller the velocity at the center dropped sharply, due to the hub, and varied in an irregular manner over the rest of the section. The flow at this point was so turbulent and so varying in direction that the measurements of velocity may contain considerable errors.

In order to make direct observations on the sharpness of definition of the edge of the stream in the experimental chamber and to determine the general nature of the flow in the chamber an observer got inside and sounded the flow with a thread. It was evident that the air in the whole chamber was much stirred up, and that the flow near the nominal edge of the stream was extremely turbulent, except in the immediate neighborhood of the entrance. Even in the farthest corners of the chamber, at a distance from the center of the stream equal to more than three times its nominal diameter, there was still a distinct movement of the air. The motion everywhere was very unsteady, the direction of flow at a given point changing 60 degrees or more almost instantaneously. The best defined part of the circulation was near the small end of the exit cone, where two strong vortices rotating in opposite directions existed in the corners of the chamber. The examination of the flow was not extended to points above and below the stream in this neighborhood, so it is not certain whether or not a complete vortex ring, surrounding the opening into the exit cone, existed. The results of this examination of the flow in the experimental chamber made it clear that the balance would have to be shielded in some way from the air currents if any accurate work was to be done. In Eiffel's tunnel, partial shielding of the balance is accomplished by placing it on a platform which, however, extends across only a small proportion of the width of the room, and can hardly act as a complete protection from air-currents for the measuring instruments.

The power curve is plotted in figure 14 (curve No. 1) and the curve of speed against revolutions per minute in figure 15 (curve No. 1). The energy ratio varied too little and too irregularly to make it worthwhile to plot a curve. Its mean value was 0.90, making no allowance for propeller losses. If the propeller efficiency be assumed to be 57 per cent (the value calculated by the Drzewiecki method), the energy ratio for the tunnel proper becomes 1.58.

In view of the irregularities of flow found in the experimental chamber it was decided to try next the effect of inclosing the stream in a cylindrical tube during its passage across the experimental chamber. No attempt was made to make the tube air-tight, the static pressure inside the tube being equal to that in the experimental chamber, which was carefully made air-tight. Curve No. 2 in figure 14, and also in figure 15, correspond to this case, and the traverse of the stream at points corresponding with those taken for the original model are plotted in figure 16.

Comparing these traverses with those in figure 13, it is seen that the nature of the distribution in the entrance cone is practically unaffected. The velocity at point C was a little less regular than for the case of the unconstrained stream, showing an increase near the walls similar to that which characterized the entrance cone. At D and E, however, the velocity was much more even with the inclosing tube than without it, being constant within 1 per cent over 75 per cent of the diameter. Evidently, from the standpoint of steadiness of flow, the inclosed type of tunnel is superior to the Eiffel type.

In the exit cone the effect of surrounding the stream with a definite boundary was still more apparent. At F the velocity three-quarters of the way from the center to the walls was 94 per cent of that at the center, as against 67 per cent in the original model. At G the corresponding figures were 82 per cent and 40 per cent. At H there was, as in the first case, a minimum at the center and two maximums, the distribution of velocity being reasonably uniform across the outer 70 per cent of the blade, which is the most effective portion.

It is reasonable to suppose, in view of the better filling of the exit cone and of the generally improved velocity distribution, that the energy ratio would be increased by inclosing the stream, and this supposition was fully justified by the power measurements. For a given rate of rotation of the propeller the wind speed was increased while the power consumption for a given wind speed was decreased just about 50 percent. The energy ratio with the inclosing tube was 1.83 for the whole installation, or, making due allowance for the propeller losses, 3.20 for the tunnel alone.

It is evident that the enclosure of the stream improves the results in every way. The experiments, so far as power consumption is concerned, check very well with those obtained in some similar experiments on model tunnels, carried out by Lieut. Castellazzi. Lieut. Castellazzi found that the efficiency was decreased 40 per cent by the use of an open experimental chamber. The experimental chamber used in his experiments was round in cross section and was twice as large in diameter as the entrance and exit cones where they entered the chamber, and the slightly greater loss in efficiency found in the experiments conducted at Langley Field may be accounted for by the larger size and more irregular form of the experimental chamber there employed.

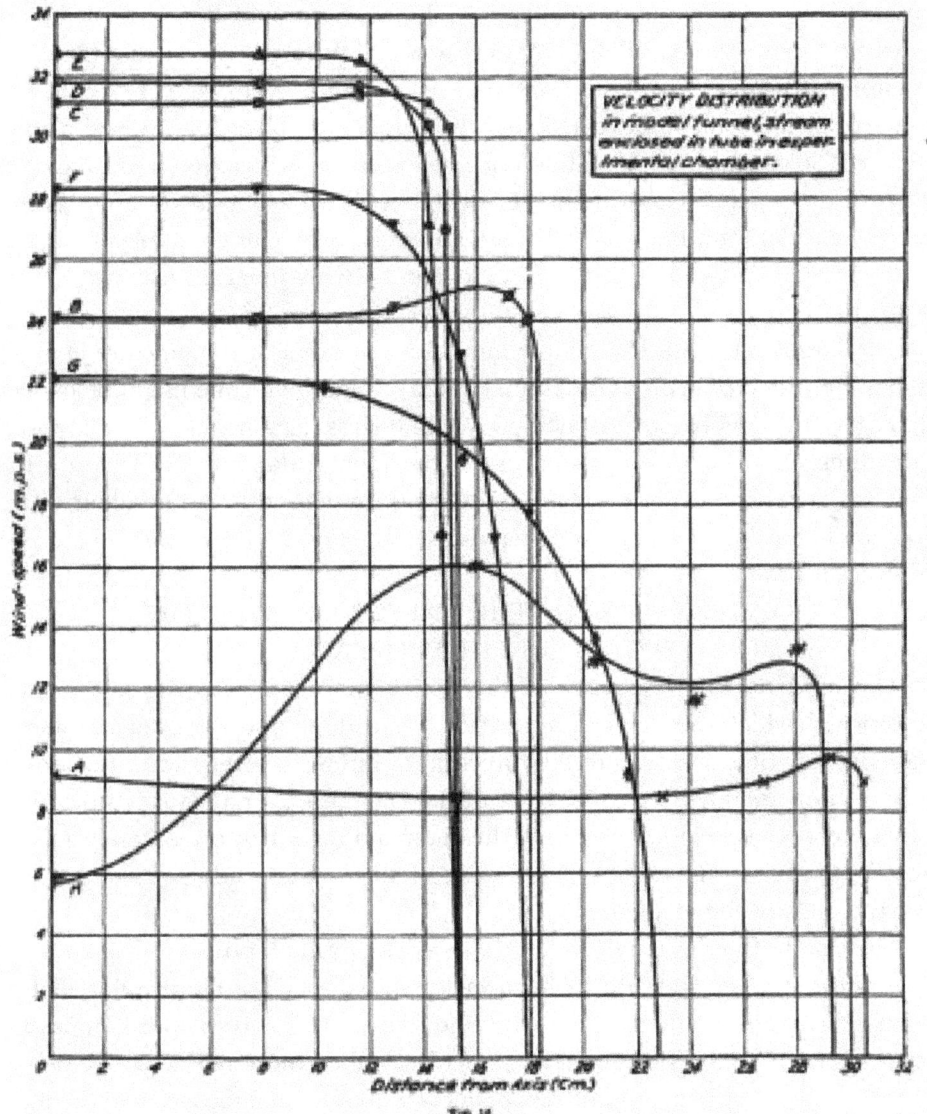

Fig. 16.

EFFECTS OF VARIATION IN EXIT CONE FORM.

The next series of experiments dealt with the effect of alterations in exit cone form. It was originally the intention to make a number of cones of different forms, but this plan was abandoned after two had been tried, and the experiments cover only the parabolic and straight forms of cone. These are as widely different from each other in respect of their acceleration curves as are any two forms which would be likely to be used.

The curves of power and revolutions per minute with the straight cone are plotted as curve No. 3 in figures 14 and 15. The mean energy ratio is 1.83 for the combination of tunnel and propeller, or 3.20 for the tunnel alone, values identical with those for the parabolic cone. It is evident from the curves that the effect of changing the exit cone from a parabolic to a straight form was very slight. The parabolic form seems to have a slight advantage at high values of VD and to be inferior at low values, but the difference between the two curves is in no case in excess of the possible experimental error. In view of these results it appears that the efficiency of a tunnel is not affected appreciably by exit cone form or by the nature of the acceleration in the cone, but only by its length, mean angle, and total expansion ratio.

The large acceleration suddenly imposed on the air at the juncture between the parallel-sided portion of the tunnel and a straight exit cone might be expected to cause turbulence, so that the flow would be less regular than with a parabolic or other smoothly curving form. No experimental data are available on this point as yet, as the experiments were temporarily halted by an accident to the propeller before traverses and investigations of the flow had been carried out with the straight cone.

OBSERVATIONS OF THE NATURE OF THE FLOW THROUGH THE PROPELLER.

The most noticeable feature of the flow behind the propeller is the great rapidity with which the slip stream spreads. Instead of contracting, as in the case of an airplane propeller, where the direction of inflow is unrestricted, the stream expands immediately on passing clear of the cone, the air changing its direction so that there is a strong movement of the air, in a direction approximately at right angles to the axis of the tunnel, at a distance of 30 cm. back and 50 cm. out radially from the edge of the exit cone.

The flow in the throat and cones was very steady at all points except near the edges of the stream. The velocity head varied with a total amplitude of oscillation of about 2 per cent of the head and a period of from 20 to 40 seconds. On passing the propeller the pulsations of velocity became much more marked. The period of the pulsations close behind the propeller was about half a second, and the maximum velocity was estimated to be about 50 per cent greater than the minimum, although no means of measuring and making a continuous record of a rapidly varying velocity were available. On going farther away from the propeller along the lines of flow of the air, the pulsations steadily increased in violence and the period lengthened, until at a distance of about 80 cm. to the rear of the propeller, the flow consisted of a violent gust about every second, the velocity in the intervals between these gusts being so low as to be hardly perceptible. These observations on the nature of the flow and its variations held in a general way for all the models tried, but the pulsations of velocity were much more marked for the case where the experimental chamber was left open than for that where it was inclosed in a tube.

EXPERIMENTS ON THE EFFECT OF DISKS AND SPINNERS ON THE PROPELLER.

In order to secure some idea of the effect of enlarging the hub of a propeller or of attaching a spinner, some experiments were made with disks of wall board attached in front of and behind the propeller, and also with a paper cone projecting from the propeller into the exit cone. The results of these tests do not fairly represent what might be secured with a good spinner and a propeller especially designed for it, as the propeller pitch should be increased when a spinner is incorporated or the hub is enlarged, but they will give some idea of the effect.

The effect of placing a disk in front of the central portion of the propeller, the rear not being covered and the blades not being housed in any way, was to decrease the wind speed and increase the power consumption. The inner parts of the blades acted as a centrifugal blower, taking air in from the rear and throwing it our radially. The increase in power, with a disk half the diameter of the propeller, was 9 per cent, the decrease of speed with the same disk 19 per cent. With a disk only one-fifth the diameter of the propeller the speed was decreased 5 per cent. These measurements were made at a speed of 10 meters per second and with the parabolic exit cone. The relative loss by the addition of a disk was greater with the straight cone and at high speeds, the addition of a disk four-tenths the diameter of the propeller causing an increase of 28 per cent in power and a decrease of 19 per cent in speed at a speed of 34 meters a second with the straight exit cone. The energy ratio was decreased 59 per cent. All subsequent tests were made with the straight cone, and the losses would probably be less with other forms.

The addition of another disk of equal size behind the propeller, so preventing any flow in from the rear and out toward the tips, improved the performance as compared with the single disk in front of the propeller, but remained inferior to the original case with no shielding at all. The power was increased only 6 per cent as compared with the original case without nay disks, but the speed was decreased 16 per cent and the energy ratio fell off 44 per cent. When the rear disk alone was in place, so that any air thrown radially outward had to come from inside the exit cone, the power was increased 6 per cent, the velocity decreased 5 per cent, and the energy ratio decreased 19 per cent, using the model without disks as a standard in all cases. The disk behind the propeller therefore gave better results than did complete sheathing, either in the form of disks or faired by a cone in front.

The addition of a cone, having a diameter equal to two-fifths the diameter of the exit cone at its large end and an altitude of one and a quarter times its own diameter, in front of the propeller decreased the power about 2 per cent and increased the speed 7 per cent as compared with the values for the disks alone, but the energy ratio was still 30 per cent lower than for the original case. It seems strange at first that the entire blocking off of a considerable portion of the blades should increase the power consumption for a given number of revolutions per minute, but the phenomenon can be accounted for by the higher air speed past the propeller when the area of the exit cone is constricted by enlarging the hub. The theory of the effect of an enlarged hub or spinner has been discussed in another section of this report.

It appears that the addition of a spinner or the enlargement of the hub caused serious loss in every case where it was tried with the straight cone. The loss with a parabolic cone is much less, and it is likely that, with a propeller properly designed to allow for the increased velocity due to the blocking off of part of the area of the exit cone by the spinner, results as good as those in the original case could be obtained. It may even be that they could be materially improved on, but this does not seem very probable in view of the uniformly poor results shown in these experiments, where the presence of the spinner can hardly have decreased the propeller efficiency more than 10 per cent (a loss which, as already noted, could be prevented by the adoption of a propeller designed especially for the new conditions). The loss in propeller efficiency, therefore, would not be sufficient entirely to account for the decrease of energy ratio. The principal value of a very large hub is to increase the power coefficient of the propeller and make possible the reduction of the peripheral speed for a given wind speed.

Document 2-12(b), F. H. Norton and Edward P. Warner, The Design of Wind Tunnels and Wind Tunnel Propellers, II, *NACA Technical Report No. 98, 1920*.

SUMMARY.

This report is a continuation of National Advisory Committee for Aeronautics Report No. 73, and was undertaken at the Langley Memorial Aeronautical Laboratory for the purpose of supplying further data to the designer of wind tunnels. Particular emphasis was placed on the study of directional variation in the wind stream. For this purpose a recording yawmeter, which could also be used as an air speed meter, was developed, and gave very satisfactory results. It is regrettable that the voltage supplied to the driving motor was not very constant, due to varying loads on the line, but as this motor was of a lightly loaded induction type, the variation in speed was not as large as the variation in voltage. The work was carried on both in a 1-foot model and the 5-foot full-sized tunnel, and wherever possible a comparison was made between them. It was found that placing radial vanes directly before the propeller actually increased the efficiency of the tunnel to a considerable extent. The placing of a honeycomb at the mouth of the experimental portion was of the greatest aid in improving the flow, but, of course, somewhat reduced the efficiency. Several types of diffusers were tried in the return air, but only slight improvement resulted in the steadiness of flow, they not being nearly as effective as the honeycomb.

APPARATUS.

The efficiency of the tunnel and the slip of the propeller were determined by the same method as described in Report No. 73, but to better record the fluctuations in velocity and direction a recording instrument was constructed. This instrument, as shown in Figs. 1 and 2, consists of a thin mica diaphragm whose movement rotates a very light spindle containing a small silvered mirror. Light from an illuminated slit is transmitted by a lens to this mirror and the reflected beam is then focused on a moving photographic film so that any movement of the mica diaphragm is recorded as a continuous curve. By this method any small and rapid variation in the air flow of the tunnel is indicated and recorded by means of a Pitot-static tube which is connected to the two compartments separated by the diaphragm, and any change in direction is recorded in the same way by connecting the sides of a yawhead to the compartments on opposite sides of the mica diaphragm. The Pitot and the connecting tubes are made comparatively large so that any rapid fluctuation in velocity can be immediately transmitted to the diaphragm without damping or lag. Over 50 records were taken but only a few

FIG. 1.—RECORDING AIR SPEED METER.

FIG 1.—RECORDING AIR SPEED METER WITH ILLUMINATING AND RECORDING APPARATUS.

typical ones are reproduced here. Numerous experiments on the efficiency of tunnels and on speed fluctuation have previously been made in England.

EFFICIENCY AND SLIP WITH NEW PROPELLER.

In order to give a more even flow of air in the exit cone a new propeller was designed for the model tunnel having a larger pitch at the tip so that the air in this portion would be drawn through with a relatively greater velocity. In every other respect this propeller is very similar to the propeller used in the test described in Report No. 73, which, owing to a piece of wood being dropped in the running tunnel, was completely destroyed. In Fig. 3 is shown the efficiency of this propeller when working in a parabolic cone and in a straight cone. It will be noted that in the same way as with the first propeller the straight cone is considerably more efficient at high speed than the parabolic cone. In Fig. 4 is shown the slip of this propeller in the parabolic cone and in the straight cone and it is noted that the slip is less at high speed for the straight cone. It is then evident that the straight cone is aerodynamically superior to the parabolic cone, in addition to being easier to build.

EFFECT OF SIZE OF THE ROOM.

All the test runs described in Report No. 73 were conducted in a large room, approximating to free-air conditions. In the tests described in this report a temporary room was built around the tunnel, representing to scale the building provided for the 5-foot N.A.C.A. wind tunnel; and all runs except those shown in Figs. 3 and 4 were made in this model room. The cross section of the model room and the

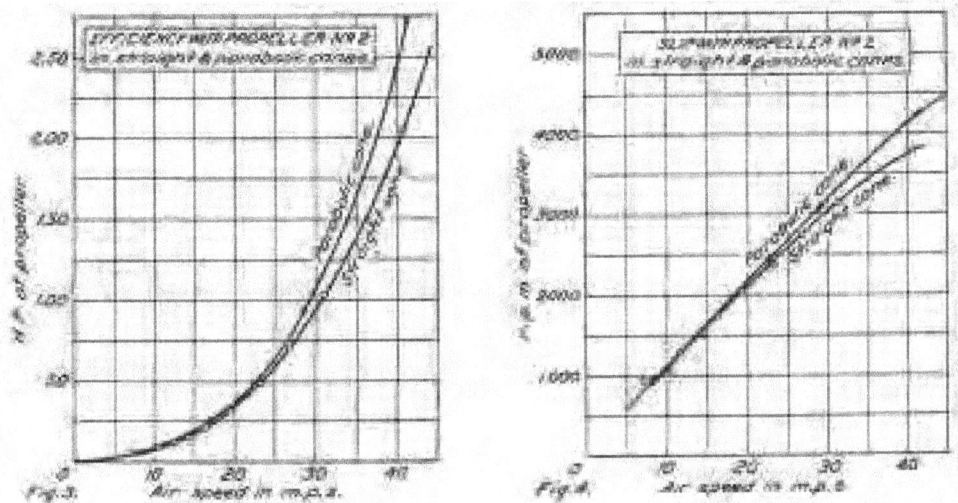

wind tunnel are shown in Fig. 5. For the same power this room decreased the air speed from 69 to 59 miles per hour or a decrease of 14.5 per cent. In the small room the maximum variation of speed was ±7 per cent and the maximum variation in direction was ±10°. The air speed records show that for the first 20 seconds after starting, in the large room, and for the first 10 seconds in the small room, the air speed is very steady, and that the fluctuations suddenly appear at a definite time and will be indicated on the record. This appearance of sudden fluctuations seems to indicate that the large part of the speed fluctuations are due to the disturbed air from the propeller as it returns through the room to the entrance cone.

EFFECT OF RADIAL VANES.

Eight radial vanes 3 mm. thick and 450 mm. deep were placed symmetrically in the exit cone immediately before the propeller. These vanes joined in the center in a stationery spinner which was of the same diameter as the propeller base. (Fig. 6.) These vanes actually increased the speed of the air in the tunnel for the same power by 5 per cent, but the fluctuations in direction and velocity remained unchanged. In order to determine what part of the vane gave the increased efficiency, 25 mm. was cut off of the outer end of each vane and the run repeated which gave a 3 per cent increase in speed for the same power over the tunnel with no vanes. Again the vanes were cut off on the end 75 mm. and in this case the same speed was obtained as with the tunnel without vanes. This seems to show that it is the whole area of the vane which acts as a straightener for the air flow and that no particular part is especially valuable in increasing the efficiency of the tunnel. Eight additional vanes 3 mm. thick were then placed along the inner surface

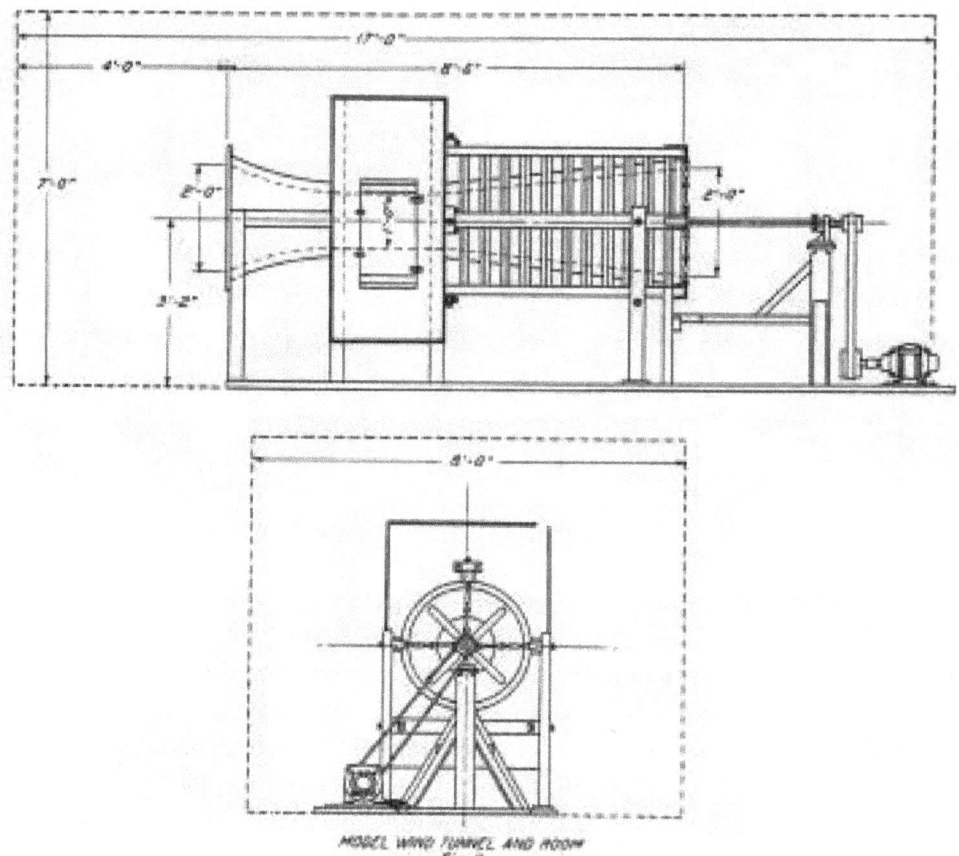

MODEL WIND TUNNEL AND ROOM
Fig. 3

of exit cone, each vane being 75 mm. wide. This distribution of vanes decreased the speed by 12 percent for the same power and the variation in speed was ±6 percent and the variation in direction was ±10°. The same vanes were then placed in the entrance cone, as shown in Fig. 7, and in this case the speed was decreased by 8 percent and the variation in direction was ±8°. With this type of vane in both the exit and entrance cone the speed was decreased by 20 percent for the same power and the variation in direction was ±8°. It is evident from these tests that the narrow vanes in either the exit or entrance cones are of little value in any way.

EFFECT OF PLACING SCREEN ACROSS THE TUNNEL.

A section of chicken wire of 25 mm. mesh was placed across the exit cone 45 centimeters ahead of the propeller. The use of the chicken wire decreased the speed by only 3 per cent, so it does not seem that this distribution of screen would be of any great harm to the efficiency of the tunnel and it is of great use in preventing small objects from being drawn into the propeller. A piece of window screen placed at the beginning of the straight portion of the tunnel decreased the

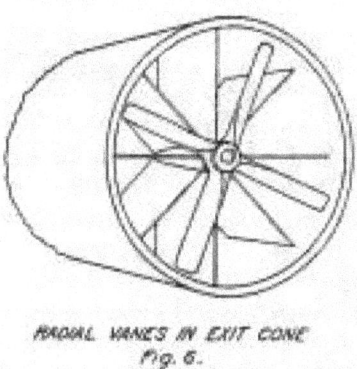
RADIAL VANES IN EXIT CONE
Fig. 6.

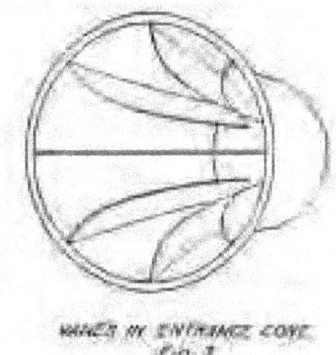
VANES IN ENTRANCE CONE
Fig. 7.

speed by 14 per cent and the fluctuation in speed was -12 per cent and was -10 per cent in direction, showing that the screen in no way helps the steadiness of flow for the particular condition of this test. Screens have been used to advantage in other tunnels. With window screen at the mouth of the entrance cone the speed was decreased by only 7 per cent.

EFFECT OF PLACING SPINNERS BEFORE THE PROPELLER.

A spinner 75 mm. in diameter and 450 mm. long was supported by steel wires before the propeller, as shown in Fig. 8. The use of this spinner seemed to have no material effect on the air flow.

THE EFFECT OF EXTENDING THE EXIT CONE BEYOND THE PROPELLER.

By extending the exit cone as shown in Fig. 9, there was no change of the air flow inside the tunnel, but the tangential flow, which had been noticed before with the propeller, was somewhat straightened out, and the air flow was more directly to the rear through the extension of the cone. A cylinder was then attached to the propeller end of the tunnel as shown in Fig. 10, which decreased the air speed

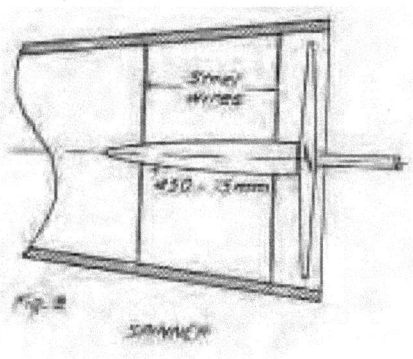
Fig. 8
SPINNER

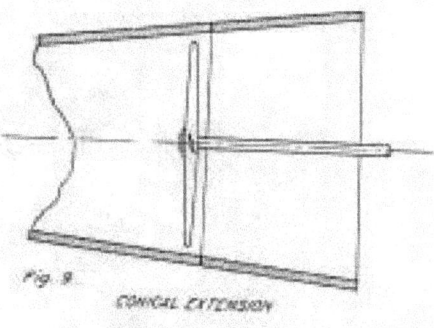
Fig. 9
CONICAL EXTENSION

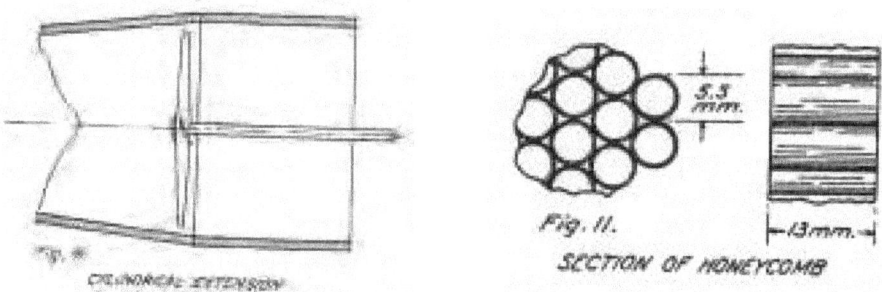

Fig. II.
SECTION OF HONEYCOMB

about 5 per cent, the air issuing from the tunnel at a considerably higher velocity and in a more compact stream, the borders of the stream still being sharply defined at a distance of 20 feet. As extensions of this kind mean a larger and longer building for the wind tunnel there would certainly be no advantage in using them.

EFFECT OF HONEYCOMBS.

A honeycomb was constructed as shown in Fig. 11 and was placed at the entrance to the straight portion of the tunnel. Owing to the difficulty in obtaining thin-walled metal tubing and to the expense of constructing honeycombs of this type, only this one was tried. It is quite evident, however, even from this one test that the honeycomb is of the greatest importance in straightening out the flow. The speed is reduced 19 per cent and the energy ratio 45 per cent by this honeycomb, but the maximum speed variation was only ±2 per cent and the variation in direction was reduced to ±0.5°.

In order to show more clearly the great increase in steadiness of flow, a curve taken with a recording yawmeter is shown for the open tunnel and for the tunnel

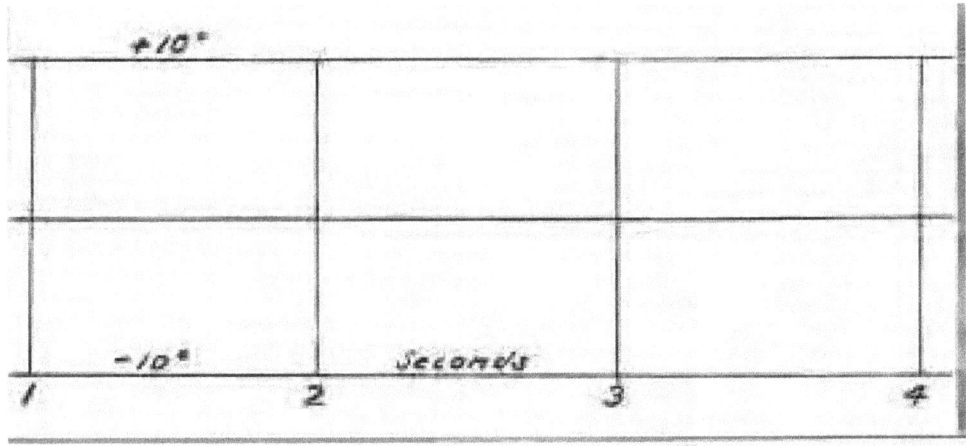

FIG. 13. VARIATION IN DIRECTION IN THE MODEL TUNNEL WITH A HONEYCOMB

containing the honeycomb. (Figs.12 and 13.) It is evident from these how great is the advantage of the honeycomb. As the length diameter ratio in the tubes of this honeycomb are only 2 ½ it is quite possible that by using longer tubes the flow would be even better and the reduction in speed should not be appreciable. There seems to be no doubt from these tests that the honeycomb is absolutely essential in most wind tunnels.

EFFECT OF DIFFUSERS.

The first diffuser tried is shown in Fig. 14 and consists essentially of a cubical box of which both sides are perforated with small holes, whose diameter is equal to the thickness of the wall of the box and whose spacing between centers is about twice that of the diameter of the hole. This box was connected rigidly to the rear

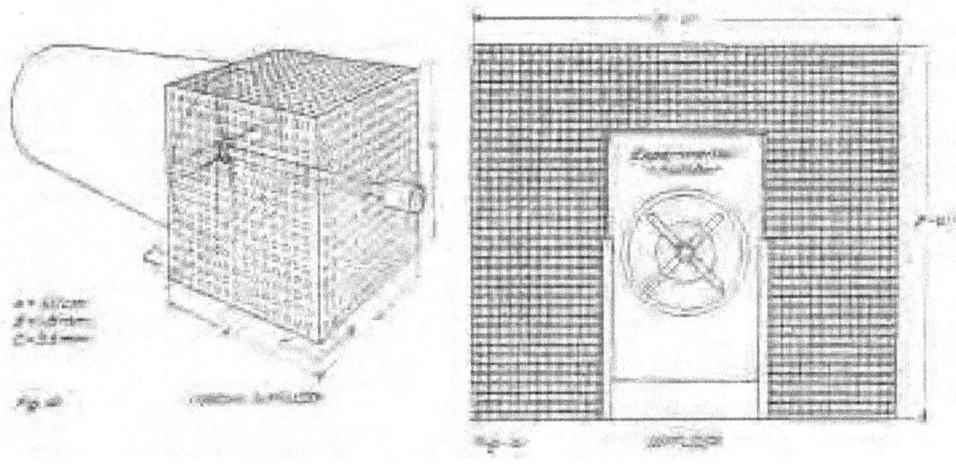

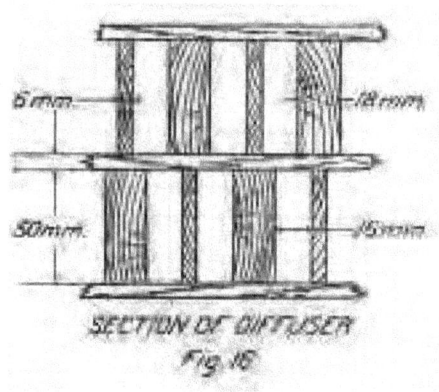

SECTION OF DIFFUSER
Fig. 16

of the exit cone so that all the air passing through the propeller must escape through these small holes. It was hoped in this way to break up any pulsations which would originate from the propeller. This arrangement decreased the speed of the tunnel by 7 per cent and the maximum variation of speed was ±6 per cent and the direction variation was ±5°, so that it would seem that the flow is slightly straightened, but nowhere near as much as with the honeycomb. A second diffuser was tried as shown in Fig. 15, which consists of a latticework across the tunnel room at the experimental chamber consisting of 50 mm. square cells having a 6 mm. wall with a length 2 1/4 times their diameter. This diffuser only reduced the speed of the tunnel by 2 per cent, and the maximum variation was ±7 per cent, and the variation in direction was ±5°. Although this diffuser has very little effect on the efficiency of the tunnel, at the same time it does not much improve the steadiness of flow. A third diffuser was constructed as shown in Fig. 16 and placed in the same position as the last. This diffuser decreased the air speed for the same power about 5 per cent, the variation in velocity was ±5 per cent, and the variation in direction was ±4°, showing only a slight improvement over the open room. It seems strange that these diffusers did not improve the air flow more, as the British have found that diffusers greatly improve the flow in their tunnels. The results of these tests would not, however, justify the use of a diffuser in a full-sized tunnel because of the rather large expense of construction of such a piece of apparatus.

EFFECT OF PERFORATING THE STRAIGHT PORTION OF THE TUNNEL.

In order to determine the effect on air flow of opening the doors in the cylindrical portion of the tunnel and in using small holes for the introduction of apparatus, various tests were made on the model in order to see how this would effect the efficiency and steadiness of flow. Also the velocity of the air in the experimental chamber was determined by a small anemometer. A slot was first cut in the cylinder parallel to its axis and one-fifteenth of the diameter wide, running the whole length of the experimental chamber. The air flow extended out about the width of the slot from the walls of the cylinder, and beyond this there was no flow in the chamber and the efficiency of the tunnel was not appreciably affected. This slot was then increased in width to one-sixth of the diameter of the tunnel, thus decreasing the efficiency of the tunnel very slightly, and the flow of air extended about one-sixth of the tunnel diameter into the experimental chamber nearest the exit cone, but this air flow was less marked as the distance to the entrance cone was decreased. When the width of the slot was increased to three-eighths of the tunnel diameter the efficiency was decreased about 15 per cent and the air flow extended two-thirds of the width of the slot into the experimental chamber, near the exit cone, but there was no flow elsewhere in the experimental chamber.

TESTS IN FULL-SIZED TUNNEL.

A few tests were made in the large tunnel in order to afford a comparison with the model. In Fig. 17 is shown the slip in the large tunnel. In comparing this with a similar condition in the model tunnel (Fig. 4) it is seen that for the same air speed the revolutions per minute is 5.7 times as large in the small tunnel as in the large one. Theoretically, the ratio should be exactly 5, but the fact that the model test was run in a proportionately larger room would account for this difference.

As the exact efficiency of the driving motor in the large tunnel is unknown, a curve of horsepower supplied to the motor is plotted against air speed, but to give some idea of the power supplied to the propeller a dotted curve is drawn from the estimated motor efficiency. (Fig. 18.)

In comparing this curve with the one obtained in the model, it is seen that the full-sized tunnel is slightly more efficient, so that results may be taken from models to safely predict the performance of the full-sized tunnels. It is also interesting to notice that the power does not increase as rapidly as the cube of the speed but more nearly as $V^{2.5}$, although, as the efficiency of the motor is not exactly known, the value of the exponent can not be determined very closely.

Records were taken in the full-sized tunnel of variations in velocity, and these are reproduced in Figs. 19 and 20. In the first figure the wind-tunnel motor was connected to a gasoline driven generator of 25 kilowatts and records taken at several

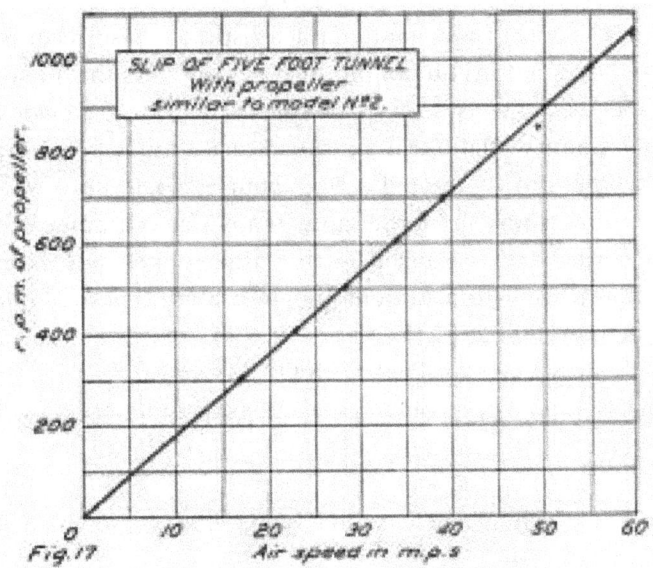

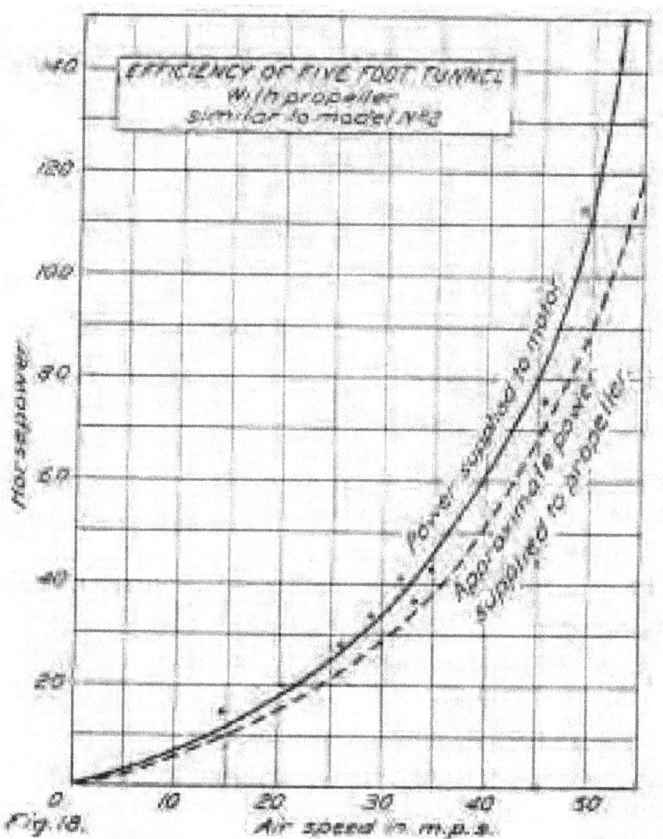
Fig. 18.

speeds. In Fig. 20 the motor was connected to a 300-kilowatt generator driven by a Liberty motor. The most important characteristic of these records is that the magnitudes of the fluctuations do not increase as rapidly as the air speed, so that at the higher speeds, quite contrary to expectations, the velocity is relatively steadier. The maximum variation in air speed at 90 miles per hour was about ±1.5 per cent, whereas in the model it was about ±2 per cent, so that it would seem that the steadiness was about the same in any size of tunnel.

Yawmeter records were also taken in the large tunnel, but were not reproduced, as they show practically a straight line, indicating that the honeycomb was satisfactorily straightening out the flow.

NATURAL PERIODS OF TUNNEL.

A wind tunnel acts as an open organ pipe and its natural period will be given by:

$$t = \frac{V}{l}$$

where l is the length of the tunnel in feet, and V is the velocity of sound, or 1,040 ft./sec.

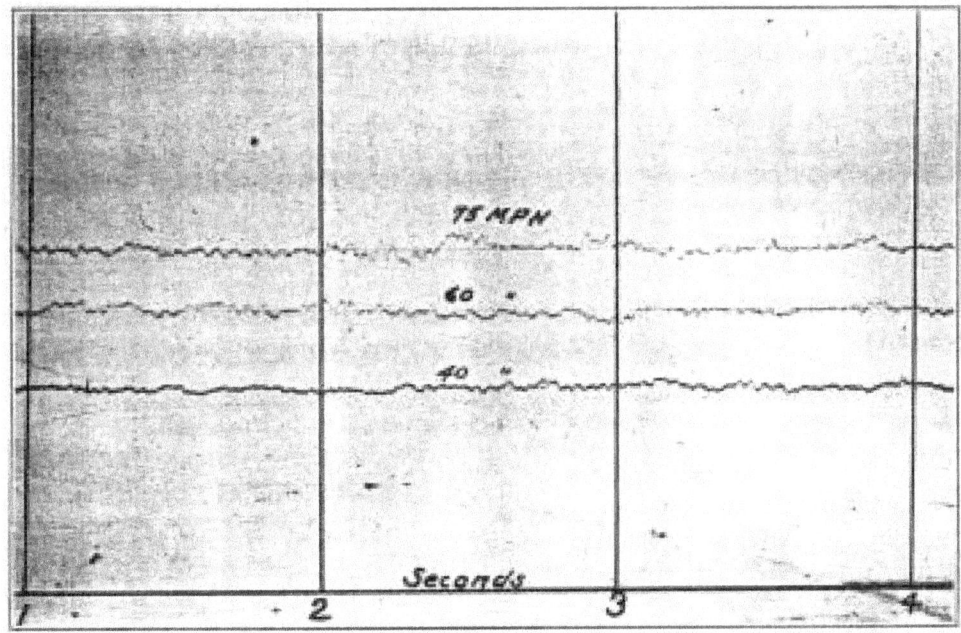

FIG. 19—VELOCITY VARIATIONS IN LARGE TUNNEL WHEN THE DRIVING MOTOR WAS CONNECTED TO A 25 K. W. GASOLINE GENERATING SET.

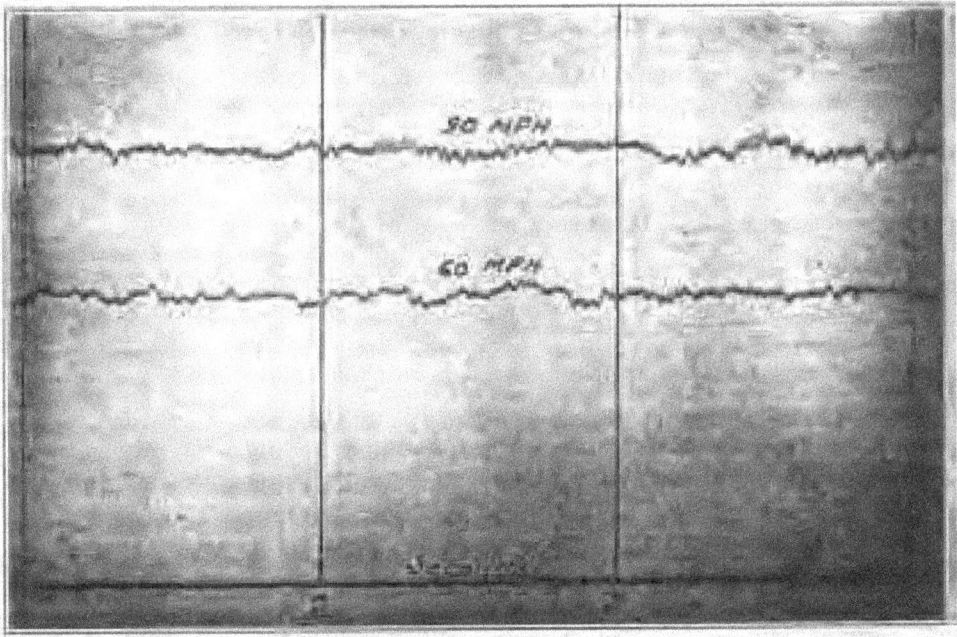

The model tunnel would then have a period of 0.03 seconds and the large tunnel a period of 0.15 seconds. Vibrations of this nature are very evident audibly in the tunnels at certain speeds, but do not seem to be present on the records, as the pitot tube is very nearly at the node of the vibration. The honeycomb has a considerable influence in damping, these vibrations, which are more of a curiosity than of any practical interest.

AUTOMATIC REGULATORS.

As it is not practical to supply a constant voltage to a wind tunnel, although some tests have been made with storage batteries where an extremely constant speed was required, it is either necessary to keep the voltage constant as nearly as possible by hand regulations or use some type of automatic regulator. In small tunnels it is quite easy to regulate the wind by hand, but in larger tunnels the inertia of the moving parts is so great that there is considerable amount of lag between the change in regulation and the response of the air speed, making hand regulation very difficult. A very complicated regulator has been constructed at Göttingen (N.A.C.A. File No. 5346-10) and seems to hold the velocity quite constant. There are also numerous electrical devices for maintaining a constant motor speed, and some of these regulators will hold the speed within 0.1 per cent. It seems probable, however, that even if the revolutions per minute of the propeller is constant that there will still be fluctuations in the air speed, so that a successful regulator must be actuated by the air flow. There is a great deal of work to be done on such regulators, and the N.A.C.A. intends to carry on work of this kind in the near future.

CONCLUSIONS.

The qualities that should be aimed at in wind-tunnel design in order of their importance are:

1. Constant direction of flow.
2. Constant velocity of flow.
3. Uniform velocity across section.
4. Efficiency.
5. Ease of working around tunnels.
6. Simplicity and cheapness of construction.

A good many of these qualities are contradictory, and the best compromise must be made between them and the type of work that is to be undertaken. For example, a tunnel for testing instruments should have a high efficiency, but need not have a very steady flow. On the other hand, a tunnel for testing wings should have its efficiency somewhat lowered in order to obtain a steady flow. It is quite possible to so arrange the honeycomb and diffusers that they may be removed when it is desired to obtain the highest speed. It would also be of value to make it possible to open the ends of the building, as there are many days when the wind would have little effect on the steadiness, and the efficiency would apparently thus be considerably increased. This arrangement would also make it possible to cool off the air in the building in a very short time, an advantage that would be greatly appreciated in hot weather.

This work seems to show conclusively that a straight exit cone is more efficient than a curved one, and it is certainly cheaper to construct. Diffusers affect the air flow very little, and they do not seem to warrant the expense of construction. Honeycombs, however, are of the greatest value and should be placed in every tunnel.

Document 2-13(a–c)

(a) "Report of Committee on Aerodynamics," *NACA Annual Report for 1919* (Washington, DC), pp. 28–31.

(b) "Report of Committee on Aerodynamics," *NACA Annual Report for 1920* (Washington, DC), pp. 22–28.

(c) "Report of Committee on Aerodynamics," *NACA Annual Report for 1921* (Washington, DC), pp. 31–36.

In its annual reports, the NACA published reports on the activities of its committees (actually subcommittees, as the NACA itself was a committee). These reports outlined the current state of the art and the investigations in progress. This series of reports (1919 to 1921) by the committee on aerodynamics identifies the aerodynamic problems of greatest concern to the NACA during that formative era, as well as the rudimentary state of aerodynamic knowledge. The reader will note how the committee on aerodynamics defined its own role as it proceeded with its research programs during these years, and how the NACA researchers developed cooperative working relationships with McCook Field and the Washington Navy Yard.

Document 2-13(a), "Report of Committee on Aerodynamics,"
NACA Annual Report for 1919.

REPORT OF COMMITTEE ON AERODYNAMICS.

The committee on aerodynamics is a consolidation, made in April 1919, of the three committees on aircraft design, navigation of aircraft and aeronautic instruments, and free flight tests.

The committee on aerodynamics recommended that in the part of the work at Langley Field within its domain the emphasis should be placed mainly on—

(*a*) Studies, in the laboratories, of propellers from the aerodynamic point of view;

(*b*) Studies, in the laboratories, of the flow of air around parts of airplanes and airships, separately or assembled, and of the forces brought to bear on those parts by the air;

(*c*) Free flight tests on fullsized machines; and

(*d*) The development of instruments, equipment, and methods for making such studies and tests.

This policy has been followed during the year.

The work at Langley Field under these four heads, carried on under the general direction of Edward P. Warner, has been largely one of development and design, although some research work complete in itself has been carried on. The wind tunnel has been under construction throughout most of the year and is now nearing completion, and a number of experiments on model tunnels have been made to determine the best form to adopt for the 5-foot tunnel now under construction. The balance for this tunnel has also been designed and constructed at Langley Field, and is ready for use. As an accessory to the design of the balance, a theoretical investigation of the errors to which such balances are subject and the considerations governing their design has been made and is published as a technical appendix to this report. The most important experimental work complete in itself has dealt with free flight tests, researches having been conducted throughout the summer on two airplanes furnished by the Air Service. These researches have dealt chiefly with the determination of the lift and drag coefficients and with the balance of the airplanes, thus giving data for purposes of comparison with those obtained in model tests. Several minor investigations on the performance of airplanes and on stresses in the airplane have also been undertaken, and two of them are published as appendixes to this report.

The development of autographic recording instruments for free flight tests has continued at the Bureau of Standards under the general direction of John F. Hayford and Lyman J. Briggs for the committee. Of the six new instruments designed five are now complete, namely, the air-speed meter, the angle-of-attack meter, the recording tachometer, the torque meter, and the thrust meter. The first three named have passed through extensive laboratory tests and calibrations, and are substantially ready for use in the air. The torque meter and thrust meter must be subjected to extensive laboratory tests and possible modifications before they will be ready for use. The stable zenith instrument, to be used for recording continuously the angle between the wing chord and the horizontal, has been redesigned as a result of considerable experience with this type of instrument. The redesigned instrument is being built and is nearly complete.

Special investigations have been made and are still in progress at the Bureau of Standards, under the recommendation of the committee, on the Parker variable camber wing for airplanes. This is a promising line of attack of the problem of reducing the landing speeds of an airplane without reducing its flying speed. Incidentally, some valuable information is being secured in regard to the possibility of using a streamline wing and the ordinary type in a biplane combination.

Researches on airplane propellers are being continued at Stanford University under the general direction of Prof. W. F. Durand.

The Bureau of Standards has undertaken four investigations for the aerodynamics committee:

1. Development of open scale altimeter.
2. Diaphragms and elastic fatigue.
3. Altitude correction for air speed indicators.
4. General report on aeronautic instruments.

The first three of these are definite experimental problems while the fourth consists not only of a summary of recent experimental investigations of instruments carried out in cooperation with the committee at the Bureau of Standards, but will go further and embrace a general survey of the state of this subject at the close of the war both in the United States and abroad.

1. The work on the open scale altimeter has progressed to a point where a working model of the essential parts of the altimeter, exclusive of the indicating mechanism, has been assembled and tested. By proper theoretical design of the spring and diaphragm elements an instrument has been made whose performance depends on the material of the steel spring almost entirely and only to a slight degree on the material of the diaphragm. It has been possible to secure steel nearly free from elastic fatigue effects, although such is not the case with the alloys used for the thin flexible diaphragm. As a result this instrument has been shown by test to have less than one-third of the fatigue effects (i.e., discrepancy between increasing and decreasing readings) permitted by the Bureau of Standards specifications for altimeters. The object of this work, which will be continued until completed, is to provide a precision altimeter suitable for altitude determination in aircraft performance tests.

2. Thin metallic diaphragms, usually corrugated for flexibility, form a necessary element in a great variety of engineering instruments, particularly in aneroid barometers for altitude measurement, in air-speed indicators, certain forms of statoscopes and rate-of-climb indicators, balloon manometers, and aviator's oxygen control apparatus. Such diaphragms are never perfectly elastic but show what are known as fatigue effects, failing to recover instantly from the deformations undergone in the normal operation of the instrument; hence, the importance of experiments to select the most promising alloys, to determine the most effective thermal and mechanical treatment in the process of manufacture, and to establish the most efficient geometrical design for the diaphragms when used either singly or in combinations. Up to the present, this investigation has resulted in the development of measuring appliances for detecting the small changes in question by micrometric methods, in the preparation and use of suitable shop equipment for spinning sample diaphragms at the bureau in considerable numbers, in a preliminary study of a variety of alloys, to select those which warrant more detailed study, and in a special study of mechanical seasoning processes. This last phase of the work has led to the conclusion that diaphragms can be seasoned mechanically; that is, artificially aged and thus brought into a permanent state where they will repeat

their performance in successive tests under the same conditions. This is done by repeated deformation of a suitable amount several thousand times, and is done automatically by a mechanism designed for the purpose. Some such seasoning process appears to be a necessary preliminary to the comparative measurement of the effects of different processes of heat treatment, different compositions of the alloy, and different mechanical design. The seasoning process, it will be understood, is not intended to eliminate elastic fatigue, although it does always reduce it somewhat. The object of the seasoning is to secure definite and uniform results, so that those factors which will diminish the fatigue can be analyzed quantitatively.

3. It has hitherto been taken for granted that Venturi tubes, when used for air-speed indicators, will follow a familiar law which states that the suction produced is directly proportional to the density of the air and to the square of the speed.

The object of this investigation is to determine by direct experiment whether this law does apply to Venturi tubes or whether on the contrary, the compressibility and viscosity of the atmosphere may cause some effects which will complicate the correction of these instruments for different altitudes. The experiments are conducted in a so-called vacuum wind tunnel (that is, a very small air-tight wind tunnel in which reduced pressures corresponding to high altitudes may be secured). The conclusion has been reached that the instruments examined are free from the effect of compressibility but not entirely free from the effect of viscosity. These experiments are to be completed and brought to a conclusion which can be expressed numerically for the purpose of correcting such instruments when used in aircraft performance tests.

4. The general report on aeronautic instruments presents the results of investigations made during the war by the Bureau of Standards to determine the characteristic sources of error of the various types of instruments. Tachometers, for example, for measuring the revolutions per minute of the propeller shaft are built in a variety of different types, operating on diverse physical principles. There are the chronometric, centrifugal, magnetic, electric, air viscosity, air-pump, and liquid types, each of which has its own characteristic sources of error. Aside from the ordinary errors met in engineering instruments, such as incorrect calibration, parallax, looseness of friction in the mechanism, elastic hysteresis, and secular changes, those used on aircraft may be further influenced by the physical conditions peculiar to aviation, viz, (1) extreme drop of pressure, (2) extreme change of temperature, (3) vibration, (4) acceleration or inclination. Besides the above results, this report, which is nearly completed, gives a description of the instruments collected by the Bureau of Standards in cooperation with this committee during the war, including those of British, French, Italian, Russian, Danish, and German construction.

In light of that report it will be evident that the objects toward which instrument development work should chiefly be directed in the immediate future may be summarized as follows:

1. Open scale instruments for performance testing of aircraft. These need not necessarily be so compact, light, or rugged as service instruments, and hence offer freedom for such design as will insure the highest accuracy.

2. Instruments for long distance navigation, including an absolute or ground speed indicator, and such a form of gyroscopic stabilizer as may be needed for mounting the instruments.

3. Instruments to guide the pilot in flying through fog, such as more reliable gyro turn indicators and compasses.

4. Better materials for springs and diaphragms and a more systematic determination of the thermal and elastic constants of the materials.

The Bureau of Standards has developed (primarily for the Air Mail Service) a field-marking radio device which enables a pilot to steer directly to the center of his landing field, although it may be obscured by clouds, rain, snow, or fog. This apparatus utilizes the same transmitting equipment for the ordinary radio direction finding signals and for the landing signals. The landing signals are projected vertically as an electromagnetic cone of great intensity, which can be heard satisfactorily at an altitude of three to four thousand feet. The device enables the pilot to first find the approximate vicinity of a landing field and then fly directly to its center, thus making a safe landing in a fog or in the dark. As elevated aerial systems are manifestly dangerous to air navigation, the Air Mail Service experimented extensively in radio transmission with antennas only 20 feet in height, highly directional, and admitting of sharp tuning. The installation of high-powered stations in the vicinity of flying fields is therefore made possible. Efforts are being made to provide and perfect a practical visual signal to take the place of the present audible signal requiring an audibility of 10,000 to overcome engine ignition interference noises. Such a signal will greatly enlarge the field of operation. A new type of gyroscopic nonmagnetic compass, intended to overcome the unreliability of the magnetic compass caused by vibrations and other disturbing influences of an airplane in motion, is now being developed by the Air Mail Service. This compass is now in a usable form for operation on land or sea, and only requires such changes as will adapt it to use on airplanes. It consists essentially of a solid metal ball floating on a film of compressed air and rotating coordinately in fixed relation to the earth's rotation.

Document 2-13(b), "Report of Committee on Aerodynamics," NACA Annual Report for 1920.

REPORT OF THE COMMITTEE ON AERODYNAMICS.

Following is a statement of the organization and functions of the committee on aerodynamics:

ORGANIZATION.

Dr. John F. Hayford, Northwestern University, chairman.
Dr. Joseph S. Ames, Johns Hopkins University, vice chairman.
Maj. T. H. Bane, United States Army.
Dr. L. J. Briggs, Bureau of Standards.
Maj. V. E. Clark, United States Army.
Commander J. C. Hunsaker, United States Navy.
Franklin L. Hunt, Bureau of Standards.
Prof. Charles F. Marvin, Chief Weather Bureau.
Edward P. Warner, Massachusetts Institute of Technology, secretary.
Dr. A. F. Zahm, United States Navy.

FUNCTIONS.

1. To aid in determining the problems relating to the theoretical and experimental study of aerodynamics to be experimentally attacked by governmental and private agencies.

2. To endeavor to coordinate, by counsel and suggestion, the research and experimental work involved in the investigation of such problems.

3. To act as a medium for the interchange of information regarding aerodynamic investigations, in progress or proposed.

4. The committee may direct and conduct research and experiment in aerodynamics in such laboratory or laboratories as may be placed (either in whole or in part) under its direction.

5. The committee shall meet from time to time, on call of the chairman, and report its actions and recommendations to the executive committee.

The committee on aerodynamics by reason of the representation of the Bureau of Standards, the Army, the Navy, technical institutions, and the industry, is in close contact with aerodynamical research and development work being carried on in the United States. Its representation enables it, by counsel and suggestion, to coordinate the experimental research work involved in the investigation of aerodynamical problems, and to influence the direction of the proper expenditure of energy toward those problems which seem of greatest importance.

The committee has direct control of aerodynamical research conducted at the Langley Memorial Aeronautical Laboratory and also directs propeller research conducted at Leland Stanford Junior University under the supervision of Dr. W.F. Durand, and through its membership it keeps in close touch with the work being carried on at the Bureau of Standards, at McCook Field by the engineering division of the Army Air Service, and at the Washington Navy Yard by the Bureau of Construction and Repair, United States Navy.

Two new wind tunnels have been completed and put in operation in the United States within the past year. A new 5-foot wind tunnel at the Langley

Memorial Aeronautical Laboratory has gone into service and has already run at speeds slightly in excess of 110 miles per hour. It is anticipated that speeds of 140 miles per hour will be attained with a new propeller which will be better suited to the characteristics of the electric motor employed. The other new wind tunnel of the year is that constructed by the Curtiss Engineering Corporation at Garden City and is of the true Eiffel type.

The committee on aerodynamics, in directing the research work at the Langley Memorial Aeronautical Laboratory, has adopted a definite policy with reference to research work to be conducted at this laboratory. The policy adopted confines the work to three general problems, and, in order to obtain results which will be of general use, experiments are to be conducted in such a manner that general conclusions and, if possible, general theories may result from them. The following three general problems covering the work of the aerodynamical laboratory for the coming year have been adopted:

(a) Comparison between the stability of airplanes, as determined from full-flight test and as determined from calculations based on wind tunnel measurements.

The committee will endeavor to determine the characteristics and peculiarities of certain existing airplanes, and attempt to account for these by calculations based on wind tunnel work. The matter of control will also fall under this heading. The first work conducted will probably be confined to the explanation of the theory of small oscillations and its verification with full-scale work. Later, a study of maneuverability and controllability will follow, as it is felt that in the present state of the art there is not available to airplane designers a rational method of predicting the maneuverability of airplanes from the drawings of the airplanes or from wind tunnel experiments with models.

(b) Similar comparison between the performance of airplanes full-scale and the calculations based on wind tunnel experiments.

A great deal of attention has been given by the British to the prediction of performance based on aerodynamic data, but there is still a gap between model and full-scale results which can not be bridged until we have more information. The performance is intimately connected with the propeller, and it is the intention of the committee to have all propeller research conducted at the Aerodynamical Laboratory of Leland Stanford Junior University under the direction of Dr. Durand. An effort will be made to tie in the results obtained at Leland Stanford with the performance work being done at Langley Field. Experiments will also be conducted on models of well-known airplanes to better understand the landing and starting characteristics of airplanes and to determine exactly what it is that makes certain airplanes require a long run.

(*c*) General aerofoil problem, including control surfaces, with particular reference to thick sections and combinations and modifications of such sections.

The committee is to undertake a systematic investigation of thick wing sections, after a thorough analysis of what has been done in this matter, and to duplicate some of the experiments already performed. After the determination of what properties of thick wing sections are of interest, work will then be carried along with a view to systematic variation of the variables which determine the aerodynamic properties of a series. Determination will also be made of the relation between aerodynamic properties of such standard aerofoils and aerofoils of similar profile but of different aspect ratio and taper. It is also desirable to know biplane and other interference effects when the aerofoils are used in combination. A careful study will also be made of recent work, by which it appears possible to predict from a knowledge of the lift coefficient the properties of aerofoils in combination and of different aspect ratio, as well as the influence of a boundary.

Such problems arising in connection with the Army and Navy programs of development as fit in logically with the above program will be referred to the committee on aerodynamics, and the research work covering the problems will be conducted at the Langley Memorial Aeronautical Laboratory.

At the Langley Memorial Aeronautical Laboratory a large number of experiments have been carried on with model wind tunnels in the past year to determine the best form for steadiness of flow and efficiency of operation. The effect of various shapes of cones, experimental chambers, and types of propellers, honeycombs, and diffusers were thoroughly studied. A special recording air-speed meter and recording yaw meter were designed in order to study the steadiness of flow, and it was found that the tunnel with a continuous throat was superior to the open or Eiffel type of tunnel both in efficiency and steadiness of flow. It was also demonstrated that a honeycomb placed in the entrance cone is of the greatest value in straightening the air flow, but a diffuser placed in the return circuit was apparently of little value.

The National Advisory Committee's 5-foot wind tunnel was completed in the spring of 1920 and has been in continuous operation since. This tunnel is designed from the data obtained in the model experiments and is very satisfactory both in efficiency and steadiness of flow. The 10-foot four-bladed propeller is driven by a 200-horsepower variable-speed electric motor. The power for this motor is obtained from gasoline-driven generating sets, and the control system is very convenient, the motor being started and stopped simply by pushing a button in the experimental chamber, and the speed being controlled by a rheostat from the same place.

The balance used in this tunnel is of the modified N.P.L. type, and was constructed in the shop of the National Advisory Committee at Langley Field. Unlike the usual balance, the weight is supported on a half bearing socket, rather than a conical pivot, as this device considerably reduces the friction and will carry a much larger load. It is also possible with this balance to simultaneously read the lift, drag, and pitching moment. As the N.P.L. type of balance is not suited to holding tapered wings, and as a large amount of work of this kind is planned for the future, a simple wire type of balance is being constructed at the present time, similar to that used in the wind tunnel at Göttingen.

It has been the practice in the past when setting up a model to align the chord of the wing with the wind by placing a thin wooden batten on the wing and comparing this batten with a straight line on the floor of the tunnel. But as this method is rather laborious and inaccurate, a new type of aligning apparatus has been designed for this tunnel, consisting essentially of a mechanism for reflecting a beam of light from a plain mirror which is attached parallel to the chord of the wing, so that by rotating the wing the reflected beam of light is brought to a cross line on a small target on the side of the experimental chamber. In this way a wing can be lined up with an accuracy of $0.01°$ in a very few seconds. As the air speeds used in this wind tunnel are considerably higher than those usually encountered, a special type of manometer was constructed to obviate the necessity of having an extremely long inclined tube. This gauge changes the head of liquid and at the same time the inclination of the tube, so that the fluctuations of the liquid are approximately equal at any speed. A multiple manometer has also been constructed for pressure distribution work on models, containing 20 glass tubes, the inclination of which can be adjusted to any desired angle.

A thorough investigation has been made of the problem of spindle interference and the best manner of protecting the spindle by a fairwater. Different types and lengths of fairwater were tested in order to determine which condition would give the least total interference. An accurate determination of the effective resistance of the spindle was made for various lengths of spindle and for various air speeds so that a complete set of data is available for use on any model tests for the future. In order to provide data for stability calculations a wing was tested through an angle of 360 degrees, and a model of an airplane was tested in the same way. In order to determine the scale corrections for model airplanes a model of the JN4H was constructed with great accuracy, and all details of the airplane were reproduced in the model, including the radiator and motor, but the wires were omitted as it was thought that their resistance could be determined better from tests of the full-sized wires. This model was tested at speeds of 30, 60, and 90 miles per hour in order to determine the corrections that must be applied to it in order to give the full-flight performance which was carefully determined on the full-sized machine.

FREE FLIGHT.

The machines available for the committee's use at the Langley Memorial Aeronautical Laboratory consist of two JN4H training machines and one DH4.

During the summer the machines have been in the air about 60 hours. Numerous small changes have been made on these machines during the different tests, including changing the stagger, changing the angle of the tail plane, and changing the position of the center of gravity by adding weight at the front or rear of the fuselage. A large number of special instruments have been designed and constructed at Langley Field for research in full flight. An accelerometer has been developed for obtaining the loads on an airplane during stunts and landings, and satisfactory results have been obtained with it, which are of considerably greater accuracy than those obtained by other types of instruments. Instruments were also developed for recording the position of and the force on all three controls of the airplane, and variable results have been obtained with these instruments. For obtaining the pressure distribution on the tail of the full-sized machine a special multiple manometer was constructed having 110 glass manometer tubes, all of which could be photographed at one time by an automatic film camera placed in the fuselage. As this instrument will only determine accurately the pressure distribution in steady flight, another manometer is now being constructed consisting of a large number of small diaphragm gauges which will record continuously on a moving film so that the rise and fall of the pressure at various points on the tail surfaces can be recorded during any stunt maneuver.

An air-speed meter and yaw meter have been constructed, working on the optical recording principle, having the actual period of the instrument high and its friction small, so that air-speed records can be obtained of any small or high period fluctuations in the wind velocity. To determine the angular rotation of the airplane during flight, in order to study its stability properties, a kymograph was constructed consisting of a narrow slit which focused the image of the sun on moving bromide paper, and another instrument of the same type has been constructed working on the gyroscopic principle. For obtaining the full-flight lift and drag coefficient a special longitudinal inclinometer was constructed which would give a large scale deflection and would be convenient and accurate to read.

The investigations undertaken consist of the determination of the lift and drag coefficients of the JN4H in free flight, and it is found possible by careful piloting to flay the machine at or slightly beyond the burble point. A thorough experimental investigation has been made of the static longitudinal stability of the airplane and a great many factors have been altered on the full-sized machine, such as changing the angle of the tail plane, changing the center of gravity of the machine, changing the section of the tail plane, and inclining the angle of the propeller axis. A study was made of the angle of attack and the air speed at the wing tips during spins and loops. This was accomplished by placing vanes and air-speed meters at the wing tips and photographing them during the maneuver by means of a camera gun and then plotting the curve of angle and speed against time from the photographs so obtained.

A very extensive investigation of the pressure distribution over the tail of an airplane in free flight has been undertaken. The pressure at 110 points on the left and right hand sides of the tail have been taken independently and the total pressure determined from these two curves. By means of photographic recording methods the time taken for making this investigation in the air is brief, but the computation and plotting of the results are laborious and require a long time for their completion. Runs were made with three positions of the center of gravity and two angles of setting of the stabilizer, as well as one run with celluloid over the crack between the stabilizer and the elevator. In all cases the pressure found over the tail was extremely low and in steady flight the load on the tail would be found very small compared with the load resulting from accelerated flight. A large number of records have been taken with the recording accelerometer designed by the N.A.C.A., these records being taken in the JN4H and several other machines during various stunts and landings. It was found that the maximum acceleration experienced in any stunt was during a roll, where the acceleration reached a maximum of 4.2 g. In order to determine the characteristics of an airplane during circling flight a record of the forces on all three controls was made doing banks of various angles up to 60 degrees and side slips up to 20 degrees of yaw.

The wind tunnel at Leland Stanford Junior University has again been occupied entirely with propeller tests. The results of the research work conducted this year are contained in technical report No. 109. Preparations are being made for tests on propellers at large angles of yaw, which will give data for the analysis of helicopters traveling horizontally.

Dr. George de Bothezat, aerodynamical expert of the National Advisory Committee for Aeronautics, has carried on at McCook Field, with the cooperation of the Engineering Division of the Army Air Service, a special investigation for the measurement of aerodynamic performance. The report on this investigation has been completed and approved as technical report No. 97, entitled "General Theory of the Steady Motion of an Airplane." This investigation involved the design and construction of a new type of barograph. Also in connection with his investigation of airplane performance, Dr. de Bothezat has designed a torque meter and a rate-of-climbmeter, which are under construction. The torque meter is a very simple design, and present indications are that it will be a most serviceable and efficient instrument. The rate-of-climbmeter is not based on a new principle; it is simply a new construction and design embodying the experience obtained in the use of other instruments.

The research work conducted by the Bureau of Construction and Repair of the Navy Department is carried on at the aerodynamical laboratory of the Washington Navy Yard and at the naval aircraft factory, Philadelphia Navy Yard. At the Washington Navy Yard two wind tunnels are in operation, and during the year a large number of airplanes and seaplane models have been given routine tests, and tests on many new aerofoil sections have also been made. Special attention has been given to testing streamline forms and struts. Yawing tests were conducted on the EP and the IE envelopes, which are formed from mathematical curves and have very low resistance. The tests indicate that the yawing moment about the center of gravity of a bare streamlined form varies but little from one shape to another. In connection with the tests on struts, it was shown that the Navy I strut has approximately 15 per cent less resistance than that given for the "Best" strut by the National Physical Laboratory. Wind tunnel tests were also conducted on two airship cars, one of faired contour and the other with facets of the same general contour, the results of which show the great value of fairing. The resistance of the faired car was 15 per cent less than that of the unfaired.

In connection with the wind tunnel at the Washington Navy Yard, a new aerodynamic balance of great interest has been developed. The balance is so designed that all adjustments of weights to bring the balance into equilibrium are automatic, and the time required for testing and the number of skilled operators are thus much decreased.

The Bureau of Construction and Repair has also undertaken the development and construction of the following instruments:

> A precision recording barograph intended for use in airplane trials, and especially for measuring the landing angle of airplanes, for which no wind tunnel test is available. This instrument will have a range of from 0 to 5,000 feet, and will incorporate the desirable features of the present Bureau of Standards precision altimeter.
>
> Two thermometer altimeters and density indicators. These instruments will combine a thermometric element with a pressure element in such a manner as to show at all times the altitude corrected for temperature.
>
> Two instruments intended to measure quantitatively the permeability of gas cells of envelopes without the removal of samples. The construction of these instruments has been suggested by the technical staff of the Bureau of Standards. This instrument is to take the form of a cup of suitable area which is pressed against the envelope at the point where the permeability is to be determined. A current of air is either sucked or driven through the cup, sweeping it out at a known rate. The mixture of gas and air from the cup is then passed through a thermal conductivity cell, and the proportion of hydrogen contained in the mixture is determined from the thermal conductivity of the mixture.

In the high-speed wind tunnel at McCook Field, which is operated under the direction of the technical staff of the Army Air Service, work has been continued along the same general lines as those indicated in technical report No. 83. During the year it is contemplated that tests will be conducted to determine the flow around a sphere and around biplane combinations. It is hoped thus to determine how nearly the action of the visible vapor particles indicate the true air flow about a body, and to visualize the flow around combinations of more than one supporting surface so as to determine the nature of the interference between the upper and lower surfaces should be of the greatest interest. It is also hoped to photograph the vapor action about a sphere over as large an air-speed range as possible. The sphere is to be supported in a manner to produce a minimum disturbance due to the support, and the photographs obtained are to be compared with existing photographs of flow about spheres and with the theoretical streamlines.

It is also hoped that tests will be conducted to determine the effect of rake and tapered wing tips on air flow, as this information may make it possible to further improve the airplane form and nature of taper in wings.

Performance tests are also conducted at McCook Field, and the committee on aerodynamics has requested that special tests be made on longitudinal stability to obtain an index of the dynamic longitudinal stability of the various airplanes used by the Army. The work already done by the staff of the National Advisory Committee for Aeronautics at Dayton with the cooperation of the Engineering Division of the Air Service on five airplanes is but a beginning of longitudinal stability investigation. It is desirable to obtain readings of stick forces and elevator angles on every type of machine in the Army's possession, and to have curves plotted in the same way as in National Advisory Committee's report No. 96.

The investigations carried on at the two wind tunnels of the Bureau of Standards under the direction of Dr. L.J. Briggs have consisted largely in instrument calibration and testing. The principal research has been in connection with the resistance of spheres and projectiles.

The work of the Aeronautic Instruments Section of the Bureau of Standards comprises the investigation, experimental development, and testing of aircraft instruments; also the development of methods of testing, fundamental researches on the physical principles involved in such instruments, and the study of their behavior in actual service.

The more important investigations which have been undertaken by the section during the past year are as follows:

An investigation has been completed and prepared for publication through the National Advisory Committee for Aeronautics on the effect on the performance of Venturi tube air-speed indicators of changes in atmospheric pressure. The results show that in certain instruments commonly used a correction should be applied for the viscosity of the air, a factor which has not hitherto been taken into account. This is of special interest in dirigible work where the air speeds may be low, and also in aircraft performance tests where exceptional precision is required.

An altimeter of exceptional accuracy designed and made at the Bureau of Standards has been completed and submitted to the Army. Another model with additional improvements has recently been designed and is under construction.

At the request of the National Advisory Committee for Aeronautics a fundamental investigation of the factors determining the behavior of flexible diaphragms as used in aeronautic instruments has been undertaken. The irreversible effects which cause the lag in diaphragm instruments has been formulated mathematically. The relation between force and deflection for diaphragms of different sizes, thickness, and materials has been studied graphically, practical methods for spinning diaphragms and building up diaphragm boxes have been investigated, and the possibilities of mechanical seasoning by repeated stress considered.

An improved rate of climb indicator, which indicates directly the rate of climb of aircraft in hundreds of feet per minute, has been completed and tested, and specifications have been prepared for the Army to use in the manufacture of a number of these instruments.

Information regarding instruments available for aerial navigation in cloudy weather or at night or for long-distance flights has been compiled at the request of the National Advisory Committee for Aeronautics and the Air Mail Service by the Aeronautic Instruments Section. This work will be continued and the development of new instruments undertaken.

Other investigations have been the development of a motion-picture apparatus for recording instrument readings during the flight of an airplane; a study of the errors in instruments used for determining the direction of aircraft, such as gyroscopic and liquid inclinometers and banking indicators, gyroscopic and magnetic compasses and turn indicators, a systematic investigation of commercial sphygmomanometers; a paper on the results of investigations on German instruments; a statistical study of the causes of failure in aeronautic instruments.

Assistance has been given the Air Service, the Aero Club of America, and others interested, during the past year, in the world's altitude competition for airplanes. Instruments have been calibrated and the best procedure for determining the altitude attained formulated.

Document 2-13(c), "Report of Committee on Aerodynamics," NACA Annual Report for 1921.

ORGANIZATION.

The Committee on aerodynamics is at present composed of the following members:
Dr. John F. Hayford, Northwestern University, chairman.
Dr. Joseph S. Ames, Johns Hopkins University, vice chairman.
Maj. T. H. Bane, United States Army.
Dr. L. J. Briggs, Bureau of Standards.
Commander J. C. Hunsaker, United States Navy.
Dr. Franklin L. Hunt, Bureau of Standards.
Maj. H. S. Martin, engineering division, McCook Field.
Prof. Charles F. Marvin, Chief Weather Bureau.
C. I. Stanton, Air Mail Service.
Edward P. Warner, Massachusetts Institute of Technology, secretary.
Dr. A. F. Zahm, United States Navy.

FUNCTIONS.

The functions of the committee on aerodynamics are as follows:

1. To determine what problems in theoretical and experimental aerodynamics are most important for investigation by government and private agencies.

2. To coordinate by counsel and suggestion the research work involved in the investigation of such problems.

3. To act as a medium for the interchange of information regarding aerodynamic investigations and developments in progress or proposed.

4. The committee may direct and conduct research in experimental aerodynamics in such laboratory or laboratories as may be placed either in whole or in part under its direction.

5. The committee shall meet from time to time on the call of the chairman and report its actions and recommendations to the executive committee.

The committee on aerodynamics by reason of the representation of various organizations interested in aeronautics is in close contact with all aerodynamical work being carried out in the United States. In this way the current work of each organization is made known to all, thus preventing duplication of effort. Also all research work is stimulated by the prompt distribution of new ideas and new results which adds greatly to the efficient conduction of aerodynamic research. The committee keeps the research workers in this country supplied with information on all European progress in aerodynamics by means of a foreign representative who is in close touch with all aeronautical activities in Europe. This direct information is supplemented by the translation and circulation of copies of the more important foreign reports and articles.

The Aerodynamic Committee has direct control of the aerodynamical research conducted at Langley Field, the propeller research conducted at Leland Stanford University under the supervision of Dr. W.F. Durand, and some special investigations conducted at the Bureau of Standards and at a number of universities.

WIND TUNNEL.

The committee's wind tunnel at Langley Field has recently had several changes made in it which have considerably improved the steadiness of flow. The most important of these is a new electrical system consisting of a synchronous motor-generator set which furnishes power direct to the wind tunnel motor. The speed of the wind tunnel motor is kept at a constant value within ±0.2 of a per cent by means of automatic voltage regulators. The air flow has also been considerably improved by placing a series of vanes around the end of the exit cone so that the air escapes radially. A wire type of balance is now used in this tunnel for all speeds between 30 and 60 meters per second.

It has long been felt that the tests made in the wind tunnel with a model varying much from the usual type are unreliable because of the uncertainty of the scale correction. For this reason the committee is now constructing at Langley Field a compressed-air wind tunnel with a throat diameter of 1.6 meters, a maximum speed of 25 meters per second, and a working pressure of 20 atmospheres. This wind tunnel will give a Reynolds number which is the same as for a full-sized airplane, and although the difficulties of supporting the model are great, the use of a comparatively low velocity and a high pressure have overcome the mechanical difficulties.

There are being constructed at the present time in the United States four other wind tunnels. At the Massachusetts Institute of Technology there are being erected a 1.25 meter and a 2.50 meter tunnel of the open-circuit type and with continuous throats. At McCook Field there is being constructed a high-speed 1.6 meter wind tunnel of the open-circuit type, which is designed for a velocity of 200 miles per hour. The Bureau of Standards at Washington is constructing a wind tunnel with a throat diameter of 3.25 meters. This wind tunnel is novel in that it is built in the open without any housing. The wind tunnel is well surrounded by trees and hills to prevent as far as possible the atmospheric conditions affecting the air flow.

The three-dimensional balance designed by Dr. A.F. Zahm for the Washington Navy Yard wind tunnel has proved very satisfactory. The weights of this balance are automatically actuated by electrically driven lead screws, and the time of making a test is much shorter than with other types of balances.

FREE FLIGHT.

The committee now has in use for aerodynamic research at Langley Field five airplanes; three *JN4H's*, one *VE-7*, and one Thomas-Morse *MB3*. The *JN4H* has been used by the committee extensively in experimental work, mainly because of its strength and the economy of operation. During the past year the flying time of the airplane has been 110 hours, representing 260 flights. Fifty-two per cent of the flying time has been used in actually making measurements in the air. No accidents of any kind have occurred with the committee's airplanes. One forced landing was made due to the sticking of the carburetor float during violent stunting, but the airplane was brought down without damage to itself or the instruments which it contained. Although complete airplanes have not as yet been constructed by the committee, a number of parts, such as wings, tail surfaces, etc., have been designed and constructed at Langley Field for use in free flight research.

INSTRUMENTS.

A number of new pieces of apparatus have been constructed for the wind tunnel, including a machine for forming plaster wings, a new micromanometer, a light balance for measuring the moments of control surfaces, and an instrument for measuring the rolling velocity of wings. It has become more and more evident, as the discrepancies between free flight and model tests have been discovered, that it is necessary to produce in the wind tunnel a slip stream comparable with that on the full-sized airplanes. A very small flexible shaft has been developed which is able to drive the model airplane propeller up to speeds of 30,000 revolutions per minute, which corresponds to the normal speed of a full-sized propeller. The flexible shaft is so small that it disturbs the air flow inappreciably and in this respect is superior to an electric motor or a turbine.

It is realized that all free-flight data must be obtained by recording instruments, first, because events happen so rapidly that observations are difficult to make, and secondly, because the observer is under rather a nervous strain and can not take observations as accurately as he could in the laboratory. For this reason the committee has designed and constructed a considerable number of standardized recording instruments, electrically driven and synchronized, for taking records on interchangeable film drums. With these instruments the only duty of the observer is to change the drums at the end of the record, for the pilot can start and stop all of the instruments with a single switch. The following instruments have been constructed and used during the year:

(1) A new accelerometer more compact and accurate than the previous model.

(2) A recording air speed meter with a high natural frequency and small friction.

(3) A new model of a kymograph.

(4) A multiple manometer which will record on a moving film 30 simultaneous records of varying pressures. The natural frequency of this instrument is very high and the volume does not change appreciably with changes in pressure, which is a very important fact when recording pressures through long tubes.

(5) An instrument for recording annular velocities about a single axis.

(6) A control position recorder for three controls.

(7) A balance for measuring the forces on a trailing wing in flight.

The aeronautic instruments section of the Bureau of Standards has been engaged in an extensive program of research and development work on aircraft instruments in cooperation with the National Advisory Committee for Aeronautics, the Army, the Navy, and to a more limited extent with other Government agencies and private concerns. In addition to the experimental investigations and the development of new instruments a considerable amount of work has been carried out in connection with the routine testing of service instruments.

The investigation of the altitude effect on air speed indicators undertaken at the request of the National Advisory Committee for Aeronautics has been continued and extended. The experiments have been conducted in an improved wind tunnel with a 16-inch throat and mounted in one of the Bureau of Standards altitude chambers. With this apparatus valuable data have been obtained at speeds up to 100 miles per hour and under conditions of pressure and temperature corresponding to altitudes up to 30,000 feet.

Research concerning the action of diaphragms and Bourdon tubes undertaken at the request of the National Advisory Committee for Aeronautics has been continued with the purpose of determining the laws of deflection and of obtaining essential information of value in the design of instruments involving the use of diaphragms and Bourdon tubes.

A series of eight reports dealing with the various aeronautic instruments has been prepared for the National Advisory Committee for Aeronautics and will be found in the Seventh Annual Report.

At the request of the Army and the Navy, the development of the following instruments has been undertaken:

      An improved aircraft sextant.
      An improved compass.
      An improved precision barometer.
      A precision altimeter compensated for air temperature.
      A precision barograph.
      An improved rate of climb indicator.
      An improved rate of climb recorder.
      A combined statoscope and rate of climb indicator.
      A synchronizing type ground speed indicator.
      An astronomical position finder.
      A horizontal angle indicator.
      An improved centrifugal tachometer.
      An air speed indicator for dirigibles.
      A ballonet volume indicator for dirigibles.
      Standard testing sets for field use.

Pursuant of the policy of following the latest developments in aeronautic instruments in foreign countries, a member of the aeronautic instruments section was detailed to investigate the recent developments in England, France, Italy, and Germany. This work was carried on in cooperation with the National Advisory Committee for Aeronautics representative in Europe and our military, naval, and commercial attachés, and much valuable information has been obtained.

A carbon pile tensiometer is being developed for the Navy which allows the accurate recording of tensions at a distance. An instrument has been devised by the Navy for the measurement of the ground speed of an airplane at frequent intervals of time on taking off or landing.

AEROFOIL TESTS.

During the past year the committee has conducted a large number of aerofoil tests in its 5-foot wind tunnel at Langley Field. The main object of these tests was to study the properties of thick aerofoils suitable for internal bracing. The tests were made at 35 meters per second, and in some cases as high as 60 meters per second as it was found that thick wings improve in efficiency with the speed more rapidly than thin wings. Some of the sections developed had at all angles a higher efficiency than the R.A.F. 15 section tested under the same conditions, and yet were more than three times as thick as that section in the center, while the maximum lift coefficients were approximately the same. A number of wings were tested which tapered in plan form, and it was found, contrary to expectations, that heavily tapered wings had the same center of pressure travel and practically the same efficiency as wings of uniform section.

The distribution of pressure was studied over 12 thick aerofoils of various types in order to determine the loading along the spars when the section varied along the span. A new method was devised for constructing pressure distribution models with comparatively little expense.

The effect of placing an aerofoil close to a flat surface representing the earth was thoroughly investigated both at the Massachusetts Institute of Technology and at the Washington Navy Yard. It was found that there was a remarkable increase in efficiency of the wing when close to a flat surface, which accounts for the fact that certain airplanes float for such long distances before landing.

Work has been continued in the McCook Field wind tunnel on various aerofoils at very high speeds, and a further study of vortex motion has been made.

Perhaps the most interesting work which has been carried out on aerofoils is that done by the committee in the testing of large aerofoils when suspended beneath a flying airplane. The aerofoil, constructed in the same way as an ordinary airplane wing of wood and fabric, is pulled up against the lower side of the fuselage in taking off, and when in the air is lowered down by means of a windlass to a distance of 20 or 30 feet or as far as is necessary to got out of the influence of the downwash. The magnitude of the resultant force is measured by a balance in the fuselage and the angle at which the wing trails back from the vertical measures the angle of the resultant. From these figures the lift and drag can be easily computed. At present only small wings of 6 feet span have been tested in this way, but it is evident from the great steadiness with which they trail beneath the airplane that accuracies probably as great as those obtained in the wind tunnel can be reached, although it is necessary to fly in smooth air for this kind of work. The results from the present apparatus although only of a preliminary character show such a good agreement with high speed wind tunnel tests of the same section that it is proposed to use a large bombing machine and trail wings of 30 feet span beneath it. Tests of this nature have not only the same Reynolds number, but also the same velocity, the same size, and the same amount of turbulence as the full-sized airplane, so that the results can be used by designers with perfect confidence.

A number of aerofoil sections have been tested, among which were several of the Göttingen series. The Washington Navy Yard tests check the Göttingen tests as closely as could be expected, the general types of the characteristic curves being very similar in every case. The Göttingen aerofoils tested were: Nos. *173, 255, 256,* and *822.*

Tests for scale effect have also been made on the *R.A.F. 15* and *R.A.F. 19* aerofoils.

STRUTS.

An interesting investigation has been made at the navy yard wind tunnel in Washington in the distribution of pressure over a strut. It is concluded that the total drag of the strut is the small difference between the upstream and downstream drag, so that a small error in measuring these will cause a huge error in the total.

STABILITY.

A very complete investigation has been made of the oscillations in flight of the *VE-7* and *JN4H,* the latter airplane with a special tail plane to make it statically stable. The results on the whole are in poor agreement with the theory, due mainly, it is believed, to the fact that the oscillations are large, often over 60 degrees, and that the slip stream has a considerable influence.

Considerable work has been done on static stability and it is becoming more and more evident that the aspect ratio of the tail plane has by far the greatest influence on the stability. It has also been found in actual flight that complete static stability may be obtained when the load is positive upon tail surfaces at all times. A study of the distribution of pressure over the tail surfaces of this surface in steady flight has given valuable information as to the functions of this surface in producing stability.

The lateral stability derivatives Y_v, L_v, and N_v have been determined in free flight for the *JN4H* and comparison has been made with the results from wind tunnel tests. On the whole the agreement is good, the discrepancies being mainly due to the influence of the slip stream and to the fact that in the model the control surfaces were assumed to be in a neutral position, whereas actually they were at a considerable angle.

A mechanical device has been constructed which will illustrate in every particular the dynamic and statical stability of an airplane. By the adjustment of weights the effect of changing the mass, the moment of inertia, the damping, etc., can be produced at will. As yet it has not been possible to obtain any quantitative value for stability with this instrument, but it is hoped that it may be used for quickly finding the stability properties of a new airplane from its known characteristics.

Tests have been made by the Navy on a series of balanced control surfaces with various types of balance. The characteristics of the type in which the axis is placed aft of the leading edge of the movable surface have been investigated at some length.

STRESSES IN FLIGHT.

The distribution of pressure was determined over the horizontal tail surfaces of a *JN4H* during all types of maneuver. In no case did the maximum loading on the tail exceed 6 pounds per square foot, and contrary to the usual expectations this load was in an upward direction. A theory has been devised which will give the loading on the tail surfaces in close agreement with the actual measurements.

The distribution of pressure over the rudder and fin have also been investigated on the same airplane and it was found that the heaviest loads occur in a roll where the loading may go as high as 10 pounds per square foot. It is interesting to notice from the standpoint of fuselage design that the maximum load on the horizontal tail surfaces, the maximum load on the rudder and fin, and the maximum load on the wings may all occur at the same time.

The recent development of very high speed airplanes has shown that very large unexpected loads may occur on the wing surface, several instances causing the stripping of the fabric or crippling of the trailing edge. A Thomas-Morse single seater which has a speed of over 160 miles an hour has been fitted up for measuring the distribution of pressure over the wing surfaces. It is hoped to determine the pressures both in steady flight and during violent maneuvering, and for this purpose the wings have been especially strengthened.

CONTROLLABILITY.

The measurement of and the design for controllability are very important problems and ones which have received but scant attention. In fact, the very definition of controllability is at the present time stated vaguely. The committee is now making an attempt to find some accurate quantitative means of measuring the controllability of various airplanes and to find the effect on controllability of various changes in control surfaces.

The desire for high speed has led many designers to eliminate the external bracing on the horizontal tail surfaces and for this reason a number of airplanes have been constructed with rather thick sections for the tail surfaces. Several airplanes of this type have been found by pilots to be extremely sluggish in responding to the controls; that is, for a certain range about the neutral position the controls have no effect. This condition was investigated in the wind tunnel on a tail plane of this type, and it was found that the elevator must be moved several degrees on either side of its neutral position before the force on the tail is appreciably changed, due to the fact that the elevator seemed to be in the shadow of the thicker portion of the tail surfaces and could have no effect until it was turned out into the free air stream.

The angular velocity and angular accelerations have been measured on a *JN4H* during all types of maneuver, in order to provide designers with data which will be of use in construction of airplanes.

The subject of control, especially lateral control, at low flying speeds has received some attention. It is evident, however, that different airplanes, although varying only slightly in external characteristics, vary tremendously in the amount of lateral control which they have at the stalling speed, and an explanation of this would be of great value. The Navy has recently devised an entirely new type of lateral control which in wind tunnel tests shows great promise.

AIRSHIPS.

Several types of external-pressure pads developed by the Navy have been tested upon the wind of an airplane at Langley Field in order to assure that such opening when cemented to the outside of the wing will give the same reading as a flush hole. One type of pad has proved to be very successful. The possibilities have been considered of measuring the pressure over the surface of an airship during accelerated flight, and as yet no satisfactory method has been devised for entirely eliminating the rather large errors due to the forces acting upon the air column in the long connecting tubes which are necessary in this experiment. The investigation, however, has not be abandoned, and it is believed that the difficulties will be overcome.

Extensive tests have been made on two models of the rigid airship *ZR-1*. These tests were made on the hulls, bare and with six types of control surfaces.

Tests have been conducted at the navy yard wind tunnel in Washington on the effect of fineness, ratio, and length of parallel middle body on airship forms.

PROPELLERS.

Experiments have been conducted in the wind tunnel to measure the drag of various propellers under various degrees of yaw and with different amounts of braking. The drags of propellers are rather small so that the possibility of the save vertical descent of the helicopter without power does not look very probably if the usual type of propeller is used. Test have also been conducted upon a helicopter propeller having blades which are automatically set at a constant angle of attack by means of individual tail planes.

An extensive investigation has been carried out at Leland Stanford University on the properties of propellers at angles of yaw. The results look very promising in connection with the horizontal travel of helicopters, as a considerable horizontal thrust may be obtained with no more power than is required in ordinary flight.

BOMBS.

The Bureau of Standards has been conducting a very extensive investigation of bombs and projectiles not only in their 150-mile wind tunnel but also in a 12-inch air stream from a high-power compressor where speed can be obtained above the velocity of sound. Some very interesting conclusions have been reached in connection with stream lining at very high speeds.

Document 2-14

J. C. Hunsaker, memorandum to NACA, "Recommendations for Research Program—Comparison of Wing Characteristics in Models and Free Flight," 10 November 1920, RA file J, Historical Archives, NASA, Langley.

One of the NACA's most important early research programs began in 1920 after Jerome Hunsaker recommended that the committee pursue a systematic program comparing the wing characteristics of wind tunnel models with those of identical, but full-size, wings in free flight. This proposal, which the NACA embraced and carried out at Langley, led to the establishment of new research methods and produced some of the first reliable data comparing actual and model airplanes. Interestingly, the agreement between model and actual wings was generally good, but the fuselage comparisons revealed considerable discrepancies. Understanding why would require several more years of experiments and analysis and would ultimately help justify larger wind tunnels at Langley. (Note: The duplication of paragraphs numbered "5" in the report is from the original.)

Document 2-14, Jerome Hunsaker to NACA,
"Recommendations for Research Program," 1920.

November 10, 1920

From: J. C. Hunsaker, Commander (C.C.), U.S.N.
To: National Advisory Committee for Aeronautics, Room 2722, Navy Building.
Subject: Recommendations for Research Program—Comparison of Wing Characteristics in Models and Free Flight.

1. It is recommended that the contemplated research on comparison between model and full-scale airplanes be made as thorough as possible in order to furnish accurate and complete data on which we may base performance estimates. A research of this nature should include a study of at least three aerofoil sections which are widely separated in their characteristics. The sections recommended by Mr. Norton, i.e., Curtiss (JN-4), R.A.F.-15, Albatross, and U.S.A.T.S.-5 seem to answer all requirements. The problem would be somewhat simplified if a Parasol type of monoplane were available, but perhaps more information may be obtained from a study of a biplane such as the JN4H now available.

2. It is suggested that very accurate scale models of the component parts of this airplane be constructed and tested in the tunnel at various speeds, singly and in combination, with a view to determine the probable scale effect and the interferences.

The wing combinations should be tested carefully and investigated analytically by means of the Gottingen equations (Prandtl and Munk). These equations give very satisfactory results and it is thought that additional constants should be determined for all cases in common use. The German tests have not undertaken a study of tapered wings and, in view of the probable importance of this feature in internally braced designs, it is suggested that a study be made of taper in monoplane and biplane combination as a parallel work. Reference is made to Munk's article in Technische Berichte II-2, for the method of determining the constants.

3. The resistance derivatives should be determined for this machine by experiments on the model and also by calculations based on considerations of the design. An attempt should be made to arrive at some conclusion in regard to the best way to determine each derivative, it being well known that certain derivatives can not be obtained with a necessary accuracy from a direct test. In this respect it is recommend that the findings of Mr. O. Glauert as reported in "Aircraft Engineering" during 1920, be given careful consideration.

4. Additional wind tunnel tests will be suggested by the results of these investigations and should be made in view of the specific requirements.

5. In the full-scale, or free-flight, tests there are several outstanding problems such as:

(a) Determination of the variation of lift and drag coefficients with angle.

(b) Pressure distributions on wings and perhaps on tail surfaces (in addition to the research now under way).

(c) Efficiencies of control surfaces.

(d) Forces on tail surfaces and effect of down wash from different wings on balance and stability.

(e) Stability.

5. The variation of lift and drag with angle should be determined by the usual method of timing a horizontal steady flight over a measured course and checked by gliding flights with the propeller stopped (see Br. A.C.A. R&M Nos. 541 and 603). A method of analyzing the performance in a climb may be found in Zeitschrift F.u.M. June 30, 1920, and it is suggested that a similar scheme be used in this research.

6. It seems desirable to actually determine the effectiveness of the control surfaces by introducing a known moment to be counteracted. This has been done in some German tests reported in Zeitschrift F.u.M. for November 15, 1919, and November 29, 1919. The results apparently justify further experimentation.

7. The study on efficiencies of and forces on control surfaces should determine the variations due to thickness and plan forms of these surfaces. Some very good work has been done in this field and the present research should be made with the view of completing such work.

8. Free flight stability testing has been limited to determinations of period, damping, etc. The previous work of the Advisory Committee furnishes a very good foundation for the present investigation.

9. A comparison of the various reports on free flight tests is sufficient to emphasize the importance of accurate data. Special attention should be given to the determinations of velocity, density, thrust, weight, and center of gravity location. This research will require an immense amount of work, but the results will justify the efforts providing care is taken in planning and executing every test.

[signed] J. C. Hunsaker

Document 2-15(a–i)

(a) Joseph S. Ames, vice chairman, committee on aerodynamics, NACA, letter to A. [F.] Zahm, L. J. Briggs, E. B. Wilson, W. F. Durand, E. N. Fales, J. G. Coffin, H. Bateman, and F. H. Norton, 23 August 1920, RA file 70, Historical Archives, NASA, Langley.

(b) F. H. Norton, acting chief physicist, NACA Langley, response to Joseph S. Ames, 26 August 1920, RA file 70, Historical Archives, NASA, Langley.

(c) A. F. Zahm, Washington Navy Yard, response to Joseph S. Ames, 17 September 1920, RA file 70, Historical Archives, NASA, Langley.

(d) Joseph G. Coffin, Curtiss Aeroplane & Motor Corporation, response to Joseph S. Ames, 18 September 1920, RA file 70, Historical Archives, NASA, Langley.

(e) W. F. Durand, Leland Stanford University, response to J. S. Ames, 24 September 1920, RA file 70, Historical Archives, NASA, Langley.

(f) Ludwig Prandtl, University of Göttingen, letter to William Knight, NACA Technical Assistant (translation), the NACA, 1 May 1920, RA file 70, Historical Archives, NASA, Langley.

(g) "International Standardization of Wind-Tunnel Results," *NACA Annual Report for 1922* (Washington, DC), p. 36.

(h) G. W. Lewis, director of aeronautical research, NACA, memorandum to Langley Memorial Aeronautical Laboratory, 1 April 1925, RA file 70, Historical Archives, NASA, Langley.

(i) Aerodynamics Department, The National Physical Laboratory [Great Britain], "A Comparison between Results for R.A.F. 15 in N.P.L. Duplex Tunnel and in the N.A.C.A. Compressed Air Tunnel," (summary), n.d. [1925], RA file 70, Historical Archives, NASA, Langley.

As the number and variety of wind tunnels around the world increased, researchers became more concerned that measurements taken in one tunnel might not be reproducible in other tunnels. In 1920, the British Advisory Committee for Aeronautics (ACA) spearheaded an effort to run a series of tests to compare the performance of the world's major wind tunnels. The NACA Committee on Aerodynamics readily endorsed the concept, but it was not sure just what should be measured or how the program should be structured. In an effort to obtain broad input and achieve some consensus in the American aeronautical community, the aerodynamics committee's vice chairman, Joseph S. Ames (for whom the Ames Laboratory would later be named), wrote to eight leading aerodynamics figures, asking them "to outline a program of tests to be made in the wind tunnels of this country and of Europe" that would enable researchers to "connect" data from different tunnels. Responses from F. H. Norton, A. F. Zahm, J. G. Coffin, and W. F. Durand, reproduced herein, uniformly expressed interest in such an effort, but there are interesting differences in their suggestions for the program. Ames received advice and comments from other noted well-known aeronautics personages as well, including Ludwig Prandtl, who offered Ames his suggestions earlier in the year.

The NACA prepared specifications for the American tests and, as the "International Standardization of Wind-Tunnel Results" section from the *NACA Annual Report for 1922* shows, participated fully in the project. Although the standardization tests were initially expected to be a one-time, straightforward proposition, the project rapidly grew in complexity, and the NACA was an active participant in tests for many years. Research Authorization (RA) 70, the in-house authority for the program and the source file for most of these documents, became one of the longest-running projects on the committee's books.

The NACA and ACA published numerous reports that described the various tests and analyzed results. The final document in this section is the summary from a British report by the Aerodynamics Department of the National Physical Laboratory comparing results from the tests of a standard wing section in the N.P.L. Duplex tunnel and the new Variable Density Tunnel (VDT) at Langley. The British expression of confidence in the American VDT is of particular interest in this document.

In the final analysis, however, standardization remained a dream that was never fully realized because of the widely varying capabilities of the world's wind tunnels and the complexities inherent in wind tunnel testing. In another sense, however, the standardization project was a success in that it focused the greatest minds in aerodynamics around the world on the fundamental questions of wind tunnel testing and stimulated an unprecedented degree of cooperation and information exchange.

Document 2-15(a), Joseph S. Ames, letter regarding standardization of wind tunnels, 1920.

August 23, 1920

My dear Sir:

At the last meeting of the Subcommittee on Aerodynamics of the National Advisory Committee for Aeronautics, I was requested to ask you if you would be kind enough to outline a program of tests to be made in the wind tunnels of this country and of Europe with a view to securing what one might call standardization, that is, information which would enable one to connect the data published, as obtained in these different wind tunnels.

This request is being sent to several others, and after replies are received from all, which I trust will be at an early date, I will see that a comprehensive program is prepared for submission at the next meeting of the committee.

Sincerely yours,
NATIONAL ADVISORY COMMITTEE FOR AERONAUTICS
[signed] Joseph S. Ames
Vice-Chairman, Committee on Aerodynamics

Copies to A. H. Zahm, L. J. Briggs, E. B. Wilson, W. F. Durand, E. N. Fales, J. G. Coffin, H. Bateman, and F. H. Norton.

Document 2-15(b), F. H. Norton, reply to Joseph S. Ames, 1920.

August 26, 1920.

From: Langley Memorial Aeronautical Laboratory.
To: National Advisory Committee for Aeronautics.
 (Atten. Vice-Chairman, Com. on Aerodynamics)
Subject: Comparison of wind tunnels.
Reference: (a) NACA Let. Aug. 23, 1920, and enclosures.

1. It is evident that for a given model, a given angle of incidence and a given velocity there is only one correct value for L_c and D_c. Tests with the same model in various wind tunnels will not necessarily give these correct values but they will show any large error for any one particular tunnel.

2. The purpose of these tests should be to determine factors for converting past, present and future results, at least approximately, to agree with similar results obtained in any one tunnel. These tests should also determine the merits of the various types of wind tunnel.

3. The main classes of errors which may arise in wind tunnel work are:—First, errors in speed of wind; Second, errors in measuring the forces; and Third, errors in determining the true resistance of the model exclusive of its supports.

4. It is advisable in tests of this nature to make the models as simple and as few in number as possible, and it would seem to me that a standard type of aerofoil would be best suited for this purpose, and I believe that a single test would show everything that would be shown by a more extensive series of tests. In order that the aerofoil may be used in the smaller tunnels it should be made with a span of 18", and a chord of 3" and should be run at a speed of 40 feet per second. It has been suggested that it might be desirable to make tests on a complete model but I do not believe that any more information would be obtained in this way than from the simple aerofoil and the model would have the disadvantage that even a very slight misalignment of the wings would introduce a considerable amount of error in the results.

5. It will be very desirable to determine the steadiness of flow both in direction and velocity for various tunnels. A method for accomplishing this in a rough way has been suggested by using a sphere or cylinder for the test but this method can not give quantitative results. It is suggested that a very complete and valuable record would be obtained by using a recording air speed meter and recording yawmeter in order to obtain records of the actual variation in the flow of each tunnel. Such an instrument has been used very successfully at Langley Field and it is suggested that the same instrument should be used in all tunnels.

[signed] F. H. Norton
F. H. Norton,
Acting Chief Physicist.

Document 2-15(c), Albert F. Zahm, reply to Joseph S. Ames from Washington Navy Yard, 1920.

September 14, 1920

Dear Prof. Ames:

Your letter of August 23 arrived while I was away on vacation.

One of the main objects of wind-tunnel research is to determine the action of the air on and about a model in a stream of indefinite extent flowing uniformly without a pressure gradient, except that caused by the model.

I would suggest first that a very few of the ablest theoretical aerodynamists, such as Prof. Prandtl, be invited to discuss the mathematical theory of the flow in a wind tunnel—not too ideal for practical use—both when empty and when containing a model; and to indicate what corrections should be made to render the wind-tunnel data applicable to a model in a uniform infinite stream, or to a model moving uniformly through an infinite still atmosphere. Prof. Prandtl has written somewhat on this subject, and Dr. de Bothezat has expressed a wish to do so.

After that I would suggest that a few laboratories be invited to determine the action of the air on and about a few very simple models in their tunnels, and thence, after their own peculiar corrections, to derive the air action on the same models in a uniform infinite stream. If the final results agree, the methods of standardization are provided.

The careful testing of a variety of identical models in a variety of tunnels differently manned would probably yield results a little more consistent than those already available. But unless the tests were guided by adequate theory, furnished before hand, it seems improbable that all the observations and precautions would be taken that are necessary to make wind tunnel data strictly comparable.

In case the comparative tests are to begin at once, I would suggest as models a sphere and a thin circular disc in normal presentation to the wind, preferably with its edge champfered [sic] on the back; both to be of specified dimensions. In each case I would have the velocity and pressure distributions, for various fixed wind speeds, mapped throughout the working part of the tunnel both when vacant and when containing the model. The resultant air force on the model, and the pressure distribution all over its surface should be determined. It should be required also to report the temperature and moisture of the air in the tunnel during the test, so that the density and viscosity may be known. Finally, correction should be made to derive the resultant air force on the model in a uniform infinite stream.

In both cases a number of spheres and discs, as nearly geometrically similar as may be feasible, should be used as a check in finding the V1 or scale effect.

If these tests are to be made to establish trustworthy doctrine, rather than to expose the defects of busy routine tunnels, I would suggest that they be limited to research laboratories, with ample equipment, and supervised by high-grade men with sufficient leisure.

Very truly yours,
[signed] A. F. Zahm.

Document 2-15(d), Joseph G. Coffin, reply to Joseph S. Ames from Curtiss Aeroplane Corporation, 1920.

September 18, 1920

Dr. Joseph S. Ames,
National Advisory Committee for Aeronautics,
Washington, D.C.

Dear Sir:—

Your communication of August 23rd must have gone astray as I have received but that of September 14th.

I am extremely interested in the question of standardization tests or rather comparison of tests in various wind tunnels. The following suggestions show in a general way the kind of tests that in my opinion would be useful.

1. Tunnels should show uniform air flow over experimental section as tested by standard type of Pitot.

2. Tests on any standard rings should give identical curves (within experimental area) when reversed.

3. Balance should be checked up for absolute forces.

If these three conditions are satisfactorily met, then:

1. Comparison of results should be made on a standard airfoil. This airfoil should be a species of primary standard sent around from place to place somewhat as an invariable pendulum is used in the determination of "g". It should be of metal to withstand high air speeds.

2. Tests should be made on similar airfoils of geometrical constructions such as shown in sketch below, for example, which can be constructed mechanically and checked up at the various laboratory shops.

3. Comparison tests of an airplane model of "invariable" all-metal type construction.

4. Resistance at zero yaw of streamline airship model which is of a fairly large size, (volume). This brings in the pressure gradient correction, standard surface and shape.

5. No uniformity can be expected unless great and careful attention is paid to spindle and attachment location corrections. These are extremely troublesome and in my opinion deserve the greatest amount of attention at the present time.

6. Tests on a sphere and a cylinder.

7. Attention must be paid to nature of surface. Surface has a model characteristic.

I have had constructed a very carefully made aluminum wing with a RAF-6 upper camber and an absolutely flat under camber. It is provided with holes for end spindle attachment at either end and also for crank spindle attachment at the center. Comparison results with this wing have already been made by M.I.T. and the Bureau of Standards. The results are very interesting.

By sending around an invariable standard wing we eliminate possible <u>surface differences</u>, spindle attachment location differences, as well as differences due to the spindles themselves. Special spindles accompany the model. I believe the Curtiss Company would be glad to cooperate by allowing the use of this standard.

8. Tests on pressure gradients along the tunnel should be made.

9. Tests should be repeated at various air speeds.

After all, what is desired is that any given tunnel, when properly used, should give identical test results. This standardization would go a long way toward attaining this ideal.

I would like to say also that a new type of balance should be developed. I have in mind such a balance which would eliminate to a great extant the difficulties with spindle corrections.

Very truly yours,

[signed] Joseph G. Coffin

Document 2-15(e), W. F. Durand, reply to Joseph S. Ames from Stanford University, 1920.

24th September 1920.

Dr. J. S. Ames
Chairman, Aerodynamic Com.
N.A.C.A.
Washington, D.C.

My Dear Dr. Ames,

I have just returned from a summer trip to the Hawaiian Islands, and find your letter in regard to experimental work in aerodynamical laboratories looking toward a standardization of results. While in the Islands I was beyond reach of mail, and this must be my excuse for the long delay in my reply.

The subject of your inquiry is one in which I have felt much interest and it has seemed to me that the general program of standardization involves two principal features:

(1) The adoption of a system of standard dimensionless units and the expression of results of laboratory research in terms of such units. Already a good start has been made by the N.A.C.A. in this respect, in the work of the technical data division, in reducing to standard terms and forms, results of miscellaneous laboratory research work in this country and abroad.

(2) Some program of test or standardization which would, in effect operate as a test on laboratories and laboratory equipment. Thus a test is made in laboratory A on a series of airfoil sections specified in terms of a drawing or a series of ordinates. Laboratory B desires to check up on Laboratory A and attempts to reproduce the same section. It is very sure that the reproduction will not be exact and may even differ to such a degree as to seriously compromise the comparison. Similarly with

propeller tests. In order to make our results strictly comparable with those made in other laboratories, notable by Eiffel and at the N.P.L., there should in effect be a standardization of laboratory equipment—balance, dynamometers, wind speed measures, etc.

Just how to accomplish such a standardization or comparison in the most effective manner is not entirely clear, but I believe that something could be accomplished by carrying out a series of tests on a series of carefully selected type models, made of metal and therefore invariable in form, and sent bodily from one laboratory to another. Thus if a persistent difference develops as between two laboratories in measuring the same thing, it must obviously be inhere in the laboratory equipment. Or again, if in a dozen laboratories all but one or two are in sensible agreement regarding certain measurements, the presumption is that in the divergent cases some error traceable to the equipment or mode of carrying on the measurements has been introduced. A search for this source of error should then serve to definitely clear up the matter and establish all laboratories on a uniform basis at least as regards this particular feature.

Of course in all laboratory equipment, time changes develop and certain standard forms should be requested from time to time in order to make sure that no secular change of importance is in progress.

The particular features regarding which such laboratory standardization should be carried out are clearly the following:

(a) Air foil sections:
Lift Force.
Drift Force.
(b) Models of planes:
Forces as above.
Tipping movements.
(c) Propellers:
Torque.
Thrust.
R.P.S.
(d) Wind Speed:
This is fundamental in all tests.

Regarding (d), I have thought that a special form of wind speed measuring device might be devised, which could be sent bodily from one laboratory to another, and which, through it's indications in comparison with the method or device used at each laboratory, might serve to give a comparison or a comparative calibration of one laboratory in terms of another.

Regarding (a), (b), and (c), standard metal forms and a standard program of test might be devised, the carrying out of which in each laboratory would thus serve to check one against another and to develop cases which might be specially divergent from the others.

There is much in the way of detail here which would have to be carefully studied, but I believe that something might be done, especially through the agency of a central body such as our Committee.

Hoping that something may develop along these or equivalent lines, and with assurances of our desire to co-operate in all ways practical, I am,

Sincerely yours,

[signed] W. F. Durand

P.S. I have not attempted to discuss this matter in detail, in the thought that a general understanding of the field to be covered and of the general grand strategy would first be desirable.

W.F.D.

Document 2-15(f), Ludwig Prandtl, letter to William Knight from Göttingen, Germany (translation), 1920.

May 1st 1920.

To: Mr. William Knight, Technical Assistant, U.S. Commission for Aeronautics, American Commission, Wilhelmsplatz 7, Berlin.

My Dear Mr. Knight:

I hope you reached Berlin safely. I want to make definite proposals at once for the comparisons to tests made at various laboratories. The following five experiments appear to be, in my opinion, the right ones to propose:

1) Measurement of resistance of a flat circular metal plate, set at right angles to the air flow in order to test the measurement of velocity in the laboratory.

2) Measurement of resistance of a smooth sphere, diameter 20 cm., made as exact as possible, to test the eddying of the air current.

3) Measurement of resistance of a metal wing of about 20 x 100 cm., equipped with the proper fastenings for all the laboratories, which will be sent from one laboratory to another.

4) Measurement or resistance of a wing as constructed from the same drawing by each laboratory, with its own manufacturing means. The drawing may be identical with the one used in experiment 3.

5) Measurement of resistance of a dirigible balloon body model, which may be made of polished wood, to test the uniformity of the air flow lengthwise. (If speed increases down stream, the measurement of resistance will be too large, and vice-versa.)

I shall be obliged if you will negotiate with the various laboratories on the basis of this program, if you agree to it, and collect the statements of the separate laboratories as to the fastenings they require (hooks, screw sockets, etc.) for each model. The models to be turned out in common can then be produced in accordance with those specifications.

I have looked over my old studies, and find that a reprint of only one of the papers you are concerned with is missing; i.e., a paper given at the International Congress of Mathematicians at Heidelberg, 1904. It was a particularly important paper, at least from a historical point of view. If you attach importance to it, you can still probably purchase the Proceedings of that Congress (1905) published by Teubner, Leipzig.

Some other important studies done at my old laboratory deserve your attention. They appeared in the Jahrbuch der Motorluftschiff-Studeingesellschaft 1910/11, 1911/12, and in the Jahrbuch der Luftfahrzeugegesellschaft 1912/13 (Springer, Berline). Everything else of consequence is to be found in the Technische Berichte, or you can get reprints of the papers in questions.

I send you my best regards and hope to hear from you.
[signed] L. PRANDTL.

Document 2-15(g), "International Standardization of Wind-Tunnel Results," NACA Annual Report for 1922.

INTERNATIONAL STANDARDIZATION OF WIND-TUNNEL RESULTS.

During the past year, the committee has entered into an agreement with the Aeronautical Research Committee of Great Britain, through the National Physical Laboratory, to arrange for the conduct of certain definite tests in the wind tunnels of the world. The tests are to be made on standard airfoil and airship models which have been designed and constructed by the National Physical Laboratory. The National Advisory Committee or Aeronautics undertook to arrange for the test in the wind tunnels of the United States. In September 1922, the committee received from the National Physical Laboratory two airship models for comparative tests. These models have been tested under the direction of Dr. A. F. Zahm, at the aeronautical laboratory of the Washington Navy Yard.

The National Advisory Committee has further authorized the testing of standard models in the United States, the models consisting of three cylinders, having length-diameter ratios of 5 to 1, and four models of the U.S.A. 16 wing section, each having an aspect ratio of 6 and the length varying from 18 to 36 inches. The tests on both cylinder models and wing models are to be made over as wide a range of V/L as possible, and to include determinations of lift, drag, and pitching moments every 4° from -4° to +20°. The streamline airship models to be tested will have the proportions of the Navy "C" class airship described in a recent report of the Washington Navy Yard wind tunnel. Four streamline airship models, of 4,

6, 9, and 12 inches diameter, respectively, are to be tested, and are to be supported by spindles of lenticular form, the least diameter of spindle being one-twentieth the diameter of the model, and the fineness ratio of the spindle being 3. After completion of the test in the wind tunnels of the United States, the models will be sent to laboratories in European countries and to Canada for test.

Document 2-15(h), George W. Lewis, memorandum to Langley Memorial Aeronautical Laboratory, 1925.

From: National Advisory Committee for Aeronautics.
To: Langley Memorial Aeronautical Laboratory.
Subject: Results of tests of standard R.A.F. airfoil.

1. The Committee is in receipt of a letter from the Aeronautical Research Committee of Great Britain suggesting that each country publish independently the results of the wind tunnel tests on the standard N.P.L. models. It is further suggested that the results of the tests on the standard R.A.F. airfoil be published as soon as possible, as the Aeronautical Research Committee desires to issue at some future date a complete memorandum comparing the results in the different countries.

2. Before a meeting is called of the representatives of the various laboratories in this country in which the airfoil was tested, it seems desirable that the reports of these tests be circulated among those representatives. There is accordingly enclosed herewith a copy of the report of the Massachusetts Institute of Technology on the tests of the standard R.A.F. model. The report of the Bureau of Standards has already been sent to you, and the report of the Engineering Division of the Air Service will be forwarded as soon as received.

3. It is desired that a meeting of the representatives of the various laboratories be held during the first week in May to consider the preparation of the joint reports on the tests of this model.

G. W. Lewis,
Director of Aeronautical Research.

Document 2-15(i), Aerodynamics Department, National Physical Laboratory [Great Britain], "A Comparison between Results for R.A.F. 15 in the N.P.L. Duplex Tunnel and in the N.A.C.A. Compressed Air Tunnel" (summary), n.d. [1925].

It does not appear justifiable to use any full scale figures for further comparison, since the above results refer to a square-ended monoplane of 6:1 aspect ratio. The only important discrepancy revealed by the above comparison is that occurring at maximum lift, and this difference, at the point where the flow is critical, can not be considered surprising. The agreement, apart from this limited region is surprisingly good, when the extreme difference between the two types of tunnels is taken into consideration, and tends to establish a considerable amount of confidence in the compressed air tunnel results. The marked difference in lift at the stall needs further explanation, and future comparisons on other sections might throw some light upon it. Further information on this point is very desirable if the merits of the compressed air tunnel are to be fairly assessed, on account of the importance of a correct prediction of maximum lift to the designer.

Document 2-16

"Summary of General Recommendations,"
NACA Annual Report for 1921 (Washington, DC) pp. 4–5.

By 1921, the NACA had the first portions of its Langley laboratory up and running, and the committee began to take its advisory role as seriously as it had its research role. The *NACA Annual Report for 1921* contained a fascinating section entitled "Summary of General Recommendations" wherein the committee made a number of specific recommendations for federal legislation and policies in support of aviation growth and safety. From what the committee termed "the most urgent need"—that of federal legislation to regulate all facets of aviation—to policies that would encourage and support the "aerological" (weather) service, air mail, aircraft manufacturers, helium production for airships, and "greater provision for the continuous prosecution of research on a larger scale," these recommendations became the nucleus of major federal legislation. The Air Commerce Act, enacted during the decade, and the ensuing regulations played a role in shaping aviation that was at least as important as the NACA's scientific and technical research programs.

Document 2-16, "Summary of General Recommendations,"
NACA Annual Report for 1921.

SUMMARY OF GENERAL RECOMMENDATIONS.

The more important general recommendations of the National Advisory Committee for Aeronautics are summarized as follows:

LEGISLATION FOR THE DEVELOPMENT OF AVIATION.

The most urgent need for the successful development of aviation at the present time, either for military or civil purposes, is the enactment of legislation providing for the Federal regulation of air navigation, and the establishment of airways and airdromes under Federal regulation. The Federal regulation should include the licensing of aviators, aircraft, and airdromes; the airways should consist of chains of landing fields providing supply and repair facilities and including the necessary meteorological stations, observations, and reports. If the Federal Government will establish and regulate transcontinental airways, as recommended, the committee is confident that air lines for the transportation of passengers or goods will be rapidly established by private enterprise in all parts of the country. The first national airways, however, should be carefully planned to serve military as well as civil needs. The committee reiterates its former recommendations as to the manner of accomplishing the desired results, and urgently recommends the establishment by law of a Bureau of Air Navigation in the Department of Commerce.

EXTENSION OF AEROLOGICAL SERVICE.

The committee emphasizes the importance of extending aerological service (under the Weather Bureau) along airways as established, and recommends that adequate provision of law be made for this service, which is so indispensable to the success and safety of air navigation.

POLICY TO SUSTAIN THE INDUSTRY.

Whatever may have been the faults or the shortcomings of the aircraft industry during or since the war, the fact remains that there must be an aircraft industry, and that it should be kept in such a condition as to be able to expand promptly and properly to meet increased demand in case of emergency. The Government, as the principal consumer, is directly concerned in the matter, and should formulate a policy which would be effective to sustain and stabilize the aeronautical industry and encourage the development of new and improved types of aircraft. In this respect the committee invites attention to the recommendation contained in its special report submitted to the President on April 9, 1921, published as House Document 17, and again recommends the adoption of a policy which, while safeguarding the interests of the Government, will tend to sustain and stabilize the industry.

IMPORTANCE OF MILITARY AVIATION.

Aviation is indispensable to the Army and to the Navy in warfare; and its relative importance will continue to increase. Other branches of the military services are comparatively well developed, whereas aviation is still in the early stages of its development. The demand for greatly reduced expenditures in the military and naval services should not apply to the air services. The committee recommends that liberal provision be made for the Army and Navy Air Services, not only that provision be made for the maintenance and training of personnel, but also that the funds be adequate to insure the fullest development of aviation for military and naval purposes.

SCIENTIFIC RESEARCH.

Substantial progress in aeronautical development, whether for military or commercial purposes, must be based upon the application to the problems of flight of scientific principles and the results of research. The exact prescribed function of the National Advisory Committee for Aeronautics is the prosecution and coordination of scientific research, and, while encouragement may be taken from the progress made, greater provision for the continuous prosecution of research on a larger scale is strongly recommended by the committee.

THE AIR MAIL SERVICE.

The Air Mail Service has demonstrated that airplanes can be utilized with certain advantages in carrying the mail. And it has done more than this, despite the handicap of using, military types of aircraft, poorly adapted to its work or to any civil or commercial purpose, in demonstrating that commercial aviation for the transportation of passengers or goods is feasible. There are several causes which are delaying the development of civil aviation, such as the lack of airways, landing fields, aerological service, and aircraft properly designed for commercial uses. The Air Mail Service stands out as a pioneer agency, overcoming these handicaps and blazing the way, so to speak, for the practical development of commercial aviation. As a permanent proposition, however, the Post Office Department, as its functions are now conceived, should no more operate directly a special air mail service than it should operate a special railroad mail service; but until such time as the necessary aids to commercial aviation have been established it will be next to impossible for any private corporation to operate under contract an air mail service in competition with the railroads. The National Advisory Committee for Aeronautics therefore recommends that provision be made for the continuation of the Air Mail Service under the Post Office Department.

HELIUM AND AIRSHIPS.

The United States has a virtual monopoly of the known sources of supply of helium, and these are limited. Experiments have been conducted by the Bureau of Mines with a view to the development of methods of production and storage, but as yet the problem of storage in large quantities has not been satisfactorily solved. Because the known supply is limited, because it is escaping into the atmosphere at an estimated rate sufficient to fill four large airships weekly, and because of the tremendously increased value and safety which the use of helium would give to airships, particularly in warfare, it is, in the opinion of the National Advisory Committee for Aeronautics, the very essence of wisdom and prudence to provide for the conservation of large reserves through the acquisition and sealing by the Government of the best helium-producing fields. Attention now being given to the development of types of airships to realize fully the advantages which the use of helium would afford should be continued. Such development would give America advantages, for purposes either of war or commerce, with which no other nation could successfully compete.

Document 2-17

F. H. Norton, "The National Advisory Committee's 5-Ft. Wind Tunnel," *Journal of the Society of Automotive Engineers* (May 1921): 1–7.

In spite of World War I and increasing discontent with its Langley Field landlord, the NACA continued work on its first facilities, a five-foot open wind tunnel modeled after Britain's four-foot National Physical Laboratory tunnel and a dynamometer laboratory for engine testing. The Atmospheric Wind Tunnel (AWT), which began operation in 1920, was an obsolete design even before it was built, but it was a proven commodity, and it allowed the NACA's technicians to finally get to work with their own equipment. During the dedication of the Langley Memorial Aeronautical Laboratory (LMAL) on 20 June 1920, praise was lavish. None other than former opponent David Taylor hailed the modest brick wind tunnel building as a "shrine to which all visiting aeronautical engineers and scientists will be drawn." Taylor's statement was a bit exaggerated for the time, but Langley soon began to chart its course and produce significant results. The following year, Frederick H. Norton, Langley's chief physicist, reported on the work being done in the AWT, noting that the work at Langley was "entirely research on the fundamental problems of aeronautics, such as the systematic design of airfoils, the scale effect on models, and the relation of the stability on models to that in free flight."

For one of its most important early programs, the NACA obtained a Curtiss JN4H Jenny for flight tests, and Langley model makers built two models of the plane for wind tunnel testing. One was used for tests measuring lift, drag, and moments, while the other was outfitted for pressure-distribution tests. The wind tunnel data then were compared to measurements obtained during flight tests of the JN4H. Gustav Eiffel had done the first such comparative tests a decade earlier in France, but the Langley program was more extensive and done with considerably greater precision. With such research programs, the NACA began to stake out a crucial role for itself.

Document 2-17, F. H. Norton, "The National Advisory Committee's 5-Ft. Wind Tunnel," 1921.

In the spring of 1919 work was started on a 5-ft. wind tunnel for the National Advisory Committee for Aeronautics at Langley Field, Va., and in the spring of 1920 the tunnel was completed and ready for calibration and for conducting tests. This tunnel has now been in operation for about one year and during this time new apparatus and equipment have constantly been added to increase efficiency and usefulness. While the tunnel is not as large as some that are now in use, it has, it is believed, the highest useful speed of any wind tunnel in the world, that is, it maintains a high velocity flow which is steady in speed and direction, and it possesses satisfactory means for measuring the forces on models at the highest velocities. At present, useful speeds up to 120 m.p.h. can be attained; but, as the propeller was originally designed for other conditions, it is estimated that with a new and higher-pitch propeller a maximum speed of 140 m.p.h. can be reached. Testing models at such high velocities is not a simple matter and a number of new methods and devices had to be developed to accomplish this successfully. For this reason a large portion of the time that the tunnel has been in active operation has been occupied in carefully studying the aerodynamic properties of the tunnel and in constructing apparatus for holding models at the high velocities which can be reached so that only in the last few months has the tunnel been devoted continuously to research work. From now on, the work of the tunnel will be devoted almost exclusively to tests on thick airfoils, including an investigation of their pressure distribution, and to study the stability of model airplanes.

TUNNEL AND BUILDING.

The wind-tunnel building, which is constructed substantially of brick and steel, is approximately 92 ft. long, 43 ft. wide, and 28 ft. high at the eaves. In Fig. 1 is shown a longitudinal section of the building and tunnel giving the principal dimensions of the structure. A heavy concrete foundation runs the whole length of the tunnel and a separate foundation is used for the power plant, so that any vibrations which may be set up by the propeller or motor are not directly transmitted to the tunnel or balance. The interior of the building is smooth and free from obstructions so that the return flow of air from the tunnel will be disturbed as little as possible. Large doors at one end of the building allow the circulation of outdoor air for cooling the building in summer, which is necessary because of the rising temperature due to the power loss in the tunnel at the higher speeds, although, of course, these doors can not be open while the actual tests are being carried on, owing to possible disturbances from the wind. Besides the main room for the tunnel there are several small rooms for offices in the building.

The tunnel itself is of the venturi type with a continuous throat of circular section, and there is an air-tight experimental chamber built about the working section in order that small holes may be opened into the tunnel while it is running, without disturbing the airflow; and this chamber has proved of great convenience in much of the work. The tunnel expands, as shown in Fig. 1, from the 5-ft. diameter working section to a diameter of 10 ft. at the mouth of the entrance and exit cones. This type of tunnel was selected after a considerable amount of investigation had been made upon a model tunnel 1 ft. in diameter, by measuring its efficiency and steadiness while varying many of its characteristics, especially the form of the cones, the type of the experimental chamber, the diffusers, and the honeycombs. It will be noted that, contrary to usual practice, no diffuser is used in the return flow, and this is because the gain in steadiness from its use was found to be very slight on the model, while the cost of the full-sized diffuser would be considerable. Taking into consideration the aerodynamic efficiency of the tunnel, the cost of construction, and the steadiness of the air flow, this type of tunnel was considered to be the best for the proposed investigations, although for some classes of work a larger diameter and slower speed tunnel would be more advantageous.

The cones of the wind tunnel as shown in Fig. 2 are supported from a concrete foundation by heavy steelwork and the surface of the cones is planked with cypress with the inside highly polished. This construction may first seem unduly

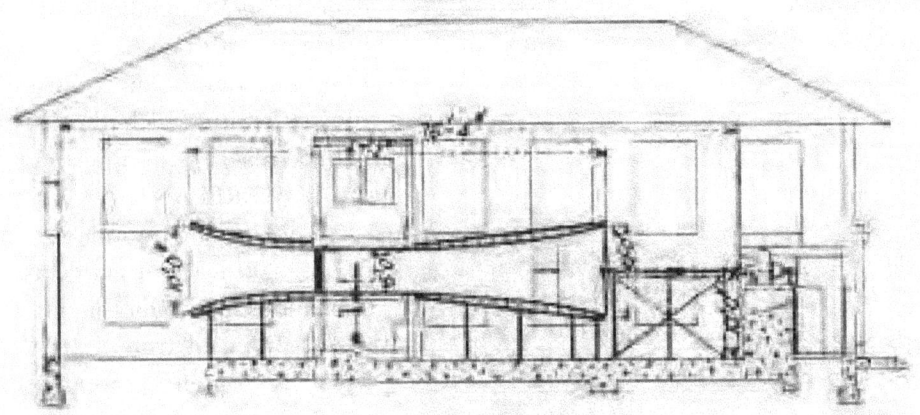

Fig. 1—Longitudinal Section of the Wind-Tunnel Building and the Tunnel.

heavy but when it is realized how great are the vibrations set up by the higher wind speeds and the necessity for having the cones remain perfectly true, it is evident that such a construction is quite justified.

The experimental chamber, which is about 10 ft. long, 14 ft. wide, and 23 ft. high, is built around four concrete columns of very massive construction to withstand the heavy pressures, in some cases as much as 80 lb. per sq. ft., that arise during the high-speed runs. The lower story of the experimental chamber contains the National Physical Laboratory balance and the controlling devices for the air speed, while the upper story contains the propeller dynamometer and wire type of balance. The chamber is entered either by a large door when the tunnel is not running, or, when it is necessary to enter with a difference in pressure, an air lock is provided which consists of two small doors with air valves. Adjustments are made on the model through large doors which can be opened in the throat of the tunnel, these doors being curved to fit

the section of the throat. In order that the model may be inspected during a test, there are three curved glass windows set flush with the inner surface of the tunnel, two in the floor of the tunnel and one in the top. Besides these inspection windows there are also four illuminating windows with electric bulbs, providing a powerful light for photographic purposes. A general view of the lower story of the experimental chamber is shown in Fig. 3, with one of the curved tunnel doors open, indicating the ease with which the model may be reached for adjustment.

POWERPLANT.

The driving motor consists of a 200-hp. direct-current, adjustable-speed motor, with a speed range of from 1 to 1000 r.p.m. At present, power is supplied to this motor at 250 volts from a dynamo driven by a Liberty engine in an adjacent building, but as this powerplant is expensive and inconvenient it is hoped that a more suitable source of supply will soon be available. The propeller is directly connected to the motor by a 3-in. shaft supported on a steel framework and mounted in three ball bearings, while the propeller itself is 10 ft. in diameter and has four blades. As this propeller was designed to be driven by a Liberty engine at 1400 r.p.m. it does not at present absorb all the power which the electric motor can deliver and a new propeller with a larger number of blades and a higher pitch is soon to be used which will increase the air speed obtainable in the tunnel to a considerable extent. The main switchboard is at the entrance of the building, and to obviate the necessity of running heavy wires down to the experimental chamber, the control is by automatic push buttons, one for starting and one for stopping, while a field rheostat is used for speed control above the normal rate of 250 r.p.m. and a series rheostat or potentiometer for speeds below this. The rheostats are placed outside of the experimental chamber to eliminate the heat which would raise the temperature in the small chamber to an uncomfortable degree.

In Fig. 4 is shown a curve of power input to the driving motor plotted against the air speed in the throat of the tunnel. It should be noted that the efficiency of the motor with the very small field excitation which occurs at the higher speeds is exceedingly low and the actual power supplied to the tunnel is much less than that supplied to the motor. An approximate energy factor for the tunnel and propeller of 1.90 is obtained, showing an efficiency considerably higher than that for either the National Physical Laboratory or the straight Eiffel type of tunnel. The reason for the power curve showing a sudden increase at the basic speed is because the potentiometer rheostat is connected in at this speed and absorbs a constant amount of power which is much more than the motor itself absorbs. In Fig. 4 is also shown a curve of the propeller speed plotted against the air speed, that is, the effective propeller slip; and, as would be expected, the slip is constant for all air speeds.

The steadiness of velocity in this wind tunnel compares favorably with that of other tunnels, the maximum variation of the velocity from the mean at any air

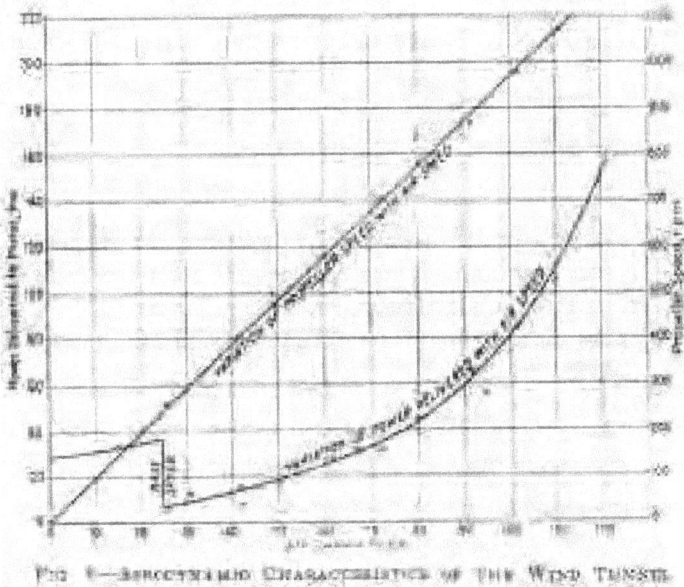

speed not being greater than 1 per cent at a given point in the tunnel. No manual operation of the controls is necessary during a test, except at long intervals, to compensate for the changing resistance of the motor due to a temperature rise. It was at first planned to construct a speed regulator similar to the one used at Göttingen, but with the proper adjustment of the governor on the generating set, such satisfactory results were obtained that the regulator was found unnecessary. The maximum variation of wind direction in the throat as determined by a recording yawmeter is ±0.5 deg., and this variation is of such a high period that it does not appreciably affect the readings of the balance.

BALANCES.

The balance mainly used in this wind tunnel is of the modified National Physical Laboratory type, designed and constructed at the Committee's laboratory, a cross-section of which is shown in Fig. 5. The distance from the center of the model to the pivot point on this balance is 54 in., while the distance from the pivot to the end of the weighing arm is 27 in., so that the weights are actually twice as heavy as they are marked. The balance was designed to measure forces on the model up to 50 lb., while the weight of the moving parts was kept down to 46 lb. by the use of aluminum alloys and high-tensile steel. While the National Physical Laboratory type of balance is convenient and satisfactory for small tunnels and low wind speeds, it is felt that an entirely different type of balance must be designed if it is desired to measure forces any larger than 30 or 40 lb., as it is found when the maximum forces are used on this balance that a great amount of trouble is introduced by deflection of the various members and especially by the vibration which is set up when a model is turned to an angle near its burble point. Another objection to the National Physical Laboratory balance for large forces is the heavy weights that must be used; that is, for balancing a weight of 50 lb. on the model, weights to the amount of 100 lb. must be lifted onto the

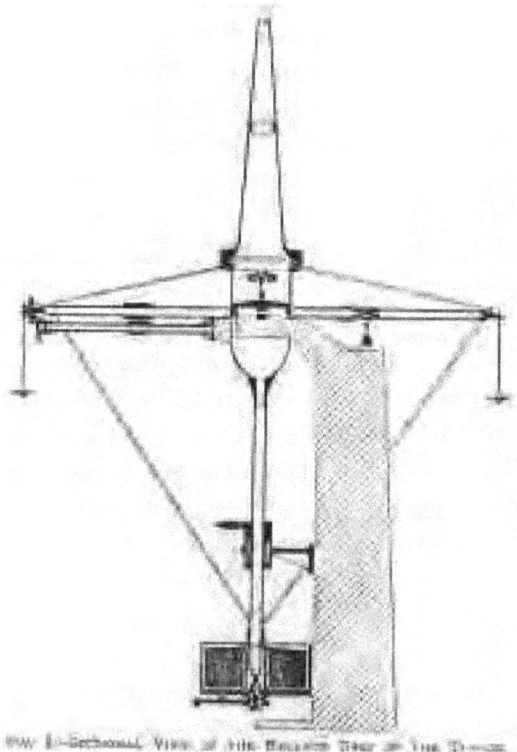

arms, which is very inconvenient when rapid tests are being made. The principal changes besides weights and dimensions which have been made in the original National Physical Laboratory balance are as follows:

(1) A ball-bearing pivot is used in place of the usual conical pivot as the latter gave considerable friction under large loads and also gave trouble through a shifting of its position so that the zero reading was changed during a run.

(2) A weighing arm is used for measuring moments instead of the former torsion wire, and in this way lift, drag and pitching moments can be measured simultaneously.

(3) The weighing arms are made of light steel tubing, and to prevent deflection they are trussed up with tie-rods, thus greatly diminishing the weight of the arm, while at the same time increasing its stiffness.

(4) A pinion is used for turning the head of the balance to make small adjustments more accurately and a prism is used for reflecting the horizontal graduations in a convenient direction.

(5) An improved locking device is used on the lower balance tube.

(6) The lift is measured very satisfactorily and quickly by a Toledo weightless scale, which allows direct readings to be made.

It was found necessary to use a mercury seal to prevent air from passing around the balance spindle into the tunnel, even though the doors in the experimental chamber were closed. This is due to the fact that even with the tightest possible construction there are a number of small leaks about the experimental chamber, the air from which accumulating at the crack around the balance spindle produces an air flow large enough to introduce a considerable error into the readings of the balance. This seal is made of cast iron and allows a maximum head of mercury of 2 in., the height of the mercury being at all times observable by a glass tube on the outside of the seal. Models are usually supported from the top of the balance by a tapered steel spindle, which is 1 in. in diameter at the base and tapers to 5/16 in. at the top, and this is enclosed up to within 2 in. of the model with a thin brass fair-water to reduce the spindle correction to a minimum. As the bending moment in this spindle is very large at high speed and as there is considerable trouble with the model vibrating, a method has been devised by D. L. Bacon for supporting the top of the model rigidly and at the same time reading the forces with accuracy and great rapidity on the balance.

This method is shown in Fig. 6 and consists of a set of small wires, one extending from the top of the model, one from the lifting arm of the balance and a diagonal to the top of the experimental chamber, the direction of this diagonal being such that if it were projected downward it would pass exactly through the center of rotation of the moving parts of the balance. It is also necessary that the plane of these

488   *Chapter 2: Building a Research Establishment*

wires shall be exactly in the plane of the spindle and the lift arm so that there will be no component of lift transferred to the drag arm; and this is done by carefully adjusting the top of the diagonal wire until a force on the model in the direction of the lift plane will have no effect upon the reading of the drag arm. The same method could be used on the drag arm but as the forces are not so great it

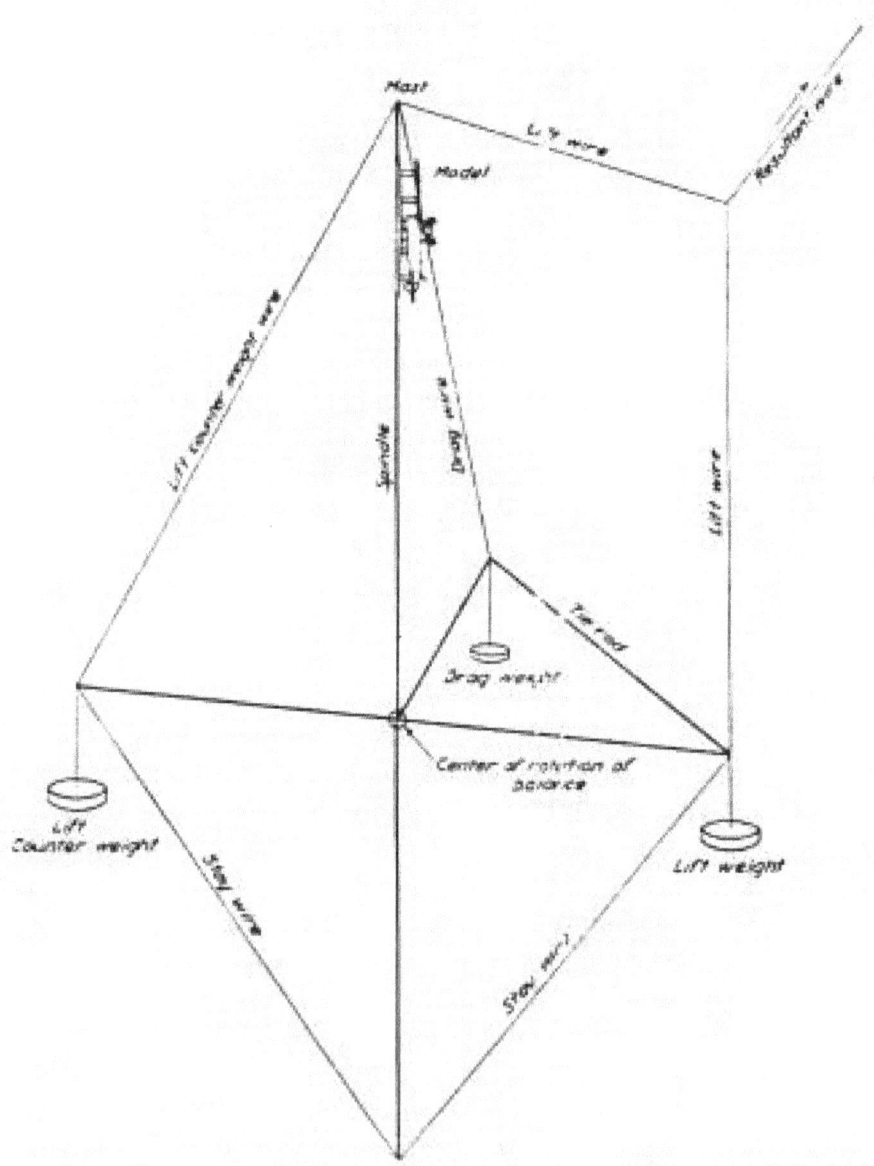

FIG. 6—METHOD OF SUPPORTING MODELS ON THE BALANCE AT HIGH SPEED

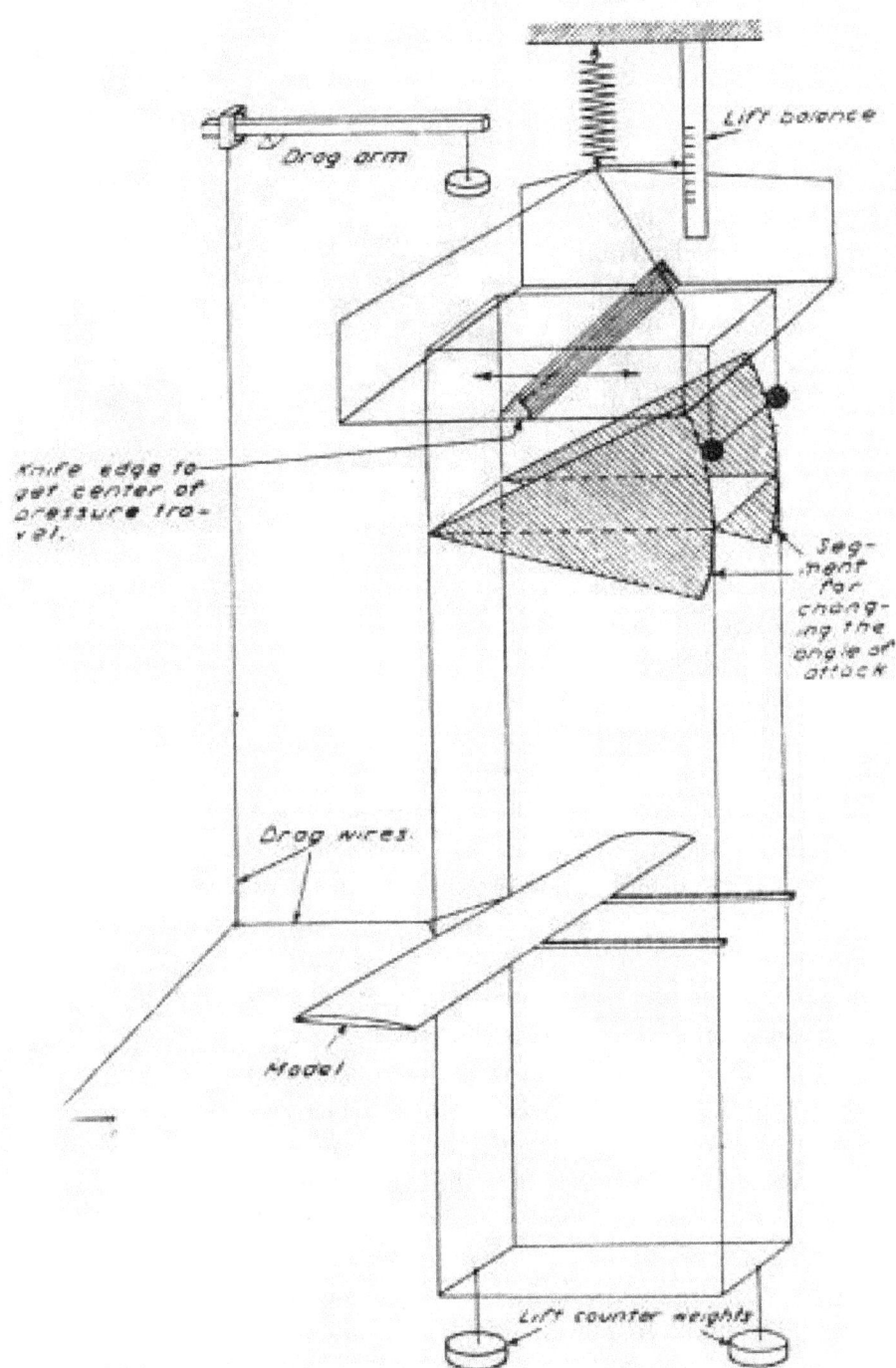

Fig. 7—Diagrammatic View of the Wire Balance

was found sufficient to run a diagonal wire directly down to the end of the arm. These wires pass through small holes in the tunnel walls which are large enough to allow the free play of the tunnel wires and yet due to the air-tight experimental chamber there is very little air passing through them. By this method it has been possible to test a $\frac{1}{24}$ size model of the Curtiss JN4 up to 100 m.p.h. without excessive vibration or deflection and this speed was the limit only because the model itself, even though made of metal, was not sufficiently rigid to stand a heavier load.

As it is believed that the important development in aeronautics of the future centers about the internally braced wing, the Committee's policy is to conduct extensive researches on this type of airfoil. The National Physical Laboratory type of balance is unfortunately unsuited to tests of this nature, especially where the wings are tapered down to a thin section at the tip, as it is practically impossible to support such a model by a spindle attached to the end of the wing. For this reason it has been necessary to design and construct another type of balance, which supports the wing nearly at its center by wires, as shown in Fig. 7. The lift and drag can be measured on this balance directly, the lift on a Toledo scale and the drag on a small balance connected to the wing by a parallelogram of wires. The center of pressure is determined directly by finding the point about which the wing is in equilibrium when balanced on knife edges, as shown in Fig. 8. This method is very convenient and accurate and eliminates the large amount of computation which was necessary in finding the center of pressure travel by the usual methods. While this balance does not take the place of the National Physical Laboratory balance for the majority of the tests, still it is a necessity when it is required to support the model by its center and for wings of high aspect ratio where it would be impossible to support them steadily by an end spindle.

SPECIAL APPARATUS.

The air speed in the tunnel is originally determined by a pitot tube and the micro-manometer shown in Fig. 9. This manometer can measure a head of water

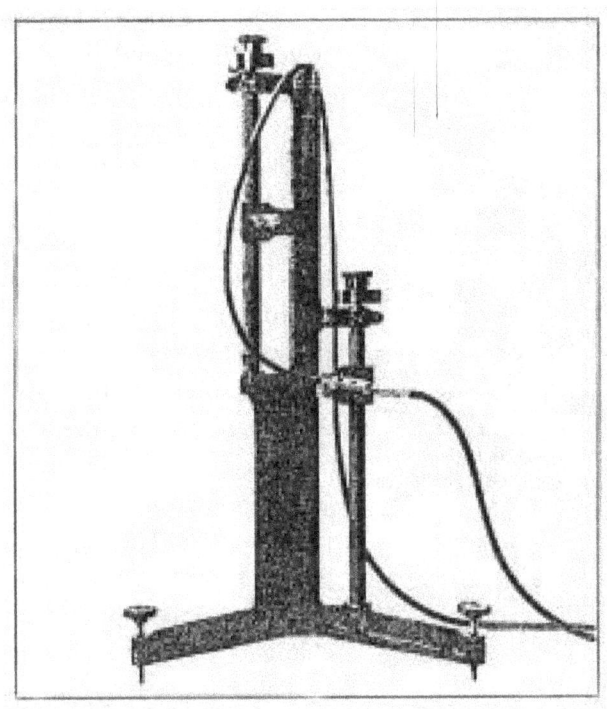

FIG. 9—THE MICRO MANOMETER

to 0.001 in., which is sufficiently accurate for any work required in the wind tunnel, and is very much more convenient than the Chattock gage generally used in wind tunnels, as its sensitivity can be changed by altering the slope of the glass tube, and its range can be extended to a head of 18 in. without difficulty.

For usual running, the air speed in the tunnel is determined by the difference between the static pressure in the side of the tunnel and the outside air in the building, this difference in pressure being measured by the manometer shown in Fig. 10. This manometer is arranged so that its sensitivity will be inversely proportional to the head measured so that at the higher speeds where the fluctuations are naturally greater than at the lower speeds, the variations shown by the liquid will be proportionately the same as at the low speed, which is a necessity when running at the highest velocities. This gage also obviates the necessity of having a very long inclined tube, which would mean that the meniscus must change its position by several feet, so that the operator would have to stand in different positions for various velocities in the tunnel. This gage is, of course, used as a secondary instrument and is calibrated from the readings of the pitot tube, but actual heads can be easily read on it by measuring the angle of the tube and knowing the distance from the center of the reservoir to the meniscus of the liquid.

A multiple manometer is used to a large extent in determining the distribution of pressure over models, and one containing 20 tubes is shown in Fig. 11. The two outside tubes are connected directly to the top of the reservoir so that they will read the height of the liquid at all times, while the other 18 tubes are connected to the pressure holes on the model. The height of the liquid can be adjusted by raising or lowering the reservoir, while the sensitivity can be changed by varying the inclination of the glass tubes.

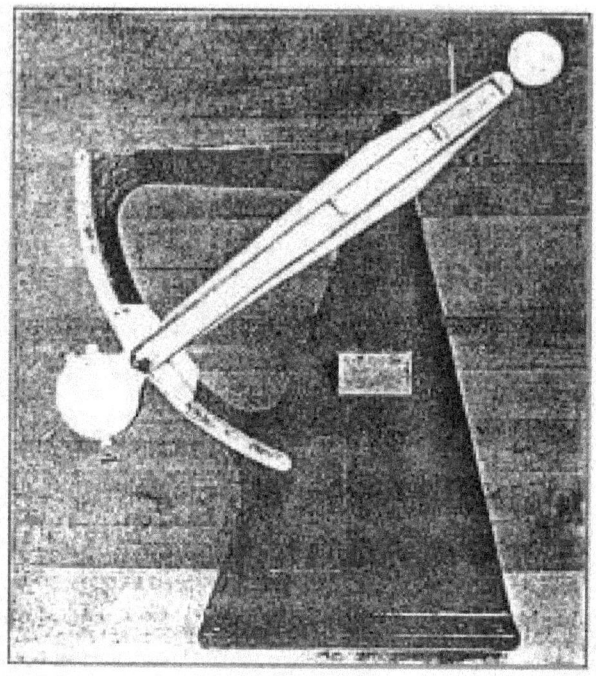

Fig. 10—The Tilting Manometer

For the study of fluctuations in velocity and direction in both the model and full-sized tunnel, a number of high-period recording airspeed meters have been constructed, the most recent one being shown in Fig. 12. This instrument was designed to be portable, requiring only a small battery for the light and the motor, and the film is carried in light-tight drums which are used like plate holders. The sensitivity and position of the zero of this gage can be changed easily, making it available for a large number of uses, while its natural period is so high and the friction is so small that it can easily record the highest period fluctuations that will occur in any wind-tunnel work.

Because of the inconvenience and inaccuracy of the old method of aligning wings by attaching a batten to them and then aligning this batten with a parallel line on the floor of the tunnel, a new method is used consisting essentially of a projector which throws a parallel beam of light upon a small plane mirror temporarily attached parallel to the chord of the wing, and this beam of light is reflected back to the cross line of a white target on the wall of the experimental chamber, Fig. 13. To align the model all the operator has to do is turn the head of the balance until the light spot falls on this cross line, which he can see from any part of the experimental chamber, so that one man can line up a wing to within 0.02 deg. in a few seconds, thus greatly reducing the time and increasing the accuracy compared with the older methods.

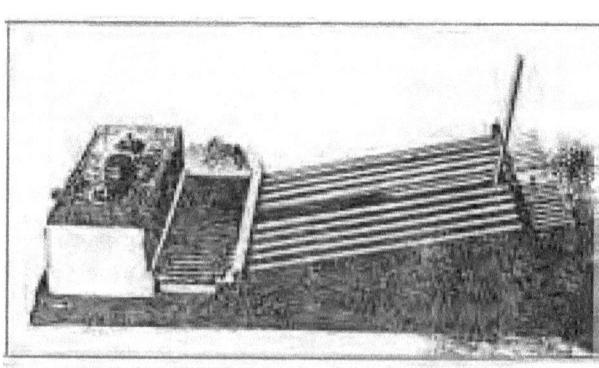

Fig. 11—The Multiple Manometer

This method is also applicable for determining the angular deflection of a wing or model during actual test. The optical method of aligning has been in use for a considerable length of time and has proved very satisfactory, entirely eliminating those errors in alignment which are bound to creep in with the older method, due to unskilled operators or curvature in the batten.

One of the chief causes of dissimilarity between a model test and a free-flight test is the lack of a slipstream in the model, and to produce this effect in the wind tunnel a small propeller is driven before the model by a belt from a high-speed electric motor above the tunnel, the propeller being supported by steel wires from the walls of the tunnel as shown in Fig. 14. The wires, the belt, and the propeller mounting undoubtedly cause a somewhat different air flow from that occurring in full flight, but this interference is probably very small and at least gives us a much closer approximation of actual conditions than has been obtained before. While it has not been possible as yet to drive a model propeller at a proportional speed to the full-size propeller, it has been possible by using a model with a pitch slightly

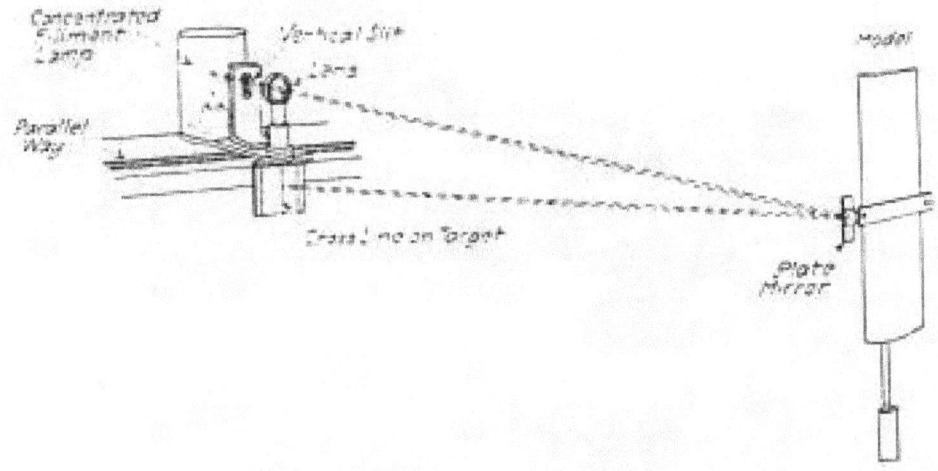

FIG. 13—METHOD OF ALIGNING THE WIRES

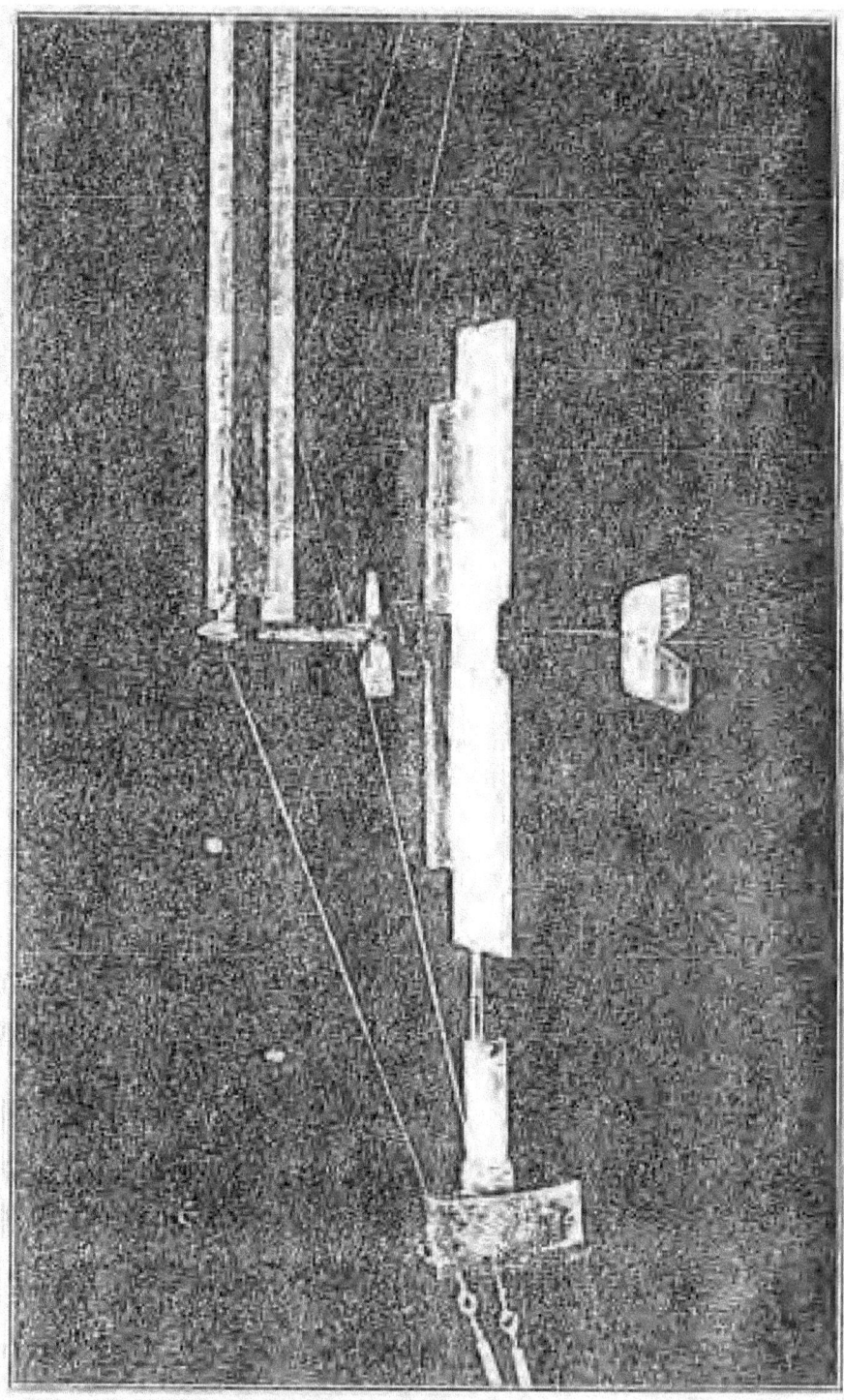

FIG. 14.—METHOD OF DETERMINING THE SLIP STREAM EFFECT

larger than the full-scale propeller to get a slipstream of the same characteristics as the slipstream in the full-sized machine, which should give identical results as far as the interference of the model is concerned. Work is being carried out on the design of a very small high-pressure turbine to drive the propeller at a higher speed and so that a smaller mounting can be used through reducing the interference.

MODELS.

While the models tested in the wind tunnel are, strictly speaking, not a part of the equipment, still those models of such standard form that they are used repeatedly in tests can be considered as such. The Committee has constructed two models of the JN4H airplane, one of them being the $\frac{1}{24}$ scale model illustrated in Fig. 15, which is constructed with the greatest accuracy to reproduce all parts that might affect the air flow, that is, the engine, radiator, and wind shields, the only omission being the wires and fittings which can be more accurately calculated than tested. The other model of the same machine has been constructed to $\frac{1}{15}$ model size for pressure distribution tests on the tail surfaces.

All of the medium or thick wings that are tested in the tunnel are constructed of maple, as this material can be worked with proper precautions to within an error of 0.002 in. The wings are cut upon a special machine shown in Fig. 16, which not only will cut the usual constant-section type, but will also cut wings tapering in plan form and thickness, as, for example, a wing with a depth-to-chord ratio falling off toward the tip as a parabolic function, and at the same time with an elliptical plan form. A much heavier and more precise machine of the same type has been designed for cutting and grinding aluminum or steel wings. Without a machine of this type the Committee's extensive program on thick tapered wings would be impossible because of the great expense of making the models by hand.

FIG. 15.—A WIND TUNNEL MODEL OF AN AIRPLANE

The wind-tunnel work of the National Advisory Committee at Langley Field, in contrast to the work of most of the tunnels in this country, is entirely research on the fundamental problems of aeronautics, such as the systematic design of airfoils, the scale effect on models and the relation of the stability on models to that in free flight. Fortunately, it is possible to carry this work on most efficiently because of the close cooperation that can be had with the free flight investigations which are being conducted by the Committee at the same time. It is, however, realized that with only one wind tunnel, the various model investigations can not be carried on simultaneously, and so the work can not be conducted as rapidly as desired, but it is expected that a second and larger tunnel will be constructed by the Committee in the near future, which will greatly extend the range and amount of investigation possible.

FIG. 16—A SPECIAL MACHINE FOR CUTTING AIRFOILS

Document 2-18(a–j)

(a) Max M. Munk, letter to J. C. Hunsaker,
7 October 1920, Hunsaker Collection, Box 4, File M.

(b) J. C. Hunsaker, letter to Lester D. Gardner,
3 May 1921, Hunsaker Collection, Box 2, Folder 4, File G.

(c) F. H. Norton, Chief Physicist, NACA Langley,
memorandum to NACA Executive Officer (G. W. Lewis),
5 August 1921, RA file 44, Historical Archives, NASA Langley.

(d) W. Margoulis, excerpts from "A New Method of Testing
Models in Wind Tunnels," *Aeronautics* [Britain] XIX,
No. 373 (New Series), (9 December 1920): 412–413.

(e) Edward P. Warner, Massachusetts Institute of Technology,
letter to G. W. Lewis, Director of Aeronautical Research,
NACA, 5 March 1921, NASA Record Group 255, General
Correspondence File, Box 12, National Archives, Washington.

(f) W. Margoulis, excerpts from "A New Method of
Testing Models in Wind Tunnels," NACA Technical
Note No. 52, [1921].

(g) Max M. Munk, "On a New Type of Wind Tunnel,"
NACA Technical Note No. 60, May 1921.

(h) F. H. Norton, memorandum to G. W. Lewis,
"Design of compressed air wind tunnel," 6 October 1921,
NASA Record Group 255, General Correspondence
Numeric File 21-5, Box 80, National Archives, Washington.

(i) "Compressed Air Wind Tunnel," *NACA Annual Report
for 1921* (Washington, DC), pp. 29–30.

(j) Max M. Munk and Elton W. Miller, *The Variable Density Wind Tunnel of the National Advisory Committee for Aeronautics*, NACA Technical Report No. 227 (Washington, DC, 1925).

Langley's study comparing a Curtiss JN4H in flight to models of the same airplane in the Atmospheric Wind Tunnel confirmed that wind tunnel testing was a valid predictor of aircraft performance and behavior, but it also pointed the NACA researchers toward another well-known, but poorly understood, problem: scale. It had long been known that the forces generated by a model were not proportional to the model's scale. A $\frac{1}{20}$-scale model of a plane in an air stream $\frac{1}{20}$ as fast as the actual plane flies generated considerably less than $\frac{1}{20}$ of the actual lift and drag forces. Early aerodynamicists had developed empirical coefficients to "scale up" the data, but the comparison tests showed just how unreliable these coefficients were. The problem was that the air itself could not be "scaled-down" to model size, and its properties, such as density and temperature, were almost the same in the wind tunnel as they were around a full-size airplane. To gain the maximum value from wind tunnel testing, an answer to the scaling problem had to be found.

The answer to the scaling problem had its roots in the work of a nineteenth-century British scientist, Osborne Reynolds. Reynolds showed that the forces a moving fluid exerts on a body, or vice versa, depended on the fluid's velocity, density, and viscosity, and on a key dimension of the body, such as length or diameter. He combined these parameters into a mathematical expression where all of the dimensions cancelled one another out. The dimensionless result, known as the Reynolds number, was a key to understanding scale factor. Because it was dimensionless, the Reynolds number could be used to compare fluid-flow forces around similarly shaped, but differently sized objects. By varying different parameters, such as increasing the velocity or decreasing the density, to obtain the same Reynolds number for different tests—a condition known as dynamical similarity—an excellent correlation between model tests and aircraft performance was possible. But this was easier said than done. When a $\frac{1}{20}$-scale model was tested in the AWT, the Reynolds number was about $\frac{1}{10}$ that of the corresponding actual flight. Larger models could theoretically be used, but wingspans greater than about 3 $\frac{1}{2}$ feet were not useful in the five-foot-diameter tunnel due to interference by tunnel walls. The AWT lacked the power to run at the high speeds necessary to generate the required Reynolds numbers; furthermore, such speeds were not practical in open wind tunnels anyway.

Two possible solutions to the problem emerged almost simultaneously in the early 1920s. In both cases, the intent was to increase the density/viscosity term in the Reynolds number calculation. Wladimir Margoulis, a Russian-born aerodynamicist and former director of the Eiffel laboratory in France, proposed using carbon dioxide instead of air in a sealed wind tunnel, because carbon dioxide's density was over one-and-one-half times that of air. Max Munk, a brilliant and quixotic German student of Prandtl, suggested that a wind tunnel be built inside a pressure vessel so that tests could be run under pressure, thus increasing the density of the air as much as twentyfold. The NACA employed both experts, bringing Munk to the United States and retaining Margoulis as an agent in Paris. Debate continues over which man was first to suggest using higher densities, but the NACA chose to pursue Munk's concept.

The result was the Variable Density Tunnel (VDT), Langley's second wind tunnel, which went into service in 1922. Externally, it appeared to be little more than a large cylindrical tank with spherical ends. A closed five-foot-diameter wind tunnel was mounted inside, such that air flowed through the central test section, past the fan, and returned via an annular passage. The entire tank could be pressurized to 300 pounds per square inch (twenty atmospheres), sufficient to produce Reynolds numbers for tests of $\frac{1}{20}$-scale models that were equivalent to full-scale flight. A wide variety of tests were run in the VDT, including studies of several model airplanes to validate the high-pressure concept, but the most significant investigations involved airfoils. Through extensive use of the VDT, Munk and other NACA researchers drew accurate performance curves for the commonly used airfoils of the era, and then extended the investigation to develop families of airfoils with similar characteristics. The VDT provided unique capabilities, and the airfoil data produced with it put the NACA and its Langley Laboratory "on the map" of first-class aeronautical laboratories in the mid-1920s. (Document 2-15(i) in the preceding section on international standardization of wind tunnels is a 1925 report from Britain's National Physical Laboratory comparing results from their Duplex wind tunnel and the NACA VDT. It notes that the favorable comparison "tends to establish a considerable amount of confidence in the compressed air tunnel results.")

The documents included herein illustrate how Munk's idea of a compressed air wind tunnel moved from a concept to a reality. In a 7 October 1920 letter to Jerome Hunsaker, Munk mentioned that he had "finally found a perfectly new manner for increasing Reynold's [sic] number," but he declined to provide any details prior to employment. Once on the NACA's payroll, he readily furnished the theoretical basis for his proposal, which the committee published as Technical Note No. 60, "On a New Type of Wind Tunnel." Once the VDT was in operation, the NACA published a thorough description of the tunnel in Technical Report No. 227, *The Variable Density Wind Tunnel of the National Advisory Committee for Aeronautics.*

A secondary theme in this section concerns some of the problems surrounding Max Munk himself. From the start he was embroiled in controversy, both national (because he was German), and personal (because he was unbearable to work with). Documents in this section hint at the difficulties Langley personnel had with Munk. As shown in the 1921 letter from Hunsaker regarding reactions to employing German experts at the time, Munk faced an uphill battle from the start. Readers should note that just to bring Munk into the United States and employ him in federal service took no less than two special Presidential orders.

Frederick Norton's 5 August 1921 letter regarding the use of German airfoils is another case in point; Norton was incensed that the NACA would even consider allowing the use of a foreign—especially a German—airfoil when a number of good American designs were available. While such national parochialism is perhaps understandable so soon after the end of World War I, these documents suggest that acceptance at Langley during this period would have been very difficult for any German immigrant, not just Munk.

Norton's letter of 6 October 1921, dealing this time with VDT issues, shows that the situation had continued to deteriorate. In his case, however, Munk managed to retain George Lewis's support, leading Norton to resign two years later. Max Munk's triumph was not long-lived, however. The recollections of Fred Weick, included as Document 2-20(c) in the section on the success of the Propeller Wind Tunnel (PRT), recount some of the continuing difficulties encountered in working with the brilliant but temperamental German aerodynamicist. This challenge culminated in a 1927 "revolt" by Langley personnel against Munk that resulted in his departure from the NACA.

Document 2-18(a), letter from Max M. Munk in Germany to Jerome Hunsaker, 1920.

Oct. 7th. 1920

To: Commander J. C. Hunsaker, U.S.N.
Assistant for Aeronautics to the Chief of the Bureau of Constr. and Repair of the
U.S. Navy Department
Washington, Navy-Building

Sir,

hoping [sic] to support my request of July 1920 at Göttingen, I beg leave to inform you, that I made an important invention concerning wind-tunnel tests. You told me then to be greatly discouraged by the present model-tests, on account of the small value of their Reynold's number and of the want of security in applying them on large bodies, following from it. I quite agree with you in this matter: the law of the lift of wings being found, there remain only investigations about details in the present wind-tunnels.

Meditating on these things, I finally found a perfectly new manner for increasing Reynold's number to about eightfold or even the elevenfold (but then with a little complication surmountable however) without increasing the expenses of the construction or the tests. The new method is important for such investigations where Reynold's number is to be considered. But then, you either get the eight-fold value of this number or otherwise you can save seven eights of the costs.—

Not even knowing, whether your address is correct, I dont [sic] like to explain the new method in this letter. Indeed I do not like to do this at all, as long as I have not yet any answer to my request. I prefer reserving my idea to my next employer. I shall however thank you very much for passing on the matter to the Committee and for helping my next employer to be in the U.S.

Trusting you to excuse my once more troubling you in this matter, I shall be very much obliged for the favor of an early reply.

Believe me Sir to be Yours most respectfully

[signed] M. Munk.

Document 2-18(b), letter from J. C. Hunsaker to Lester D. Gardner, 1921.

May 3, 1921.

Mr. Lester D. Gardner,
The Gardner, Moffat Co., Inc.
225 Fourth Avenue
New York, N.Y.

Dear Gardner:

I note with interest the row you appear to have got into over the report to the President on the aeronautical situation, and so long as you are convinced you are doing good I suppose your policy is correct from your point of view. However, you ought to be more careful about using the term "minority report" since there wasn't any, and since the gentleman who wrote a special letter to the President had already agreed in committee that [there] would be no minority report.

The information you wanted about Dr. Munk is that he was got from the University of Göttingen by the National Advisory Committee for Aeronautics on my recommendation as knowing more about German aerodynamics and theoretical questions generally than [anybody] else they could get, and that the best way to bring ourselves up to date with the research work at the Göttingen laboratory, and especially the unpublished portions of it, was to import the man who had done most of it himself. He has not been turned over to the Navy or anybody else but is employed in Lewis' office in translating German data and preparing dope [intelligence information]. I don't think you can get anywhere by attacking the Committee for having employed a German in order to make available in this country what the Germans know.

With kind regards,
Very truly yours,
[signed] J. C. Hunsaker

Document 2-18(c), F. H. Norton, memorandum to George W. Lewis, 1921.

August 5, 1921.

From: Langley Memorial Aeronautical Laboratory
To: National Advisory Committee for Aeronautics
(Attention Executive Officer.)
Subject: Tests of Gottingen wing section.
Reference: (a) NACA Let. 54-6(44/21)—18648, Aug. 4, 1921.

1. Under no circumstances would I consider it advisable to use a German wing section on a Committee helicopter as I am sure our own wing sections are fully as good as Germany's and if Dr. Munk can not find one which we have already tested which has the properties he desires I will gladly undertake to design and test one if he will furnish me with more particulars.

2. I think you will see that we should use every means to advertise the Committee and the Committee's work and the use of a German wing section in its own helicopter should certainly not be advisable from this point of view.

F. H. Norton
Chief Physicist

**Document 2-18(d), W. Margoulis, excerpts from
"New Method of Testing Models in Wind Tunnels," Aeronautics [Britain], 1921.**

In forecasting the conditions of flight of aeroplanes by the results of model tests made in existing laboratories, serious errors are inevitably made, owing to the fact that in the laboratory it is impossible to observe the laws of similitude requiring the equality of Reynolds numbers and the equality of the ratios of the velocities to the velocity of sound (Law of Bairstow and Booth). The first of these conditions is due to viscosity, and is of special importance at low speeds; the second is due to the consideration of compressibility, and should be observed at high speeds.

Now, the Reynolds numbers attained in existing laboratories are from 15 to 25 less than those reached by machines in flight, whilst the velocities of the airstream remain from two to three times lower than the speed in free flight. Thus, when a model aeroplane is tested in the laboratory, the streamline wires resist relatively twice as much, and the struts of the rigging and landing chassis five times as much, whilst the wings carry up to 30 per cent less on the model than on the aeroplane.

It will be the same in the large laboratories (1,000 to 1,500 h.p.) now being planned, which, though realizing higher velocities, will not attain the seventh part of the true value of Reynolds number.

WE WILL SHOW, HOWEVER, THAT IT IS POSSIBLE TO HAVE WIND TUNNELS GIVING HIGHER VALUES OF REYNOLDS'S NUMBER AND OF THE RATIO OF THE VELOCITY TO THE

VELOCITY OF SOUND THAN THOSE REACHED BY FULL SCALE MACHINES, AND THAT WITH LESS OUTLAY FOR INSTALLATION AND UPKEEP THAN IS REQUIRED FOR THE LABORATORIES NOW BEING PLANNED.

THIS RESULT WILL BE ATTAINED BY EMPLOYING SOME GAS OTHER THAN AIR, AND ESPECIALLY CARBONIC ACID, AT PRESSURES AND TEMPERATURES WHICH ARE SUITABLE AND GENERALLY VERY DIFFERENT FROM THOSE OF THE SURROUNDING ATMOSPHERE.

Let v and d be respectively the velocity and diameter in the working section of a wind tunnel; μ—the coefficient of viscosity; $\rho°$—the density at $1 \text{ kg}/\text{cm}^2$ and $273°$; p—the pressure, and T the absolute temperature of the fluid circulating in the flue. The units employed are the kg., the metre, and the second.

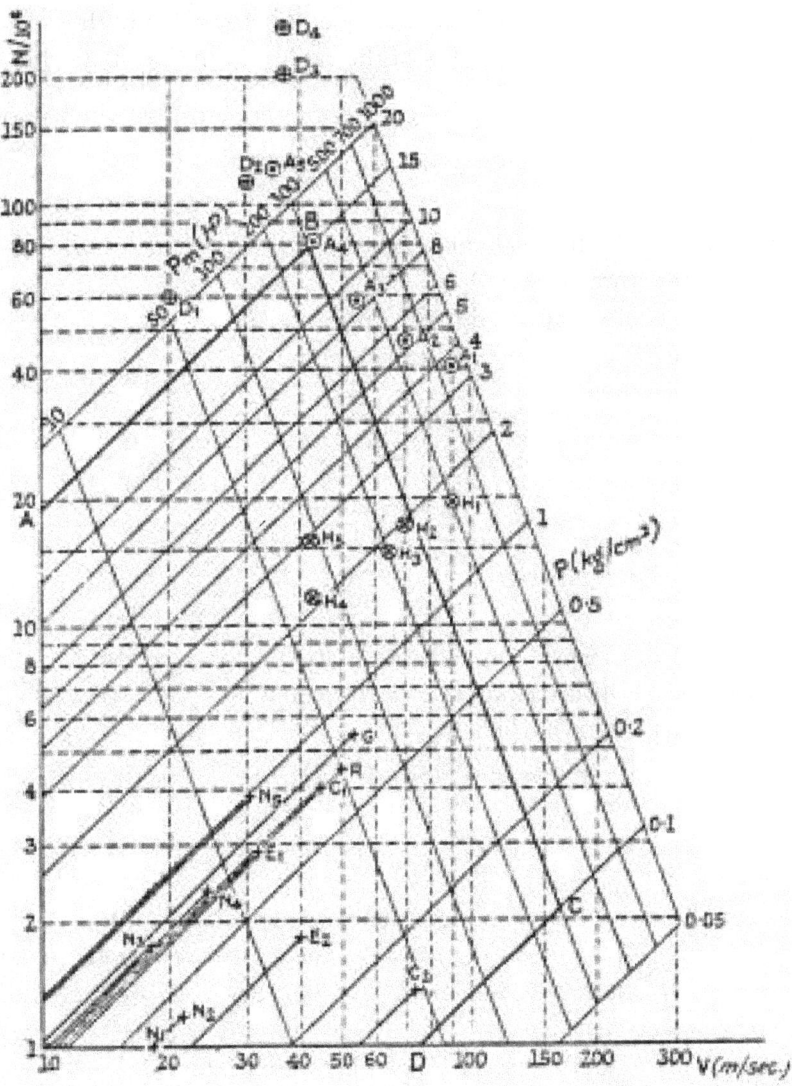

The motive power Pm required for working the fan of a closed circuit wind tunnel of the Crocco type is:

$$Pm = 0.47 v^{2.75} d^{1.75} \mu^{1.25} \rho_0^{0.75} p^{1.75} T^{-0.75} \qquad (1)$$

If the span of the model aeroplane is equal to 6/10 of the diameter of the flue the Reynolds number will be:

$$N = 0.6 \frac{vd}{\nu} = 0.0164 v d \mu^{-1} \rho_0 p T^{-1} \qquad (2)$$

Where ν is the kinematic coefficient of viscosity; $\nu = \mu/\rho$.

The velocity of sound in a fluid being equal to $\sqrt{\gamma \, p/\rho}$, γ being the ratio of the specific heats, the condition of equality of the ratios of the velocity to the velocity of sound requires that:

$$v = 1.1086 \rho_0^{-0.5} \gamma^{0.5} T^{0.5} V \qquad (3)$$

V being the speed of the full scale machine.

We will examine three cases:

1st Case.—Tests of model aeroplanes and dirigibles; N is given.

From formulas (1) and (2) we deduce:

$$\qquad (4)$$

and

$$\qquad (5)$$

THE USE IN A WIND TUNNEL OF CARBONIC ACID AT 15 KG./CM² AND 253° REDUCES THE POWER REQUIRED FOR REALIZING A GIVEN REYNOLDS' NUMBER WITH A GIVEN DIAMETER OF FLUE IN THE RATIO OF 1088 TO 1.

2nd Case.—Tests of model propellers; V is given.

We have:

$$\qquad (6)$$

the true speed (V) in the tunnel being determined by formula (3).

THE USE OF CO_2 AT 0.5 KG./CM² AND 253° REDUCES THE POWER REQUIRED BY 70 PER CENT. AND IF WE REDUCE THE PRESSURE TO 0.1 KG./CM², THE POWER REQUIRED WILL BE REDUCED BY 90 PER CENT.

3rd Case.—Any model whatever: V, N, and d are given.

We have:

$$P_m = \ldots \frac{\mu^? \ldots}{r^3} \ldots \quad (7)$$

and

$$p = \ldots \quad (8)$$

the value of v being given by formula (3).

THE USE OF CO_2 AT 253° AND AT A PRESSURE DETERMINED BY FORMULA (8), REDUCES THE POWER REQUIRED BY 55 PER CENT.

We would point out that the terms characterizing the nature of the fluid and the conditions of temperature and pressure in formulas (4) and (7) are independent of the experimental value of n the expression giving the power drop in tunnels:

$$\Delta = a \cdot \mu^n \rho_0^{1-n} \cdot p^{1-n} T^{n-1} \cdot v^{2-n} d^{-1-n}$$

It is the same for all the terms of formulas (3), (5), and (8).

As an application of the method, we will consider a closed circuit tunnel 2 m. in diameter, 300 horse-power, utilizing carbonic acid. Model aeroplanes will be tested at a pressure of 15 kg./cm² and at 253°; we shall thus realize at 30 m./sec. a Reynolds number of 81.10^6, corresponding to that of an aeroplane with a span of 26 m. at 150 km./h. or of a racing plane at 650 km./h.

If the propeller tests are made at 0.5 kg./cm², the speed attained will be 76 m./sec., equivalent to a speed in free flight of 103 m./sec., that is 370 km./h, whilst at 0.1 kg./cm2 the speed will be 570 km./h.

FOR REALIZING THESE CONDITIONS AN ORDINARY WIND TUNNEL 2 METRES IN DIAMETER WOULD REQUIRE 326,000 H.P. IN THE FIST CASE, AND 3,000 H.P. IN THE SECOND CASE.

During the manipulations of the model, the working section can be isolated by means of two doors, the carbonic acid contained in the working section having been previously collected in a special tank. The measuring devices, registering the stresses automatically, will be placed in an airtight cabin fixed on the wall of the working section. . . .

[*The author here discusses a graph and calculations, omitted here, that "represents the functioning of a tunnel of our system (diameter 2 m, carbonic acid at 253°)."*]

We thus see that for all aeroplanes, except giant planes (42 m. span), we realize the condition of equality of Reynolds's number, that for dirigibles (24 m. in diameter) we attain the half of Reynolds's number which is sufficient to give a good approximation, and that for all propellers we realize the conditions imposed by the consideration of compressibility and viscosity. . . .

Remark.—In the application just given of our system to a tunnel of 2m., we assumed, in order to treat a general case, that the carbonic acid was cooled to $-20°$. Practically this cooling leads to complications of installation and functioning which are not justified by the slight gain of power (see formulas 4, 6 and 7) which results.

We therefore consider it preferable to work at the temperature of the surrounding atmosphere. If we assume a mean temperature of $+10°$, formulas (7) and (8) show that for the same values of V and N the power is increased by 11 per cent and the pressure by 6 per cent, with respect to those corresponding to a temperature of -20°.

Document 2-18(e), letter from Edward P. Warner, Massachusetts Institute of Technology, to George W. Lewis, 1921.

March 5, 1921.

Mr. G.W. Lewis,
National Advisory Committee for Aeronautics
Washington, D.C.
Dear Mr. Lewis:

I have recently received a letter from Mr. Margoulis relative to his new system of high-pressure windtunnel employing carbon dioxide, and I am really quite favorably impressed with such a tunnel for some classes of work. For determining the speed and density effects on air speed meters, for example, such a tunnel would be favorably used and also in connection with the test of streamlined bodies.

Margoulis is anxious to be retained as consulting engineer to construct such a tunnel here, and while I do not see the necessity of this, as it would be easy to secure the necessary talent for securing a high-pressure tunnel in America (incidentally I think that it would be best to use air at high pressure in preference to carbon dioxide sacrificing some of the theoretical advantages but considerably simplifying the construction and obviating the necessity of absolute leak-proofness), I do recommend that Mr. Margoulis be retained by contract for not over 2500 francs to prepare a 12,000-work report on "European Windtunnel Practice" and on suggestions for windtunnel design. He has shown himself possessed of many excellent schemes and we would avail ourselves of his knowledge at least to that extent. I further recommend that Margoulis be retained to prepare a general report on the use of nomograms and alignment charts in aeronautics. It is a subject which we have too much neglected and which he is better qualified by experience to treat than any one else in the world, as he is the originator of almost all the very ingenious nomograms used at the Eiffel laboratory.

Yours sincerely,
[signed] Edward P. Warner

Document 2-18(f), W. Margoulis, excerpts from "A New Method of Testing Models in Wind Tunnels," NACA Technical Note No. 52, 1921.

We know that the two essential conditions of the application of the law of proportionality of pressure to the product of density, the square of the linear dimensions and the square of the speed to the results of model tests are:

1) THE EQUALITY OF REYNOLDS' NUMBER.

$$\aleph = \frac{vl}{v_1} = \frac{Vl}{v_1}$$

v being the velocity of the airstream in the tunnel,
V being the speed of the machine in free flight,
l and L respectively one of the principal linear dimensions of the model and of the full-scale airplane.

v and v_1 respectively the kinematic coefficients of viscosity of the fluid circulating in the tunnel and of the air in which the machine flies.

2) THE EQUALITY OF THE RATIOS OF THE SPEED TO THE VELOCITY OF SOUND (Law of Bairstow and Booth).

$$M = \frac{v}{w} = \frac{V}{W}$$

w and W being respectively the velocities of sound in the tunnel and in the air.

The first of these conditions is due to viscosity and is important especially at low speeds (tests of model airplanes); the second condition is due to the consideration of compressibility and must be observed at high speeds (tests of model propellers).

Now, in existing laboratories utilizing a horsepower of 100 to 300, the models are generally made to a $\frac{1}{10}$ scale and the speed is appreciably lower than the speeds currently attained by airplanes; the Reynolds' Number realized in the laboratories is thus from 15 to 25 times smaller than that reached by airplanes in free flight, while the ratio M varies between the third and three-fourths of the true ratio.

Thus, when a model airplane, for instance, is tested in such a laboratory, the streamline wires resist relatively twice as much, the struts of the rigging and undercarriage five times as much, while the wings carry 30% less on the model than on the airplane, so that RESULTS OBTAINED IN EXISTING LABORATORIES CAN NOT BE PRACTICALLY UTILIZED.

We cannot appreciably increase Reynolds' Number by increasing either the diameter (d) of the tunnel, or the velocity (v) of the airstream, for the motive power required for working the fan producing the airstream is proportional to d1.75 v2.75 and such increase would therefore lead to installations much too costly both as to establishment and upkeep.

Thus wind tunnels are now being planned having a diameter of 3 to 5 m., speeds of 60 to 75 m/sec., and horsepower of 1000 to 1500; but the Reynolds' Numbers attained in such tunnels will still be 8 times less than those of existing large airplanes.

We will show, however, that it is possible to have wind tunnels in which the Reynolds' Number will be greater than that now attained by airplanes, and in which the ratio of the velocity to the velocity of sound will also be greater than that realized in practice, and we will show that this can be done with an outlay for installation and upkeep much below that required by the laboratories now being planned.

In order to attain this result we have only to employ a gas other than air, at a pressure and temperature different from those of the surrounding atmosphere.

[. . .]

2.—1st CASE—REYNOLDS' NUMBER IS GIVEN; TESTS OF MODEL AIRPLANES AND AIRSHIPS.

Eliminating v from equations (1) and (2) [*same as (1) and (2) in Document 2-18d*] we have:

$$P_m \approx 37\pi x \frac{\mu^3}{\rho_0^2} \cdot \frac{1}{p^2} \cdot \frac{T^2}{d} \qquad (4)$$

and

$$v = 61 \cdot \frac{\mu}{\rho_0} \cdot \frac{1}{p} \cdot \frac{T}{r} \qquad (5)$$

Formula (4) shows that the power P_m is proportional to the term μ^3/ρ_0^2 characterizing the fluid, and to the term T^2/p^2 characterizing the conditions of temperature and pressure.

THE GAS WHICH IS PRACTICALLY MOST SUITABLE TO THE DIFFERENT CONDITIONS LAID DOWN BY THE ABOVE FORMULAS, AND BY THOSE WHICH FOLLOW, IS CARBONIC ACID (CO_2), AS A GAS HAVING A LOW COEFFICIENT OF VISCOSITY, HIGH DENSITY, AND A LOW RATIO OF SPECIFIC HEATS.

For air, $\mu^3/\rho_0^2 = 301/10^{18}$ for CO_2, $\mu^3/\rho_0^2 = 72.4/10^{18}$,

[Footnote:] (Other gases, such as chloride of methyl (CH_3Cl) and Xenon would give better results but could not be practically employed. Thus μ^3/ρ_0^2 is equal to $571/10^{18}$ for water, to $15.6/10^{18}$ for CH_3Cl and to $33/10^{18}$ for Xe. It is evident that in the formulas we must always assume for water $p = 1$ kg/cm² and $T = 273°$.)

WE THUS SEE THAT FOR A GIVEN REYNOLDS' NUMBER AND WITH EQUAL DIAMETER OF FLUES, THE USE OF CARBONIC ACID AT 273° AND 1 KG/CM^2 WILL ECONOMIZE 3/4 OF THE POWER REQUIRED WITH ATMOSPHERIC AIR AND THAT COMPRESSION TO 15 KG/CM^2 AND COOLING TO 253° WILL REDUCE THIS POWER IN THE RATIO OF 1088 TO 1, WHICH CERTAINLY CONSTITUTES A REMARKABLE RESULT.

THE SPEED WILL BE REDUCED IN THE RATIO OF 1.85 TO 1 IN THE FIRST CASE AND IN THE RATIO OF 30 TO 1 IN THE SECOND CASE.

[*Margoulis here includes two additional cases that are essentially the same as the 2nd and 3rd case in Document 2-18(d).*]

WE THUS SEE THAT IF WE WISH TO ESTABLISH A LABORATORY FOR WELL DETERMINED VALUES OF V AND N, THE USE OF CARBONIC ACID AT 15 KG/CM^2 AND 253° REDUCES THE POWER REQUIRED IN THE RATIO OF 48.5 TO 1 AND REDUCES THE DIAMETER IN THE RATIO OF 22 TO 1.

IN A LABORATORY ALREADY BUILT, THE USE OF CARBONIC ACID AT 253° REDUCES THE POWER REQUIRED IN TEE RATIO OF 2.2 TO 1; THE PRESSURE SHOULD THEN BE EQUAL TO 68/100 OF THE PRESSURE REQUIRED WITH AIR.

[...]

REMARK I.—When, in the formulas giving the value of the power in the three cases considered, we compare the values of the terms characterizing the nature of the fluid, we find that water forms the least advantageous fluid for use in a laboratory, the more so as, being incompressible, its density can not be varied.

We may remark, however, that heating water to 100° reduces its coefficient of viscosity in the ratio of 6 to 1; in the 1st case it then becomes more advantageous than air at atmospheric temperature and pressure. On this subject we may say that we may consider the use in wind tunnels not only of gas (that is, of fluids for which the temperature of saturation at 1 kg/cm2 is below 273° absolute) but also of vapors, which must be heated so that their temperature is above the temperature of saturation at the pressure at which they are utilized. Thus we may consider using water vapor, although its characteristics ($\mu = 0.89 \times 10^6$, $\rho_0 = 0.079$) are not favorable.

REMARK II.—In our theory of wind tunnels, we have called the coefficient of utilization of a tunnel (rs) the ratio of the kinetic energy of the fluid stream in the working section to the power of the engine running the fan.

For the closed circuit tunnels which we are studying, this coefficient takes the very simple form:

$$r = \frac{v \cdot \Omega \cdot \rho v^2}{T \rho v^3 \cdot N} \quad [\text{illegible}]$$

$N_s = vd/v = N/0.6$.

5.—PRELIMINARY PROJECT OF A WIND TUNNEL.

As a practical application let us consider a closed circuit tunnel, 2 m. in diameter, utilizing carbonic acid.

For tests of MODEL AIRPLANES we use carbonic acid compressed to 15 kg/cm² and cooled to 253° and we fix a speed of 30 m/sec. (corresponding, according to formula (3) to a speed in the air of 41 m/sec.).

The power required given by formula (1) will be 300 HP and the Reynolds' number realized will be 81.10^6 and equal to that realized by the largest existing airplanes (N is greater for large airplanes going slowly than for small planes flying at a high speed). This Reynolds' number will correspond to that of a racing plane flying at 650 km/hr.

The large laboratories now being planned will realize Reynolds numbers 8 times smaller with powers about 4 times greater. If in one of these laboratories having a diameter of 3 m. we wished to attain $N = 81.10^6$, we should require a power of $300 \times 1088 \times 2/3 = 217{,}000$ HP. [*Footnote:*] (This figure must be considered rather as a proof of the impossibility of realizing this Reynolds' number in an ordinary tunnel than as an exact value of the required power. As a matter of fact, the speed in this case should reach 900 m/sec. and we have not the right to apply our formulas to such speeds, for which moreover the phenomena of compressibility would completely distort the results.)

For high speed tests, and especially for PROPELLER tests, the pressure must be below 1 kg/cm². Thus formula (1) shows that with carbonic acid at 0.5 kg/cm² and 253°, the speed realized with the same power of 300 h.p. would be:

$$305 \left[\frac{1.5}{0.5} \right] = 30 \times 2.53 = 76 \text{ m sec.}$$

equivalent to a speed in the air of $76/0.74 = 103$ m/sec., that is 370 km/hr.

If the pressure is reduced to 0.1 kg/cm², the equivalent speed in the air would be 570 km/hr., with an economy in power of 90%; to attain such a speed an ordinary tunnel of 3 m. would thus require

$$300 \times \left[\frac{5.1}{1} \right] \times 10 = 60{,}000 \text{ h.p.}$$

Lastly, we would say a few words on the realization of this wind tunnel.

There should be a closed circuit flue with continuous wall formed of thick sheet metal and protected from over-heating. Two doors sliding perpendicularly to the axis of the flue will isolate the working section while the model is being handled. The measuring devices will be placed in an airtight cabin on the wall of the working section, so that the rods of the model supports can traverse the wall of the flue by joints which should not be airtight. The measurements will be registered automatically by apparatus installed either in the cabin and visible from outside, or actually installed outside the cabin. In the latter case the apparatus will consist of manometers connected with dynamometric capsules placed in the cabin. The propeller-fan will have adjustable blades so that it can be adapted to the density of the fluid used in the tunnel.

[. . .]

Document 2-18(g), Max M. Munk, "On a New Type of Wind Tunnel," NACA Technical Note No. 60, 1921.

Introduction.

The difficulties involved in conducting tests on airplanes and airships in actual flight, difficulties greater in the early years of aviation than now, and the matter of expense also, induced investigators to seek information through tests upon models. The first of such tests was made by moving the model through stationary air either by means of a whirling arm or in a straight line. Later the method adopted was to suspend the model in a current of air flowing in a large tube. <u>Wind tunnels</u> of this type have become of increasingly great importance. At first the tunnels were only small pieces of physical apparatus in a laboratory, but at last they require an entire building. The latest wind tunnel of the Zeppelin Company in Germany provides a current of air ten feet in diameter, which has a velocity of 110 mi/hr. and absorbs 500 H.P.

The results obtained with this type of wind tunnel are of very great value and at the present time they are the chief source of information for the aircraft designer. However, there are certain critics who declare that the results of wind tunnel tests are valueless for purposes of design. Indeed, justification for such opinions is not wholly lacking. There is, in fact, no necessary and exact connection between the motion of air around a small airplane model and that around the full-sized airplane. Sometimes the results of the tests on models agree well with those observed with the airplane itself, but important cases are known where the two do not agree. Further, there are questions the answers to which it is most important for the designer to have, and yet the answer deduced from tests of models in wind tunnels would be absolutely wrong. There is always an uncertainty connected with such tests, because one is never quite sure whether or not the results thus obtained may be applied to full-sized bodies.

In spite of this uncertainty, wind tunnels have been of the greatest use in the development of aeronautics. Tests upon models led to the construction of streamlined bodies having small resistance, and of aerofoils of good section. Experiments in wind tunnels led to the discovery of the theorems referring to the lift of aerofoils and to the effect of combining several aerofoils. A wind tunnel is still the most important means available for scientific tests. It can not be denied, however, that it is becoming more and more difficult to find a problem suitable for study by a wind tunnel, which can be immediately applied in aeronautics. Many tests of a theoretical character can be suggested, but it is difficult to interpret them. There are many important and urgent tests with respect to the design of aircraft which should be performed, but the results would be worthless if they were carried out in a wind tunnel of the present type. The theory of non-viscous motion is almost complete, the tests referring to it have been made, and the field of investigation lying between non-viscous motion and actual motion in the air is cultivated so intensely that it is difficult to find a new problem.

For all these reasons, the author believes that his proposition to make use of compressed air in a new type of wind tunnel comes at the right moment. Tests in such a tunnel will give information concerning those questions which could not be investigated with the present tunnels because of the exaggerated effect of viscosity. The new type of tunnel is free of the uncertainty characteristic of the older type, and will indicate clearly what problems may be undertaken with the latter. It will make unnecessary many full-flight tests, and will mark a step in advance in aeronautics.

Let us then consider this new type of wind tunnel; its advantages, the difficulties attendant upon its use, and the special methods required.

I. PRINCIPLE OF THE PROPOSED WIND TUNNEL.

The main difference between the new type of wind tunnel and the ones now in operation is the use of a different fluid. The idea is to diminish the effect of viscosity. It would not be surprising if any other fluid were better than air in this respect. However, there does not seem to be such a fluid. Water, the liquid most easily obtained, has, indeed, a comparatively small viscosity; that is, the ratio of its viscosity to its density is only the 13th part of the similar ratio for air. The density of water, however, is so great that it is hardly possible to afford the horsepower required to force water through a large tunnel. But, even supposing that such a current of water could be obtained, e.g. by using a natural waterfall, it would be quite impossible to make tests in it. A model could not be made sufficiently strong to withstand the enormous forces acting on it, nor would it be possible to hold the model stationary. The same difficulty would be met in using any other liquid. As for gases other than air, carbonic acid is the only one which has a ratio of viscosity to density less than that of air, but the difference is so small that it would not pay to use it. It is less expensive to build a larger wind tunnel than to construct one for using carbonic acid gas, which has to be sealed and requires gasometers and other contrivances for holding the gas; and, further, the difficulties of operation would all be increased.

The fact that there is still another way of changing the fluid did not occur to any one for many years. Air may be used, but, if it is compressed, it becomes a fluid with new properties,—a fluid which is the best suited for reliable and exact tests on models. When air is compressed, its density increases but its viscosity does not. The increased pressure, it is true, requires strong walls for the tunnel to withstand the pressure and to prevent the air from expanding, but the increase of effectiveness secured for the tests is so great that it will pay to make the necessary changes and to replace the light walls of existing tunnels by heavy steel ones.

Before discussing this point we must first convince ourselves that the increase of pressure greatly increases the range and value of wind tunnel tests.

II. THE REYNOLDS NUMBER.

We are inclined naturally to compare small objects with large ones, with the assumption that all the qualities are independent of the size of the object, and that therefore the effects will be correspondingly smaller or larger. Coming at once to our problem, we are disposed to think that useful information for the designer of a flying machine may be obtained by observing the shapes of a butterfly or of various insects. In fact, this is the idea underlying tests on models. The absolute size of bodies is, it must be noted, a concept devoid of exact meaning. There is no <u>absolute</u> length; the length of any object can only be compared with that of another. Imagine all scales to have been destroyed, and let us not be conscious of the dimensions of our own bodies. Then we would not be able to decide whether our physical world should be called a dwarf one or a giant one—we would have no basis of comparison. We may therefore reasonably expect that a world on a different scale than ours would not differ essentially from ours if the same physical laws are valid in both.

This does not mean that all numerical ratios would be the same in both. It is not necessary that the same physical laws produce the same motion of a fluid, i.e. a geometrically similar motion, around two similar bodies. For the streamlines of a fluid around an immersed solid are not related to its shape by geometrical relations but by those derived from the laws of mechanics. It is possible, however, to derive the condition for obtaining such similar motions by extending our general considerations, without using mathematical processes.

We picture two phenomena, independent of each other; in particular we presuppose that no scale is carried from the seat of one phenomenon to that of the other. We consider separately two geometrically similar solids, each immersed in its own fluid, and endeavor, under these conditions, to see if we can detect any difference between them. If we can not, it would be absurd to expect two different motions, for one of the absolute truths, of which everyone is convinced, is that equal causes have equal effects. Further, where we can not find a difference, we believe, there is equality.

The two solids being supposed to be geometrically similar, no difference can be found between them, since we do not have a scale. By selecting any particular length of the body, its dimensions can provide us only with a standard length for the investigation of the relation between the body and the space qualities of the fluid.

For the same reason we can not detect any difference between the densities of the two fluids. Instead of considering density as the second standard unit—length being the first—we will obtain a more useful one and one to which we are more accustomed if we combine the concepts of volume and of density, and consider, for instance, the mass of a cube of unit volume filled by the fluid as our standard unit of mass.

The velocity of the fluid relative to the immersed body and at a great distance from it may be considered as a third standard unit.

It is essential to realize that it is not possible to find any relation between these three quantities. Neither do any two of them mean the same physical thing, nor can any two of them be combined in such a way that the third appears. If, therefore, the qualities mentioned were sufficient to determine all the features of the phenomenon, the flow around similar bodies would always be similar also; we would not be able to detect any difference. This is the actual case if the fluid is non-viscous, and therefore motions around similar bodies immersed in perfect fluids are similar.

The viscosity of a fluid is characterized as follows: consider a unit cube of the fluid, so chosen that in any plane parallel to one of its faces the fluid has a constant velocity; let the velocity of the fluid increase uniformly as one passes from this face across to the opposite one; then, if this change in velocity equals the unit of velocity, the force of friction on the face of the cube is called the coefficient of viscosity of the fluid. This appears to be a complicated concept, so we shall try to combine it with the two standard units of length and of mass, so that we obtain a velocity characteristic of the viscosity of the fluid, in combination with the other two qualities. Let us imagine now a unit cube of the fluid and any difference of velocity on the two opposite sides. There is a force of friction on each such face. If this force were to act on a unit cube of the fluid, i.e., on a unit mass, it would produce an acceleration, and in the course of being moved through a unit distance this cube would have its velocity increased from 0 to a definite value.

We may imagine the conditions of velocity on the two opposite faces of the unit cube varied until the force of friction is such that the resulting velocity of the second cube equals the difference in velocity at the two faces of the first cube. Half this velocity may be called the "Reynolds velocity." It is characteristic of the viscosity of a fluid whose density is known, the dimensions of a solid body immersed in it being known, so as to furnish a unit of length. It can be determined for one of the two phenomena considered without reference to the other.

Therefore the ratio of the velocity of a fluid to this Reynolds velocity can be determined without reference to another phenomenon; it is an absolute number, called the Reynolds number. It may be the same in the case of two phenomena, or it may be different. If it is not the same, here is an essential difference between the phenomena, which may be observed and stated; and it would be most remarkable if, in spite of this difference, the fluids should have the same motions; it would in fact be impossible. But if, on the other hand, the two numbers are equal, we describe the motions of the two fluids as identical, taking viscosity into account, too. We may seek other differences; if there are none, it would be absurd to expect different motions.

Before extending our general considerations, we shall express the Reynolds number in terms of the quantities ordinarily used. Let r be the density of the fluid; m be the coefficient of viscosity; B be the characteristic length of the immersed solid. The mass of a cube of the fluid of length B on each edge is B3r; the force of friction on a face of area B2 is mBV1, when V1 is the difference of velocity at the two opposite faces; the work performed by this force if acting through a distance B is mB2V1, which equals the kinetic energy gained by the (second) cube of mass rB3 —i.e. ½ ρB³V₁². Hence, if $V_1 = V_2$

$$\mu B V_1 = \frac{1}{2} \rho B^3 V_1^2 \quad \text{or} \quad V_1 = \frac{2\mu}{\rho B}$$

the Reynolds velocity is one-half of this, i.e.

$$V_K = \frac{\mu}{\rho B}.$$

Writing V for the velocity of the fluid at a great distance from the solid, we have, by definition, the Reynolds number

$$\frac{V}{V_K} = \frac{V B}{\mu} \rho$$

If this has the same value in two phenomena of flow, they are alike in all respects. This may be called the Reynolds Law.

III. DEDUCTIONS FROM THE REYNOLDS LAW.

In the preceding section an attempt has been made to derive the expression for the Reynolds Law in as elementary a manner as possible. Only by knowing the basis of the law can one grasp its complete meaning and obtain the absolute confidence in it which is required for one to apply it safely. A mathematical proof was not given although it would have been shorter, for it would at the same time have been poorer of content.

We considered only the viscosity of the air, and did not discuss the other differences which exist between the tests on models and those on full-sized objects. The next stop is to investigate whether these differences do not introduce such errors that it would not be worthwhile simply to get rid of a possible error due to viscosity. Before doing this we must consider the deductions from the Reynolds Law so far as wind tunnel tests are concerned.

Let the span of the wing of a model be 3 ft., and the air velocity be 60 mi/hr. (= 88 ft./sec.). The kinematical viscosity of air at 0°C and normal pressure is 0.001433, i.e., about $1 \text{ ft.}^2/700 \text{ sec.}$. Hence the Reynolds number, regarding the span as the characteristic length is

$$\frac{3\text{ft} \times 88\text{ft. s.}^{-1}}{\text{1ft.}^2} = 185,000$$
$$\overline{700\text{s.}}$$

That is, the velocity of the air in the tunnel would be almost <u>two hundred thousand times</u> the velocity called the Reynolds velocity. The full-sized airplane may have a span ten times as great, and the velocity of flight may be 1 ½ times as great; so that its Reynolds number is

$$10 \times 1.5 \times 185{,}000 = 2{,}775{,}000.$$

The magnitude of these numbers is surprising. The viscosity of the air is so small that in the neighborhood of the wings of an airplane the velocities produced by the forces of friction are only about <u>three millionth</u> of the velocity of flight. Equation (1) shows that the kinetic energy is proportional to the square of the velocity, while the work performed by the frictional force is proportional to the velocity. Hence the work performed by the frictional force is a minute fraction of the kinetic energy, $10/185{,}000$ in the model test referred to and $10/2{,}775{,}000$ in the case of the airplane. It seems surprising that any effect of friction can be detected, since it increases or decreases the kinetic energy by such a small fraction.

However, in the calculation of the Reynolds number one quantity is chosen arbitrarily. An arbitrary length occurs in the formula, and the magnitude of the number depends upon the choice of this length. Indeed, within a range of a dimension like the span of wings, the viscosity has almost no influence, but the smaller the range considered, the greater is the effect of viscosity, provided there are in this range the same differences of velocities as in the other. It must be noted that great differences of velocity occur within very small ranges. Near the surface of the wing velocities almost zero occur close to velocities of the magnitude of the velocity of flight. The character of the motion depends upon the stability of flow near the surfaces, and therefore, upon phenomena within small ranges. Within these the Reynolds number and the ratio of the acceleration to the viscosity is less than the number commonly used for comparison.

In any case, tests show that there are considerable differences of motion at the Reynolds numbers of the test and the flight. There is even instability, changing the character of the motion near the largest airships, on increasing its velocity, when flying at normal velocities.

These facts are not contradictions of the Reynolds Law, but, on the contrary, are in agreement with it. The surprising fact that, even when the Reynolds number is large, its influence is considerable, does not furnish the least reason-for doubting the correctness of a law based upon such elementary considerations.

Doubts about the Reynolds Law are based upon a different fact. In spite of the convincing proof, it happens that model tests at the same Reynolds number sometimes give quite different results. Now the Reynolds Law does not mean that at the same Reynolds number only one particular motion of the air is possible. It states that there is no difference between two phenomena with the same number. It may be that two or more motions are possible, but then they are possible in any case of the same Reynolds number.

There must be some reason, however, why the one or the other motion occurs. The reasons may be different. Sometimes there is a kind of hysteresis, the fluid remembered, as it were, what happened before this particular motion began; and the motion is different, for instance, if the angle of attack was larger or smaller immediately before. If such a phenomenon occurs with the full-sized body, it can be investigated by a model test at the same Reynolds number. Sometimes there is no such hysteresis, but the motion is very sensitive and is changed by the least change of the shape of the body, or of the smoothness of its surface, or with a change of the turbulence of the air. In such cases the motion around the full-sized body will be sensitive at the same Reynolds number as in the model tests. In this case it will be difficult to obtain the exact shape of the model and the right smoothness of its surface in order to have the same motion. At the same time other differences between the model test and the actual flight will produce differences in the results; but in such cases it is very doubtful whether two airplanes which are apparently identical have the same qualities. There does not exist a definite motion around the body at that particular Reynolds number. The careful investigator will observe this fact. Then the model test has shown all there is to be shown, and the method is not to be blamed for revealing phenomena which are surprising to the designer but true nevertheless.

IV. ERRORS DUE TO OTHER CAUSES.

There are still other differences between the tests on models and in actual flight, which will cause errors. It is necessary to realize that these, other than the one due to viscosity, do not affect seriously the value of the results of the tests. The new type of wind tunnel may, then, be expected to give reliable results.

The best evidence of the insignificance of these errors due to other causes is obtained by comparing tests made in different wind tunnels. It may be stated that there is found a certain agreement, but only with the same value of the Reynolds number. Reynolds himself deduced the law called by his name from experiments upon water flowing through pipes. In the two wind tunnels at Göttingen very careful investigations were made on aerofoils, over a large range of Reynolds numbers, and under very different conditions. Most results at the same Reynolds number agree well; even the results which can not be plotted on a curve against the Reynolds number appear much more regular when so plotted than when plotted in any other way. The results of these tests show that full-sized tests are much better than model ones, and provide the designer with clear, reliable and useful information. It is not sufficient, however, to compare the results of several tests in a perfunctory manner; care must be taken.

The Göttingen tests were not made under conditions geometrically similar; the two tunnels are not equally good. There are many tunnels which have more turbulence than is necessary, the designer having only taken care to obtain a uniform velocity. The older wind tunnel at Göttingen was exceedingly turbulent. The surfaces of the models were different purposely. Only the results obtained in good wind tunnels should be compared, the model having a proper surface, and the test being thoroughly laid out with reference to its influence. Then the differences would be smaller, and the reliability and usefulness of tests at the full-sized Reynolds number would appear more distinctly.

The matter may be considered also from another point of view. The tests show that under particular conditions the results of different tests agree very well; in certain cases only is good agreement lacking. Now it is not evident that the results may be expected to agree. It may be and is very probable that the motions which do not agree with each other are such sensitive motions as were described in the previous section. Of course this sensitiveness appears exaggerated if the differences in the test conditions are.

Theoretical reasons are not wanting, however, as to why the character of the motion depends almost exclusively on the ratio of the velocity of air to the Reynolds velocity, and not upon other ratios, e.g. the ratio of the Reynolds velocity to the velocity of sound in the medium, the latter being characteristic of its compressibility. It is not at all sufficient to state that this ratio is small, the Reynolds number (or its inverse) being small too. But the ratio of velocity to the velocity of sound has only one meaning; there is no arbitrary quantity used in forming it—such as B in the Reynolds number. It does not matter whether this ratio is calculated for a wide range or for a small one. There is no discontinuity if the range or the compressibility passes to zero. In this case the fluid acts, with respect to its compressibility, like a perfect fluid. If the ratio of the velocity of flight to the velocity of sound is small, there is no physical reason for expecting a large influence. So much the less is the influence of a difference of compressibility in the tests on the model and in flight. Stated mathematically, any coefficient is a function of the two ratios; but, when both are small, the function is continuous with respect to the one and irregular with respect to the Reynolds number.

The same deduction is valid for the other errors; whether the cause be the contrivance used for supporting the model, the turbulence of the air, the variation of pressure or of velocity, or the finite distance of the walls of the tunnel or the boundaries of the current of air, the error is small provided the cause is. Their influence can be made as small as is necessary and customary in any technical test. Not only is the error small, it is regular, it can be compensated for, and it does not impair the comparison of different tests, as would the error due to viscosity.

V. THE DIMENSIONS OF A COMPRESSED AIR WIND TUNNEL.

In a tunnel filled with compressed air it is possible to obtain a Reynolds number much larger than in the tunnels now in use. But the range is limited in several respects, and its features must harmonize with each other in order to secure good results and also a low cost of operation.

The size of the tunnel is limited by the size of the models. It is not possible to make correctly shaped models if they are too small. The velocity of flow, on the other hand, must not be too great, lest the contrivances for supporting the model become so large that they disturb the motion. The stresses in the model must also be considered. This condition is duly respected if the dynamical pressure of the air does not exceed a particular value.

Hence the velocity must be the smaller the greater the density. This is desirable also with respect to the power required, to the increase of temperature produced, and to the dimensions of the fan and its shaft. The designer must also consider the time required to fill the tunnel with a compressor of proper dimensions. The pressure is limited only by questions of construction.

Let D be the diameter of the section where the model is placed, V be the velocity of the air and P be the maximum pressure. Then

Reynolds number	$R \alpha DVP$
Power required	$P \alpha D^2 V^3 P$
Heat produced per unit of surface	$\alpha V^3 P$
Dynamical pressure	$q \alpha V^2 P$
Weight of tunnel walls	$\alpha D^3 P$

Energy required to fill tunnel	$\alpha D^3 P^{1.25}$
Shaft diameter/diameter of tunnel	$\alpha V P^{1/3}$

(velocity of circumference of fan constant)

The designer, in the first place, must choose the dynamical pressure he can permit without the supports of the model introducing too great an error. Then he may calculate the pressure needed for the Reynolds number desired, and the smallest diameter he considers proper. If he selects too high a pressure, the diameter must be made greater. Generally this will increase both the cost of operation and other difficulties. The Reynolds number and the dynamical pressure being given, the diameter and the velocity may be expressed as functions of the pressure.

If $R=aDVP$ and $q=bV^2P$ then $V=AP^{-1/2}$ [and] $D=BP^{-1/2}$ where a, b, A and B are constant coefficients. Substitutions may then be made in the expressions for the different quantities. It appears:

Power absorbed	$\alpha\ P^{-3/2}$
Heat produced per unit of surface	$\alpha\ P^{-1/2}$
Weight of tunnel walls	$\alpha\ P^{-1/2}$
Energy required to fill tunnel	$\alpha\ P^{-1/4}$
Shaft diameter/diameter of tunnel	$\alpha\ P^{-1/6}$

That is to say, all the quantities mentioned are more favorable the higher the pressure. This advantage must be compared with the difficulty of construction in consequence of high pressure, and the disadvantage of a smaller diameter. A theoretical limit for the pressure is the critical point where the air ceases to be a "perfect gas." In the neighborhood of this point the viscosity increases and therefore it is of no advantage to increase the pressure; but reason of construction would prevent this point being reached. The critical point of carbonic acid gas is, however, much lower, especially if it is cooled.

We can not close this chapter without considering the most interesting question, whether it would be possible to build a wind tunnel for tests of models of airships, having a Reynolds number equal to flight conditions. Let the length of the actual ship be 655 ft., and its velocity be 95 mi/hr. In a tunnel designed for tests of ship models only, the dynamical pressure could be increased to 2000 lbs./ft.² The pressure could be 100 atmospheres (200,000 lb./ft.²). Then the velocity would have to be just 95 mi/hr., and happens to be "full sized." The scale would be 1:100; the diameter could be 2 ft., and the power about 1000 HP. We think this tunnel could be made. It would give the designer information long desired.

The results of tests in a compressed air wind tunnel would be applied in the same way as is the practice with existing tunnels. The tunnel would give the ordinary coefficients, and the right ones. The Reynolds number could be calculated from the observed temperature and pressure.

The results would be, first of all, for the information of the designer of aircraft, giving him the true values of the coefficient required for any problem. The tunnel could also be used with advantage for scientific investigations. The differences in the Reynolds numbers which could be realized in such a tunnel are much greater than can be obtained in existing tunnels. At the same time, the pressures and the forces on the model vary only as the Reynolds number, if the same model is used, whereas in existing tunnels they vary as the square of this number.

Document 2-18(h), F. H. Norton, memorandum to G. W. Lewis, 1921.

October 6, 1921

From: Langley Memorial Aeronautical Laboratory
To: National Advisory Committee for Aeronautics
(Attention Executive Officer)
Subject: Design of compressed air wind tunnel

1. There are at present three Draftsmen spending almost all of their time on this work as it takes most of Mr. Morgan's time to supervise Mr. Pratt and Mr. McAvoy. Mr. Morgan feels that they are working very inefficiently and are getting very few results, as Dr. Munk does not seem to have any clear idea as to what he wishes in the engineering design excepting that he is sure that he does not want anything that Mr. Griffith or myself suggest. At the end of this week I must take Mr. Morgan and Mr. Pratt entirely away from this work and let Mr. McAvoy struggle along as best he can. For this reason it is requested that Mr. McAvoy be ordered to Washington so that he can be directly under Dr. Munk's supervision. I am getting so disgusted with the way the whole thing is being carried out that I would like to keep as much of it in Washington as possible.

[signed] F. H. Norton
F. H. Norton
Chief Physicist.

Document 2-18(i), "Compressed Air Wind Tunnel," NACA Annual Report for 1921.

THE LANGLEY MEMORIAL AERONAUTICAL LABORATORY.

In previous annual reports the committee described the progress made in the development of its field station at Langley Field, Va., for the prosecution of scientific research in aeronautics. The station now comprises three principal units, namely, an aerodynamical laboratory or wind tunnel, an engine dynamometer laboratory, and a research laboratory building, the latter including administrative and drafting offices, machine and woodworking shops, and photographic and instrument laboratories. The research laboratory and the wind tunnel building are of permanent brick construction; the engine dynamometer laboratory is housed in a temporary four-section steel airplane hangar.

The committee has recently completed the construction of a factory type building of brick and steel, designed to house the new compressed-air wind tunnel. It is expected that the new wind tunnel will be in operation about July, 1922.

The Langley Memorial Aeronautical Laboratory occupies a plot of ground known as plot 16, Langley Field, Va., the plot having been set aside for the committee's use by the Chief Signal Officer of the Army in 1916, at the time the site was selected as a proposed joint experimental station and proving ground for the Army and Navy air services and the advisory committee. The use of that plot of ground was officially approved by the Acting Secretary of War on April 24, 1919. The four buildings at present constituting the Langley Memorial Aeronautical Laboratory have been erected by the committee pursuant to authority granted by Congress.

COMPRESSED AIR WIND TUNNEL.

On June 9, 1921, the executive committee of the National Advisory Committee for Aeronautics authorized the construction at the Langley Memorial Aeronautical Laboratory of a compressed air type of wind tunnel designed by Dr. Max Munk, technical assistant of the committee.

The utility of the present type of wind tunnel is limited by the fact that owing to a "scale effect" the results of tests on the small models, which are usually about $\frac{1}{20}$ scale, are not immediately applicable to the full-size machine. Obviously it is very desirable to obtain, if possible, test results which are strictly proportional to those obtained in free flight. This condition may be realized by the use of a wind tunnel in which the air is compressed to about 20 atmospheres or more in order to compensate for the difference in the "scale" or Reynolds number for the model and for the full-size airplane.

The wind tunnel under construction has a diameter of 5 feet, the wind tunnel proper being placed within a steel cylinder 15 feet in diameter and 34 feet long. The steel cylinder has been tested for an internal pressure of 450 pounds per square inch and is designed for an average working pressure of 300 pounds per square inch.

The design of the cylinder further provides for a large door at one end and means for observing and operating the balance and setting wing angles from without the cylinder. The design of the balance has been carefully considered and due provision is made for the large forces to be measured.

The wind tunnel motor is 300 horsepower and the Reynolds number will be controlled by changing the air density rather than by changing the air speed. The air compressing units consist of two 300-horsepower compound compressors which compress the air to 115 pounds per square inch. The air is compressed into a receiving chamber and is then compressed by a 175-horsepower duplex booster compressor to the desired pressure in the test chamber. With the compressor units selected it will require approximately one hour to fill the chamber with air at a pressure at 300 pounds per square inch and every provision is being made in the design to make it unnecessary to open the chamber until the model is completely tested. Provision is also being made to maintain constant density so as to take care of temperature variations.

This tunnel when in operation will test models with a span of about 2 feet, but the results will be strictly comparable to similar data for a full scale machine, with a span of 30 feet, flying at 100 miles per hour. The construction of the models will therefore require special study and care.

Document 2-18(j), Max M. Munk and Elton W. Miller, The Variable Density Wind Tunnel of the National Advisory Committee for Aeronautics, NACA Technical Report No. 227, 1925.

SUMMARY

This report contains a discussion of the novel features of this tunnel and a general description thereof.

PART I
FUNDAMENTAL PRINCIPLES
By Max M. Munk

All the novel features of the new variable density wind tunnel of the National Advisory Committee for Aeronautics were adopted in order to eliminate the scale effect. The leading feature adopted was the use, as the working fluid, of highly compressed air rather than air under normal conditions.

It is not at once obvious that the substitution of compressed air eliminates the scale effect with aerodynamic model tests, although the necessary theoretical discussion has been available for some years. The idea of using compressed air must have occurred, in all probability, to many. It was not, however, till early in 1920 that the thought came to the writer; and in what follows is given his own line of reasoning, expressed in as simple language as possible.

In a paper entitled "Similarity of Motion in Relation to the Surface Friction of Fluids," by T. E. Stanton and J. R. Pannell, Philosophical Transactions A, volume 214, pages 199–224, 1914, will be found an excellent treatment of the subject, with references to the earlier discussions by Newton, Helmholtz, and Rayleigh.

Proceeding at once to the motion of a rigid body immersed in a fluid, the aim of the investigation is to obtain information concerning the fluid forces on such a body. Everything in connection with the problem has to be studied to that end, and has to be included in the investigation, whether this latter be analytical or, as we suppose now, experimental. There are the properties of the immersed body, its shape, its direction of motion, eventually the character of its surface. Even more important is the action of the fluid brought into play by these properties. Every detail of the motion of the fluid, together with the physical properties of the fluid, is immediately connected with the kind and magnitude of the forces created. We can only attain to a full knowledge of the forces created by regarding their cause, the fluid motion. All velocity components at all points of the flow are important and characteristic details of the cause of the forces on the body immersed in the fluid.

Then, why do investigators think that they can learn about what will occur on a large scale by observing what occurs on a small scale? Not from any intuitive feeling, inexpressible in words because devoid of thought; not from any vague metaphysical argument difficult to explain. There is a definite, extremely sound and simple reason why we expect to obtain reliable information from model tests. It is because we expect the two cases when compared with each other will perfectly, at all points, conform to each other, point by point. We do not mentally confine the geometrical similarity to the bodies immersed and to the dimensions of the entire arrangement, leaving as an unsolved and uninteresting question what the fluid does in the two cases. We do not expect that, for some mysterious reason, the fluid forces will correspond to each other in accordance with some simple rule. On the contrary, we include the flow patterns in our conception of "model." Any two corresponding portions of the flow, however small, are supposed to be similar with respect to shape and direction of the streamlines and with respect to the magnitude of velocities. The ratio of the lengths of a pair of corresponding portions of a streamline is supposed to be constant throughout the flow, and so is the ratio of two velocities corresponding to each other. We are under the impression that with respect to every detail the entire small-scale experiment is an exact replica of what occurs on a large scale, and we believe that the smallest quantity, whatever it is, occurs in a numerically corresponding way with the same conversion factor throughout the entire flow. In such a case, and only then, are we entitled to expect a simple relation between the fluid forces of the model test and those on the large-scale experiment. Such forces are the integrals of the elementary forces, and hence they stand in a constant ratio if the elementary forces do. This constant ratio can furthermore be expected to be a simple algebraic expression of the ratios between the characteristic quantities of the two arrangements.

Not only the model but the entire flow is the replica. There is a good illustration. It sometimes occurs in aerodynamics that the same body moved in the same way in the same fluid gives rise to different configurations of flow. The air forces are then also different.

The question, "Can we learn from aerodynamic model tests?" is thus reduced to the equivalent question, "Can flow patterns be geometrically similar?" If so the boundaries of the flow in general, and the immersed bodies in particular, have to be similar, but this alone is no sufficient reason why the similarity should extend to every streamline. The question whether a test is really a model test in the strict meaning, the question whether the small-scale flow is similar to the large-scale flow, requires a special examination. This examination will decide whether we can obtain reliable information from the test. If the flows are not exactly similar, but only approximately, the information also will only be approximately correct and not wholly reliable. There will exist a "scale effect."

Two configurations of aerodynamic flow are created in different fluids under conditions geometrically similar. We wish to know whether the flow patterns are geometrically similar. We imagine a small-scale flow to exist exactly similar to the large-scale flow really existing, and we ask whether this imagined small-scale flow is compatible with the general laws of mechanics and hence identical with the actual small-scale flow. More particularly, we examine whether each particle of the imagined small-scale flow is in equilibrium, remembering that the corresponding particle of the large-scale flow is.

We assume first that no physical properties of the fluids, nor differences of such properties, have any influence on the shape of the flow pattern or on the fluid forces, except the density of the fluids. We dismiss also any external influence, like that of gravity. Then the only type of force brought into action by the motion of the fluid is the mass force of all the particles, and they are equalized by means of a variable pressure. The pressure distribution is only the natural reaction against changes of mutual positions of all the fluid particles, which changes must be compatible with the continuity conditions of the fluid. Each particle has the natural tendency to move straight ahead with constant velocity. This tendency is in conflict with the other tendency of each fluid particle to claim its own space, not to share its space with any other particle. These two conflicting tendencies lead to a distribution of varying pressure and to mass forces on the particles due to their motion along curved paths and with varying velocities. The pressure distribution gives rise to an elementary force on each particle, and the flow arranges itself in such a configuration that this pressure force is in equilibrium with the mass force.

Let us consider now the case when the linear dimensions are diminished in the ratio l_2/l_1, all velocities diminished in the ratio V_2/V_1, and the density ρ_2, bears the ratio ρ_2/ρ_1 to the original density.

The mass forces are expressed mathematically by a type of term occurring in Euler's or Bernouilli's equation. Per unit volume, they are of the type

$$\frac{\text{Density} \times \text{Velocity}^2}{\text{Length}}$$

and hence resultant mass forces of corresponding portions of the flow are of the type

(1) Density × Length² × Velocity³

Such forces are in equilibrium with the pressure forces, and this determines the latter. Hence a change of density, scale, and velocity gives rise to a change of all elementary forces and hence of all resultant forces in the ratio

$$\frac{\rho_2 \, l_2^2 \, V_2^2}{\rho_1 \, l_1^2 \, V_1^2}.$$

The equilibrium of the particles remains unimpaired by the change of scale, and we conclude that corresponding flow patterns are necessarily similar. Hence, if the density of the fluid were the only property influencing the fluid paths and hence the fluid forces, all aerodynamic model tests would be interpreted correctly by the application of the so-called "square law." Corresponding fluid forces would be proportional to the fluid density, to the square of the velocity, and to the square of the linear scale. Accordingly, the absolute coefficients generally in use for expressing the magnitude of fluid forces would not only be absolute, but also constant for similar shapes and arrangements.

Experience has shown that the "square law" does not strictly hold, but that the air-force coefficients vary, sometimes slightly and sometimes in a very pronounced way. This is due to the influence of other properties of fluid, neglected before. There arises the question which other property of air is the principal cause of variations of flow patterns under conditions otherwise geometrically similar. All men who have devoted much thought to this problem agree that viscosity has such an effect, greatly in excess of that of other properties. The point is that the forces taken care of by the introduction of such properties of the fluid are very small when compared with the mass forces, which latter alone are governed by the "square law." This holds true at all points of the flow and with respect to all fluid properties, except with viscosity, where it only holds at most points. Viscous forces are proportional to the rate of sliding of adjacent layers of fluid, and are expressed by terms of the type,

$$(2) \quad \mu \frac{\partial u}{\partial y} dxdz$$

Here the constant quantity m is called the modulus of viscosity. u, a velocity, is at right angles to y, a Cartesian coordinate, together with x and z. Hence $\delta u/\delta y$ has the physical dimension of an angular velocity, $1/\text{Time}$. Now, this rate of sliding is small throughout an aerodynamic flow except near the boundary. There it may assume a very large magnitude. So, in spite of the small value of the modulus of friction of air, μ, the friction $\mu \, \delta u/\delta y$ can assume a very large value and can become dominating at certain points of the flow. It can then produce essential changes of the entire flow pattern. Very little in detail is known about these things, and it seems useless to carry the discussion on at this point. Experience has shown that proper attention to the viscosity brings system and regularity into results of tests otherwise obscure and contradictory. It is for this reason that the elimination of the effect of viscosity for many years was thought desirable in the first place as a fundamental improvement of aerodynamic model tests, resulting in the elimination of the scale effect.

There has been some controversy as to whether these arguments are sufficient for the final decision that viscosity is the all-important fluid property. No arguments whatsoever will definitely decide that, but only final success. The separation of the physical effects to be taken into consideration for any practical purpose from those which may be neglected is a mental step which can not be accomplished by mere logics.

Granted, now, that viscosity is of practical importance, the question arises, Are similar flows possible in viscous fluids; and if so, under what conditions will the flows be similar? It is understood now that the arrangements are geometrically similar, that only the density r and viscosity m of the fluid have to be considered in addition to the linear scales of the arrangement and the ratio of the velocities.

The answer to the last question depends again upon the result of the examination whether each particle of an imagined small-scale flow, similar to an actual large-scale flow, is in equilibrium or not. Now, in viscous fluids the mass forces are not in equilibrium with the pressure forces, but in equilibrium with the combination of both the pressure forces and the viscosity forces. We have now three types of forces in equilibrium with each other, and that gives rise to a variety of possibilities. Two forces in equilibrium are, of necessity, numerically equal, hence if one of them be changed in a given ratio the other will too. With three forces, all three may be changed in a different ratio and still the equilibrium maintained.

The criterion for the similarity of flows is, therefore, that two of the three forces be changed in the same ratio. Then the third, in equilibrium with the two, will be changed in this same ratio and needs no special examination.

We compare the ratio of chance of the mass forces and of the viscosity forces with each other. We have seen already (1) that the mass forces are changed in the ratio $\frac{\mu_2 l_2^3}{\mu_1 l_1^3}$.

The viscous forces being of the type $\frac{\mu v l^2}{\mu^* l^*}$

are seen to be changed in the ratio $\frac{\mu_2 v_2 l_2}{\mu_1 v_1 l_1}$.

Now, the two flow patterns will be similar and the test will be a strict model test only if the mass forces and the viscosity forces are changed in the same ratio. Hence we obtain, as the condition of an exact model test,

$$\frac{\mu_2 l_2^3}{\mu_1 l_1^3} = \frac{\mu_2 v_2 l_2}{\mu_1 v_1 l_1}$$

or, written in a different way,

$$[3] \quad \frac{V_1 l_1 \rho_1}{\mu_1} = \frac{V_2 l_2 \rho_2}{\mu_2}.$$

The expressions on either side of equation (3) are generally called "Reynolds numbers," from Osborne Reynolds, who was the first to emphasize their importance. Since V and l are certain velocities and lengths in the two flows, corresponding to each other, but otherwise arbitrarily chosen as "characteristic" velocity or length, the value of one special Reynolds number in one single case has as little meaning as the scale of one single object. The equality of the Reynolds numbers of two arrangements, different but geometrically similar, expresses the dynamic equivalence of the two flows compared.

If the ratio of the two Reynolds numbers is different from unity the value of this ratio can be considered as a kind of relative scale between these two tests, not of the geometric scale but one which may be called dynamic scale. The ratio of the Reynolds numbers indicates differences in the relative importance of the mass forces and of the viscosity forces. A single Reynolds number, together with the definition of the characteristic velocity and length, is only an identification number, not much more than the street number of a house. Comparison of Reynolds numbers of flows where the conditions are not geometrically similar have hardly any meaning.

The preceding discussion has led us to the condition under which a wind tunnel will have no scale effect due to viscosity, and probably not any scale effect of practical importance. This condition is not equal velocity in model test and in flight. Full velocity is only of value for investigating certain original airplane parts and original flight instruments. The test with a model of diminished scale but at the velocity of flight is by no way distinguished from tests at other wind-tunnel velocities. On the other hand, if there is no scale effect expected, the Reynolds number being equal in both model test and free flight, the dynamic scale being 1, and if there are still arguments raised doubting the validity of such tests, such arguments hold with equal right or wrong against all other model tests, more particularly against such tests in ordinary atmospheric wind tunnels. For the principal difference between the variable density tunnel and atmospheric tunnels is the elimination of one source of error, of the one moreover, which is believed by most experts to be the most serious.

The fact is, then, that in general model tests in atmospheric wind tunnels are made at a Reynolds number smaller than in free flight. The linear dimensions of the model are largely diminished, and nothing is done to make up for this; the velocity is at best the same as in flight and the ratio μ/ρ is the same, the same fluid being used in test and in flight.

It is neither practical nor sound to make up for the diminution of the model by correspondingly increasing the velocity so as to obtain the original value of the product Vl as required in equation (3). It is not practical because such a wind tunnel would consume an excessively high horsepower, and because the air forces on the model would become excessive to such an extent as to make the test practically impossible. Such a method would also be unsound. For the differences in air pressure, which amount only to little more than 1 per cent in flight and in ordinary wind tunnels, would increase rapidly with velocities approaching the velocity of sound. Thereby the influence of the compressibility would be rapidly increased, and thus another error, now negligible, would make the results unsuitable for the desired purpose.

There remains then only the diminution of the ratio μ/ρ often denoted by v, in order to make up for the diminution of l in equation (3). This means the choice of another fluid. The use of water instead of air has been seriously proposed. With water $v = \mu/\rho$ is indeed seven times as small as with air. The problem of the large power consumption could eventually be solved, either by using a natural stream or by towing the model. However, water is about 800 times as dense as air, and hence the forces produced at the same velocity are 800 times as large, giving rise to stresses 800 times enlarged. It is practically impossible to make ordinary model tests with forces on the model 800 times as large as they are now.

What we need is a fluid which may be denser than atmospheric air at sea level, but only so to a moderate degree. Its dynamic viscosity modulus $v = \mu/\rho$ should be distinctly smaller than that of air, in order to make up for the scale of the model and eventually for the diminished velocity necessary for bringing down the pressure on the model and the absorbed horsepower. No such fluid is known under ordinary atmospheric conditions. Further consideration showed that a high pressure transforms air (or another gas) into a fluid suitable for wind-tunnel work giving results without scale effect. This fact depends on the physical property of air of keeping the same viscosity modulus m under all variations of pressure. This has been confirmed by experiments and is mentioned in treatises on physics. It is in keeping with the molecular theory, with denser air the average free paths are proportionally shorter. The viscosity modulus μ remains the same, but the density increases when the pressure increases. Hence the ratio $v = \mu/\rho$ varies inversely with the pressure (the temperature remaining unchanged). Hence we have

> Kinematic viscosity ~ Pressure^{-1}
> Model pressure ~ Pressure x Velocity^2
> Absorbed horsepower ~ Pressure x Velocity^3

Assuming a model scale of say 10, we want a kinematic viscosity at least 10 times as small as with air. With pressure of 20 atmospheres we could get

Test velocity = ½ flight velocity.

Resultant model pressure = $20(½)^2$, 5 times actual pressure.

Horsepower consumption of the tunnel = $20(½)^3$, 2.5 that of an atmospheric tunnel of the same size and operating at full scale velocity.

Reynolds number = Reynolds number in free flight. These figures seemed practical. On them the design of the variable density wind tunnel of the National Advisory Committee for Aeronautics has been based.

More generally it can be seen that the principle of compressing the air allows any Reynolds number, even with a small model, if only the pressure can be produced and maintained. For keeping the Reynolds number constant and increasing the pressure in the ratio A, decreases the resultant pressure on the model as A^{-1} and the required horsepower as A^{-2}.

The throat diameter of 5 feet was chosen in order to be able to use the same models as in the atmospheric wind tunnel of the National Advisory Committee for Aeronautics. A small diameter would require smaller models, and it becomes increasingly difficult to construct such models accurate enough.

Furthermore, 5 feet is the smallest diameter for a closed tunnel where a man can walk and work without exceeding discomfort. The choice of the smallest diameter suitable was necessary in view of the large costs and difficulties for procuring a large enough housing strong enough to withstand an internal pressure of 25 atmospheres.

The same restriction of space decided the choice of a closed (not free jet) type of tunnel.

All other novel features can be traced back to the particular features of this tunnel, the large inside pressure and the larger resultant force on the model. They are described in the second part of this paper.

PART II
DESCRIPTION OF TUNNEL
By Elton W. Miller.

In the pages which follow a description is given in some detail of the tunnel and the methods of operation. The purpose in preparing this report is to make clear the testing methods employed, in order that the technical reports now in preparation may be better understood. The building of this tunnel was first suggested by Dr. Max M. Munk in 1921 (Reference 1). The writer has assisted Doctor Munk and Mr. David L. Bacon in the design and development of the mechanical features of the tunnel.

The tunnel is shown in sectional elevation in Figure 1, and consists briefly in an experiment section, E, 5 feet (1.52 meters) in diameter, with entrance and exit cones housed within a steel tank 15 feet (4.57 meters) in diameter and 34 feet 6 inches (10.52 meters) long. The air is circulated by a two-blade propeller, returning from the propeller to the entrance cone through the annular space between the walls of the tank and an outer cone, C_o. The balance, which is of novel construction, is mounted in the dead, or noncirculating, air space between the walls of the experiment section and the outer cone. The balance is operated electrically, and readings are taken through peepholes in the shell of the tank. Figures 2 and 3 are general views of the tunnel. Figure 4 is a plan of the building showing the tunnel and compressors.

The tank, which was built by the Newport News Shipbuilding & Dry Dock Co., of Newport News, Va., is capable of withstanding a working pressure of 21 atmospheres. It is built of steel plates lapped and riveted according to the usual practice in steam boiler construction, although, because of the size of the tank and the high working pressure, the construction is unusually heavy. There is a cylindrical body portion of 2 ⅛-inch (53.98 millimeters) steel plate with hemispherical ends 1 ¼ inches (31.75 millimeters) in thickness. Entrance to the tank is gained by an elliptical door K 36 inches (914 millimeters) wide by 42 inches (1,066 millimeters) high. The tank, which with its contents weighs about 100 tons (90.7 metric tons), is supported by a foundation of reinforced concrete.

The walls of the experiment section and cones are of wood; those of the experiment section consist of a series of doors which may be unbolted and removed to gain access to the balance. The cross-sectional area at the large end of the exit cone is substantially twice that of the experiment section, and the cross-sectional area of the return passage at its largest part is about five times that of the experiment section. Two honeycombs, H_p and H_s, are provided for straightening the air flow. Honeycomb H_p is of 2-inch (50.8 millimeters) round cells, while honeycomb H_s is of 1 ¼-inch (31.75 millimeters) square cells. The latter honeycomb is made removable to permit access to the experiment section; it is suspended from a removable trolley track by which it may be rolled to one side of the entrance cone. In order that the honeycomb may be returned to exactly the same place each time, it is made to seat on three conical points where it may be securely locked. Arrangements have also been made for adjusting the position of the honeycomb, as shown in Figure 5.

The propeller is driven directly by a synchronous motor of 250 horsepower (253.5 metric horsepower), which runs at a speed of 900 revolutions per minute. The synchronous motor has an advantage over the usual direct-current motor in that no complicated devices are necessary for maintaining a constant speed of revolution. Such variations in dynamic pressure as are made in the ordinary atmospheric tunnel by changing the air velocity are here made by changing the density of the air. It is therefore not necessary to vary the air velocity. Fluctuations of a fraction of a per cent occur, due to variations in the frequency of the electric current supplied to the motor; otherwise the velocity is constant for a given tank pressure. There is a slight increase in air velocity with an increase in tank pressure, as shown in Figure 16, but this is not objectionable.

The propeller, which is 7 feet (2.14 meters) in diameter, is mounted on a ball-bearing shaft which passes through one end of the tank. The stuffing box through which this shaft passes is only loosely packed, and air leakage is reduced to a minimum by means of oil which is fed by gravity from a reservoir above. The oil which is carried through the stuffing box is returned to the reservoir by a motor-driven pump.

Air compressors for filling the tank with air are shown in Figure 4. The air is compressed in two or three stages, according to the terminal pressure in the tank. A two-stage primary compressor is used up to a terminal pressure of about seven atmospheres. For pressures above this a booster compressor is used in conjunction with the primary compressor. The booster compressor may be used also as an exhauster when it is desired to operate the tunnel at pressures below that of the atmosphere. The primary compressors are driven by 250-horsepower synchronous motors and the booster compressor by a 150-horsepower squirrel-cage induction motor.

A diagrammatic drawing of the balance is shown in Figure 6. It consists essentially in a structural aluminum ring (1) which encircles the experiment section, two lever balances (2) and (3) for measuring lift, and a third lever balance (4) for measuring drag. The ring as it looked before assembly in the tunnel is shown in Figure 7. An assembly view in the tunnel is seen in Figure 8. The doors which surround the experiment section have here been removed, exposing the balance to view. The model is attached to the ring, by wires or other means, and all forces are transmitted to the ring and thence to the lever balances. The ring is suspended from lever balances (2) and (3), Figure 6, by the vertical members (9), of which there are four, two on each side. Cross shafts and levers are employed in order to carry the full weight of the ring to the two lever balances. The drag forces are transmitted by horizontal members (10) to bell cranks and thence by vertical members (11) to lever balance (4). Hanging from the ring are bridges which carry coarse weights (5) and (6). Any desired number of coarse weights may be added or removed by means of motor-driven cam shafts. A similar bridge carrying coarse weights (7) is hung from lever balance (4).

The sliding weights are moved by motor-driven screws to which are geared revolution counters; these may be read through peepholes in the shell of the tank. At the end of each beam is a pair of electrical contact points by which the beam may be made to balance automatically. The sliding weights may also be controlled by a manually operated switch. The lift balances are sensitive to plus or minus 10 grams and the drag balance to plus or minus 1 gram.

It is possible with this balance to measure any three components; for instance, lift, drag, and pitching moments. The lift is first approximately counterbalanced by increasing or decreasing the number of coarse weights hanging from the two weight bridges. The remainder is then counterbalanced by moving the sliding weights on the two lever balances. The drag is measured similarly. The total lift is the sum of the readings of the two lift balances; the pitching moment is the algebraic sum of the three balance readings multiplied by their respective lever arms.

The model may be supported in the tunnel by wires only, or by a combination of wires or struts and a spindle. In the latter case the spindle is attached to a vertical bar (12) which may be raised or lowered by appropriate gearing thus changing the angle of attack of the model. The angle of attack is indicated by an electrically controlled dial on the outside of the tank. The vertical bar (12) is protected from the air flow by a fairing (13).

Round wires of about 0.040 inch (1 millimeter) diameter have been used for supporting models, this much larger diameter being necessary because of the large forces, but streamlined wires of much larger section have been found preferable. These wires are attached to the balance ring below and to the model above, thus serving as struts or free columns to support the weight of the model when the air stream is not on. The struts may be attached to the wheels of the model as shown in Figure 9 or to threaded plugs screwed into the wings as in Figures 10 and 11. The advantage of the streamline wires over the round wires is illustrated in Figure 12. The wire and spindle drag for two airfoils and one airplane model have been reduced to a per centage of the gross minimum drag of the model with wires and plotted against Reynolds number.

All the various operations required within the tunnel while running, such as the shifting of balance weights and the setting of the manometers, are performed by small electric motors. It has been necessary, therefore, to carry a large number of electric wires through the shell of the tank. These wires pass through a suitable packing gland and are attached to terminal boards inside and out. The outside terminal board may be seen in Figure 3.

The airspeed is measured by static plates, one of which is located in the wall of the experiment section and the other in the wall of the other cone. The static plates are calibrated against Pilot tubes placed in the experiment section. A micromanometer designed especially for use in this tunnel is shown in Figure 13. Alcohol is the liquid used, and a head up to 1 meter may be measured. This manometer is similar in principle to that described in National Advisory Committee for Aeronautics Technical Note No. 81, but is different in that the index tube is stationary and the reservoir is raised or lowered by a motor-driven screw. A revolution counter geared to the motor indicates the head to 0.1 millimeter. It is possible to determine the dynamic pressure to an accuracy of plus or minus 0.2 per cent.

The dynamic-pressure distribution in the experiment section is represented by contour lines in Figure 14. This survey was made by using a number of Pitot tubes mounted on a bar which could be revolved in the tunnel. Observations were thus made at a large number of points. The dynamic pressure will be seen to vary in the region occupied by the model within a range of plus or minus 2 per cent. This survey was made at one and two atmospheres only. We know from check runs that the same flow condition holds for other pressures. The horizontal static pressure gradient in the tunnel at various pressures is shown in Figure 15. Pressures are given with reference to a static plate located in the wall of the experiment section. It will be noted that the curves which are plotted on semilog paper are parallel, indicating that the pressure gradient is proportional to the density. Operating data of general interest, as the time required for raising pressure in the tank, the time required to exhaust the tank, the power consumption of the compressors and drive motor, are shown in Figure 16. The velocity change with change of tank pressure is also shown. The energy ratio of the tunnel for various tank pressures is shown in Figure 17.

The building of this tunnel and the development of its various mechanical devices to a point where routine testing may be done has required the solution of a number of mechanical problems. This development period has passed, and the results now being obtained in the tunnel are believed to be as consistent and reliable as those obtained in any other wind tunnel. Two airplane models and thirty-seven airfoils have so far been tested. Tests of a Sperry Messenger airplane model provided with eight different sets of wings are now in progress.

The variation of the aerodynamic characteristics of an airplane model with change of scale is shown in Figure 18. This figure gives the polar curves of the

Fokker D-7 airplane model tested at various tank pressures. The minimum drag and the lift/drag ratio for this model, and also for a Sperry Messenger model, are plotted against the Reynolds number in Figure 19.

CONCLUSIONS

The underlying theory of the variable density tunnel has been discussed, the mechanical construction of the tunnel has been described, and some typical results obtained on an airplane model have been given. The tunnel is in continuous operation, and there is every reason to believe that the results obtained at the higher densities are truly representative of full-scale conditions.

REFERENCES

Reference 1.—"On a New Type of Wind Tunnel," by Dr. Max M. Munk. N. A. C. A. Technical Note No. 60.

"Abriss der Lehre von der Flüssigkeits und Gasbewegung," by Dr. L. Prandtl. Handwörterbuch der Naturwissenschaften, vol. 4.

"Experimental Investigations," by O. Reynolds, Phil. Trans. 174 (1883), and "On the Dynamic Theory of Viscous Fluids," Phil. Trans. A. 186 (1894).

"Similarity of Motion in Relation to the Surface Friction of Fluids," by T. E. Stanton and J. R. Pannell, Phil. Trans. A, vol. 214, pp. 199-224, 1914.

Document 2-19(a–c)

(a) Ludwig Prandtl, letter to J. C. Hunsaker (translation), 30 March 1916, Hunsaker Collection, Box 4, File 4.

(b) Joseph S. Ames, NACA Headquarters, letter to J. C. Hunsaker, Navy Department, 15 October 1920, Hunsaker Collection, Box 1, File A.

(c) Ludwig Prandtl, *Applications of Modern Hydrodynamics to Aeronautics, Part 1*, NACA Technical Report No. 116 (Washington, DC, 1921).

It is difficult to overstate the degree of difference between the American and German approaches to understanding aerodynamics. While the American researchers opted for a practical, problem-solving approach, the Germans chose to pursue a rigorous, mathematically based understanding of the supporting theory. Ludwig Prandtl was the leader of the German school of thought, and he influenced many of the leading aerodynamicists of the twentieth century, including Max Munk and Theodore von Kármán. Most of Prandtl's publications were in German, but the NACA commissioned and published a seminal paper, *Applications of Modern Hydrodynamics to Aeronautics*, in 1921. Much of this extensive paper involved mathematical derivations, but portions are included herein to illustrate what Prandtl called "the leading ideas" and their experimental confirmation.

Not all Americans disdained the theoretical approach, however. Jerome Hunsaker and Joseph Ames, both of whom held graduate degrees in science or engineering, understood what a sound theoretical footing could add to the experimental work of the NACA, and both men were well acquainted with the German developments in this area. Hunsaker had known Ludwig Prandtl since his 1913 European inspection trip, and the latter's 30 March 1916 letter to Hunsaker, noting that "it has pleased me very much to see that you have agreed with my ideas regarding the boundary lamina of viscous fluids," indicates they shared a belief that the complex nature of fluid flow could be understood and predicted. Ames, too, realized the importance of the German theoretical work. Ames's enthusiasm for having obtained several copies of the "Technische Berichte," a collection of German wartime research reports, is readily apparent in his 15 October 1920 letter to Hunsaker, where he states, "The importance of the information . . . can not be over-estimated."

Such an attitude provides a stark contrast to the parochial views of Frederick Norton noted elsewhere in this chapter, and the NACA was fortunate to have the services of such strong and well-respected personalities to argue the merits of theoretical methods and press for the employment of Prandtl protégés. Through the immigration of Prandtl's students, especially von Kármán, theoretical aerodynamics came to the United States, and it gradually claimed a leading role in the saga of aeronautical progress.

Document 2-19(a), letter from Ludwig Prandtl to J. C. Hunsaker (translation), 1916.

[Göttingen, Germany, 30 March 1916]
<u>TRANSLATION</u>

Mr. J.C. Hunsaker
Instructor in Aeronautics, Massachusetts Institute of Technology
Boston, Mass.
Dear Sir:

Please accept my heartiest thanks for the gift of your survey of the Hydrodynamical Theory in its application to experimental aerodynamics as well as for the interesting collection of the results of the experiments of the aerodynamical laboratory at the Smithsonian Institute. I am forwarding under separate cover a small acknowledgement and hope that you will receive it as well as I received yours. (This may be translated, "hope that you will value it as highly as I valued yours").

With reference to the survey of the "Hydronamical Theory" it has pleased me very much to see that you have agreed with my ideas regarding the boundary lamina of viscous fluids. This is the first publication in the English language which has taken notice of it. The exceptional collection of original references has been a great pleasure as it has also been of great value to me. It contains some things with which I was not yet acquainted.

With greatest respect, I am, Yours very truly,
[Prof. Dr. L. Prandtl]

**Document 2-19(b), letter from Joseph S. Ames
to J. C. Hunsaker from NACA Headquarters, 1920.**

[October 15, 1920]

Commander J. C. Hunsaker, U.S.N.
Navy Department
Washington, D.C.
Dear Commander Hunsaker:

It is with pleasure that I am informing you that the National Advisory

Committee for Aeronautics has been successful in obtaining a number of sets of the "Technische Berichte" and we are mailing you under separate cover volumes No. 1, 2, and 3. The Committee is also forwarding a carefully prepared translation of the index of the first three volumes together with a list of symbols used.

The "Technische Berichte" consists of separate memoranda issued as confidential material during the course of the war by the Aeronautical Supply Department of the German Air Force.

These reports contain practically all of the official German aeronautical information resulting from research conducted during the period of the war. The research work carried on in German covers a wide range of subjects, including the study of propellers, the pressure distribution on control surfaces, resistance of struts, methods of calculating performances, analysis of the performance and design of airplane engines, and systematic tests of approximately 350 wing sections.

The importance of the information contained in the "Technische Berichte" can not be over-estimated and it is the desire of this Committee that all research laboratories and individuals interested in aeronautical research should become familiar with the results of the aeronautical research carried on in Germany during the War.

Respectfully,
National Advisory Committee for Aeronautics
[signed] Joseph S. Ames
Chairman, Committee on Publications and Intelligence.

Document 2-19(c), Ludwig Prandtl, Applications of Modern Hydrodynamics to Aeronautics, Part 1, NACA Technical Report No. 116, 1921.

PREFACE.

I have been requested by the United States National Advisory Committee for Aeronautics to prepare for the reports of the committee a detailed treatise on the present condition of those applications of hydrodynamics which lead to the calculation of the forces acting on airplane wings and airship bodies. I have acceded to the request of the National Advisory Committee all the more willingly because the theories in question have at this time reached a certain conclusion where it is worthwhile to show in a comprehensive manner the leading ideas and the results of these theories and to indicate what confirmation the theoretical results have received by tests.

The report will give, in a rather brief Part I, an introduction to hydrodynamics which is designed to give those who have not yet been actively concerned with this science such a grasp of the theoretical underlying principles that they can follow the subsequent developments. In Part II follow then separate discussions of the different questions to be considered, in which the theory of aerofoils claims the greatest portion of the space. The last part is devoted to the application of the aerofoil theory to screw propellers. [*Note: Only Part I is included in this chapter.*]

At the express wish of the National Advisory Committee for Aeronautics I have used the same symbols in my formulae as in my papers written in German. These are already for the most part known by readers of the Technische Berichte. A table giving the most important quantities is at the end of the report. A short reference list of the literature on the subject and also a table of contents are added.

PART I.
FUNDAMENTAL CONCEPTS AND THE MOST IMPORTANT THEOREMS.

1. All actual fluids show internal friction (viscosity), yet the forces due to viscosity, with the dimensions and velocities ordinarily occurring in practice, are so very small in comparison with the forces due to inertia, for water as well as for air, that we seem justified, as a first approximation, in entirely neglecting viscosity. Since the consideration of viscosity in the mathematical treatment of the problem introduces difficulties which have so far been overcome only in a few specially simple cases, we are forced to neglect entirely internal friction unless we wish to do without the mathematical treatment.

We must now ask how far this is allowable for actual fluids, and how far not. A closer examination shows us that for the interior of the fluid we can immediately apply our knowledge of the motion of a nonviscous fluid, but that care must be taken in considering the layers of the fluid in the immediate neighborhood of solid bodies. Friction between fluid and solid body never comes into consideration in the fields of application to be treated here, because it is established by reliable experiments that fluids like water and air never slide on the surface of the body; what happens is, the final fluid layer immediately in contact with the body is attached to it (is at rest relative to it), and all the friction of fluids with solid bodies is therefore an internal friction of the fluid. Theory and experiment agree in indicating that the transition from the velocity of the body to that of the stream in such a case takes place in a thin layer of the fluid, which is so much the thinner, the less the viscosity. In this layer, which we call the boundary layer, the forces due to viscosity are of the same order of magnitude as the forces due to inertia, as may be seen without difficulty. [*Footnote*] (From this consideration one can calculate the approximate thickness of the boundary layer for each special case.) It is therefore important to prove that, however small the viscosity is, there are always in a boundary layer on the surface of the body forces due to viscosity (reckoned per unit volume) which are of the same order of magnitude as those due to inertia. Closer investigation concerning this shows that under certain conditions there may occur a reversal of flow in the boundary layer, and as a consequence a stopping of the fluid in the layer which is set in rotation by the viscous forces, so that, further on, the whole flow is changed owing to the formation of vortices. The analysis of the phenomena which lead to the formation of vortices shows that it takes place where the fluid experiences a retardation of flow along the body. The retardation in some cases must reach a certain finite amount so that a reverse flow arises. Such retardation of flow occurs regularly in the rear of blunt bodies; therefore vortices are formed there very soon after the flow begins, and consequently the results which are furnished by the theory of nonviscous flow can not be applied. On the other hand, in the rear of very tapering bodies the retardations are often so small that there is no noticeable formation of vortices. The principal successful results of hydrodynamics apply to this case. Since it is these tapering bodies which offer specially small resistance and which, therefore, have found special consideration in aeronautics under similar applications, the theory can be made useful exactly for those bodies which are of most technical interest.

For the considerations which follow we obtain from what has gone before the result that in the interior of the fluid its viscosity, if it is small, has no essential influence, but that for layers of the fluid in immediate contact with solid bodies exceptions to the laws of a nonviscous fluid must be allowable. We shall try to formulate these exceptions so as to be, as far as possible, in agreement with the facts of experiment.

2. A further remark must be made concerning the effect of the compressibility of the fluid upon the character of the flow in the case of the motion of solid bodies in the fluid. All actual fluids are compressible. In order to compress a volume of air by 1 per cent, a pressure of about one one-hundredth of an atmosphere is needed. In the case of water, to produce an equal change in volume, a pressure of 200 atmospheres is required; the difference therefore is very great. With water it is nearly always allowable to neglect the changes in volume arising from the pressure differences due to the motions, and therefore to treat it as absolutely incompressible. But also in the case of motions in air we can ignore the compressibility so long as the pressure differences caused by the motion are sufficiently small. Consideration of compressibility in the mathematical treatment of flow phenomena introduces such great difficulties that we will quietly neglect volume changes of several per cent, and in the calculations air will be looked upon as incompressible. A compression of 3 per cent, for instance, occurs in front of a body which is being moved with a velocity of about 80 m./sec. It is seen, then, that it appears allowable to neglect the compressibility in the ordinary applications to technical aeronautics. Only with the blades of the air screw do essentially greater velocities occur, and in this case the influence of the compressibility is to be expected and has already been observed. The motion of a body with great velocity has been investigated up to the present, only along general lines. It appears that if the velocity of motion exceeds that of sound for the fluid, the phenomena are changed entirely, but that up close to this velocity the flow is approximately of the same character as in an incompressible fluid.

3. We shall concern ourselves in what follows only with a nonviscous and incompressible fluid, about which we have learned that it will furnish an approximation sufficient for our applications, with the reservations made. Such a fluid is also called "the ideal fluid."

What are the properties of such an ideal fluid? I do not consider it here my task to develop and to prove all of them, since the theorems of classical hydrodynamics are contained in all textbooks on the subject and may be studied there. I propose to state in what follows, for the benefit of those readers who have not yet studied hydrodynamics, the most important principles and theorems which will be needed for further developments, in such a manner that these developments may be grasped. I ask these readers, therefore, simply to believe the theorems which I shall state until they have the time to study the subject in some textbook on hydrodynamics.

The principal method of description of problems in hydrodynamics consists in expressing in formulas as functions of space and time the velocity of flow, given by its three rectangular components, u, v, w, and in addition the fluid pressure p. The condition of flow is evidently completely known if u, v, w, and p are given as functions of x, y, z, and t, since then u, v, w, and p can be calculated for any arbitrarily selected point and for every instant of time. The direction of flow is defined by the ratios of $u, v,$ and w; the magnitude of the velocity is $\sqrt{u^2 + v^2 + w^2}$. The "streamlines" will be obtained if lines are drawn which coincide with the direction of flow at all points where they touch, which can be accomplished mathematically by an integration. If the flow described by the formulas is to be that caused by a definite body, then at those points in space, which at any instant form the surface of the body, the components of the fluid velocity normal to this surface must coincide with the corresponding components of the velocity of the body. In this way the condition is expressed that neither does the fluid penetrate into the body nor is there any gap between it and the fluid. If the body is at rest in a stream, the normal components of the velocity at its surface must be zero; that is, the flow must be tangential to the surface, which in this case therefore is formed of stream lines.

4. In a stationary flow—that is, in a flow which does not change with the time, in which then every new fluid particle, when it replaces another particle in front of it, assumes its velocity, both in magnitude and in direction and also the same pressure—there is, for the fluid particles lying on the same stream line, a very remarkable relation between the magnitude of the velocity, designated here by V, and the pressure, the so-called Bernouilli equation—

$$p = \frac{V^2}{2} + p = \text{const.} \tag{1}$$

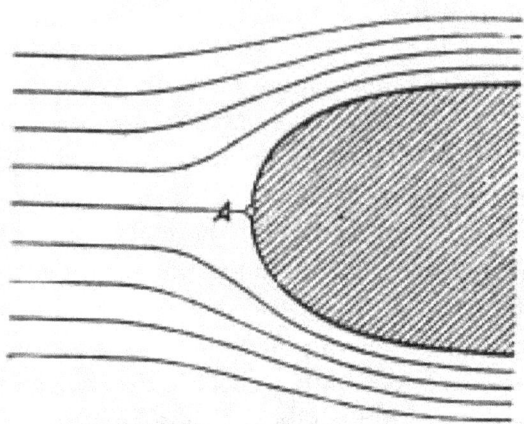

Fig. 1.—Flow around a blunt body.

(ρ is the density of the fluid, i.e., the mass of a unit volume). This relation is at once applicable to the case of a body moving uniformly and in a straight line in a fluid at rest, for we are always at liberty to use for our discussions any reference system having a uniform motion in a straight line. If we make the velocity of the reference system coincide with that of the body, then the body is at rest with reference to it, and the flow around it is stationary. If now V is the velocity of the body relative to the stationary air, the latter will have in the new reference system the velocity V upon the body (a man on an airplane in flight makes observations in terms of such a reference system, and feels the motion of flight as "wind").

The flow of incident air is divided at a blunt body, as shown in figure 1. At the point A the flow comes completely to rest, and then is again set in motion in opposite directions, tangential to the surface of the body. We learn from equation (1) that at such a point, which we shall call a "rest-point," the pressure must be greater by $\rho/2 V^2$ than in the undisturbed fluid. We shall call the magnitude of this pressure, of which we shall make frequent use, the "dynamical pressure," and shall designate it by q. An open end of a tube facing the stream produces a rest point of a similar kind, and there arises in the interior of the tube, as very careful experiments have shown, the exact dynamical pressure, so that this principle can be used for the measurement of the velocity, and is in fact much used. The dynamical pressure is also well suited to express the laws of air resistance. It is known that this resistance is proportional to the square of the velocity and to the density of the medium; but $q = \rho/2 V^2$; so the law of air resistance may also be expressed by the formula

$$W = c \cdot F \cdot q \qquad (2)$$

where F is the area of the surface and c is a pure number. With this mode of expression it appears very clearly that the force called the "drag" is equal to surface times pressure difference (the formula has the same form as the one for the piston force in a steam engine). This mode of stating the relation has been introduced in Germany and Austria and has proved useful. The air-resistance coefficients then become twice as large as the "absolute" coefficients previously used.

Since V^2 can not become less than zero, an increase of pressure greater than q can not, by equation (1), occur. For diminution of pressure, however, no definite limit can be set. In the case of flow past convex surfaces marked increases of velocity of flow occur and in connection with them diminutions of pressure which frequently amount to $3q$ and more.

5. A series of typical properties of motion of nonviscous fluids may be deduced in a useful manner from the following theorem, which is due to Lord Kelvin. Before the theorem itself is stated, two concepts must be defined. 1. The circulation: Consider the line integral of the velocity $\int V \cos(V, ds) \cdot ds$, which is formed exactly like the line integral of a force, which is called "the work of the force." The amount of this line integral, taken over a path which returns on itself is called the circulation of the flow. 2. The fluid line: By this is meant a line which is always formed of the same fluid particles, which therefore shares in the motion of the fluid. The theorem of Lord Kelvin is: In a nonviscous fluid the circulation along every fluid line remains unchanged as time goes on. But the following must be added:

(1) The case may arise that a fluid line is intersected by a solid body moving in the fluid. If this occurs, the theorem ceases to apply. As an example I mention the case in which one pushes a flat plate into a fluid at rest, and then by means of the plate exerts a pressure on the fluid. By this a circulation arises which will remain if afterwards the plate is quickly withdrawn in its own plane. See figure 2.

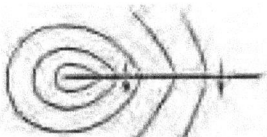

Fig. 2 – Production of circulation by introduction and withdrawal of flat plate.

(2) In order that the theorem may apply, we must exclude mass forces of such a character that work is furnished by them along a path which returns on itself. Such forces do not ordinarily arise and need not be taken into account here, where we are concerned regularly only with gravity.

(3) The fluid must be homogenous, i. e., of the same density at all points. We can easily see that in the case of nonuniform density circulation can arise of itself in the course of time if we think of the natural ascent of heated air in the midst of cold air. The circulation increases continuously along a line which passes upward in the warm air and returns downward in the cold air.

Frequently the case arises that the fluid at the beginning is at rest or in absolutely uniform motion, so that the circulation for every imaginable closed line in the fluid is zero. Our theorem then says that for every closed line that can arise from one of the originally closed lines the circulation remains zero, in which we must make exception, as mentioned above, of those lines which are cut by bodies. If the line integral along every closed line is zero, the line integral for an open curve from a definite point O to an arbitrary point P is independent of the selection of the line along which the integral is taken (if this were not so, and if the integrals along two lines from O to P were different, it is evident that the line integral along the closed curve OPO would not be zero, which contradicts our premise). The line integral along the line OP depends, therefore, since we will consider once for all the point O as a fixed one, only on the coordinates of the point P, or, expressed differently, it is a function of these coordinates. From analogy with corresponding considerations in the case of fields of force, this line integral is called the "velocity potential," and the particular kind of motion in which such a potential exists is called a "potential motion." As follows immediately from the meaning of line integrals, the component of the velocity in a definite direction is the derivative of the potential in this direction. If the line-element is perpendicular to the resultant velocity, the increase of the potential equals zero, i. e., the surfaces of constant potential are everywhere normal to the velocity of flow. The velocity itself is called the gradient of the potential. The velocity components u, v, w are connected with the potential Φ by the following equations:

$$u = \frac{\partial \Phi}{\partial x}, \quad v = \frac{\partial \Phi}{\partial y}, \quad w = \frac{\partial \Phi}{\partial z} \tag{3}$$

The fact that the flow takes place without any change in volume is expressed by stating that as much flows out of every element of volume as flows in. This leads to the equation

$$\frac{\partial u}{\partial x} + \frac{\partial v}{\partial y} + \frac{\partial w}{\partial z} = 0 \tag{4}$$

In the case of potential flow we therefore have

$$\frac{\partial^2 \Phi}{\partial x^2} + \frac{\partial^2 \Phi}{\partial y^2} + \frac{\partial^2 \Phi}{\partial z^2} = 0 \tag{4a}$$

as the condition for flow without change in volume. All functions $\Phi\,(x, y, z, t)$, which satisfy this last equation, represent possible forms of flow. This representation of a flow is specially convenient for calculations, since by it the entire flow is given by means of the one function Φ. The most valuable property of the representations is, though, that the sum of two, or of as many as one desires, functions Φ, each of which satisfies equation (4a), also satisfies this equation, and therefore represents a possible type of flow ("superposition of flows").

6. Another concept can be derived from the circulation, which is convenient for many considerations, viz, that of rotation. The component of the rotation with

reference to any axis is obtained if the circulation is taken around an elementary surface of unit area in a plane perpendicular to the axis. Expressed more exactly, such a rotation component is the ratio of the circulation around the edge of any such infinitesimal surface to the area of the surface. The total rotation is a vector and is obtained from the rotation components for three mutually perpendicular axes. In the case that the fluid rotates like a rigid body, the rotation thus defined comes out as twice the angular velocity of the rigid body. If we take a rectangular system of axes and consider the rotations with reference to the separate axes, we find that the rotation can also be expressed as the geometrical sum of the angular velocities with reference to the three axes.

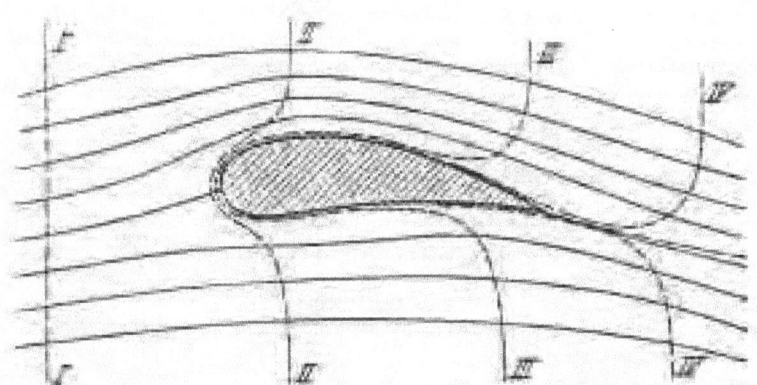

FIG. 3.—Successive positions of a field line in flow around a solid body.

The statement that in the case of a potential motion the circulation is zero for every closed fluid line can now be expressed by saying the rotation in it is always zero. The theorem that the circulation, if it is zero, remains zero under the conditions mentioned, can also now be expressed by saying that, if these conditions are satisfied in a fluid in which there is no rotation, rotation can never arise. An irrotational fluid motion, therefore, always remains irrotational. In this, however, the following exceptions are to be noted: If the fluid is divided owing to bodies being present in it, the theorem under consideration does not apply to the fluid layer in which the divided flow reunites, not only in the case of figure 2 but also in the case of stationary phenomena as in figure 3, since in this case a closed fluid line drawn in front of the body can not be transformed into a fluid line that intersects the region where the fluid streams come together. Figure 3 shows four successive shapes of such a fluid line. This region is, besides, filled with fluid particles which have come very close to the body. We are therefore led to the conclusion from the standpoint of a fluid with very small but not entirely vanishing viscosity that the appearance of vortices at the points of reunion of the flow in the rear of the body does not contradict the laws of hydrodynamics. The three components of the rotation ξ, η, ζ are expressed as follows by means of the velocity components u, v, w.

$$\xi = \frac{\partial w}{\partial y} - \frac{\partial v}{\partial z}, \quad \eta = \frac{\partial u}{\partial z} - \frac{\partial w}{\partial x}, \quad \zeta = \frac{\partial v}{\partial x} - \frac{\partial u}{\partial y} \tag{5}$$

If the velocity components are derived from a potential, as shown in equation (2), the rotation components, according to equation (5) vanish identically, since

$$\frac{\partial}{\partial y}\frac{\partial \varphi}{\partial z} = \frac{\partial}{\partial z}\frac{\partial \varphi}{\partial y}, \text{ etc.}$$

7. Very remarkable theorems hold for the rotation, which were discovered by V. Helmholtz and stated in his famous work on vortex motions. Concerning the geometrical properties of the rotation the following must be said:

At all points of the fluid where rotation exists the direction of the resultant rotation axes can be indicated, and lines can also be drawn whose directions coincide everywhere with these axes, just as the stream lines are drawn so as to coincide with the directions of the velocity. These lines will be called, following Helmholtz, "vortex lines." The vortex lines through the points of a small closed curve form a tube called a "vortex tube." It is an immediate consequence of the geometrical idea of rotation as deduced above that through the entire extent of a vortex tube its strength—i. e., the circulation around the boundary of the tube—is constant. It is seen, in fact, that on geometrical grounds the space distribution of rotation quite independently of the special properties of the velocity field from which it is deduced is of the same nature as the space distribution of the velocities in an incompressible fluid. Consequently a vortex tube, just like a stream line in an incompressible fluid, can not end anywhere in the interior of the fluid and the strength of the vortex, exactly like the quantity of fluid passing per second through the tube of stream lines, has at one and the same instant the same value throughout the vortex tube. If Lord Kelvin's theorem is now applied to the closed fluid line which forms the edge of a small element of the surface of a vortex tube, the circulation along it is zero, since the surface inclosed is parallel to the rotation axis at that point. Since the circulation can not change with the time, it follows that the element of surface at all later times will also be part of the surface of a vortex tube. If we picture the entire bounding surface of a vortex tube as made up of such elementary surfaces, it is evident that, since as the motion continues this relation remains unchanged, the particles of the fluid which at any one time have formed the boundary of a vortex tube will continue to form its boundary. From the consideration of the circulation along a closed line enclosing the vortex tube, we see that this circulation—i.e., the strength of our vortex tube—has the same value at all times. Thus we have obtained the theorems of Helmholtz, which now can be expressed as follows, calling the contents of a vortex tube a "vortex filament": "The particles of a fluid which at any instant belong to a vortex filament always remain in it; the strength of a vortex filament throughout its extent and for all time has the same value." From this follows, among other things, that if a portion of the filament is stretched, say, to double its length, and thereby its cross section made one-half as great, then the rotation is doubled, because the strength of the vortex, the product of the rotation and the cross section, must remain the same. We arrive, therefore, at the result that the vector expressing the rotation is changed in magnitude and direction exactly as the distance between two neighboring particles on the axis of the filament is changed.

8. From the way the strengths of vortices have been defined it follows for a space filled with any arbitrary vortex filaments, as a consequence of a known theorem of Stokes, that the circulation around any closed line is equal to the algebraic sum of the vortex strengths of all the filaments which cross a surface having the closed line as its boundary. If this closed line is in any way continuously changed so that filaments are thereby cut, then evidently the circulation is changed according to the extent of the strengths of the vortices which are cut. Conversely we may conclude from the circumstance that the circulation around a closed line (which naturally can not be a fluid line) is changed by a definite amount by a certain displacement, that by the displacement vortex strength of this amount will be cut, or expressed differently, that the surface passed over by the closed line in its displacement is traversed by vortex filaments whose strengths add up algebraically to the amount of the change in the circulation.

The theorems concerning vortex motion are specially important because in many cases it is easier to make a statement as to the shape of the vortex filaments than as to the shape of the stream lines, and because there is a mode of calculation by means of which the velocity at any point of the space may be determined from a knowledge of the distribution of the rotation. This formula, so important for us, must now be discussed. If Γ is the strength of a thin vortex filament and ds an element of its medial line, and if, further, r is the distance from the vortex element to a point P at which the velocity is to be calculated, finally if α is the angle between ds and r, then the amount of the velocity due to the vortex element is

$$dv = \frac{\Gamma \cdot ds \cdot \sin\alpha}{4\pi r^2}, \qquad (6)$$

the direction of this contribution to the velocity is perpendicular to the plane of ds and r. The total velocity at the point P is obtained if the contributions of all the vortex elements present in the space are added. The law for this calculation agrees then exactly with that of Biot-Savart, by the help of which the magnetic field due to an electric current is calculated. Vortex filaments correspond in it to the electric currents, and the vector of the velocity to the vector of the magnetic field.

As an example we may take an infinitely long straight vortex filament. The contributions to the velocity at a point P are all in the same direction, and the total velocity can be determined by a simple integration of equation (6). Therefore this velocity is

$$v = \frac{\Gamma}{4\pi} \int \frac{ds \cdot \sin\alpha}{r^2}.$$

As seen by figure 4, $s = h \operatorname{ctg} \alpha$, and by differentiation,

$$ds = -\frac{h}{\sin^2 \alpha} d\alpha \quad \text{Further } r = \frac{h}{\sin \alpha}; \text{ so that}$$

$$v = \frac{\Gamma}{4\pi h} \int \sin\alpha \, d\alpha = -\frac{\Gamma}{4\pi h}|\cos\alpha|\Big|_0^\pi = \frac{\Gamma}{2\pi h}. \quad (6a)$$

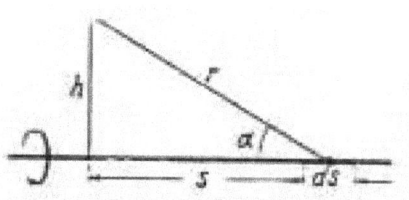

FIG. 4.—Velocity-field due to infinite rectilinear vortex.

This result could be deduced in a simpler manner from the concept of circulation if we were to use the theorem, already proved, that the circulation for any closed line coincides with the vortex strength of the filaments which are inclosed by it. The circulation for every closed line which goes once around a single filament must therefore coincide with its strength. If the velocity at a point of a circle of radius h around our straight filament equals v then this circulation equals "path times velocity" $= 2\pi h \cdot v$, whence immediately follows $v = \Gamma/2\pi h$. The more exact investigation of this velocity field shows that for every point outside the filament (and the formula applies only to such points) the rotation is zero, so that in fact we are treating the case of a velocity distribution in which only along the axis does rotation prevail, at all other points rotation is not present.

For a finite portion of a straight vortex filament the preceding calculation gives the value

$$v = \frac{\Gamma}{2\pi h}(\cos\alpha_1 - \cos\alpha_2), \qquad (6b)$$

This formula may be applied only for a series of portions of vortices which together give an infinite or a closed line. The velocity field of a single portion of a filament would require rotation also outside the filament, in the sense that from the end of the portion of the filament vortex lines spread out in all the space and then all return together at the beginning of the portion. In the case of a line that has no ends this external rotation is removed, since one end always coincides with the beginning of another portion of equal strength, and rotation is present only where it is predicated in the calculation.

9. If one wishes to represent the flow around solid bodies in a fluid, one can in many cases proceed by imagining the place of the solid bodies taken by the fluid, in the interior of which disturbances of flow (singularities) are introduced, by which the flow is so altered that the boundaries of the bodies become streamline surfaces. For such hypothetical constructions in the interior of the space actually occupied by the body, one can assume, for instance, any suitably selected vortices, which, however, since they are only imaginary, need not obey the laws of Helmholtz. As we shall see later, such imaginary vortices can be the seat of lifting forces. Sources and sinks also, i.e., points where fluid continuously appears, or disappears, offer a useful method for constructions of this kind. While vortex filaments can actually occur in the fluid, such sources and sinks may be assumed only in that part of the space which actually is occupied by the body, since they represent a phenomenon which can not be realized. A contradiction of the law of the conservation of matter is avoided, however, if there are assumed to be inside the body both sources and sinks, of equal strengths, so that the fluid produced by the sources is taken back again by the sinks.

The method of sources and sinks will be described in greater detail when certain practical problems are discussed; but at this point, to make the matter clearer, the distribution of velocities in the case of a source may be described. It is very simple, the flow takes place out from the source uniformly on all sides in the direction of the radii. Let us describe around the point source a concentric spherical surface, then, if the fluid output per second is Q, the velocity at the surface is

$$v = \frac{Q}{4\pi r^2}; \qquad (7)$$

the velocity therefore decreases inversely proportional to the square of the distance. The flow is a potential one, the potential comes out (as line-integral along the radius)

$$\Phi = \text{const.} - \frac{Q}{4\pi r}. \tag{7a}$$

If a uniform velocity toward the right of the whole fluid mass is superimposed on this velocity distribution while the point source remains stationary—then a flow is obtained which, at a considerable distance from the source, is in straight lines from left to right. The fluid coming out of the source is therefore pressed toward the right (see fig. 5); it fills, at some distance from the source, a cylinder whose diameter may be determined easily. If V is the velocity of the uniform flow, the radius r of the cylinder is given by the condition $Q = \pi r^2 \cdot V$. All that is necessary now is to assume on the axis of the source further to the right a sink of the same strength as the source for the whole mass of fluid from the source to vanish in this, and the flow closes up behind the sink again exactly as it opened out in front of the source. In this way we obtain the flow around an elongated body with blunt ends.

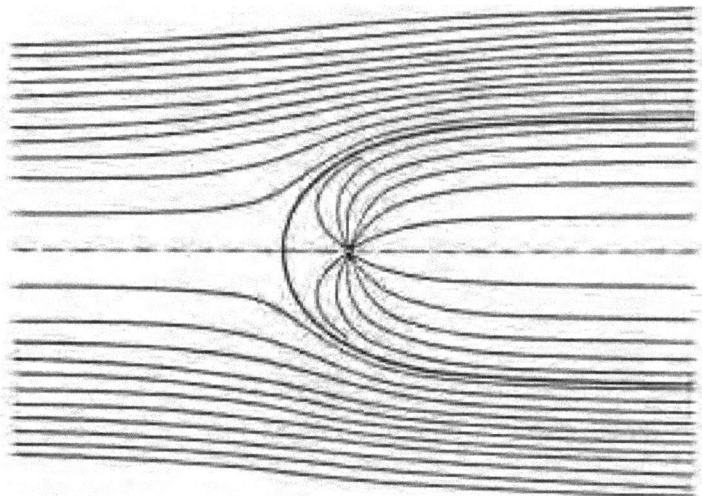

FIG. 5.—Superposition of uniform flow and that caused by a source.

10. The special case when in a fluid flow the phenomena in all planes which are parallel to a given plane coincide absolutely plays an important role both practically and theoretically. If the lines which connect the corresponding points of the different planes are perpendicular to the planes, and all the streamlines are plane curves which lie entirely in one of those planes, we speak of a uniplanar

flow. The flow around a strut whose axis is perpendicular to the direction of the wind is an example of such a motion.

The mathematical treatment of plane potential flow of the ideal fluid has been worked out specially completely more than any other problem in hydrodynamics. This is due to the fact that with the help of the complex quantities ($x + iy$, where $i=\sqrt{-1}$, is called the imaginary unit) there can be deduced from every analytic function a case of flow of this type which is incompressible and irrotational. Every real function, $\Phi(x, y)$ and $\Psi(x, y)$, which satisfies the relation

$$\Phi + i\Psi = f(x + iy), \tag{8}$$

where f is any analytic function, is the potential of such a flow. This can be seen from these considerations: Let $x + iy$ be put $= z$, where z is now a "complex number." Differentiate equation (8) first with reference to x and then with reference to y, thus giving

$$\frac{\partial \Phi}{\partial x} + i\frac{\partial \Psi}{\partial x} = \frac{\partial f}{\partial z} \cdot \frac{\partial z}{\partial x} = \frac{\partial f}{\partial z}$$

$$\frac{\partial \Phi}{\partial y} + i\frac{\partial \Psi}{\partial y} = \frac{\partial f}{\partial z} \cdot \frac{\partial z}{\partial y} = i\frac{\partial f}{\partial z} = i\frac{\partial \Phi}{\partial x} - \frac{\partial \Psi}{\partial x}.$$

In these the real parts on the two sides of the equations must be equal and the imaginary parts also. If Φ is selected as the potential, the velocity components u and v are given by

$$u = \frac{\partial \Phi}{\partial x} = \frac{\partial \Psi}{\partial y} \;::\; v = \frac{\partial \Phi}{\partial y} = -\frac{\partial \Psi}{\partial x}. \tag{9}$$

If now we write the expressions $\delta\Phi/\delta x + \delta v/\delta y$ (continuity) and $\delta\Phi/\delta x - \delta v/\delta y$ (rotation) first in terms of Φ and then of Ψ, they become

$$\left. \begin{array}{c} \frac{\partial u}{\partial x} + \frac{\partial v}{\partial y} = \frac{\partial^2 \Phi}{\partial x^2} + \frac{\partial^2 \Phi}{\partial y^2} = \frac{\partial^2 \Psi}{\partial x \partial y} - \frac{\partial^2 \Psi}{\partial y \partial x} \\ \frac{\partial u}{\partial y} - \frac{\partial v}{\partial x} = \frac{\partial^2 \Phi}{\partial y \partial x} - \frac{\partial^2 \Phi}{\partial x \partial y} = \frac{\partial^2 \Psi}{\partial x^2} + \frac{\partial^2 \Psi}{\partial y^2} \end{array} \right\} \tag{10}$$

It is seen therefore that not only is the motion irrotational (as is self-evident since there is a potential), but it is also continuous. The relation $\delta^2\Phi/\delta x^2 + \delta^2\Phi/\delta y^2 = 0$ besides corresponds exactly to our equation (4a). Since it is satisfied also by Ψ, this can also be used as potential.

The function Ψ, however, has, with reference to the flow deduced by using Φ, as potential, a special individual meaning. From equation (8) we can easily deduce that the lines Ψ = const. are parallel to the velocity; therefore, in other words, they are streamlines. In fact if we put

$$d\Psi = \frac{\partial \Psi}{\partial x} dx + \frac{\partial \Psi}{\partial y} dy = 0, \text{ then } \frac{dy}{dx} = \frac{\frac{\partial \Psi}{\partial x}}{\frac{\partial \Psi}{\partial y}} = \frac{v}{u}$$

which expresses the fact of parallelism. The lines Ψ = const. are therefore perpendicular to the lines Φ = const. If we draw families of lines, Φ = const. and Ψ = const. for values of Φ and Ψ which differ from each other by the same small amount, it follows from the easily derived equation $d\Phi + id\Psi = \frac{df}{dz}(dx + idy)$ that the two bundles form a square network; from which follows that the diagonal curves of the network again form an orthogonal and in fact a square network. This fact can be used practically in drawing such families of curves, because an error in the drawing can be recognized by the eye in the wrong shape of the network of diagonal curves and so can be improved. With a little practice fairly good accuracy may be obtained by simply using the eye. Naturally there are also mathematical methods for further improvement of such networks of curves. The function Ψ, which is called the "stream function," has another special meaning. If we consider two streamlines $\Psi = \Psi_1$ and $\Psi = \Psi_2$, the quantity of fluid which flows between the two streamlines in a unit of time in a region of uniplanar flow of thickness 1 equals $\Psi_2 - \Psi_1$. In fact if we consider the flow through a plane perpendicular to the X-axis, this quantity is

$$Q = \int_{}^{} u\, dy = \int \frac{\partial \Psi}{\partial y} dy = \int d\Psi = \Psi_2 - \Psi_1.$$

The numerical value of the stream function coincides therefore with the quantity of fluid which flows between the point x, y and the streamline $\Psi = 0$.

As an example let the function

$$\Phi + i\Psi = A(x + iy)^n$$

be discussed briefly. It is simplest in general to ask first about the streamline $\Psi = 0$. As is well known, if a transformation is made from rectangular coordinates to polar ones r, φ, $(x + iy)^n = r^n (\cos n\varphi + i \sin n\varphi)$. The imaginary part of this expression is $ir^n \sin n\varphi$. This is to be put equal to $i\Psi$. $\Psi = o$ therefore gives $\sin n\varphi = o$, i.e., $n\varphi = o, \pi, 2\pi$, etc. The streamlines $\Psi = o$ are therefore straight lines through the origin of coordinates, which make an angle $\alpha = \pi/n$ with each other, the flow is therefore the potential flow between two plane walls making the angle α with each other. The other streamlines satisfy the equation $r^n \sin n\varphi = $ const. The velocities can be obtained by differentiation, e.g., with reference to x:

$$\delta\Phi/\delta x = i\,\delta\Psi/\delta x = u - iv = An (x + iy)^{n-1} = Anr^{n-1} \{\cos (n-1)\,\varphi + i \sin (n-1)\,\varphi\}.$$

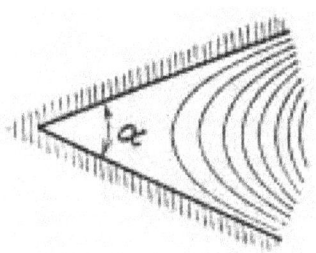

Fig. 6.—Uniplanar flow between plane walls making an angle $\alpha = 45°$ with each other.

For $r = o$ this expression becomes zero or infinite, according as n is greater or less than 1, i.e., according as the angle α is less or greater than $\pi (= 180°)$. Figures 6 and 7 give the streamlines for $= \pi/4 = 45°$ and $3/2\pi = 270°$, corresponding to $n = 4$ and $2/3$. In the case of figure 7, the velocity, as just explained, becomes infinite at the corner. It would be expected that in the case of the actual flow some effect due to friction would enter. In fact there are observed at such corners, at the beginning of the motion, great velocities, and immediately thereafter the formation of vortices, by which the motion is so changed that the velocity at the corner becomes finite.

It must also be noted that with an equation

$$p + iq = \varphi (x + iy) \qquad (11)$$

the x-y plane can be mapped upon the p-q plane, since to every pair of values x, y a pair of values p, q corresponds, to every point of the x-y plane corresponds a point of the p-q plane, and therefore also to every element of a line or to every curve in the former plane a linear element and a curve in the latter plane. The transformation keeps all angles unchanged, i. e., corresponding lines intersect in both figures at the same angle.

By inverting the function φ of equation (11) we can write

$$x + iy = \chi(p + iq)$$

and therefore deduce from equation (8) that

$$\Phi + i\Psi = f[x(p+iq)] = \Phi(p+iq) \qquad (12)$$

Φ and Ψ are connected therefore with π and θ by an equation of the type of equation (8), and hence, in the p-q plane, are potential and stream functions of a flow, and further of that flow which arises from the transformation of the Φ, Ψ network in the x-y plane into the p-q plane.

This is a powerful method used to obtain by transformation from a known simple flow new types of flow for other given boundaries. Applications of this will be given in section 14.

11. The discussion of the principles of the hydrodynamics of nonviscous fluids to be applied by us may be stopped here. I add but one consideration, which has reference to a very useful theorem for obtaining the forces in fluid motion, namely the so-called "momentum theorem for stationary motions."

We have to apply to fluid motion the theorem of general mechanics, which states that the rate of change with the time of the linear momentum is equal to the resultant of all the external forces. To do this, consider a definite portion of the fluid separated from the rest of the fluid by a closed surface. This surface may, in accordance with the spirit of the theorem, be considered as a "fluid surface," i. e., made up always of the same fluid particles. We must now state in a formula the change of the momentum of the fluid within the surface. If, as we shall assume, the flow is stationary, then after a time dt every fluid particle in the interior will be

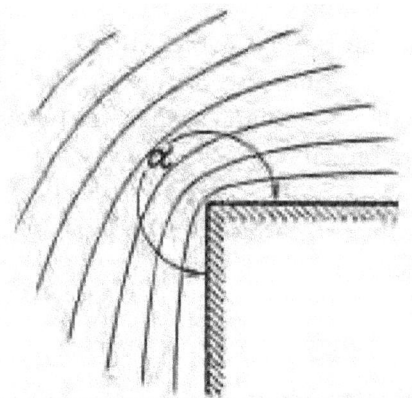

FIG. 7.—Unipolar flow around plane walls making an angle 270° with each other.

replaced by another, which has the same velocity as had the former. On the boundary, however, owing to its displacement, mass will pass out at the side where the fluid is approaching and a corresponding mass will enter on the side away from which the flow takes place. If dS is the area of an element of surface, and v_n the component of the velocity in the direction of the outward drawn normal at this element, then at this point $dm = \rho dS \cdot v_n\, dt$. If we wish to derive the component of the "impulse"—defined as the time rate of the change of momentum—for any direction s, the contribution to it of the element of surface is

$$dJ_s = v_s (dm/dt) = \rho dS \cdot v_n v_s. \qquad (13)$$

With this formula we have made the transition from the fluid surface to a corresponding solid "control surface."

The external forces are compounded of the fluid pressures on the control surface and the forces which are exercised on the fluid by any solid bodies which may be inside of the control surface. If we call the latter P, we obtain the equation

$$\sum P_s = \iint p \cdot \cos(n,s) \cdot dS + \rho \iint v_n v_s dS \tag{14}$$

for the s component of the momentum theorem. The surface integrals are to be taken over the entire closed control surface. The impulse integral can be limited to the exit side, if for every velocity v_s on that side the velocity v_s' is known with which the same particle arrives at the approach side. Then in equation (13) dJ is to be replaced by

$$dJ - dJ' = (v_s - v_s')(dm/dt) = \rho dS\, v_n\, (v_s - v_s'). \tag{13a}$$

The applications given in Part II will furnish illustrations of the theorem.

Document 2-20(a–c)

(a) Max M. Munk, memorandum to George W. Lewis, 10 March 1925, RA file 44, Historical Archives, NASA Langley.

(b) Fred E. Weick and Donald H. Wood, *The Twenty-Foot Propeller Research Tunnel of the National Advisory Committee for Aeronautics*, NACA Technical Report No. 300 (Washington, DC, 1928).

(c) Fred E. Weick, excerpts from *From the Ground Up* (Washington, DC: Smithsonian Institution Press, 1988), pp. 46–67.

While the VDT provided unprecedented capabilities for airfoil research at high Reynolds numbers, it was not well suited for certain other investigations, such as propeller testing. No practical means to utilize model propellers existed, because they did not deflect during operation the same way full-size ones did; this introduced significant error. After a promising propeller design had been tested for structural integrity and to determine its basic characteristics, an evaluation of its actual performance on an airplane required expensive and sometimes dangerous flight tests. However, accurate measurements in flight were difficult at best. Langley had a dynamometer to get reliable data on engine performance, and the NACA had sponsored propeller research in Stanford University's Eiffel-type tunnel since 1917, but no valid laboratory method of studying the entire propulsion system and its interaction with the aircraft fuselage existed in the mid-1920s.

The NACA's response was to construct the first very-large wind tunnel, the Propeller Research Tunnel (PRT). The PRT, completed in 1927, was a closed tunnel with an open test section, often referred to as an open-jet type. The nozzle supplied a twenty-foot-diameter air stream with velocities up to 110 miles per hour. A tunnel this size required more electrical power than was available at LMAL, so the PRT used two 1,000 horsepower diesel engines that had been removed from a navy submarine. The balance could support an airplane or a substitute "test fuselage" that featured an onboard dynamometer to directly measure engine torque. Technicians developed optical methods to measure blade deflection during operation. The PRT quickly proved its worth for propeller testing in conditions that were very close to those in actual flight, but since it was the first tunnel large enough to accommodate full-size airplanes (except for their full wing spans), its role soon expanded to include drag studies of aircraft components such as landing gears, radiators, tail planes, and—perhaps most significantly—radial engine cowlings. The PRT also served as the model for even larger wind tunnels that the NACA would build over the next two decades.

As the 10 March 1925 letter from Max Munk to George Lewis shows, it was Munk who conceived this large tunnel, and it was he who convinced Lewis that such a tunnel would be not only feasible, but also tremendously valuable to the NACA. As he had with the VDT, Munk correctly assessed a current research problem—propeller performance tests that were not possible in the VDT—and devised a solution that was at once bold and theoretically sound. Unfortunately, he ran into serious non-technical problems when Lewis put him in charge of the PRT's construction.

While Max Munk brought a keen theoretical mind to America in 1920, he also brought a large ego and an unremitting belief in the German organizational model, in which a leader's ideas were unquestioningly accepted and practiced. Munk was clearly brilliant, but he did not—perhaps could not—adapt to the more egalitarian mode of operation at Langley. The other Langley researchers and engineers considered his style overbearing and excessively autocratic, and they finally revolted against him. No single document fully portrays the situation, but Fred Weick recalled several incidents in his autobiography, *From the Ground Up*, which gives a good picture of the mismatch between Munk and other Langley personnel. In addition, a letter from Fred Norton to NACA Director of Aeronautical Research George Lewis, one of many written by Langley managers between 1921 and 1927, illustrates his frustration with the situation. However, these letters met blind eyes, as Lewis was Munk's biggest supporter. Even after the mass resignation of LMAL section heads in 1927, Lewis tried, unsuccessfully, to find some way to retain Munk in the Washington office. Munk, however, would not accept this loss of "his" laboratory and chose to resign from the NACA. Although Munk maintained some visibility in aeronautics for a while, notably as the author of a series of articles on basic aerodynamics in *Aero Digest* during the 1930s, his resignation from the NACA effectively marked the end of his serious contributions to the field.

Document 2-20(a), Max M. Munk, memorandum to George W. Lewis, 1925.

March 10, 1925.

MEMORANDUM for Mr. Lewis.
Subject: Laboratory for Testing Full Size Propellers.

1. At the present, our aerodynamic free flight research has not yet been brought forward enough to separate the performance of an airplane into that of

the engine, that of the propeller, and that of the remaining airplane. Nor has it been possible to obtain any definite comparisons between the results of model tests and free flight tests.

The first step taken to logically and systematically adapt our research work to the requirements of the practice was the construction of the variable density tunnel in order to get rid of the scale effect. This step was successful as far as could be expected. The model results in the new tunnel agree better with free flight experiments than with results from ordinary tunnels. The influence of the propeller slipstream will be included in the variable density tunnel tests by providing apparatus for driving and observing a model propeller.

The model propeller in the variable density tunnel will be free from the scale effect proper. It will, however, experience pressures and hence elastic (and permanent) deformation different from those of the full size propeller. Further, the influence of the compressibility of the air will not properly be taken care of. It will be very difficult and expensive to make an exact model of the propeller and of the portion of the airplane in its vicinity. At last, the variable density tunnel is fully occupied for the next years by tests not directly referring to the propeller.

2. I wish, therefore, to recommend that the tests referring to propeller performance and to the influence of the other portions of the airplane on the performance of the propellers be excluded from the research assigned to the variable density tunnel. I wish further, to recommend that the mentioned part of our research program be turned over and assigned to a special laboratory to be constructed for this purpose.

A similar laboratory was built during the last war at Fishamend (near Vienna, Austria) by Prof. Th. v. Karman. No published description has ever come to my attention. As far as I know, this laboratory was destroyed under the terms of the Peace Treaty of Versailles. It may be possible to obtain some information about it from Dr. v. Karman, now at Aachen.

3. The tests to be made in the laboratory would comprise:
> (a) Determination of Torque and Thrust of actual propellers (up to 10 ft. diameter, say) at different ratios U/V of the propeller tip velocity U to the velocity of motion and at actual speeds.

The tests would be run in air of sea level pressure and would be correct as to deformation of the propeller, compressibility of the air, and would not involve any scale effect.
> (b) The same tests with the propeller in front of any airplane fuselage, and in the presence of the wings and of other parts of the airplane.

The results will then include the forces on the several parts of the airplane and will give the actual merits of the propeller under these conditions,
> (c) Propeller tests with the axis of the propeller slightly tilted. This will correspond to the use of propellers under different angles of attack.

(d) Determination of the engine performance. The propeller can be driven by any airplane engine and the tests can then be made to include a test of that engine. In particular, the merits of the cooling system can be examined.

(e) The tests will further give information of the merits of different shapes of fuselages or engine cars taking into account the slipstream produced by the propeller under different conditions.

(f) In special cases, the laboratory would be available as the largest wind tunnel in the world.

4. The laboratory is contemplated to consist of a 20-ft. wind tunnel with open jet test chamber and with the return channels being under compression. At first glance, the cost of a 20-ft. tunnel would seem to be prohibitive, but a closer examination will show that such is not the case. Our present atmospheric 5-ft. tunnel could be built much cheaper and must not be taken as a basis for the costs of the new laboratory. Our variable density tunnel is costly on account of the high compression of the air used. This new 20-ft. tunnel is supposed to be built especially for the purpose indicated and will be comparatively cheap on account of the following items:

1. The power plant will consist of [two] 400 H.P. combustion engines, say Liberties, and not of electric motors. A third engine will be required to drive the propeller.
2. The building will be made chiefly of wood frame construction, and internal steel rod bracing used where required.
3. No high priced special balances will be built but only ordinary balances will be used as can be purchased on the market.

No estimate of the costs has been made yet, but I believe that the entire cost will not be in excess of the cost of an ordinary good wind tunnel. The results will be of much greater value, being free from any important error and giving plenty of information needed and wanted by the practice for many years. The research program will not be exhausted for many years, but on the contrary will probably grow larger and larger as the development of aeronautics goes on.

5. The need for a laboratory as the one in question is demonstrated by the many unsuccessful attempts to require the information to be obtained in other ways.

There have been constructed giant whirling arms, towers on railroad cars, dynamometer hubs on airplanes, airplanes gliding along cables in an airship shed, and the Fishamend Laboratory mentioned above. (The latter was possible with open return.)

[signed] Max M. Munk.

Document 2-20(b), Fred E. Weick and Donald H. Wood, The Twenty-Foot Propeller Research Tunnel of the National Advisory Committee for Aeronautics, *NACA Technical Report No. 300, 1928.*

SUMMARY

This report describes in detail the new propeller research tunnel of the National Advisory Committee for Aeronautics at Langley Field, Va. This tunnel has an open jet air stream 20 feet in diameter in which velocities up to 110 M.P.H. are obtained. Although the tunnel was built primarily to make possible accurate full-scale tests on aircraft propellers, it may also be used for making aerodynamic tests on full-size fuselages, landing gears, tail surfaces, and other aircraft parts, and on model wings of large size. [Italics appeared in original.]

INTRODUCTION

The need of an accurate means for making aerodynamic measurements on full-size aircraft propellers has been realized for some time. Tests on model propellers in wind tunnels are not entirely satisfactory because the deflection of the model is different from that of a similar full-scale propeller, which introduces a rather large error in some cases. The difference in scale and tip speed between the model and full-scale propeller is also a cause of error. Full-scale flight tests on propellers are made, of course, under the correct conditions, but at the present time they can not be made with sufficient accuracy.

In the spring of 1925 the design and construction of a propeller research wind tunnel to fill this need for full-scale tests was started by the National Advisory Committee for Aeronautics. It was completed during the summer of 1927 and testing has been carried on since that time. The tunnel is of the open-jet type with an air stream 20 feet in diameter. This is large enough to permit the mounting of a full-sized airplane fuselage with its engine and propeller. The open-jet type is particularly suitable for testing propellers because no corrections are required for tunnel-wall interference. (References 4 and 5.) Also, since with the open-jet type the inside of the experiment chamber is free from restricting walls, the installation of the objects to be tested is relatively simple.

This wind tunnel makes it possible for the first time to make aerodynamic tests with laboratory accuracy on full-scale aircraft propellers and also on full-scale fuselages, engine cowlings, cooling systems, landing gears, tail surfaces and other airplane parts. Full-scale tests of wings are not, of course, possible in a 20-foot air stream, but large model wings (12 feet in span) can be tested at comparatively high values of the Reynolds number.

Dr. Max M. Munk is responsible for the general arrangement of the propeller research tunnel, and the detail design and construction were carried out under the direction of Mr. E. W. Miller of the laboratory staff.

DESCRIPTION OF THE TUNNEL
GENERAL

The propeller research tunnel of the National Advisory Committee for Aeronautics is located at Langley Field, Va., on a plot adjacent to the committee's other research equipment. Figure 1 is a diagrammatic sketch indicating the general arrangement of the tunnel and Figure 2 illustrates the exterior appearance.

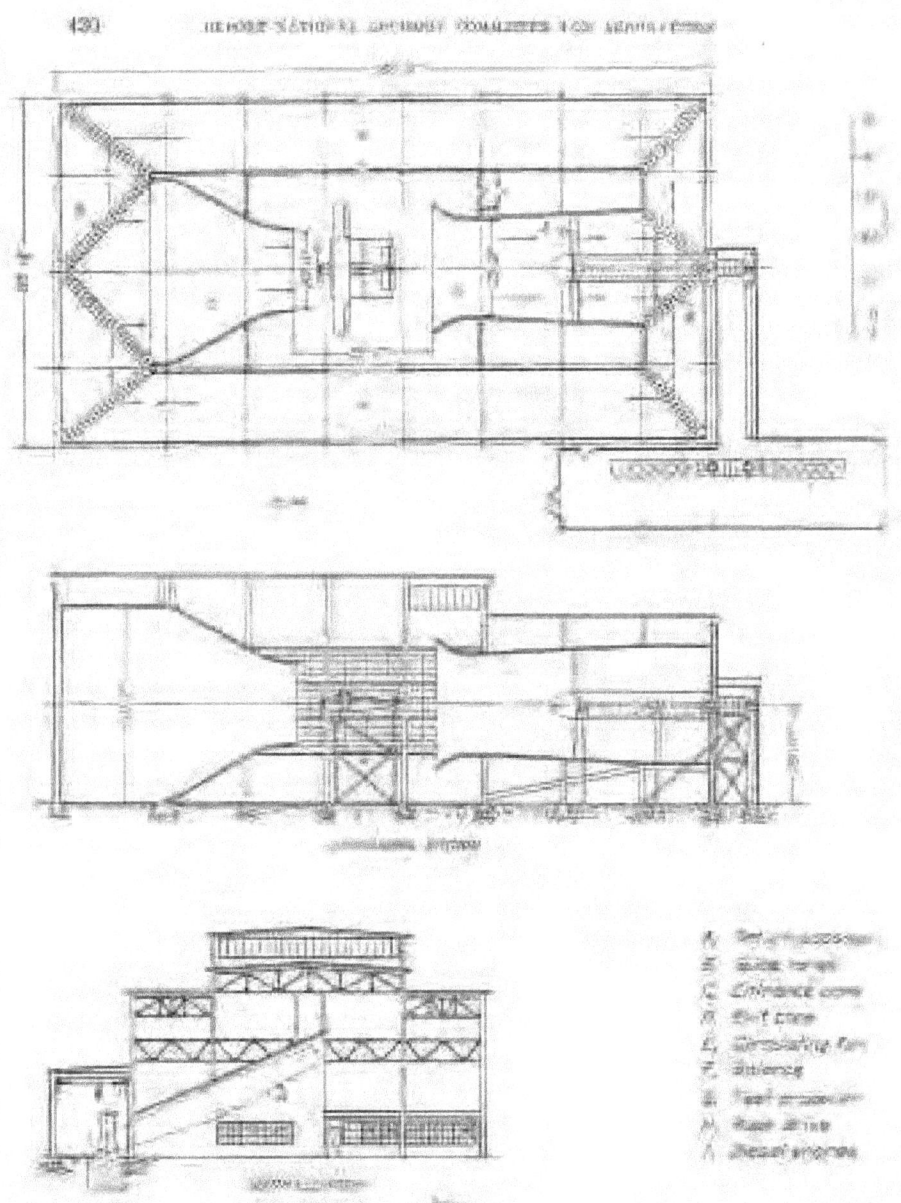

The tunnel proper is a wood walled steel-framed structure 166 feet long and 89 feet wide, having a maximum height of 56 feet. The walls are of 2 inch by 6 inch tongued and grooved pine sheathing attached to steel columns with wooden nailers. Except for the fact that the walls are on the inside of the framing only and that the heights vary from point to point, standard structural practice is followed.

The tunnel (fig. 1) is of the open-throat, closed test chamber, return passage type. The direction of the air flow is indicated by arrows. The air is drawn across the test chamber into the exit cone by a propeller fan. After passing through the fan the air column divides, passes through successive sets of guide vanes at the corners, and returns through the side passages to the entrance cone. The areas of the passages are varied in the case of the exit cone by varying the diameter, and of the return passages by sloping the roof and floor, so that the velocity of the moving air is gradually decreased at the large end of the entrance cone to about one-eighth that through the test chamber. It is then rapidly accelerated in passing through the entrance cone.

TEST CHAMBER

The test chamber is about 50 by 60 by 55 ft., located, as shown in Figure 1, near the center of the tunnel structure. Large windows in the east and west walls afford ample light. Doors open out of the west wall to permit the movement of material to and from the test chamber. An electric crane traveling along a roof truss is useful in lifting loads about the chamber and onto the balance. Electrical outlets for light and power are provided at convenient points.

ENTRANCE CONE

The entrance cone (fig. 3) is of 50 ft. square section at the large end, changing to 20 ft. diameter in its length of 36 ft. It is constructed of a double layer of ¾ in.

by 2 in. sheathing bent, fitted, and nailed to wood forming rings. These, in turn, are bolted to angle clips riveted to I-beams bent to proper shape. A built-up wood ring forms the end of the cone. At the large end the cone runs into the return passage on a gradual curve.

FIG. 2

EXIT CONE

The exit cone (fig. 4) is similar in construction to the entrance cone. It is circular in section from the mouth of the bell in the test chamber to the fan. The cone has a diameter of 33 ft. at the mouth of the bell, reducing to 25 ft. at the test chamber wall and then increasing with a 7∞ included angle to 28 ft. in diameter at the fan. From the fan a gradual change is made to 30 ft. square at the return passage. The total length of the exit cone is 52 ft.

GUIDE VANES

Guide vanes (fig. 5) are located, as shown in Figure 1, at each point of change of direction of the air stream. These consist of metal covered wood framed curved

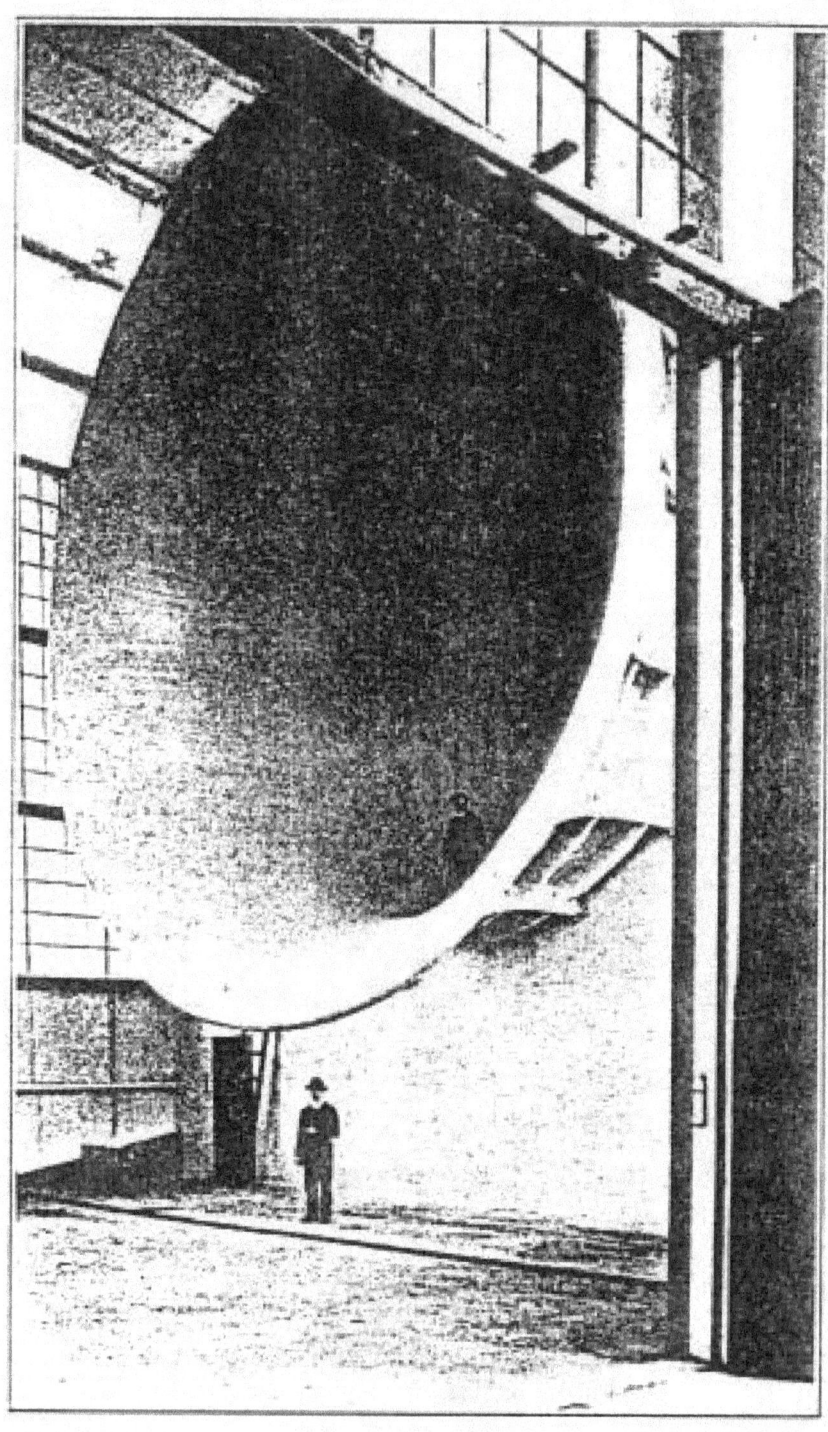

Fig. 4

shapes built up in sections 5 ft. long. Rounded leading edges and pointed trailing edges are of wood. The vanes are so proportioned that the free area between them is about a mean of the passage areas before and behind them. Streamlined wood separators run diagonally across the corners and act as stiffeners and supports for each tier of vane sections. It may also be noted in Figure 5 that cross bracing in the return passages is streamlined in the direction of flow.

FIG. 5

FAN

Circulation of air is accomplished with a 28 ft. diameter propeller type fan. (Fig. 6 and fig. 1.) It consists of eight cast, heat-treated, aluminum-alloy blades screwed into a cast steel hub and locked in place by means of wedge rings which are forced between the blade shanks and the hub. This makes it possible to change the pitch to adapt the fan to the driving engine characteristics or to secure different air speeds with the same engine speed. At present 100 M. P. H. is obtained with 330 R. P. M. of the engines and fan. The weight of each blade is 600 pounds and the total weight of the fan is about 3 ½ tons.

Fig. 1

A steel framed sheet aluminum spinner 7 ft. in diameter is attached to the hub. This fairs into the cylindrical propeller shaft housing.

DRIVE SHAFT

The fan hub is keyed on to the tapered end of an 8-in. solid steel shaft running back through the exit cone and return passage. This shaft is supported on four plain, collar oiled, bearings and one combination plain radial bearing and deep groove ball thrust bearing. The latter is located at the end of the shaft opposite the propeller. The bearings are supported, in turn, on steel I-beam A frames resting on spread footings in the ground below the exit cone.

The shaft and bearing bracing are surrounded by a cylindrical sheet steel fairing on wood formers of the same diameter as the fan spinner. The legs of the A frames are also suitably faired.

POWER PLANT AND TRANSMISSION

Because of local conditions it was found advisable to use Diesel engines rather than electric motors to furnish power for circulating the air through the tunnel. Two Diesel engines, which had been removed from a submarine, were furnished by the Navy Department.

These engines are full Diesel M.A.N. type, 6 cylinder, 4 cycle, single acting, rated at 1,000 H.P. each at 375 R.P.M. After due consideration, it was decided to install these end to end as they had been in the submarine, using the existing flywheels and clutches, spacing them far enough apart to allow the installation of a driving sheave between. The location of the engine room is shown in Figure 1, with the engine and sheave position indicated. The auxiliary machinery is arranged on the opposite side of the room from the engines. Figure 7 is a general view of the engine room.

Power is transmitted from the driving sheave to a similar sheave located forward of the thrust bearing on a part of the fan shaft extending through the main tunnel wall. Forty-four "Texrope" V-belts are used with two adjustable grooved idler pulleys located as shown in the end view, Figure 1. The transmission ratio is 1 to 1. The belt pull is carried on a suitable steel structure and the whole framing, is roofed over and sided with a protected corrugated metal. This same material is a covering for the engine room proper, rendering this part of the installation practically fireproof.

BALANCE

The testing of full size airplane fuselages necessitated the design of a new type balance. This, as shown in Figures 8 and 9, consists essentially of a triangular frame **A** of steel channels and gussets resting on tubular steel posts **B**, which in turn bear on the platforms of ordinary beam scales **C**. Double knife edges are provided at both ends of these posts. The rear post of the frame is on the longitudinal center line of the balance and the forward posts are at equal distances (5 ft.) on either side. The sum of the net readings on all three balances is the lift. The pitching moment is computed from the sum of the front balance readings and from the rear balance reading. Since the rear balance is on the longitudinal axis, the rolling moment is computed from the net readings on the front balances.

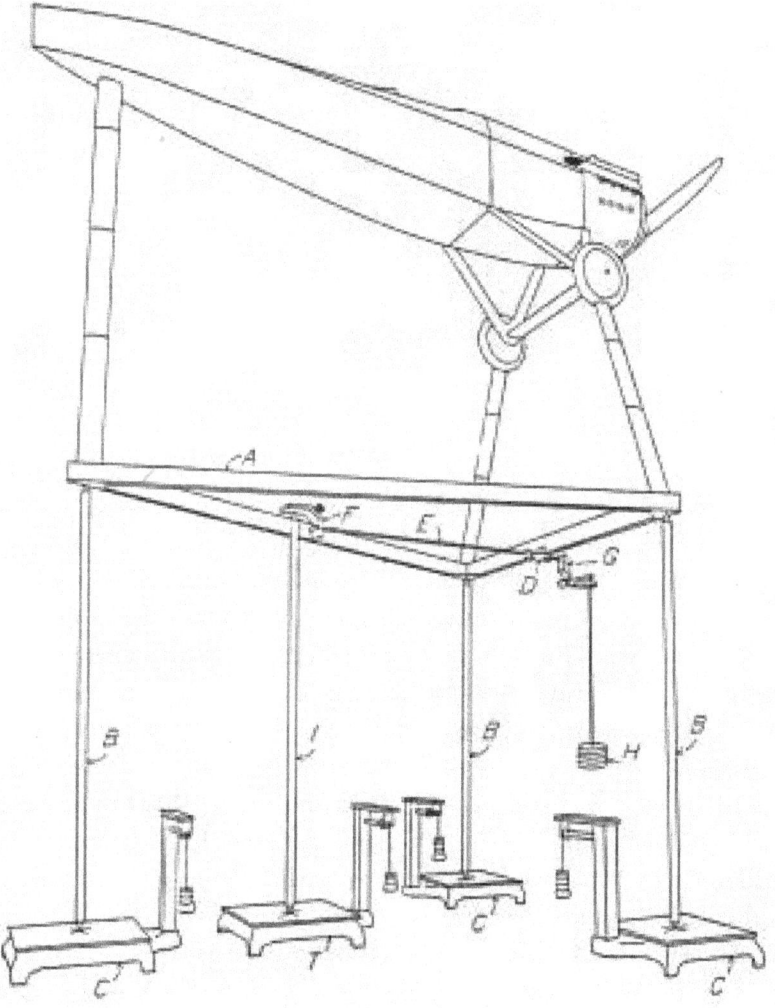

Fig. 8

At **D** are located knife edges connected to tie rods **E** running forward to a bell crank **G** and a counterweight **H**, and aft to a bell crank **F** and a post **I** resting on the scale **T**. A forward pull or thrust on the frame produces a down force on the post or an increase in load on the scale T. The counterweight H produces an initial load on the scale T and consequently a drag or backward force is measured as a diminution of load on the scale. The counterweight consists of several 50-lb. units and can be easily adapted to the range of thrusts and drags expected during any one test.

Fig. 6

TORQUE DYNAMOMETER

The fixed knife edges on the bell cranks are seated on blocks bolted to a rectangular steel frame rigidly fastened to the floor, as shown in Figure 9. In addition, this frame is provided with knife edges, links, and counterweights which hold the triangular frame in a fixed lateral position. Screws are also provided for raising the triangular frame from the knife edges while working on the attached apparatus. A stairway at the rear and a grating floor facilitate work on the supports and apparatus mounted on the balance.

At each corner of the triangular frame are ball ended steel tubes, adjustable in length and angle, which support the body under test. The forward tubes, in the case of a fuselage with landing gear, have a fitting at the upper end which clamps the axle of the landing gear. The rear post has a ball-and-socket attachment to the fuselage. The drag of these supports is reduced by streamline fairings which also serve to cover wires and fuel and water lines running to the fuselage.

TORQUE DYNAMOMETER

As the engine power is one of the major variables determining the propeller characteristics, a test fuselage has been developed which allows the engine driving the propeller to be mounted on a dynamometer and the torque to be measured directly.

As shown in Figure 10, this is a heavy angle and strap steel frame so shaped that it can be slipped inside a standard airplane fuselage and supported by suitable

PROPELLER BLADE DEFLECTION

ENGINE STARTER

blocking. At its forward end a steel casting is fixed carrying two large ball bearings and an extension shaft and plate. An airplane engine can be mounted on this plate. Its torque, which is carried through the plate and shaft and a special linkage, is read on a dial scale mounted farther back in the fuselage. This dial reads directly in lb. ft. up to 2,000 lb. ft. and a total of 4,000 lb. ft. may be obtained with a counterweight. A double link system renders the operation independent of the direction of engine rotation.

Figure 11 shows a VE-7 airplane mounted on the test fuselage with an E-2 engine on the plate. The radiator is mounted independently of the engine and is not used for cooling. Cooling water is supplied and returned through rubber hose running back through the fuselage and down the rear post to the floor. Figure 12 shows the dial in the rear cockpit.

To reduce the fire hazard and to simplify installation, fuel is supplied from a small tank located on the outer wall of the tunnel, feeding by gravity to the engine carburetor. The gravity tank is filled from a large storage tank by an electric gear pump which is started and stopped by an automatic float switch in the gravity tank.

PROPELLER BLADE DEFLECTION

Propeller blade deflections are measured as follows. A telescope with cross hairs, in conjunction with a prism, is mounted on a lathe bed beneath the propeller

being tested. One blade of the propeller at a time is painted black and a black background is painted on the ceiling. Two lights are arranged so that their beams strike the propeller blade. On sighting through the telescope no image will be seen when the black blade passes the black background; but when the white or bright metal blade passes, a line of the leading or trailing edge will appear. By locating, the cross hairs successively on these lines and reading, the distance moved it is possible to compute the angular deflection of the propeller blade at any given radius. Further development of this apparatus is in process.

MANOMETER

For routine testing, velocities are calculated from the readings of an N.A.C.A. micro-manometer, one side of which is connected to plates set in the walls of the return passage and entrance cone, and the other side open to the air in the test chamber.

SPEED REGULATOR

An air-speed regulator has been developed to insure a uniform dynamic pressure, but to date its use has not been found necessary.

ENGINE STARTER

For starting an airplane engine mounted on the balance, an electric starter is secured to the entrance cone shown in Figures 12 and 13. A hollow shaft with a pin meshing with a dog on the propeller shaft is driven by means of a chain from an electric motor. The whole unit is arranged to swing down clear of the air stream during a test.

CALIBRATIONS

A velocity survey has been made over the entire cross section of the air stream at a point about 6 ft. back of the entrance cone edge. Seventy-nine points were taken at 2 ft. intervals. The velocity without a honeycomb or air straightener was found to be constant within 1 per cent over the test area. This is attributed to the large reduction of area in the entrance cone. Large variations of velocity at the

entrance to the cone are greatly reduced by the rapid acceleration through it. In consequence, while provision was made in the structure for the installation of a honeycomb, none has been deemed necessary.

The wall plates and manometer are calibrated from time to time against a group of Pitot tubes set in the air stream. These are attached to a movable frame to which one or more Pitot tubes may be attached and the velocity at any point in the air stream determined without a special installation. In particular, this apparatus is used to measure the velocities in the plane of the airplane propeller.

The tunnel was designed to give a velocity of 100 M.P.H. with an energy ratio of 1.2 based on the power input to the fan. A velocity of 110 M.P.H. has been obtained indicating an energy ratio higher than that assumed.

Figure 13 is a view in the test chamber during a standard propeller test. Balances, manometer and deflection apparatus are shown in operation. An observer stationed in the fuselage to control the engine and read the torque scale does not appear in this view.

SOME RESULTS

A considerable amount of testing has already been accomplished since operation began in July, 1927. Figure 14, taken from Technical Note No. 271 (Reference 1), indicates the proportional drag of various parts of the Sperry Messenger airplane fuselage. The propeller research tunnel is particularly adapted to full-scale tests of this nature. Figure 15 shows the characteristics of Propeller I, previously tested in model form at Stanford University, and in two separate flight tests. (References

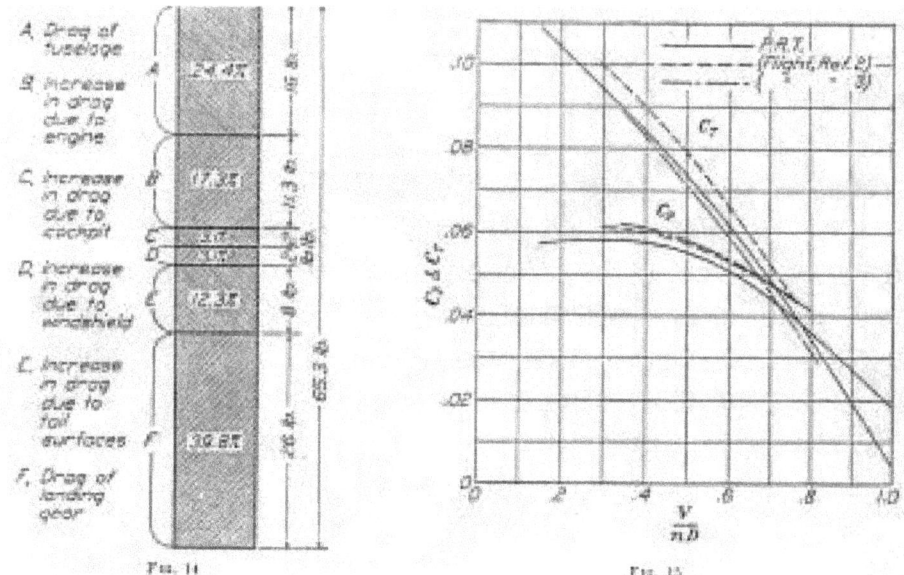

Fig. 14   Fig. 15

2 and 3.) Curves from these tests are given for comparison. Attention is called to the inaccuracy of flight data mentioned in the introduction to this paper. Tests of wings of 12 ft. span have also been made at speeds up to 100 M. P. H. A comparatively high Reynolds number is thus attained. Figure 16 is a view of a wing set up for test. A comprehensive program of tests to determine the effect of propellers on air-cooled engines operating in front of various types of fuselages with several shapes of cowling is now in progress. The effect of these bodies on the propeller is also being determined.

ACCURACY

Dynamic pressure, thrust, torque, and R.P.M. are measured with an accuracy of from 1 to 2 per cent. Computed data are, therefore, correct to approximately plus or minus 2 per cent and final faired curves through computed points to about plus or minus 1 per cent. This compares favorably with other engineering measurements. The beam thrust balance is to be replaced with a dial scale which will increase the accuracy and will enable the observers to read more quickly and more nearly simultaneously. A change in the linkage of the torque scale is contemplated which will increase the accuracy of that reading. When these changes are in effect it is hoped that computed points will be correct to plus or minus 1 per cent.

CONCLUSION

The propeller research tunnel fulfills a long-felt want in aerodynamic research. Propellers can be tested full scale, and with actual engines and bodies in place, with an accuracy not attained in flight tests. The components of the airplane, fuselage, landing gear, and tail surfaces can be tested full scale. While full size

FIG. 18
ACCURACY

wings can not be accommodated, a stub wing can be installed which is sufficient to study the effects of all parts of the airplane on the propulsive system, and vice versa. Tests thus far made are consistent and reliable and it is increasingly evident that the propeller research tunnel is a useful addition to the extensive research facilities of the National Advisory Committee for Aeronautics.

LANGLEY MEMORIAL AERONAUTICAL LABORATORY,
NATIONAL ADVISORY COMMITTEE FOR AERONAUTICS,
LANGLEY FIELD, VA., June 2, 1928.

REFERENCES

Reference 1. Weick, Fred E.: Full Scale Drag on Various Parts of Sperry Messenger Airplane. N.A.C.A. Technical Note No. 271, 1928.

Reference 2. Durand, W. F., and Lesley, E. P.: Comparison of Tests on Air Propellers in Flight with Wind Tunnel Model Tests on Similar Forms. N.A.C.A. Technical Report No. 220, 1926.

Reference 3. Crowley, J. W., Jr., and Mixson, R. E.: Characteristics of Five Propellers in Flight. N.A.C.A. Technical Report No. 292, 1928.

Reference 4. Glauert, H., and Locke, C. N. H.: On the Advantages of an Open Jet Wind Tunnel for Airscrew Tests. B.A.C.A. R.& M. 1033, 1926.

Reference 5. Durand, W. F.: Experimental Research on Air Propellers. N.A.C.A. Technical Report No. 14, 1917.

Document 2-20(c), Fred E. Weick, excerpts from From the Ground Up, 1988.

In response to the requests of Dr. Lewis about how NACA might help the Bureau of Aeronautics, I kept mentioning the need for fullscale propeller tests at high tip speed, where compressibility losses became evident. Compressibility is the physical property by which the volume of matter decreases to some extent as pressure is brought to bear on it. The British had made high-tip-speed tests on two-foot models in a wind tunnel, but the Reynolds number was so low that the results were questionable when applied to full scale. Their compressibility losses were much greater than those indicated by our meager flight tests in this country. At that time, however, I could not see any practical way of making full-scale tests other than in flight.

One day in early 1925 Dr. Lewis called me into his office and asked me how I would like to see a wind tunnel capable of making the full-scale propeller tests. But, I said, in order to make full-scale tests on a 10-foot propeller, the diameter of the tunnel's throat would have to be at least 20 feet, or four times the size of NACA's largest wind tunnel at Langley up to that time. "Yes, you're right," Dr. Lewis said. "But," I said, "in order for it to be practical, the tunnel's airflow would have to reach at least 100 miles an hour, and to achieve that, you'd have to have an immense amount of power—probably a couple of thousand horsepower." "Yes," Dr. Lewis said again. "I've been talking it over with Dr. Munk and we think that such an arrangement might be practical." I was astonished: a 20-foot tunnel would require a structure of 4^3 or sixty-four times the volume! Neither the Hampton nor Newport News power plant was large enough to supply electric power, but NACA, Dr. Lewis informed me, had arranged to get two navy surplus diesel engines of 1,000 horsepower each, taken from a T-2 submarine. "If we can get this tunnel built," Dr. Lewis asked me, "would you like to come down to Langley Field and run it for us?" Without hesitation, I said that I would indeed.

When Dr. Lewis made the suggestion that I be transferred from the navy to NACA, the assistant chief of the Bureau of Aeronautics, Capt. Emory S. Land, absolutely refused. The captain could not see losing me to another government organization when I had just become useful in his propeller department. Some time passed without approval for my transfer, until the line officer in charge of the propeller department, Lt. Stanton Wooster, who had replaced my old boss Lieutenant Shoemaker, advised Captain Land that I would probably do the Navy Department at least as much good doing research in NACA's new propeller tunnel at Langley. Moreover, Wooster told him he didn't exactly like standing in the way of a possible improvement for me. So after this polite badgering, Captain Land finally said, "All right, all right," and the transfer was made.

NACA at that time had an annual appropriation of roughly $250,000 with which to pay a staff of one hundred twenty-five, buy buildings and equipment, and take care of operating expenses. As I remember it, the total cost of the bare structure for the propeller research tunnel was about $70,000. In order to get it started as soon as possible, NACA had skimped and saved $35,000 out of its FY appropriation, which ended on June 30, until the early 1970s the end of the government's fiscal year. On June 30 NACA had entered into a contract with the Austin Company to construct the tunnel's outer shell for $35,000. Then on the next day, July 1, it had entered into another contract to construct the internal structure, entrance and exit cones, and return passages. Construction started at once. This all happened before my transfer and was the main reason I was itching to move from Washington down to Langley.

[. . .]

The distance by air from Washington, D.C., to Hampton, Virginia, the town nearest Langley Field, is only about 120 miles, but by road through Richmond it is about 175 miles. In the 1920s these roads were surfaced with gravel and often badly rutted; smooth ribbons of concrete were not to be found in rural Virginia. In our Model T roadster, packed to overflowing, it took us all day to make the trip.

The engineer in charge of the Langley Memorial Aeronautical Laboratory at that time, a Californian by the name of Leigh M. Griffith, appeared unhappy with the idea that I had been placed under him from above; in fact, Griffith must have been generally unhappy with his situation at Langley, for he left within the month. He was replaced by Henry J.E. Reid, an electrical engineer who had been in charge of the laboratory's instrumentation. Reid remained the engineer-in-charge until he retired from the National Aeronautics and Space Administration (NASA) in 1960. The lab had a flight research division headed by test pilot Thomas Carroll, a power plant division headed by Carlton Kemper, and two wind tunnel sections, one, the 5-foot or atmospheric wind tunnel (AWT) section headed by Elliott G. Reid, and, the other, the variable-density tunnel (VDT) section headed by George J. Higgins. There was also an instrument shop, model shop, technical service department, and a clerical and property office headed by Edward R. ("Ray") Sharp. The new 20-foot propeller research tunnel (PRT) was being constructed under the supervision of Elton W. Miller, a mechanical engineer who had previously been in charge of the construction of the variable-density tunnel. I was placed under Miller until the tunnel was ready for operation.

By the time I started at Langley, the outer shell of the new tunnel had been completed but work on the entrance and exit cones and guide vanes was still going on. The tunnel, which had been laid out by Dr. Max Munk in Washington, was of the open-throat type then most suitable for testing propellers. My first job was to design and get constructed a balance arrangement that measured the aerodynamic forces on the model and the model's reaction to them. This balance had to support an airplane fuselage, complete with engine and propeller, 25 feet above the floor in the center of the tunnel's 20-foot-in-diameter airstream. All of the pertinent forces, such as drag, thrust, and moments, were to be measured down below by four small and simple beam scales.

Since 1921 Dr. Munk had been holed up in a little office at NACA headquarters in Washington, where he had been turning out excellent theoretical work. Munk had studied under Ludwig Prandtl at the University of Göttingen in Germany and had been brought to this country by the NACA in 1921. His entry into this country required two presidential orders: one to get a former enemy into the country, and another to get him a job in the government. And I guessed this helped him to appreciate his importance.

Without question, Munk was a genius, and, without question, he was a difficult person to work with. In early 1926 he decided on his own that, since Langley laboratory was where all the real action was taking place, that was where he should be. NACA headquarters must have agreed, because it made him the lab's chief of aerodynamics; this put him in charge of the flight research division and the two wind tunnel sections. My boss, Elton Miller, now reported to Munk, and all of my work ultimately had to be approved by him. I had known Munk in Washington and had great respect for his abilities. On the other hand, I did not want my balance design turned down at the last minute; so I had taken the pain to take each detail of design, mostly on cross-section paper, up to Munk to get his approval, and I got his initials on every single one of them. This, I thought, would certainly assure his final approval.

The movable parts of the balance supporting the airplane were supported by a structural steel framework about 12-feet high, 12-feet wide, and 16-feet long. In place of adjustable cables, steel angles ¼ by 2 ½ by 2 ½ inches provided the diagonal bracing.

A couple of days before we expected to try out the balance using a little Sperry Messenger airplane with its 60-horsepower engine running, Munk made an unannounced visit to the PRT building. Just as he walked into the bare-walled 50-foot cubicle that housed the test section, a loud horn squawked, calling someone to the telephone. This sent Dr. Munk into a tantrum, and I immediately had one of my mechanics disconnect the horn. Before he had entirely calmed down, he walked over toward the balance structure and put his hands on the long diagonal braces. These were fairly flexible, and he found he could move them back and forth a bit. Visualizing the entire structure vibrating to the point of failure and the whole airplane and balance crashing to the ground, the perturbed Munk ordered me to tear down the balance entirely and to design a new foundation and framework for it. He then turned and went back to his office a couple of blocks away.

Naturally I, too, was perturbed. Munk, after all, had approved every detail of my balance design. Not knowing what to do, I waited for some time to give him an opportunity to cool down. Then I went to his office and, as calmly as I could manage, mentioned that I thought the natural frequencies of the long diagonal members would be so low that vibrations would not be incited by the more rapid impulses from the engine and propeller. But mainly I suggested that, inasmuch as all the parts were made and ready to be put up, why not wait a couple of days before tearing it down and make a careful trial using the Sperry Messenger, starting at low speed, gradually increasing it, before dismantling the apparatus. Munk finally agreed, but demanded to be present when the test was made.

I did not like the idea of his presence one iota. To start the engine, the Messenger's propeller had to be cranked by hand from a balloon ladder that was put up in front of the propeller 25-feet above the floor. (A balloon ladder was like a fireman's ladder but its base was attached to a pair of weighted wheels, which permitted it to be "leaned" out into space. At its base there was a "protractor" that told you how far it could be angled without tipping over.) This sweaty business often took some time. It was not the kind of operation I wanted the excitable Munk to watch. Moreover, since no one else in the PRT section had ever started an airplane engine by turning the propeller, I was the one who was going to have to do it.

I brought my problem to Elton Miller, my boss, and to Henry Reid, the engineer-in-charge. Together, we decided that the only thing to do was to make an end-run around Munk and check out the tunnel balance system in his absence. This was easily done, as Munk worked on theoretical problems in his room at a Hampton boarding house every afternoon. We set up the test run and after a bit got the engine started without any difficulty. We then experimented with it until we could start it easily and felt ready for the final trial.

The problem of convincing Munk remained. We could not simply tell him about the successful test, so we agreed to arrange another "first test" for Munk to witness. Engineer-in-charge Reid escorted Munk to the tunnel the next morning. I casually said, "Good morning," clambered up the ladder, and pulled through the Messenger's prop. Luckily, the engine started on the first try. We then moved the ladder away, ran the engine through its entire range with no vibration difficulty, and then shut it down. Now, I wondered, what sort of explosion will we have? I needn't have worried. Munk walked toward me with his hand outstretched and congratulated me on the success of the operation. Everything had turned out all right. The balance system of the PRT operated satisfactorily with engines of up to 400 horsepower into the late 1930s, when it was replaced by a new and better one.

In 1926 Dr. Munk gave a number of lectures on theoretical aerodynamics to a select group of young Langley engineers. I was very happy to learn these things from him. Ever since graduation from the University of Illinois, I had thought about taking some graduate courses in aeronautical engineering. While working for Tony Yackey, I had read in a magazine article about the graduate courses in aeronautics offered at Massachusetts Institute of Technology. I had written MIT for information and had received a letter back from Professor Edward P. Warner, who had been Langley's first chief physicist in 1919 and would later become assistant secretary of the navy for aeronautics, editor of the magazine Aviation, and finally president of the International Civil Aviation Organization (ICAO), which continues to coordinate the rules and regulations for aeronautical activities throughout the nations of the world. I had hoped still to find a way to work in some graduate courses even after reporting to work at the Bureau of Aeronautics. But Dr. Lewis had talked me out of the idea on the basis that formal aeronautical engineering education was inferior to what I could learn if I went to work for the NACA at Langley. I guess he was probably right in regard to the aeronautical courses per se, but on occasion, in later years, I sorely missed the extra mathematics and physics that would have been obtained in school.

As mentioned earlier, the power plant for the new PRT consisted of two 1,000-horsepower, six-cylinder in-line diesel engines taken from a T-2 submarine. These engines were located end-to-end with crankshafts connected to a large sheave or pulley between them. This sheave carried forty-four Tex-rope V-belts to a similar sheave on the shaft of the propeller fan that drove the air through the tunnel. The shaft of the propeller fan was 25 feet above the ground, and the two sheaves were 55 feet apart center to center. Because we were concerned that some destructive vibrations might occur in the crankshaft-sheave assembly, we decided that a theoretical analysis of the torsional oscillations should be made, with Dr. Munk outlining the problem and a new man, Dr. Paul Hemke, to work out the solution. As a junior engineer, my assignment was to give the measurements and sizes that I would get from the drawings of the engines and sheaves.

I had no difficulty giving them the measurements, but Dr. Hemke was never able to get the gist of the torsional pendulum problem as described by Munk. This went on for some time with no results being obtained. Finally, I looked into my mechanical engineers' handbook and into a couple of textbooks and found that considerable work had been done on the problem and that the solution was not too difficult. I made the computation myself, coming out with a natural frequency of 312 RPMs. Later on, after the tunnel was in operation, some men came down from the navy shipyard in Brooklyn with equipment to measure the torsional oscillations; they found exactly the same natural frequency as I had computed. Hitting it exactly, of course, was a matter of luck, but it helped give me a good reputation, whether I deserved it or not. The success put me in good with Munk, but unfortunately Dr. Hemke was never able to work satisfactorily with him. A short time later he left the NACA. Hemke later joined the faculty of the U.S. Naval Academy, after holding a prestigious Guggenheim Fellowship for research under B. Melville Jones at Cambridge.

Another problem I helped to solve was the design of the 28-foot propeller fan that was to circulate the air in the propeller research tunnel. This fan needed to have eight blades of normal width. The exact energy ratio of the tunnel was not known in advance, so I desired to have blades that could be adjusted so that the pitch could be set exactly right after trial runs. Aluminum-alloy blades therefore seemed the best choice, but the blades we wanted were too large to be forged in the manner of the aluminum-alloy propeller blades then being manufactured. Fortunately, the propeller was to turn at only 375 revolutions per minute, which meant that the stresses would be very low in comparison even with airplane propellers having large diameters. This gave me the idea that a cast aluminum alloy might be used successfully, which it was.

I arranged with the Aluminum Company of America to cast the blades in their plant at Cleveland, Ohio. Before the large blades were cast, however, the company made two blades for a small ten-foot model that I then took to McCook Field in Dayton, where they were tested by Army Air Service engineers on their propeller whirl rig. This test showed the blades to be sufficiently strong.

[. . .]

We had gotten the diesel engines to run satisfactorily in a short time, but the long Tex-rope drive, with its forty-four different V-belts, were always getting so tangled up that they could not stay on the sheaves satisfactorily. It took several months of experimenting with idling sheaves in various locations before the operation became satisfactory. The entire drive, with Tex-ropes and sheaves, had been purchased from the Allis Chalmers Company and had been guaranteed to operate satisfactorily. After spending much money and time on this problem, the NACA sued Allis Chalmers for a sizable rebate. My daily log of operations was used as part of the testimony, and after one futile operation I had put down in disgust, "No soap." Although this was a generally used term in the Midwest, where I came from, indicating failure, no one in the room, all of whom were from the East, knew what I had meant. So I had to explain it in detail, an amusing interlude in an otherwise dull trial.

[. . .]

Finally, in 1927, the Tex-rope problem was solved and the PRT was ready for actual testing. The tunnel personnel included Donald H. Wood, a mechanical engineer from Renssalaer Polytechnic Institute who was about my age and who had been with the PRT section from the start; Melvin N. Gough, a young engineer who had come to Langley directly from Johns Hopkins; William H. Herrnstein, Jr., an engineer who had come directly from the University of Michigan; and John L. Crigler and Ray Windier, also engineers. The power plant and shop work under Ted Myers, who was a little older than the rest of us, included George Poe and Marvin Forrest. There were two or three others whose names I have forgotten. At any rate, we had a good team and we all worked together very well.

Just before testing actually started in the PRT, Langley experienced a rather sad affair: a revolt against Dr. Munk, the head of our aerodynamics division. Munk was a wonderful theoretical aerodynamicist, but, as the story of my design of the PRT's balance structure illustrates, he was also an extremely difficult supervisor, not just for me but for all the section heads working directly under him. Eventually all the section heads, including Elton Miller, decided they couldn't work with Munk any longer and handed in their resignations. Munk was then relieved of his job, which I feel was a great loss. If Dr. Lewis could only have kept him holed up in his little office in Washington, Munk could have produced a great deal more of his useful theoretical work.

With Munk's departure, Elton Miller became the chief of the aerodynamics division, and I became head of the PRT section.

In the first months of PRT operation, just to get experience, we merely tested the Sperry Messenger and obtained drag data with various parts of the airplane removed. This was then written up and published as an NACA technical note. Before we could actually make propeller tests, though, we had to design and build a dynamometer that would support the engine and propeller up in the airstream and measure the torque of the engine while it was operating. We mounted the dynamometer in a long structural steel frame of small cross-section. Any engine up to about 500 horsepower could be mounted ahead of it, and any fuselage form could be put on over it. The engine torque we measured directly in foot pounds on a dial-type Toledo scale.

For the first propeller tests, we slipped this dynamometer inside the fuselage of a Vought VE-7 airplane. The engine was a 180-horsepower Wright E-2 liquid-cooled unit, similar to the old Hissos. These tests were made primarily to compare with the propeller data we had gotten from our flight testing and with the small-model data acquired previously from wind-tunnel testing at Stanford University. Three full-size wooden propellers of the same type previously used on the same VE-7 airplane, and models of which had also been investigated in the Stanford tunnel, were tested. The results agreed as well as could have been expected, considering the difference in Reynolds number (the nondimensional coefficient used since Osborne Reynolds's pioneering experiments at the University of Manchester in the 1880s as a measure of the dynamic scale of a flow) between flight and tunnel testing.

An interesting sidelight about the accuracy of aerodynamic testing appeared soon after we began PRT testing. In our final plots, the points of our wind-tunnel data for the full-scale tests had substantial scatter about the curves, whereas the small-model tests had the points right on the curves. I worried about this for a while, until I found the answer. Our results at NACA were plotted with fine points and fine lines at a large scale, and the model tests at Stanford were in printed form with heavy lines and large points. When we plotted our results in the same manner as the model test results, our accuracy was at least as good as Stanford's. I reported these results in NACA Technical Report (TR) 301, "Full-Scale Tests of Wood Propellers on a VE-7 Airplane in the Propeller Research Tunnel," in 1928. TR 300, describing the tunnel and testing equipment, had been prepared just previously by Donald H. Wood and me.

After the technique of making satisfactory propeller tests in the propeller research tunnel had been worked out satisfactorily, one of the first things that I wanted to investigate was the effect of high propeller tip speeds; after all, this had been one of the main reasons for having the tunnel built in the first place. The original propeller-testing setup with a 180-horsepower engine was not powerful enough to cover the range desired in high tip speeds, but we did what we could with it. The tests had to be made with a very low pitch setting that corresponded to angles of attack of the cruising or level range of flight but not of climb or take-off conditions. The range of the tests was from 600 to 1,000 feet per second, about 0.5 to 0.9 times the velocity of sound in air, or in modern terminology, from Mach 0.5 to Mach 0.9. Within the range of these tests, the effect of tip speed on the propulsive efficiency was negligible, and there was no loss due to the higher tip speeds. The results are given in my NACA Technical Report 302 (June 20, 1928). Later on, with a more powerful engine, we were able to test a whole series of propellers at tip speeds of up to 1,300 feet per second, which is well above the speed of sound. The results of these tests are recorded in NACA TR 375 (November 1930) by Donald H. Wood.

Document 2-21(a–c)

(a) Elliot G. Reid, "Memorandum on Proposed Giant Wind Tunnel," 3 April 1925, National Archives, 171.1, Washington, D.C.

(b) Arthur W. Gardiner, "Memorandum on Proposed Giant Wind Tunnel," 22 April 1925, National Archives, 171.1, Washington, D.C.

(c) Smith J. DeFrance, *The NACA Full-Scale Wind Tunnel*, NACA Technical Report 459, Washington, D.C., 1933.

By 1930, the VDT had proven that wind tunnel testing of models at high Reynolds numbers produced data that accurately predicted the performance of airfoils, and the PRT had shown the value of full-scale testing in determining the drag characteristics and synergistic effects of integrating the various components of an airplane. Yet, neither tunnel totally reproduced the conditions experienced by a complete airplane in flight, so the NACA decided to build a still larger wind tunnel, one capable of testing full-size aircraft. This was not a brand new idea. Shortly after Max Munk proposed the twenty-foot-diameter PRT in March 1925, others at Langley began to call for a ten-foot increase in its diameter to enable the tunnel to test large- and full-scale aircraft components. Arthur W. Gardiner and Elliot G. Reid, assistant aeronautical engineers at Langley, each prepared a "Memorandum on Proposed Giant Wind tunnel" that applauded the idea of a facility suitable for propeller research and suggested how such a "giant wind tunnel" could do much more than test propellers. Their ideas had little effect on the design of the PRT, which was built with a twenty-foot-diameter nozzle, but good ideas rarely die.

The Full-Scale Tunnel (FST), completed in 1931, was gigantic in all respects, and it loomed over every other structure at the LMAL. Like the PRT, the FST was a closed-circuit tunnel with an open test section and return ducts whose outer walls formed the building walls. With its larger size, the FST design included dual return ducts. The elliptical nozzle—a brilliant solution to the problems associated with very large circular or rectangular designs—delivered an air stream that measured thirty by sixty feet and attained velocities of up to 118 miles per hour. Two thirty-five-foot-diameter propellers, each driven by a 4,000 horsepower electric motor, circulated almost 160 tons of air through the 838-foot-long circuits. The FST could handle airplanes or large-scale models with wingspans of up to forty-five feet. While the FST's speeds were not extraordinary, they were sufficient to enable measurements that could be confidently extrapolated to cover the aircraft's entire speed range because no scale factor was involved. In addition to lift and drag studies, the FST supported stability analysis and research into the interferences between the various parts of an airplane. With the FST joining the VDT and PRT in service, the NACA's Langley Laboratory possessed the most extensive and versatile research facilities in the world. As with its earlier facilities, the NACA published a technical report discussing the FST and its capabilities.

Document 2-21(a), Elliot G. Reid, "Memorandum on Proposed Giant Wind Tunnel," 1925.

MEMORANDUM ON PROPOSED GIANT WIND TUNNEL.

I have been thinking over the possibilities of a tunnel of the size and type suggested and have become very enthusiastic over the project. As requested, I am outlining my ideas of the value of such an apparatus.

Below are listed a number of problems which I consider particularly suitable for investigation in the large tunnel:

(A) Propeller research:
- (a) Tests of full size propellers, isolated, for determination of characteristics under the conditions of deformation imposed by air loads and centrifugal forces.
- (b) Investigation of the propeller slipstream, including confirmation or rejection of the inflow theory and investigation of the phenomena accompanying tip speeds greater than that of sound.
- (c) Tests of propellers with axes inclined to the wind.
- (d) Tests of propellers on actual airplane fuselages for determination of full scale interference characteristics.
- (e) Pressure distribution tests on propellers.
- (f) Investigation of the magnitude and causes of full size propeller deflections.
- (g) Tests of adjustable and reversible propellers and their operating mechanisms under flight conditions.

(B) Testing of full size airplane parts which cannot be accurately reproduced to the scale of present models.
- (a) Tests of radiators, air-cooled engine installations, etc.
- (b) Tests of fuselages with and without slipstream for development of better forms.
- (c) Pressure distribution over full size wings under conditions either very difficult or dangerous to obtain in flight. (Conventional airplane practice, i.e., spars, ribs, and usual covering to be used in construction).

(C) Strength and stiffness of wing cellules.
  (a) Tests of large models or full size half cellules for the purpose of measuring air load deflections and, if it is found possible to determine aerodynamic characteristics under these conditions, improve the structures in such ways as to eliminate dangerous or inefficient conditions.
(D) Power plant testing.
  (a) Tests of air-cooled engines under flight conditions. To be made by installing engine in fuselage and running in tunnel under propeller load.
  (b) Development of cowling and cooling controls for air-cooled engines.
  (c) Heat dissipation tests on full-scale radiators.
(E) Testing of very large models of wings and complete airplanes.

Probably the one line of research to which the large tunnel would be the greatest boon is propeller testing. It is known, of course, that models could be tested in the Dense Air Tunnel [VDT] at Reynolds numbers equal to those existing in the flight operation of full size propellers. This, however, is not the criterion for complete dynamic similarity. Dr. Durand's excellent analysis of the necessary relations for the establishment of complete dynamic similarity, as given in N.A.C.A. Technical Report No. 14, Part II, shows that dimensional homogeneity can be fulfilled only by making the model of the same size as the original. While he does not take up the possibility of testing in a different medium—as compressed air—the fulfillment is still incomplete because the factors involving medium density are different.

It has been the feeling for some time that the deflections of full-scale propellers were considerably different from those of their smaller prototypes and that numerous failures to predict full-scale propeller performance arose from this fact. Mr. Lesley was convinced that such a condition did exist and supported his belief with the statement that model and full scale results were in better agreement in the case of stiff bladed propellers than for limber ones.

Therefore it would seem not only desirable, but almost imperative to have some means of testing full-size propellers and the large wind tunnel seems the only possible solution.

With regard to tests of propellers with axes inclined to the wind, it would seem that only small angles could be explored if the closed throat type of tunnel be used. This is more particularly the case for propellers mounted on full size fuselages. The wall effect would probably so modify these results as to make them misleading.

Interference and slipstream studies could be carried out with an accuracy never before possible. The possibility of using a full size radiator directly behind the propeller solves one of the most troublesome problems of such tests.

Fuselage forms have received very little study up to the present. The Variable Density Tunnel provides a means of eliminating the scale effect factor but the difficulty of producing accurate replicas of fuselages is great and to study their characteristics under the action of a slipstream is almost out of the question with this tunnel.

Static testing of wing cellules will give very good and reliable information on the ultimate strength but only very approximate data concerning deflections. It is of course known that wing characteristics may undergo large changes if any deflections are introduced and yet very little attention has been paid to this matter. By the use of a very large tunnel, it would be possible to test complete semi-span cellules of the smaller machines and very large scale models of large ones. By the use of proper strength relations, the deflections in full size cellules could be very accurately predicted from tests of models sufficiently large to permit use of the same type of construction. It would thus be possible to investigate the conditions which have caused inefficient flight in some cases, failure of controls in others and accidents in some few. The failure of monoplane wings in service during a dive is one of the latter.

The testing of aero engines under actual flight conditions is a thing which can not be accurately done at present. It is known that the ideal conditions of the electric dynamometer test do not prevail in actual operation and just how much the performance suffers is a matter of conjecture. As it would be possible to put an entire plane, minus the outer portions of the wings, into the proposed tunnel, the conditions of flight could be exactly simulated and the propeller testing equipment utilized to measure the engine output.

Likewise the cooling of the air-cooled engine could be studied with an accuracy and completeness quite out of the question in flight and it should be possible, as a result of such testing, to devise much more efficient (from the aerodynamic standpoint) cowling for such engines.

The testing of large-scale models of wings and airplanes is particularly desirable. We feel quite sure that the much feared scale effect becomes almost vanishingly small as the VL product rises to the range covered by actual airplanes and that in this range the effects of turbulence are also practically negligible. If the maximum value of this product could be brought to three or four times as large as that obtainable in present tunnels, it should be quite safe to predict full-scale performance directly from such test data. While the Variable Density Tunnel can, theoretically, accomplish this very thing, there always exists the difficulty of reproducing the airplane with sufficient accuracy in the small size necessary. Then, too, any extensive pressure distribution investigation can be carried out only with an enormous expenditure of time, apparatus, and labor.

The large tunnel would eliminate a number of these outstanding difficulties. Models could easily be built to the required accuracy with only a little more pains than is usually used in normal airplane construction. Their cost would be relatively small and thru their use it would be possible to study a large number of problems which are important but can not be undertaken for one reason or another at the present.

The questions of size and general arrangement will naturally involve much thought and planning. A few remarks on these subjects are appended.

The suggested 20 ft. throat diameter seems ample except for one possibility which was not mentioned in the list above. It would be a marvellous asset to be able to test scaled-down models of large airplanes and having the models capable of actual flight. This could be done in many cases if the tunnel were to have a throat of about 30 ft. diameter, but any model capable of carrying a pilot would be too large for a 20 ft. tunnel. The comparison of flight and tunnel tests on the same model would be just about the ne plus ultra of aerodynamic investigation.

The use of return ducts would effect such a great saving of power and make possible a maximum speed so much higher, with any given power, that the open circuit type of tunnel is completely out of the question.

The outstanding advantages of the Eiffel chamber would seem to overbalance any possible aerodynamic advantage inherent to the closed throat tunnel. Installation and operation of measuring apparatus have few terrors in an Eiffel chamber but, as a result of my experience in the atmospheric tunnel, I would expect many troublesome and expensive complications if the throat were to be closed.

With the use of an open balance room, the shape of throat is pretty well limited to a polygon or circle, as a free jet of air having a square or rectangular cross section would be rather difficult to handle, particularly in the matter of collecting after having traversed the balance room. The circular jet would eliminate several difficulties attendant upon the propeller installation.

A single large propeller might be difficult to build but its operation would doubtless be more satisfactory than any other system that could be devised. It is thought that a multiblade fan of the inserted blade type, somewhat similar to the one in use in the 5 ft. tunnel at McCook Field, offers the best possibilities for adaptation to very large installations.

[signed] Elliott G. Reid
Elliott G. Reid,
Assistant Aeronautical Engineer
Langley Field, Va.
April 3, 1925.

Document 2-21(b), Arthur Gardiner, "Memorandum on Proposed Giant Wind Tunnel," 1925.

MEMORANDUM ON PROPOSED GIANT WIND TUNNEL.

As requested, I submit herewith a brief memorandum on the proposed giant wind tunnel indicating certain investigations for which such a wind tunnel would seem to be peculiarly well adapted.

To my mind, a wind tunnel of the proposed proportions would fulfill a very definite need in aeronautic research, and, if built, would have a threefold purpose: (1) to aid in correlating the wind tunnel testing of models with the flight testing of full-size airplanes, i.e., to duplicate, after a fashion, the function of our variable density tunnel, but attacking the problem from the standpoint of increased linear dimension, approaching or equalling full-scale, rather than from the standpoint of increased air density; (2) to take over certain phases of flight research, thereby substituting more accurate methods for the somewhat unreliable flight test methods, and, at the same time, making it possible to exercise control of the operating conditions, and (3) to enable some very pertinent research problems to be investigated which are not being undertaken at present due to the lack of suitable equipment.

To utilize a giant wind tunnel to fulfill the purpose mentioned under (1) above might appear, at first thought, to be an unnecessary duplication of effort. However, full-size, or nearly full-size, model testing would seem to have many advantages over small model testing in a variable density tunnel, as full-size models could be made in our own shops according to standard airplane construction practice, thus effecting a great saving in time and enabling a greater variety of model building. Also, it would seem that considerable time could be saved in the actual testing of the model. The advantages accruing from either the direct application of wind tunnel test data from a giant tunnel or the application of the test data with but a small correction for scale effect would be highly desirable in predicting flight performance.

The purpose mentioned under (2) above might also appear to be a duplication of effort, but here again it would appear that certain phases of flight research could be conducted in a giant wind tunnel much more expeditiously and with greater accuracy than possible with flight testing methods. In addition, there are cases where a greater range of investigation is possible in wind tunnel testing than is permissible in flight testing. The fact that wind tunnel tests can be conducted independent of weather conditions would also make for a saving in time.

The purpose to be fulfilled under (3) above is by far the most important. The fact that a giant wind tunnel would enable certain investigations to be made, which are not being undertaken at present due to the lack of suitable equipment, might very well, in itself, be a sufficient reason for taking the pioneer step in constructing a wind tunnel of the proposed type. Some pertinent investigations that could be conducted might include tests to determine: full-scale propeller characteristics, either with a free mounting or when mounted adjacent to full-size slipstream obstructions, the data obtained including all scale and installation factors; performance of engine-propeller units mounted in full-size fuselages; the cooling capacity of, and the heat distribution throughout the cylinders of air-cooled engines when mounted with cowling in full-size fuselages; full-scale radiator performance under all conditions of mounting in or near a full-size fuselage (nose, wing, and retractable mountings; drag characteristics of full-size landing gears assembled to fuselage, the investigation to include a study of the retractable type looking toward its general adoption; comparative performance of direct-driven and geared propellers, gear efficiency included in results; the efficiency and general operating characteristics of adjustable and variable pitch propellers; drag characteristics of full-size fuselages with a view to increasing the efficiency of design; parasite resistance of full-size airplane parts either independently or in assemblies; effect on airplane characteristics of such equipment as spoiler gears, variable camber wings, etc., and the behavior of empenage assemblies (reference is made to recent tests with the MO-1 tail unit).

One of the greatest benefits to be derived from the construction of a giant wind tunnel would be its utilization in connection with the investigation of certain power plant problems.

In connection with our analysis of airplane performance, the outstanding need at present is a means for determining the power output during flight of normal, over-compressed and throttled, supercharged and geared engines. All of these types of engines either are being investigated at the present time or are included in our immediate future program of tests. As the purpose of the present and proposed tests is to arrive at a definite answer as to the most efficient power plant unit, it is extremely essential that means be provided for determining power output during flight. In lieu of a suitable torque-meter, propellers, calibrated as installed in the airplane, would serve the purpose. A giant wind tunnel would provide a means for supplying the required propeller calibrations.

The determination is a giant wind tunnel of the performance characteristics of engine-propeller units mounted in full-size fuselages, complete with radiator installation, et al, would aid materially in arriving at the efficiency of a given power plant installation. There is some question as to whether or not accurate torque and thrust measurements could be made with the vibrations of the engine imposed on the measuring devices, but if the effect of engine vibrations could be reduced to a negligible amount, the proposed method of engine-propeller calibration would seem to have many possibilities. It is thought that the problem of handling the engine exhaust would not be serious.

A giant wind tunnel would serve directly as a means for studying air-cooled engines. If the required accuracy of measurement could be secured, a great advantage would accrue from the direct testing of air-cooled engine-propeller units mounted with cowling in full-size fuselages. We have felt the need for investigating such problems as heat distribution throughout air-cooled engine cylinders, effect of fuel mixture on cylinder temperature, effect of cylinder temperature on power, etc., in connection with our tests with the supercharged Lawrance J-1 engine, but have been unable to make any such tests due to lack of equipment. A giant wind tunnel might serve to supplant the immediate necessity for securing special equipment for testing air-cooled engines.

The study of adjustable and variable pitch propellers is tied up directly with the power plant problem, especially with the supercharged engine problem. In light of present knowledge, it would seem that variable pitch propellers must be developed into safe and efficient units if the ultimate advantages of the supercharged engine are to be realized. In our tests with the Roots supercharger we have felt a definite need for the variable pitch propeller, especially for high altitude operation. These two types of propellers could be carefully studied in a giant wind tunnel, whereas their characteristics could not be investigated efficiently, if at all, in a small model due to the difficulty of reproducing them to a small scale.

In conclusion, I concur with the general opinion that the tentative size (20 ft.) of the proposed tunnel be increased to such a size as to enable certain of the smaller size airplanes, or specially built small size airplanes, being tested therein in toto, and to permit accurate testing of large propellers (we are at present using a propeller having a diameter of 13 ft.) such as are used with geared engines.

[signed] Arthur W. Gardiner
Arthur W. Gardiner,
Assistant Aeronautical Engineer
Langley Field, Va.
April 23, 1925.

Document 2-21(c), Smith J. DeFrance, The NACA Full-Scale Wind Tunnel, *NACA Technical Report No. 459, 1933*.

SUMMARY

This report gives a complete description of the full-scale wind tunnel of the National Advisory Committee for Aeronautics. The tunnel is of the double-return flow type with a 30 by 60 foot open jet at the test section. The air is circulated by two propellers 35 feet 5 inches in diameter, located side by side, and each directly connected to a 4,000-horsepower slip-ring induction motor. The motor control equipment permits varying the speed in 24 steps between 25 and 118 miles per hour. The tunnel is equipped with a 6-component balance for obtaining the forces in 3 directions and the moments about the 3 axes of an airplane. All seven dial scales of the balance system are of the recording type, which permits simultaneous records to be made of all forces.

The tunnel has been calibrated and surveys have shown that the dynamic-pressure distribution over that portion of the jet which would be occupied by an airplane having a wing span of 45 feet is within ± 1 1/2 per cent of a mean value. Based on the mean velocity of 118 miles per hour at the jet, the ratio of the kinetic energy per second to the energy input to the propellers per second is 2.84. Since it is generally recognized that a long open jet is a source of energy loss, the above figure is considered very satisfactory.

Comparative tests on several airplanes have given results which are in good agreement with those obtained on the same airplanes in flight. This fact, together with information obtained in the tunnel on Clark Y airfoils, indicates that the flow in the tunnel is satisfactory and that the air stream has a very small amount of turbulence.

INTRODUCTION

It is a generally accepted fact that the aerodynamic characteristics of a small model can not be directly applied to a full-sized airplane without using an empirical correction factor to compensate for the lack of dynamic similarity. Two methods have been used to overcome this difficulty. One is to compress the working fluid and vary the kinematic viscosity to compensate for the reduction in the size of the model. This method is used in the variable-density wind tunnel where tests can be conducted at the same Reynolds number as would be experienced in flight. The other method is to conduct tests on the full-scale airplane.

The variable-density wind tunnel offers a satisfactory means for testing the component parts of an airplane and is particularly suitable for conducting fundamental research on airfoil sections and streamline bodies. However, this equipment has its limitations when the aerodynamic characteristics of a complete airplane are desired, especially if the effect of the slipstream is to be considered. It is practically impossible to build a model of the required size that is a true reproduction of a complete airplane. This difficulty is increased by the requirement that the model withstand large forces.

It is apparent that the most satisfactory method of obtaining aerodynamic characteristics of a complete airplane is to conduct a full-scale investigation. Heretofore such investigations have been conducted only in flight. Because of the variation in atmospheric conditions, it has been necessary to make a large number of check flights to obtain enough data to average out the discrepancies. Furthermore, in flight testing the scope of experiments is often limited by the fact that the possible alterations that can be made are restricted to those that do not seriously affect the weight or airworthiness of the airplane. In order to provide a means of full-scale investigation by which the conditions can be controlled and alterations made without serious limitations, the full-scale wind tunnel has been erected. Of course, only the steady-flight conditions can be readily investigated in the wind tunnel, but the execution of this work in the tunnel will facilitate full-scale testing and allow the flight-research personnel of the laboratory to concentrate on those problems possible of solution only in flight.

The full-scale wind tunnel may be used to determine the lift and drag characteristics of a complete airplane, to study the control and stability characteristics both with and without the slipstream, and to study body interference. In addition, equipment has been installed to determine the direction and velocity of the flow at any point around an airplane. Aircraft engine cooling and cowling problems can also be investigated under conditions similar to those in flight.

The design of the full-scale wind tunnel was started in 1929. Since this was to be the first wind tunnel constructed with an elliptic throat and with two propellers mounted side by side, a 1/15-scale model was constructed to study the flow problems. Very satisfactory flow conditions were obtained in the model tunnel. This piece of equipment is now being used for small-scale testing. Construction of the full-scale wind tunnel was started in the spring of 1930; it was completed and operated for the first time in the spring of 1931.

DESCRIPTION OF TUNNEL

The general arrangement of the tunnel is shown in figure 1 and an external view of the building is given in figure 2. The tunnel is of the double-return flow type with an open throat having a horizontal dimension of 60 feet and a vertical dimension of 30 feet. On either side of the test chamber is a return passage 50 feet wide, with the, height varying from 46 to 72 feet. The entire equipment is housed in a structure, the outside walls of which serve as the outer walls of the return passages. The overall length of the tunnel is 434 feet 6 inches, the width 222 feet, and the maximum height 97 feet. The framework is of structural steel and the walls and roof are of 5/16-inch corrugated cement asbestos sheets. The entrance and exit cones are constructed of 2-inch wood planking, attached to a steel frame and covered on the inside with galvanized sheet metal as a protection against fire.

Entrance cone.—The entrance cone is 75 feet in length and in this distance the cross section changes from a rectangle 72 by 110 feet to a 30 by 60 foot elliptic section. The area reduction in the entrance cone is slightly less than 5:1. The shape of the entrance cone was chosen to give as far as possible a constant acceleration to the air stream and to retain a 9-foot length of nozzle for directing the flow.

Test chamber.—The test chamber, in which is located the working section of the jet, is 80 by 122 feet. The length of the jet, or the distance between the end of the entrance cone and the smallest cross section of the exit-cone collector, is 71 feet. Doors 20 by 40 feet located in the walls of the return passage on one side provide access for airplanes. In the roof of the test chamber are two skylights, each approximately 30 by 40 feet, which provide excellent lighting conditions for daytime operation; eight 1,000-watt flood lights provide adequate artificial illumination for night operation. Attached to the roof trusses and running across the test chamber at right angles to the air stream and also in the direction of the air stream are tracks for an electric crane which lifts the airplanes onto the balance.

Exit cone.—Forward of the propellers and located on the center line of the tunnel is a smooth fairing which transforms the somewhat elliptic section of the

Chapter 2: Building a Research Establishment

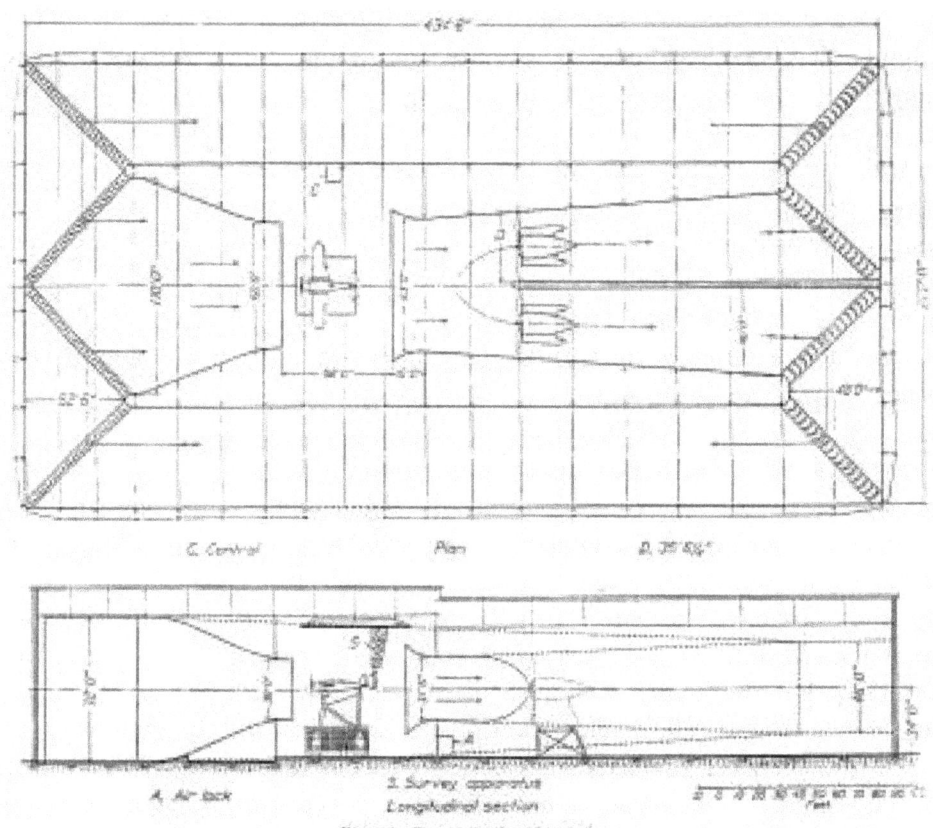

single passage into two circular ones at the propellers. From the propellers aft, the exit cone is divided into two passages and each transforms in the length of 132 feet from a 35-foot 6 1/2-inch circular section to a 46-foot square. The included angle between the sides of each passage is 6°.

Propellers.—The propellers are located side by side and 48 feet aft of the throat of the exit-cone bell. The propellers are 35 feet 5 inches in diameter and each consists of four cast aluminum alloy blades screwed into a cast-steel hub.

Motors.—The most commonly used power plant for operating a wind tunnel is a direct-current motor and motor-generator set with the Ward Leonard control system. For the full-scale wind tunnel it was found that alternating current slip-ring induction motors, together with satisfactory control equipment, could be purchased for approximately 30 per cent less than the direct-current equipment. Two 4,000-horsepower slip-ring induction motors with 24 steps of speed between 75 and 300 r.p.m. were therefore installed. In order to obtain the range of speed one pole change was provided and the other variations are obtained by the introduction of resistance in the rotor circuit. This control permits a variation in air speed from 25 to 118 miles per hour. The two motors are connected through an automatic switchboard to one drum-type controller located in the test chamber. All the control equipment is interlocked and connected through time-limit relays, so that regardless of how fast the controller handle is moved the motors will increase in speed at regular intervals.

The motors are provided with ball and roller bearings, which reduce the friction losses to a minimum. Roller bearings of 8.5- and 11.8-inch bores are provided at the slip-ring and propeller ends respectively, while the thrust of the propellers is taken on a ball bearing at the rear end of each motor shaft. The motors are mounted with the rotor shafts centered in the exit-cone passages. The motors and supporting structure are enclosed in fairings so that they offer a minimum resistance to the air flow.

Guide vanes.—The air is turned at the four corners of each return passage by guide vanes. The vanes are of the curved-airfoil type formed by two intersecting arcs with a rounded nose. The arcs were so chosen as to give a practically constant area through the vanes.

The vanes at the first two corners on back of the propellers have chords of 7 feet and are spaced at 0.45 and 0.47 of a chord length, respectively. Those at the opposite end of the tunnel have chords of 3 feet 6 inches and are spaced at 0.41 of a chord length. By a proper adjustment of the angular setting of the vanes, a satisfactory velocity distribution has been obtained and no honeycomb has been found necessary.

Balance.—The balance, which is of the 6-component type, is shown diagrammatically in figure 3. Ball and socket fittings at the top of each of the struts **A** hold

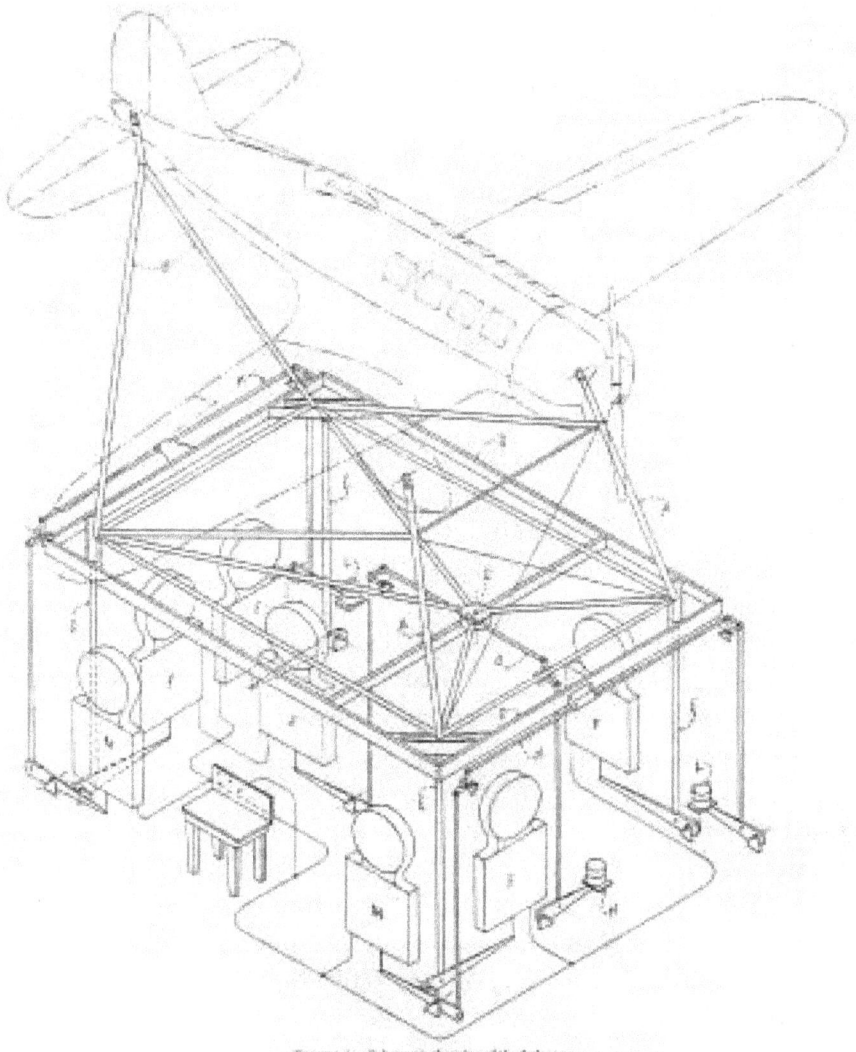

FIGURE 2—Schematic drawing of the balance.

the axles of the airplane to be tested; the tail is attached to the triangular frame **B**. These struts are secured to the turntable **C**, which is attached to the floating frame **D**. This frame rests on the struts **E**, which transmit the lift forces to the scales **F**. The drag linkage **G** is attached to the floating frame on the center line and, working against a known counterweight **H**, transmits the drag force to a scale **J**. The cross-wind force linkages **K** are attached to the floating frame on the front and rear sides at the center line. These linkages, working against known counterweights **L**, transmit the cross-wind force to scales **M**. In this manner forces in three directions are measured and by combining the forces and the proper lever arms, the pitching, rolling, and yawing moments can be computed.

The scales are of the dial type and are provided with solenoid-operated printing devices. When the proper test condition is obtained, a push-button switch is momentarily closed and the readings on all seven scales are recorded simultaneously, eliminating the possibility of personal errors.

The triangular frame **B** is caused to telescope by electrically operated screws which raises and lowers the tail of the airplane and thereby varies the angle of attack. By a similar mechanism the turntable **C** can be moved so as to yaw the airplane from 20° left to 20° right.

The entire floating frame and scale assembly is enclosed in a room for protection from air currents and the supporting struts are shielded by streamlined fairings which are secured to the roof of the balance room and free from the balance. In figure 4 it can be seen that very limited amount of the supporting structure is exposed to the air stream. The tare-drag measurements are therefore reduced to a minimum.

Survey equipment.—Attached to the bottom of the roof trusses is a 55-foot structural steel bridge (fig. 5), which can be rolled across the full width of the test chamber; mounted on this bridge is a car which can be rolled along the entire length. Suspended below the car is a combined pitot, pitch, and yaw tube which can be raised or lowered and pitched or yawed by gearing with electrical control

on the car. This arrangement permits the alinement of the tube with the air flow at any point around an airplane. The alinement of the tube is indicated by null readings on the alcohol manometers connected to the pitch and yaw openings in the head and the angle of pitch or yaw is read from calibrated Veeder counters connected to the electric operating motors. This equipment is very valuable for studying the downwash behind wings and the flow around the tail surfaces of an airplane.

CALIBRATIONS AND TESTS

The velocity distribution has been measured over several planes at right angles to the jet, but the plane representing approximately the location of the wings of an airplane during tests was most completely explored. The dynamic-pressure distribution over the area that would be occupied during tests by an airplane with a wing span of 45 feet is within $\pm 1?$ per cent of a mean value. It is possible to improve the distribution by further adjustment of the guide vanes. However, tests already conducted in the tunnel indicate that the present distribution does not detrimentally affect the results. This fact has been shown by the excellent agreement which has been obtained between the tunnel and flight results.

A survey of the static pressure along the axis of the tunnel showed that the longitudinal pressure gradient is small, as evidenced by the fact that between 11 and

36 feet from the entrance cone the variation of the static pressure is within ± 1 per cent of the mean dynamic pressure at the test section.

Two wall plates with static orifices are located in each return passage just ahead of the guide vanes at the entrance-cone end of the tunnel. The orifices are connected by a common pressure line, which is led to a micromanometer on the control desk in the test chamber. The other side of the manometer is left open to the test-chamber pressure. This installation has been calibrated against the average dynamic pressure determined by pitot surveys of the jet at the test location and it is used to determine the dynamic head during tests.

A series of Clark Y airfoils of the same aspect ratio, but with spans of 12, 24, 36, and 48 feet, have been tested at the same Reynolds number to determine the jet-boundary correction. Tests have also been made to determine the blocking effect of an airplane in the jet. The results of the complete investigation will be presented in a separate report.

Using the mean velocity across the jet of 118 miles per hour for computing the kinetic energy per second at the working section and dividing this by the energy input to the propellers per second gives an energy ratio for the tunnel of 2.84. This ratio, considering the length of the open jet, compares very favorably with the most efficient open-throat tunnels now in operation and exceeds the efficiency expected when the tunnel was designed.

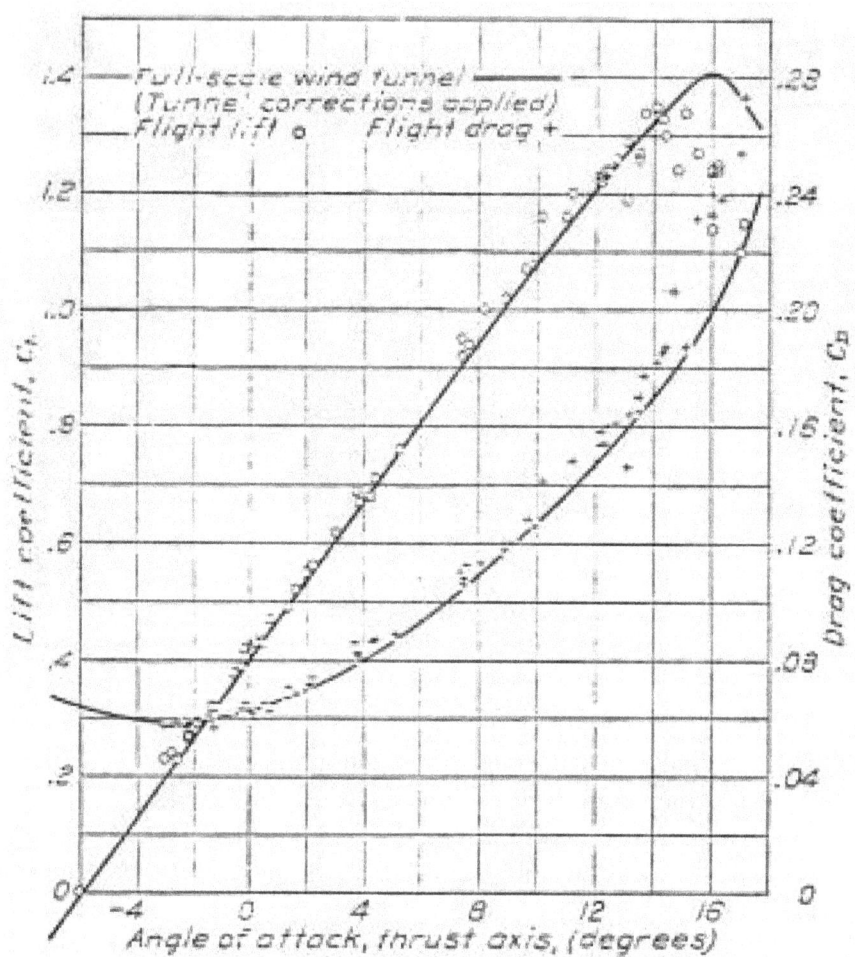

FIGURE 7.—Lift and drag characteristics of the F-22 airplane.

Before force measurements are made on an airplane, the airplane is suspended from the roof trusses by cables and held within one half inch from the balance supports while the tare forces are measured. The tare-drag coefficient determined in the above manner has been of the order of 25 per cent of the minimum drag coefficient of the airplanes tested.

When testing airfoils the airplane supports are replaced by those shown in figure 6. The angle of attack is changed by displacing the rear support arms and rotating the airfoil about pins in the top of the main supports. The rear support arms are moved by linkages, which are connected to long screws on the back of the main supports, and the screws are operated by hand cranks inside the balance house. The tare drag of this support system is exceptionally small and amounts to only 3 per cent of the minimum drag of a 6 by 36 foot Clark Y airfoil.

The lift and drag characteristics have been measured in the tunnel on several airplanes which had been previously tested in flight and their polars determined. These tests were conducted to obtain a check between the tunnel results and those from flight tests. A comparison of the results from the two methods of testing for one of the airplanes, the Fairchild F-22, is shown in figure 7. The wind-tunnel results are shown by the solid lines and the flight results are presented by the experimental points. These curves are representative of the results obtained with the different airplanes.

The agreement that has been obtained between the flight and full-scale tunnel results, together with the consistent manner in which measurements can be repeated when check tests are made, has demonstrated the accuracy and value of the equipment for aeronautical research.

    LANGLEY MEMORIAL AERONAUTICAL LABORATORY,
    NATIONAL ADVISORY COMMITTEE FOR AERONAUTICS,
        LANGLEY FIELD, VA., March 13, 1933.

Document 2-22

Max M. Munk, memorandum on "Recommendations for New Research," 16 November 1926, AV400-1, LaRC Correspondence Files, NARA, Philadelphia, PA.

This letter from Max Munk, chief of the Aerodynamics Division at Langley, to George Lewis, the NACA's director for research in Washington, furnishes an intriguing glimpse into Munk's vision for himself and the Langley Laboratory. Five days earlier, Lewis had issued instructions that all recommendations for new research projects from the Langley technical staff be forwarded to his office, via the engineer-in-charge, for consideration by the appropriate subcommittee. While this appeared to be an innocuous request to streamline the process, Munk took decided issue with a process that could bypass him. Noting that Langley was "at present pretty well filled up with problems; we are really overstocked," Munk downplayed the need for the Langley staff to make such recommendations. He expressed concern over his staff's need to instead focus "the fullest amount of thought and interest" on current projects, and suggested that a staff member should only propose projects "derived directly from the problem he is engaged in at the time." Such suggestions from subordinates would, of course, be reviewed by Munk. Given Munk's uncompromising faith in himself and his commanding methods, one senses that this may have been more of an effort to retain control over the research program at Langley than an expression of real concern over workload and a possible compromise of research quality.

Document 2-22, Max M. Munk, memorandum on "Recommendations for New Research," 1926.

November 16, 1926

Comment of Dr. Munk.
Subject: Recommendation for new research.
Reference: NACA Let. Nov. 11, 1926. 21-1.

1. The type of wings of reference has been discussed at the last meeting of the Subcommittee on Aerodynamics.

2. As a comment to the memorandum of Mr. E. G. Reid, and to Mr. Lewis' letter, I wish to call to Mr. Reid's and to Mr. Lewis' attention, that we are at present pretty well filled up with problems, we are really overstocked. Each problem should receive the fullest amount of thought and interest and should be carried through as far as can be. Otherwise, we might degenerate into a mere test factory. From

this point of view it is desirable to have only as many problems being turned over from outside as absolutely necessary. It is further desirable that each staff member propose chiefly such new problems as are derived directly from the problem he is engaged in at the time. Otherwise, the conclusion can not be avoided that he does not concentrate his entire mind on his problem; and furthermore, he is less prepared to know about the desirability of his proposed problem, if it does not belong to his present work in investigating.

3. To sum up, we need on the side of our staff members the serious will and the intense interest necessary to solve problems, rather than reflecting on new problems to be solved by somebody else.

[signed] Max M. Munk
Technical Assistant

Document 2-23(a–b)

(a) Daniel Guggenheim, letter to Secretary of Commerce Herbert Hoover, 16 January 1926, reprinted by Reginald M. Cleveland in *America Fledges Wings* (Chicago: Pittman Publishing Co., 1942), pp. 3–7.

(b) California Institute of Technology, "Development of Aeronautics," August 1926, Historical Files, Folder A1.1, California Institute of Technology Archives, Pasadena, CA.

While the Federal government's investment in aeronautical research facilities grew throughout the 1920s, other wind tunnels were built with private funding. While "airmindedness" and a general interest in the possibilities of commercial aviation were on the rise during this period, no one could be sure of its future. Flying remained a risky business, and aircraft manufacturers found it difficult to attract investment capital until some of the safety issues could be resolved. There was a pressing need for trained aeronautical engineers in 1925, but only five American schools offered aeronautical-engineering courses, and among them only MIT and the University of Michigan awarded degrees in aeronautical engineering. Such programs required wind tunnel laboratories for instruction and research, but the colleges were either private or state institutions and Federal funding was generally unavailable in the mid-1920s.

Into this void stepped Harry Guggenheim and his father, Daniel. Daniel Guggenheim was an American businessman who, along with his brothers, had acquired a large fortune managing mining operations started by his father. Harry, who had served as a pilot during World War I and maintained a keen interest in aviation after the war, determined that a committee of wealthy men should underwrite research that would lead to safer airplanes. Over lunch one day, the younger Guggenheim heard of a plan by New York University to raise $50,000 to endow a new aeronautical engineering program. Rather than a public appeal, he proposed that New York University Chancellor Elmer E. Brown write his father—Harry would personally deliver the letter—and request a gift from the Guggenheim family. Rather than involve his brothers, the elder Guggenheim decided to make the donation himself, and NYU opened its Guggenheim School of Aeronautics in 1925.

Daniel Guggenheim soon began to think beyond NYU, however, and in a letter to Commerce Secretary Herbert Hoover he offered to underwrite a larger, nationwide program, the Daniel Guggenheim Fund for the Promotion of Aeronautics, and endow it with $2,500,000, a substantial sum for the day. Established in 1926, the

Guggenheim Fund relied on a board of distinguished aviation personalities to determine the fund's distributions. Between 1926 and 1930, the Guggenheim Fund awarded major grants totaling $1,693,000 to seven engineering schools: the California Institute of Technology (Caltech), Leland Stanford University, the University of Michigan, MIT, the University of Washington, the Georgia Institute of Technology (Georgia Tech), and the University of Akron. A large portion of these grants went into construction of wind tunnel laboratories, but, significantly, some funds were earmarked for the hiring of exceptional professors and researchers. A Guggenheim grant allowed Clark and Robert Millikan to woo Theodore von Kármán, Germany's outstanding theoretical aerodynamicist (and Prandtl protégé), into moving to California and assuming the directorship of Caltech's Guggenheim Aeronautical Laboratory (GALCIT), where he would play a dominant role in shaping the growth of theoretical aeronautics in the United States. While the Guggenheim wind tunnels were generally unremarkable, exhibiting no significant technical innovations, they were a vital part of programs that began to supply the professional engineering talent that American aviation so badly needed. By midcentury, over 90 percent of the nation's leading aeronautical engineers were graduates of Guggenheim-funded colleges.

Document 2-23(a), letter from Daniel Guggenheim to Herbert Hoover, 1926.

Honorable Herbert Hoover
Secretary of Commerce
Washington, D. C.
My Dear Mr. Secretary:

Under your general direction the United States Government has made substantial progress in the promotion of civil aviation. I am venturing to advise you, therefore, by this letter, of my purpose to establish a Fund which will cooperate with you and with all agencies of the Government and the public generally in advancing the art and science of aeronautics and aviation.

This action is taken particularly in view of the very wise endorsement by the President of the United States of the recommendation by the National Advisory Committee for Aeronautics that a Bureau of Air Navigation be established in the Department of Commerce. President Coolidge stated that:

"The outstanding weakness in the industrial situation as it affects national defense is the inadequacy of facilities to supply air service needs. The airplane industry in this country at the present time is dependent almost entirely upon Governmental business."

The Department of Commerce, in studying the need for commercial aviation, reported:

"There are indications of a great change in the last few months which has given an impetus to plans for developing American civil aeronautics that is bound to produce permanent results. The extent of our country, its physical characteristics and its intimate contact with Canada, Mexico, and the West Indies are such as to make air service highly desirable. The success of the transcontinental air mail service operated by the Post Office Department and the general approval which has greeted its operations indicate that the choice of a method for developing civil aeronautics in the United States is the question demanding immediate solution. There is every reason to hope that before the end of the present year, civil aviation in the United States will have taken a long step forward toward a position of permanent security."

You, yourself, have pointed out that a Government public service must be provided to cooperate with aviation in a manner comparable to the cooperation our Government now gives to shipping, and you have very properly pointed out that without such service "aviation can only develop in a primitive way." There is undoubtedly a function in this situation which the Government alone can perform.

I have also been much impressed by the question raised and the answer given, in the report of the President's Aircraft Board which, as of November 30, 1925, said:

"How can the civilian use of aircraft be promoted? This . . . may well be the most important question which aviation presents in its far-reaching consequences to our people. A great opportunity lies before the United States. We have natural resources, industrial organization, and long distances free from customs barriers. We may, if we will, take the lead in the world, in extending civil aviation."

Such considerations as the foregoing have convinced me that there is a function which can only be performed by private enterprise aside from the proper function of the Government. So much remains to be done before civil aviation can realize the possibilities before it, that everyone must recognize that there intervenes a period of necessary study and experimentation.

In these circumstances I have decided to establish the Daniel Guggenheim Fund for the Promotion of Aeronautics and to place at its disposal the sum of $2,500,000. The Fund will be administered by a Board of Trustees composed of men of eminence and competence.

I shall place the sum of $500,000 immediately in the hands of the trustees to defray the expense of their studies and any work they may decide immediately to undertake. In addition, I will hold myself in readiness to supply any additional sum, up to a total of a further $2,000,000 as and when the judgment of the trustees may indicate that the money can be used wisely to promote the aims of the Fund. The trustees will have unrestricted power to do anything which in their judgment may develop aeronautics, the only condition being that the Fund shall not be a profit-making enterprise. Any earnings the Fund may realize from its efforts will go back into the Fund to carry on the work for which it was created.

The trustees are to have the power to spend the principal sum thus contributed and there is no purpose to establish a permanent foundation. The thought is, rather, that, the whole art and science of aeronautics and aviation being now in its infancy, it will be possible with the sums thus contributed, to bring about such an advance in the art that private enterprise will find it practicable and profitable to "carry on" and thus render a continuous and permanent endowment for this purpose unnecessary.

You will perhaps recall that last year I established a School of Aeronautics at New York University, my desire in making that gift being more quickly to realize for humanity the ultimate possibilities of aerial navigation, and to give America the place in the air to which her inventive genius entitles her. This school is already making gratifying progress, and the studies that have been made in connection with it have indicated clearly the enormous field of opportunity which unfolds itself to the pending development of air transportation. The establishment of additional schools such as that at New York University may well be warranted in the future.

Among the most important objects which I would now like to see accomplished at the earliest possible moment is the development of opportunities for new fields of employment for American young men. My family, as you know, has long been identified with exploration beneath the earth. We have tried to assist in development which would make mining more safe as well as more profitable, and therefore, of the greatest economic and value. Not the least desirable results which have followed from this effort have been the opportunities for the profitable employment of able engineers and workmen generally. My hopes, therefore, are that, through the impetus which the Daniel Guggenheim Fund for the Promotion of Aeronautics will give, attractive opportunities for men to work and serve in the air may develop far more rapidly than would otherwise be the case.

The general purposes to which I trust the new Fund will devote itself may be broadly defined as follows:

1. To promote aeronautical education both in institutions of learning and among the general public.

2. To assist in the extension of fundamental aeronautical science.

3. To assist in the development of commercial aircraft and aircraft equipment.

4. To further the application of aircraft in business, industry, and other economic and social activities of the nation.

I am hopeful without desiring in any sense to restrict their own freedom of judgment, that the Trustees of the Fund will govern themselves as far as possible by the following principles:

1. Restrict the work to civil activities.
2. Avoid duplication of effort with other aeronautical organizations.
3. Avoid work which is properly a Government function.
4. Plan carefully to concentrate effort and to carry an investigation or project through to definite conclusions.
5. Maintain a simple, inexpensive directing organization depending on outside established agencies wherever possible, to carry out the aims of the Fund.

I have confidence that the Fund can serve an important purpose. Recent events in the United States have stimulated much discussion of aviation. The time is ripe for action. There is urgent need in our country for immediate, practical, and substantial assistance to aviation in its commercial, industrial, and scientific aspects. No less urgent is the need to awaken the American public, especially our business men, to the advantages and possibilities of commercial aircraft—in a word, to make the American public in a very real sense, "air-wise."

In closing, Sir, may I express my delight at the intelligent and constructive interest you yourself are manifesting in the development of aeronautics as one of the most important new agencies of civilization. In making my deed of gift to the trustees of the Fund, I shall accordingly request that the trustees cooperate with your Department in every possible manner.

With best wishes, I am, Mr. Secretary,
Yours very truly,
Daniel Guggenheim
January 16, 1926

Document 2-23(b), California Institute of Technology, "Development of Aeronautics," 1926.

DEVELOPMENT OF AERONAUTICS

It has been announced that the Daniel Guggenheim Fund for the Promotion of Aeronautics has made a gift of $300,000 to the California Institute of Technology, and an equivalent gift to Stanford University. News of this gift was received at the California Institute in the following telegram from Harry F. Guggenheim, President of the Daniel Guggenheim Fund:

"It gives me great pleasure as President of the Daniel Guggenheim Fund for the Promotion of Aeronautics to advise you that the Trustees of the Fund have authorized a grant amounting altogether to approximately $300,000 for the erection of a permanent building at the California Institute of Technology to be devoted to the study of aeronautics and including a provision of fifteen thousand dollars a year for a term of years for the conduct of study and experiments in this rapidly developing science and art. May I remark that this gift is made in recognition not merely of the opportunities for study and research which the climatic and other conditions in California make possible, but also as a tribute to the distinguished work in science and education of yourself and associates, and because of our belief that you are developing in Southern California an institution which is destined to make very great contributions to the progress not only of our own country but of the whole world."

On the basis of this gift the California Institute has established the Daniel Guggenheim Graduate School of Aeronautics at the California Institute of Technology, and there will be constructed immediately a new Aeronautics Building containing a new ten-foot High-speed wind tunnel, the total construction involving an expense of approximately $200,000.

The California Institute has just announced the programs:

1. The extension of the Institute's theoretical courses in aerodynamics and hydrodynamics, with the underlying mathematics and mechanics, taught by Professors Harry Bateman, Edward T. Bell, and Paul S. Epstein.

2. The initiation of a group of practical courses conducted by the Institute's experimental staff in cooperation with the engineering staff of the Douglas Airplane Company, with the aid of the facilities now being provided at the Institute combined with those of the Douglas plant.

3. The initiation of a comprehensive research program on airplane and motor design, as well as on the theoretical bases of aeronautics.

4. The immediate perfection of the new stagger-decalage, tailless airplane recently developed at the Institute, primarily by one of its instructors in aeronautics, A. A. Merrill, a radical departure from standard aeronautical design, which in recent tests has shown promise of adding greatly to the safety of flying.

5. The establishment of a number of research fellowships in aeronautics at the California Institute.

6. The building and testing, not only of models for wind tunnel work, but also of full-size experimental gliders and power planes for free flight work.

It is considered of especial importance that the facilities of the Douglas Airplane Company in Santa Monica, with its large corps of engineers, will be added to those of the Institute for both instructional and research purposes in the effort to make a center of the first importance in Southern California for the development of both the theoretical and practical phases of aeronautics.

The Daniel Guggenheim Fund for the Promotion of Aeronautics was established by Mr. Daniel Guggenheim last February with provision for supplying $2,500,000 as needed by the Fund, the Fund being unique among the great foundations of the country in that its founder did not contemplate a permanent foundation but merely provision for sums which would make possible experimentation and development in the field of aeronautics and aviation during the infancy of this art and science as a civil enterprise.

Last year Mr. Guggenheim established the School of Aeronautics at New York University, with a principal fund of $500,000; and now one of the first acts of the new Guggenheim Fund is to make financial provision of equal amount in order that the scientific experience and the equable climatic conditions of California may be utilized to supplement the work of eastern and European institutions in the study of this important subject.

A further announcement was made today that Dr. von Karman of Aachen, Germany, one of the foremost mathematical physicists and one of the most outstanding aeronautical engineers of Europe, has accepted the invitation of the Daniel Guggenheim Fund to visit the aeronautical centers of this country in the near future. He will go immediately to Pasadena for the sake of advising with the aeronautical staff of the Institute, both as to the best type of aeronautical installation and as to the design of new planes already being perfected there.

Document 2-24(a–b)

(a) C. G. Grey, excerpts from "On Research," *Aeroplane* 35, No. 21 (21 November 1928): 837–840.

(b) C. G. Grey, excerpts continued from "On Research," *Aeroplane* 35, No. 22, (28 November 1928).

Before the end of the 1920s, the American model for aerodynamics research had become the standard by which much of the industrialized world judged its own efforts. While the empirical and theoretical contributions of other nations, particularly Great Britain and Germany, were substantial, critics condemned their governmental agencies and longed for research organizations and facilities like those in the United States. The criticism in England is especially interesting since that nation's Advisory Committee for Aeronautics served as the model for the American NACA. In 1928, *Aeroplane* Editor C. G. Grey penned a two-part editorial in response to comments by C. W. Brett, managing director of Barimar Ltd. Brett had called for an aeronautical research laboratory independent from the government agencies, which he felt could not address the needs of independent inventors. He proposed that it be funded and managed by a consortium of three existing British aviation organizations. Grey agreed that such a facility was needed, but he doubted that Brett's organization plan would work and instead called for a wealthy benefactor to underwrite the facility, much like Daniel Guggenheim had done in America. Such a laboratory never came to pass in England, but this editorial called attention to the fact that "ad hoc" research—research focused on solving practical problems, such as that practiced in American universities and the NACA—had produced some outstanding results.

Document 2-24(a), C. G. Grey, excerpts from "On Research," 1928.

AS THINGS ARE.

The official reply to Mr. Brett's suggestion, if a Question were asked about it in the House of Commons, would be that any inventor who has an idea, whether patented or not, for the improvement of anything aeronautical, has only to take it to the Air Ministry, and that if the official experts judge it to be of any value it will be tested impartially on behalf of the Air Ministry either at the Royal Aircraft Establishment at Farnborough or at the National Physical Laboratory at Teddington.

But the trouble is, as Mr. Brett suggests, that the Air Ministry grant for purely experimental work is in fact too small,—in spite of the half million or so pounds per annum spent by the R.A.E. at Farnborough. And so an undue amount of time passes before experiments can be made.

The N.P.L. at Teddington is always far behind in its experimental work, owing to pressure of work and likewise lack of money. Moreover it is not primarily interested in aeronautics. And so aeronautical experiments have to take their turn with all other branches of engineering.

As for the R.A.E., there is always the suspicion that if an inventor submits a really good idea his invention may be held up for months, and even for years, while the bright brains at Farnborough, who are in fact paid for producing new ideas themselves, try to invent some better way of doing the same thing as the inventor has done. Even when there is no suspicion of that sort the R.A.E. is so busy experimenting with its own inventions that it does take a long time to test anybody else's. For example, the Pobjoy engine, which recently passed its type-test successfully, was actually sent to Farnborough last March, and was only put through its tests a few weeks ago.

The really wily inventor can, certainly, get his invention tested in other ways. If he happens to have a friend on the spot he may have tests made at South Kensington or at Cambridge, provided that his invention comes within the scope of the experts at those places. Or if he is a good enough salesman (which few inventors are) he may induce one or two manufacturers to test his invention in their own experimental shops, or on their machines.

All these methods have their objections, and at the [finish] the inventor who manages to get his invention properly tested within a reasonable time must be something of a specialist in the gentle art of wangling,—without which in peace as in war, achievement of a desired end is difficult.

Therefore there is a great deal to be said for Mr. Brett's suggestion that a central and completely equipped laboratory and experimental station which exists entirely by, with and from aviation is desirable. The only question is, who is going to pay for it?

The Aircraft Industry certainly has not the money. And if it had the money, individual members of the industry would certainly not cooperate in anything which might help general progress and so help their rivals. The British Taxpayer obviously could not be asked to fork out money for the purpose so long as there are doles and pensions to be paid.

There remains then only the dim hope that some British millionaire, possibly of extraneous origin, such as Sir Basil Zaharoff, who has already founded the Chair of Aviation of the London University, might be induced to put his hand in his pocket for the purpose much in the same way as that good hundred-per-cent American, Mr. Daniel Guggenheim, has done in establishing the Daniel Guggenheim Fund for Aeronautics.

[. . .]
WHAT RESEARCH MEANS.

Research is of two kinds, that which is known as basic research, and that which is known as ad hoc research. In the former kind seekers after truth dig around blindly, working on more or less established, scientific facts, in the hope that something may turn up. In the latter kind deliberate research is made to discover a certain thing,—*ad hoc* meaning "*to this*," the word "*object*" or "*end*" being understood.

Occasionally in *ad hoc* research for one thing quite a different thing, or a different application of the same thing, may be discovered accidentally,—as, for example, the usefulness of the Handley Page slot in assuring control below stalling point, when in fact search was being made for a method of improving lift and slowing landing speeds. Basic research may, by sheer luck, discover something which revolutionizes the whole industry, or science, with which it is concerned. But, on the whole, the cheapest and quickest way of making progress is by means of *ad hoc* research.

That is just where the Americans have scored over us so far. Over and over again their practical aeronautical engineers have started out with one definite end in view, and they have pursued that end until they have got it.

The present improved performances of aircraft practically all over the World is due primarily to the search for the proper stream-line shape of fuselages and engine housings for high-speed racing machines in the Curtiss Company's wind tunnel. The Curtiss designs were brought to this country by Mr. C. R. Fairey five years ago, and those lines dominate the design of the machines with the highest performances all over the World.

Further search for improved performance, and consequent economy of running, caused the invention of the high-pressure wind tunnel (300 lbs. to the square inch) by a German engineer who was imported to the States just after the War. And the high-speed wind tunnel with its air-stream of 100 miles an hour, and 20 feet diameter, which the American National Advisory Committee for Aeronautics have at Langley Field was built because the N.A.C.A. wanted to experiment with engines and airscrews on fuselages under flying conditions.

The N.A.C.A. has had its high-pressure wind tunnel for five years or so, and it has discovered all kinds of things about interference effects as between fuselages and planes and struts and undercarriages and tail units, of which we in this country know but little. We are just beginning to build a high-pressure wind tunnel.

If only we had a completely equipped laboratory and full-scale experimental station, such as Mr. Brett has suggested, and if we only had the money to run it, regardless of expense, and if we only had the man at the top and a staff to spend that money efficiently and to get results from it, we certainly could make as much progress in the next five years as we are likely to make in the next twenty-five.

This is said without belittling in any way the splendid work which is being done by the Department of Supply and Research at the Air Ministry. There everything possible is done, even to the point of being unconstitutional, to hasten experiments and to urge research. But there are limits to what any Government Official can do. And private enterprise has no limits except those imposed by finance.

Nor are there any limits to the things to be discovered, if Aviation is to progress as it should. Next week we will consider some of these.—C.G.G.

(To Be Continued.)

Document 2-24(b), C. G. Grey, excerpt continued from "On Research," 1928.

THINGS TO BE DISCOVERED

One of the interesting facts which has been discovered in the N.A.C.A.'s high-speed tunnel is that by totally enclosing a radial engine in cowling the speed of a machine can be increased beyond anything which mere calculation would show to be possible, and yet the engine can be kept cool. A machine of the popular cabin-monoplane type which, with a 200 h.p.Whirlwind engine and normal cowling covering about three-quarters of it, has a speed of 125 m.p.h., has had its speed up to up to 133 m.p.h. with complete cowling. And, incidentally, as photographs show, the appearance of the machine is improved.

Figures quoted in THE AEROPLANE recently from an American paper gave the improvement as being from 118 m.p.h. to 137 m.p.h., and a scientific critic proved that the higher speed was impossibly high. The true figures, as herein quoted on reliable authority, are startling enough. No doubt they also could be proved by pure or applied mathematics to be impossible. And quite possibly the higher figure may be reached by further improvements.

Experiments have gone that far, but there is still more to be done in the way of discovering what shape of cowling best suits particular forms of fuselages. Also the shape of spinner which best suits particular forms of cowling has to be discovered,—and whether, with certain airscrews and cowling, any kind of spinner is worthwhile has to be proved.

Then there are experiments to be made as to the best shapes of fuselage which can be made big enough to include a cabin. And there is the question of the best way of fixing a wing in relation to a cabin fuselage.

Naturally all our own experimentalists have been and still are engaged on these problems. Some of them try to get there with slide rules and the Prandtl Theory. Some of them try to do it in wind-tunnels with such low speeds that the results may be quite misleading. And some of them try to do it on actual aeroplanes, in which all sorts of errors are likely to occur because a combination of undercarriage, wing, fuselage, engine, and cowling which gives a certain performance may have that performance entirely upset by altering any one of the components, though altering other components to fit in with the first alteration might give a vastly improved performance.

Document 2-25(a–c)

(a) Frank A. Tichenor, "Why the N.A.C.A.?"
Aero Digest 17, No. 6 (December 1930): 40, 124–134.

(b) Frank A. Tichenor, "The N.A.C.A. Counters,"
Aero Digest 18, No. 2 (February 1931): 50, 122–126.

(c) Edward P. Warner, "Speaking of Research,"
Aviation 30, No. 1 (January 1931): 3–4.

While the NACA's continued expansion of its research programs and facilities was generally applauded, both at home and abroad, it was not without controversy and criticism. The NACA's founding legislation had charged the committee to "direct the scientific study of the problems of flight, with a view to their practical solution," but this mandate was interpreted differently by different people. By 1931, the NACA tended to emphasize the last phrase, and its efforts were largely directed toward finding practical solutions to well-known problems. Critics, including *Aero Digest* Editor Frank Tichenor, argued that Congress intended something else entirely. In two of his "Air—Hot and Otherwise" columns, Tichenor blasted the NACA for its failure to continually generate new, purely scientific knowledge, something he claimed its charter demanded. The NACA's chief of aerodynamics, Elton Miller, prepared a response for both his boss, Henry Reid, and *Aviation* Editor (and NACA Committee on Aeronautics member) Edward Warner, who based a rebuttal editorial on it. Although glossing over the fundamental science issue, they countered that the NACA saw its development work as science with a purpose, but legitimate science nonetheless. In truth, the argument was really over the difference between science and engineering, a distinction frequently missed in the era, even within ranks of the NACA. Tichenor's criticism had little long-term effect on the NACA, and Warner received numerous compliments for his stand, but the exchange did serve to illustrate that aeronautical research had come to play a visible, important, and growing role in aviation development, even if it was not purely, or even primarily, one of fundamental science.

Document 2-25(a), Frank A. Tichenor, "Why the N.A.C.A.?" 1930.

AIR—HOT AND OTHERWISE
Why the N.A.C.A.?
By Frank A. Tichenor

HERE is a matter of such vital importance to the industry that we can not write of it save with plain words of considerable solemnity. It is a matter to which we respectfully would call the attention of the President. Indeed, we do so explicitly and respectfully, refraining from anything except such a statement as will make facts clear.

In this period of industrial readjustment, particularly in the aviation industry, our thoughts turn to a very important basis of technical enterprise, experimental aeronautical research. A young industry is more dependent on research, and at the same time less able to provide for it, than older and better established industries. Because the Government has been well aware of this situation, nearly all aeronautic research in this country has been financed and carried on by the Federal Government. Foremost in this activity has been the National Advisory Committee for Aeronautics, for which Congress has provided funds. The N.A.C.A. has obtained from Congress funds for the largest, the most splendidly equipped and the most modern laboratories, and facilities for aeronautic research. To all practical purposes aeronautic research in America means N.A.C.A. research. Our thoughts turn in this hour to this research activity, and with full concern for conditions in the aeronautic industry, we ask ourselves whether the N.A.C.A. has discharged its duty well, whether it has given to the industry the full return to which it is entitled for these appropriations.

How greatly aeronautic progress depends upon research has indeed been fully realized by those in charge of N.A.C.A. work, as is indicated in the annual report of the N.A.C.A. for 1921 (page 5) :

"Substantial progress in aeronautical development must be based upon the application to the problems of flight of scientific principles and the results of research."

Research activity of the N.A.C.A. has been going on for more than ten years. The first appropriation for a wind tunnel having been made in 1917, this tunnel was reported to have been completed in 1918. Experts tell us that a year is ample time to build an ordinary small wind tunnel. Nevertheless, although the wind tunnel was completed, it was not then put into operation. In 1919, the tunnel was again reported not yet in operation. Finally, in 1920, the same tunnel originally reported as finished in 1918, was once more reported as finished. The year 1920, therefore, we are entitled to consider as the beginning of research activity, particularly inasmuch as an engine laboratory and free flight test facilities had been announced as completed in 1919.

This fact is important because the results of research can not be judged from the activity of one day, or one month or even one year. After ten years of uninterrupted activity, however, with continuous liberal financial support, the N.A.C.A. can be judged according to the results derived from its research work and an estimate can be made of what we have a right to expect in the future. Let us, therefore, review these results and ascertain what the N.A.C.A. has achieved.

The standard by which the results of research should be appraised is defined by the N.A.C.A. itself. Repeatedly, its annual reports have stressed scientific research as of paramount importance. For instance, almost all reports close like that of 1927 (page 76) : "Further substantial progress is dependent largely upon the continuous prosecution of scientific research," and farther below on the same page, "its (the N.A.C.A.'s) work in the fields of pure and applied research on the *fundamental* problems of flight." The latest report, that for 1929, states (page 87) : "The most important active influence upon aeronautics has been the farsighted and constructive policy of the Federal Government, liberally supported by Congress and the President, in providing for the continuous prosecution of organized scientific research." In the 1926 report we find (page 69), "The more *fundamental* investigations are undertaken by the Committee in its own laboratory," and (page 68), "to conduct investigations of a truly *scientific* character." (The italics are mine.)

We could easily quote other passages from N.A.C.A. publications to the same effect. The N.A.C.A. is not an aircraft factory; it is not interested in the properties or the development of any particular airplane. More general scientific investigations are its domain. It is charged with the responsibility of furnishing information concerning aeronautics as a science.

Nor do the annual reports of the N.A.C.A. leave any doubt about what is meant by "scientific research." That of 1922 (page 48), defines the term clearly:

"By scientific research is meant the investigation by trained men in a properly equipped laboratory of the *fundamental phenomena of nature*. . . . All progress depends upon the acquisition of knowledge, of new knowledge. This can be obtained only by long continued investigations *directed by men who know the problems and the methods used for their solutions.*"

Perhaps the best standard by which to judge the results of ten years of N.A.C.A. research is in terms of returns for the funds spent. Even with a small appropriation there is no upper limit to what can be obtained in the way of research if that research is directed "by men who know. . . . " There is, however, a lower limit to what ought to be obtained for a given amount of money. It stands to reason that we can expect more for an expenditure of $2,500 than for one of $250, and more for one of $25,000 than for one of $2,500.

The N.A.C.A. has spent on each of its research items undertaken more than $100,000, and we have a right to count on important results from $100,000 researches. This average expenditure for each problem investigated is computed by dividing the sum of the money spent by the number of problems undertaken. Thus far the N.A.C.A. has received $4,936,370 in appropriations. Approximately $4,800.000 has been spent (presuming the expenditure of the whole sum of $1,508,000 appropriated for 1930). The results of its research are laid down in eighty-eight Technical Reports. All other N.A.C.A. Technical Reports contain information obtained from outside sources, the N.A.C.A. acting only as publisher. This means that more than $30,000 has been spent for each report on a research project. It means much more per research, for at least four reports are always issued on the same research. This would give $200,000 per research item. Allowing for those research projects not yet completed for which no reports have yet been published and allowing also deductions for other expenses of the N.A.C.A., we are certainly justified in estimating that more than $100,000 has been spent for each research undertaken. Since 1925, and until 1930, the annual appropriation for the National Advisory Committee for Aeronautics has been approximately $500,000. This year it was increased to $1,508,000. No one can claim that during any one of the last four years more than five research problems have been finished and the results made available to the public. One hundred thousand dollars per research is perhaps too moderate an estimate.

It is pertinent to ask whether really useful scientific results have been obtained, and if not, to inquire about the reasons why research so liberally supported failed to furnish an adequate return. This sum can not be considered exorbitant if valuable results have been obtained from it.

If we make a more detailed analysis of the N.A.C.A.'s research of the past ten years, we find that it can be classified into wind tunnel research, free flight research on actual airplanes, and engine laboratory research.

In the engine laboratory, tests have been conducted with a view to improving the efficiency of gasoline aircraft engines by the choice of the best compression ratios, richness, and mixtures, and the like. That work would be valuable if important results had been obtained, but we doubt whether, lacking this research, any one existing engine would be worse. To say the least, this study and experiment has not been of a scientific nature. In addition, the Diesel engine was studied, likewise not a scientific or new phenomenon, and no tangible results were achieved, except possibly in the case of the spray research with solid injection.

The free flight researches gave valuable information concerning the maximum accelerations and maximum pressures occurring in maneuvers. Also some practical information regarding the ice hazard and similar subjects was obtained. Apparently the only fact demonstrated in the study of the supercharger was that such a device increases the available horsepower, and that was known before. This can hardly be considered an outstanding success. On the whole it can, nevertheless, be said that the free flight research has been the most beneficial conducted by the N.A.C.A. At the same time it can be said that no free flight test has been a scientific test nor dealt with investigation of fundamental phenomena of nature. Test flights conducted over a period of ten years, with the aid of good instruments, can not but yield some valuable information, especially at a time when flying is new, but they are not likely to advance fundamental science.

The class of wind tunnel research should correspond most to the description "scientific." Therefore, we ought to consider it in more detail in order to find there at least some of the promised scientific work. In this category the pressure distribution work of the N.A.C.A. showed only that wings should be rounded at the tips, which was known before, and which could be and was demonstrated in the course of natural industrial development. Merely to make pressure distribution measurements is not scientific. We are sometimes inclined to believe that it would be better for wind tunnel research if it were more difficult to do this kind of work; an abundance of patience is necessary but not much creative mental effort. The results are not of great practical value, because they are made under steady wind tunnel conditions, whereas the largest pressures occur under unsteady flight conditions. For this reason, the pressure measurements made in flight tests are much more valuable.

In addition there have been wind tunnel tests on complete airplane models, and drag measurements on airplanes and airplane parts. This research can not yield new results of general value, and is therefore outside the scientific research the N.A.C.A. is charged to undertake.

During all of the ten years, much time and effort has been spent on a series of tests undertaken to standardize wind tunnels throughout the world. This work showed merely that different wind tunnels give slightly different results and that these differences can not be predictedæwhich facts we knew before. Tests referring to wind tunnel technique are secondary anyhow. Someone has claimed that all wind tunnels could continue to do research even if no airplanes existed. They could, but we would not accept such work as useful unless science had been advanced.

Propellers have been investigated and found to possess a certain thrust and torque. Interesting, but again not scientific progress, not even technical progress.

We come at last to the research having most of the scientific element in itæthat dealing with the rotating cylinder. This stirred the imagination when the first tests were made and showed undreamed-of lifts. Right now, a very prominent manufacturer is making experiments along that principle. Unfortunately, the first tests along this line were not made by the N.A.C.A. On the contrary, the N.A.C.A. refused a suggestion in 1921 to measure this phenomenon. Several years later, it did repeat measurements made abroad without adding one new thought or result.

The Autogiro is the most painful subject in connection with the N.A.C.A. research. The N.A.C.A. had the priority in this new and perhaps most important invention of recent years. Autogiro models were investigated in 1922. It is hard to believe, but nevertheless true, that these tests were never published in a Technical Report. Five years later, after the practical value of the Autogiro had been demonstrated abroad, the results were published in mimeographed form, giving evidence of an opportunity to contribute to scientific progress which was woefully neglected.

In the investigation of auto-rotation of wings, it was demonstrated that, in a wind tunnel, wings can be made to rotate like windmills. This has hardly any bearing on or connection with the spinning of airplanes. It can hardly be called a research, but rather only making pretense of research. No airplane designer gives any attention to such tests, and science rejects them entirely.

A study of boundary layer control is on the program of the N.A.C.A., according to its statement, but no report has appeared in print on the results and we have not been apprised of any progress. This should be the most important subject of the work, but in fact hardly anything seems to have been done except the repetition of some work abroad.

Finally there is the wing section research. This is the only line in which the N.A.C.A. has contributed to aeronautics by way of its own experimental research. The M wing sections were developed by the N.A.C.A., in its wind tunnel, and at least two of them have been adopted in practice, being considered superior to older ones. Accordingly, the N.A.C.A. report for 1924 (page 50) says: "satisfactory progress has been made in the science of aerodynamics during the past year. . . . One important result of wind tunnel investigations has been the development of a number of remarkably efficient wing sections of adequate thickness for economical structures. *It is desirable that this development continue substantially along the present course*."

This was indeed desirable for the investigation was intended only as the first and preliminary step of a more systematic research. Much better wing sections were expected from the next series of tests, as the report indicates (page 59), "It is believed that a fruitful field for research lies in the determination of these sections which have a stable flow with good aerodynamic properties." In the interim, however, there has been no evidence of further work and the M section research, so admirably begun, has never been, continued.

We do not believe that we have overlooked a major research item of the N.A.C.A.: we are certain we have not overlooked a successful one. The N.A.C.A. was officially awarded the congressional medal for its low drag cowling. Apparently, even the friends of the N.A.C.A. consider this the most outstanding of the research projects completed. Yet, in the true sense, this cowling work was a development rather than an original work. Moreover, because it had reference to special airplanes and engines, it can not be regarded as having general value. Therefore, it can not be considered scientific work. It does not involve the study of new and fundamental phenomena of nature. Its doubtful value in this connection is clearly contrasted with the research of similar aim though along entirely different lines carried on at the same time in England. The Townend Ring is definitely superior to the N.A.C.A. cowling. It is the outcome of strictly scientific research carried on with scientific spirit, involving the systematic exploration of new and fundamental phenomena, and incurring relatively little expense. It represents more brain and less expenditure than for the N.A.C.A. cowling research.

The results of the N.A.C.A. experimental research are not, in our opinion, an adequate return for the money spent. There is hardly one research project of scientific value, and only a few of technical value. There is an enormous gap between the principles of research laid down and those applied.

It can not be denied that there is keen feeling of disappointment throughout the industry about the outcome of the N.A.C.A. research. Every year the industry gathers at Langley Field to acquaint itself with the latest results of the research going on, but every year it is presented with stone rather than with bread. New laboratories and instruments are exhibited but no new results worth speaking of.

Responsibility for the N.A.C.A.'s failure to make substantial contributions to aeronautic science does not rest entirely on the organization itself. General supervision of the research undertaken is in the hands of committees, which are composed of members serving without compensation. Under these circumstances, they can not give much time to this research; and after all, they are not to be blamed for its shortcomings. Scientific knowledge can not he amassed by a committee any more than an opera can be written by a committee. The capable and patriotic members of the several research committees feel that they can give best service by keeping their hands off, by assisting with advice and suggestion only, without showing too much initiative.

The real responsibility would seem to rest, therefore, upon the director of research. Is he one who knows "the problems and the methods used for their solution"? We fear not. But then it must be remembered that this director exercises the direction of the research from a distance of 200 miles, and as an auxiliary duty only. His primary duty, is that of an executive. In the first place he must practice diplomacy and exercise organizing talent: only secondarily need he exhibit any scientific spirit. Most of his direction of the research is done over the long-distance wire, or on occasional visits. These facts, together with his normal duties which stand in distinct contrast to the duty of research supervision, and require entirely different capabilities, make it plausible to believe that the director of research is not in a position properly to discharge his duty. As one important reform that will improve the present conditions, we suggest that the Langley Field laboratory be separated entirely from the Washington political office of the N.A.C.A. and be put in charge of a capable research engineer who would be fully responsible for the research and for it only.

As it is, the true initiative must come from the local head of the laboratory, and from the heads of the single divisions. We expect most from the aerodynamic sections. It is now a fact that both positions, the head of the L.M.A.L. and of the aerodynamics division, have been occupied in recent years by men who are decidedly not research engineers at all. Neither of them has ever contributed anything to science, and neither of them expects to do so. They are mere routine engineers, and hardly that; they are mere bureaucrats, signing letters and unwrapping red tape.

This brings us to the question of the N.A.C.A. staff. Friends of the N.A.C.A. have claimed that the staff has suffered great losses because the industry has induced its best men to leave by offering them lucrative positions. This does not sound probable. In the first place, a capable research engineer does not leave his work if he has found favorable working conditions, and is progressing satisfactorily in his work. The fact that nearly all good research engineers have left the N.A.C.A. constitutes in itself a reproach to the management. From inside information we know that most engineers left of their own initiative, because they were dissatisfied with the management. They are now employed in industry, and most of them did not leave as friends of the Committee. During these ten years, the head of the laboratory at Langley Field has changed four times, and two and a half years is about the average time the engineers used to stay. There must be a reason for this state of flux in the personnel. Most of the research engineers are young graduates, and the few older men who have stayed with the organization are for the greatest part less capable than those who left. Jealousy and petty politics have always played too great a part in the activities at Langley Field. The spirit of research and scientific work was never really encouraged by the management. Nobody can carry on research work successfully if he is compelled to devote a great part of his time to fighting for the cooperation of others to which he has a right, and fighting off the aggressiveness of his colleagues. The failure of the National Advisory Committee for Aeronautics is the failure typical of so many public organizations. There is no effective check on what is accomplished. If the results of the N.A.C.A. could be computed according to their worth in dollars and cents, the Committee would long ago have been bankrupt. But it is not a money-making organization; it is a money-spending organization. That leaves much energy free, and unfortunately the conditions in such a case are favorable to the survival of those most unsuitable for carrying on scientific research.

The activity of the N.A.C.A. has become a mere building of new laboratories without distinct ideas of what to do with them after they are built, and it has become a mere weighing and measuring of less value than the weighing of a grocery clerk. No concerted efforts are made to advance science; no efforts are made to apply the results of the tests to any logical system, to digest them, and to interpret their significance in the sum of general knowledge. The truth is that the tests can not be interpreted that way because the program has not been guided by scientific reasoning. Weighing for weighing's sake is not scientific research, but at the best a kind of indoor golf.

We urge that radical changes in the management be made with the view to improving the conditions to the end that real and honest talent be attracted to the N.A.C.A. Only then will there be some prospect of an intelligent use of the research equipment and a reasonable return on the money spent.

Let's devote a period of thought to wondering if these large appropriations devoted to the N.A.C.A. have served, are serving, or will serve the industry.

Let's hope that Congress, yes, and even the President of the United States, will give consideration to the self-same subject.

Let us spend money, certainly—no detail of aviation should be stinted but let us have men in charge of its expenditure who will see to it that the money which we spend shall count.

Document 2-25(b), Frank A. Tichenor, "The N.A.C.A. Counters," 1931.

AIR—HOT AND OTHERWISE
The N.A.C.A. Counters
Frank A. Tichenor

IN these columns in December, I reviewed the conditions prevailing in the National Advisory Committee for Aeronautics which prevent it from functioning in a manner useful to the best interests of the industry it purports to serve. The discussion has disclosed evidence of widespread interest in this question. It becomes ever more apparent that the points touched on have for some time been a subject of concern to many. The importance of a wise and honest expenditure of public funds appropriated specifically for scientific research and not for a cheap substitute for it, is generally recognized. The conditions which urged us to stress this vital phase of aeronautical development have found sympathetic response among all who have at heart the good of aviation and the country.

The comments of those who concur in our contentions contain little that has not already been said in these columns. As a matter of fact, little more can be said concerning these conditions which have lasted so long and about which informed public opinion fairly agrees. It was with interest and curiosity, however, that we awaited replies of defenders of the present N.A.C.A. management. We had hoped that some bright spots might be brought to shine on the otherwise dark picture, that perhaps things are not quite as bad as they appeared to us. Unfortunately we were disappointed in this hope. Although the management of the N.A.C.A. has an advocate who apparently tries to defend its policy, actually he only concurs in the broad picture which we painted. Indeed, we fear the picture is even darker after his defense than it would have been without it. We are disposed to believe now that conditions are even worse than we at first suspected, and that the trouble is not only lack of ability, but also absence of an honest attempt to accomplish the laudable task which the N.A.C.A., through its presiding body, has assigned itself.

This seems to be another case in which a feeble attempt to defend a weak cause serves only to render its defects more vulnerable. It almost looks as though the defender of the N.A.C.A. management in his own heart agrees with us; and although he finds it expedient to depreciate our criticism, he writes as though he himself would like to see reform effected. He does not call attention to one successful research, nor one scientific advancement which can be credited to the N.A.C.A., nor even one technical advance which we may have overlooked. Nor does he suggest that such advances can be expected in the near future. Indeed, there is not even the assurance that they will eventually be forthcoming or that anything is being done to hasten that day. Our principal criticism, the absence of scientific research, is tacitly admitted. Such research, he contends, is the proper sphere of universities, not of the N.A.C.A.

Now, we have not, merely as the result of our own judgment, specified scientific research as the task of the N.A.C.A.; we quoted this as the N.A.C.A.'s task from the Committee's own annual reports. The defender of the N.A.C.A. can not logically ignore this point altogether, as he does, for it is the most important consideration, the keynote of the N.A.C.A.'s shortcomings. This is not a question of opinion only: rather, it is far more a question of keeping faith, of loyalty to duties defined by the supervising body of the N.A.C.A. The policy of conducting scientific research was adopted ten years ago by the presiding Nominee Committee, made up of the foremost experts of the country. In all annual reports since then, it has been recorded as the accepted policy of this body. It has been pleaded for in hearings before congressional committees. It has formed the basis for public appropriations. Does the defender of the N.A.C.A. mean to imply that there is one policy for obtaining appropriations and for general advertising and publicity purposes and quite another one for the actual service and activity within the walls of the N.A.C.A.?

We are referring chiefly to a reply to our article expounded under the sponsorship (and probably under the personal authorship) of one who is himself a member of the supervising body of the N.A.C.A. Having been appointed by the President to serve as a member of the group prescribing the policy of the N.A.C.A., he certainly can not plead ignorance of that policy. His words would indicate that he is opposed to the policy of scientific research, that he prefers something easier and cheaper. If that is so, he can well argue that everyone has a right to an opinion of his own. It is, however, one thing to advocate a change of policy and quite another to advocate disobedience to a policy established by authority, laid before Congress and the nation, for the execution of which policy appropriations have been made. No ordinary citizen should advocate disobedience to rules established by authority;

he ought rather to satisfy himself by urging and pleading for reforms. How much more shocking is it that a member of the Committee should defend the management's activities when they are in direct contrast to the policies promulgated by the presiding body! Our opponent's failure to acknowledge the authorized policy of the Committee has greatly intensified our conviction that something is wrong with the N.A.C.A.

We expected the defender of the N.A.C.A. to give us evidence (or at least to try to give us evidence) that we had erred, and that the N.A.C.A. after all is true to its avowed ideals. Instead he denounces these ideals, admitting thereby the justification of our criticism. He joins us so far as the facts are concerned; he concedes the point that the N.A.C.A. has made no "contribution to pure science" and merely tries to make that failure appear insignificant.

This thing is not insignificant, however. If money is appropriated for scientific research, can we consider it of no consequence that those funds are spent for something else? The presiding body of the N.A.C.A., Congress, the industry—all of us expect the N.A.C.A. to be a bright torch of science; we have a right to expect that. We know that a great national organization is needed as a guiding light for a prosperous aviation industry, not as a satellite which merely reflects faintly the illumination of more brilliant bodies. This point is not one for which compromise can suffice. Either there is scientific research or there is not, and authority has decreed that the N.A.C.A. should conduct scientific research.

Such is the defense the management of the N.A.C.A. summons, a defense worse than none. It begins with mild mocking on the degree of courage necessary for attacking scientists—as if any scientists were attacked. The absence of scientists on the staff of the N.A.C.A. was attacked, and the politicians who occupy positions rightfully belonging to scientists were attacked. Certainly it takes more courage to defend their presence there than to criticize the absence of scientists. It takes more courage to disclaim the policy laid down by authority of office and public consent than to plead for it. It takes more courage to advocate a use of public funds other than that for which they were appropriated. It takes more courage to sacrifice the interest of the industry and to advocate administrative measures foreign to American spirit than to ask for a clean policy. Pleading weakness and inability to defend oneself in a case like this, where the defense would be easy if the criticism were unjust, is inadvertently an admission of guilt. Where are the advances of science made by the N.A.C.A. at Langley Field? Why not enumerate them, if there are any, instead of lapsing into disputes about personal courage?

Let us review, in short, what results the champion of the N.A.C.A. is able to point out. As the most conspicuous achievement, he mentions the Townend Ring and the fuselage cowling, admitting that the N.A.C.A. did not invent the latter. With the evolution of the cowling, the N.A.C.A. had nothing whatsoever to do. It is nevertheless seriously suggested that neither cowling nor ring would have been adopted by the industry had it not been for the N.A.C.A. The industry is alleged to be so timid that the information about improvements available is not sufficient to induce it to adopt them; the industry needs the guiding hand of the N.A.C.A.; the industry does not trust and has no confidence in its own speed tests made by its own pilots. The implication is that, instead, it waits until the N.A.C.A. measures in pounds and ounces the diminishment of the drag in consequence of some improvement and then computes the increase in the speed. The industry, it is seriously alleged, has more confidence in such computed speed gain than in a speed gain directly observed. How grotesque! We really have cause to admire the courage of one who advances such opinions. He dignifies this business of measuring by calling it "determination of relative merit," which must have something to do with the theory of relativity. At least we must confess that we fail to understand it fully.

As further useful research of the N.A.C.A., our attention is called to the pressure measurements of floats. There is certainly some use in such measuring. A tailor taking down the measurements of his customer also does something useful. Measuring pressures is but a secondary duty of the N.A.C.A.; its chief duty is to measure pressures (if any) in such a connection that scientific theory is advanced thereby. This was not done in the case mentioned.

As a third citation of successful activity of the N.A.C.A., another pressure measurement is mentioned, one which likewise failed to advance science—the pressure measurement over the dirigible *Los Angeles*. Although genuine contribution to science is in fact not claimed for this work, it is contended that these measurements have proved indispensible for practical purposes, and that the new Goodyear Navy airships will be stronger and lighter in consequence of them. We doubt whether the design staff or the Goodyear-Zeppelin Company will agree. The strength and weight of a dirigible depend upon structural improvements and aerodynamic progress. The air force loadings assumed for the structure of a dirigible are based chiefly on theoretical aerodynamic developments (to which no research at Langley Field has ever contributed), together with one numerical factor, based on operation experience. It is a curious fact that the Goodyear-Zeppelin Company, after its experience with more than 120 airships, arrived at a different factor than the U.S. Navy with its experience with two airships. A commendable research project would have been an attempt to clear up that discrepancy. Instead, the N.A.C.A. made some pressure measurements, leaving the whole question as it was before. No harm done; dirigibles will continue to be improved without the assistance of the N.A.C.A. But, after all, why all the measurements?

It is obvious, therefore, that our arguments have been strengthened by those of the advocate of the N.A.C.A. We stand in line with the supervising body of the N.A.C.A. and insist with it that scientific research is the proper and chief domain of N.A.C.A. activity. We advocate that capable scientists be encouraged to join, not to leave, the N.A.C.A. staff. We heartily agree with the able and patriotic chairman of the N.A.C.A. that the N.A.C.A., to borrow his own words, "undertake investigations directed by men who know the problems and the means of their solution" and that such changes be effected as are necessary to bring this about. We insist that is not the function of the N.A.C.A. to gape and squint at what others are doing, like a loafer standing on the street, and do at best a little measuring to see whether the work was done right, but that it should be the purpose of the N.A.C.A. to do things itself. Aeronautics has not yet reached its goal. The final shape of airplanes will eventually be quite different from what we have now. We want that development hastened. We want a critical and scientific survey, an exploration of all known possibilities. It may be possible (it probably is possible) to increase the specific lift to ten times what we have now, and we want a central institution of research to give us light on that. It may be possible to reduce the specific drag to one-tenth what we have now; the theory of air motion producing drag is still entirely in the dark. Friction of air, as such, does not account for more than one-twentieth of actual drag. We want to have some light on that too. We want knowledge concerning boundary control, concerning the effect of rotating cylinders, of vibrating surfaces, of lubrication, of autogiros, of Flettner cylinder, of jet action, of shooting action, of sound wave action, and of chemical action. Indeed the possibilities are without limit. We want a national agency to explore these unexplored regions, and to do so with scientific spirit, systematic thought, and honest endeavor. We are not satisfied with useless pressure measurements and with the building of wind tunnels which will never be really usefully employed. Build small laboratories and do big things in them; not the other way. Only then will the nation attain high rank in world aviation.

Our conviction as to the failure of the N.A.C.A. is not an original idea. The conditions existing have been recognized by none other than General John J. Pershing, who states in the third chapter of the story of his war experiences, appearing in the New York Times, Wednesday, January 14, "...we had some fifty-five training planes in various conditions of usefulness—all entirely without war equipment. Of these planes it is amusing now to recall that the National Advisory Committee for Aeronautics, which had been conducting an alleged scientific study of the problem of flight. . . . " General Pershing was in a position to know and no one can question his sincerity.

Document 2-25(c), Edward P. Warner, "Speaking of Research," 1931.

Nothing is easier, and nothing demands less courage, than attacking scientific work for the benefit of a non-scientific audience. The general public, and even most of us who are engaged in applying science to industry, have little understanding of scientific theories or of laboratory work, and it is immensely consoling to feel that people who pretended to be so much wiser than ourselves, and claimed to "understand all that stuff," were wrong all the time.

To assail the scientist is the safest of pursuits, for he has neither the inclination nor the equipment for rebuttal. Reasoned and orderly discussion in reply to a vindictive assault on the caliber of work done in a laboratory is quite impossible before an audience that has no background or experience of its own to give it an understanding of the nature of the controversy.

The National Advisory Committee for Aeronautics has had its share of vilification. Perhaps no answer, and not even passing comment, is necessary. The steady growth of interest in the industry's pilgrimages to Langley Field, annual since 1926, and the lively discussions that take place there, give evidence of the esteem in which manufacturers and operators hold the N.A.C.A. Nevertheless it is worth recalling how much influence that body's activities have already had on American aeronautical development. Let us take an example.

The venturi and ring cowlings are the most conspicuous of recent contributions to aerodynamic efficiency. They are the exclusive invention of no individual or group. Their genesis can be traced back at least to 1920. They have appeared in various forms here and abroad,—but the most important single step in practical application was the direct quantitative proof of the reduction in resistance they permitted. Without the tests that were made at Langley Field, the new forms of cowl might have been suggested and argued and perhaps tried in a small way, but their general acceptance would have been a matter of many years. Even in England, the original home of the ring, its adoption as a standard feature of airplane design has been enormously accelerated by the American laboratory work, which has given generalized in place of specialized results, and definite measurements of resistance in place of speculations based on a measurement of maximum speed. The most important function of the Committee, in short, was not to invent a new type of cowling, but to determine the relative merits of all available types and to make the determination on a scale which no other laboratory in the world was prepared to duplicate.

It is a good general rule that there are three types of research work, and they are adapted to three different types of organization. First, and in the very long run most important, is contribution to pure science and underlying theory, perhaps most generally the product of members of the staffs of educational institutions or of laboratories endowed by private capital or by especially farsighted industries. At the other extreme is the study directed to solve a particular and specialized problem of particular design, or to lead directly to the invention of a new proprietary device, and that is the proper sphere of the research department of a corporation.

It is a sphere, be it said in passing, not as yet sufficiently exploited by American airplane manufacturers, for in the face of the great affluence of the industry in 1928 and 1929 the absence of research departments in most airplane factories remained a proper ground for surprised comment by foreign visitors and for shame-faced acknowledgment by American engineers.

Between the two extremes there is a third class of work, the conduct of "practical" studies, general but immediate in their application. That is the particularly fitting task for a government laboratory.

To produce a theory of heat flow which will make it possible to calculate the cooling characteristics of an air-cooled cylinder is the function of an individual mathematical physicist, and it is on a college faculty that he will most often be found. To determine the cause of repeated spark-plug failure in the XYZ engine is the responsibility of the XYZ Engine Company. But to find out by actual measurement on a group of typical engines how temperatures are typically distributed in cylinders and how they are affected by changing conditions of flight, information that can be applied to the XYZ engine or any other, is work most profitably to be undertaken for the general good, and for general dissemination of the results, by such a body as the Advisory Committee.

The determination of design data may not be inspiring or spectacular. It does not appeal to the imagination as does an invention, but it is extraordinarily important. Without it, design does not progress.

The structural design of every airplane built in the United States today is dependent to some degree upon the N.A.C.A. work on pressure distribution and air loads in flight. The structure of seaplanes has gained a rational foundation for the first time through the studies made on pressure distribution on float bottoms. The airship being built at Akron will be a stronger and a lighter craft than would be possible without the N.A.C.A. measurements on the Los Angeles. Examples can be multiplied without number.

Science is a term that covers a multitude of widely different things. Not the least important among them is the skillful devising of means for accumulating data upon which the designers of engineering material may lean.

The Advisory Committee has acquired the material and the personnel to do that work for American aviation, and has been doing it. Aeronautical engineers in Europe are quick to express their envy of their American colleagues' good fortune in having at their disposal an institution of such resources.

Document 2-26

Joseph S. Ames and Smith J. DeFrance, remarks at the dedication of the NACA Full-Scale Wind Tunnel in "Report of Proceedings of Sixth Annual Aircraft Engineering Research Conference," Langley Field, Virginia, 27 May 1931, NACA, pp. 20–26, Historical Archives, NASA Langley.

As the NACA's research capabilities, physical plant, and budget grew, the committee clearly understood that continued public, and especially industry, support would be crucial. The committee also knew that it needed a way for the emerging civil aviation industry to bring its problems to the attention of the NACA. In a brilliant stroke, Director of Research George Lewis and Executive Secretary John Victory devised a conference where civilian and military leaders would be brought to Langley for a first-hand look at the facility and an opportunity to talk directly with NACA people. The first conference, with 38 invited guests in 1926, proved popular, and it quickly grew into the Annual Aircraft Engineering Conference. By 1931, the conference had become a well-staged and -attended event hosting over a hundred guests—including a number of prominent aviation journalists—to ensure publicity, and the NACA used the event to show off its latest technology. In 1931, the latest technology was the Full-Scale Tunnel, a thirty- by sixty-foot monster of a wind tunnel, and the committee made the dedication of the FST the highlight of that year's conference. This excerpt from the conference report outlines the carefully scripted dedication program. The dedication opened with a brief review of the history of aeronautical research by Joseph Ames, followed by a report by tunnel chief engineer Smith J. DeFrance, and culminated with a live demonstration of the new tunnel.

While the actual number of research projects that resulted from industry inputs at these conferences was small—less than two dozen in all—the conferences succeeded in projecting the NACA's best image to the aviation industry and the public, and they helped ensure essential congressional support.

Document 2-26, Joseph S. Ames and Smith J. DeFrance, remarks at the dedication of the NACA Full-Scale Wind Tunnel in "Report of Proceedings of Sixth Annual Aircraft Engineering Research Conference," 1931.

DEDICATION OF FULL-SCALE WIND TUNNEL.

The members of the conference then proceeded to the new full-scale wind tunnel, where a Vought Corsair airplane was mounted on the balance in position for test.

DR. JOSEPH S. AMES.

The tunnel was dedicated by Dr. Ames, who presented the following statement:

Before putting this new wind tunnel into operation I would like to say a few words with regard to the history of wind tunnels and something with regard to the design and study of equipment of this type. The problem is to find the resistance offered to the passage of a solid body through the air. Early experiments along this line were attempts to study the effect of wind on the human body by determining how far a man could jump with the wind and how far against the wind.

The beginning of the science of aerodynamics probably dates back to the year 1661, when Hooke read a paper before the Royal Society of London, on the resistance of the air. The material in this paper was based on experiments Hooke had conducted by throwing different shaped bodies horizontally from the top of a tower and observing the time they remained in the air before striking the ground. Similar experiments were made by Sir Isaac Newton in 1710, with spherical bodies. In 1746, Benjamin Robins, an Englishman and a distinguished mathematician, developed an early form of whirling table and accumulated considerable data on air resistance and on the motion of bodies projected into the air.

Although these investigations were concerned chiefly with ballistics and had little bearing on the aerodynamics of flight, they led to the statement of the pressure velocity law by Charles Huttin about 1790, that pressure varies as the square of the velocity, and to the importance of aspect ratio.

The whirling table, or whirling arm, and straight-away towing arrangements, were used for the investigation of problems in aerodynamics until the beginning of the twentieth century. These devices were supplemented by the actual flight of models, man-carrying gliders, and in some cases engine-driven airplanes. The era of gliding, from the time of Lilienthal, did much to lay the foundation of flight and the principles of stability.

Langley, early in his career, conducted research on the sustentation of bodies by inclined planes with a view to determining the fundamental data as to lift and drag of airfoils and the probable efficiency of air propellers.

About 1890, Sir Hiram Maxim constructed a wind tunnel with a three-foot square throat.

In 1901, Orville and Wilbur Wright set themselves to solve various problems of flight and started a lengthy series of experiments to check previous data on wind resistance and lift of curved surfaces, besides problems on lateral control. They built a wind tunnel at their home in Dayton, which had a 16-inch square throat and was 6 feet long. In this tunnel they measured the lift and drag of over 200 miniature wings. In the course of these tests they produced comparative results on the lift of square and oblong surfaces, with the result that they rediscovered the importance of aspect ratio.

It was not until 1909 and 1910 that organized research was undertaken, and wind tunnels were constructed at the National Physical Laboratory in England, at the Eiffel Laboratory in France, and at the Aeronautical Research Institute at Göttingen.

In the United States, wind tunnels were constructed at the Washington Navy Yard, at the Massachusetts Institute of Technology, and at the Bureau of Standards, and the first wind tunnel constructed by the National Advisory Committee for Aeronautics was completed in 1919.

In 1927 the Committee placed in operation the present propeller research wind tunnel. This wind tunnel was designed and constructed largely for the purpose of investigating the characteristics of full-sized airplane propellers. A large number of propeller investigations have been completed in this wind tunnel, but the major portion of the operating time has been taken up in the study of other than propeller problems. The investigation of large wing models, the study of the cowling and cooling of engines, the study of engine-nacelle-wing arrangements, the investigation of different forms of airship models, and many other projects have so filled this program that in 1928 the Committee submitted to the Director of the Bureau of the Budget a request for authority to expend $5,000 for the development and procurement of a design of a wind tunnel suitable for research on full-sized airplanes. With the approval of the Bureau of the Budget, authority was obtained from Congress for the expenditure of $5,000 for the study and design of a full-scale wind tunnel on May 16, 1928.

Preliminary designs and estimates were prepared and submitted to the Bureau of the Budget, and by act approved February 20, 1929, the Congress provided an appropriation to extend over a period of two years the construction of a full-scale wind tunnel.

The contract was awarded for the construction of this wind tunnel on February 12, 1930, and we are here today to place in operation this important equipment. With the completion of this equipment we now have available a means for actually studying a full-sized airplane under flight conditions.

The ideal method of investigating the stability and control characteristics of an airplane would be to place on the airplane in flight means of measuring the lift, drag, and moment characteristics. This, of course, would be very difficult to accomplish so as to obtain accurate measurements. In this wind tunnel we have practically done the same thing by placing the airplane on a balance capable of measuring all the changes, and providing the air stream flowing past the airplane.

The completion of this wind tunnel opens up a new vista of important problems, the solution of which I am confident will mean much toward increasing the safety and efficiency of aircraft. The Committee has received many suggestions for research problems from the military services and from aeronautical engineers, which will provide a research program that will keep this piece of equipment in continuous operation for a long period of time.

The Executive Committee in 1928 authorized the preparation of a preliminary design, and Mr. Smith J. DeFrance, of our technical staff, was placed in charge of this project. Mr. DeFrance, with his assistants and other members of our technical staff, has been responsible for the design and construction of the completed equipment. I wish to add at this point the appreciation of the members of the Committee to the staff of the Committee and to Mr. DeFrance for the excellent manner in which they have carried through this enterprise.

Mr. SMITH J. DeFRANCE.

The Chairman then called on Mr. DeFrance, who described the operation of the full-scale wind tunnel, referring, as he spoke, to diagrams of the plan of the tunnel and balance. His remarks were as follows:

After what Dr. Ames has said I am sure that you are all familiar with the principle of the wind tunnel. You have seen small wind tunnels in operation. This tunnel is in many respects similar to a small tunnel but of course much larger. The cross-sectional area of the throat is five times that of the next largest wind tunnel in the world, the propeller research tunnel.

This is the first wind tunnel ever constructed for the purpose of testing complete full-sized airplanes, and as such it will fill a very important place in the field of aeronautics. Its principal use will be in the determination of the lift and drag characteristics of an airplane. Previously it has been necessary to do this from glide tests in flight, and sometimes the tests have been very lengthy because of inability to control test conditions. Here we will be able to control the test conditions, and to obtain the polar of an airplane in approximately one hour whereas it might take a month in flight.

In this tunnel we will be able to study control, especially control at low speeds and at high angles of attack; and the drag of air-cooled engines, and of water-cooled engines with radiators, under practically the same as flight conditions.

The dimensions of the tunnel are 30 by 60 feet at the throat, and, as may be seen from the chart, the tunnel is of the double-return flow type. The velocity of the air stream may be varied up to 115 miles per hour. This stream is produced by two 35 1/2-foot propellers, each directly connected to a 4,000-horsepower motor of the slip ring induction type. The speed of the motors may be varied between 75 and 300 r.p.m. in twenty-four equal steps.

The airplane is mounted on a six-component balance through tubular struts. The chart shows the arrangement of the balance and struts, which are at present streamlined by fairings. Eventually the fairings will be separately supported and merely serve as shields for the tubes, thereby reducing the support drag to a minimum. The tubes in turn are secured to a floating structural steel framework, which is connected by linkages to six recording scale heads. The lift is taken on two scales forward and one in the rear, the drag on one scale, and the side-wind force on two scales. From the lift readings it will be possible to compute the pitching and rolling moments, and from the side-wind readings the yawing moments.

Mounted on top of the floating frame is a turntable by means of which it will be possible to turn the airplane from 15 degrees left to 15 degrees right while the tunnel is in operation. We will also be able to vary the angle of attack from 5 degrees to +20 degrees, thereby taking the airplane through the stalled condition and making it possible to obtain data so difficult to obtain in flight.

The scales are equipped with electrical recording devices which are operated from a single control stand. The amounts of the forces are printed on cards from all scales simultaneously, thereby eliminating errors which may arise from readings. The cards are moved from time to time as readings are made.

Because of the amount of power required to operate the tunnel and the small capacity of the local power plant, we are compelled to take the power on off-peak load, or between midnight and 6:00 a.m. The amount of power permitted during the day is 750 kilowatts, which will give an air speed of 55 miles an hour. This afternoon we are operating at that speed. Before the tunnel is started, the pilot will climb aboard the airplane and after the air stream has been started he will start the airplane engine. Readings will be taken on the scales, and you will be notified by placards when the cards are moved and when the angle of attack and the angle of yaw of the airplane are changed.

The pilot will now go aboard the airplane. Dr. Ames, I ask you, as Chairman of the National Advisory Committee for Aeronautics, to dedicate this full-scale wind tunnel.

Dr. Ames.

This Committee started its work in 1915. I regard this moment as probably the most important moment in the history of the Committee, because it is to set in operation a piece of apparatus which promises to give in the shortest time the most important information desired in the development of aerodynamics; an instrument which is unique in the world, and which we owe to the ingenuity of our engineers and to a Congress and a Budget Committee who understood our problem and were willing to cooperate with us.

The pilot having climbed aboard the airplane, Dr. Ames pressed the button and propellers were rotated, starting the air stream. The pilot started the engine of the airplane and readings were taken on the balances, the members of the conference being notified by placards as to the variation of the attitude of the airplane with respect to the air stream.

Document 2-27

Minutes of the Second Technical Committee Meeting, United Aircraft and Transport Corporation, 5 December 1929, pp. 522–531, Boeing Company Archives, Seattle, Wash.

Once the merits of wind tunnel testing had become apparent, as they had by the late 1920s, American aircraft builders began to think seriously about constructing major new tunnels for their own research and development work. The Curtiss Aeroplane & Motor Company had been operating a private wind tunnel for two decades with some notable results, including designs for Schneider Cup racers, but most manufacturers were content with the data coming from the NACA and university laboratories. Aircraft building was a risky endeavor in 1920, and the fledgling firms engaged in it did well to merely survive. As the decade drew to a close, however, the future of aviation seemed much more assured, thanks to a combination of federal regulation, greater public acceptance and enthusiasm, and technological progress. The aircraft industry, especially the larger companies formed through consolidation, could also support greater investments in research facilities.

The United Aircraft and Transport Corporation established a technical committee made up of prominent individuals in aviation to identify technical trends and recommend actions for the company to take. At the committee's second meeting in December 1929, the minutes recorded a discussion about wind tunnels. These minutes provide unique insight into the thoughts of several influential people in the aeronautical community concerning the role of private versus government laboratories, the state of intercompany cooperation, and the overall understanding of wind tunnels. While no two situations can be exactly alike, these minutes, and especially the debates over the desired size and costs, reveal much about the decision process for governing and managing boards.

Document 2-27, minutes of the Second Technical Committee Meeting, United Aircraft and Transport Corporation, 1929.

CHAIRMAN MEAD: Another question which has been brought up a number of times was that in the group we have no aerodynamic facilities in which to deposit all the information we have gotten so far, and the next decision is whether with as much money as we have in aviation we can rely on the outside sources of information whenever we need such data.

I have no brief either way, but it just seems peculiar not to have some facilities of that kind in our own outfit, and I would like to hear what you think about that.

I have incidentally heard a good deal of criticism of having some work done outside because of the leaks which occur; everybody knows about it before you do.

MR. MONTEITH: Before you go into the aerodynamic research, I think that everybody in the group who is doing flight testing of production ought to be equipped with decent barograph and calibration instruments.

CHAIRMAN MEAD: Tomorrow, we want to have you, if you will, get your gang together on this flight testing and see if you can not come to some conclusions as to both how to do it and what equipment is required. Then we could get the various units so equipped, because that certainly is most important.

MR. SIKORSKY: I think that is very important, and I would even add the suggestion that maybe we can order some entirely identical instruments, because it will help if we can know the instruments; it will simplify the comparison of the test.

MR. McCARTHY: We bought a barograph recently which is the same as the Navy uses at Anta Costa. It was quite expensive, of course.

MR. SIKORSKY: As a central unit, it seems to me a simple method to install a method of correcting these instruments.

MR. CHATFIELD: The operation for calibrating an ordinary barograph is not very elaborate, ten or fifteen dollars will buy one. We use mercuro-chrome with a bell jar. Apparently the temperature areas are not important enough to worry about.

MR. McCARTHY: I think we ought to talk a little about wind tunnels before we go away. It seems to me that the United can well afford to buy one wind tunnel somewhere in the East for the use of our outfit.

MR. MONTEITH: Why the East?

MR. McCARTHY: Well, it would certainly have to be in the East or West to be of any use to anybody.

—You will have to get your own facilities out there. I don't see how we could have one wind tunnel to serve everyone.

But, for a fairly modest sum we can put up about an eight-foot tunnel that would serve all the Eastern units and probably excluding Stearman.

MR. SIKORSKY: I am very much in favor of it. I think it is simply a necessity for us to arrange our own research laboratory of such size as would give reasonable service. As was stated here, the very important fact of secrecy,—and we know that it is almost impossible to get it,—is sufficient.

Besides this, simply the service is sometimes very hard to get. The data is not so reliable and not comparable with each other, and I believe the wind tunnel will simply pay back its cost in one year or so to everyone of us, besides the special work for every unit the wind tunnel would do, and the general work everyone would be interested in.

Again, it is interesting both so to speak in a positive and a negative way. Today, for example, we asked the question about these new wheels; who knows accurately what the data is? With our own wind tunnel, we could test it correctly and have reliable information available.

The same thing holds true with the new shape of stresses which come out, the results would be the actual tests of such new refinements, ideas and so on which may come out. Because, probably one of the strongest things which United may have is to keep leading the industry, and to do it accurately, I believe a research laboratory would be of considerable value, it is certainly worth spending twenty or thirty thousand dollars on, or whatever it would cost.

CHAIRMAN MEAD: There is quite a variance in opinion as far as I can gather as to what kind of a wind tunnel it should be, and how large.

MR. McCARTHY: I don't think you want to go below eight feet.

MR. CHATFIELD: I think the central tunnel should be,—I favor going a little larger than eight feet, with the idea that possibly some of the individual plants may have smaller tunnels and would like a larger one in which to take those problems which can not very well be handled in a four or five foot tunnel.

CHAIRMAN MEAD: I think we want to interrupt Chat just a second—I realize that this gang around here is being looked upon to provide the equipment which will return a good profit on the investment in United, and therefore it is a much different picture than those of us individually faced before when we have all pinched the pennies here because we have had to. Now, by pooling,—everyone giving a bit to this project, we certainly can afford better equipment than we could ever think of having ourselves, and I don't think we ought to look at this thing in too niggardly fashion.

What we really should have is desirable equipment, or at least ask for it, and if the money is not forthcoming that is too bad and we will have to trim our sales accordingly.

I feel that if you want a wind tunnel, I certainly want it to be a good one whatever size that might be. I don't know anything about wind tunnels so I can't advise as to size.

MR. CHATFIELD: I would like to have Mr. Weick's opinion on that point.

MR. WEICK: Well, I think that size is more important than velocity for instance. You can put money into a wind tunnel in two ways, one is in size and the other is in power to obtain velocity, and while of course, you want to get as near to full scale results as you can in both cases, the size when you are dealing with all sorts of models is more important than just velocity, and the power goes up very quickly with increased velocity. I think that Mr. Sikorsky's point is worth emphasizing that one wind tunnel, if you are going to rely on wind tunnel work which apparently you are to some extent, one wind tunnel in which all of these various models can be tested in by uniform methods under uniform conditions would be greatly valuable, because as it is with some testing in one tunnel and some testing in another you can not prepare those models with any degree of exactness, you can't expect to be able to compare them.

CHAIRMAN MEAD: We can take for definite example, Boeing Company developed something here in the wind tunnel that might exist at the University of Washington. The result of that test might go readily to Hartford to be checked, or wherever this other tunnel would be, so that our data would all be alike.

MR. WEICK: That would be very well.

MR. MONTEITH: I am pessimistic about shipping wind tunnel models. We used to ship them to Atlantic, or M I T, and the Railroad Company used them like they were cord wood.

MR. CHATFIELD: We have had fairly good luck recently at M I T with models coming in.

CHAIRMAN MEAD: We have not answered the question of how big or how fast here.

MR. McCARTHY: You would not want to answer it in one gulp, would you?

MR. MONTEITH: It depends on how much money you have.

CHAIRMAN MEAD: Well, let's put down two or three operations here, and let's see then how much money we can get; we will go out and canvas the crowd.

MR. MONTEITH: I think you have only two operations. You have either the full sized wind tunnel, or a reasonably sized tunnel like eight or ten feet; eight feet preferably. I think the four foot tunnel is absolutely out of the question except for very minor tests.

CHAIRMAN MEAD: Your optional one is how big then? Would you say this big fellow?

MR. MONTEITH: It is twenty feet like NACA.

MR. WEICK: That is very good because you can put all full scale bodies into it, and propellers and so forth, and the only thing you can't get in full scale is the wings.

CHAIRMAN MEAD: How low speed in a twenty-foot tunnel is going to be any earthly good?

MR. MONTEITH: Sixty miles an hour anyhow.

MR. WEICK: You ought to have at least I would say eighty. You see the tunnel there has only one hundred ten as an absolute maximum at the present time, and one hundred is what we call top speed for testing, and that was quite satisfactory for almost all conditions.

CHAIRMAN MEAD: Can I tell again what is the top limit on this thing?

MR. MONTEITH: I don't think you should go above one hundred.

MR. McCARTHY: What did it cost to put in that unit down there?

MR. WEICK: It is hard to say exactly, because the power units were put in by the Navy with a couple old submarines.

MR. McCARTHY: I think the eight or ten foot tunnel is about the limit.

MR. WEICK: The only thing I can say about the ten-foot tunnel is what someone here said the other day about the cost of the Berliner-Joyce tunnel. They have one which cost $37,000.00.

MR. CHATFIELD: California Tech's actually cost $60,000.00.

CHAIRMAN MEAD: And what velocity in a ten-foot tunnel?

MR. WEICK: I would say you want one hundred miles an hour.

CHAIRMAN MEAD: Well, now, that is the way to go at it and we can ask them for what we feel we need. Is there any use of having an intermediate size here? That is, the cost of these things seem to go up as the sixth power.

MR. MONTEITH: No use going beyond ten feet if you can't go to the full size.

CHAIRMAN MEAD: Well, isn't NACA apt to go at things in rather an expensive manner? As long as you are out of it, Weick, you can perhaps feel this is not criticizing, and we could not perhaps build a big tunnel for much less expense. As an example, Cline's tunnel down here, he admits for certain reasons that it is very expensive and apparently could be well cut down to $30,000.00 or $35,000.00.

Do you think there is any chance of the NACA's being very elaborate,—very much more than would be necessary?

MR. WEICK: I don't think the NACA tunnel could be classed elaborate in any sense of the word. It is built of only a single thickness of board on a steel framework for the walls, and it was sort of a factory job in construction all the way through with no trimmings whatever.

CHAIRMAN MEAD: I would think from what little I could see that the cost is in the machinery and not in the tunnel itself.

MR. McCARTHY: They got the machinery for nothing.

MR. MONTEITH: How much power have they there?

MR. WEICK: 2000 horsepower is all.

CHAIRMAN MEAD: Now, wouldn't it be worthwhile, we seem to be interested in wind tunnels, to make Chat the dog again and have him go around again and see or get information on what it really might cost to build these tunnels based on actual cost of other places and examination of NACA's, and discussion with them as to production and costs, and so on, and then we can put up a figure which looks reasonable and let the Executive Committee decide what they want to do about it.

MR. WEICK: I think that is a good idea.

Incidentally, they are now building another tunnel at NACA which would give better information than our tunnel, but you see when you double the linear dimension of a tunnel, which is done in that one, then you go up as a cube in volume and the price of your building goes up to beat the band.

MR. McCARTHY: That type of tunnel is not usable for a lot of things. The small tunnel is too.

CHAIRMAN MEAD: We might find a grandpa to give us these things, who knows?

MR. SIKORSKY: I believe a ten-foot tunnel would do all right, because in a big tunnel simply the speed with which you can make big tests and actually put models in,—I think it may be a little too big.

CHAIRMAN MEAD: Of course, we could do this, we could start off in a modest fashion and take the ten-foot, then if we kept growing and there was any need for a bigger one and we found it was desirable we could build it perhaps.

MR. WEICK: I would certainly recommend starting with a ten-foot tunnel.

CHAIRMAN MEAD: What else in the way of equipment do we have to have,—or is the desirable equipment to have, is the better way perhaps to put it?

Here is one wind tunnel—

MR. CHATFIELD: (Interrupting) There is a point in the operation, I think a great many wind tunnels are limited in their usefulness, not in their ability to conduct the test, but in their ability to get them written up afterwards.

To get the full use of a tunnel, I think it ought to have a larger personnel than many wind tunnels have so as not to have the delay in the making of the tests and the reporting of the results.

MR. McCARTHY: I think that is a secondary question. If you get the Executive Committee to approve of the construction of a wind tunnel, you merely go out and hire people to run it, that follows naturally.

Document 2-28(a–b)

(a) A. L. Klein, "The Wind-Tunnel as an Engineering Instrument," *S. A. E. Journal* 27, No. 1 (July 1930): 8–90.

(b) A. L. Klein, letter to V. E. Clark, 14 August 1934, Klein Collection, Folder 3.2, California Institute of Technology Archives, Pasadena, California.

At a Los Angeles meeting of the Society of Automotive Engineers (a leading professional society for aeronautical as well as automotive engineers) in 1930, A. L. Klein, an aerodynamics professor at Caltech, presented "The Wind-Tunnel as an Engineering Instrument," which briefly discussed some of the ways wind tunnels could be used, with particular emphasis on research to enable high-speed planes to land on short fields. The paper itself is mildly interesting, but the discussion that followed—moderated by former Langley Laboratory Engineer-in-Charge Leigh M. Griffith—ventured into several contemporary issues concerning wind tunnel design and operation. The discussion concerning the problems with small, high-speed electric motors with model propellers explained part of the rationale for building the PRT and FST at Langley, and other comments, primarily by Griffith, noted both problems and successes with the VDT. Considering his background with the NACA, two of his remarks are especially interesting. Griffith opened the discussion with, "The wind-tunnel is more or less of a mystery to many who are otherwise well versed in aeronautics. I have personally had some experience with it, and it is still a mystery to me." His ensuing comments belie that self-effacing remark, and his final comment at the end of the discussion sums up his belief that "It is only a question of time, I think, when we shall be able to design aircraft upon the basis of tunnel tests and not miss the computed performance on the full-scale machine by more than 2 or 3 per cent." Time would prove him right.

While Klein's 1930 S. A. E. paper shows his interest and mature understanding of wind tunnels and research techniques, a letter he wrote to Virginius E. Clark in 1934 went into greater detail concerning many of the practical considerations for wind tunnel work. Klein's letter, the result of years of experience with Caltech's tunnel, outlined a careful procedure for testing new aircraft designs and included an itemized estimate of the costs associated with wind tunnel testing. The dollar amounts appear miniscule today, but at a time when an engineer might earn $15 per week, spending $800 to $1,200 on a test series was not something to be taken lightly, and the planning of test programs grew increasingly complex to ensure the greatest possible benefits.

Document 2-28(a), A. L. Klein,
"The Wind-Tunnel as an Engineering Instrument," 1930.

Of the many problems that arise in the design of an airplane, those in connection with the wings can be most easily investigated in the wind-tunnel. The determination of the mean aerodynamic chord of an unorthodox wing cellule is one of the most obvious types of wind-tunnel problems. It is highly desirable that the wing cellule of the airplane be investigated independently, as only by determining the polars of the wing cellule alone and then repeating the measurements with the fuselage, nacelles and other parts in place, can the interference between them be measured. By following this procedure and then trying different types of filleting, marked improvements in the characteristics of the complete airplane can be obtained. Muttray [Assistant Professor of Aerodynamics, California Institute of Technology] has shown that an improperly filleted fuselage can have a marked effect upon the wing-fuselage interference. [See National Advisory Committee for Aeronautics Technical Memorandum No. 517.] A badly designed fuselage and fillets cause a great decrease in the equivalent span of the airplane. The interference drag can be almost completely eliminated by correct design. It is well known that anything attached to the upper surface of a wing has a very detrimental effect. If protuberances can not be avoided, a model of sufficient scale should be constructed and their design worked over so that they will have the least effect. All re-entrant angles and small gaps, especially up above the wing, should be avoided. The recent work of the National Advisory Committee for Aeronautics has shown that a properly mounted wing-engine has only one-sixth the drag of the present normal type of engine nacelle. [See National Advisory Committee for Aeronautics Technical Note No. 320.]

Miscellaneous Drag Problems

The present type of landing-gear has very large drag, principally because of acute angles between the struts. Landing-gears can be tested at full-scale in a large wind tunnel or at half scale in a smaller wind-tunnel. The N.A.C.A., in its brilliant development of the Venturi cowling, has pointed the way for a more scientific attack upon drag. The British townend ring, though different in principle from the N.A.C.A. cowl, produces similar results.

A newer type of drag problem has arisen in connection with very high-speed airplanes. This type of plane is necessarily so clean that its gliding angle is very flat. This characteristic, combined with some of the ways in which the tendency toward low-wing monoplanes, has produced airplanes having great floating tendencies. These planes are very difficult to land in small fields over obstructions. There are three possible methods of landing them, all unsatisfactory: (a) The pilot may glide into a field steeply, picking up speed all the time and then floating a long distance before making contact; (b) he may glide in at his minimum gliding-speed and touch the ground at approximately the same distance from the obstruction as before; or (c) use the last method which is to squash into the field and pull out just before contact, thus making a short landing.

The first two methods are impossible in small fields, while the last method requires great skill and is very dangerous in bumpy air. Side-slipping a high-speed airplane is not very effective, as the fuselage used in this type is a very good streamline body at any ordinary angle of yaw. A few calculations will show that enough flat-plate area to decrease the lift-drag ratio to a reasonable value will be almost impossible to obtain in a safe and controllable manner. The usual form of spoiler is likewise inadvisable as it decreases the lift markedly, thus increasing the sinking speed. The only reasonably safe way to decrease the lift-drag ratio is to use some form of interference-drag device that will not spoil the lift and yet will produce a large increase in drag. This is an ideal wind-tunnel problem, and the polar and pitching-moment curves of any contemplated device can be easily determined.

High-lift devices should always be investigated to determine their effectiveness. Enough tests have been made with models and with the corresponding full-scale airplane to prove the reliability of the wind-tunnel methods.

Dynamical Stability Difficult To Test

All of the foregoing tests can be made with the ordinary three-component wind-tunnel balance. To investigate the complete airplane with its six degrees of freedom, a six-cylinder balance is necessary. The model of the complete airplane can be tested for stability and control, the effectiveness of the controls measured and the statical stability investigated. The problem of stability, power on, is more difficult. The statical stability of the airplane can be determined without the slipstream, and after the coefficients have been determined the moments of the gravity forces and the propeller thrust can be added in. To work with the slipstream a small high-speed motor is necessary. To date no satisfactory power unit has been developed, although a number of laboratories have designed or purchased apparatus for this purpose.

The study of dynamical stablity is very difficult, as it requires an entirely different type of set up than any of the foregoing tests. The model must be free to oscillate and it must be dynamically as well as geometrically similar to the full-scale airplane to be investigated.

The problem of wing or tail flutter is very difficult to investigate, as the model must be constructed so as to have the same geometrical shape, structural rigidities and mass distribution as the airplane. Work of this nature has been done and more will be done in the future. The surface texture of airfoils is now being worked on in the world's laboratories, and definite data on the effect of corrugations, rivet heads and the like will be available in the near future.

It is well to mention that propellers are being constantly tested in the laboratories that specialize in this work and our present remarkably high propeller efficiency has been achieved as a result of their efforts.

The writer does not believe that all of the tests mentioned are necessary for the design of a conventional airplane, but every plane should have its polars, moment curves and static stability determined. The rules now extant for dynamic stability give satisfactory results and those for the prevention of dangerous spinning characteristics are sufficient for the designer in most cases.

THE DISCUSSION

CHAIRMAN L. M. GRIFFITH [M.S.A.E.—Vice-president, general manager, Emsco Aero Engine Co., Los Angeles]:— The wind-tunnel is more or less of a mystery to many who are otherwise well versed in aeronautics. I have personally had some experience with it, and it is still a mystery to me. One thinks the building of a tunnel is a simple sort of job. He sees a tunnel running, notes its character, gets the dimensions and drawings and builds one like it. If he has had no experience, he says, "In two months we will have the tunnel finished and start making tests." But after the tunnel is completed in the two months, usually a year or two years is required to find out whether it is a good tunnel or not. There seem to be many things to contend with when one deals with air at high velocity through a wind-tunnel; the air does not follow the nice, smooth lines that were laid down on the drawing-board. Information resulting from wind-tunnel tests of all kinds, however, forms the real basis of our aerodynamic advance. We discover many things with full-size machines but can not very conveniently measure them. The quantities involved can not be determined readily, as we found at Langley Field; therefore we are dependent for much of our information upon the results of tests in wind-tunnels on models and parts of airplanes.

DR. A. L. KLEIN:— I can echo what Mr. Griffith has said, since, after the tunnel at the Institute was built, we found that one place inside of its perfectly conical body the air was moving upstream. We were much astounded; then we did a few things and got the air to go in the same direction over all of the tunnel.

Full-Scale Application of Tunnel Results

CHAIRMAN GRIFFITH:— Many airplane designers who have had a little experience are prone to think that the wind-tunnel is suitable only for the use of research men working on problems that have no bearing on the actual airplane. On the other hand, there may be one or two designers who have implicit faith in any result that comes from a wind-tunnel. Somewhere between these two views is a happy medium where the work of the designer is guided, not controlled, by wind-tunnel results. All such results are subject to interpretation and modification as necessary to suit the actual full-scale design, taking into consideration the difference in operating conditions between the flight of the full-size airplane and the passage of air around the small model in the wind-tunnel.

STANLEY H. EVANS [Aeronautic Engineer, design staff, the Douglas Co., Santa Monica, Calif.]:— What model airscrew speed can you get in the tunnel, Dr. Klein?

DR. KLEIN:— Our largest models will be of 6-ft. span, and an airscrew of the same proportional size as that used in an airplane would be approximately 18 in. in diameter. To run that propeller at the same V/D ratio as the actual propeller would require a speed of 10,000 to 15,000 r.p.m. Great difficulties have been experienced with the small electric motors at such speeds because of overheating. We hope soon to have a high-frequency generator to drive a three-phase motor at any speed up to 20,000 or 25,000 r.p.m. and to be able to control its speed by controlling the speed of the motor generator. Such apparatus is very expensive and the sets built to date have not been very satisfactory.

MR. EVANS:— I assume you could use a much larger propeller and only a small portion of the airplane model.

DR. KLEIN:— That could be done, but we were thinking of running the propeller in the stability tests of the airplane as a whole. If you were developing nacelles, you could make a model of just the parts of the structure adjacent and use a larger propeller; this would require more horsepower. The only successful work of this type has been done in England and Germany, and one of the aerodynamical laboratories in this Country received a duplicate of one of these motors and found that it ran red hot.

Trouble with High-Speed Electric Motors

WELLWOOD E. BEALL [Jun. S.A.E.—Assistant chief engineer, Walter M. Murphy Co., Pasadena, Calif.]:— The motor to which Dr. Klein refers was imported from Germany by Prof. Alexander Klemin of New York University. It was about 2 ½ in. in diameter and about 9 in. long. It operated on 500 cycles and required a special converter, also of German manufacture. This small motor developed, as I recall, about 1 ¾ hp. and was similar to one the Navy experimented with some time ago.

This motor was intended to be mounted in a wind-tunnel model and drive a propeller so that conditions approximating actual powered flight could be simulated. It was designed for operation at 40,000 r.p.m. with the propeller geared down to a suitable speed. However, operation at this speed was found to be impracticable, due to the motor overheating. It was then adjusted to turn at 36,000 r.p.m. With the motor mounted by itself in the laboratory and when it was turning a small propeller which threw considerable air upon it, this speed proved to be practicable. However, as soon as it was mounted inside a model for test, where no air current could strike it, it immediately became hot and after 45 sec. of running became too hot to operate.

The propeller reduction-gears were mounted on the motor in such a way that the torque reaction, and consequently the power delivered to the propeller, could be measured. This reduction-gear train was carried by a frame that pivoted in such a way that the torque reaction tended to rotate it. This rotation was restrained by a calibrated spring and the torque was indicated by a long, thin arm. This torque indicator operated satisfactorily in still air but, when placed in the slipstream of the propeller or in the wind-tunnel, it became inoperative due to the impact of the wind on it. This prevented the indication of the torque and consequently the calculation of the power.

This motor was also equipped with a revolution counter consisting of a worm-gear train and a small disc about an inch in diameter with one mark on its circumference. To obtain the speed of the motor, it was necessary to watch this disc, count its revolutions and calculate the result. This method is suitable for obtaining the speed of the motor before it is mounted in the model but very inconvenient when mounted in the model and in the tunnel. The reasons for this are obvious.

Although this motor was rather disappointing, it did arouse considerable interest and at least has provided a start in obtaining data for predicting the effect of the propeller slipstream and wash by means of wind-tunnel tests. The motor, I believe, has been sent back to its manufacturers to be rewound and rebuilt to operate at lower temperatures under load. A new system of distance-type indicating devices for the speed and torque is also being devised. With this rebuilt motor it is hoped that many valuable data may be obtained.

CHAIRMAN GRIFFITH:— The difficulty of running small-motor tests in the wind-tunnel is one of the factors that led to the present large tunnel at Langley Field and is leading the National Advisory Committee for Aeronautics to plan the construction of a much larger tunnel. I understood that the size of this was to be in the neighborhood of 30 ft. high and 40 ft. wide, but Dr. Klein tells me it has been increased to 30 x 60 ft. It is interesting to note that the 20-ft. tunnel takes about 2000 hp. to drive it.

With reference to Dr. Klein's comment about the detrimental effect of protuberances on top of the wing, I have been curious to know how much the Dornier-X speed might be below that of a similar airplane having the engines mounted within the wing itself.

DR. KLEIN:— German engineers have made some tests but the results have not been completely published. They showed that locating the propeller completely above the wing does not interfere with the wing. Before he built the flying-ship, Dr. Dornier expected to get a considerable increase in lift at take-off, because the slipstream would be entirely above the wing and increase the circulation about the wing.

QUESTION:— Has any work been done on an airplane which has some variable-drag device to increase the drag on landing so as to reduce the landing speed?

DR. KLEIN:— I do not know of any that has been done. We expect to try several devices of our own and of other people for this purpose. I think personally that the only feasible means is to use some interference-drag device; any other way is open to objections on the ground of reduction of controllability.

Variable-Density Wind Tunnels

QUESTION:— What is the situation in regard to increasing the air density, using a closed pressure-system?

DR. KLEIN:— That is one way of achieving a large Reynolds Number, which is our criterion of scale effect. A tunnel 5 ft. in diameter that can be pumped up to 20 atmospheres has been built at Langley Field and has been very successful. The British are contemplating building a similar tunnel. The Langley Field tunnel was exceedingly expensive. I imagine a high-pressure tunnel would cost about five times as much as an open tunnel of the same size. Our tunnel has a 10-ft. diameter, and we get an increased scale-factor by running at air-speeds up to 200 m.p.h. I think that our tunnel, without the building, cost approximately $75,000. We did not expect to get such high speeds but are pleased that we can get them. We build wind-tunnels and get astonishing results; nobody has very clear-cut ideas as to what the ideal wind-tunnel is.

CHAIRMAN GRIFFITH:— An interesting item about the variable-density wind-tunnel at Langley Field is the tank in which the tunnel was placed. It was a very good piece of ship-plate work. The shell is 15 ft. in diameter and about 30 ft. long, with hemispherical ends, and weighs 43 tons. The side plates are $1\frac{1}{4}$ in. thick. This tank was tested to a pressure of 450 lb. per sq. in. and showed very little leakage. It cost $24,000, and $1,200 more was spent to get it from the place where it was built to the site of the tunnel. When we got it there, we began to figure how much more money we would have to spend. We had enough to put up a building and, through the cooperation of the Navy Department, used Navy equipment that originally cost about $80,000 and had been used in the helium plant at Fort Worth, Texas. Consequently, we were able to do a relatively big job for a small sum of money.

Plans for Huge Tunnel at Langley Field

A MEMBER:— Is that very large tunnel at Langley Field actually being constructed and is it possible to make a guess as to the power that will be required?

DR. KLEIN:— I believe that the National Advisory Committee for Aeronautics obtained from the Congress an appropriation of $900,000 for it and expects to use about 8000 hp. The Committee was considering the larger tunnel very seriously and was debating how to build it.

CHAIRMAN GRIFFITH:— This wind-tunnel problem is really very interesting. When we built the 5-ft. variable-density tunnel at Langley Field we thought we would be in an excellent position to investigate all kinds of aerodynamic problems in the tunnel. Shortly after that tunnel was finished, we started the 20-ft. tunnel with the idea that we would then be able to make tests at the same Reynolds number but with different air densities and model scales. The interior of the high-pressure tunnel burned out several times and we found that wood was not a suitable material at 20 atmospheres, or a pressure of about 300 lb. per in., as combustion is extremely energetic at that pressure and air velocity.

All this time the Committee was carrying on full-scale work with airplanes with about 17 different varieties of recording instrument and found the limitations of that method. Having completed the 20-ft. tunnel, it is now building one 30 x 60 ft. To the industry the world over it looks as if it were going to be worth a lot of money. I really believe that, with proper coordination between full-flight tests with all the instruments that can be crowded into the cockpit, tests of the model airplane in a variable-density tunnel and tests in an ordinary tunnel, we can produce a mass of coordinated data that will tie up rather closely the various testing means.

In any case we can look forward to the increasing use of the wind-tunnel and to its influence being reflected in greater aerodynamic efficiency of aircraft. That is very definite. We know that the wind-tunnel has given us the basis on which we have built most of our aerodynamic progress and is going to be the main instrument for further development. It never will take the place of free-flight development, but it is coming closer to it. It is only a question of time, I think, when we shall be able to design aircraft upon the basis of tunnel tests and not miss the computed performance on the full-scale machine by more than 2 or 3 per cent.

Gerald Vultee [Chief engineer, Lockheed Aircraft Co., Burbank, Calif.; now with Detroit Aircraft Corp., Detroit]:— The result of my experience in flying airplanes is that, if one sometimes could be sure of hitting within 25 per cent of calculations, he would feel much better.

Present Landing-Speeds Seem Safe

JOHN K. NORTHROP:—Will the new landing rules of the Department of Commerce necessitate the use of variablelift devices, or will it be possible to bring a full-scale machine to a landing much slower than the theoretical figures would indicate?

MR. VULTEE:—As I remember the rule, unless the theoretical landing-speed was below 60 m.p.h. the Department required special flight-tests to prove the practicability of the design. If those were passed, and they were not particularly stringent tests, the design was approved. However, I believe from the experience we have had that the plane can be brought in at a considerably lower speed than would appear from theoretical considerations.

It is natural to assume that most manufacturers' performance figures on landing-speeds are somewhat optimistic; checking the maximum lift–coefficients of existing commercial planes against the performance that is claimed for them, the maximum lift–coefficients are found to run in the neighborhood of 0.0040 or higher. These are rather high figures. We encountered something similar to that in trying to reduce the landing-speed we had. We found we could not get any assurance of being able to reduce the landing-speed appreciably with any normal wing-section, as by actual flight-tests we already had a maximum lift-coefficient of about 0.0040. However, I believe that a good pilot can bring a plane in rather more slowly than the theoretical figures indicate.

Perhaps we have become used to seeing planes coming in at 60 and 65 m.p.h. and it looks like 45. The landing-speeds we are using now seem to be satisfactory as regards safety. The planes get in and out of fairly small fields and do not average a large per centage of crack-ups on landing; therefore, as an increase in landing-speed will make possible a greater increase in high speed, it seems that we should go a little slowly in drawing conclusions regarding specifying slower landing-speeds. Planes designed for a landing-speed of about 40 m.p.h. would look rather queer compared with the planes that are being built at present. All the builders, I believe, are making a little increase in the allowable landing-speed in their designs except for airplanes that are built for special purposes, such as training, for which a lower landing-speed is absolutely necessary.

Document 2-28(b), letter from A. L. Klein to V. E. Clark, 1934.

August 14, 1934

Mr. V. E. Clark
30 Rockefeller Plaza
New York City
Dear Mr. Clark:

You will please find enclosed a copy of our wind tunnel specifications together with some additional blueprints and addenda to bring them up to date.

In answer to the first point in your letter, I would state that the maximum span of models which we have tested in our tunnel is $7\ 3/4$ ft.; any size up to this will be satisfactory. We usually try to pick a model span of some simple ratio to the full scale airplane, in order to simplify the computation of the model dimensions. The wind tunnel corrections are now in such good order that we have found that this size of model, $77\ 1/2\%$, gives perfect satisfaction. The only reason we do not try to go to a larger model is because we are afraid that our velocity distribution will not be satisfactory very much nearer the walls. This size of model results in approximately a $1/7$th scale model of a modern single engine transport.

Your question 2 is covered, I think, in the wind tunnel model specifications.

Our normal procedure in testing airplane models is to test the wing alone first, in order to find its characteristics, then to build up the model part by part, putting on the fuselage first and filleting it with wax in order to get the minimum detrimental effect, then adding the engine and landing gear, etc. Finally, having made a complete test of the model less horizontal tail surfaces, we add the horizontal tail surfaces and take measurements with three positions of the stabilizer in order to determine the optimum stabilizer setting. After making the stabilizer tests with the elevator fixed, we work with the elevator set at various angles for its effectiveness, and similarly with the ailerons and rudder. If control tabs are used on the elevator, we make elevator free tests, in which case the elevator must be mounted on ball bearings and counterweighted so that it is in static balance. These tests are made at several tab settings in order to get the tab effectiveness.

We have found from experience that the following gives a fairly accurate estimate for experiments on normal airplane models:

a) A fixed cost of about $120 for preparing the model for the tunnel, making the preliminary calculations for reducing the data, etc.

b) A running cost of approximately $28 per Run, a normal Run consisting of a series of three or six component measurements at about 15 angles of attack and at one air speed.

Our costs are based on the following items:

a) 1.5 times our labor cost.

b) A wind tunnel charge of $6.00 per hour for all time in which the tunnel is tied up for the investigation.

c) Electrical power used at 1.2 cents per K.W.H. (about $2 or $3 per Run).

We prefer to base our charges directly on the above costs, but if you prefer, we will make a definite bid based on a detailed list of exactly the tests you desire.

In order to quote on a single set of lift, drag and pitching moment curves corrected for Reynolds' Number and turbulence, it will be necessary for us to make at least three runs at various speeds in the wind tunnel and then to extrapolate to full scale. This will cost in the neighborhood of $200 to $225 on account of the large overhead of one test. The unit cost of a small number of runs is rather large. Our costs have been as small as $20 per run for investigations that were of some length, i.e. from 60 to 100 runs on the model. We have found that the normal cost of a complete test on a model as mentioned is from $800 to $1200.

I hope that the foregoing will give you all of the necessary information which you desire.

Sincerely yours,

A. L. Klein

Document 2-29

Starr Truscott, Aeronautical Engineer, memorandum to Engineer-in-Charge [Henry J. E. Reid], "Work in connection with special aerodynamic tests for Bureau of Aeronautics which has been requested by Mr. Lougheed," 5 April 1932, RA file 210, Historical Archives, NASA Langley.

As noted elsewhere in this chapter, there is a close link between hydrodynamics and aerodynamics, and the latter discipline frequently benefited from theoretical and experimental work in the former. The NACA constructed two model-towing tanks at Langley for the express purpose of investigating the behavior of seaplanes during takeoff and landing, but some rather unusual tests were performed in the tanks on occasion. One such test involved a study of seagull flight characteristics for the U.S. Navy's Bureau of Aeronautics (BuAer). BuAer's Victor Lougheed devised a program to capture several seagulls, freeze them with outstretched wings, and test their aerodynamic characteristics by towing them submerged through one of the seaplane tanks. Always cooperative, the NACA approved the project but, as this memo shows, the Langley staff wanted to be sure that the navy assumed most of the responsibility and risk for such an unorthodox program. The tests were run, but there is no evidence that they contributed anything of significance to aeronautical engineering.

Document 2-29, Starr Truscott, "Work in connection with special aerodynamic tests for Bureau of Aeronautics which has been requested by Mr. Lougheed," 1932.

April 5, 1932
MEMORANDUM For Engineer-in-Charge.
Subject: Work in connection with special aerodynamic tests for Bureau of Aeronautics which has been requested by Mr. Lougheed.

1. On the morning of Wednesday, March 30, Mr. Lougheed appeared at the tank. The special dynamometer for use in making the tests in which he is interested had been received the day before. Under Mr. Lougheed's direction the dynamometer was unpacked and a rack made especially to suit it. He also explained the operation in detail.

2. In view of the general delicacy of the device and the ease with which the small spring hinge "knife edges" could be put out of adjustment, it was decided to leave the balance exactly as it was received, with all motions locked, until the work of assembling it to its support and on the carriage had been completed. The

support has not yet been received from the Norfolk Navy Yard, although it is understood that it is completed.

3. After discussing the installation and the precautions to be observed in this work and in the operation of the balance, Mr. Lougheed gave an outline of the further work which would be required to prepare for and carry out the tests.

4. A working platform across the fan end of the tank is required. On this is to be carried the freezing box for freezing the birds, two protractors for measuring the angle of attack of the wings, a projecting walk for reaching the balance on the spur projecting from the carriage, a receptacle for liquid air, two scales, two scale pans, and fine shot for calibrations of the balance.

5. The freezing box must be about 8 or 10 feet long and 4 feet wide. In plan form it should taper from the center to the ends so as to reduce the volume to be filled with liquid air. The depth of the box should be as little as will accommodate the birds, for the same reason. This box can be made of insulite or some other insulating board with wood batten stiffening.

6. The two protractors can be made or bought. They can be relatively crude in construction but must include blades 12 inches to 18 inches long to extend under the wings to measure the slope of the wing chord relative to some level line. Two are required so that one can be held while the other is being used to set the other wing.

7. The projecting walk is required to provide access to the calibrating screw which lies under the balance and some 4 feet back from the tip of the spike on which the models are supported—or on which the birds are impaled.

8. The receptacle for the liquid air will probably be the one in which it is received, but it must be supported in such a manner as to make access easy and replacing simple.

9. The two scales are required for weighing the birds, or models, and for weighing the calibrating loads. This balance has no calibrated springs or lever balances. On each test run the device is brought to a null point and left there. After the run the model is removed and the forces required to restore the null are measured by dead loading with shot.

10. One scale should read up to 5 pounds by $\frac{1}{4}$ ounce; the other should have a capacity of $\frac{1}{2}$ to 1 pound and should read to 5 grains.

11. With the balance there have been supplied stirrups for supporting scale pans in which the shot may be placed, but the scale pans must be supplied. Mr. Lougheed suggests paper cups as easy and light.

12. In addition to the material to be used in connection with the tests, there will also be required a supply of birds for testing. Mr. Lougheed suggested 5 gulls and 5 turkey buzzards. These could be obtained by local trapping he thought and probably boys might be interested in getting them.

13. For keeping these birds a small menagerie will be required. The cages must be large enough to permit the birds to spread their wings to preen them. If the birds do not have sufficient space they will disarrange and break feathers by pushing them against the walls of the cages. Mr. Lougheed estimated that a cage about 12 feet long, 5 feet high, and 5 feet deep would suffice. This should be divided by a solid partition into a section 7 feet long and one 5 feet long. The larger section is for the buzzards, the smaller for gulls. Perches should be provided.

14. The birds will require to be fed to keep them in condition. The gulls get fish and the buzzards meat in the form of spoiled meat or carrion.

15. Mr. Lougheed now has one gull and one buzzard in the zoo at Boston which will be shipped down here.

16. I called his attention to the protection against killing or taking which is given these birds by State and Federal laws. He said he would get the necessary permits issued to himself and supply copies of the original to anyone who undertook to catch birds for this job.

17. Discussing the balance and its method of operation, Mr. Lougheed referred to the adjusting of speed, while the carriage was in motion, to suit the model or bird. I called his attention that we could not do it from the carriage. He replied that he understood that, but that by fitting two colored lights which could be seen by the operator at the desk and which would indicate plus or minus speed, it ought to be possible to do it. These lights would be operated by contacts on the balance. I recalled to him that we had told him in the beginning that it was not possible to change speed while running. He said the speed wasn't critical and if it didn't work out it would simply mean more runs!

18. He requested that we install the balance on the balance support which is still at the Navy Yard. This would require installing the wiring from the contacts, the fitting of lamps, with condensers (if required) and batteries, for signalling the operator on the carriage, and the fitting of signal lamps for the operator at the desks.

19. After we had the balance installed we should play with it, mounting discs or other objects on the spike and determining their resistance, etc. A box of wing models which could be fitted was also on its way to the laboratory.

20. Mr. Lougheed inquired about a local supply of liquid air. I told him I had no information but doubted that it could be obtained locally or even in Norfolk. He said it could be obtained in Washington and how it could be transported. I suggested the Ludington Line airplanes if it was urgent. He replied that probably the boat would be all right.

21. When the recital of this list of things to be done began I, of course, acquiesced in the items of the freezing box and the access platform. The first was relatively simple, requiring only a few sheets of insulite and a little carpenter work, while the second we have in the form of our small portable bridge.

22. However, as the list kept increasing in length—and obviously in cost—I thought it might be better to let Mr. Lougheed complete his tale before commenting. When this involved the provision of the menagerie and keepers, to say nothing of trappers, I decided that it would be better to make no objections but simply to get the picture well in mind, study it, and then make proposals as to what we should do and what the bureau should provide.

23. Accordingly, I suggest the following:

A. The Bureau of Aeronautics should be informed that Mr. Lougheed has visited the laboratory in connection with this work and has discussed the equipment required in addition to the balance and also the method of taking the birds. Certain items of this material can be supplied by the laboratory but others should be supplied by the bureau because they are of types which are not used at or easily available to the laboratory. These items are:

(1) The birds required for experiment

(2) the liquid air for freezing them

(3) the protractors for measuring angle of attack of wings

(4) two scales, one reading 5 pounds to $\frac{1}{4}$ ounce, the other reading 1 pound to 5 grains

The bureau should also be reminded that in the course of the conferences before these tests were authorized, the representatives of the bureau were informed that the speed of the carriage was not controllable from the carriage and hence could not be varied during the run. From the recent discussion with Mr. Lougheed it has been learned that the operation of the balance which has been provided requires that the speed of the carriage shall be adjusted while running. This will require the fitting of some small items of equipment not contemplated in the original plan and the adopting of a method of operation which will make heavy demands on the control operator. The laboratory is willing to attempt this method of operation, but of course can not promise certain success.

B. The laboratory will assemble the balance to its support, provide all incidental fastenings, wiring, lights, extensions of control rods, batteries, and other items connected with the operation of the balance, and will install the balance on the carriage in the manner described by Mr. Lougheed on his recent visit. It will also provide and have ready the means of access to the balance, the freezing box, and one or more cages for the birds. When the birds are received it will provide for their feeding. (It is assumed that requisitions covering the necessary meat and fish will get by the Comptroller.)

24. It will be noted that this calls for the bureau to supply the birds. It seems to me they should do this because it will require quite a bit of arranging. Mr. Lougheed spoke of putting an advertisement in the papers. Such a thing would only start trouble. Some group or person either well intentioned or seeking notoriety would protest and the resulting troubles would keep us busy.

25. A further thought is that Mr. Lougheed warned me that the gulls might easily put out an eye for anyone who handled them, as they always strike for it, while the buzzards can bite off a finger without trouble, thanks to their "tinsnip" jaws. I would just as soon others should handle such birds while catching them. Feeding them will be bad enough.

26. A further thought is that if this work is to be kept at all confidential the purpose of the birds must not be advertised. This would surely occur if the birds ere taken locally. It would be much better to take them around Washington and ship them here. It might even be feasible to handle the taking at the Naval Air Station.

Starr Truscott,
Aeronautical Engineer

Document 2-30(a–b)

(a) Edward P. Warner, "Research to the Fore," *Aviation* 33, No. 6 (June 1934): 186.

(b) "Research Symphony: The Langley Philharmonic in Opus No. 10," *Aviation* 34, No. 6 (June 1935): 15–18.

Aviation Editor Edward P. Warner was clearly a big fan of the NACA. He extolled the virtues of the committee, its research methods, and the products of that research in numerous columns, but Warner, who served on the NACA's Committee on Aeronautics, knew what he was talking about. Thus, his comments and conclusions are worthy of consideration as more than simply those of an apologist.

The final documents presented in this chapter both come from the pages of *Aviation*; the first a Warner editorial, and the second an article likely written—and certainly approved—by him as well. They were published about the time of the NACA's 1934 and 1935 Annual Aircraft Engineering Research Conferences at the Langley Laboratory, and both dealt with the rise of aeronautical research to a leading role in aircraft design. In 1934's "Research to the Fore," Warner observed that "every group in the aircraft industry" had come to "a new appreciation of the vital importance of the scientific fundamentals of aircraft design. Research has ceased to be the servant of aeronautical development, and has become its guide." Journalistic hyperbole aside, Warner's observation had some solid evidence, in the form of recently built wind tunnels and new aircraft designs, to back it up.

Aviation took a whimsical tone in its report on the tenth (1935) conference, titling it "Research Symphony: The Langley Philharmonic in Opus No. 10" and using musical analogies to categorize the research work being done under the baton of "Conductor Joseph S. Ames" and his able "Concert Master George W. Lewis." Nevertheless, this "brief recapitulation of some of the principal movements" included a wealth of information on the current state of aeronautical research and its impact on the aviation industry. Reading it, one will quickly sense how far aviation technology had come in the two decades since the NACA's founding, thanks in no small part to that organization's increasingly sophisticated research program.

Document 2-30(a), Edward P. Warner, "Research to the Fore," 1934.

Research to the Fore

About this time each year the aircraft industry prepares to move in force on Langley Field and the laboratories of the N.A.C.A. This springtime visit has become an annual habit, and out of the annual attendance of 200 or 300 at the N.A.C.A. field day the majority are veterans who have acquired the habit so thoroughly that they never think of missing the trip. In the last two or three years those old timers have been sensing a change in the atmosphere of the meeting, for an interest in the detail of research equipment, method, and results that was once concentrated among scientists has spread to every group in the aircraft industry. The representatives of the builders of aircraft for the military market attend in steadily growing numbers, and the manufacturers of light commercial craft and the operators of airlines, though they came comparatively late to the roster of the meeting, are finally beginning to play a part.

All this is quite a compliment to the National Advisory Committee, but it is much more than that,—it is evidence of a new appreciation of the vital importance of the scientific fundamentals of aircraft design. Research has ceased to be the servant of aeronautical development, and has become its guide. One need not go to the very remote past to find that research, like God and the doctor in the ancient jungle, was valued by the man of strictly practical interests principally in time of trouble, when he had run into an unexpected obstacle and needed to have it removed in a hurry. Now he has learned not only to avoid the obstacle by a sufficiently intense preliminary study, but to make certain that he is really getting the best possible result from his product, and not merely a passably good one, by trying out the whole range of possible alternatives under laboratory conditions. On a modern high-speed transport the difference between a wing-fuselage fillet casually faired in to look about right and one determined as the ideal through a long series of studies in the wind tunnel may be 3 m.p.h. in maximum speed. On a 12-passenger twin-engined transport that means a saving of about $2,500 a year in operating cost on a single plane. On an order for twenty such planes, the saving in a single year would be enough to pay the cost of building and equipping a first-class wind tunnel in which to do the work. That fact has made itself felt, and whereas no more than four or five years ago it was rather an extraordinary thing to have any extensive wind-tunnel testing done before building a new ship it has now become the general rule. Not only the wind tunnel, but the seaplane channel as well, has become an accepted and an almost necessary instrument of the designer in his preliminary planning of a new type.

Aviation has suffered at all times from a delusive belief, which one may still encounter here and there, that research and analysis in airplane design are futile frills and that what is needed is to have a good practical man with an extended experience as a pilot and a good eye for line draw a picture of the new airplane and build it accordingly. There was a time when that went so far that the very making of engineering drawings for an experimental machine was considered to fall under the head of "frills," and one pioneer builder used to boast "he could start in the middle and work outwards and not decide what anything on the plane was going to be like until he came to it." That certainly is one way of doing it, but not the best way. How far it is from being the best becomes apparent when the practical man scornful of theory and of research has done his best and when another designer unhampered by any such scorn, but possessed of wind tunnels and believing their results, takes the same set of specifications and produces a machine to be put into the competition. The airplane built by inspiration and by seasoned judgment is prone to look extremely foolish, under those conditions, as against the ship in which judgment is backed by careful application of science and of laboratory technique.

Already it is true that a majority of America's foremost airplane builders either have wind tunnels of their own or have access to the tunnels of neighboring universities. We predict that within another three or four years the company that fails to own and operate its own tunnel will be quite out of the running, and that an aerodynamic section attached to every engineering staff, with its personnel concerning themselves exclusively with aerodynamic research and with the analysis of aerodynamic problems handed over to them by the designers, will be no questionable extravagance but quite as much of a necessity as the stress analysis group is today. The compass points that way, and the wise management will make its plans accordingly.

Document 2-30(b), "Research Symphony: The Langley Philharmonic in Opus No. 10," 1935.

Research Symphony
The Langley Philharmonic in Opus No. 10

With conductor Joseph S. Ames insisting on inflexible adherence to tempo and Concert Master George W. Lewis getting brilliant performance from his individual group and solo performers, the Tenth Annual Engineering Research Conference of the National Advisory Committee for Aeronautics proceeded with all the smoothness of a major symphonic ensemble in action. The audience was the aviation industry in numbers that strained the facilities of the isolated Tidewater Peninsular at Old Point Comfort almost to the limit. It is impossible within our space limitations to reproduce each nuance in the detailed development of every research theme. That we leave to the extensive literature of the committee. All we can do here is to give a brief recapitulation of some of the principal movements.

Spinning Song

Most fascinating number of this year's performance was the demonstration of the brand new tunnel where model airplanes may spin freely in a vertically rising air jet of controllable velocity. Most remarkable are the models themselves, miniature airplanes that are not only geometrically similar to their full scale counterparts, but also must have identical mass distribution and similar dynamic characteristics. To add wonder to wonders, into each model has been built delicate timing machinery connected to controls to reset rudder, elevators, ailerons at predetermined intervals during spinning to promote recovery from the spin, or to change its characteristics. To watch these intricate models swimming in the air stream like goldfish in a bowl, reproducing well-known spinning maneuvers as though under the control of a miniature human Pilot, was an experience that few of this year's visitors will soon forget.

Thanks to Public Works Administration funds which made possible the installation of the new tunnel and its accessory equipment, the 800 factors which contribute to spins outlined some five years ago by Fred E. Weick, may now be subjected to exacting laboratory tests, reducing by many times the expense and the danger involved in full-scale spinning research.

Wings; Crescendo

Long before the Department of Commerce undertook a program of private flying encouragement through equipment purchases, the NACA had been at work on the fundamentals of what makes flying simpler. Five years of work on variable lift devices for lowering landing speed and on surer lateral controls had gradually split the boundaries of knowledge, and the conference revealed for the first time a new series of results.

Particularly important was the report of flight trials on designs so far tried only in the wind tunnel. Flaps and variable area wings, generally considered only as a means of reducing minimum speed, are established on a new footing as proven aids to performance in getting off and climbing. On a standard Fairchild monoplane, the installation of a Fowler wing (sliding a flap out of the lower surface of the wings to rear and at the same time pulling it down so that both the area and the camber are increased) reduced take-off distance from 500 ft. to 330 ft. in still air. The total distance to clear a 50-ft obstacle came down from 910 ft. to 720, the calculated distance to accomplish the same take-off and climb with a heavily loaded twin engine ship, from 1,500 ft. to 840. On the twin engine take-off, the use of a simple split flap without area increasing features would reduce the space needed for take-off only from 1,500 ft. to 1,100 ft.

A trend toward variable lift devices of higher efficiency than the plain hinged flap, a readiness to accept as necessary whatever mechanical complications their virtues might involve, were plainly indicated in the report and their discussions by the engineers in attendance. Area variation, for example, is clearly a matter of practical interest for the near future.

Most elaborate and most effective of devices so far known is boundary-layer control, sucking off the air from the surface of the wing into its interior through slots parallel to the span along the upper surface. The committee's studies show that with a fixed wing with a single slot halfway back on the cord, burbling can be eliminated, a steady flow and a steadily increasing lift be maintained up to an angle of attack of over 50 deg., a maximum lift coefficient of 3.0 be attained with an application to the blower of less than 3 per cent of the engine power. That would make it possible to land at 55 m.p.h. with a wing loading of 23 lb. per sq. ft. Even with a thin tapered wing well suited for high speed use the same lift could be secured with expenditure of about 6 per cent of the total power.

Interesting experiences with the handling of flap-equipped airplanes came out in the course of discussion. The committee's pilots have found that if the flaps are pulled down suddenly to steepen tile glide path in coming in over an obstacle, the immediate effect is exactly the opposite of what is wanted. The increase of lift by the flaps sets the ship to climbing above its original course, and with a light airplane fully 850 ft. had to be covered before the flight-path dropped below the level that would have been reached if the flaps had not been used at all. To overcome any such reversal of effect it was suggested that the flaps be pulled down to very large angles, as much as 80 deg., where there will be a pure air-brake action with no further increase of lift. One experimental machine of private-owner type has been fitted with such a control for trial.

Another way of increasing the lift, well known to test pilots trying to meet a minimum speed specification but seldom made the subject of research, is by pulling back into a full stall and then opening the throttles wide to blow the slip-stream across the wings and to carry a part of the weight directly on the propeller thrust. Measurement on a typical plane in the propeller research laboratory at Langley Field showed an effective increase of 0.3 in the lift coefficient from such a maneuver, a possible reduction of about 10 per cent in minimum speed.

The effect of large flap angles on stability has proved to be bad throughout, even when the flap extends over only a part of the span. The Fairchild with the Fowler flap became longitudinally unstable both with free and with fixed controls at all speeds above 70 m.p.h. with the flap clear down. Fortunately the stability characteristics prove to be best in the part of the speed range where the flap is most likely to be wanted, but at low speeds there is an extreme sloppiness of rudder control that requires as much as 11 deg. of rudder to hold a straight course. In some cases, in fact, the machine could not be flown at all at minimum speed with full flap effect because of the impossibility of keeping it straight even with full rudder. Mr. McAvoy of the Committee's technical staff and Temple N. Joyce debated flap landing technique and agreed that it differed from normal practice in that the machine need never be brought anywhere near a stalled attitude. The drag being so large that the nose call be put down sharply without picking up much speed, the angle of attack can be kept small until the very last instant of flattening out.

For that reason, the abrupt collapse of the lift coefficient at angles beyond the burble point that characterizes all flap arrangements makes no trouble in practice.

The NACA slot lip control is a new and most promising addition to the long list of lateral controls developed especially for use in conjunction with flaps and to be effective beyond the stall. A combination of slot and spoiler, it is a small flat plate lying flush with the upper surface and hinged at its forward edge. So placed as normally to block very largely but not entirely the exit of air from a slot through the wing just forward of mid-chord, its raising cuts down the lift with none of the time lag that marks the action of an ordinary spoiler.

Drag; Diminuendo

Few of the committee's programs have been as extensive or as fruitful of practical results as the relentless pursuit of drag. First notable contribution to aerodynamic efficiency was the familiar NACA radial engine cowling, then a long series of investigations on nacelle position, interference, correlated with more highly theoretical studies of air flow, scale effect and turbulence.

In the course of recent research the laboratories have dipped deeply into fundamentals without losing sight of visual manifestations of flow phenomena. For example, the smoke streamer studies of flow separation from airfoil bodies have revealed that a turning propeller in optimum position (tractor or pusher installation) improves the airflow in normal flight and at high angles of attack.

Optimum position for engine installation has been found to be within the structure or in leading edge nacelles. An ideal installation would have the engine completely enclosed in the structure fitted with extension shaft. Rear extension of the shaft gives best net efficiency at low speeds but, at 200-300 m.p.h. is worse than one with propeller ahead of the leading edge.

Tests on in-line engine installation in wing-nacelle combinations have yielded slightly higher drag figures than for radials, although it appears probable that with higher power concentration in a given nacelle, conditions may be reversed.

Bombshell for retractable landing gear advocates was the news that clean fixed landing gears had only slightly lower drag than fully retracting types. In terms of top speed, the difference was approximately 3 per cent. Airline engineers pointed out in conference that 6 m.p.h. at the upper end of the range was the equivalent to 10 per cent power, asked for study of fixed landing gear effect on stability and spinning qualities. Investigation of the take-off characteristics of airplanes with normal retracting gear extended as compared with fixed faired gears was suggested by T. P. Wright.

To run down the influence of various fuselage-wing combinations on drag and therefore on speed, a series of tests was run in the variable density tunnel to determine optimum wing position with respect to fuselage. Beginning with a bare fuselage with an uncowled radial engine in the nose and a rectangular wing in the best low-wing position, refinements were added successively. Engine cowling, wing root fillets, complete housing of engine in the wing, adaptation of a symmetrical airfoil (rectangular then tapered), and finally a shift of the wing up to a high mid-wing position, boosted the potential high-speed of the combination (for constant engine power) from 145 up to 205 m.p.h. The addition of a trailing edge flap naturally did not add to the top speed but gave, as would be expected, a greatly increased speed range.

New sources of efficiency were promised by Eastman N. Jacobs, whose airfoil family increases yearly. Ideal offspring, NACA 23012, is symmetrical, has maximum camber relatively far forward. In such airfoils, maximum camber may be moved forward beneficially to 15 per cent of chord length, harmlessly to 5 per cent.

Stresses, Giros, Boats; Miscellany

Heretofore the study of gust loads has been confined to effects on wings. Recent flight research has indicated, however, that gust loads on tail surfaces are much greater than has been June, suspected. Accelerometer readings give an average for the action of a gust over the entire wing span but since the wave length of the gust may be much shorter than the wing span, the peak loads imposed by the gust may be much higher than the average recorded for the entire wing. Where the wing may extend beyond the boundaries of a single gust wave, the shorter span tail surfaces may take the full gust impact. On an O2H machine, for example, where an average wing gust of 13 ft. per second was recorded, simultaneous readings on the horizontal and vertical tail surfaces showed 33 and 45 ft. per second gusts respectively. An extensive investigation of tail loading to supplement the results so far obtained from acceleration readings on transport planes in actual service is now under way.

Work with rotating wing systems, although apparently not as active as a year ago, still has a place in the research program. Recent investigations have covered the selection of airfoil sections for rotor use, also plan form modifications for maximum values of L/D. Thin cambered sections show higher efficiency than the symmetrical types; for example, NACA 4412 was found to be some 13 per cent more efficient than NACA 0018. Although it was suspected that greater rotor efficiencies might be obtained by cutting out portions of the effective blade area near the hub, experiment soon indicated not only that maximum L/D's were obtained with a full span blade, but also that greater efficiencies might be expected from tapered plan forms where the chord at the blade root was considerably greater than at the tip.

The ability of an autogiro to take off vertically on energy stored in an over speeded rotor was demonstrated with a 10 ft. electrically driven model. During initial rotation the blades were held at zero lift position then suddenly released to a high angle of attack position. The model rose vertically to a height of 20-25 ft. It was shown that the vertical distance attained during the initial jump is a function of disk loading. Tests on the model showed that, with a 392 ft. per second tip speed and a disk loading of 2 lb. per sq. ft., the initial rise is only 1 ft. Cutting the disk loading in half, however, the initial jump goes up to 15 ft.

The research program which led to the discovery of the beneficial effects of pointed main steps for flying boat bottoms (announced in 1934) was extended to cover steps of varying depths, and to study the effects of changes in the angle of dead rise and after-body keel. Shallow steps were shown to perform better at low speeds, deeper steps at higher speeds. Relatively flat after-body angles (the range between 0 and 8 deg. has so far been investigated) appear more advantageous at low speeds, low loadings. Dead rise angles did not seem critical, however, for the characteristics of similar hulls with bottom angles of 15, 20 and 25 deg, were essentially alike.

Most important result announced concerned the effect of various shapes of rivet heads in bottom plating. By towing metal planing surfaces with a standard pattern of dimpled, brazier and button headed rivets, the advantage of keeping bottoms as smooth as possible became evident. The resistance of dimpled rivets was 5 per cent over the smooth plate, brazier heads 12 per cent, button heads 17 per cent.

Power plants; Energico

Some twenty forms of NACA cowl with varying ratios of nose opening to overall diameter and rear gap area are being studied at full scale in the 20 ft. tunnel. Charts will shortly be available from which designers will be able to select cowl characteristics for all desired engine and flight conditions. As suggested by airline operators who experience cooling troubles at normal angles and speeds of climb, the program will be extended to include rang of angles of thrust in line to flight path of from 0 to 10 deg.

Preliminary test results indicate that cylinder cooling at constant power output is independent of altitude as long as the mass movement of air over the cylinder is constant. The cooling effect per unit of cooling surface seems to vary approximately as the square root of the airflow in terms of pounds per second per square feet of area covered. Studies of variation of cylinder temperatures with changes of cooling air temperatures indicate that the variation is linear, independent of brake mean effective pressure, mass flow.

Cooling fins should run at least eight to the inch, should be closely jacketed to force airflow to follow the fin and the wall closely. The shape of the jacket and the size of the intake and outlet openings are being investigated not only for free air flow cooling but also for blower cooling.

Interest in completely housed-in engines for drag reduction prompts research in blower cooling. Of chief interest is the cost of blower operation in horsepower. Some 30 to 35 per cent of total horsepower goes into the cooling of a bare engine, 13 to 16 per cent to cool the same engine with a properly designed

NACA cowl. Calculations based on skin friction of the average radial engine indicate that the absolute minimum of power required is about 1? per cent of the total horsepower. Therefore, the range in which blower designs must work is between 1? and 13 per cent of the total horsepower. On the basis of 65 per cent blower efficiency, the cooling loss should not be over 5 per cent.

Where long range performance is required, the low specific fuel consumption of the compression ignition engine put is independent of altitude as long as is very attractive since 40 to 45 per cent of initial useful load of an airplane may be required for fuel alone. Work has progressed far enough to indicate that it is possible to obtain the same power for the same displacement and r.p.m. for carburetor and for compression ignition engines with reduction of specific fuel consumption of some 20 per cent.

Shape of combustion chamber affects the efficiency of compression ignition engines. Spherical or disk type mixture chambers were found not as effective as a new displacer type recently developed by the laboratories. A solid boss cast on top of the piston projects into the mixture chamber at top dead center, causing a high velocity air flow in the narrow clearance between boss and sidewall and great turbulence in the chamber at the instant of fuel injection.

Among questions submitted for consideration of the committee: Best cooling arrangement for six-cylinder in-line engines? How can airflow inside an NACA cowl be restored to the outside flow most efficiently?—through an open slot or through louvers?—is it permissible to support the skirt of the cowl on the fire wall? Are internal guide vanes permissible? How may the accessories on an in-line engine best be cooled? What is the effect of a propeller spinner on the cooling of an in-line engine? How best to cool a flat engine—where to take in the air, where to discharge it? What is the drag of openings such as the ends of exhaust stacks, facing aft,

Propellers; Vibrato

Resonance with frequencies originating in the engine rather than aerodynamic flutter is now recognized as the cause of vibrations frequently leading to fatigue failures in propellers. It is demonstrable that true flutter can occur only at speeds far in excess of the speed of sound. The natural frequency of a rotating propeller is quite different from one at rest, due to the blade tension induced by centrifugal forces. For full scale propellers, therefore, vibration analysis by accoustical [sic] methods becomes the only practical method.

A method of scaling down propeller vibration effects for study was shown. A propeller of full-scale diameter was prepared with the aluminum alloy blades reduced to one-tenth normal width and one-tenth normal thickness. It is mathematically demonstrable that the vibration characteristics of such a propeller when rotated at one tenth normal revolutions per minute are exactly similar to those of the normal propeller at normal speeds. It is therefore possible by rotating the propeller (with blades completely enclosed in steamline tubes to eliminate all air effects) and imposing at the same time axial vibrations of known frequencies on the hub, to determine the resonant frequencies of the modified blades. Results scale up to full size by simply multiplying all data by ten.

It was stated that undesirable resonances had been eliminated in some cases by changing the mounting of a propeller from a position parallel to the engine crank throw to one 90 deg. away, but the desirability of such a method of correction was seriously questioned by propeller and engine manufacturers.

A new light on the composition of propeller noise was obtained by filtering out certain frequencies (or combinations of frequencies) from the sound picked up by microphone 50 ft. away from an electrically driven full scale propeller. It was evident that the troublesome ranges were (1) the higher frequency harmonics and (2) the vortex noises which produce the characteristic tearing sounds emitted by a propeller at high speed. It was shown that the basic frequency filtered out from all the super-imposed harmonics and vortex noises was a musical note of low pitch. This is the note normally heard by an observer at a relatively great distance. The unpleasant frequencies are filtered out by the intervening atmosphere. This effect was produced electrically to conclude the demonstration.

NASA History Series

Reference Works, NASA SP-4000:

Grimwood, James M. *Project Mercury: A Chronology.* NASA SP-4001, 1963.

Grimwood, James M., and C. Barton Hacker, with Peter J. Vorzimmer. *Project Gemini Technology and Operations: A Chronology.* NASA SP-4002, 1969.

Link, Mae Mills. *Space Medicine in Project Mercury.* NASA SP-4003, 1965.

Astronautics and Aeronautics, 1963: Chronology of Science, Technology, and Policy. NASA SP-4004, 1964.

Astronautics and Aeronautics, 1964: Chronology of Science, Technology, and Policy. NASA SP-4005, 1965.

Astronautics and Aeronautics, 1965: Chronology of Science, Technology, and Policy. NASA SP-4006, 1966.

Astronautics and Aeronautics, 1966: Chronology of Science, Technology, and Policy. NASA SP-4007, 1967.

Astronautics and Aeronautics, 1967: Chronology of Science, Technology, and Policy. NASA SP-4008, 1968.

Ertel, Ivan D., and Mary Louise Morse. *The Apollo Spacecraft: A Chronology, Volume I, Through November 7, 1962.* NASA SP-4009, 1969.

Morse, Mary Louise, and Jean Kernahan Bays. *The Apollo Spacecraft: A Chronology, Volume II, November 8, 1962–September 30, 1964.* NASA SP-4009, 1973.

Brooks, Courtney G., and Ivan D. Ertel. *The Apollo Spacecraft: A Chronology, Volume III, October 1, 1964–January 20, 1966.* NASA SP-4009, 1973.

Ertel, Ivan D., and Roland W. Newkirk, with Courtney G. Brooks. *The Apollo Spacecraft: A Chronology, Volume IV, January 21, 1966–July 13, 1974.* NASA SP-4009, 1978.

Astronautics and Aeronautics, 1968: Chronology of Science, Technology, and Policy. NASA SP-4010, 1969.

Newkirk, Roland W., and Ivan D. Ertel, with Courtney G. Brooks. *Skylab: A Chronology.* NASA SP-4011, 1977.

Van Nimmen, Jane, and Leonard C. Bruno, with Robert L. Rosholt. *NASA Historical Data Book, Volume I: NASA Resources, 1958–1968*. NASA SP-4012, 1976, rep. ed. 1988.

Ezell, Linda Neuman. *NASA Historical Data Book, Volume II: Programs and Projects, 1958–1968*. NASA SP-4012, 1988.

Ezell, Linda Neuman. *NASA Historical Data Book, Volume III: Programs and Projects, 1969–1978*. NASA SP-4012, 1988.

Gawdiak, Ihor Y., with Helen Fedor, compilers. *NASA Historical Data Book, Volume IV: NASA Resources, 1969–1978*. NASA SP-4012, 1994.

Rumerman, Judy A., compiler. *NASA Historical Data Book, 1979–1988: Volume V, NASA Launch Systems, Space Transportation, Human Spaceflight, and Space Science*. NASA SP-4012, 1999.

Rumerman, Judy A., compiler. *NASA Historical Data Book, Volume VI: NASA Space Applications, Aeronautics and Space Research and Technology, Tracking and Data Acquisition/Space Operations, Commercial Programs, and Resources, 1979–1988*. NASA SP-2000-4012, 2000.

Astronautics and Aeronautics, 1969: Chronology of Science, Technology, and Policy. NASA SP-4014, 1970.

Astronautics and Aeronautics, 1970: Chronology of Science, Technology, and Policy. NASA SP-4015, 1972.

Astronautics and Aeronautics, 1971: Chronology of Science, Technology, and Policy. NASA SP-4016, 1972.

Astronautics and Aeronautics, 1972: Chronology of Science, Technology, and Policy. NASA SP-4017, 1974.

Astronautics and Aeronautics, 1973: Chronology of Science, Technology, and Policy. NASA SP-4018, 1975.

Astronautics and Aeronautics, 1974: Chronology of Science, Technology, and Policy. NASA SP-4019, 1977.

Astronautics and Aeronautics, 1975: Chronology of Science, Technology, and Policy. NASA SP-4020, 1979.

Astronautics and Aeronautics, 1976: Chronology of Science, Technology, and Policy. NASA SP-4021, 1984.

Astronautics and Aeronautics, 1977: Chronology of Science, Technology, and Policy. NASA SP-4022, 1986.

Astronautics and Aeronautics, 1978: Chronology of Science, Technology, and Policy. NASA SP-4023, 1986.

Astronautics and Aeronautics, 1979–1984: Chronology of Science, Technology, and Policy. NASA SP-4024, 1988.

Astronautics and Aeronautics, 1985: Chronology of Science, Technology, and Policy. NASA SP-4025, 1990.

Noordung, Hermann. *The Problem of Space Travel: The Rocket Motor.* Edited by Ernst Stuhlinger and J. D. Hunley, with Jennifer Garland. NASA SP-4026, 1995.

Astronautics and Aeronautics, 1986–1990: A Chronology. NASA SP-4027, 1997.

Astronautics and Aeronautics, 1990–1995: A Chronology. NASA SP-2000-4028, 2000.

Management Histories, NASA SP-4100:

Rosholt, Robert L. *An Administrative History of NASA, 1958–1963.* NASA SP-4101, 1966.

Levine, Arnold S. *Managing NASA in the Apollo Era.* NASA SP-4102, 1982.

Roland, Alex. *Model Research: The National Advisory Committee for Aeronautics, 1915–1958.* NASA SP-4103, 1985.

Fries, Sylvia D. *NASA Engineers and the Age of Apollo.* NASA SP-4104, 1992.

Glennan, T. Keith. *The Birth of NASA: The Diary of T. Keith Glennan.* J. D. Hunley, editor. NASA SP-4105, 1993.

Seamans, Robert C., Jr. *Aiming at Targets: The Autobiography of Robert C. Seamans, Jr.* NASA SP-4106, 1996.

Garber, Stephen J., editor. *Looking Backward, Looking Forward: Forty Years of U.S. Human Spaceflight Symposium.* NASA SP-2002-4107, 2002.

Project Histories, NASA SP-4200:

Swenson, Loyd S., Jr., James M. Grimwood, and Charles C. Alexander. *This New Ocean: A History of Project Mercury.* NASA SP-4201, 1966; rep. ed. 1998.

Green, Constance McLaughlin, and Milton Lomask. *Vanguard: A History.* NASA SP-4202, 1970; rep. ed. Smithsonian Institution Press, 1971.

Hacker, Barton C., and James M. Grimwood. *On the Shoulders of Titans: A History of Project Gemini.* NASA SP-4203, 1977.

Benson, Charles D., and William Barnaby Faherty. *Moonport: A History of Apollo Launch Facilities and Operations.* NASA SP-4204, 1978.

Brooks, Courtney G., James M. Grimwood, and Loyd S. Swenson, Jr. *Chariots for Apollo: A History of Manned Lunar Spacecraft.* NASA SP-4205, 1979.

Bilstein, Roger E. *Stages to Saturn: A Technological History of the Apollo/Saturn Launch Vehicles.* NASA SP-4206, 1980, rep. ed. 1997.

SP-4207 not published.

Compton, W. David, and Charles D. Benson. *Living and Working in Space: A History of Skylab.* NASA SP-4208, 1983.

Ezell, Edward Clinton, and Linda Neuman Ezell. *The Partnership: A History of the Apollo-Soyuz Test Project.* NASA SP-4209, 1978.

Hall, R. Cargill. *Lunar Impact: A History of Project Ranger.* NASA SP-4210, 1977.

Newell, Homer E. *Beyond the Atmosphere: Early Years of Space Science.* NASA SP-4211, 1980.

Ezell, Edward Clinton, and Linda Neuman Ezell. *On Mars: Exploration of the Red Planet, 1958–1978.* NASA SP-4212, 1984.

Pitts, John A. *The Human Factor: Biomedicine in the Manned Space Program to 1980.* NASA SP-4213, 1985.

Compton, W. David. *Where No Man Has Gone Before: A History of Apollo Lunar Exploration Missions.* NASA SP-4214, 1989.

Naugle, John E. *First Among Equals: The Selection of NASA Space Science Experiments.* NASA SP-4215, 1991.

Wallace, Lane E. *Airborne Trailblazer: Two Decades with NASA Langley's Boeing 737 Flying Laboratory.* NASA SP-4216, 1994.

Butrica, Andrew J., editor. *Beyond the Ionosphere: Fifty Years of Satellite Communication.* NASA SP-4217, 1997.

Butrica, Andrew J. *To See the Unseen: A History of Planetary Radar Astronomy.* NASA SP-4218, 1996.

Mack, Pamela E., editor. *From Engineering Science to Big Science: The NACA and NASA Collier Trophy Research Project Winners.* NASA SP-4219, 1998.

Reed, R. Dale, with Darlene Lister. *Wingless Flight: The Lifting Body Story.* NASA SP-4220, 1997.

Heppenheimer, T. A. *The Space Shuttle Decision: NASA's Search for a Reusable Space Vehicle*. NASA SP-4221, 1999.

Hunley, J. D., editor. *Toward Mach 2: The Douglas D-558 Program*. NASA SP-4222, 1999.

Swanson, Glen E., editor. *"Before this Decade Is Out . . .": Personal Reflections on the Apollo Program*. NASA SP-4223, 1999.

Tomayko, James E. *Computers Take Flight: A History of NASA's Pioneering Digital Fly-by-Wire Project*. NASA SP-2000-4224, 2000.

Morgan, Clay. *Shuttle-Mir: The U.S. and Russia Share History's Highest Stage*. NASA SP-2001-4225, 2001.

Leary, William M. *"We Freeze to Please": A History of NASA's Icing Research Tunnel and the Quest for Flight Safety*. NASA SP-2002-4226, 2002.

Mudgway, Douglas J. *Uplink-Downlink: A History of the Deep Space Network 1957–1997*. NASA SP-2001-4227, 2001.

Center Histories, NASA SP-4300:

Rosenthal, Alfred. *Venture into Space: Early Years of Goddard Space Flight Center*. NASA SP-4301, 1985.

Hartman, Edwin P. *Adventures in Research: A History of Ames Research Center, 1940–1965*. NASA SP-4302, 1970.

Hallion, Richard P. *On the Frontier: Flight Research at Dryden, 1946–1981*. NASA SP-4303, 1984.

Muenger, Elizabeth A. *Searching the Horizon: A History of Ames Research Center, 1940–1976*. NASA SP-4304, 1985.

Hansen, James R. *Engineer in Charge: A History of the Langley Aeronautical Laboratory, 1917–1958*. NASA SP-4305, 1987.

Dawson, Virginia P. *Engines and Innovation: Lewis Laboratory and American Propulsion Technology*. NASA SP-4306, 1991.

Dethloff, Henry C. *"Suddenly Tomorrow Came . . .": A History of the Johnson Space Center*. NASA SP-4307, 1993.

Hansen, James R. *Spaceflight Revolution: NASA Langley Research Center from Sputnik to Apollo*. NASA SP-4308, 1995.

Wallace, Lane E. *Flights of Discovery: 50 Years at the NASA Dryden Flight Research Center.* NASA SP-4309, 1996.

Herring, Mack R. *Way Station to Space: A History of the John C. Stennis Space Center.* NASA SP-4310, 1997.

Wallace, Harold D., Jr. *Wallops Station and the Creation of the American Space Program.* NASA SP-4311, 1997.

Wallace, Lane E. *Dreams, Hopes, Realities: NASA's Goddard Space Flight Center, The First Forty Years.* NASA SP-4312, 1999.

Dunar, Andrew J., and Stephen P. Waring. *Power to Explore: A History of the Marshall Space Flight Center.* NASA SP-4313, 1999.

Bugos, Glenn E. *Atmosphere of Freedom: Sixty Years at the NASA Ames Research Center.* NASA SP-2000-4314, 2000.

General Histories, NASA SP-4400:

Corliss, William R. *NASA Sounding Rockets, 1958–1968: A Historical Summary.* NASA SP-4401, 1971.

Wells, Helen T., Susan H. Whiteley, and Carrie Karegeannes. *Origins of NASA Names.* NASA SP-4402, 1976.

Anderson, Frank W., Jr. *Orders of Magnitude: A History of NACA and NASA, 1915–1980.* NASA SP-4403, 1981.

Sloop, John L. *Liquid Hydrogen as a Propulsion Fuel, 1945–1959.* NASA SP-4404, 1978.

Roland, Alex. *A Spacefaring People: Perspectives on Early Spaceflight.* NASA SP-4405, 1985.

Bilstein, Roger E. *Orders of Magnitude: A History of the NACA and NASA, 1915–1990.* NASA SP-4406, 1989.

Logsdon, John M., editor, with Linda J. Lear, Jannelle Warren-Findley, Ray A. Williamson, and Dwayne A. Day. *Exploring the Unknown: Selected Documents in the History of the U.S. Civil Space Program, Volume I, Organizing for Exploration.* NASA SP-4407, 1995.

Logsdon, John M., editor, with Dwayne A. Day and Roger D. Launius. *Exploring the Unknown: Selected Documents in the History of the U.S. Civil Space Program, Volume II, Relations with Other Organizations.* NASA SP-4407, 1996.

Logsdon, John M., editor, with Roger D. Launius, David H. Onkst, and Stephen J. Garber. *Exploring the Unknown: Selected Documents in the History of the U.S. Civil Space Program, Volume III, Using Space*. NASA SP-4407, 1998.

Logsdon, John M., general editor, with Ray A. Williamson, Roger D. Launius, Russell J. Acker, Stephen J. Garber, and Jonathan L. Friedman. *Exploring the Unknown: Selected Documents in the History of the U.S. Civil Space Program, Volume IV, Accessing Space*. NASA SP-4407, 1999.

Logsdon, John M., general editor, with Amy Paige Snyder, Roger D. Launius, Stephen J. Garber, and Regan Anne Newport. *Exploring the Unknown: Selected Documents in the History of the U.S. Civil Space Program, Volume V, Exploring the Cosmos*. NASA SP-2001-4407, 2001.

Siddiqi, Asif A. *Challenge to Apollo: The Soviet Union and the Space Race, 1945–1974*. NASA SP-2000-4408, 2000.

Monographs in Aerospace History, NASA SP-4500:

Launius, Roger D. and Aaron K. Gillette, compilers, *Toward a History of the Space Shuttle: An Annotated Bibliography*. Monograph in Aerospace History, No. 1, 1992.

Launius, Roger D., and J. D. Hunley, compilers, *An Annotated Bibliography of the Apollo Program*. Monograph in Aerospace History, No. 2, 1994.

Launius, Roger D. *Apollo: A Retrospective Analysis*. Monograph in Aerospace History, No. 3, 1994.

Hansen, James R. *Enchanted Rendezvous: John C. Houbolt and the Genesis of the Lunar-Orbit Rendezvous Concept*. Monograph in Aerospace History, No. 4, 1995.

Gorn, Michael H. *Hugh L. Dryden's Career in Aviation and Space*. Monograph in Aerospace History, No. 5, 1996.

Powers, Sheryll Goecke. *Women in Flight Research at NASA Dryden Flight Research Center, from 1946 to 1995*. Monograph in Aerospace History, No. 6, 1997.

Portree, David S. F. and Robert C. Trevino. *Walking to Olympus: An EVA Chronology*. Monograph in Aerospace History, No. 7, 1997.

Logsdon, John M., moderator. *Legislative Origins of the National Aeronautics and Space Act of 1958: Proceedings of an Oral History Workshop*. Monograph in Aerospace History, No. 8, 1998.

Rumerman, Judy A., compiler, *U.S. Human Spaceflight, A Record of Achievement 1961–1998*. Monograph in Aerospace History, No. 9, 1998.

Portree, David S. F. *NASA's Origins and the Dawn of the Space Age*. Monograph in Aerospace History, No. 10, 1998.

Logsdon, John M. *Together in Orbit: The Origins of International Cooperation in the Space Station*. Monograph in Aerospace History, No. 11, 1998.

Phillips, W. Hewitt. *Journey in Aeronautical Research: A Career at NASA Langley Research Center*. Monograph in Aerospace History, No. 12, 1998.

Braslow, Albert L. *A History of Suction-Type Laminar-Flow Control with Emphasis on Flight Research*. Monograph in Aerospace History, No. 13, 1999.

Logsdon, John M., moderator. *Managing the Moon Program: Lessons Learned From Apollo*. Monograph in Aerospace History, No. 14, 1999.

Perminov, V. G. *The Difficult Road to Mars: A Brief History of Mars Exploration in the Soviet Union*. Monograph in Aerospace History, No. 15, 1999.

Tucker, Tom. *Touchdown: The Development of Propulsion Controlled Aircraft at NASA Dryden*. Monograph in Aerospace History, No. 16, 1999.

Maisel, Martin D., Demo J. Giulianetti, and Daniel C. Dugan. *The History of the XV-15 Tilt Rotor Research Aircraft: From Concept to Flight*. NASA SP-2000-4517, 2000.

Jenkins, Dennis R. *Hypersonics Before the Shuttle: A Concise History of the X-15 Research Airplane*. NASA SP-2000-4518, 2000.

Chambers, Joseph R. *Partners in Freedom: Contributions of the Langley Research Center to U.S. Military Aircraft in the 1990s*. NASA SP-2000-4519, 2000.

Waltman, Gene L. *Black Magic and Gremlins: Analog Flight Simulations at NASA's Flight Research Center*. NASA SP-2000-4520, 2000.

Portree, David S. F. *Humans to Mars: Fifty Years of Mission Planning, 1950–2000*. NASA SP-2001-4521, 2001.

Thompson, Milton O., with J. D. Hunley. *Flight Research: Problems Encountered and What They Should Teach Us*. NASA SP-2000-4522, 2000.

Tucker, Tom. *The Eclipse Project*. NASA SP-2000-4523, 2000.

Siddiqi, Asif A. *Deep Space Chronicle: A Chronology of Deep Space and Planetary Probes, 1958–2000*. NASA SP-2002-4524, 2002.

Merlin, Peter W. *Mach 3+: NASA/USAF YF-12 Flight Research, 1969–1979*. NASA SP-2001-4525, 2001.

Renstrom, Arthur G. *Wilbur and Orville Wright: A Bibliography Commemorating the One-Hundredth Anniversary of the First Powered Flight on December 17, 1903.* NASA SP-2002-4527, 2002.

No monograph 28.

Chambers, Joseph R. *Concept to Reality: Contributions of the NASA Langley Research Center to U.S. Civil Aircraft of the 1990s.* SP-2003-4529, 2003.

Peebles, Curtis, editor. *The Spoken Word: Recollections of Dryden History, The Early Years.* SP-2003-4530, 2003.

Jenkins, Dennis R., Tony Landis, and Jay Miller. *American X-Vehicles: An Inventory-X-1 to X-50.* SP-2003-4531, 2003.

Index

Ader, Clément, 14, 180, 181
Adlershof. See Deutschen Versuchsanstalt für Flugwesen, Adlershof.
Advanced Group for Aeronautical Research and Development (AGARD), lxi
Aerial Experiment Association, Hammondsport, N.Y., 230
Aerial Transport Company. See Henson and Stringfellow airplane.
Aero Club of America, 230, 453
Aeronautical Society of Great Britain, 18, 58, 59, 84, 126, 196, 232, 241
Ahlborn, Friedrich, 300
Air Commerce Act, 478
Air Force Aeronautical Systems Center (ASC), U.S., xlvi
Air Force Flight Dynamics Laboratory, U.S., xlvi
Air Force Flight Test Center History Office (AFFTC/HO), U.S., xlvii
Air Force Historical Collection, U.S., xxxvii
Air Force Historical Research Agency (AFHRA), U.S., xlvii
Air Force History Support Office, U.S., xlvi
Air Force Institute of Technology (AFIT), U.S., 265
Air Force Museum, U.S., xlvii
Air Force, U.S., xviii, xliii, xlvi, xlvii, lxvi
Air Mail Service, U.S., 444, 453, 479, 480
Air Production Board (World War I), 249
Air University, Maxwell AFB, Ala., xlvii, lvi
Albatross airplane, 463
Allegheny Observatory (Penn.), 131, 133, 241
Allen, James, 225, 229
Allis Chalmers Co., 581
Ambrosian Library, Milan, Italy, 5
American Association for the Advancement of Science (AAAS), 17, 137, 147
American Astronomical Society, 233
American Institute of Aeronautics and Astronautics (AIAA), lvi, lvii
American Mathematical Society, 233
American Nautical Almanac Office, 233

American Society of Civil Engineers (ASCE), 17
American Society of Mechanical Engineers (ASME), lvii
Ames Aeronautical Laboratory. See Ames Research Center, NACA/NASA.
Ames Collection, Milton, xl, xli, xlii
Ames, Joseph S., 247ill., 249, 253ill., 255, 360, 445, 454, 465, 467, 468, 469, 470, 472, 536, 538, 631, 632, 635, 657, 659
Ames Research Center, NACA/NASA, xxxix, xli, xliv, xlv, 467
Anderson, John D., Jr., xviii, xxiii, lviii, lix, 3, 5, 7, 23, 240
Archimedes, 3
area rule, xli-xlii
Aristotle, 3
Army Air Service, U.S., 264, 581
Army, U.S., 225, 247, 248, 251, 265, 294, 306, 310, 362, 385, 441, 445, 450, 451, 452, 453, 457, 476, 479
Arnold Engineering Development Center (AEDC), lxix
Arnot, Matthias, 149
Associated Press, 221, 223
Atmospheric Wind Tunnel, NACA Langley, 249, 250ill., 428, 445, 481, 484ill., 498, 577
Atmospheric Wind Tunnel, 7 x 10-Foot, NACA Langley, 261ill., 261, 262, 263, 435
Auburn University, xii, xiii, xxxv, lvi, lvii, xlvii, lxvii, lxx
Austin Co., 577
Auteuil, France, 245, 403
autogiro, 620-21
Automobile Club of America, 289
Avery, Norm, lxvi
Avery, William, 20, 210, 211
Avion III (Ader), 14

Baals, Donald D., and Corliss, William, lxiv
Bacon, David L., 487, 531
Baden-Powell, Baden Fletcher Smyth, 126
Badin Double Venturi Head, 371, 372, 372ill.

Bairstow, Leonard, 312, 316
Bane, Thurman H., 247ill., 362, 374, 445, 454
Barimar Ltd., 611
Barlow, Jewel B., and Rae, William H., Jr., lxiv
Bateman, Harry, 466, 468, 609
Battle of Britain (World War II), 339
Bauer, Eugene E., lxvi
B. E. 2C airplane, 319, 324, 344
Beall, Wellwood, E., 646
Becker, John V., lxiv
Bell aircraft:
   Model 204/VH-1, xlvii
   X-1, xxxiii, xxxviii, xlii, xliv, xlvii, lxii
   XS-1. See X-1.
   X-series, xliv
   XV-15 Tilt Rotor Research Aircraft, xxxiv, lxi
Bell, Alexander Graham, 10, 130, 145, 146, 230
Bendemann, Friedrich, 300, 304
Berlin, Don, lxvi
Berliner-Joyce wind tunnel, 640
Bernouilli, Daniel, 6
Bernouilli's theorem, 398, 399, 400, 401, 417
Betz, Albert, 391
Bilstein, Roger (historian), lx, 234
Board of Ordnance and Fortification, U.S. Army, 226, 227
Boeing Co., xxxv, l, li, lii, lxv, xli
   Historical Archives, xlix, l, 636
   Historical Services, lxv
Boeing aircraft:
   Boeing 247, xxxii, l
   Boeing 707, xxxiii, l
   Boeing 747, l
   B-17, xxxii, l
   B-29, xxxiii, l
   B-47, xxxiii, l
   B-52, l
   Model 299, l
   Model 367-80, l
Boeing Field, xxxviii
Boeing North American, lii
Bolling AFB, D.C., xlvi
Bostonis, Ronald G., lxiii
Boulton & Watt steam engine, 35, 49
Boushey, Homer A., 268ill.
Bowers, Peter M. (historian), lxvi

Braslow, Albert L., lxiii
Brett, C. W., 611, 612, 613
Brewster aircraft:
   F2FA, xxxiii
Briggs, Lyman J., 441, 445, 452, 454, 465, 468
Brinkley, W. C., 222
British Advisory Committee for Aeronautics (ACA), 302, 305, 316, 318, 467, 475, 476
British Aircraft Production Technical Department, 366
British Air Ministry, 614
British National Physical Laboratory, 244, 246, 248, 249, 286, 297, 300, 302, 311, 312, 313, 316, 317, 319, 379, 380, 387, 391, 405, 411, 412, 416, 417, 447, 448, 451, 467, 473, 475, 476, 481, 483, 485, 486, 487, 490, 499, 611, 612, 633
Brown, Elmer, 604
Bryan, George Hartley, 392
Bugos, Glenn (historian), lxi
Bureau of Aeronautics, U.S. Navy, xlii, 271, 576, 652, 655
Bureau of Aircraft Production, 376
Bureau of the Budget, 633
Bureau of Construction and Repair, U.S. Navy, 309, 445, 450, 451
Bureau of Mines, 385, 480
Bureau of Navigation, U.S. Navy, 286, 605
Bureau of Standards, xxvi, 289, 306, 309, 441, 442, 443, 444, 451, 452, 453, 455, 457, 462, 471, 476, 633
Burnham, Walter E., lxvii
Busemann, Adolf, xxv
Busteed, Mr., Squadron Commander, 353

California Institute of Technology (Caltech), xviii, xix, lv, lxv, 265, 604, 605, 608, 609, 640, 642
Caltech wind tunnel, 609
Cambridge Electric Light Co., England, 315
Cambridge Scientific Instrument Co., England, 316
Cambridge University, 580
Caproni Co., 384
Caquot, Albert, 381
Carnegie, Andrew, 273, 284

Cartwright, Edmund, 36
Castellazi, Lt., 400, 403, 412, 422
Cathers, Richard, liii
Catholic University of America, 244, 248
Cayley, George, xi, xx, xxi, 6, 7, 8, 12, 15, 33, 57, 79, 188, 240, 241, 268
Cayley glider of 1804, 7, 8
Centennial of Flight Commemoration Act (Public Law 105-389), xxvii
Century Series fighters, xliv, xlvii
Chambers Report (U. S. Navy), 287
Chambers, Washington Irving, 246, 286, 287, 288, 296
Champs de Mars, France, 245
Chanute-Herring glider of 1896, 21ill., 23
Chanute, Octave, 15, 17, 17ill., 18, 19, 22, 107, 126, 147, 148, 149, 163, 171, 179, 187, 189, 195, 196, 197, 198, 200, 203, 208, 225, 229, 244, 273, 275, 282, 284
Chapman, William, 35, 36
Chatfield, James S., 637, 638, 639, 640, 641
Chattock gauge, 491
Churchill, Winston, 384
Civil Aeronautics Authority. See Federal Aviation Administration.
Clark, Kenneth (historian), xx
Clark, Virginius E., 445, 642, 650
Clark Y airfoil, 592, 599, 601
Cleveland, Reginald M., 604
Clinton, William J., xxvii
Coffin, Joseph G., 466, 467, 468, 470, 472
Cogswell, John, Dr., 223
Cold War, xliv, xlvii, l, liii
Collier Trophy, Robert J., xlii, lxiv, 258
Columbus, Christopher, 4
Commission for Aeronautics, U.S., 474
Committee for the Scientific Survey of Air Defense, 339
Cone, Hutchinson, 264ill.
Congress of Mathematicians, Heidelberg (1904), 475
Congress, U.S., xxvi, 238, 246, 296, 305, 306, 307, 361, 616, 617, 624, 626, 633, 635, 648
Consolidated B-24 Liberator, xxxiii
Consolidated Vultee, liii
Convair, liii

Convair aircraft:
  Convair 240, liii
  Convair 340, liii
  Convair 440, liii
  Convair 880, liii
  Convair 990, liii
  B-36, liii
  B-58, xxxiii
  XFY-1 Pogostick, liii
  XF2Y Sea Dart, liii
  F-102, xxiv, xxxii, xxxiii, xliv
  F-106, xxiv, xxxi, xxxiii, xliv
  XF-92, xxiv, xxxi, xxxiii
Cook, William, li
Cooper, Bertram, 355
Copernicus, Nicolas, 7
Corn, Joseph J. (historian), lxvi
Council of National Defense, 376
cowling, NACA, xxxii, xli, 258, 614, 626, 627, 643, 662, 664, 665
Crigler, John L., 581
Crocco, Antonio, 405, 412
Crouch, Tom D. (historian), lix, 10, 23, 25, 148, 241, 243
Crowley, John W., 575
Curtiss Aeroplane and Motor Co., 470, 472, 613, 636
Curtiss aircraft:
  AT-5 airplane, xxxii
  JN4H Jenny, xxxii, 251, 448, 449, 450, 455, 459, 460, 461, 463, 481, 490, 495, 498
  NC-4, xxxii
Curtiss Engineering Corp., 446
Curtiss, Glenn H., 149
Curtiss wind tunnels, 403, 411
Curtiss-Wright Corp., liv, lxvi
Cygnet, 230

Daedalus, 1
Daniel Guggenheim Fund for the Promotion of Aeronautics, lv, 264ill., 265, 266, 605, 606, 607, 609
Daniel Guggenheim Graduate School of Aeronautics at the California Institute of Technology, 609, 612
Daniels, John T., 222, 223

Daniels, Josephus, 306
Darwin, Charles, 7
Da Vinci, Leonardo, xi, xix, xx, 1, 4, 5, 239, 240, 245
Davison, F. Trubee, 264ill.
Dawson, Virginia P. (historian), lxviii
Dayton Press, 221
Dayton-Wright RB-1, xxxii
DeBothezat, George, 373, 374, 450
DeFrance, Smith J., 584, 591, 631, 634
Degen, Jakob, 34, 36, 41, 49, 52
DeHavilland DH.4 airplane, 397, 448
De Houthulst, Willy Coppens, xxxv
Department of Commerce, 605, 660
Department of Transportation, xxvi
Detroit Aircraft Corp., 649
Deutschen Versuchsanstalt für Flugwesen, Adlershof, 300, 302–304, 395
Dines, W. H., 342
Dobson, B. Palin, 356
Dornier, Claude, 647
Douglas aircraft:
  A-1 Skyrocket, liv
  A4D Skyhawk, liii
  A-20, liii, liv
  A-26, liii, liv
  C-17 Globemaster, liii
  C-47, liii
  C-54, liii
  D-558, liii, xliv, liv, lxii
  DC series, xxiv, xxxi, lii, lxvi
  DC-1, xlvii
  DC-2, xxxii, xlvii
  DC-3, xxxii
  F4D Skyray, liv
  O2H, 663
  SBD Dauntless, liii, liv
  TBD Devastator, liii
  X-3, liii
Douglas Aircraft Co., lii, liii, liv, xlvii, 609, 646
Douglas, Donald W., liii
Dough, W. S., 222
"Dragonfly" engine, 378
Dryden Flight Research Center (NACA/NASA), xxxix, xl, xliii, lxii, lxiii
Dryden, Hugh L., lxii

Drzewiecki method, 314, 421
Drzewiecki theory of propeller action, 401, 407, 408
Durand, William F., 250, 264ill., 306, 360, 376, 400, 441, 445, 446, 455, 466, 467, 468, 472, 474, 575
Du Temple, Felix, 14

Edson, Lee, xvii, xxx, lvii
Edwards Air Force Base, Calif., xxxviii, xxxix, xliii, xlvii
Egtvedt, Claire, li
Eiffel, Gustav A., 245, 251, 253, 268, 287, 291, 294, 296, 297, 298, 303, 318, 381, 405, 412, 422, 446, 473, 481, 485, 499, 556, 588
Einstein, Albert, 263
EMSCO Aero Engine Co., Los Angeles, 645
Engelhardt, Lloyd, liv
Engineering Research Corp. (ERCO), xlii
Eole (Ader), 14, 14ill.
Ercoupe, xlii
Erie Railroad, 17
Etheridge, A. D., 222
Euclid, 4
Euler, Leonhard, 6
Evans, Stanley H., 646
Ewing, Susan E., xlviii
Experimental Techniques Branch, NASA Langley, xliii

Fairchild F-22 airplane, 600, 601
Fairey, C. R., 613
Fales, Elisha N., 466, 468
Farren, William S., 355
Federal Aviation Administration (FAA), xxvi
Ferri, Antonio, xxv
Fifth International Aeronautic Congress, 289
Fishamend Laboratory, Vienna, Austria, 558, 559
Fliegner, Albert, 417
Flight Research Center. See Dryden Flight Research Center.
Floyd L. Thompson Technical Library, NASA Langley, lvi, xliii
Focke, Heinrich, 297
Fokker D-VII airplane, 535

Fokker Tri-Motor 5-AT airplane, xxxii
Ford, Henry, liv
Ford Motor Co., liv
Ford Trimotor airplane, liv
Forrest, Marvin, 581
Fort Meyer, Va., 226, 228
Fowler flap, 661
Fowler wing, 660
Francillon, René, lxvi
French Academy of Sciences, 39, 56
French Aeronautical Technical Service, 380, 381
Freud, Sigmund, xx
Fuhrmann, H., 297, 304
Full-Scale Tunnel, NACA/NASA Langley, xxxviii, xliii, 258ill., 259ill., 259, 260ill., 260, 261, 584, 585, 591, 592, 593, 594ill., 595, 596ill., 597ill., 598ill., 599ill., 631, 634, 635, 642

Galilei, Galileo, 3, 5
Gallaudet, Edson Fessenden, 23
Gardner, Lester D., 501
Gardiner, Arthur W., 584, 588, 591
Gates Learjet Model 55, xxxiv
Geer, Mary Wells, lxv
General Aviation Collection, University of Texas at Dallas History of Aviation Collection, lv
General Aviation Manufacturing Corp., li, lii
General Dynamics aircraft:
    F-16, xxxiv
    F-111, xxxiii, xlii
Georgia Institute of Technology, 265, 266, 605
German Society of Mechanical Engineers, 300
German Wright Co., 297
Gillette, Aaron K., lxiii
Glauert, Heinrich, 464, 575
Glazebrook, R. T., 312
Glenn Research Center at Lewis Field, NASA, xxxix, lxvii
Goin, Kenneth, lxiv
Goodyear Navy airships, 627
Goodyear Zeppelin Co., 627
Gorn, Michael (historian), lxii, lxv, 270
Göttingen airfoils, 459, 502
Gough, Melvin N., 581

Gray, George W., lx
Greenwood, John T. (historian), lix
Grey, C. G., 237, 264, 611, 614
Griffin, Robert S., Admiral, 385
Griffith, Leigh M., 259, 521, 577, 642, 645, 647, 648
Grumann aircraft:
    X-29, xxxiv
Guggenheim Aeronautical Laboratory at the California Institute of Technology (GALCIT), xviii, xxxvii, liii, lv, 265, 270, 605
Guggenheim, Daniel, 264ill., 265, 604, 610
Guggenheim, Harry, 264ill., 604, 608, 612
Gull, Lloyd Dake, xxxvi, xxxvii
Gwynn-Jones, Terry (historian), lix

Halley, Edmond, 363
Hallion, Richard P. (historian), lxii, lxiii, 265
Handley-Page Co., 378
Handley-Page wing slot, 613
Hanle, Paul A. (historian), lxv, 269
Hansen, James R. (historian), xi, xii, xxxv, xl, xliii, lvii, lix, lx, lxi, lxxi, 239, 257, 258, 263
Hargrave, Thomas/James, 149, 162, 163, 164, 180, 185
Hartman, Edwin P., lxi
Harvard University, 233
Hayford, John F., 247ill., 360, 441, 445, 454
Hebbert, C. M., 398
Heinemann, Edward H., lii, liii, liv
Helium Board, 385
Hemke, Paul A., 580,
Henderson, Paul, 247ill.
Henry Ford Museum and Greenfield Village Archives, Dearborn, Mich., liv
Henson-Stringfellow airplane, 12, 13
Henson, William, 12
Herring, Augustus M., 15, 19, 19ill., 20, 147, 148, 149, 156, 173, 190, 196, 209, 210, 211
Herring glider of 1896. See Chanute-Herring glider of 1896.
Herrnstein, William, Jr., 581
Hicks, Frederick C., 383, 384
Hicks, Harold, liv
Higgins, George J., 577

Highly Maneuverable Aircraft Technology (HIMAT), xxxiv, xliv
High-Speed Civil Transport (HCST) program, liii
High-Speed Flight Station (NACA). See Dryden Flight Research Center.
History of Aviation Collection at the University of Texas at Dallas, xxxvii
Holden, Henry M., lxvi
Hooke, Robert, 632
Hoover, Herbert, 604, 605
Houston, Robert S., lxiii
Huffaker, Edward C., 24ill., 187, 200, 202, 283
Huffman Prairie, Dayton, Oh., xxxviii, 27, 28
Hughes H-1 aircraft, xxxiii
Hunley, J. D. (historian), lxii
Hunsaker, Jerome C., 245, 246, 250, 251, 255, 270, 286, 287, 296, 301, 306, 311, 317, 359, 383, 445, 454, 463, 465, 497, 499, 500, 501, 536, 537
Hunt, Franklin L., 445, 454
Huttin, Charles, 632
Huyghens, Christian, 5
Hyper-X (X-43), NASA/Industry, xliv

Icarus, 1
Ide, John Jay, 266, 375, 376, 377, 380, 382
Indiana Dunes, Ind., 17, 19, 20, 147
Institute Aerotechnique de St. Cyr, France, 302, 303, 304, 387, 392
International Aeronautic Commission, 289
International Civil Aviation Organization (ICAO), 580
Internet, xxxvi, xxxvii
Issy-les-Moulineaux, France, 375, 380
Jacobs, Eastman N., 663
Jakab, Peter (historian), lix, 22
Jamestown Exposition of 1907, 230
Jefferson, Thomas, 57
Jenkins, Dennis R., lxiii
Johannistal flying grounds, Germany, 301
Johns Hopkins University, 233, 581
Jones and Paterson method of airfoil design, 322
Jones, B. Melville, 580
Joukowski, Nicolai, xxv, 253, 270, 272
Joyce, Temple N., 661

Junkers Co., 394
Junkers laboratory at Dessau, Germany, 394
Junkers wing, 395

Kaiser Foundation (Germany), 297, 303
Kaiser Prize (Germany), 300, 304
Katydid (Chanute glider), 19, 20, 147
Kelvin, Lord, 10, 125, 128, 543, 546
Kemper, Carlton, 577
Kilgore, Robert, xliii
Kindelberger, James H., "Dutch," lii
Kinney, Jeremy (historian), lxxi
Kitty Hawk, N.C., viii, xxvii, xxviii, xxx, 1, 2, 20, 23, 24, 26, 28, 130, 187, 192, 198, 199, 200, 204, 208, 212, 221, 222, 223, 231, 244, 284
Klein, Arthur L., lv, 359, 642, 643, 645, 646, 647, 648, 650, 651
Klemin, Alexander, 646
Knight, Montgomery, 265
Knight, William, 416, 474
Krell manometer, 419

Laboratoire Aerodynamique Eiffel, 302, 499, 506
Lamar Files, William, xlvi
Lanchester, Frederick William, xxv, 270, 392
Land, Emory S., 576
Langley Aerodromes:
    Great Aerodrome of 1903, 28, 29ill., 30ill., 130
    No. 0, 134
    No. 1, 135
    No. 3, 136
    No. 4, 136, 137
    No. 5, 10, 11ill., 130
    No. 6, 9ill.
Langley Aeronautical Laboratory. See Langley Research Center, NACA/NASA.
Langley Field, Va., 237, 247, 248, 249, 306, 400, 403, 410, 411, 417, 418, 422, 441, 446, 447, 455, 456, 458, 461, 469, 481, 496, 522, 560, 561, 575, 588, 591, 601, 613, 623, 626, 627, 629, 631, 648
Langley Memorial Aeronautical Laboratory. See Langley Research Center, NACA/NASA.
Langley Research Center, NACA/NASA, xxiv, xxxi, xxxviii, xxxix, xlii, lvii, lx, 29, 256

Langley Research Center, NACA/NASA (continued), 257, 258, 259, 260, 261, 263, 265, 266, 267, 270, 272, 362, 375, 376, 398, 426, 445, 446, 448, 468, 478, 481, 502, 521, 522, 556, 575, 576, 577, 580, 585, 601, 602, 622, 623, 642, 657
Langley, Samuel P., 9, 10, 12, 22, 26, 28, 31, 89, 107, 126, 127, 128, 130, 131, 141, 145, 149, 179, 180, 181, 187, 205, 217, 221, 238, 241, 283, 284, 293, 632
Lapresle, Charles, 303
Launius, Roger D. (historian), xxvii, lxiii
Law of Bairstow and Booth, 502
Lawrance J-1 engine, 590, 591
Le Bris, Jean Marie, 16
Lee, J. Lawrence (historian), lxxi
Lee, William D. B., 57
Les Hanuadières, France, 31ill.
Lesley, Everett P., 575, 586
Lessing, Sam, 148
Leviathon (Maxim biplane of 1894), 13, 14, 13ill., 127
Lewis, George W., 271, 466, 476, 497, 502, 506, 521, 556, 557, 576, 580, 602, 631, 657, 659
Lewis Research Center, NACA/NASA, xli
Liberatore, Eugene K., lxii
Liberty engine, 436, 484
Library of Congress, xxxvii
Lilienthal, Gustav, 101-02
Lilienthal, Otto, xi, xxi, xxii, 10, 15, 16, 17, 18, 22, 25, 89, 91, 92, 101, 102, 103, 104, 105, 106, 126, 153, 154, 155, 158, 159, 166, 172, 174, 177, 182, 184, 185, 187, 188, 189, 192, 193, 194, 195, 196, 197, 198, 199, 200, 201, 202, 203, 204, 207, 209, 212, 214, 215, 242, 273, 275, 280, 283, 284
Lindbergh, Charles A., 264ill.
Link, Edwin A., lv
Link trainers, lv
Lippisch, Alexander, xxv
Littlewood, William, 68, 248
Locke, C. N. H., 575
Lockheed aircraft:
 Air Express, xxxii

Lockheed aircraft (continued):
 F-104, xxxiii, xliv
 F-117, xxxiv
 L1049/C-69, xxxiii
 P-38, xxxiii, xlvi
 P-80, xxxiii
 U-2, xxxi
 Vega, xxxii
 YF-12/SR-71, xxxiii
Loftin, Laurence K., Jr., xxxi, xxxii, lix
Logsdon, John M., xxvii
London University, 612
Lougheed, Victor, 652, 653, 654, 655, 656
Luddington Line airplanes, 654
Luftfahrzeug Gesellschaft, 297
Luftwaffe, 538

Mach number, 583
Macintosh screw propeller, 72
Mack, Pamela E. (historian), lxiv, 258
Maclaurin, Richard, 246, 311
Malina, Frank, 268ill.
Mansfield, Harold, lxv
Marey, Étienne-Jules, 196
Margoulis, Wladimir, xxv, 253, 497, 499, 506, 507
Mariotte, Edme, 5
Martin B-10 airplane, xlvii
Martin, H. S., 454
Marvin, Charles F., 247ill., 275, 282, 309, 445, 454
Massachusetts Institute of Technology (MIT), 242, 246, 265, 266, 300, 311, 360, 361, 455, 458, 471, 476, 497, 537, 580, 604, 605, 633, 639
Maurain, Charles, 303
Maxim, Hiram, 13, 14, 107, 126, 128, 135, 149, 179, 180, 181, 217, 242, 302, 632
Maxwell AFB, Ala., xlvii
McAvoy, William H., 661
McCarthy, Charles, 637, 638, 639, 640
McCook Field, Oh., xlvi, 248, 249, 263ill., 264, 265, 362, 363, 373, 374, 403, 404, 411, 445, 450, 451, 452, 455, 458, 581, 588. See also Wright Field and Wright-Patterson AFB.

McDonnell aircraft:
  F-4, xxxiii
  F-101, xliv
McDonnell Douglas Corp., lii, liii, liv, xlix, lxvi
McDonnell Douglas aircraft:
  DC-10, liii
  MD-11, liii
  MD-90, liii
McFarland, Marvin W., lvii, 187, 273
Mead, George, 636, 637, 638, 639, 640, 641
Meiler, John, 148
Merrill, A. A., 609
Meurthe, Deutsch de la, 303
Middle Ages, The (European), 2
Millar, G. H., 354
Miller, Elton W., 256ill., 498, 523, 531, 577, 578, 579, 582
Miller, J. W., 264ill.
Miller, Ronald, and Sawers, David (historians), lix
Millikan, Clark B., lv, 265, 268ill., 605
Millikan, Robert, 264ill., 265, 270, 605
Missouri Historical Society Archives, liv
Mixson, Ralph E., 575
Moedebeck, Hermann, 214, 275
Moffett Field, Calif., xxxix
Moffett, William A., 247ill.
Montgolfier brothers, xxi
Montgomery, John Joseph, 16
Montieth, Charles N., 637, 639, 640
Moore, Johnny, 222
Morgan, William C., 521
Motorluftschiff-Studien-Gesellschaft, 296, 297, 303
Mouillard, Louis-Pierre, 16, 187
Mozhaiski, Alexander, 13, 14
Muenger, Elizabeth A. (historian), xliv, lxi
Munk, Max M., xxv, lxv, 253, 255, 257, 266, 267, 267ill., 269, 270, 271, 272, 391, 395, 396, 464, 497, 498, 499, 500, 501, 502, 511, 521, 522, 523, 535, 556, 557, 559, 577, 578, 580, 582, 602, 603
M (Munk) wing sections, 621
Muttray, M., 643
Myers, Ted, 581

NACA 4-digit airfoils, 663
NACA annual manufacturers conferences, 260ill., 631, 657, 659
NACA Report Server, lvi
NACA slot lip, 662
NACA Wind Tunnel No. 2. See Variable-Density Wind Tunnel.
NASA Center for Aerospace Information (CASI), lvi
NASA Marshall History Office, lxviii
NASA Technical Report Server, lvi
National Air and Space Act of 1958, xxxix
National Air and Space Museum (NASM), Washington, D.C., xxxv, xxxvi, xxxvii, xlviii, lix
National Archives and Records Aministration (NARA), xxxv, xliv, xlv, xlvi, lxviii
National Bureau of Standards. See Bureau of Standards.
National Physical Laboratory at Tedington. See British National Physical Laboratory.
National Research Council, 376
National Signal Bureau, 212
Naval Academy, U.S., 580
Naval Aircraft Factory, Philadelphia, Pa., 450
Naval Appropriations Act of 1916, 306, 385
Naval Historical Center at the Washington Navy Yard, Washington, D.C., xlviii-xlix
Naval Observatory, Washington, D.C., 233
Navy Bureau of Construction and Repair, U.S., 250
Navy, U.S., xliii, xlvii, xlviii, 246, 251, 265, 286, 294, 309, 311, 359, 361, 385, 445, 457, 458, 460, 461, 479, 537, 627, 646, 648
Nayler, J. L., 379, 380
Newcomb, Simon, 233, 234, 235
Newcomen, Thomas, 132
Newport News (Va.) Shipbuilding & Dry Dock Co., 531
Newton, Isaac, xx, xxi, 5, 6, 7, 55, 71, 240, 241, 525, 632
New York Public Library, xxxvii
New York University Guggenheim School of Aeronautics, 604, 607, 610
Nieuport "Nighthawk," 377, 378
Norfolk (Va.) Navy Yard, 653, 654
North American Aviation, Inc., lii, lxvi

North American Aviation aircraft:
  AT-6, li
  B-25, li
  BT-9, li
  F-86, xxxiii, li
  F-100, xliv, li
  F-107, xliv
  FJ-1, li
  Model NA-16, li, lii
  Model NA-73, lii
  P-51, xxxiii, xlii, xlvi, li, lxvi
  X-15, xlii, li, lxiii, lxiv
  X-15A, xxxiii
  XB-70, xxxiii, xliv, li
  XFJ-1, lii
  XP-51, xl
North American Rockwell Corp., xlix, li, lxv
Northrop aircraft:
  HL-10, xxxiv
  N-1M, xxxiii
Northrop, John K., "Jack," li, lii, 649
Norton, Frederick H., 266, 271, 398, 426, 463, 466, 467, 468, 469, 481, 497, 500, 502, 521, 537

Office of Aeronautical Intelligence, NACA, 253, 266, 375, 376, 377

Pannell, J. R., 525, 535
Parseval airship, 296, 298, 299
Pascal, Blaise, 5
Pénaud, Alphonse, 8–10, 133, 134, 140, 150, 184, 188, 209
Pennell, Maynard, li
Pershing, John, 628
Phillips, Horatio, lviii, 59, 115, 116, 180, 181, 242, 268
Pilcher, Percy, 126, 128, 177, 182, 184, 190, 191, 196, 197, 200, 209
Piper aircraft:
  Cub, xxxii
  Pawnee, xliii
Piper Aircraft Corp., xlii
Pitcairn PCA (Cierva), xxxii
Pitot tube, 313, 363, 364ill., 365, 365ill., 367, 371, 419, 427, 471, 573

Pobjoy engine, 612
Poe, George, 581
Pope, Alan, lxiv
Post Office, U.S., 480, 606
Powers, Sheryll Goecke, lxiii
Prandtl, Ludwig, xxv, lxv, 244, 245, 249, 253–255, 257, 269-70, 272, 287, 288, 296, 297, 303, 386, 387, 391, 394, 466, 467, 469, 470, 474, 475, 499, 535, 536, 537, 538, 578
Propeller Research Tunnel (PRT), NACA Langley, 256, 257, 258, 259, 260, 261, 271, 376, 500, 556, 557, 560, 561, 561ill., 562ill., 563ill., 564ill., 565ill., 566ill., 567ill., 568ill., 569ill., 570ill., 571ill., 572ill., 573ill., 575ill., 576, 577, 579, 580, 581, 582, 584, 642
Public Works Administration (PWA), 660
Pupin, Michael I., 247ill., 360

Quittner, P., Dr., 299

Radio Direction Finding (RADAR), 339
R.A.F. airfoil sections:
  R.A.F. 6, 324, 471
  R.A.F. 15, 458, 459, 463, 467, 476
  R.A.F. 19, 459
  R.A.F. Mark IV airspeed indicator, 371, 372
Rathbun, Richard, 187
Rawdon, Herb, lxvii
Rayleigh, Lord, 10, 125, 302, 524
Raymond, Arthur E., lv
Reed, R. Dale, lxii
Reid, Elliot G., 265, 584, 585, 588, 602
Reid, Henry J. E., 577, 579, 616, 652
Renaissance, The (European), xix, 2, 239
Renard, Paul, 289
Republic aircraft:
  F-105, xliv
  P-47, xxxiii
  P-47B, xlii
  XP-47 Thunderbolt, 78
Reynolds Law, 515, 517
Reynolds number, 252, 255, 257, 259, 455, 459, 498, 499, 500, 501, 502, 505, 507, 508, 509, 510, 513, 515, 516, 517, 518,

Reynolds number (continued),
519, 520, 521, 522, 523, 528, 529, 530,
533, 556, 574, 584, 586, 592, 599, 648, 651
Reynolds, Osborne, 251, 252, 498, 518, 528,
535, 582
Reynolds velocity, 514, 518
Riabouchinski, Dimitri, xxv
Robins, Benjamin, 37, 47, 240, 632
Robinson Collection (University of Texas at
Dallas History of Aviation Collection), lv
Rockwell International, li, lii
Rockwell International aircraft:
B-1, xxxiv, li
X-31, li
Rockwell Space Transportation System
(STS), xxxiv
Rockwell Standard Corp., li
Roland, Alex (historian), lx, 239
Root, Elihu, Jr., 264ill.
Roots supercharger, 591
Rosendahl Collection (University of Texas at
Dallas History of Aviation Collection), lv
Royal Aeronautical Society (British), lvii, 59, 84
Royal Aircraft Establishment (RAE) at
Farnborough, England, 378, 379, 391,
611, 612
Royal Aircraft Factory (British), 302, 340, 397
Royal Air Force (RAF), 339, 342, 355, 356,
377, 384, 385
Royal Flying Corps (Canadian), 339, 340
Royal Society of London, 240, 632
Ryan aircraft:
Dragonfly YO-51 STOL, liii
Fireball FR-1, liii
NYP Spirit of St. Louis, xxxii, liii
SV-5A, liii
SV-5B, liii
Vertiplane V/STOL, liii
X-13 Vertijet, liii
Ryan Aircraft Co., liii
Ryan, John D., 264ill.

Sabine, W. C., 376
San Diego Aerospace Museum, liii
Sarah Clark Collection (National Archives),
lxv, lxviii

Schmid, Edgar, lxvi
Schultz, James, lxi
Science Technology Corp., Hampton, Va., xxvii
Scientific Revolution, The, xx, 4, 8
Scott, Catherine (historian), xxxvii
Scriven, George P., 305, 306, 307, 309, 310
Seely, General, 384
Selfridge, Thomas E., 226, 230
Serling, Robert J., lxv
Seversky, Alexandre, xxv
Sharp, Edward R., 577
Shoemaker, Eugene, 576
short takeoff and landing (STOL), liii
Siddeley Co. of Coventry, 377
Siddeley "Siskin" aircraft, 377
Signal Corps, U.S. Army, 225, 226, 227, 228,
229, 231, 232, 369, 522
Sikorsky aircraft:
S-42, xxxii
S-60/64, xxxiii
VS-300, xxxi
Sikorsky, Igor, xxv, li, 637, 638, 641
Silberman, Paul E., xlviii
Smeaton, John, 37, 203, 240, 273
Smith, Richard K. (historian), lix, 221
Smithsonian Institution, Washington, D.C.,
10, 22, 31, 91, 127, 130, 131, 145, 147,
187, 188, 246, 287, 288, 293, 301, 305, 36,
310, 311, 386, 395
Society of Automotive Engineers (SAE), lvii, 642
sound barrier, xli, lxii
Space Shuttle (NASA), xxxiv, lxii, lxiii, lxiv
Sperry M-1 Messenger airplane, xxxii,
256ill., 534, 573, 575, 578, 579, 582
Sperry Venturi speed meter, 372
Spratt, George A., 200, 202, 244, 273, 274
Stack Collection, John, NASA Langley, xlii
Stack, John, xl, xlii, lxiv
Stanford University, 247, 250, 257, 266, 306,
389, 403, 411, 412, 441, 445, 446, 450,
455, 461, 472, 556, 582, 605, 608
Stanton, C. I., 454
Stanton, T. E., 525, 535
Stearman Co., l
Stearman, Lloyd, 637
Stillman, Wendell H., lxiii

Stout Metal Airplane Co., liv
Stratton, Samuel W., 376
Stringfellow, John, 12
supersonic transport (SST), l, liii, xxxix, xlii
Sundry Civil Act of 1920, 385
Sydenham, Lord, 353, 358

TACT, xxxiii
Tactical Fighter Experimental (TFX). See General Dynamics F-111.
Talay, Theodore A., lix
Tate, Dan, 24ill., 200
Tate, W. J., 200
Tatin, Victor, 180
Taylor, David W., 246, 247ill., 248, 287, 383, 384, 386, 481
Taylor, D. Bryan (historian), lxxi
Taylor, Geoffrey I., 318
Taylor Cub, xxxii
Technische Hochschule, Aachen, Germany, 270, 298, 383, 386, 392
Teddington, England. See National Physical Laboratory.
Teledyne Ryan, liii
Texas A&M University, xlii
TFX (Tactical Fighter Experimental). See General Dynamics F-111.
Theater of the Royal Society of Arts, London, 353
Theodorsen, Theodore, xi, xxv
Thomas-Morse MB3 airplane, 455, 460
Thompson, Milton O., lxii
Tichenor, Frank, 272, 616, 617, 624
Tizard, Henry T., 339, 340, 353, 354, 355, 358
Torricelli, Evangelista, 5
Toussaint-Lepère airspeed meter, 368
Townend ring cowling, 622, 629, 643
Travel Air aircraft:
    Model R Racers, lxvii
    Mystery Ship, lxvii
Travel Air Co., lxvii
Treaty of Versailles, 558
Triaca, Albert C., 230
Truscott, Starr, 652, 656
Tsien, Hsue-tsen, xxv
Tunison, Frank, 221

United Aircraft and Transport Corp., xlii, li, 636
University of Akron, 265, 605
University of Dessau (Germany), 383, 386, 394, 395
University of Göttingen (Germany), lxv, 244, 250, 253–255, 286, 287, 296, 297, 298, 300, 301, 302, 303, 304, 383, 386, 389, 390, 394, 396, 438, 448, 474, 500, 501, 518, 537, 578, 633
University of Illinois, 579
University of Manchester (England), 252, 582
University of Michigan, 265, 581, 604, 605
University of Paris (France), 296, 303
University of Texas at Dallas History of Aviation Collection, lv
University of Washington, 265, 605
U.S. Army (U.S.A.) 16 wing section, 475
U.S.S. Los Angeles (dirigible), 627, 630

Variable-Density Wind Tunnel (VDT), NACA, xli, 254ill., 255, 256, 257, 258, 259, 260, 267, 267ill., 271, 467, 476, 497, 498, 499, 523, 535, 556, 559, 577, 584, 586, 587, 592, 642, 648
Venturi cowling, 629
Venturi, Giovanni Battista, 5
Venturi tube, 364ill., 365, 367, 372, 372ill., 443
Verein Deutschen Ingenieure, 304
Versuchsanstalt für Flugwesen der königliche technische Hochschule, Berlin, 299
Vertical/short takeoff and landing (V/STOL), xxxix, xlii, lxi
Vertical Wind Tunnel, 5-Foot, NACA, 261, 262ill., 263
Vertol aircraft:
    CH-46, l
    CH-47, l
Vertol Aircraft Co., l
Vickers Sons, 302, 379
Victory, John F., 631
Vincenti, Walter G., lix
Volta Congress on High-Speed Aeronautics, xxvi
Von Helmholtz, Ludwig, 525, 546, 547, 549
Von Parseval, Major, 299, 300
Von Kármán, Theodore, xi, xvii, xviii, xix, xx, xxi, xxii, xxv, xxx, lv, lvii, lxv, 268ill.

Von Kármán, Theodore (continued), 269, 270, 392, 393, 536, 558, 610
Von Zeppelin, Ferdinand, 183, 298
Vought aircraft:
   F8-U Crusader, xxxiii, xliv
   O3U-1 Corsair, 259ill., 631
   VE-7, 455, 459, 571, 582, 583
Vultee, Gerald, 649

Wagner, Ray (historian), lxvi
Walcott, Charles D., 247ill.
Walker, Lois E., and Wickam, Shelby E., lxv
Wallace, Lane E. (historian), lxii
Wallops Island, Va., xli
Ward Leonard motor control system, 595
Warner, Edward P., 253, 272, 383, 386, 398, 426, 441, 445, 454, 497, 506, 580, 616, 628, 657
Washington Navy Yard, Washington, D.C., 246, 247, 248, 249, 250, 265, 287, 295, 305, 306, 309, 440, 445, 450, 451, 455, 458, 466, 469, 475, 633
Watson (James) and Crick (Francis), 7
Watson-Watt, Robert A., 339
Weather Bureau, U.S., 295, 306, 309, 479
Weick, Fred E., xl, xlii, xliii, li, lvii, 257, 271, 500, 556, 557, 560, 575, 576, 638, 639, 640, 641, 660
Weick W-1A airplane, xlii
Wells, A. J., 242, 266
Wenham, Francis H., lviii, 25, 58, 59, 77, 78, 80, 83, 84, 85, 87, 241, 268
Western Society of Engineers, 192, 208
Westland "Wagtail" (British), 377
Whitcomb, Richard T., xli
Whitcomb winglet, xxxiv
Whitter Aviation Library and Archives of the San Diego Aerospace Museum, N. Paul, liii
Wichita State University, lxvii
Williams, H. M., 359
Wilson, E. B., 466, 468
Wilson, Woodrow, 255
Windler, Ray, 581
Wood, Donald H., 556, 560, 581, 583
Wooster, Stanton, 576

World War I, xxxv, 20, 234, 244, 245, 261, 264, 287, 306, 383, 500
World War II, xli, xlii, xliv, xlvi, liii, 339, 383, 386
Wright brothers, xi, xxv, xxvii, xxx, xxxviii, xlii, lvii, lix, 1, 6, 8, 10, 12, 15, 22, 23, 24, 26, 28, 31, 89, 130, 148, 149, 202, 203, 208, 212, 215, 218, 225, 226, 230, 231, 232, 233, 234, 237, 238, 239, 242, 244, 245, 268, 286
Wright brothers' aircraft:
   Flyer of 1903, xxxii, 23, 27ill., 29, 192, 223, 224, 238
   Flyer of 1905, 28
   Glider of 1900, 23, 201
   Glider of 1901, 24
   Glider of 1902, 25, 26ill., 26, 192
Wright Brothers Memorial Trophy, xlii
Wright Centennial Anniversary, xi, xxviii
Wright-Curtiss patent dispute, 287
Wright drag balance, 243ill.
Wright Field, Oh., xlvi
Wright, Katherine, 244
Wright, Milton, 1, 188, 192, 193, 223
Wright, Orville, 1, 21, 24, 26, 27, 91, 183, 187, 188, 192, 208, 211, 221, 222, 223, 225, 226, 229, 238, 242, 243, 264ill., 632
Wright Patterson Air Force Base, xlvi, xlvii, lxviii, 27, 265
Wright State University, xxvii
Wright, Theodore P., 662
Wright Whirlwind engine, 614
Wright, Wilbur, 1, 10, 21, 24, 26, 27, 31ill., 91, 126, 130, 183, 187, 188, 189, 192, 193, 208, 211, 218, 221, 222, 223, 225, 229, 238, 242, 244, 273, 274, 275, 282, 284, 286, 632
Wright wind tunnel, 239ill.

X-series aircraft, xlvii
X-43. See Hyper-X.
XYZ Engine Co., 630

Yackey, Tony, 580
Young, Pearl I., 18

Zaharoff, Basil, 612
Zahm, Albert F., 244, 246, 248, 250, 265, 268, 270, 286, 287, 288, 301, 305, 445, 454, 455, 465, 467, 468, 469, 470, 475
Zahm Pitot-Venturi tube, 369
Zeppelin Airship Co., 385, 395, 396, 511
Zeppelin Versuchsanstalt, Friedrichshafen, 298, 386, 389, 395, 396

www.ingramcontent.com/pod-product-compliance
Lightning Source LLC
Chambersburg PA
CBHW080919100426
42812CB00007B/2321